THE SAGA OF THE NERVE GROWTH FACTOR

Preliminary Studies, Discovery, Further Development

World Scientific Series in 20th Century Biology

Published

Vol. 1 Selected Papers of Frederick Sanger (with Commentaries)
edited by F. Sander and M. Dowding

Vol. 2 A Portrait of the Immune System: Scientific Publications of Niels Kaj Jerne
edited by I. Lefkovits

World Scientific Series in 20th Century Biology – Vol. 3

THE SAGA OF THE NERVE GROWTH FACTOR

Preliminary Studies, Discovery, Further Development

Rita Levi-Montalcini

Istituto di Neurobiologia, Italy

World Scientific
Singapore • New Jersey • London • Hong Kong

Published by

World Scientific Publishing Co. Pte. Ltd.

P O Box 128, Farrer Road, Singapore 912805

USA office: Suite 1B, 1060 Main Street, River Edge, NJ 07661

UK office: 57 Shelton Street, Covent Garden, London WC2H 9HE

The author and publisher would like to thank the following publishers for their assistance and permission to reproduce the articles found in this volume:

Academic Press, Inc. (*Dev. Biol.*, *J. Ultrastruc. Res.*, *Exp. Cell Res.*), American Physiological Society (*Physiol. Rev.*), Bioscience Ediprint Inc. (*Int. J. Tiss. Reac.*), Birkhäuser Verlag (*Agents Actions*, *Encyclopedia of Neuroscience*), Cancer Research, Inc. (*Cancer Research*), Elsevier Publ. Co. (*Prog. Brain Res.*, *Brain Res.*), Elsevier Science Publ. B. V. (*TINS*, *J. Neurol. Sci.*), Elsevier Scientific Publ. Ireland Ltd. (*Neurosci. Lett.*), Flammarion Médecine-Sciences (*Eur. J. Clin. Biol. Res.*), Gordon and Breach Publ. Group (*Growth Factors*, *Autoimmunity*), IMPR. de L'Académie (*Arch. Biol.*), Macmillan Magazines Ltd. (*Nature*), National Academy of Sciences (*Proc. Natl. Acad. Sci. USA*), Press of the Wistar Institute of Anatomy and Biology (*J. Morphol.*, *J. Exp. Zool.*), Rapid Communications of Oxford Ltd. (*NeuroReport*), Raven Press (*Histochem. Cell Biol. of Auton. Neurons, SIF Cells, and Paraneurons*), Springer-Verlag (*Rheumatol. Int.*), The American Association for the Advancement of Science (*Science*), The American College of Rheumatology (*Clin. Exp. Rheumatol.*), The New York Academy of Sciences (*Ann. N. Y. Acad. Sci.*), The Royal Society of Medicine (*Proc. Royal Soc. Med.*), The Williams & Wilkins Co. (*Pharmacol. Rev.*), Thieme Medical Publ., Inc. (*Prog. NeuroEndocrinImmunol.*).

While every effort has been made to contact the publishers of reprinted papers prior to publication, we have not been successful in some cases. Where we could not contact the publishers, we have acknowledged the source of the material. Proper credit will be accorded to these publishers in future editions of this work after permission is granted.

British Library Cataloguing-in-Publication Data
A catalogue record for this book is available from the British Library.

THE SAGA OF THE NERVE GROWTH FACTOR — Preliminary Studies, Discovery, Further Development

ISBN 981-02-2604-7

Printed in Singapore.

This book is dedicated with love to Viktor Hamburger,

Rita

FOREWORD

The discovery in the early fifties of the Nerve Growth Factor (NGF) brought to light the existence of an entirely new class of endogenous products identified in proteins collectively designated as Polypeptide Growth Factors (PGFs).

PGFs are distinct from other, vital diffusible substances such as hormones and cytokines, since their main action occurs during terminal differentiation of previously determined responsive cells, or as yet undetermined embryonic cells.

In the adult, PGFs play important roles in the survival, repair and proper functioning of the cells receptive to their action. Mutated versions of PGFs, of their receptors, or over-expression of the former, have been shown to cause transformation of these cells with acquisition of a neoplastic phenotype.

This volume contains a series of articles which outline the discovery and life history of NGF. The most attractive and perhaps the most unusual feature of this long series of investigations which started almost half a century ago, is that each finding has signaled a turning point and opened up new perspectives. The history of NGF is more like a detective story than a scientific enterprise, which usually unfolds according to well-defined rules along a route paved by previous findings.

A brief comment on another feature of the NGF saga: the nerve cell which first brought NGF to our attention is the sympathetic adrenergic neuron, a third-rate citizen in the galaxy of billions of cells which constitute the nervous system. And yet it was this humble cell that revealed the humoral nature of the transmission of the nerve impulse, which is valid for "simple" nerve cells as well as for those endowed with the highest functions in the Central Nervous System (CNS). Another third-rate citizen in the immense and diversified cell population of the immune system is the mast cell, which suddenly sprung to the forefront of NGF studies.

In his 1993 article, "New concepts about the mast cell", Stephen J. Galli, one of the foremost experts in the study of the structure and functional properties of these cells, outlined their life history from the time of their discovery by Paul Erlich in the last century to the present day. Erlich coined the term *Mastzellen* — i.e., fattened or well-fed cells, on the assumption that the granules that filled their cytoplasm contained nutrient substances which were made available, when needed, to the adjacent cells.

A series of subsequent studies disproved this hypothesis. It was demonstrated that the granules' content, synthesized and stored in these cells, consists of a variety of mediators including cytokines and NGF which, upon activation of these cells by a whole variety of stimuli, are released and influence a vast array of important biological processes. "For this reason" states Galli, "I have suggested, only partly in jest, that "mast cell" can also stand for "master cell" ".[1] This term seems today even more appropriate in view of the recently discovered specific link-function that mast cells

[1]Galli S.J., New concepts about the mast cell, *N. Eng. J. Med.* **328**, 257–265 (1993).

carry out among the three homeostatic systems. A similar, much broader NGF role came to our attention when the studies moved from the cellular level to that of the organism.

Unexpectedly, functions of mast cells and NGF which had previously appeared to be unrelated were instead discovered to be closely interconnected as described in Part IV of this volume. These findings and the prospected modulatory role of NGF on cells belonging to the neuro-endocrine-immune triad, opened an entirely new line of basic and clinical research.

Lastly, I apologize for listing only very few selected citations from the thousands of excellent articles which have been published, mainly in these past two decades, on NGF. I also wish to acknowledge and thank individuals and groups who have shared with me the moments of excitement and frustration which are the rule, rather than the exception, in any scientific adventure.

The first part of my work was performed in the Department of Zoology of Washington University in St. Louis, chaired by Prof. Viktor Hamburger, who not only invited me to join his group and offered me generous hospitality, but also took active part in the initial phases of this research. Stanley Cohen, the discoverer of the chemical nature of the NGF, shared with me the Nobel Prize in 1986. A close friend, the late Pietro Angeletti, was a prominent figure in the early discoveries. Other dear and life-long friends participating in our research are Pietro Calissano and Luigi Aloe. The latter, along with his group of young co-workers in our laboratory (L. Bracci-Laudiero, A. Micera, P. Tirassa, M. A. Tuveri and E. Vigneti), deserve most of the credit for the discovery of the role played by NGF in the interaction between the nervous, immune and endocrine systems. Other investigators from our institute who made contributions to the study of the mechanism of action of NGF are S. Biocca, A. Cattaneo, A. Levi and F. Tirone. I wish to extend particular gratitude to Lifegroup of Padova, headed by F. della Valle, and to F. della Valle, A. Leon and co-workers for their important discovery of NGF-mast cell interaction, described in Part IV of this volume. I also wish to thank in particular some of the many investigators from different laboratories who work in collaboration with our own and continue to make most valuable contributions in this field: E. Alleva, R. Bradshaw, L. Calzà, C. Cuello, L. Giardino, M. Goedert, I. Hendry, R. Hogue Angeletti, E. Johnson, G. Kollias, T. Lundeberg, W. Mobley, B. Oderfeld-Nowak, L. Olson, U. Otten, E. Shooter, M. G. Spillantini, L. Swanson and H. Thoenen. Finally, I wish to express my gratitude to L. C. Delucchi for her help in reviewing the English and to G. Tripodi for her assistance in the laborious task of selecting the articles and typing the manuscript.

Rita Levi-Montalcini

CONTENTS

PART I

EXPERIMENTAL NEUROEMBRYOLOGY IN THE FIRST HALF OF THIS CENTURY: PERSONAL CONTRIBUTIONS

Ever since the science of experimental neuroembryology emerged from the more general field of experimental embryology, most effort has been directed towards obtaining information on the mechanisms which control growth and differentiation of nerve cells in the vertebrate nervous system and on the factors which interact in the building of the immensely complex wiring network of the Central Nervous System (CNS). It is fortunate that the first researchers in this field did not realize the magnitude of the problem they were facing and the impossibility of solving it with the primitive tools and techniques then available, otherwise they would have invested no further efforts along this line.

The first attempts to obtain some knowledge of the control mechanisms of the nervous system date back to the beginning of this century. They took their start from, and were in fact part of, the more general area of investigation known as experimental embryology, which explores growth and differentiation of organs and tissues of the developing embryo under normal and experimental conditions. The latter consisted in ablation and implantation of end organs, permutation and translocation of different segments of the CNS and exchange of limb buds between embryos of closely related species of different size with microsurgical instruments devised for this purpose. The results of a long series of investigations performed at first in amphibian larvae and then in chick embryos, gave evidence for the all-important role played by peripheral tissues in the differentiation of their innervating nerve centers.

In 1934, Viktor Hamburger (a student of Hans Spemann, the 1935 Nobel Laureate who discovered the "organizer-induction" phenomenon) explored the effect of limb bud extirpation, in 3-day-old chick embryos, on spinal cord motor neurons in charge of the innervation of the limb.[1] Deprived of their fields of innervation, motor nerve cells in the segment of the spinal cord innervating the limb failed to undergo growth and differentiation. This finding suggested to the author that, under normal conditions, the pioneer nerve fibers, after reaching their target limb tissues, send signals to their centers which at this stage consist mainly of still undifferentiated nerve cells. Upon receiving the "signal", they would be induced to undergo differentiation. Their number, under normal conditions, would be dictated by the requirements of the target area to be saturated with the needed innervation. The 1934 article by Hamburger on this phenomenon came to my attention many years later, in 1940. I

[1] Hamburger V., The effects of wing bud extirpation on the development of the central nervous system in chick embryos, *J. Exp. Zool.* **68**, 449–494 (1934).

read it with great interest, but some doubts were raised in my mind regarding the hypothesis submitted by the author. My background, very different from his own, suggested other possible explanations for this effect, which I was eager to put to the test.

As a medical student at the University of Turin, Italy, in the early thirties, I had the fortune to have as professor the foremost authority in neurohistology: Giuseppe Levi. By the second year I had mastered, under his guidance, the specific techniques for selectively staining nerve cells and their fibers. At the same time I became acquainted with the monumental work of Santiago Ramon y Cajal who, between the last decades of the past century and the first decade of this century, uncovered the histological structure of all nerve cells and their interconnections in virtually every part of the nervous system of vertebrates, from the lowest species to man.

His success in this formidable task was due not only to his extraordinarily intuitive mind, but also to the application of the histological techniques discovered by the Italian anatomist Camillo Golgi, which consisted in immerging small fragments of the nervous system in a silver-chromium solution. The darkly-stained nerve cells, in thin sections of the imbued tissue, stood out in all their detail against the golden-yellow background of non-stained surrounding cells. Cajal, in turn, developed a new silver-reduced technique which, likewise, selectively stains nerve cells from the earliest embryonic stages to maturity. His two stupendous volumes entitled "Histologie du systéme nerveux de l'homme et des vertebrés",[2] originally published in Spanish in 1904, are to the present day recognized as the foundation of the field of neurobiology. To unravel the forbidding complexity of the fully developed nervous system, he resorted to the strategy of exploring these structures, from their earliest inception in developing lower and higher vertebrate embryos to all subsequent stages.

Inspired by Cajal's work, I decided to follow his same approach. Even before receiving the M.D. degree in 1936, I undertook a series of studies of the nervous system at the histological, electrophysiological, pharmacological and behavioral levels.[3,4] The chosen object of study was the chick embryo from the third day of incubation to hatching.

This successful research project came to an abrupt end on July 14, 1938, when the promulgation of the fascist racial laws deprived Italian Jewish citizens of all civil rights, including access to public schools and universities. I spent the year 1939 at the Institute Neurologique in Brussels, but in December of the same year the threatened invasion of Belgium by the Nazi armies prompted my return to Italy.

---

[2]Cajal R.S., *Historie du Système Nerveux de l'homme et des Vertébrés*, Paris, Maloine, **1** (1904).

[3]Levi-Montalcini R. and Visintini F., Le prime vie nervose dimostrabili con metodi meurofibrillari nell'embrione di pollo all'epoca dei primi movimenti, *Boll. Soc. Ital. Biol. Sper.* **XIII**, 983–985 (1938).

[4]Visintini F. and Levi-Montalcini R., Relazione tra differenziazione strutturale e funzionale dei centri e delle vie nervose nell'embrione di pollo, *Arch. Suisses Neur. et Psych.* **44**, 119–150 (1939).

Back in Turin, I created a small laboratory in my bedroom in order to resume, at least in part, the investigations pursued in previous years on the differentiation of the nervous system in the chick embryo. Space and equipment limitations prevented the continuation of electrophysiological studies. The only path still open to me was the study, at the histological level, of the developing nervous system of the chick embryo, in normal and in experimentally modified conditions. In retrospect, this was a most advantageous forced restriction. I was, in fact, forced to concentrate all my attention on the exploration of the intrinsic and extrinsic causal factors involved in the development of the vertebrate nervous system. My extensive training in developmental neurology and the experience acquired in previous investigations offered me the possibility to approach the problem of elucidating the role of peripheral tissues in the differentiation of their nerve centers from a neurological rather than from an experimental embryological "a la Spemann" perspective.

I reinvestigated the effect of the surgical extirpation of a limb bud in three-day-old chick embryos. My mentor, Giuseppe Levi, joined me in this research project. A large number of operated specimens survived until the end of the incubation period. Twenty-four hours after the operation, we began to selectively sacrifice the embryos every six hours, to fix them, to stain them using the silver-specific technique, to section them and, finally, to examine them with the optical microscope. The results suggested an entirely different interpretation from the "induction-recruitment hypothesis" formulated by Hamburger. According to our observations, based on the inspection of hundreds of experimental cases, limb ablation resulted in regressive changes followed by massive degeneration of motor and sensory neurons deprived of their peripheral field of innervation.

Unlike Hamburger, who focused his attention on motor neurons in the spinal cord, I explored the sequence of events in sensory nerve cells of dorsal root ganglia (DRG). On the basis of my observations, I advanced the hypothesis that the failure of sensory neurons to establish synaptic contact with their end organs was responsible for the massive death of these cells. Since the racial laws barred the publication of articles by non-Aryan authors, we sent this and other manuscripts to the Pontificia Academia Scientiarum which did not discriminate racially. A brief account of our findings was also published in the Belgian Archives de Biologie (paper 1).

The neuroembryological studies that had been interrupted during the period of occupation of Italy by the Nazi armies (1943–1945) were resumed in 1946 after the defeat of the Germans and restoration of civil rights to non-Aryan Italian citizens. In the Spring of that year I received an invitation by Prof. Hamburger, Chairman of the Department of Zoology of Washington University in St. Louis, Missouri, to reinvestigate together with him the effects of limb bud ablation on its innervating nerve cells. He had read the brief article published in the Belgian journal and the more extensive Italian publication of the Pontificia Academia Scientiarum.[5]

[5]Levi-Montalcini R. and Levi G., Correlazioni nello sviluppo tra varie parti del sistema nervoso, *Commentationes Pontifica Accademia Scientiarum* **VIII**, 527–575 (1944).

Almost half a century later, in 1992, Hamburger gave a vivid and most accurate report of our different interpretations of the effect exerted by end organs on the differentiation of their innervating nerve cells.

"To an experimental embryologist of the Spemann school" he wrote, "the idea that we might be dealing with the death of neurons would have been hardly conceivable. Embryonic induction, on the other hand, was a familiar theme, and I had formulated the recruitment hypothesis after the fashion of assimilative induction which plays a role in the activity of the organizer. But the neurologist of the Levi school was not encumbered by the mindset of the experimental embryologist. Dr. Levi-Montalcini had no difficulty imagining a loss of neurons during development. Her background in neurology had another advantage: she mastered the silver impregnation technique of Cajal-De Castro which enabled her to distinguish between undifferentiated cells and neurons with axons. This played an important role in later investigations".[6]

In the joint paper published two years after my arrival in St. Louis, the Levi-Montalcini and G. Levi hypothesis which had been reported in the previous, 1942–1944 articles, was fully confirmed.[7]

While pursuing this investigation, another most interesting phenomenon came to my attention. Cervical and thoracic ganglia of 5–6 day-old untreated chick embryos exhibited the same massive cell degeneration of sensory neurons which takes place in brachial and lumbo-sacral ganglia upon surgical ablation of the limb bud. In both normal and experimental cases, the degenerative process occurred in the early differentiated neurons located in the Ventro-Lateral (VL) section of the ganglia, described in our earlier papers (not listed), and tentatively identified as exteroceptive neurons. The process of nerve cell degeneration was rigidly time-patterned and coincided with the arrival of sensory nerve fibers at their cutaneous target cells, as evidenced in my previous electrophysiological studies with F. Visintini (not included).

In the 1950 Hamburger and Levi-Montalcini article we stated that, in the experimental situation, the reduction of the peripheral area was definitely responsible for the process of degeneration. It was possible that the same mechanism operated in the case of normal cervical and thoracic ganglia. This would imply that in early stages, cervical and thoracic VL cells send out more neurites than the periphery can support. The excess neurons would break down at the stage in which the VL cells are highly dependent upon some factor released from target tissue.

A much more striking instance of normally occurring cell death in the developing nervous system is described in my 1950 article (paper 2) dealing with early developmental processes of the spinal cord in chick embryos. This study, begun in my private laboratory in Turin and completed after my arrival in St. Louis, describes the

[6]Hamburger V., History of the discovery of neuronal death in embryos, *J. of Neurobiol.* **23**, 1116–1123 (1992).

[7]Hamburger V. and Levi-Montalcini R., Proliferation, differentiation and degeneration in the spinal ganglia of the chick embryo under normal and experimental conditions, *J. Exp. Zool.* **111**, 457–502 (1949).

development of regional differences in the spinal motor column in chicks from the 3rd to the 11th day of incubation. Regional changes take place after 4.5 days. After that amount of time, a spectacular event occurred in the cervical region. Within half a day (between days 4.5 and 5), a massive degeneration removed the greater part of the motor column. The number of degenerating cells was so large as to obscure the presence of intact cells. Before their demise, the cells had been tentatively identified as preganglionic sympathetic neurons. I advanced the hypothesis that we were dealing with an abortive preganglionic sympathetic system which, I suggested, may have been functional in ancestral forms.

This is, to my knowledge, the first well-documented instance of normally occurring neuronal death (paper 2). Another, even more startling instance will be reported in Part II of this volume. However, it was only many decades later that the biological significance of the cellular and molecular events which result in the sudden disappearance of thousands of cells in their early differentiative stages was elucidated. This phenomenon, designated with the term "apoptosis" and also indicated with the expression "programmed cell death", has in recent years acquired a most prominent role among the developmental processes explored at the supracellular, cellular and molecular levels.[8–11]

The neurogenetic events which intervene in the segregation and differentiation of neuronal cell populations in the developing brain and spinal cord are described in paper 2 and in a later, more extensive and detailed review (paper 3). The chick embryo again served as an invaluable object for the inspection, in silver stained slides, of the temporal and topographic sequence of these morphogenetic events.

Another equally impressive and, at that time, unknown developmental event which came to my attention was the migration of nerve cells in an incipient stage of differentiation from the germinal layer lining the brain ventricles and the central canal of the spinal cord to the sites where they undergo their final differentiative phases. Migratory movements also occurred in different segments of the cerebro-spinal axis, suprasegmental brain structures, and in the thoracic segment of the spinal cord (paper 3). Recently the phenomenon of cell migration during early neurogenetic events has become the object of extensive analysis.[12,13]

The scenarios I observed through the eyepieces impressed me not only because of their dramatic and dynamic character, but also because they showed that the nervous system employed a strategy completely different from that which had, until

---

[8]Oppenheim R.W., Cell death during development of the nervous system, *Ann. Rev. Neurosci.* **14**, 453–501 (1991).

[9]Raff M.C., Social controls on cell survival and cell death, *Nature* **356**, 397–399 (1992).

[10]Schwartz L.M., The role of cell death genes during development, *Bioassays* **13**, 389–395 (1991).

[11]Singh N. and Anand S., Cell death by apoptosis, *Indian J. of Exp. Biol.* **32**, 843–847 (1994).

[12]Komuro H. and Rakic P., Modulation of neuronal migration by NMDA receptors, *Science* **260**, 95–97 (1993).

[13]Rakic P., Principles of neural cell migration, *Experientia* **46**, 882–891 (1990).

then, been attributed to it. It had been commonly believed, in fact, that quantitative and qualitative differences among cell populations in the central nervous system were the result of the different proliferative activities of the various segments of the cerebrospinal axis. It now appeared clearly evident that the nervous system instead makes use of a different tactic, providing for the different functions carried out by nerve cells in one of several ways: by drastic elimination within the ranks of excess and unnecessary cells, as was the case in the cervical spinal cord; through the active migration of cell populations in charge of different functions such as somatic and visceral activities; or by separating cell populations with the same function but destined to innervate different peripheral tissues (papers 2 and 3).

Arch. Biol. **LIII**, 537–545 (1942)

Les conséquences de la destruction d'un territoire d'innervation périphérique sur le développement des centres nerveux correspondants dans l'embryon de Poulet

PAR

RITA LEVI-MONTALCINI et GIUSEPPE LEVI (Turin)

Les recherches expérimentales, dont nous résumons ici d'une façon tout à fait sommaire les résultats, ont eu pour but d'analyser les conditions qui déterminent l'arrêt du développement des centres nerveux, après la destruction de l'ébauche du territoire que ces centres sont destinés à innerver. Des recherches assez nombreuses ont déjà été effectuées sur ce sujet ; nous nous bornerons à rappeler celles de DÜRKEN, de DETWILER, d'A. WEBER, de MAY sur les Amphibiens, de SHOREY (1) et d'HAMBURGER (2) sur le Poulet.

Pourtant les résultats de ces importants travaux, tout en démontrant d'une façon indiscutable que l'absence du territoire périphérique entraîne une réduction du nombre des neurones sensitifs correspondants chez les Amphibiens, et des neurones sensitifs et moteurs chez le Poulet, n'ont pas réussi à éclaircir la cause de l'arrêt dans la différenciation des neurones qui auraient été destinés à innerver, chez l'embryon normal, le territoire artificiellement supprimé. En effet SHOREY (1) et HAMBURGER (2), les seuls qui ont travaillé sur l'embryon de Poulet, ont examiné les embryons opérés,

(1) SHOREY, M. L. — *Journ. exp. Zool.*, 7, p. 25-63, 1909.
(2) HAMBURGER. — *Journ. exp. Zool.*, 64, p. 71, 1932.

par les méthodes de coloration histologique ordinaire, lesquelles ne permettent pas de distinguer d'une façon sûre les neurones différenciés des éléments qui n'ont pas de caractères spécifiques. De plus ces auteurs, au lieu de suivre la marche du processus qui détermine l'arrêt dans le développement des centres nerveux chez des embryons à différents stades, se sont bornés à examiner un seul stade.

Nous avons travaillé sur le même matériel qu'HAMBURGER, c'est-à-dire sur l'embryon de Poulet à la fin du 3e jour d'incubation ; dans la plupart des expériences, nous avons extirpé l'ébauche du membre postérieur qui, à ce stade, apparaît sous forme d'une éminence peu prononcée, fusiforme et assez allongée dans le sens cranio-caudal. A ce stade, on reconnaît dans la moelle quelques neurones qui se colorent par l'argent (neurones radiculaires et cellules associatives) ; la racine antérieure est ébauchée. Les ganglions spinaux n'ont pas encore de limites bien nettes et on n'y retrouve pas d'éléments qui se colorent par l'argent.

Le développement des embryons opérés se produit d'une façon tout à fait normale, si on fait abstraction de l'absence du membre postérieur, laquelle était complète dans la plupart des expériences. Les embryons ont été fixés à des temps différents après l'opération : depuis le 4e jour d'incubation jusqu'au 20e. Nous avons utilisé la méthode à l'argent réduit de Cajal (formule De Castro), laquelle donne des résultats excellents dans tous les stades. Un embryon seulement a été fixé par le sublimé picrique et coloré par la méthode de Nissl.

Nous avons constaté un arrêt dans le développement de la partie lombo-sacrée de la moelle du côté opéré et dans les ganglions spinaux correspondants, à partir de 5 jours et demi d'incubation, c'est-à-dire trois jours après l'opération ; cet arrêt, assez faible dans les stades précoces, était d'autant plus prononcé que le développement de l'embryon était plus avancé. Dans les embryons plus âgés, il y avait une réduction assez sensible du ganglion de la chaîne sympathique correspondante.

Les cordons postérieurs étaient très réduits à partir du 6e jour ; leur réduction était progressive, atteignant un degré extraordinaire à partir du 15e jour d'incubation. La réduction des cellules des cornes postérieures n'était bien évidente que dans les embryons avancés. Au contraire, celle des cellules radiculaires motrices était bien manifeste déjà au 6e-7e jour; au 14e-19e jour, presque tous les neurones radiculaires étaient disparus.

Nous nous proposons, pour le moment, d'exposer les transformations qui se produisent dans les ganglions spinaux, dans lesquels il est plus aisé, pour des raisons bien connues à chaque histologiste, d'entreprendre des recherches quantitatives. Tous les ganglions du plexus lombo-sacré étaient plus ou moins affectés (du 23e au 29e ganglion) ; c'est dans les gros ganglions de la 25e, de la 26e et de la 27e racine que nous avons retrouvé la répercussion la plus prononcée.

Nous avons déterminé dans le 25e ganglion chez des embryons à différents stades, le nombre total des cellules qui, par la présence des neuro-fibrilles dans le cytoplasme et par l'existence des neurites, devaient être considérés comme des neurones bien différenciés, et aussi le nombre des cellules qui, par leur forme et par l'aspect du noyau, avaient les caractères des éléments nerveux, mais ne se coloraient pas par l'argent, étant dépourvus de neurofibrilles et de neurites.

Dans nos déterminations quantitatives, aussi bien dans les ganglions du côté indemne que du côté opéré, nous avons groupé les neurones en deux classes distinctes. Dans la première, nous avons réuni les cellules nerveuses jeunes et dépourvues de neurofibrilles et de neurites, dans la deuxième, les neurones bien différenciés et pourvus de neurites. Nous n'avons, au contraire, pas pris en considération les cellules de taille plus petite, qui étaient évidemment destinées à former les amphicytes, les lemnoblastes, ainsi que les éléments du mésenchyme du ganglion.

Il est indispensable de résumer en peu de mots, quelques données sur la marche du processus de différenciation des neurones du ganglion spinal. Ces données ont été développées, il y a longtemps, par l'un de nous ; mais il y a sur ce sujet quelques points qui doivent encore être éclaircis, et nous nous proposons de le faire dans un autre travail. La multiplication des cellules indifférentes s'arrête, selon les observations d'OLIVO, PORTA et BARBERIS, à partir du 8e jour d'incubation ; dès ce moment, le nombre des éléments nerveux du ganglion est fixé chez l'embryon de Poulet. La différenciation des ganglions est caractérisée par une augmentation progressive du nombre des cellules à caractères spécifiques et en même temps, par un accroissement de leur taille. Parallèlement à ces deux processus, le nombre des cellules sans neurofibrilles et sans neurites décroît progressivement, évidemment parce qu'un nombre toujours plus grand de ces éléments se différencie.

Les résultats obtenus par nos numérations des cellules appartenant aux deux classes sont reproduits dans le tableau ci-après.

Pendant les deux premiers jours qui suivent l'opération, on ne remarque aucun changement dans les centres nerveux. Les neurones radiculaires moteurs du côté opéré sont indemnes et aussi les neurones sensitifs. Il faut ajouter que pendant le premier jour, les nerfs n'ont pas encore pénétré dans le membre ; c'est seulement au début du 5e jour d'incubation que les nerfs dépassent la racine du membre postérieur et atteignent rapidement la partie distale. Du côté opéré, on aperçoit, dans les embryons au 6e-7e jour, au bout distal de chaque nerf du plexus lombo-sacré, un peloton formé par des fibres entrelacées d'une façon compliquée ; ce peloton se trouve à l'extrême limite de la paroi abdominale ; on voit aussi quelques faisceaux qui se dirigent vers le mésonéphros.

C'est seulement à partir de la fin du 6e jour d'incubation qu'on commence à remarquer une répercussion de l'intervention dans les ganglions du côté opéré, et plus précisément, ce sont les éléments à caractères spécifiques bien nets qui sont affectés ; le nombre des neurones que nous avons

DÉTERMINATION DU NOMBRE DES CELLULES DU 25e GANGLION DU COTÉ NORMAL ET DU COTÉ OPÉRÉ

	Ganglion du côté normal			*Ganglion du côté opéré*			
Age de l'embryon	Cellules indifférenciées et relativement peu différenciées (1)	Cellules différenciées	Nombre total	Cellules indifférenciées et relativement peu différenciées (1)	Cellules différenciées	Nombre total	Rapport entre le nombre des cell. différ. des deux cotés
6 jours et 6 heures .	2400	1134	3534	2400	719	3119	100 : 63,4
7 jours	4387	2380	6767	4056	1686	5742	100 : 70,8
8 jours	5718	2929	8647	5190	1695	6885	100 : 57,8
9 jours	6480	5568	12048	5338	1432	6770	100 : 25,7
10 jours	4414	4179	8593	3546	929	4475	100 : 22,2
11 jours et 8 heures ..	7361	4282	11643	5838	1136	6974	100 : 26,5
12 jours	5406	2691	8097	7075	1115	8190	100 : 41,4
12 jours .... 26e ganglion (2)	7118	4924	12042	7162	1195	8347	100 : 24,2
14 jours	4521	4812	9333	4404	1002	5406	100 : 20,8
15 jours	2263	6558	8821	2306	1257	3563	100 : 19,1
17 jours	2133	5761	7894	1866	1012	2878	100 : 17,5
19 jours et 15 heures .	1373	9812	11185	5429	1821	7250	100 : 18,5

(1) A ce sujet nous renvoyons à ce que nous avons exposé dans le texte.
(2) Dans ce cas, pour des raisons qu'il serait trop long d'expliquer ici, nous avons pris en considération aussi le 26e ganglion.

réunis dans la deuxième classe, est assez sensiblement réduit, tandis que le nombre des éléments de la première classe, c'est-à-dire de ceux qui sont dépourvus de neurofibrilles et de neurites, est presque égal du côté normal et du côté opéré.

Cette différence s'accentue dans les embryons au 8e jour d'incubation, et davantage encore dans les embryons plus avancés. Il est donc évident que les cellules dépourvues de neurites ne sont affectées d'aucune façon par l'absence du territoire périphérique.

La question est moins simple à partir du 12e-13e jour d'incubation, parce que dans les stades plus avancés, on ne peut plus parler, à vrai dire, de cellules indifférenciés; tous les éléments du ganglion ont une taille assez considérable et un noyau vésiculeux. Pourtant, même à un stade assez proche de l'éclosion, il y a encore un certain nombre de cellules qui ont tous les caractères d'éléments nerveux quoiqu'elles ne contiennent pas de neurofibrilles dans le cytoplasme et qu'elles soient dépourvues de neurites; à notre avis, ces neurones sont arriérés dans le processus de différenciation ([1]).

([1]) L'objection qu'on pourrait élever contre notre interprétation de la nature de ces éléments qui ne se colorent pas par l'argent et selon laquelle il s'agirait d'une insuffisance technique, n'est pas défendable pour plusieurs raisons que nous nous proposons de développer dans un autre travail.

Nous signalerons ici que la coloration de nos embryons était parfaitement réussie, et de plus nous rappellerons un autre argument d'ordre indirect, mais très probant. Il existe un parallélisme très strict, dans le développement, même chez les embryons très avancés (nous parlons évidemment des embryons normaux), entre la diminution du nombre des cellules dépourvues de neurofibrilles et l'augmentation d'épaisseur du cordon postérieur, lequel est formé par les branches de bifurcation ascendantes et descendantes des neurites proximaux des neurones sensitifs.

En d'autres mots, il n'y a aucun doute que, quoique le nombre des neurones sensitifs soit fixé déjà au 8e jour d'incubation à peu près, il y ait un certain nombre de neurones qui restent pendant longtemps arriérés dans le développement, et que même peu avant l'éclosion, le cordon postérieur s'épaississe d'une façon considérable aux dépens des neurites centraux arriérés qui atteignent la moelle très tard.

Mais même à ces stades avancés, on remarque une différence très sensible, tout à fait comme dans les embryons plus jeunes, entre les deux classes d'éléments, c'est-à-dire qu'il y a une énorme réduction du nombre des neurones différenciés, pourvus de neurites, tandis que le nombre de neurones dépourvus de neurofibrilles et sans neurites est à peu près le même des deux côtés.

Il est donc hors de discussion qu'à tous les stades du développement, aux plus jeunes aussi bien qu'aux plus avancés, seuls les neurones dont les neurites atteignent, d'un côté, la peau et les muscles, de l'autre côté, la moelle, sont affectés par l'élimination du territoire périphérique d'innervation.

Maintenant il faudrait analyser de plus près le processus qui affecte les neurones différenciés, à la suite duquel un grand nombre de ceux-ci disparaissent, tandis que les autres s'atrophient d'une manière plus ou moins évidente. Nous nous proposons de le faire dans un exposé plus détaillé et plus documenté de nos résultats ; nous nous bornerons ici à rappeler que des mensurations de la taille des éléments des ganglions, accomplies sur un grand nombre de ceux-ci, ont démontré qu'il se produit une diminution de la taille des neurones survivants ; de même que la réduction du nombre, celle de la taille est progressive et atteint son maximum peu avant l'éclosion. De plus, en quelques cas, nous avons constaté qu'un grand nombre de neurones et parfois presque tous présentent une atrophie très prononcée, caractérisée par la conglutination des neurofibrilles dans une masse homogène et par l'atrophie du noyau.

Ce fait de la réduction de la taille des neurones du côté opéré apporte une confirmation à l'hypothèse avancée, il y a longtemps, par l'un de nous sur la corrélation qui existe entre le volume du pyrénophore du neurone et l'extension du territoire innervé par le neurite correspondant. Nous rappellerons encore qu'A. WEBER (1) est le seul auteur qui ait constaté, comme nous, mais dans un autre matériel

(1) WEBER, A. — *C. R. Assoc. Anat.* 20e Réunion (Turin, 1925), p. 391.

(larves de Bombinator), une réduction de la taille du pyrénophore des neurones à la suite de l'extirpation de l'ébauche de la patte.

Discussion

Plusieurs hypothèses ont été avancées pour expliquer l'arrêt dans le développement des centres nerveux, qui se produit à la suite de l'extirpation du territoire périphérique que ces centres sont destinés à innerver dans les conditions normales. Ce n'est pas ici le lieu de discuter ces hypothèses. Nous nous bornerons à rappeler que l'hypothèse selon laquelle l'accroissement du système moteur et sensitif est réglé par le champ périphérique, dans le sens d'un contrôle exercé sur la prolifération cellulaire des centres (Detwiler) (1), n'est pas confirmée par nos résultats ; en effet, le nombre des cellules indifférenciées chez les embryons de moins de 8 jours (et c'est la seule période où une prolifération cellulaire est possible) n'est pas réduit à la suite de la suppression du territoire périphérique.

Puisque, d'autre part, pendant une première période, le développement des centres n'est nullement affecté, et puisque la réduction du nombre et de la taille des neurones sensitifs différenciés avec neurofibrilles et pourvus de neurites, commence à se produire seulement après que les neurites de ces neurones ont atteint une certaine longueur et sont empêchés, à la suite de l'extirpation de l'ébauche, de parvenir aux organes périphériques auxquels ils sont destinés (récepteurs de la peau et des muscles), il nous semble probable que la disparition d'un nombre considérable de neurones et la réduction de la taille d'autres neurones (et aussi l'atrophie des neurones très différenciés) constatées par nous, puissent être expliquées par l'absence de synapses normales au niveau des neurites périphériques. Mais nous formulons cette

(1) Detwiler. — Neuroembryology. An experimental study. The Mamillan Com., 1936.

hypothèse de travail d'une façon provisoire et nous nous proposons de la mieux documenter et de la discuter dans l'avenir.

Résumé

1. A la suite de l'extirpation de l'ébauche du membre postérieur dans les embryons de Poulet à la fin du 3e jour d'incubation, il se produit un arrêt très prononcé dans le développement des cellules radiculaires motrices et des fibres du cordon postérieur de la moelle lombo-sacrée ainsi que des neurones différenciés des ganglions spinaux du plexus lombo-sacré. Les cellules de la corne postérieure et des ganglions de la chaîne sympathique de cette même région sont aussi affectées, quoique d'une façon moins prononcée.

2. Il résulte de la numération des cellules moins différenciées, effectuée au niveau du ganglion du 25e segment à la fois du côté indemne et du côté opéré, qu'à tous les stades, ce nombre est à peu près le même des deux côtés. L'hypothèse qui a été avancée que la multiplication serait réglée par le champ périphérique, n'a pas été confirmée par nos recherches.

3. Au contraire, le nombre des neurones bien différenciés avec neurofibrilles et pourvus de neurites, est réduit déjà deux jours et demi après l'opération, d'une façon sensible du côté opéré comparativement au ganglion du côté opposé. Cette réduction devient progressivement plus prononcée dans les stades ultérieurs, atteignant son maximum entre le 15e et le 20e jour d'incubation.

4. Dans les pyrénophores des neurones sensitifs différenciés survivants, se produit, du côté opéré, une réduction de la taille à partir du 9e jour d'incubation ; plus tard, cette réduction s'accentue beaucoup. Dans quelques cas, nous avons constaté au 16e-17e jour d'incubation, une véritable atrophie des neurones différenciés.

Reprinted from the JOURNAL OF MORPHOLOGY
Vol. 86, No. 2, March, 1950

THE ORIGIN AND DEVELOPMENT OF THE VISCERAL SYSTEM IN THE SPINAL CORD OF THE CHICK EMBRYO [1]

RITA LEVI-MONTALCINI
Department of Zoology, Washington University

TWENTY-SIX FIGURES

INTRODUCTION

The identification of the preganglionic visceral column in the chick embryo by Terni ('24) attracted the attention of the embryologists to this system. Little new information has been added, however, to the classical investigation of Terni. His description, which has been generally accepted, may be summarized briefly as follows:

In the chick the preganglionic system extends from the first thoracic to the second lumbar segment, that is between the brachial and lumbo-sacral swellings. The cells are arranged in two columns which are located in a medio-dorsal position. They are usually referred to as *Columns of Terni.* Their neurites leave the spinal cord together with the somatic efferent fibers and synapse with cells in the ortho-sympathetic chain ganglia. Since the latter extend to the cervical level, a number of preganglionic fibers grow rostrad to synapse with the cervical ganglia.

The evidence for the visceral nature of these cells is partly direct and partly indirect. Terni traced the neurites in early embryonic stages from their origin to their termination in the adjacent ortho-sympathetic ganglia. Indirect evidence was obtained from the absence of rami communicantes in the cervical level, where a preganglionic column is missing.

[1] Aided by a grant from the Rockefeller Foundation to Washington University.

The study of this system was greatly impeded because of the fact that its affinity for silver impregnation decreases considerably after the 7th day and is practically lost after 12 days. The iron hematoxylin and other histological techniques do not stain the cells differentially.

Terni's findings have been extended to reptiles and mammals. In the turtle a preganglionic column was not described but the presence of rami communicantes in the thoracic level and their absence in other levels was considered as indirect evidence of the existence of a thoracic location of the preganglionic system (Terni and Muratori, '37). In mammals, the visceral motor system is located in the same level, but in a more lateral position than in the chick embryo.

Information concerning the sacral visceral system is more scanty on account of the greater difficulty in staining this system. Terni states that, in chick embryos, this system extends between the 8th and the 14th lumbo-sacral segment, and that it is located in the same paraependymal position as the thoraco-lumbar system. However, the neurites of the cells have never been followed to the peripheral parasympathetic ganglia.

Terni attempted to trace the embryonic origin of the thoraco-lumbar preganglionic cells. He observed them in $4\frac{1}{2}$-day embryos where they are located in close contact with the ependymal layer, in a medio-dorsal position and slightly more ventrally than in later stages. He emphasizes that their embryonic origin is different from that of the somatic motor system, as is their morphology and function. He states, "from its very beginning the visceral column is sharply separated from the somatic motor column. The visceral motor cells undergo their differentiation after the 4th day, that is two days after the differentiation of the somatic efferent cells, and their origin is connected with a new episode in differentiation, independent of the origin of the somatic motor centers" ('24, p. 484). According to this version, the parent cells of the visceral neurons are located in the intermediate segment of the neural epithelium, whereas the somatic motor cells originate from the ventral part of this layer.

Our own investigations give a different picture of the embryonic origin of the preganglionic system. The analysis of the first phases of development between 2 and 4 days showed that it has a common origin with the somatic motor system, and that both start their differentiation at the same time. These results disagree with the generally accepted dualism of the two systems in the spinal cord. On the other hand, they establish a close parallelism between the origin of the spinal and the cephalic visceral motor systems, particularly with the vagus complex. The investigation of the early stages of the differentiation of the motor system was extended to the other regions of the spinal cord. The study of the cervical region gave evidence of an abortive visceral system at this level which had never been suspected before.[2]

I wish to express my appreciation of the generous advice and assistance of Dr. Viktor Hamburger throughout this investigation and for his help in editing the manuscript.

MATERIAL

The material on which the present investigation is based consists of a complete series of chick embryos ranging from 38 hours to hatching. They were impregnated with silver following De Castro's modification of Cajal's method (for details see Levi-Montalcini, '49).

The early phases of development, between 2 and 6 days, which are the most important ones for the present investigation, were analyzed in transverse, frontal and sagittal sections. A total of 30 series was used. In addition several series stained with Heidenhain's hematoxylin (transverse sections)

[2] The Lenhossek system which is found in the cervical level has been considered by many authors as a visceral system. Beccari ('42) has established definitely its somatic character.

were at our disposal. In order to obtain a better understanding of the degenerative process, the technique of vital staining with trypan blue was applied. The technique is as follows: Embryos between 4 and 4½ days of incubation were exposed by sawing a window in the shell above the embryo. A few drops of trypan blue solution, 1:20,000, were injected into the amniotic cavity, using a syringe with a fine needle. The injection was repeated 5 or 6 times, at two-hour intervals. After about 12 hours, the embryos were fixed according to Williams' modification of the technique of Lavdowski (Williams and Frantz, '48). A total of 8 cases were used. These embryos were sectioned at 5–8 μ in transverse orientation. No counter staining was applied in 5 of them. In the other three cases the sections were counterstained with neutral red (see Williams and Frantz, '48).

THE ORIGIN OF THE NUCLEUS OF TERNI

Terni ('24) has traced the origin of the preganglionic columns to the begininng of the 5th day of incubation. At that stage they are located adjacent to the neural epithelium of the basal plate, in a position which is more ventral than their final position (Terni, '24, fig. 14). At the moment when Terni recognized them for the first time they were composed of unipolar neuroblasts, which sent their neurites to the primordia of the sympathetic ganglia. The presence of these rami communicantes was the criterion he used for their identification as preganglionic cells.

The findings of Terni were confirmed by us. Yet, his contention that the earliest stage observed by him represents the beginning of the differentiation of the preganglionic system does not seem to be well substantiated. The columns as he described them exhibit the features of an already differentiated system. Their unipolar cells are in the same stage of differentiation as the somatic motor cells. Their neurites are collected in bundles and have traversed the fairly long distance between the cord and the sympathetic cell clusters.

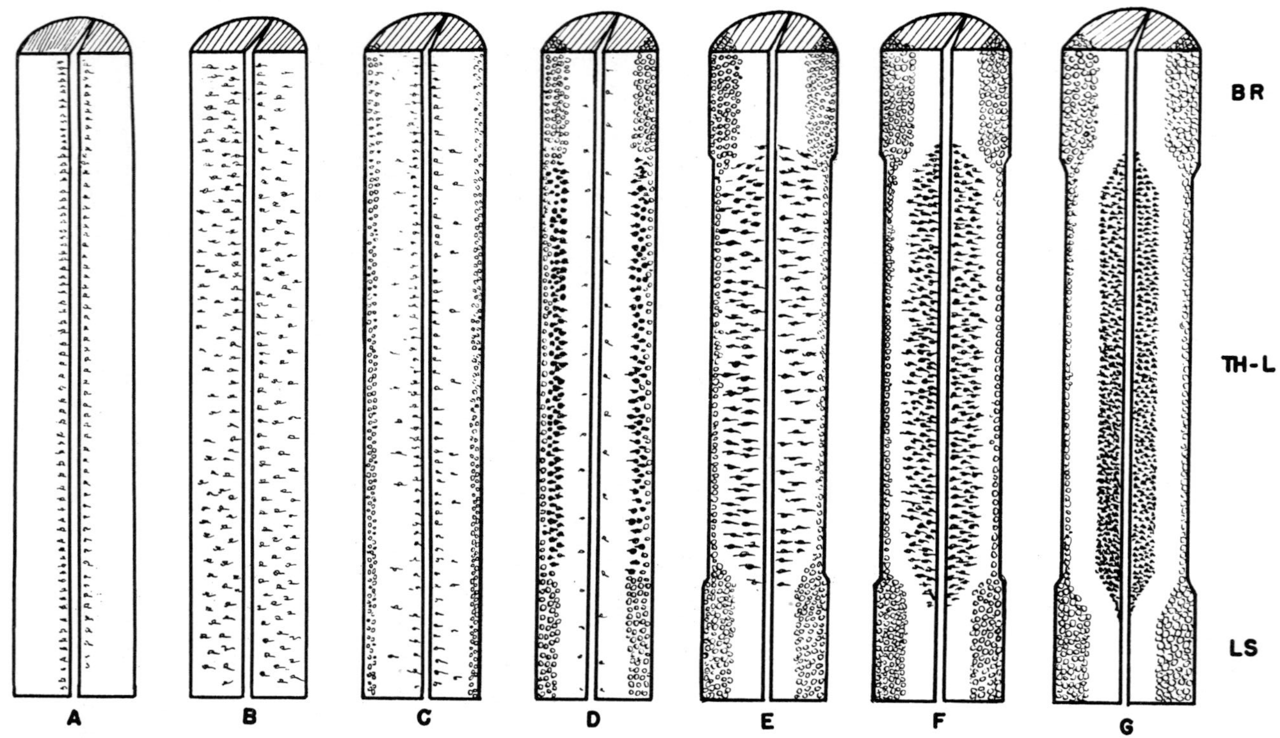

Fig. 1 Frontal sections of the spinal cord between the brachial (BR) and the lumbo-sacral (LS) swellings. TH–L = thoraco-lumbar level; A–D = origin of uniform motor column; E–G = emergence of regional patterns. A = 30–40 hours; B = migration of motor neuroblasts to periphery; C = 72 hours, beginning of formation of motor column; D = 4 days, uniform motor column formed. Open circles represent somatic elements and spindle-shaped cells represent visceral cells. F = 7 days, migration almost completed; G = 8 days, preganglionic columns of Terni formed.

One suspects that the preganglionic cells which are as advanced as the somatic motor cells at 4 days, might have started their differentiation at the same time as the latter, that is on the second day of incubation.

A careful study of early stages shows that between 3 and 4½ days the thoraco-lumbar motor column, which grows rapidly in this period, is uniform and not subdivided into two separate groups (fig. 1,A–D). Since the study of the motor column gave no clue as to the origin of the preganglionic cells, we applied the same criterion as was used by Terni, who identified the cells by their connections with the sympathetic cell clusters. We were able to trace the origin of the rami communicantes to unexpectedly early stages. At the end of the third day, single fibers were observed, emerging from the motor root, and growing toward the sympathetic primordia. On the 4th day these fibers become more and more numerous, and they are collected in thin bundles. Their size increases progressively. It is not possible to identify the cells of origin of these fibers. There can be no doubt, however, that they originate in the motor column. The possibility that we are dealing with aberrant motor fibers, which would disappear later, is ruled out by the fact that a closely timed series revealed a continuity of these fibers from their appearance at three days, to the stage described by Terni. These findings indicate that the thoraco-lumbar motor column of the 3- and 4-day embryos, though morphologically uniform is structurally heterogeneous and composed of visceral and somato-motor cells.

Crucial evidence for this assumption is derived from the study of the further development of the motor column, again using a closely timed series. In 4½-day embryos, we have observed the beginning of the formation of the Nucleus of Terni by a segregation of cells from the uniform column. Between 4½ and 5 days, this group migrates to the position described correctly by Terni ('24, fig. 14). The following

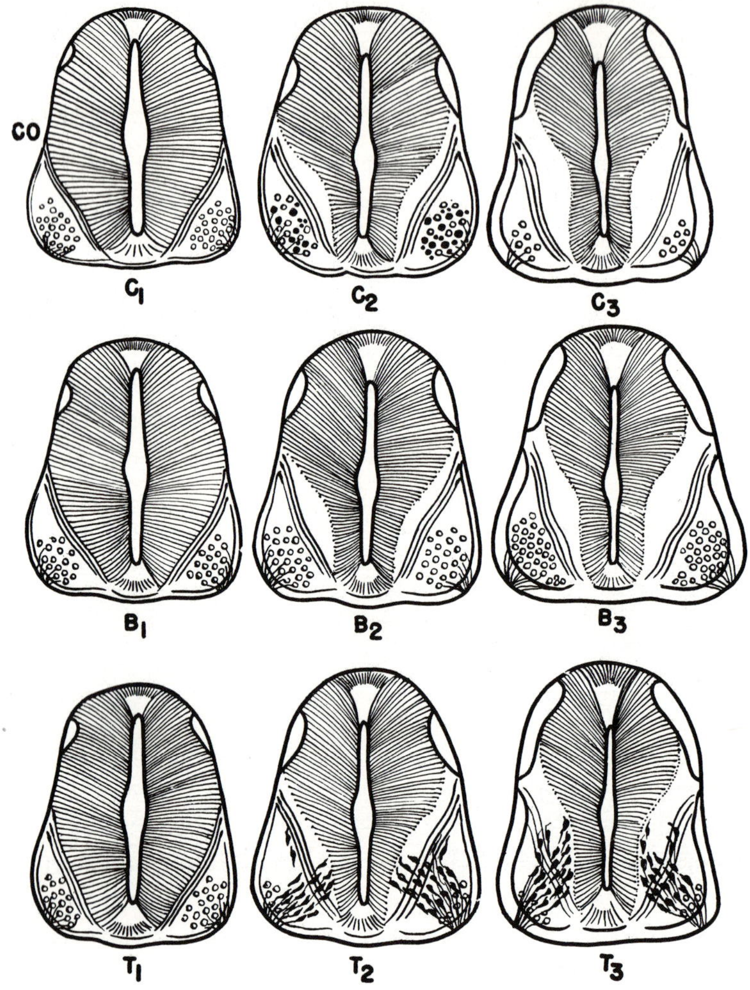

Fig. 2 A comparison of the regional differentiations in the cervical (C_1–C_3), brachial (B_1–B_3) and thoraco-lumbar (T_1–T_3) levels. C_1,B_1,T_1 = 4 days; C_2,B_2,T_2 = $4\frac{1}{2}$ days; C_3,B_3,T_3 = $5\frac{1}{2}$ days. Neural epithelium: hatched. CO = commissural fibers. C_1,B_1,T_1 = uniform motor column; C_2 = degeneration in the motor column (black dots); T_2 = beginning of migration of preganglionic neurons (solid); C_3 = the visceral elements have disappeared; B_3 = swelling of the motor horn; T_3 = advanced stage of migration. Notice the relation of the preganglionic cells to the commissural fibers and the neural epithelium.

description will show that this author observed a transitory stage of a migratory process which continues to the 8th day of incubation.

The migration of the visceral cells is foreshadowed by a loosening of the compact texture of the motor column. Following this, a large number of cells shift to a more median position. The relative position of the visceral motor cells with respect to the commissural fibers makes its certain that we are dealing with a cell migration. A comparison of figures 2,T and 3,T shows that the arcuate fibers (commissural system) which appear at three days near the external limiting membrane (fig. 4,C) and are seen at 4 and 5 days in a median

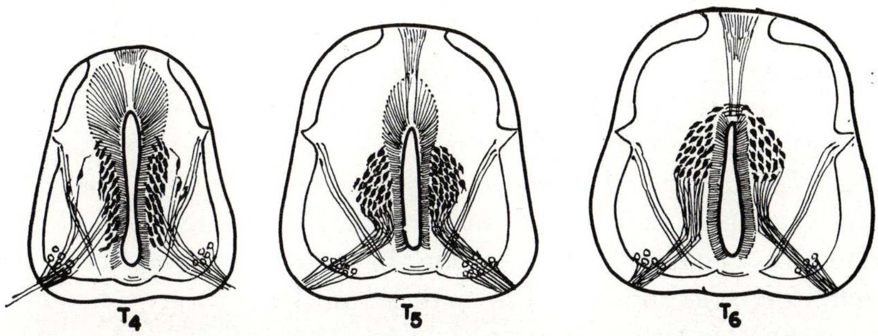

Fig. 3 Further development of the Nucleus of Terni. $T_4 = 6$ days; $T_5 = 7$ days; $T_6 = 8$ days; connection of the two columns by commissural cells.

position (figs. 2,T and 3,T), have in fact retained a static position with respect to the central canal. Measurements show that the distance between the central canal and these fibers remains practically constant between 3 and 7 days. Their shift in position is only apparent. It is due to the rapid formation of the motor column (which, at 4 to 5 days bulges out conspicuously), and to the simultaneous size reduction of the neural epithelium in the basal plate. Therefore, the arcuate fibers can be used as a landmark for the relative position of the preganglionic cells.

At 4 days the entire motor column is situated ventrolaterally to the commissural fibers, which at this stage are deeply impregnated with silver (fig. 5). At 4½ days the most medially located motor neuroblasts are interlaced with the commissural fiber tract (fig. 2,T_2). It is at this stage that the formerly uniform column segregates into two cellular groups: a small lateral group of cells which exhibit a tangential orientation, and a larger group of loosely arranged cells which shift progressively in a median direction. Referring again to the arcuate fibers, we find at 5½ days the great majority of this visceral cellular group medial to these fibers (fig. 2,T_3).

An examination of figures 2,T and 3 shows better than a detailed description that the migration is closely correlated with changes in the configuration of the neural epithelium of the basal plate. Between 3 and 5 days the latter becomes progressively narrower and transforms itself into the ependymal layer. The majority of migrating cells remain in contact with the receding outer border of the neural epithelium, until the latter has become very narrow (5 days, figs. 2,T_3 and 7). From then on, the main direction of migration shifts dorsally. The cells move along the border line between the neural epithelium and the mantle until they have reached their final position at 8 days (figs. 3,T_6 and 11). As early as 5 days, individual cells, or small groups of cells, separate from the main clusters and move in chains in a dorsal direction along the arcuate fibers (see figs. 3,T_4 and 15). At 7 and 8 days, all cells, including these independent travellers, have gathered in the definitive location forming two compact preganglionic columns (of Terni). The final phase of migration (7–8 days) coincides with the main phase of the size reduction of the central canal which is accomplished by a stretching of the roof plate in a dorso-ventral plane (Hamburger, '48, fig. 3). By the time the Nucleus of Terni is definitely established, the dorsal raphe has reached its level. The left and right centers are no longer separated by the canal but merely by fibers of the raphe. They establish connections with each other across the midline by way of associative neurons.

After the 5th day, differences between the visceral and the somatic motor cells with respect to size and to silver affinity become evident. The visceral cells are smaller than the somatic motor cells and until the 7th day they appear more intensely impregnated with silver. After this period their affinity for silver decreases very rapidly, whereas the affinity of the somatic motor cells for silver increases. At hatching the visceral motor cells are 10 times smaller than the somatic motor cells; in the adult chicken they are 20 times smaller (Terni, '24). Terni points out that these growth relations are correlated with the differential increase in their respective peripheral fields: the preganglionic fibers are connected with sympathetic ganglionic cells which enlarge only moderately, whereas the peripheral field of the somato-motor cells, the musculature, enlarges conspicuously.

The small cellular group which does not migrate in a median direction, undergoes further differentiation and shift into a medio-ventral position forming the mesial motor column. There is no doubt concerning the somatic motor character of this cell group. These cells enlarge and line up with the corresponding columns at the more rostral and more caudal levels.

To summarize. The visceral and somatic neurons are at first combined in a single motor column. They have a common origin in the ventral part of the neural epithelium and are indistinguishable up to $4\frac{1}{2}$ days. From then on, the visceral cells separate and migrate to their definitive position (fig. 1).

The evidence for the composite nature of the early motor column was derived from the observation of the rami communicantes in stages preceding the first appearance of a distinct Nucleus of Terni.

COMPARISON OF THE DEVELOPMENTAL PATTERN OF THE MOTOR COLUMN IN DIFFERENT REGIONS OF THE CORD

In the earliest stages of development (2–3 days of incubation) the spinal cord appears largest in its rostral segments, in accordance with a cephalo-caudal growth gradient of the whole embryo. These differences decrease in the following stages, as we mentioned above. We found that in embryos between the end of the third day and 4½ days, the motor column is morphologically uniform from the cervical to the lumbo-sacral region, showing no sign of regional differences (figs. 1,A–D, 19–22).[3]

In the sacral level, the motor column, and in fact the entire spinal cord, is smaller than in the more rostral regions, and the size differences persist throughout development.

The regional patterns emerge from the general pattern after 4½ days of incubation. They are distinct at 5 days and become increasingly conspicuous during the following stages. On the basis of these morphological data, we distinguish between two phases: a first phase (4–4½ days) characterized by a general or preregional pattern, and a second phase characterized by a regional pattern, beginning at 4½ days.

The preregional or general pattern in the motor column (4–4½ days of incubation)

The origin of the motor column in the spinal cord of the chick embryo has been the object of many investigations. It will not be discussed here. It may suffice to state that the migration of neuroblasts from the germinal layer of the basal plate starts on the second day of incubation, beginning in the cervical level. It extends rapidly to the more caudal levels. Bipolar neuroblasts migrate in rows through the radially arranged cells of the neural epithelium and form very early a motor column which results from the aggregation of the neuroblasts in the ventro-lateral edge of the neural tube. As soon as this position is reached, the median processes of the neuroblasts are withdrawn and their neurites perforate

[3] The Lenhossek neurons which send fibers through the dorsal roots and extend between the 1st and 7th cervical segments are located in a position medio-ventrally to the motor column. They can be recognized in the earliest stages as a system separate from the uniform motor column and are not taken into consideration in the present investigation.

the external limiting membrane and grow into the mesenchyme. It has been found that the proliferative activity in the basal plate reaches its peak during the third day of incubation (Hamburger, '48). At the same time and shortly thereafter the number of neuroblasts increases. Therefore, we must assume that the products of the cell divisions proceed immediately to neuroblast formation.

The uniformity of the motor column during the first phase is manifested in the following features:

1. The diameter of the motor column is of equal thickness in all levels. There is no brachial or lumbo-sacral swelling present at this stage (see figs. 1,A–D; 2 and 19–22).

2. All cells show complete morphological identity in hematoxylin and silver preparations at this stage, notwithstanding their functional heterogeneity in some parts of the column (see pp. 262 and 266).

3. Sagittal sections show clearly that the bundles of motor root fibers at the points of emergence from the spinal cord are of the same size from the cervical to the lumbo-sacral levels. Measurements of the diameters of the bundles at the base on one particularly well oriented section confirmed this observation. At the stages under consideration, (fig. 22) the brachial and lumbo-sacral motor roots are not thicker than the other roots.

4. The uniform size of the motor column conforms to the equal distribution of mitotic figures in the basal plate throughout the length of the spinal cord, as shown by Hamburger ('48).

The uniformity of the motor column in the stages under consideration is of a purely morphological and quantitative character. Evidence was given above that at the thoracolumbar level, this morphological uniformity conceals a duality of cell types, namely visceral and somatic cells (fig. 1,D). The question arises whether the same condition prevails in other levels. In the absence of other criteria the only approach to this problem lies in the investigation of the peripheral distribution of the motor fibers.

We have found, unexpectedly, that the ventral roots of the cervical segments also send side branches to the primary sympathetic cell clusters which are present in this level, as early as 70 hours of incubation. At three days (72–79 hours) these connections are represented by only a few scattered fibers which leave the motor roots and are directed toward the sympathetic cell clusters. One might be inclined to consider these fibers as aberrant somatic efferent fibers which have no special significance. The configuration at 4 days shows, however, that we are dealing with a regular and consistent feature. At that stage the connections are no longer individual isolated fibers but bundles which spread inside of the sympathetic clusters (fig. 25). They can be observed in practically every segment of the cervical level. These fibers have been observed before by Ganfini ('16), Tello ('25) and Terni ('31). The conflicting interpretation of their significance by these authors will be discussed on page 273.

The similarity of these fiber bundles as observed between 3 and 4½ days with the rami communicantes of the thoraco-lumbar level in the same stages is so striking as to suggest that the cervical connections are likewise rami communicantes of a visceral nature. This implies that the motor column in the cervical level like that of the thoraco-lumbar level is structurally of a composite nature, containing somatic and visceral neurons. This interpretation seems to contradict the well known fact that no preganglionic column exists at this level. The seeming paradox finds its solution in the observation that between 4½ and 5 days of incubation part of the motor column and the rami communicantes disappear (see p. 268).

Corresponding rami communicantes were not found in the brachial and lumbo-sacral levels. All motor root fibers retain the form of compact bundles over a long distance and spread in the limb buds. At 4½ days they can be traced already into the deeper parts of the limb buds.

To summarize. The motor column of the spinal cord from the cervical to the lumbo-sacral level is uniform between 4 and 4½ days of incubation. This uniformity refers to size, morphological features and mitotic pattern. In the cervical and thoraco-lumbar levels, however, the column is of a composite structure, comprising somatic motor and visceral motor neurons, whereas it seems to be structurally uniform (somatic motor) at the brachial and lumbo-sacral levels.

Regional patterns

From 4½ days on, the morphologically uniform motor column undergoes regional differentiation. The changes occurring in the thoraco-lumbar level have been described on pages 260–262. Briefly, they result in a segregation of visceral and somatic cell groups, the former migrating first mediad and then dorsad to form the Nuclei of Terni. The somatic elements shift ventro-mediad to form the mesial motor column. In the following, the cervical, brachial and lumbo-sacral levels will be considered.

1. *An abortive visceral system in the cervical level.* The condition of the cervical level was described above for stages between 2 and 4 days. Special attention was called to two features: The uniformity of the motor column, and the regular occurrence of rami communicantes in every segment of the cervical level (fig. 25). In both respects, the cervical level resembles the thoraco-lumbar level. However, the two regions are sharply different in later stages, in so far as the latter, but not the former, develops a preganglionic nucleus.

If the cervical segment of the 4-day embryo contains visceral elements, as seems to be indicated by the presence of rami communicantes, then these elements should be expected to disappear. A close inspection of the critical stages has shown that the motor column undergoes, indeed, a large-scale disintegration of cells which is strictly limited in time and space (figs. 2,C_2 and 13). It concerns the cervical level only, and it lasts only about half a day (between 4½ and 5 days). During this period the number of degenerating cells is so large as to obscure the presence of intact cells.

The process of disintegration is best studied in preparations stained with Heidenhain's hematoxylin. Heavily stained spherical structures of the size of a nucleus of a neuron are scattered throughout the motor column (fig. 13). Occasionally, smaller granules were found collected in clusters. They suggest an advanced stage of cellular disintegration. The majority of these elements is found inside of the motor column; however, they also occur regularly in the white matter adjacent to the motor column.

Besides the spherical bodies we found elements which were likewise deeply stained but exhibited at the same time structural characteristics of a cell, including the presence of a thin membrane. These latter corpuscles we interpreted as macrophages. Since the ordinary techniques do not provide criteria for the identification of macrophages, the widely used technique of vital staining with trypan blue was applied (see p. 256). The vital staining experiments gave positive evidence of the presence of macrophages among the disintegrating cells. However, it should be remembered that the vital-staining technique is not specific for macrophages. For instance, secretory cells and damaged cells take up the dye. Williams and Frantz ('48) have worked out staining techniques by which these different types of cells can be distinguished from each other. The application of these techniques gave further evidence that macrophages and degenerating cells are mingled in the lateral motor column. The presence of numerous macrophages accounts for the occurrence of deeply staining bodies in the white matter adjacent to the motor column, and for the rapid disappearance of the disintegrating cells.

At 5 days, neither degenerating cells nor macrophages are present. The cervical motor column is now represented by slender strands of cells (fig. 2,C_3) which are located in a ventro-lateral position. During the following days, they shift to their final medio-ventral position, and form the mesial column.

The same phases of development were investigated in parallel series impregnated with silver. Again, large numbers of degenerating cells were recognized in the motor column. They are characterized by their deep black color. The silver preparations revealed another point of great importance for the present considerations, that is, that the rami communicantes of the cervical level disappear at the same stages in which the cellular breakdown occurs. It seems reasonable to assume that this coincidence is not accidental, and that the fibers which disappear are the neurites of the degenerating cells. In other words, we believe that we are dealing with an abortive and short-lived cervical visceral system, homologous to the precursor of the Terni system in the thoraco-lumbar level.

It is interesting to note that simultaneously with the disappearance of the rami communicantes the sympathetic cell clusters with which they are connected move away from their original position. A great number of them migrate rostrad to form the superior cervical ganglion (Tello, '25), and others migrate dorsad to form the definitive ortho-sympathetic cervical ganglia. At the same time, the ascending preganglionic fibers of the Nucleus of Terni have reached the cervical level where their growth cones were observed.

2. *Brachial and lumbo-sacral levels.* At the stages in which the migration of the preganglionic neurons begins in the thoraco-lumbar level and the degenerative process occurs in the cervical segments, the motor column in the brachial (13 to 16 segment) and lumbo-sacral (23 to 29 segment) levels shows no sign of either degeneration or migration of cells.

No traces of rami communicantes were found, with the exception of segment 16, which demarcates the rostral end of the Nucleus of Terni. Beginning with the 5th day, the motor column enlarges considerably to form the brachial and lumbo-sacral swellings (fig. 2,B_3). In both levels, the originally uniform and compact motor column segregates into a lateral and a mesial motor column. The mesial motor columns of the brachial and lumbo-sacral levels come to lie in the same position as those of the cervical and thoraco-lumbar regions. The mesial column is the only element which retains the original uniform character of the motor system throughout the length of the cord.

ORIGIN OF THE PREGANGLIONIC SYSTEM IN THE SACRAL AND MEDULLARY LEVELS

Sacral column

A preganglionic sacral column has been described by Terni ('24) in the 7-day chick embryo. It is located in segments 30 to 36 which are the segments caudal to the lumbo-sacral swelling. The cells are located in a median position corresponding to that of the Nucleus of Terni. The author was not able to follow the neurites of these cells to the enteric ganglia, which are located at a considerable distance from the spinal cord, even at these early stages.

Our own observations are confined to the first phases of the development of this center. We have found that the origin of the sacral preganglionic neurons is identical with that of the thoraco-lumbar preganglionic neurons. Up to 5 days, the motor column at the sacral level is uniform morphologically and structurally; it differs from the more rostral levels only in that it is more slender than the latter. On the 5th day, the cells lose their compact arrangement. Part of the cells migrate in separate groups and chains in a medial and dorsal direction and, at 7 days, reach their final position adjacent to the ependyma near the middle of the central canal (figs. 16, 17 and 18). From this stage on, the cells suddenly lose their affinity for silver, and after 8 days the center is barely recognizable even in good silver preparations.

The vagus complex

It is known that the vagus complex in mammals is composed of a somatic center (nucleus ambiguus) and a visceral center (dorsal motor nucleus). The same situation is found in the chick. The high development of the nucleus intermedius in birds which is interposed between the two, however, and which seems to be concerned with the innervation of the syrinx (Kappers, Huber and Crosby, '36) is responsible for a frequent confusion of the latter with the nucleus ambiguus.

Both components of the vagus complex have a common origin from the neural epithelium of the ventral part of the basal plate. At three days, a slender column has formed which is located in a medio-ventral position. The neurites of this column turn dorsad, grow along the lateral wall of the medulla and leave it near its dorsal edge. On the 4th day, the texture of the ventral part of the medulla changes profoundly. As a result, the vagus complex appears now to be located in a relatively dorsal position. At the same time a segregation takes place between the somatic and the visceral components (fig. 23). One group of cells shifts dorsad, for a short distance, along the outer edge of the neural epithelium (fig. 24). This is the primordium of the *preganglionic column* (nucleus motoris dorsalis). The remaining cell group gives origin to the *somatic motor component* which corresponds to the nucleus ambiguus and to the nucleus intermedius of mammals. Its migration to a ventro-lateral position has been described by Kappers, Huber and Crosby ('36). We have likewise observed a migration of separate strands of cells between 4 and 7 days. Yet, some doubts may arise if we are actually dealing with the precursor of the nucleus ambiguus, or with components of the nucleus intermedius, because we have observed in later stages a large somatic motor center, pertaining to the vagus complex in the same medio-dorsal position where it started its differentiation. In other words, only part of the somatic motor component seems to migrate. A reinvestigation of the development and topography of the somatic vagus centers in the chick is needed.

It is generally contended that the *preganglionic component* of the vagus complex retains its primitive position, and does not migrate. However, the investigation of closely timed series covering the period between 3 and 7 days gave evidence of the dorsal migration of the preganglionic cells of this center.

Its further development has been studied in transverse and frontal sections. Eventually, it occupies the same relative position as the Nucleus of Terni (fig. 26), and its cells show the same small size and high affinity for silver as the spinal preganglionic cells. After 8 days, their affinity for silver decreases rapidly, as was described for the spinal visceral centers. At the same time, the somatic motor neurons increase conspicuously in size.

DISCUSSION

The main results of our investigation are the following:

1. The Nucleus of Terni does not originate in situ but it has a common origin with the somatic motor column. The cells of the Nucleus of Terni reach their final position by migration between 4½ and 8 days of incubation (fig. 1).

2. At stages preceeding the migration, the motor column is uniform in size throughout the length of the spinal cord from the cervical to the lumbo-sacral level (figs. 1,A–D and 2,C_1,B_1,T_1).

3. The fate of this column is different in different levels. In the cervical level, a considerable number of cells, presumably potential visceral cells, break down. In the thoraco-lumbar and sacral levels, the column is segregated into a somatic and a visceral component, the latter moving mediad. In the brachial and lumbo-sacral levels, neither breakdown nor migration occur, and all cells present at 5 days serve the somatic motor innervation of the limbs (fig. 2,C_2,C_3 and B_2,B_3 and T_2, T_3).

These three points require further discussion.

1. *Origin of the Nucleus of Terni.* Our version of the origin of the column of Terni is at variance with the conclusions of its discoverer. Terni ('24) states that this column originates in a position medio-dorsal to the somatic motor column, two days after the latter has become recognizable, and he

emphasizes that both systems are distinguishable from the start, with respect to their structure and function as well as embryonic origin.

We have pointed out above that he failed to recognize the earliest stages, and that the stage which he described as the beginning of the differentiation of the column represents an advanced stage of differentiation of this system. Our observation of rami communicantes in stages preceding the existence of a separate preganglionic column, gave evidence that the visceral cells are already present at three days. Our contention that the column of Terni has a common origin with the somatic motor column is based on the followings: (a) the observation, in closely timed stage series, of the segregation of the originally uniform column into two groups; (b) the migration of the main group to its final position, and (c) the progressive depletion of the lateral motor group.

2. *An abortive visceral system in the cervical level.* The concept has been developed that the cervical motor column of the $4\frac{1}{2}$-day embryo contains an abortive, potential visceral system which is homologous to the permanent visceral system of the thoraco-lumbar system. The cervical visceral system never reaches the stage of segregation from the somatic system. Its existence would have been overlooked by us, as it was by previous investigators, if we had not analyzed the earliest stages of the entire motor column and realized the striking similarity between the cervical and thoraco-lumbar levels with respect to the rami communicantes.

Our contention that we are dealing with an abortive visceral system is based on the following points. (a) The regular occurrence of fiber bundles connecting the ventral roots with the sympathetic primordia, in 3- to 4-day embryos. (b) The coincidence of two events, at $4\frac{1}{2}$ days: the sudden breakdown of a great number of cells in the cervical motor column, and the disappearance of the fiber bundles which we consider as rami communicantes. (c) Additional support for our assumption is derived from the comparison of the cervical level and the thoraco-lumbar level at these initial stages. In both regions, there exist connecting fiber bundles between motor columns and sympathetic primordia. In the thoraco-lumbar level, these bundles have been identified as the primordia of the persistent rami communicantes, and the existence of visceral motor neurons within the uniform motor column is well established (see p. 260). It is therefore reasonable to interpret the identical connecting bundles in the cervical level as rami communicantes and not as aberrant somatic fibers.

The regular occurrence of these fiber bundles in all cervical segments was observed by Ganfini ('16). He described correctly their disappearance on the 5th day. According to this author, the fibers have the function of serving as tracks for the sympathetic neurocytes which he believed to originate from the motor column. They were supposed to disappear after their task had been fulfilled. Apart from the controversial issue of the origin of the sympathetic chain ganglia, all data presented above speak against this interpretation. Tello ('25) and Terni ('31) also saw fibers connecting the ventral roots with the primordia of the sympathetic ganglia in the cervical region of 3- to 4-day embryos. However, they considered them as aberrant fibers of no particular significance, in accordance with the well established fact that no visceral center exists at this level. Our evidence against this viewpoint was presented above.

The degeneration of a number of cells in the spinal cord of 3- to 4-day embryos has been described before. Collin ('06) gave a detailed description of cell degeneration in the nervous system. He interpreted this phenomenon as a regulative mechanism by which an overproduction of neurons is corrected. Ernst ('26) considers cell degeneration as a general phenomenon occurring in all embryonic organs. As far as the spinal cord is concerned he describes cell degeneration only for the thoracic level. This contention is contradicted by our observations. Both authors failed to recognize the important fact that the cell degeneration in the motor column is limited spatially to the cervical region and is an integral part of the pattern of differentiation of the motor system.

The significance of the breakdown of differentiated neurons as part of a developmental pattern in the nervous system has been stressed in connection with the development of the spinal ganglia (Hamburger and Levi-Montalcini, '49). A large-scale degeneration of differentiated sensory neurons was found in the cervical and thoracic ganglia, during a limited period of development. No such degeneration occurs in the brachial and lumbo-sacral ganglia. The regional size differences of ganglia were therefore attributed, in part, to a patterned degeneration process, which is comparable to the present findings.

We are planning to investigate the condition of the cervical motor system in lower vertebrates. The occurrence of a more permanent preganglionic system in this region in ancestral forms would give strong support to our hypothesis.

3. *The emergence of regional differences from a morphologically uniform system.* We have described the motor column in its earliest stages as a system which is morphologically uniform and of equal width throughout the length of the spinal cord from the cervical to the lumbo-sacral level. We have listed on page 264 all features which support this statement.

In this connection, the findings of Hamburger ('48) concerning the mitotic patterns in the spinal cord, are of particular interest. He described a rather uniform mitotic activity in the basal plate, in all regions which he investigated, that is the cervical, brachial, and thoraco-lumbar levels. The peak of the mitotic activity in the basal plate was found in $2\frac{1}{2}$- to 4-day embryos. These stages precede and overlap in part the beginning of the formation of the motor column. The two sets of data are in full agreement with each other: the lack of a regional pattern in the mitotic activity is reflected in the uniformity of the motor column which persists until $4\frac{1}{2}$ days. From this uniform column the regional differences emerge in the following way: the cervical system is secondarily reduced in size by cell degeneration. The thoraco-lumbar region segregates into two components; a larger group migrates medially and dorsally to form the Nucleus of Terni, and the residual group forms the mesial motor column. The brachial and lumbo-sacral levels continue to differentiate without a loss or splitting off of cell material. In this way the apparent discrepancy between mitotic and histogenetic processes is resolved, which Hamburger ('48) stated as follows, "These findings (i.e., the uniform mitotic activity in all regions) are not in accordance with the expectation that the brachial swelling of the cord would be characterized by a peak of mitotic activity." (p. 281).

We do not contend that the morphological uniformity of the motor column during its earliest phases of differentiation reflects a structural uniformity. On the contrary, the presence of rami communicantes in the cervical and thoraco-lumbar levels of embryos of 3 and 4 days proves that in these regions the column is composed of visceral and somatic neuroblasts. The problem of the determination of regional and structural differences is not within the scope of the present investigation. The intracoelomic and flank grafts of parts of the spinal cord (Bueker, '43, '45) and the infraspinal transplantations of Wenger (unpublished) give evidence that regional differences are determined in very early stages.

A preliminary experiment of our own, in which the thoraco-lumbar segment of the spinal cord of a 25 somite embryo was transplanted between the brachial cord and the wing bud of an embryo of the same stage, gave evidence that the segregation and migration of the visceral component follows the normal pattern under these conditions of isolation. The nucleus of Terni was identified by silver impregnation.

Our observation of the migration of the visceral component in the sacral and vagus levels give evidence that all preganglionic components originate in the same way: by segregation from a common viscero-somatic motor column. The common origin of the visceral and somatic components of the vagus system has been described before by Tello ('22).

SUMMARY

1. Up to $4\frac{1}{2}$ days, a morphologically uniform ventro-lateral motor column extends throughout the length of the spinal cord. It is of uniform width in all levels except the sacral level. In it are combined all preganglionic visceral and somatic motor neuroblasts. Regional patterns emerge from this column, beginning at $4\frac{1}{2}$ days.

2. The thoraco-lumbar preganglionic system segregates from this column by cell migration in a medial and dorsal direction. The cells reach their definitive location at 8 days and form the Nucleus of Terni. The residual motor cells form the mesial motor column.

3. An abortive visceral system was discovered in the cervical level. It was identified by rami communicantes which are present on the 3rd and 4th days. Its neurons are located in the motor column, together with the somatic neurons. This system disappears on the 5th day. The degeneration of its neurons was observed. The remaining cells of the motor column form the mesial somatic column.

4. The preganglionic sacral center originates in the same way as the thoraco-lumbar center, that is by migration in a medio-dorsal direction.

5. The common origin of the visceral and somatic vagus complex was confirmed. The migration of the visceral component in a medio-dorsal direction was observed.

6. In the brachial and lumbo-sacral levels, the ventro-lateral motor column differentiates as a purely somatic motor system. It segregates into a large lateral and a smaller mesial column.

LITERATURE CITED

BECCARI, NELLO 1942 Revisione del problema del nervo accessorio. Arch. Ital. Anat. Embr., *47:* 7–71.

BUEKER, ELMER D. 1943 Intracentral and peripheral factors in the differentiation of motor neurons in transplanted lumbo-sacral spinal cords of chick embryos. J. Exp. Zool., *93:* 99–129.

——— 1945 The influence of a growing limb on the differentiation of somatic motor neurons in transplanted avian spinal cord segments. J. Comp. Neur., *82:* 335–361.

COLLIN, REMY 1906 Recherches cytologiques sur le developpement de la cellule nerveuse. Le Nevraxe., *8:* 185–309.

ERNST, M. 1926 Ueber Untergang von Zellen während der normalen Entwicklung bei Wirbeltieren. Zeitschr. Anat. u. Ent., *79:* 228–262.

GANFINI, C. 1916 Lo sviluppo del sistema nervoso simpatico negli uccelli. Arch. Ital. Anat. Embr., *15:* 91–138.

HAMBURGER, VIKTOR 1948 The mitotic patterns in the spinal cord of the chick embryo and their relation to histogenetic processes. J. Comp. Neur., *88:* 221–284.

HAMBURGER, VIKTOR, AND R. LEVI-MONTALCINI 1949 Proliferation, differentiation and degeneration in the spinal ganglia of the chick embryo under normal and experimental conditions. J. Exp. Zool., *111:* 457–502.

KAPPERS, A. U. C., G. C. HUBER AND E. C. CROSBY 1936 The comparative anatomy of the nervous system of vertebrates, including man. *1.* The Macmillan Company, New York.

LEVI-MONTALCINI, RITA 1949 The development of the acoustico-vestibular centers in the chick embryo in the absence of the afferent root fibers and of descending fiber tracts. J. Comp. Neur., *91:* 209–242.

TELLO, FRANCISCO 1922 Las diferenciaciones neuronales en el embryon de pollo, durante los cuatro primeros dias de la incubacion. Libro en honor de D. Santiago Ramon y Cajal. Madrid, *1:* 205–258.

——— 1925 Sur la formation des chaines primaires et secondaires du grand sympathique dans l'embryon de poulet. Trav. du lab. de Rech. biol. de l'Univ. de Madrid, *23:* 1–28.

TERNI, TULLIO 1924 Richerche anatomiche sul sistema nervoso autonomo degli uccelli. Arch. Ital. Anat. Embr., *20:* 433–510.

——— 1931 Il simpatico cervicale degli amnioti. "Richerche di morfologia comparata." Zeit. Anat. u. Entwickl., *96:* 289–426.

TERNI, TULLIO, AND G. MURATORI 1937 Sui rami communicanti embrionari del sistema autonomo toracolombare dei cheloni. Mon. Zool. Ital., *48:* 47–56.

WILLIAMS, W. L., AND MARTHELLA FRANTZ 1948 Histological techniques in the study of vitally stained normal and damaged cells. Anat. Rec., *100:* 547–557.

PLATE 1

EXPLANATION OF FIGURES

All sections, except figure 13, from material impregnated with silver following De Castro's modification of Cajal's technique, and sectioned 10 μ and 14 μ. The microphotos were not retouched.

4 to 11 Formation and migration of the visceral preganglionic columns of Terni between 3 and 8 days.

4 Three-day embryo. The commissural fibers (C) run along the external limiting membrane. The motor column (M) is recognizable. × 150.

5 Four-day embryo. The motor column bulges laterally to the commissural fibers (C). × 100.

6 Four and one-half day embryo. The visceral preganglionic cells start to migrate in a medio-dorsal direction. × 74.

7 Five-day embryo. Segregation of visceral cells from somato-motor cells. × 74.

8 Five and one-half day embryo. Migration in cellular strands along the outer edge of the neural epithelium. × 74.

9 Six-day embryo. The visceral cells forming two cellular columns migrate along the lateral border of the neural epithelium. × 74.

10 Seven-day embryo. Visceral preganglionic columns condensed. × 74.

11 Eight-day embryo. Visceral columns in their definitive position, connected by associative fibers (A). × 74.

ORIGIN OF VISCERAL SPINAL SYSTEM
RITA LEVI-MONTALCINI

PLATE 1

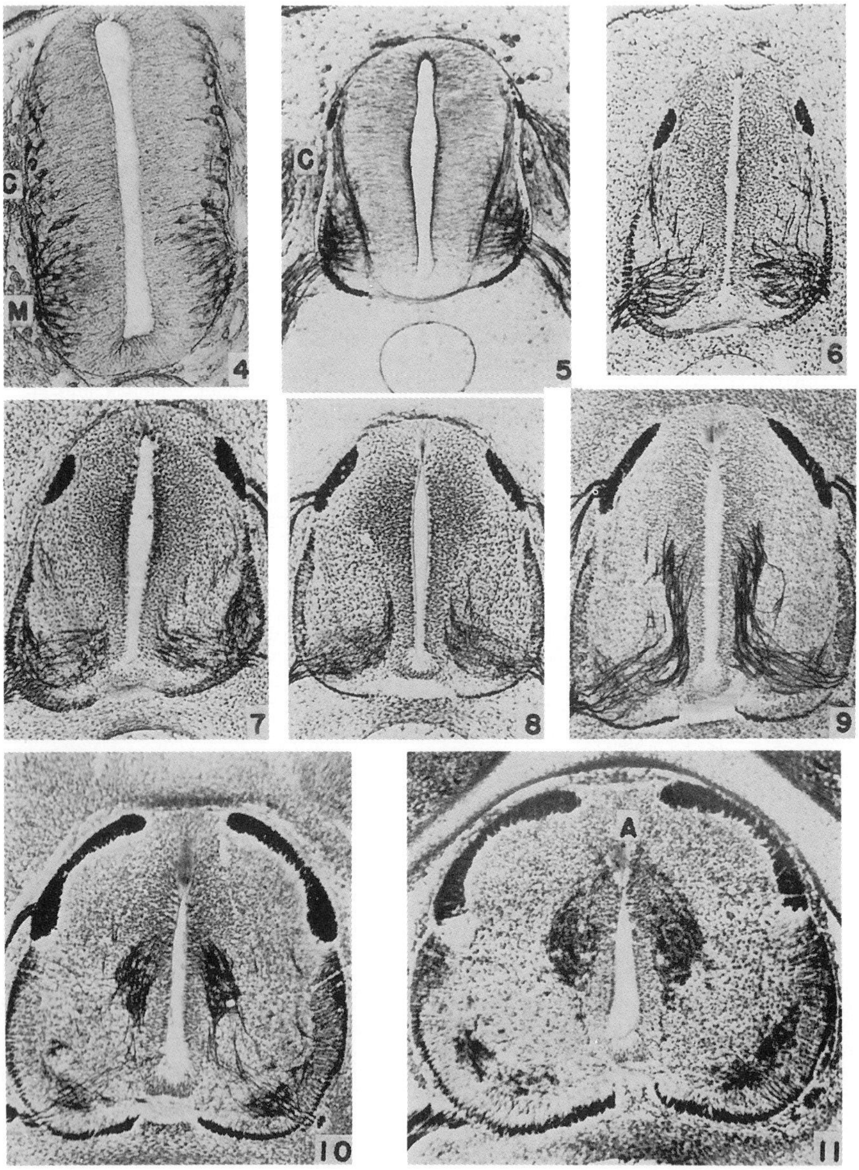

PLATE 2

EXPLANATION OF FIGURES

12 Detail of stage corresponding to figure 6. Some visceral cells in contact with the neural epithelium, medially to commissural fibers (C). × 560.

13 Four and one-half day embryo. Cervical level. Heidenhain's Hematoxylin. Degenerating cells (black spherical bodies) scattered throughout the motor column. × 620.

14–15 Two successive stages of migration of the visceral cells in the thoracic level at 6 and 6½ days. Main group in close contact with the outer edge of the neural epithelium. A few cells (V) migrate along the commissural fibers. 14: × 316, 15: × 188.

16, 17, 18 Migration of preganglionic visceral cells at the sacral level. Sixteen = 5-day embryo; 17 = 6-day embryo; 18 = 7-day embryo. × 76.

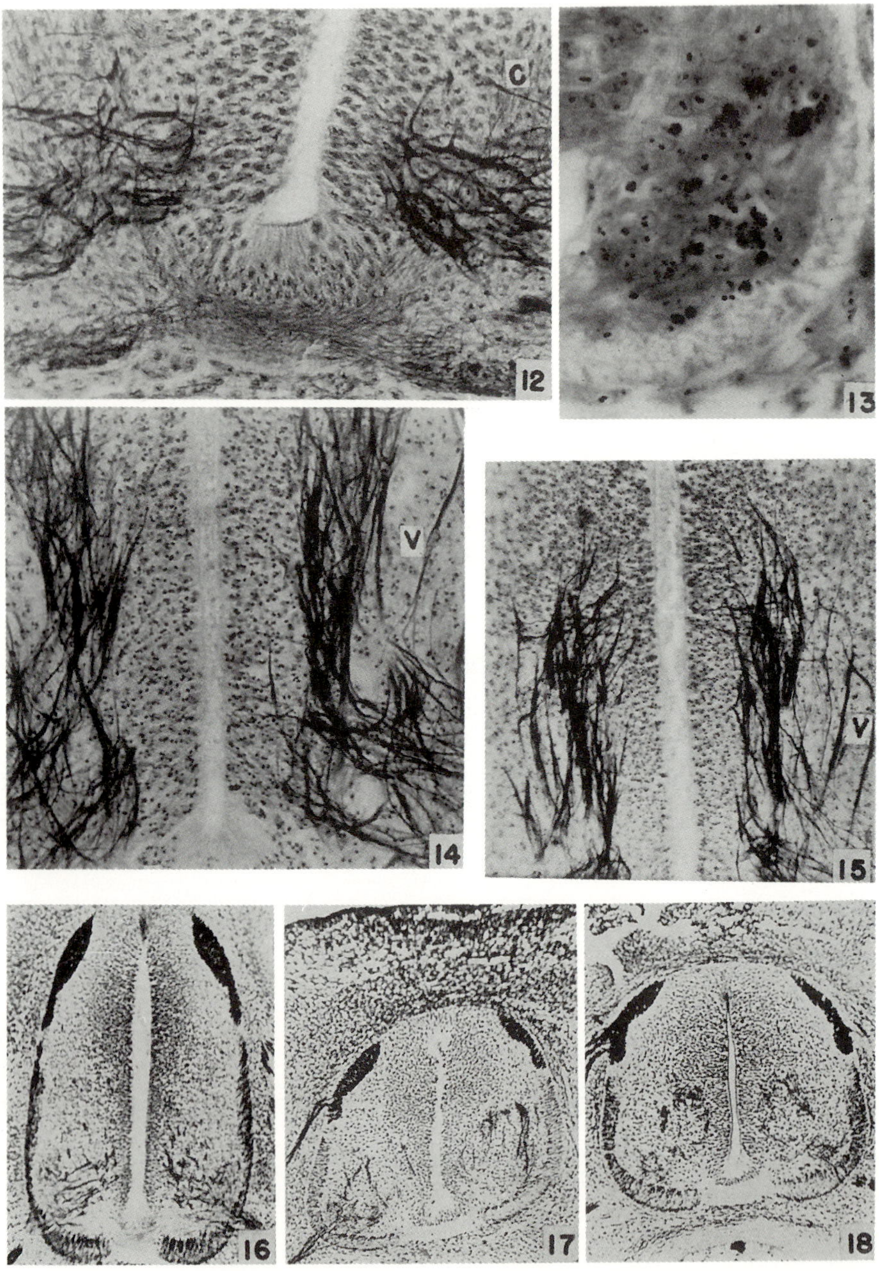
C
12
13
V
14
V
15
16
17
18

PLATE 3

EXPLANATION OF FIGURES

19, 20, 21 Transverse sections of the spinal cord of the same 4-day embryo. 19 = at cervical level; 20 = at brachial level; 21 = at thoraco-lumbar level. × 75.

22 Sagittal section of a 4-day embryo. Note the uniform size of the lateral motor column throughout and the uniform size of the motor roots at their emergence from the spinal cord. W = wing; L = leg. × 30.

23 Transverse section of the medulla of a 4½-day embryo. The dorsal motor nucleus (DM), visceral component, begins to segregate from the somatic complex (A). × 42.

24 Detail of figure 23, showing the dorso-medial migration of the dorsal motor nucleus (DM). A = somatic complex. × 180.

25 Four-day embryo. Cervical level. RC = ramus communicans. × 340.

26 Seven-day embryo. Frontal section through the medulla. DM = dorsal motor nucleus. × 40.

ORIGIN OF VISCERAL SPINAL SYSTEM
RITA LEVI-MONTALCINI

PLATE 3

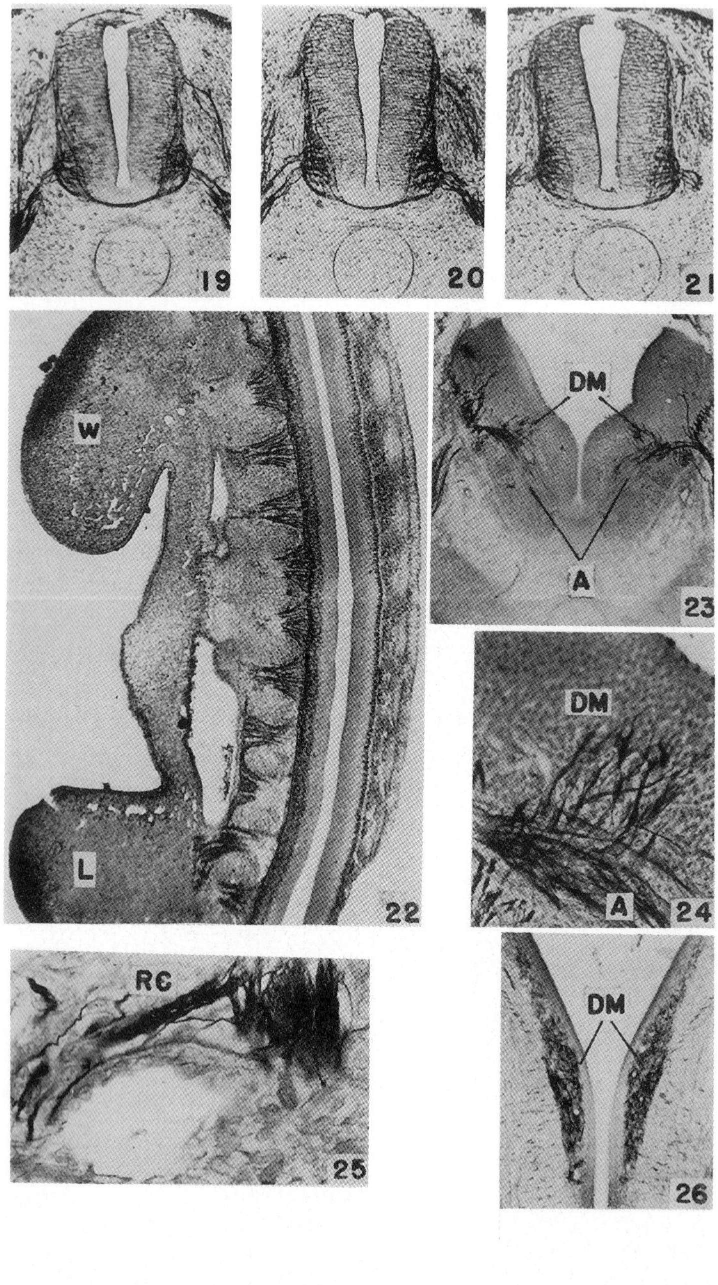

Reprinted from **Progress in Brain Research**, Vol. 4, Growth and Maturation of the Brain, ed. by D. P. Purpura and J. P. Schadé, 1–29 (1964).

Events in the Developing Nervous System

RITA LEVI-MONTALCINI

Washington University, St. Louis, Mo. (U.S.A.)

Ever since the most talented and perceptive of all neurologists, Ramon y Cajal, made available the silver technique for the selective impregnation of nerve cells, the nervous system of all phyla from celenterates to primates has become the object of an extensive and everlasting exploration for generations of neurologists. If the study of the different phyla provided a key for the understanding of the complexity of this system in higher forms, man included, it was the study of the developing nervous system which provided the Ariadne's thread to unraveling the labyrinthic complexity of nerve centers and nerve fiber tracts in the mature brain. Again it was the intuitive mind of Cajal which realized the enormous potentialities of this approach. To him and to his students we owe the most penetrating and rigorous series of investigations on the developing vertebrate nervous system. This analysis served as the basis for Cajal's fundamental work, *The Histology of the Nervous System*, which still represents today the most complete and authoritative study of this type (Cajal, 1909).

At the same time as Cajal and co-workers concentrated their studies on the embryonic nervous system, other brilliant scholars such as Edinger, Ariëns Kappers, Herrick, Crosby, and their students concentrated on the comparative analysis of the brain structures in the vertebrate phyla. Between the end of the past century and the first decades of this century, such a wealth of facts became known on the developing nervous system, as well as on its evolutionary changes, as to make one wonder if this field of investigation had not now been exhausted. It was with this outlook and a feeling of awe and curiosity for this most complex system that we first approached it two decades ago, shortly after the voice of the great master Cajal had become silent and while other outstanding scholars such as Ariëns Kappers, Herrick and Crosby were still actively pursuing their search on the evolution of brain structures.

The object of our first analysis was the chick embryo. As we became more and more familiar with this object and scrutinized its developing nervous system day after day and week after week, we became aware of the fact that this system had still not revealed more than a few of its many facets and that the essence of its developing mechanisms remained to be uncovered. Today, 20 years later, with the experience which comes from daily contact with the object of interest and with the dispassionate approach of the mature mind, the writer feels that we have still taken little more than a first glance at the endless complexity and intricacy of the differentiating nervous system. Thanks to the joined efforts of a large number of investigators, we have

References p. 25/26

possibly acquired a more precise knowledge of the sequence of events which take place during the growth and differentiation of the nervous system, but little progress has been accomplished in the understanding of the mechanisms which control these events.

It is with this feeling of inadequacy and the awareness of the number of unsolved problems still ahead of us, that I propose to discuss some of the events which take place in the developing nervous system. I will limit myself to the presentation and discussion of some of the aspects which we explored and attempted to analyze in the past years. They will be considered in reference to the time pattern of their occurrence and also in reference to the opportunity they offer to obtain some information on the agents which control and direct growth and development of nerve cells and nerve centers. Two main processes which take place in early phases of the differentiation of the nervous system will be considered, namely the *migration and degeneration of differentiating nerve cells.* We will limit our presentation to these areas not because we believe that the events to be presented may be of more interest than other events taking place in the developing nervous system, but because of our personal experience and interest in these aspects of the nervous system. Most of the observations to be reported are based on the analysis of these developmental processes in the chick embryo. They were supplemented by a few observations in mammals, specifically in the mouse embryo.

Hence when not specifically stated, we refer to the chick embryo. We will then briefly consider the results of some experimental work aimed at the analysis of the above processes.

MIGRATORY PATTERNS IN THE DEVELOPING NERVOUS SYSTEM

The capacity of differentiating nerve cells to migrate during the early stages of their differentiation has been known for a long time. Little, however, has been known about the extent and magnitude of these processes and the role they play in molding nerve centers and wiring them together. We will consider here some of these migratory movements which were the object of a close inspection in the chick embryo. The first to be described take place outside the central nervous system. They concern the cells which will form some of the cephalic ganglia and have their origin in the dorso-lateral and epibranchial placodes. Other important migratory movements occur at the same time outside of the spinal cord. These consist of the migration from the neural crest of the precursors of sensory spinal and sympathetic nerve cells. These migratory movements were considered at length during the past years and are well-known in all their aspects (Van Campenhout, 1931; Hörstadius, 1950; Tello, 1925; Yntema and Hammond, 1947); for this reason they will not be further considered here. Next we will consider the active displacements which occur in the central nervous system.

Migration from dorso-lateral and epibranchial placodes

The origin of some of the cephalic ganglia from ectodermal placodes has been ascertained through a number of investigations (Yntema, 1942, 1944; Levi-Montalcini, 1946). It is not within the aim of this presentation to report on the experimental evidence which provided the basis for the present knowledge of the origin of these ganglia. We refer the reader to the literature in this field and, in particular, to the extensive and exhaustive experimental analysis performed recently by Hamburger (1961) on the dual origin of the trigeminal ganglia in the chick embryo. Here we will consider some aspects of these migratory movements which give origin to 3 cephalic ganglia: the geniculate ganglion of the VIIth nerve, the petrosum and the nodosum which belong respectively to the IXth and Xth nerve. The observations to be reported below are based on the inspection of a large number of 2- to 6-day-old chick embryos stained with the De Castro modification of the Cajal technique (Levi-Montalcini, 1949) and sectioned serially in transversal, sagittal, and frontal sections. Although we referred many times during the past years to these studies, they were never the object of a detailed description.

The first evidence for active cell migration from the dorso-lateral and epibranchial placodes can be obtained in embryos 2½ to 3 days of age. Since the maturation of the

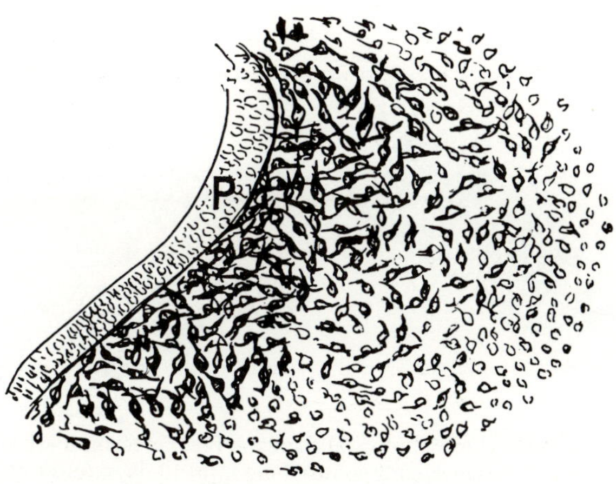

Fig. 1. Epibranchial placode (P) giving origin to the ganglion nodosum in a 3-day-old chick embryo. The migrating neuroblasts appear darkly stained and have short processes at opposite poles.

embryonic structures takes place along a rostro-caudal gradient, the migration from the dorso-lateral placode which gives origin to the geniculate ganglion precedes to some extent the migration from the 4th and 5th epibranchial placodes which give origin to the petrosum and nodosum ganglia. It was the migratory movements of these last two ganglia which offered the best condition for a close inspection.

Immediately beneath the epithelial thickening which represents the epibranchial placodes, one sees in 2½- and 3½-day-old embryos a large number of swarming cells which show intense silver affinity. They have two short opposito-polar prolongations which also appear intensely stained with silver (Figs. 1, 17).

Between the end of the 3rd and the 4th day, these cells move more and more away from the epithelial placodes where they were first seen and aggregate in two large

References p. 25/26

oval-shaped agglomerates which can be easily identified as the petrosum and nodosum ganglia (Fig. 18). At the end of the 4th day the two ganglia are well-delimited and stand out against the pale background of surrounding mesenchymal cells. One gets the impression that all neuroblasts have now assembled together and that the further size increase of these ganglia results from development and growth of individual nerve

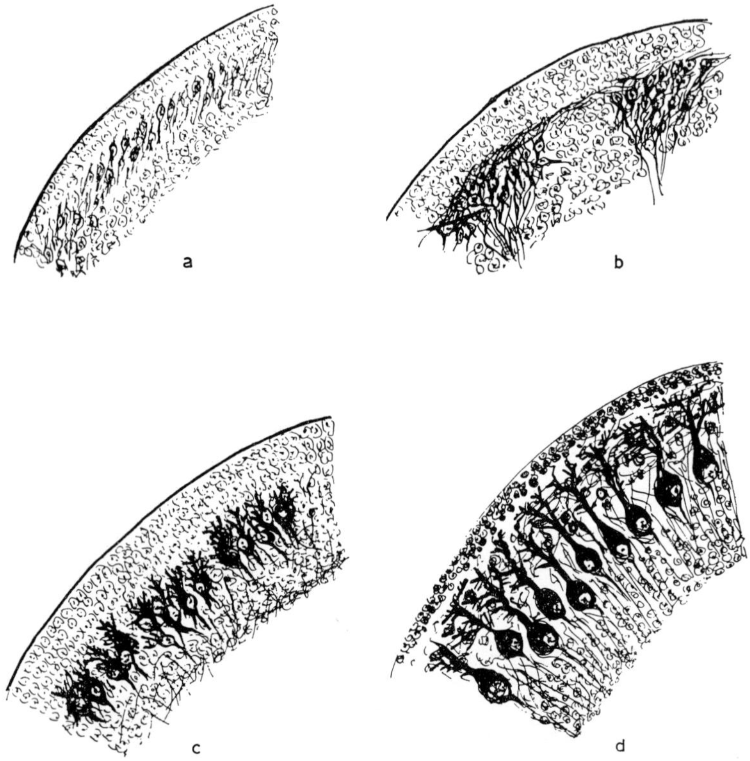

Fig. 2. The drawings illustrate the morphological changes of the cerebellar Purkinje cells in successive developmental stages of the chick embryo: (a) 8 days, (b) 10 days, (c) 14 days, (d) 17 days.

cells rather than from addition of new units. And yet all around the ganglia one sees a large number of scattered cells intensely stained with silver, with one or two short, dark filament-like prolongations. In the subsequent days the silver affinity of these cells weakens and eventually one cannot recognize them from the surrounding mesenchymal cells. The question arises as to the fate of these cells and their identity. If one has to judge on the basis of the affinity for silver, one would consider these cells as neuroblasts but this view finds no support in the observations mentioned above. Two alternatives should be considered. The first is that not all cells which react positively to silver should be identified as nerve cells. In accepting this viewpoint we would, however, deny any significance to the intense silver affinity of nerve cells as contrasted with the lack of affinity of adjacent mesenchymal cells. Also we would have to deny any value to the opposito-polar prolongations of these cells which further add to the notion that these are indeed nerve cells. The inspection of hundreds of embryos impregnated with the Cajal–De Castro silver technique confirmed us in the belief that nerve cells, but not mesenchymal cells, react positively to silver. The second alternative would be that neuroblasts are produced in excess by the ectodermal placodes. Once the ganglia have reached a given size, they become incapsulated and the neuroblasts still present in the area no longer find access to the main ganglionic

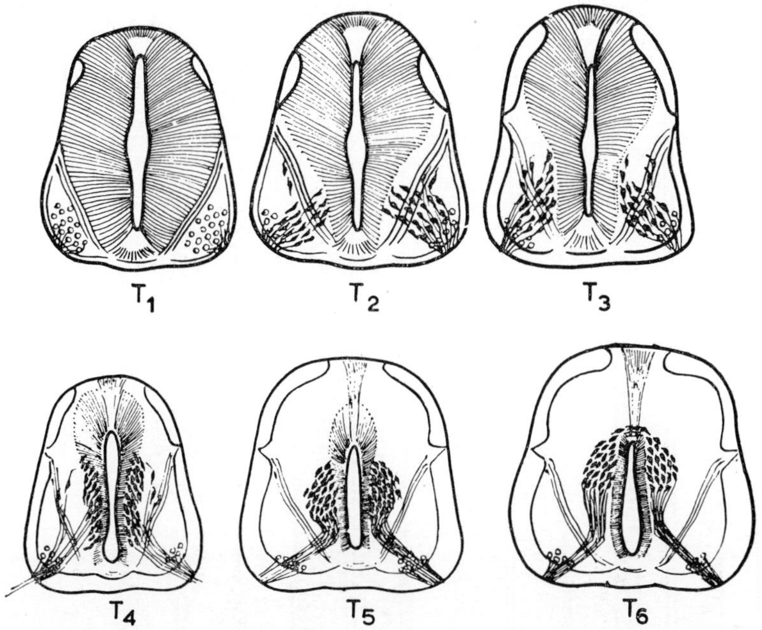

Fig. 3. Diagrammatic representation of the segregation and subsequent migration of the thoracic visceral column in the chick embryo. T_1, 3-day-old chick embryo. The somato-motor and the visceromotor cells form a compact column. T_2, $4\frac{1}{2}$-day-old embryo. Beginning of migration of preganglionic neurons. T_3, T_4, T_5, T_6 illustrate the shift of the migrating cells from a ventro-lateral to a mediodorsal position in embryos of 5 (T_3), 6 (T_4), 7 (T_5) and 8 (T_6) days of incubation. (From R. Levi-Montalcini, *J. Morph.*, **86**, 1950.)

agglomerates. They become dispersed and eventually die off, or possibly are otherwise utilized. Neither the first nor the second alternative were submitted to test and at present we must say that the problem of the identity and fate of these cells remains to be established.

Migratory movements in the central nervous system

Between the 4th and the 9th day of incubation, the developing nervous system of the chick embryo is the stage of the most extensive, elaborate and complex cell movements. While the brain vesicles and the spinal cord increase considerably in size, but do not undergo radical changes in shape, neuroblasts in the thousands engage in most sectors of the neural tube in long migrations which last for 2–3 days. To the student who is still not familiar with this object, these complex migratory movements appear hopelessly complicated and chaotic. It is only the hour after hour inspection of the develop-

References p. 25/26

ing neural tube and higher brain centers in transversal, frontal and sagittal sections which persuade the observer that these movements are not at all chaotic but on the contrary are highly organized and rigidly patterned like complicated military maneuvers displacing hundreds of moving units along different routes, under the control of the central headquarters. Like soldiers, or like armies of ants or termites, the

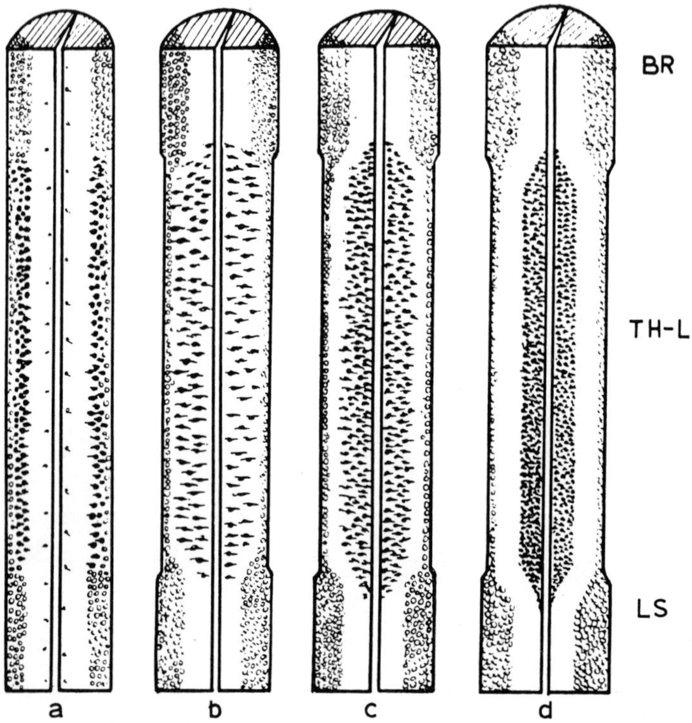

Fig. 4. Diagrammatic representation of the migratory movement of the thoracic visceral column as it appears in frontal sections of chick embryos fixed at (a) 4 days, (b) 6 days, (c) 7 days, (d) 8 days. BR, brachial segment; LS, lumbo–sacral segment; TH-L, thoraco-lumbar segment. (From R. Levi-Montalcini, *J. Morph.*, **86**, 1950.)

neuroblasts move in compact rows one after the other in long trails where each slender and spindle-shaped cell follows the other.

A characteristic common to all migrating nerve cells is their intense silver affinity and the elongated and slender shape of their bodies which stand out sharply against the pale background which forms the dense matrix of the developing nervous system. Each cell is provided with a short apical, and a long caudal filament which trails behind the body and actually represents the axon of the migrating cell. As a large number of cell units move simultaneously in different and many times opposite directions, one sees in many sectors of the brain stem or higher brain centers lines of cells crossing each other and then pursuing their own path which might bring them to entirely different areas of the developing nervous system. This intense traffic reaches its maximal expression toward the end of the 5th and the beginning of the 6th day

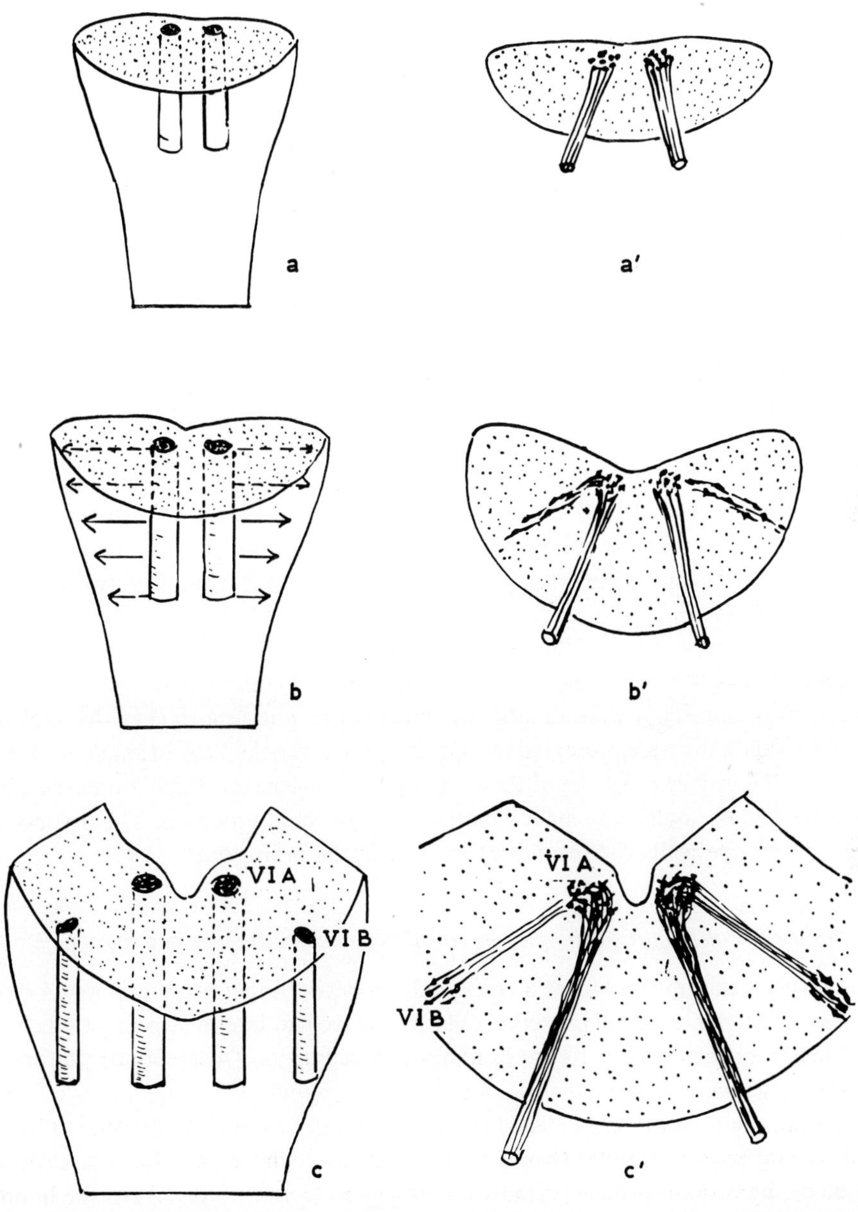

Fig. 5. Diagrammatic frontal and transversal representation of the segregation of the accessory nucleus from the main nucleus of the VIth nerve in the chick embryo. (a), (b), (c): stereographic views of the VIth nuclei at 5, 6, and 12 days. In (a), the nuclei appear as two columns; in (b), the arrows indicate the migration of neuroblasts from the main nuclei in a lateral direction; in (c), the migration is completed and the two accessory columns are formed. (a′), (b′), (c′): the same process as viewed in transversal sections. VI A, main nucleus; VI B, accessory nucleus. (From The Nature of Biological Diversity, edited by Dr. J. Allen. Copyright, 1963. McGraw-Hill Book Co. Used by permission.)

of incubation. Immediately after, most of the migrating cells have reached their destination and settle down in the area where they will be lodged for their life. As the cells become sedentary and lose the capacity for migrating, they undergo significant morphological changes. The previously elongated cells with two opposito-polar prolongations round up and the apical filament is reabsorbed or transformed into a number of short dendrites which vary in number and length depending on the cell type. The most impressive of these transformations of the short apical filament is illustrated by the cerebellar Purkinje cells. Even before their long migratory movements became known (Uzman, 1960; Levi-Montalcini, 1963) Cajal was impressed by the sudden transformation of these cells once they have settled in the cerebellar cortex (Cajal, 1929). Figs. 2, 9–11 give an idea of these complex morphological changes which transform the immature nerve cells into full fledged nerve units. Once a migratory group has reached its terminal settlement it rounds up in most instances in a compact cell agglomerate. An exception to this rule is represented by the above-mentioned cerebellar Purkinje cells (Fig. 6). These cells in fact do not aggregate, but display a distribution in a single row where each cell is parallel to the neighboring cells. In all other instances which we explored in detail, such as the formation of the preganglionic columns, the segregation of the cephalic motor nuclei, the shaping of the telencephalic centers, the cells aggregate in columns or in round or oval nuclei as mentioned above.

It is not in the purpose of this paper to present a detailed description of all, nor for that matter of a number, of the migratory movements taking place in the developing nervous system. They were already the object of previous reports (Levi-Montalcini, 1942, 1950) and they were presented in a current review article (Levi-Montalcini, 1963). Here we will consider only two of these migratory movements which are particularly impressive and exemplify the main characteristics of these processes. The first occurs in the neural tube while the second takes place in the brain stem.

Migration of the preganglionic viscero-motor column in birds and mammals

If one inspects the spinal cord of a chick embryo between the stages of 3 and 4 days, one sees a slender column of nerve cells in the ventro-lateral aspect of the tube extending from the cervical to the lumbo-sacral segments. Observations performed in previous stages showed that this column, easily identifiable as the motor column, results from the migration of cells which move laterally from their first site of origin in the ventral part of the basal plate of the ependyma which lines the central canal. At 4 days, the column presents almost the same size from the cervical to the lumbo-sacral segment of the spinal cord (Fig. 31). It is toward the middle of the 4th day that differences become evident in the cervical, brachial, thoracic, lumbar and sacral segments of the spinal cord. The developmental patterns characteristic of each segment were the object of a detailed report in a previous paper (Levi-Montalcini, 1950). Here we will consider only the thoracic segment, while in a subsequent section we will consider the differentiation of the same column in the cervical segment of the neural tube.

Until the middle of the 4th day, the motor column in the thoracic segment of the neural tube appears to consist of a rather homogeneous cell population. If one traces, however, the peripheral distribution of motor fibers which leave the neural tube in this trunk segment, one realizes the composite nature of this apparently uniform cell population. About $\frac{3}{4}$ of the nerve fibers emerging with the thoracic ventral roots make

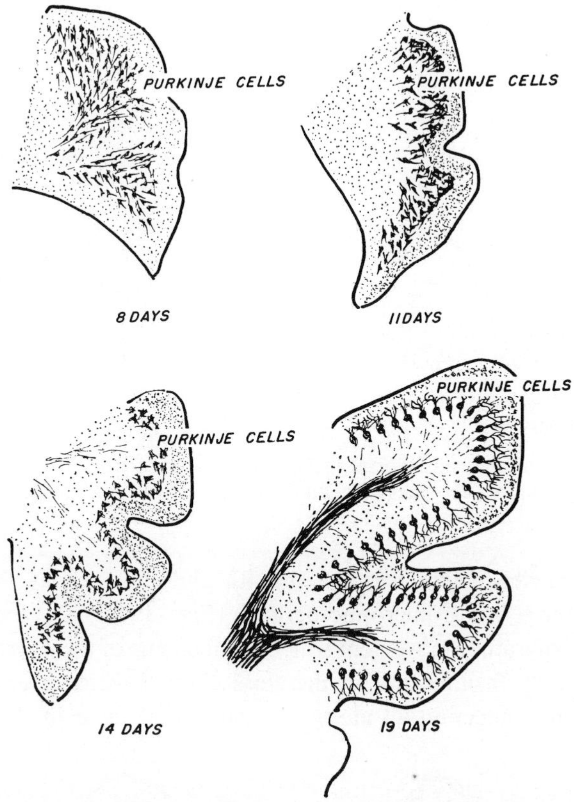

Fig. 6. Diagrammatic representation of the migratory movements of the Purkinje cells in the cerebellum of the chick embryo. (From The Nature of Biological Diversity, edited by Dr. J. Allen. Copyright, 1963. McGraw-Hill Book Co. Used by permission.)

connection with the primary sympathetic trunks which at this stage are still located in close apposition to the aorta. The residual $\frac{1}{4}$ sends fibers to the trunk muscles of the same level. Hence about $\frac{3}{4}$ of the motor nerve cells in the thoracic segment of the neural tube differ from the other $\frac{1}{4}$ in their peripheral distribution. From the above connections with the sympathetic chain ganglia they were identified as the preganglionic visceral cells. At $4\frac{1}{2}$ days these nerve cells sharply segregate from the other motor cells. The segregation of the two cell components is foreshadowed by a loosening of the compact texture of the motor column. Soon after, the $\frac{3}{4}$ of the entire group shift to a more medial position. Between the end of the 4th and the 7th day, the cells which have segregated from the residual $\frac{1}{4}$ of the motor column, move in compact cell rows in a ventro-dorsal direction. During the migration the cells are elongated

References p. 25/26

in shape and exhibit the intense silver affinity characteristic of migrating nerve cells. It is thanks to the dark, almost black shade of the neuroblasts during their entire displacement that it is possible to inspect closely each of the thousand units during the subsequent phases of their migration. Since this process presents in fact identical features in all embryos, one can reconstruct it by examining embryos fixed at slightly

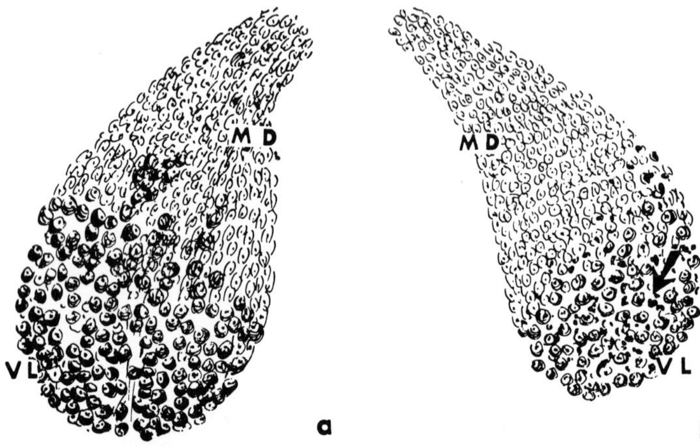

Fig. 7. The illustration of the medio-dorsal (MD) and the ventro-lateral (VL) cell populations in a brachial ganglion (a) and in a thoracic ganglion (b) of a 5-day-old chick embryo. The arrow in (b) points to some of the degenerating neurons in the ventro-lateral sector of this ganglion.

different developmental stages between the 4th and the 7th day. We refer to the drawings and to the series of microphotographs (Figs. 3, 4, 19–24) for a visualization of this impressive migratory movement which reminds one of the migration of termites or ants in its orderly fashion and in the close contact which keeps the individual units in reciprocal connection all along the route they cover in the 2½-day period.

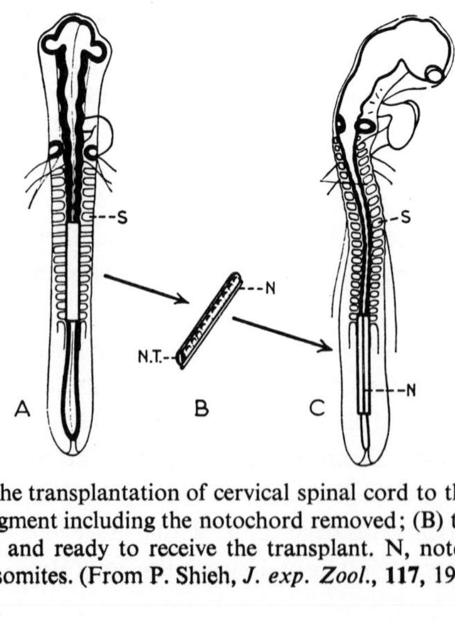

Fig. 8. The illustration of the transplantation of cervical spinal cord to the thoracic level. (A) donor with cervical spinal cord segment including the notochord removed; (B) transplant; (C) host with its thoracic segment removed and ready to receive the transplant. N, notochord; N.T., neural tube; S, somites. (From P. Shieh, *J. exp. Zool.*, **117**, 1951.)

One wonders whether any of the migrating cells acts as 'leader' of these cells or if each embarks on the same migration under the directing forces of some agent extrinsic to the cells. During the entire migration period, each cell trails behind its long axon, which has already established synaptic contact with one or the other of the sympathetic nerve cells in the chain ganglia.

Once the preganglionic motor cells have reached their permanent location adjacent to the dorsal aspect of the central canal, they gather together in two dense and compact columns. At this stage, namely at the end of the 7th day of incubation, they can be easily identified in the two columns which were first described by Terni as the preganglionic viscero-motor columns. In honor of their discoverer they are also at present known to neurologists as the 'Terni columns' (Terni, 1931).

In mammals the preganglionic motor cells undergo a similar migratory movement but their final location is in the medio-lateral aspect of the neural tube instead of the dorso-ventral position adjacent to the central canal as in birds. The mass movement of the young nerve cells was examined in mouse embryos. It is represented in Figs. 25, 26.

Migratory movements in the brain stem

The brain stem of the chick embryo between 5 and 8 days of incubation shows even more than any other sector of the central nervous system the dynamic aspects of the differentiating nerve centers. It is only through the studies of this brain segment in transversal, frontal and sagittal sections that one can reconstruct the simultaneous and many times opposite movements of nerve cell populations in their way toward their different destinations. As shown in Figs. 12–14, some of these cells form compact columns, while others are loosely arranged and detach sharply on account of their dark color on the pale background of the brain stem texture, and still others are oriented in long rows which bridge the distance between some of the cell aggregates. It was again the close inspection of this material hour after hour and day after day which revealed a number of highly organized and patterned mass movements in this apparently chaotic distribution of differentiating nerve cells. It was in fact found that nearly all the nuclei of the cephalic motor nerves do not form in the areas where they will be eventually located in the mature organism, but start to differentiate in other sectors of the brain stem and reach their end station through long and sometimes complex routes. In so doing, neuroblasts belonging to different nuclei crisscross each other and in this way one gets the impression of a chaotic and random cell distribution, while the individual units, on the other hand, move along precise and rigorously organized paths which invariably bring them to their final station.

Here we will briefly describe only one of these migratory movements which first came to our attention many years ago (Levi-Montalcini, 1942) and which presents many suggestive, although still unexplained, facets. The migratory movement under consideration takes place in the brain stem of embryos between the end of the 5th and the 8th day of incubation. It consists of the segregation of about half the population which builds the nucleus of origin of the VIth nerve and results in the formation of the

References p. 25/26

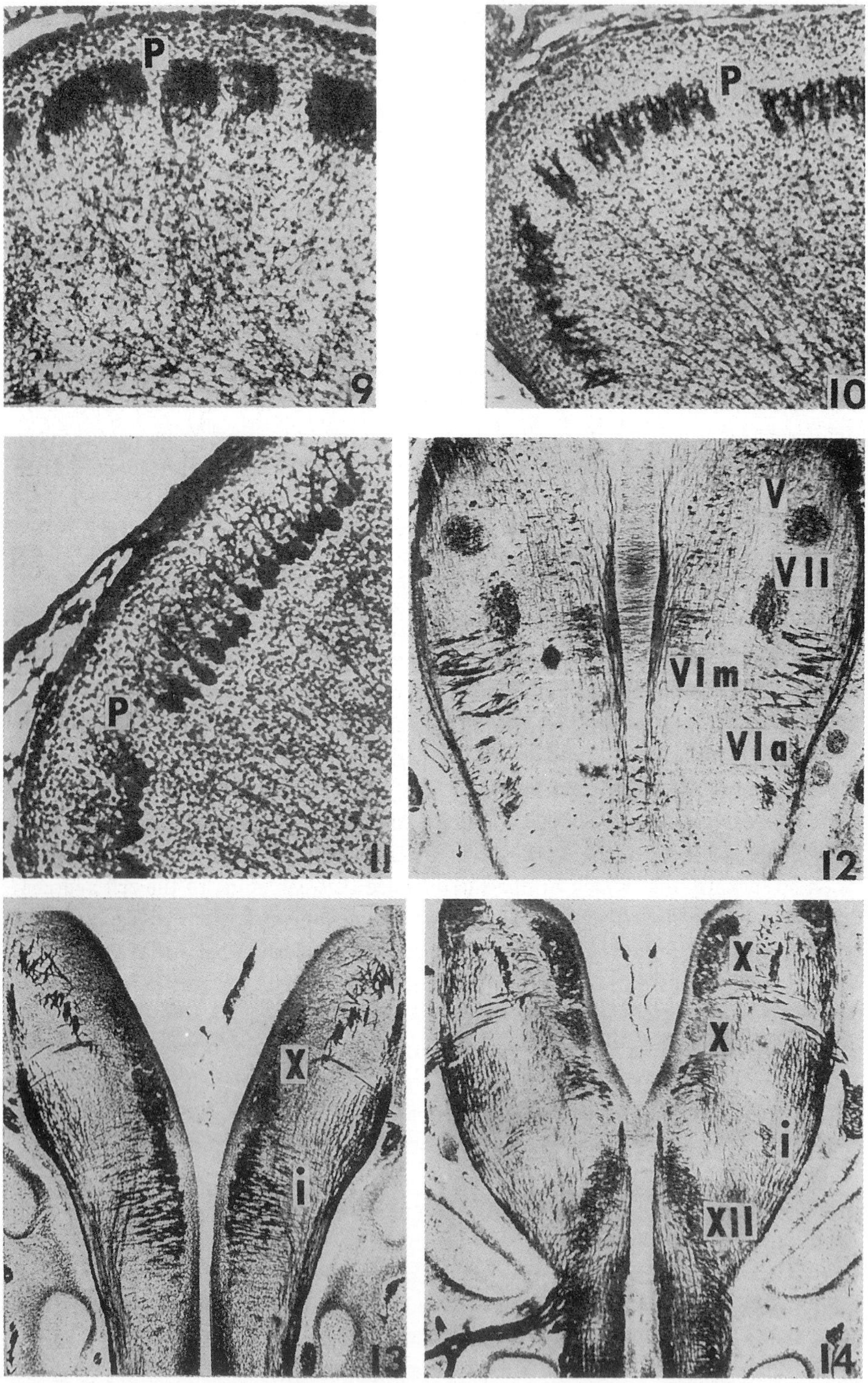

Figs. 9–14. For legend see p. 13.

accessory nucleus of the VIth nerve which provides for the innervation of the 3rd lid and of the two small muscles bursalis and quadratus. Up to the end of the 5th day, the main and accessory nuclei form a unique compact nucleus on the floor of the 4th ventricle. The right and left nuclei are located in close apposition to the two longitudinal fasciculi (Fig. 5a). They are easily identified on the basis of the axons of the cells which leave the brain stem in two compact nerve bundles immediately lateral to the midline (Fig. 5a′).

At this time one sees a few neuroblasts detach from the lateral aspect of the two nuclei and move in a ventro-lateral direction. The neuroblasts trail their axon behind and one can therefore follow the migrating cells along the long route which will bring them from the nucleus of origin to their final destination in the ventro-lateral sector of the brain stem. During the subsequent days, hundreds of darkly-stained and spindle-shaped neuroblasts detach in turn from the two main nuclei and move along the same trail followed by the first migrating nerve cells (Fig. 5b′).

When the migration comes to an end on the 8th day of incubation, one sees two compact nuclei in the ventro-lateral areas of the brain stem connected through their axons to the main nuclei which retain their position on the floor of the 4th ventricle. The axons of the main and of the two accessory nuclei leave the brain stem together in two compact nerve bundles which form the VIth nerves (Fig. 5c′).

Even before this impressive mass movement of the accessory nucleus became known, its position far away from the main nucleus had been the object of interest by neurologists such as Ariëns Kappers (1911) and Terni (1921). The latter considered the position of the accessory nuclei in the brain stem as the most convincing evidence of the validity of Ariëns Kappers' theory of neurobiotaxis. This theory, as is well-known, claims that the position of nerve centers is dictated by the influence exerted upon them by 'the greater number or the more dominant impulses which reach the cells' (Ariëns Kappers *et al.*, 1936). The location of the accessory nucleus of the VIth nerve, in close apposition to the descending root of the Vth nerve, is indeed most favorable for the discharge of impulses which, in turn, call for the closure of the lid when the eye or adjacent areas are endangered by noxic stimuli entering the field innervated

Figs. 9–14. All microphotographs from silver-stained, unretouched material, at different magnifications. Figs. 9–11 illustrate the transformation of the Purkinje cells (P) in embryos of 10 days (Fig. 9), 14 days (Fig. 10) and 17 days (Fig. 11). In Fig. 9, the neuroblasts are densely packed in cell groups, and the individual units cannot be recognized. In Fig. 10, the Purkinje cells appear in one single row. The cells show many dendrites (compare with drawing (c) in Fig. 2). In Fig. 11, the apical dendrite characteristic of the Purkinje cell is formed (compare with drawing (d) in Fig. 2).

Figs. 12–14 show 3 frontal sections of 8-day-old (Figs. 12, 14) and 7-day-old (Fig. 13) chick embryos. Fig. 12, V, main motor trigeminal nucleus, VIa and VIm, show the accessory and main nuclei respectively of the abducens nerve. The accessory nucleus has completed its migration. In subsequent stages this nucleus becomes more compact. Fig. 13 shows the dorsal motor nucleus of the Xth nerve (X) and the cells of the nucleus intermedius (I) as they move from a medial to a lateral position. Fig. 14 shows the two motor nuclei of the vagus complex (dorsal motor nucleus and nucleus ambiguus and the nucleus intermedius (I) in their final location. See also the long columnar nucleus of the XIIth nerve (XII).

References p. 25/26

by the Vth nerve. While there is no way of testing the validity of this principle as directing force in the evolution of this center, we can object to its validity as acting force during the ontogenesis. We tested this hypothesis by extirpating the rostral part of the rhombencephalon in 2-day-old embryos. In this way, the Vth nerve and its descending root were absent while the caudal part of the rhombencephalon was not affected by the operation. In these embryos lacking the descending root of the Vth

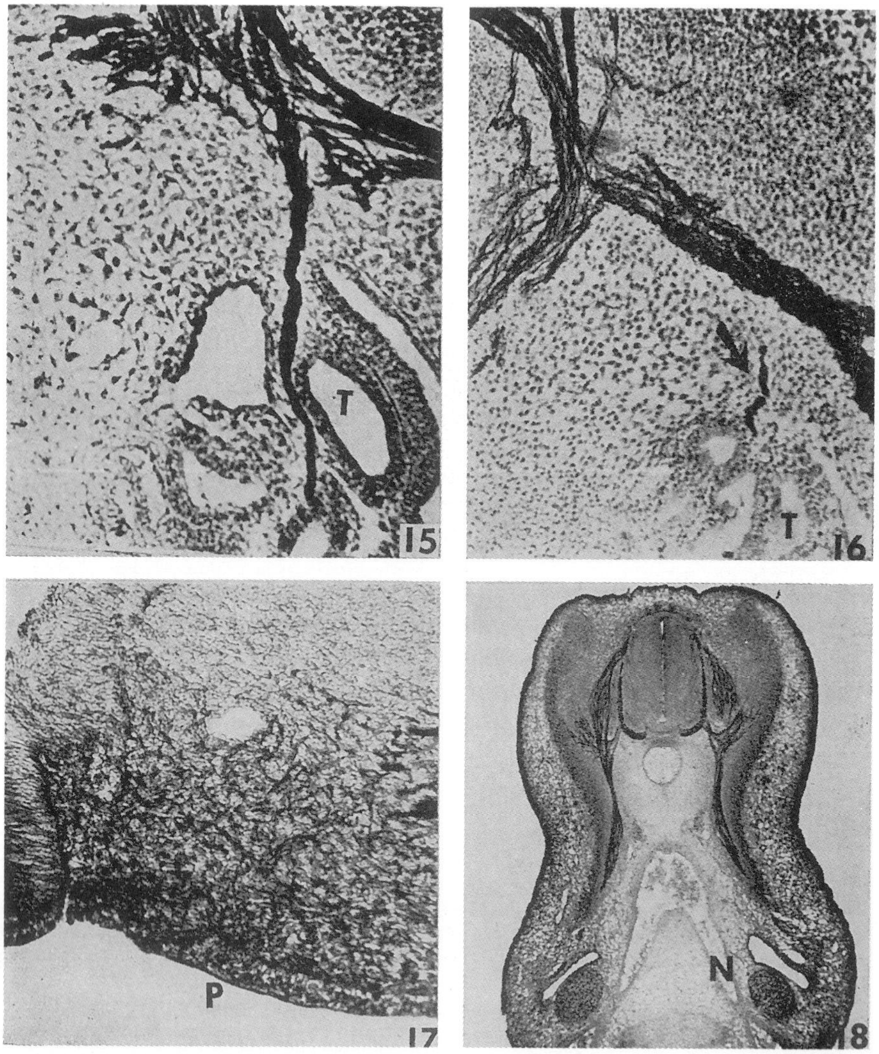

Figs. 15–18. All microphotographs from silver-stained, unretouched material, at different magnifications. Fig. 15 shows a large sensory nerve bundle entering the lateral part of the mesonephros from an adjacent sensory ganglion in a 4½-day-old chick embryo. T, mesonephric tubule. Fig. 16 shows the same nerve bundle in the process of disintegration in a 5½-day-old embryo. Arrow points to the nerve bundle. T, mesonephric tubule. Fig. 17 shows the epibranchial placode which gives origin to the ganglion nodosum in a 3-day-old chick embryo. P, placode. Compare with Fig. 1. Fig. 18 shows the ganglion nodosum (N) in a 4½-day-old chick embryo.

nerve, the mass migration of the accessory nucleus took place as in control embryos (Levi-Montalcini, 1942).

Migratory movements in higher brain centers

Although these movements are no less impressive than the ones which take place in the spinal cord and in the brain stem, they will be only mentioned here since they were already the object of extensive reports in previous publications (Jones and Levi-Montalcini, 1958; Levi-Montalcini, 1963).

Extensive mass migration of neuroblasts occurs in the cerebellum where the immature Purkinje cells cover long distances in order to reach their final location in the cerebellar cortex (Fig. 6). During the migratory phase the nerve cells exhibit the characteristic elongated spindle-shaped form which is common to all nerve cells in the way of active displacement from the original to the final location (Uzman, 1960; Levi-Montalcini, 1963).

A massive migratory movement takes place in the telencephalon (Jones and Levi-Montalcini, 1958). Also in this case we followed the displacement of thousands of nerve cells and their final aggregation in the nucleus epibasalis centralis. By way of conclusion one can say that active displacement and migration of nerve cells for long distances in the developing nervous system is the rule rather than the exception. The orderly fashion of these movements and their time pattern which is in all instances so rigorous as to permit prediction of when the first neuroblasts will start their long journey, leaves no doubt as to the existence of directing and guiding forces of these movement. It is for future work to attempt to identify these agents which are at present entirely unknown.

DEGENERATIVE PROCESSES IN THE DEVELOPING NERVOUS SYSTEM

Death of isolated units or entire cell populations is a normal occurrence in morphogenesis and is so familiar to students of developmental processes as to be considered of no more interest than all other aspects of development. During metamorphic processes cell degeneration occurs in the central nervous system (or in ganglia in invertebrates) as well as in a number of other organs and body tissues (Glücksmann, 1951). Localized degenerative processes are described in mutants as in the well-known case of mice exhibiting the Waltzer–Shaker syndrome as a consequence of degenerative changes in the cochlea and vestibular apparatus (Grüneberg, 1956). The analysis of the extensive literature on cell death led Saunders and co-workers to state that 'the onset and pace of degenerative changes during development are ultimately under genetic control' (Saunders *et al.*, 1962, p. 173) and we cannot but agree with this viewpoint which receives ample support from the numerous instances of massive cell degeneration taking place during developmental processes. We also agree with the above-quoted authors that in placing on the genes the burden for the sudden death of thousands of cells during development, we still leave unsolved the problem of why these cells should undergo early differentiation and then die when the 'death clock' is set in mot on.

References p. 25/26

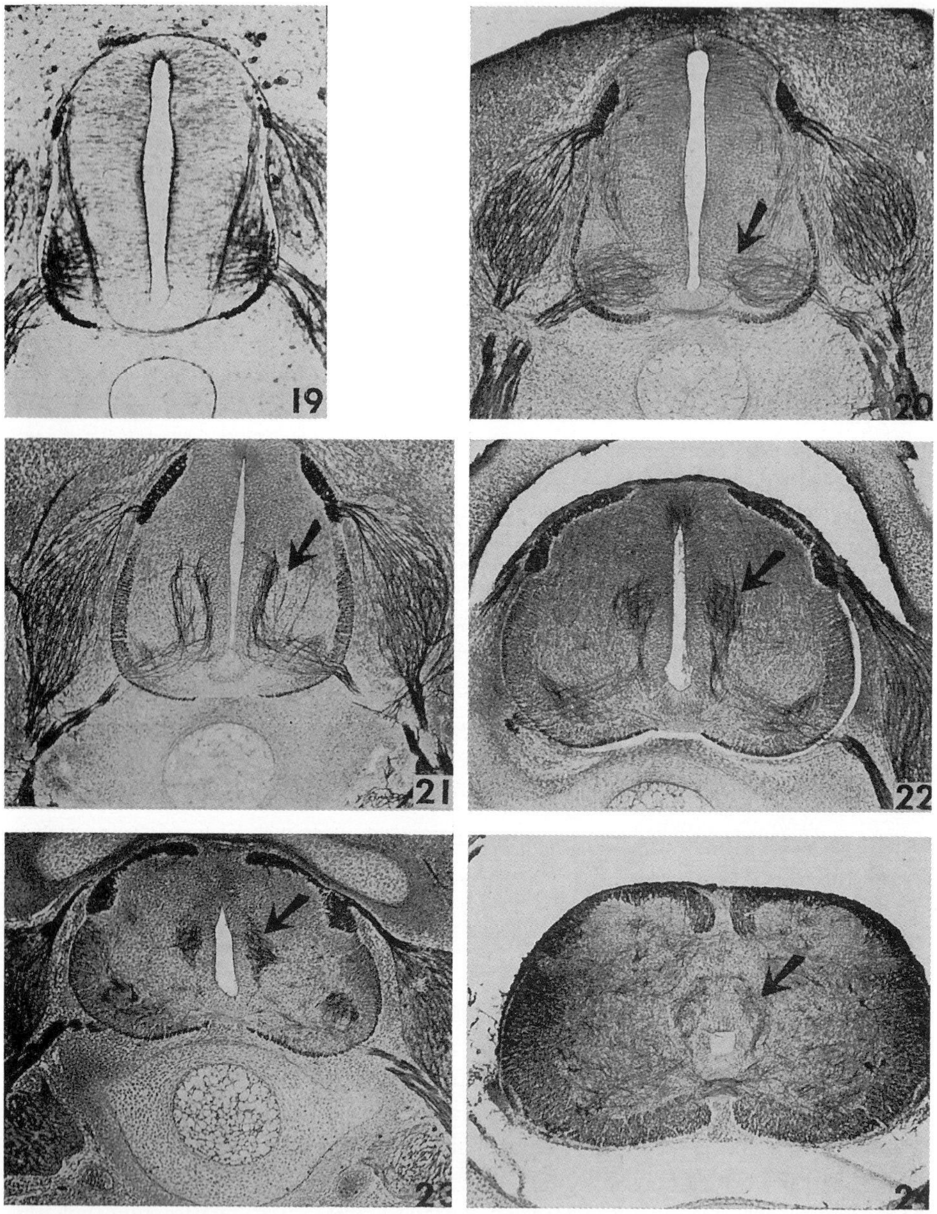

Figs. 19–24. All figures from silver stained, unretouched material, at different magnifications. Figs. 19–24 illustrate the formation and migration of the preganglionic thoracic column in the spinal cord of the chick embryo between 3 and 11 days of incubation. Transversal sections of embryos at the same trunk level. Fig. 19, 4-day-old embryo. Fig. 20, 4½-day-old embryo. Fig. 21, 6-day-old embryo. Fig. 22, 7-day-old embryo. Fig. 23, 8-day-old embryo. Fig. 24, 11-day-old embryo. Notice the progressive reduction of the central canal from 4 to 11 days. Arrows point to the preganglionic neurons.

In the following we will briefly describe one of the most striking instances of these massive degenerative processes which takes place in the developing spinal cord and was described by us some time ago. We refer the reader for a more detailed analysis of this process to our original paper (Levi-Montalcini, 1950). We will then consider similar degenerative processes which occur in some of the sensory spinal ganglia.

Mass cell degeneration in the cervical level of the spinal cord

Till the beginning of the 4th day of incubation the motor column in the cervical segment of the spinal cord appears as a rather dense and compact cell population of about the same size as in the brachial, thoracic and lumbo-sacral segments of the cord. Toward the middle of the 4th day, a massive degenerative process sets in in the cervical segment and in a few hours at least $\frac{2}{3}$ of the cell population in this region are reduced to cellular debris and pyknotic cells, scattered throughout the whole area. One sees in such preparations a large number of deeply stained round bodies scattered among the healthy nerve cells. A close inspection of these 'dark bodies' shows that they are the remnants of pyknotic or disintegrating nerve cells. During the period between $4\frac{1}{2}$ and 5 days, the number of these pyknotic cells is so large as to obscure the presence of intact cells (Fig. 32). Occasionally, smaller granules are found collected in clusters. They suggest an advanced stage of cellular disintegration. Besides the pyknotic cells, one notices a large number of macrophages. The vital staining technique used in such embryos and described in detail in our original paper (Levi-Montalcini, 1950) clearly indicates that these cells are present in large numbers in the area where degenerative processes take place. It is possibly the activity of these 'scavengers', which engulf the remnants of the damaged and dead cells, which accounts for the rapid disappearance of pyknotic cells from this segment of the neural tube. A few hours later in fact (the beginning of the 5th day) neither degenerating cells nor macrophages are detectable in this area. The cervica motor column is, however, reduced in width to less than $\frac{1}{3}$ its original size. The cells spared from death gather in a slender column, easily identifiable as the medial motor column which provides for the innervation of the axial muscles of the cervical region. We suggested in our original report that the cells undergoing early death could be part of an abortive visceral system. This suggestion was based on the observation that in stages prior to the massive degenerative processes, one sees rami communicantes in every segment of the cervical level. They are directed toward the primary sympathetic ganglia which are in the process of forming in this, as in more caudal levels of the trunk. These rami communicantes disappear at the same time as the large-scale cell degeneration takes place at $4\frac{1}{2}$ days of incubation. Experiments to be briefly mentioned in a following section gave support to the above hypothesis which will be again discussed after presentation of the results.

Degenerative processes in the cervical and thoracic spinal ganglia

While thousands of cells undergo sudden death and disappear from the cervical

References p. 25/26

segment of the developing spinal cord, an equally impressive degenerative process takes place in the cervical and thoracic spinal ganglia. In only one aspect do the degenerative processes in the neural tube and in the spinal ganglia show differences. While death hits most of the cells in the motor column of the cervical spinal cord, and dead cells are evenly scattered among the living ones in this area, in the cervical

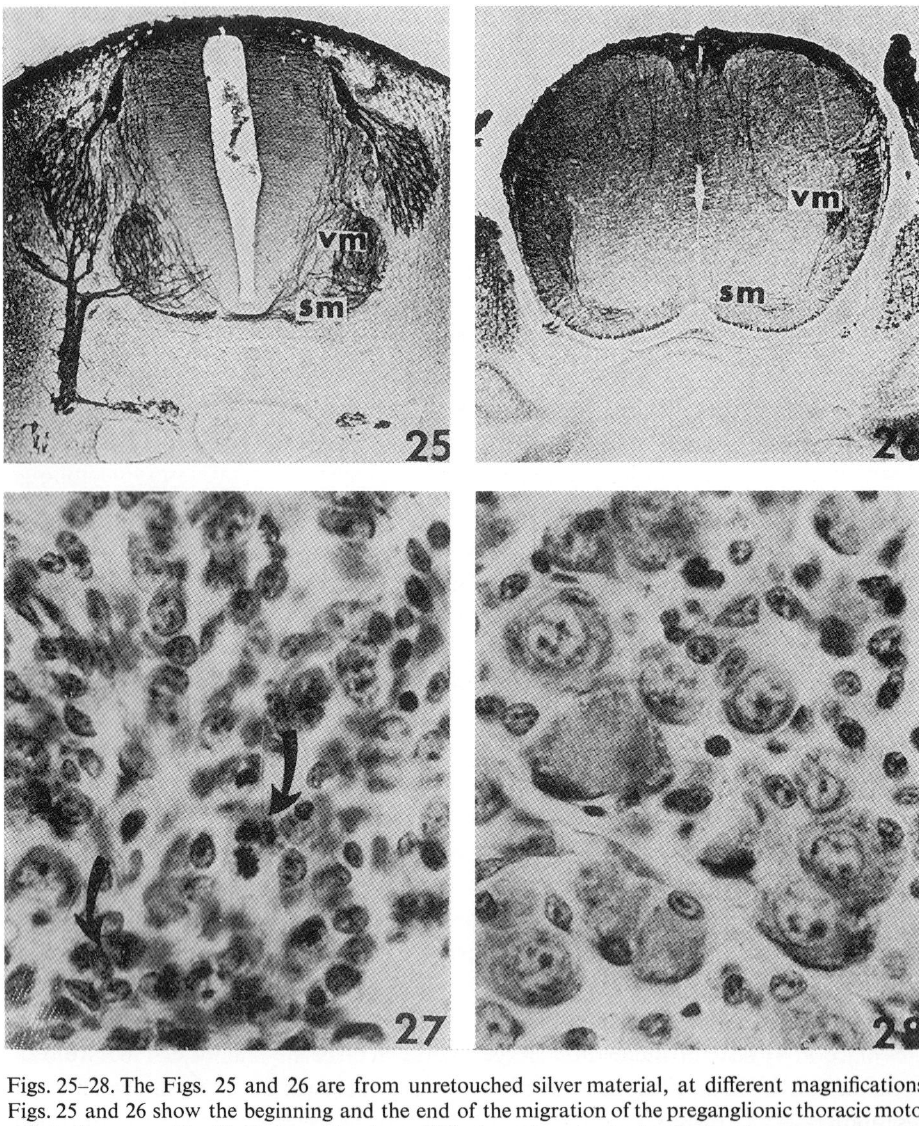

Figs. 25–28. The Figs. 25 and 26 are from unretouched silver material, at different magnifications. Figs. 25 and 26 show the beginning and the end of the migration of the preganglionic thoracic motor cells in the mouse. Fig. 25, 13-day-old embryo. This visceral-motor (vm) cells start to segregate from the somato-motor (sm) cells. Fig. 26, 17-day-old mouse embryo. The visceral-motor (vm) cells are now located in their permanent dorso-lateral position. The somato-motor cells (sm) have retained their ventral position. Fig. 27 shows the degeneration of sympathetic neuroblasts in the superior cervical ganglion of a 4-day-old mouse, following daily injections of antiserum to the NGF. Arrows point to some of the degenerating cells. Toluidin blue stain. Fig. 28, control.

and thoracic ganglia the disintegrating cells are confined to only one sector of the ganglia. It is in the ventro-lateral but not in the medio-dorsal part of these ganglia that massive degeneration takes place. The localization of pyknotic and dead cells is in fact so sharp that one could draw a line between the two parts of the ganglia where this process does or does not occur. We described in previous papers (Levi-Montalcini and Levi, 1942; Hamburger and Levi-Montalcini, 1949) the morphological differences between the cells which lodge in the dorso-medial and in the ventro-lateral sectors of these ganglia. We also reported in subsequent papers on the different behavior of these two cell types when confronted with given experimental situations (Levi-Montalcini and Hamburger, 1951). It is not the purpose of the present study to consider these differences. Here we will only mention that while the dorso-medial cells undergo normal differentiation, a large number of the ventro-lateral cells undergo sudden death and are wiped out by macrophages which invade the area (Fig. 7a,b).

What is the nature and the function, if any, of these cells doomed to die before their differentiation is fully achieved? Their number, as in the case of the motor cells in the cervical spinal cord, can be calculated in the thousands. Their early history does not differ, at least as far as one can say on the basis of their morphological analysis, from the history of neighboring cells which instead undergo normal differentiation and have a life expectancy of the same length as the organism to which they belong. Although we are not in a position to offer an answer to this question, we believe that it may be of more than passing interest to consider some parallel findings which may offer a clue to the identification of these cells and to their function.

At 4½ days, just before the degenerative processes set in in the ventro-lateral sectors of thoracic spinal ganglia, we noted that each of these ganglia contribute 2 to 3 rather large nerve bundles to the adjacent mesonephroi (Fig. 15). Since the ganglia which contribute these nerves are in the number of 6 to 7 for each side, the total number of these nerves is between 15 and 18. As the nerves reach the lateral-most part of the developing mesonephroi, they resolve themselves in individual nerve fibers and each fiber establishes contact with the epithelium lining some of the lateral tubules of the mesonephros. In the subsequent developmental stages, starting at the middle of the 4th day, the thickness of these nerve bundles decreases and toward the end of the 6th day nothing is left but remnants of the nerve bundles (Fig. 16). Also these residual nerve fibers eventually disappear altogether and no nerve can be seen to make connection with the mesonephroi in these or in subsequent developmental stages. Since the ganglia contributing nerves to the mesonephroi are exactly the same as those which show the massive cell degeneration described above, it is very tempting to correlate the two observations and identify the dying cells as the ones which establish a temporary contact through their nerve fibers with the mesonephroi. We are well aware of the fact that this observation leaves the main problem unanswered: what is the function of these nerve cells and why should the mesonephros (itself an organ doomed to a short life) receive a 'temporary' innervation from adjacent sensory ganglia? Also we must admit that we have no explanation for the equally impressive cell degeneration which occurs in the cervical ganglia. Their peripheral field was not the object of a close inspection and we do not know the termination of the axons of

References p. 25/26

these nerve cells which disintegrate between the middle of the 4th and the 5th day of incubation. As in the other instances described by Saunders, the rigid time pattern of these processes which present the same features in all specimens, clearly indicates that these massive cell degenerations are under genetic control in the same way as the mass migratory movements in other nerve cell populations.

One may now ask if these developmental processes are amenable to experimental analysis. The results to be briefly reported in the following pages indicate that this is indeed possible. In at least one instance we have evidence that nerve cells which would not, under normal circumstances, undergo migration, can be induced to migrate in a given experimental condition. We will also show that massive cell degeneration can be experimentally produced in cell populations which under normal circumstances are not doomed to die in their early differentiative stages. The similarity between normally occurring and experimentally produced cell disintegration is in some instances so striking as to raise the question whether the same or similar mechanisms might not operate in both cases. We shall consider this aspect of the problem after presentation of the results.

Mass cell migration experimentally produced in the cervical spinal column

The hypothesis that degenerative nerve cells in the cervical spinal cord could represent an abortive preganglionic visceral system was tested by Paul Shieh in our laboratorty He transplanted the cervical segment of the spinal cord of 9–25 somite chick embryos to the thoracic level of hosts of the same age which had been previously operated on for the extirpation of the thoracic segment of the spinal cord (Fig. 8). The cervical segment was then made to fit in the place of the extirpated thoracic segment. This operation is rather difficult and for this reason only a few cases survived up to the 7th–8th day. In all instances Shieh found that the transplanted segment of the cervical spinal cord underwent a differentiation similar to the differentiation which is characteristic of the thoracic segment (Shieh, 1951). In fact a slender column of cells detaches from the other motor cells of the transplant. In some instances the migration follows the same path as described in the thoracic segment of normal embryos (Figs. 33, 34). In other instances the migration is more atypical but also in these cases it resulst in the building of a cell column adjacent to the dorsal aspect of the central canal (Figs. 35, 36). Nerves emerging from these cells were traced to the adjacent paravertebral sympathetic ganglia. The author states that not all cells which normally undergo degeneration are spared this fate in the transplanted segment of the spinal cord. On the contrary, the degeneration is still well apparent in the rostral part of the transplant. It is only from the caudal-most part of the transplant that migration actually occurs.

These results, while giving support to the hypothesis that the degenerating cells in the cervical spinal cord might indeed represent an abortive preganglionic visceral column, also indicate that in a favorable environmental condition such as the one provided by the trunk region, motor nerve cells can undergo a massive migration and build a preganglionic motor column which establishes synaptic contact with adjacent sympathetic ganglia.

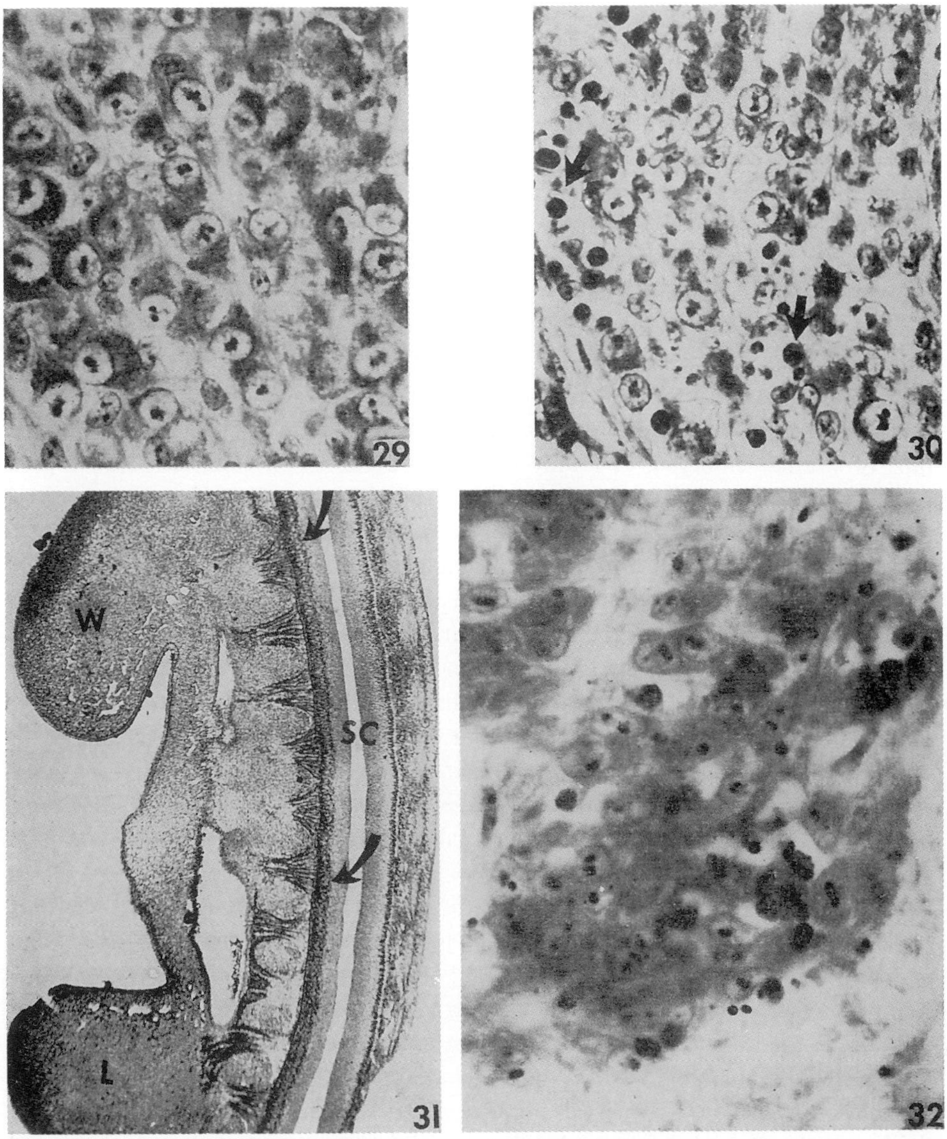

Figs. 29–32. Fig. 29 shows the ventro-lateral cells in a normal brachial ganglion of a 6-day-old chick embryo. Fig. 30 shows the same area in the brachial ganglion following removal of the adjacent wing bud at 3 days. Arrows point to disintegrating nerve cells. Heidenhain hematoxylin stain. Fig. 31, sagittal section of a 5-day-old chick embryo. Note the uniform size of the motor column (two arrows point at the column in the brachial and thoracic levels). L, leg; SC, spinal cord; W, wing. Silver stain. Fig. 32, 4½-day-old chick embryo. Transverse section at the cervical level of the spinal cord. Heidenhain hematoxylin. Degenerating cells appear as black spherical bodies scattered throughout the motor column. (Figs. 29, 30 from V. Hamburger and R. Levi-Montalcini, *J. exp. Zool.*, **111**, 1949. Figs. 31, 32 from R. Levi-Montalcini, *J. Morph.*, **86**, 1950.)

References p. 25/26

Mass cell degeneration experimentally produced in spinal ganglia and in the neural tube

Degeneration never occurs under physiological conditions in the brachial and lumbo-sacral ganglia nor does it take place in the corresponding segments of the neural tube. One can however produce a massive cell degeneration in the ventro-lateral sectors of these ganglia and in the motor column of the brachial and lumbo-sacral segments of the neural tube simply by extirpating the anterior or posterior limb buds in 2-day-old embryos (Hamburger and Levi-Montalcini, 1949). Two aspects of these experimentally produced cell degenerations are of interest: (*1*) degeneration does not follow a direct injury of motor or sensory nerve cells, nor of their axons. The extirpation of the limb buds is in fact performed long before sensory or motor nerve fibers branch in the limb bud; (*2*) the degenerative processes in the spinal ganglia and in the neural tube exhibit the same features as in the sensory ganglia and neural tube of normal embryos. The only difference between 'physiological' and experimentally produced cell degenerations is to be found in the regional areas where these processes take place. Physiological mass degeneration takes place in the cervical neural tube and in the cervical and thoracic ganglia, while the experimentally produced mass degeneration takes place in the brachial and lumbo-sacral segments of the neural tube and adjacent ganglia. We will first consider the degenerative processes which take place in the brachial motor column and were the object of a detailed analysis by Hamburger (1958). The extirpation of the anterior limb bud deprives these cells of their peripheral field of innervation and a sudden and abrupt degenerative process wipes out most of the cells of this area which provide for the innervation of the wing. Hence the extirpation of the limb bud, while not causing direct injury to these cells, has results fatal to them. In the sensory ganglia a massive cell degeneration takes place in the ventro-lateral cells and a slow and progressive atrophy sets in in the dorso-medial cells (Figs. 29, 30). Eventually both cell populations are dramatically reduced in size if the embryos are permitted to survive until the end of the incubation period (Levi-Montalcini and Levi, 1942). Evidence presented in a previous paper seems to indicate that the cells which undergo abrupt death are the proprioceptive cells, while the cells which undergo slow atrophy are the exteroceptive cells (Levi-Montalcini, 1963). While we have no explanation to offer for the behavioral differences between the two sensory cell populations, it is of interest to consider the close similarities between the degenerative processes in the neural tube in normal and experimental conditions and the similarities of these processes in the ventro-lateral cells of the thoracic and the brachial and lumbo-sacral ganglia in normal and experimental conditions. We suggested that death in the motor column of the cervical neural tube follows the failure of these cells to establish synaptic contact with the sympathetic ganglia (Levi-Montalcini, 1950). In the thoracic ganglia we noticed the sudden withdrawal of nerve fibers which had succeeded in earlier stages in establishing contact with the adjacent mesonephroi. In the experimental material, the deprivation of the peripheral field of innervation is fatal to the same types of nerve cells. As a working hypothesis one could therefore suggest that in both normal and experimentally produced mass cell degeneration, death follows the withdrawal from the medium

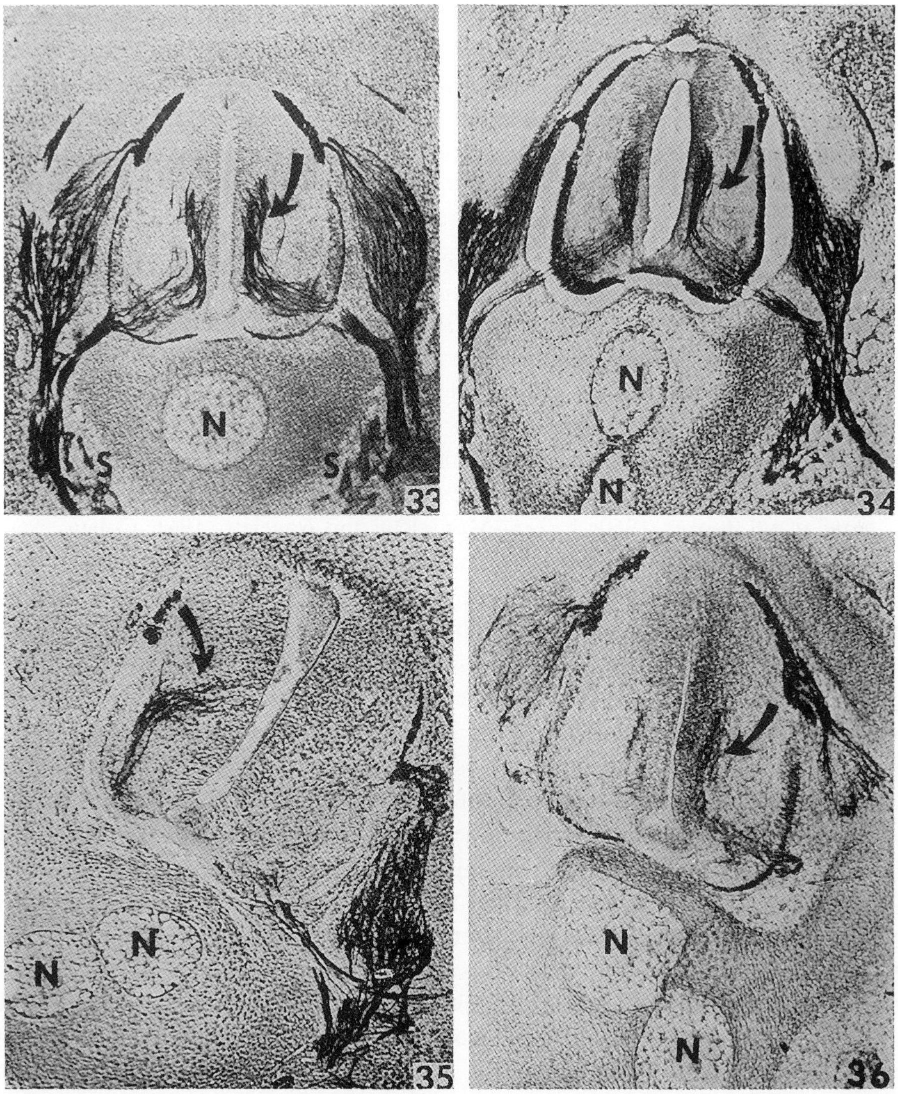

Figs. 33–36. All microphotographs from silver-stained, unretouched material, at different magnifications. Fig. 33, cross section through the medial part of normal thoracic spinal cord of 6-day-old chick embryo. Arrow points to preganglionic column in process of migration. N, notochord; S, paravertebral sympathetic ganglia. Fig. 34, cross section at the same level as Fig. 33 through transplanted cervical segment of the spinal cord. Note similar migration of motor cells as in Fig. 33. Arrow points to migrating cells. N, notochord. Figs. 35, 36 show two other experimental cases. In Fig. 35 the migration is atypical. Arrow points to the migrating cells in this and in Fig. 36. Note 2 notochords (N) from donor and host in the experimental cases. (All figures from P. Shieh, *J. exp. Zool.*, **117**, 1951.)

of some agent which is essential to the very life of these cells. The experimental results by Paul Shieh would add support to this point. In that particular instance in fact, survival and migration of nerve cells doomed to die under normal conditions, were obtained when these cells were able to establish a permanent synaptic contact with the sympathetic ganglia at the thoracic trunk level (Shieh, 1951). In the instances of

References p. 25/26

the latero-motor column of the brachial and lumbar spinal cord and of the ventro-lateral sensory cells, this factor would be normally supplied by the growing limb bud.

It might be of interest, in connection with the above problem, to consider how one can produce massive cell degeneration in other sectors of the nervous system by depleting the nerve cells of some essential factor by chemical rather than surgical procedures. We will consider these results in the following section.

Massive degeneration of the sympathetic nerve cells by an 'Anti-Nerve Growth Factor'

During recent years most of our work was devoted to the analysis of a specific protein isolated from mouse salivary glands. Upon injection of this protein in newborn and adult mammals, one obtains a striking size increase of the sympathetic nerve cells of injected animals (Levi-Montalcini and Booker, 1960a). This factor is now known as the Nerve Growth Factor (NGF). An antiserum to this NGF was produced by injecting the purified protein in rabbits. As reported in previous publications (Levi-Montalcini and Booker, 1960a,b; Levi-Montalcini and Angeletti, 1961; Cohen, 1960), the antiserum injected in newborn mammals results in the sudden and massive death of almost the totality of the sympathetic nerve cells (Figs. 27, 28). This devastating effect is similar to the massive degenerative processes described above. Only in the last instance, however, do we have precise indication as to the cause of death. This follows the destruction of a protein which plays an essential role in the life of the receptive nerve cells. It is tempting to draw a parallel between this degenerative process and the ones considered above in normal embryos and in embryos deprived of some parts of their peripheral fields of innervation. 'Physiological death' of some nerve cell populations could be due to the sudden deprivation of an essential factor. Death in other nerve cell populations follows the surgical extirpation of their field of innervation. The above considerations, if substantiated by more extensive studies, could shed light on some more general principles of nerve cell differentiation. Each cell population might be dependent for its growth and survival on some specific nerve growth factor, which is normally supplied by the environment. In only the last mentioned instance was this factor identified and isolated. It remains to future work to seek for such factors and attempt to correlate the observations reported above with the depletion of these 'essential factors'.

ACKNOWLEDGEMENTS

The work described in this paper was supported by grants from the National Institute of Neurological Disease and Blindness, Public Health Service (B 1602-C-C_6), and from the National Science Foundation (G 13946) and by a contribution from the American Cancer Society (C 1801-C-C_2) to Washington University.

SUMMARY

This review deals with two developmental mechanisms which play an important role in neurogenesis: migration and massive degeneration of large cell populations. Both processes take place during the early differentiation of the central nervous system and of the sensory ganglia. They were investigated under normal and experimental conditions in the chick embryo, which provides a most favorable material for such analysis. The significance of these mechanisms is discussed on the basis of experimentally produced migration and degeneration of nerve cell populations.

REFERENCES

ARIËNS KAPPERS, C. U., (1911); The migrations of the motor root-cells of the vagus group, and the phylogenetic differentiation of the hypoglossus nucleus from the spino-occipital system. *Psychiat. neurol. Bl. (Amst.)*, **15**, 408–427.

ARIËNS KAPPERS, C. U., HUBER, G. C., AND CROSBY, E. C., (1936); *The Comparative Anatomy of the Nervous System of Vertebrates, including Man.* Vol. I. New York, Macmillan Company (p. 76).

CAJAL, R. S., (1909); *Histologie du Système Nerveux de l'Homme et des Vertébrés.* Vol. 1. Paris, Maloine.

CAJAL, R. S., (1929); *Etudes sur la Neurogenèse de quelques Vertébrés.* Madrid.

COHEN, S., (1960); Purification of a nerve-growth promoting protein from the mouse salivary gland and its neurocytotoxic antiserum. *Proc. nat. Acad. Sci. (Wash.)*, **46**, 302–311.

GLÜCKSMANN, A., (1951); Cell deaths in normal vertebrate ontogeny. *Biol. Rev.*, **26**, 59–86.

GRÜNEBERG, H., (1956); Hereditary lesions of the labyrinth in the mouse. *Brit. med. J.*, **12**, 153–157.

HAMBURGER, V., (1958); Regression versus peripheral control in motor hypoplasia. *Amer. J. Anat.*, **102**, 365–410.

HAMBURGER, V., (1961); Experimental analysis of the dual origin of the trigeminal ganglion in the chick embryo. *J. exp. Zool.*, **148**, 91–124.

HAMBURGER, V., AND LEVI-MONTALCINI, R., (1949); Proliferation, differentiation and degeneration in the spinal ganglia of the chick embryo under normal and experimental conditions. *J. exp. Zool.*, **111**, 457–502.

HÖRSTADIUS, S., (1950); *The Neural Crest.* London, New York, Toronto, Oxford University Press.

JONES, A. W., AND LEVI-MONTALCINI, R., (1958); Patterns of differentiation of the nerve centers and fiber tracts in the avian cerebral hemispheres. *Arch. ital. Biol.*, **96**, 231–284.

LEVI-MONTALCINI, R., (1942); Origine ed evoluzione del nucleo accessorio del nervo abducente nell' embrione di pollo. *Acta Pontif. Acad. Sci.*, **6**, 335–345.

LEVI-MONTALCINI, R., (1946); Ricerca sperimentale sull'origine dei gangli del glosso-faringeo e del vago nell'embrione di pollo. *Atti Accad. naz. Lincei*, **1**, 1349–1352.

LEVI-MONTALCINI, R., (1949); The development of the acoustico-vestibular centers in the chick embryo in the absence of the afferent root fibers and of descending fiber tracts. *J. comp. Neurol.*, **91**, 209–242.

LEVI-MONTALCINI, R., (1950); The origin and development of the visceral system in the spinal cord of the chick embryo. *J. Morph.*, **86**, 253–284.

LEVI-MONTALCINI, R., (1963); Growth and differentiation in the nervous system. *The Nature of Biological Diversity.* J. Allen, Editor. New York, McGraw-Hill Book Company.

LEVI-MONTALCINI, R., AND ANGELETTI, P. U., (1961); Growth control of the sympathetic system by a specific protein factor. *Quart. Rev. Biol.*, **36**, 99–108.

LEVI-MONTALCINI, R., AND BOOKER, B., (1960a); Excessive growth of the sympathetic ganglia evoked by a protein isolated from mouse salivary glands. *Proc. nat. Acad. Sci. (Wash.)*, **46**, 373–384.

LEVI-MONTALCINI, R., AND BOOKER, B., (1960b); Destruction of the sympathetic ganglia in mammals by an antiserum to a nerve-growth protein. *Proc. nat. Acad. Sci. (Wash.)*, **46**, 384–391.

LEVI-MONTALCINI, R., AND HAMBURGER, V., (1951); Selective growth stimulating effects of mouse sarcoma on the sensory and sympathetic nervous system of the chick embryo. *J. exp. Zool.*, **116**, 321–362.

LEVI-MONTALCINI, R., AND LEVI, G., (1942); Les conséquences de la destruction d'un territoire d'innervation périphérique sur le développement des centres nerveux correspondants dans l'embryon de poulet. *Arch. Biol.*, **53**, 537–545.

SAUNDERS, J. W., JR., GASSELING, M. T., AND SAUNDERS, L. C., (1962); Cellular death in morphogenesis of the avian wing. *Develop. Biol.*, **5**, 147–178.

SHIEH, P., (1951); The neoformation of cells of preganglionic type in the cervical spinal cord of the chick embryo following its transplantation to the thoracic level. *J. exp. Zool.*, **117**, 359–395.

TELLO, J. F., (1925); Sur la formation des chaînes primaires et secondaires du grand sympathique dans l'embryon de poulet. *Trav. Lab. Rech. biol. Univ. Madrid*, **23**, 1–28.

TERNI, T., (1921); Ricerche sul nervo abducente e in special modo intorno al significato del suo nucleo accessorio d'origine. *Folia neuro-biol.*, XII, n. 2.

TERNI, T., (1931); Il simpatico cervicale degli amnioti (Ricerche di morfologia comparata). *Z. Anat. Entwickl.-Gesch.*, **96**, 289–426.

UZMAN, L. L., (1960); The histogenesis of the mouse cerebellum as studied by its tritiated thymidine uptake. *J. comp. Neurol.*, **114**, 137–160.

VAN CAMPENHOUT, E., (1931); Le développement du système nerveux sympathique chez le poulet. *Arch. Biol. (Paris)*, **42**, 479–507.

YNTEMA, C. L., (1942); Experiments on the origin of some of the sensory cranial ganglia in the chick. *Anat. Rec.*, **82**, 455.

YNTEMA, C. L., (1944); Experiments on the origin of the sensory ganglia of the facial nerve in the chick. *J. comp. Neurol.*, **81**, 147–167.

YNTEMA, C. L., AND HAMMOND, W. S., (1947); The development of the autonomic nervous system. *Biol. Rev.*, **22**, 344–359.

DISCUSSION

WAELSCH: Is there anything known about the chemical make-up of the cells before and after migration?

LEVI-MONTALCINI: Not that I know. The only difference I am aware of is that the migrating cells react much more intensively to silver stain than the stationary cells and this may well reflect biochemical differences, which would be most interesting to investigate.

WAELSCH: I am very much interested to know how these cells move.

LEVI-MONTALCINI: It is an amoeboid type of movement. The cell which is preparing to migrate becomes spindle-shaped and elongated. The cells assume a fish-like appearance and this, in our own experience, indicates that they will soon migrate.

WAELSCH: I wanted to ask something in regard to some work which Dorothea Rudnick did in our laboratory about enzyme development. An interesting relationship was found between the development of the enzymes in the yolksac and those in the embryo. They showed some kind of a mirror image. A certain enzyme which was very high in the yolksac decreased when it went up in the embryo. I wonder whether there are in the yolksac or in the embryo substances which have a blocking or inhibiting effect?

LEVI-MONTALCINI: I would not be able to answer this question.

WAELSCH: Could you say something more about the general significance of your growth factor?

LEVI-MONTALCINI: Although I am not in a position to give a satisfactory answer to your question, I must say that we do believe that the growth factor is normally present in the embryo as well as in the mature organism. It is quite possible that during development and differentiation the nerve growth factor is present in the peripheral field and acts upon the nerve centers which become associated with this periphery. This is, however, only a hypothesis and may be very difficult to prove. Also we believe that every nerve cell has its own specific growth factor. We are planning to investigate this point in detail in the future and search for other specific nerve growth factors.

SCHADÉ: In conjunction with some earlier work we have done on the development of unit activity in the cerebral cortex, I would like to ask you whether you have any information regarding the functional activity of the cells in the oculomotor nucleus. We have detected spike-activity in neurons of the prenatal rabbit cortex, which showed with routine histological techniques hardly any sizable basal dendrites and a very short apical dendrite.

LEVI-MONTALCINI: I regret to say we have no particular information regarding the early functional activity of the oculomotor nucleus.

SCHADÉ: I am curious to know whether these moving cells process any number of dendrites or just like in the early differentiation of pyramids, have a single dendrite opposite to the axon?

LEVI-MONTALCINI: They are just spindle-shaped and have a single process or dendrite opposite to the axon like the differentiating pyramid cells, which you mention. There seem to be no other dendrites growing out from the cell bodies as long as the cell is in the process of migrating.

SCHADÉ: I have one more question regarding your very interesting experiments. Looking from your slides I saw that in the sympathetic ganglia, treated with growth factor, there was also an increase of glial cells. Do you have any idea why this is?

LEVI-MONTALCINI: It may be that you are right, but I never studied this point in detail. I can, however, tell you something about the effect of the antiserum to the nerve growth factor. Upon injection of the antiserum, the glial cell population is not affected for the first few days but then suddenly decreases and then disappears almost completely. I assume that the opposite situation occurs when the growth factor is injected, namely that the glial cells increase in number. In this case the effect of the growth factor would be an indirect one, that is, it would affect the nerve cells and this, in turn, would affect the population of satellite and glial cells.

PURPURA: The assumption that nerve centers are under the influence of growth factors implies inhibiting as well as excitatory substances. Do you believe these factors are operating in the effects observed in favor of injury to cortical nerves as originally described by Cajal and confirmed in some of our recent studies?

LEVI-MONTALCINI: It could well be, but I regret to say that I am not able to answer this question.

PURPURA: Do you believe that the phenomenon of neurobiotaxis is explainable in terms of the release of different and specific growth factors by different centers?

LEVI-MONTALCINI: I would not say that we have evidence for it. It was indeed a very interesting hypothesis offered by the late C. U. Ariëns Kappers that nerve centers shift position in different phyla and become located in close apposition to nerve fibers which establish a synaptic contact with them. I must say, however, that I was not able to prove the validity of this principle as acting force during the differentiation of embryonic nerve centers. I refer to the experiment I reported on the extirpation of the rostral part of the medulla in the 2-day-old chick embryo. As a consequence of this operation, the descendant root of the Vth nerve did not form but the accessory center of the abducens underwent migration as normal, thus disproving that its migration is dependent upon a direct influence of the descendant root of the Vth nerve.

BONAVITA: You said that the growth factor can be isolated from several sources. Can you make a comparison in chemical as well as in biological terms? In other words, can you say that there is a relative species specificity of the factor?

LEVI-MONTALCINI: The factor was isolated first from mouse sarcomas 180 and 37, then from snake venom and salivary gland, and more recently from experimentally produced granuloma in several mammalian species. As found by Dr. Cohen, the factors from the snake venom and salivary gland differ in molecular weight, which is 20,000 for the snake venom and 44,000 for the salivary factor. There are also other differences as to their antigenic properties. Biologically all the above nerve growth factors call forth the same effect. That is, they all affect embryonic sensory and embryonic and mature sympathetic nerve cells.

BRIZZEE: What was the earliest developmental stage at which growth factor was employed?

LEVI-MONTALCINI: When the sarcoma 180 is implanted in the 3-day-old chick embryo, the first reaction in the sensory nerve cells is apparent at 6 days. Of the two cell populations which we described in the sensory ganglia, it is only the medio-dorsal but not the ventro-lateral which is affected by the tumor nerve growth factor. The snake venom and salivary gland factors were instead tested only in embryos 7–9 days old and in all instances the medio-dorsal sensory nerve cells and the sympathetic nerve cells showed a growth response to these agents. In subsequent developmental stages the sensory nerve cells lose their reactivity to the nerve growth factor while instead the sympathetic nerve cells remain reactive throughout all their life cycle.

BRIZZEE: Do cells in the central nervous system invariably migrate in a direction opposite to axonal localisation?

LEVI-MONTALCINI: Yes, in the central nervous system the cells migrate, as far as I could see, always in a direction opposite to the axonal outgrowth.

WENDER: I wanted to know if there are morphological or histochemical differences between the cells which migrate and those which degenerate.

LEVI-MONTALCINI: No, from a morphological point of view I was not able to see any difference in cells which would eventually migrate or degenerate.

WENDER: Is it possible to say that this migration process is a very active process?

LEVI-MONTALCINI: Yes, the process is certainly an active one and is due to amoeboid movement of individual nerve cells. In order to reach the terminal location many times they have to cross a fiber barrier or to move in a dense texture of the central nervous system. The hour after hour and day after day analysis gave us positive evidence that we are dealing with an active displacement and not a passive one.

BONAVITA: Do neurons show the silver reaction at a different degree before differentiation? In the absence of histochemical data this could tentatively indicate the existence of chemical differences at an early stage.

LEVI-MONTALCINI: Yes, neuroblasts show a different silver affinity from one group to another and also the affinity in the same cell population changes from one stage of differentiation to another. I certainly agree with you that this may reflect histochemical differences.

WAELSCH: May I ask you another question, Rita, which I wanted to ask you in the beginning. You said that in the adult your growth factor leads to an increase in size without cell division. This is probably a very important phenomenon from a biochemical point of view. Some investigators have postulated that also in the adult nervous system you can show incorporation of tritiated thymidine into the nuclei, or what is meant into the DNA of the adult nucleus. Others like Maurer and collaborators in Germany were unable to show this. Your experiment would be an ideal case to test whether without cell division in growth there would be an activity of DNA. And this would explain something which starts now to be very puzzling: does the nucleus in the adult nerve cell have a metabolic activity?

LEVI-MONTALCINI: I fully agree with you that our material could indeed be most suitable for investigation of this problem.

PART II

DISCOVERY OF THE NERVE GROWTH FACTOR (NGF): ITS CHEMICAL CHARACTERIZATION, MULTIPLE SOURCES AND EFFECTS

The papers presented in Part I of this volume provided the basic information of the structural organization of the nervous system of the chick embryo under normal and experimental conditions. In studies presented in Part II, I resorted to the same methods and experimental approach which I had previously devised.

In 1948, Elmer Bueker, a former student of Viktor Hamburger then at Georgetown University, performed a bold and ingenious experiment. Rather than examining the effects of implantation of an additional limb bud on innervating nerve centers, he implanted a fragment of neoplastic mouse tumor (sarcoma 180) into the body of a 2.5 day-old chick embryo. In his 1948 article,[1] which Hamburger brought to my attention, he reported that sensory fibers outgrowing from adjacent Dorsal Root Ganglia (DRG) branched into the tumor, while no motor nerves had gained access into the neoplastic tissue. Sensory ganglia examined 5 days later appeared somewhat larger than contralateral ganglia innervating the limb. The author concluded that the histochemical property of the tumor favored the entrance of sensory but not of motor nerves. The slight volume increase of the DRG was attributed by Bueker to the possibility of sensory nerve fibers to ramify into the fast growing tumor which provided a much broader area of innervation than the limb bud. These results were so much in line with current concepts of the interrelation between nerve centers and end organs that Bueker overlooked other, more perplexing aspects of the growth response. My reinvestigation of this effect showed that the growth response of sensory ganglia innervating the tumor was, in fact, much more pronounced than that of the same ganglia innervating a supernumerary limb. Furthermore, while the somatic motor nerve cells in the spinal segment facing the tumor transplant were not increased (they actually decreased in number), the sympathetic ganglia were tremendously enlarged and contributed to the innervation of the neoplastic tissues to a much larger extent than did sensory ganglia. Sensory and sympathetic nerve fibers branching into the tumor established no contact with the neoplastic cells but wandered at random among the cells. And yet, even conceding that the effect was much more marked than that produced by implantation of a supernumerary limb bud, and different in many respects from it, we were still not prepared in 1951 (paper 4) to see in this response any flagrant deviation from normality, so difficult it is to refute generally accepted concepts and to evaluate novel results with an unprejudiced mind.

[1]Bueker E.D., Implantation of tumors in the hind limb field of the embryonic chick and the developmental response of the lumbosacral nervous system, *Anat. Record* **102**, 369–389 (1948).

It was a spring day in 1951 when the block was suddenly removed, and it dawned on me that the tumor effect was different from that of normal embryonic tissues in that the tumor acted by releasing a growth factor of unknown nature rather than by making available to the nerve fibers a larger-than-usual field of innervation. This hypothesis, which became a certitude with me long before I obtained supporting evidence in its favor, was suggested by the observation that the viscera of embryos bearing transplants of mouse sarcomas 180 or 37 were invaded by sympathetic nerve fibers at a stage when there was still no innervation apparent in controls. Surprisingly, and against all rules of normal development, sympathetic fiber bundles even perforated the normally impermeable tunica of the large and small veins. They protruded inside their lumen, reaching, in some cases, such a volume as to block the circulation.

Another and even more startling effect (paper 5) elicited by sarcoma 180 and a tumor of the same origin known as sarcoma 37. The finding showed that para- and pre-vertebral sympathetic ganglia, too distant to gain access into the neoplastic tissue, had also markedly increased in volume. Thick nerve bundles innervating most of the viscera (thyroid, adrenal glands, sex glands, mesonephroi and other organs) did not establish contact with the cells of these organs, but branched irregularly among them, keeping close contact with the circulatory system. Sympathetic nerve fibers also gained access into the veins of the host embryo. This most atypical effect suggested to me that sarcomas 180 and 37 released a humoral factor which reached the target nerve cells not only through their fibers, but also by diffusion into the blood vessels. It seemed at the same time to be indicative of a neurotropic effect of the still unidentified tumoral nerve growth factor. Together with other waste products of the neoplastic cells, this factor could be present at a higher concentration in the veins than in the surrounding embryonic tissues. In favor of this hypothesis was also the fact that sympathetic nerve fibers entered and branched inside the lumen of the veins but not in that of the arteries, which do not carry the tumor catabolic substances. Experiments performed more than two decades later, with the much larger quantity of NGF extracted from the mouse submaxillary salivary glands (see below), provided compelling evidence for such an effect. These findings have a direct bearing on one of the most important and still unsolved problems in developmental neurobiology: assessment of the role of chemical signals in the formation of neuronal circuits in neurogenesis and in the reestablishment of fiber connections during regeneration.

Convincing evidence for a NGF neurotropic role came from *in vivo* experiments consisting of daily injections of NGF with a microcapillary glass pipette into the floor of the fourth ventricle of neonatal rats. We wanted to prove or disprove a putative effect of this protein molecule on dopaminergic neurons located in the brainstem.

The unforeseeable result was the penetration of nerve fibers, outgrowing from adjacent sympathetic ganglia, into the spinal cord of the injected pups, and their ending at the point of NGF injection. Discontinuation of the daily supply of exogenous NGF resulted in the reabsorption of these ectopic fiber bundles.[2,3] This *in vivo* approach, however, left unanswered the question whether sympathetic nerve fibers directed their course along a NGF diffusion gradient or if their entrance and route along this gradient in the cerebro-spinal axis could result from a trophic rather than a tropic effect. *In vitro* studies on single dissociated embryonic nerve fibers disproved this latter hypothesis and provided definitive evidence for the NGF neurotropic effect.[4,5]

At variance with the hypothesized NGF neurotropism, firmly established only many years later, the diffusion effect received immediate confirmation through implantation of fragments of sarcomas 180 or 37 into the yolk sack or onto the chorio-allantoic membrane. In both cases, the embryonic and neoplastic tissues shared the circulation but no direct contact was established between them. As reported in paper 5, the results were identical to those elicited by intra-embryonic tumor transplants.

These findings demonstrated that the tumor promoted an excessive growth of the sympathetic system with total disregard for the requirements of the organism, altering the sequence of events which characterize normal developmental processes. In embryos bearing transplants of sarcoma 180 or 37, the sympathetic system overtook that of other embryonic organs and tissues. The chain ganglia became eight times larger than normal. This effect was particularly impressive since the embryos were considerably smaller than untreated controls, due to the metabolic requirement of the fast growing tumor. The release into the circulation by neoplastic cells not only of the growth factor, but also of toxic agents such as a cytokine, was identified many decades later as Tumor Necrosis Factor (TNF).

The excitement of this discovery was tempered by the realization that it would be tremendously difficult to identify the tumor factor by resorting to the exceedingly laborious and time-consuming embryological experiments. Attempts to identify the hypothetical humoral neoplastic factor through injections of the tumor extract into developing embryos failed. What was needed was a much simpler and faster bioassay. This was provided by the tissue culture technique which I had learned at the University of Turin, under the guidance of Professor Giuseppe Levi.

In 1952, this technique was not yet applied to the investigation of basic biological problems. Tissue culture facilities were not available at Washington University, but a most efficient center had been established at the Biophysics Institute of the University of Rio de Janeiro, under the direction of my friend, Herta Meyer. In September of

[2]Levi-Montalcini R., Menesini Chen M.G. and Chen J.S., The neurotropic effects of the nerve growth factor in chick embryos and in neonatal rodents. In Formshaping movements in neurogenesis, *Zoom.* Vol. **8**, 201–201 (1978).

[3]Menesini M.G., Chen J.S. and Levi-Montalcini R., Sympathetic nerve fibers ingrowth in the central nervous system of neonatal rodents upon intracerebral NGF injections, *Arch. Ital. Biol.* **116**, 53–84 (1978).

[4]Gundersen R.W. and Barrett J.N., Neuronal chemotaxis: chick dorsal-root axons turn toward high concentration of nerve growth factor, *Science* **206**, 1079–1080 (1979).

[5]Letourneau P.C., Chemotactic response of nerve fiber elongation to nerve growth factor, *Dev. Biol.* **66**, 183–196 (1979).

that year, upon the invitation of the Institute Chairman, Professor Carlos Chagas, I boarded a plane for Rio carrying in my handbag a small box containing two mice, each bearing a transplant of mouse sarcoma 180 or 37 (paper 6).

The outcome of the experiments performed in that hospitable milieu marked the turning point of the investigation. The tumor factor had given a first hint of its existence in St. Louis, but it was in Rio de Janeiro that it revealed itself, and it did so in a theatrical and grandiose way, as if spurred by the bright atmosphere of that explosive and exuberant manifestation of life that is the carnival in Rio.

Sensory and sympathetic ganglia from 8-day-old chick embryos were explanted in a semi-solid medium consisting of rooster's plasma and embryonic chick extract, in close proximity to, but not in contact with, fragments of mouse sarcomas 180 or 37. In control cultures, other mouse tumors or embryonic mouse or avian tissues were combined *in vitro* with the same ganglia. The sealed cultures were incubated at 38°C and inspected at time intervals between 6 and 48 hours. At the end of the first day, all ganglia facing the sarcoma 180 or 37, but none in the control cultures, had developed a fibrillar halo which appeared particularly dense on the tumor's side (paper 7).

Upon my return to St. Louis in the winter of 1953, a young biochemist, Stanley Cohen, joined our group to investigate the biochemical nature of the factor released by the tumor. In 1954 we reported that this activity resided in a nucleoprotein moiety which became known as the Nerve Growth Promoting Factor, later to be shortened to Nerve Growth Factor (paper 8), and indicated ever since with the acronym NGF.

If the *in vitro* bioassay afforded the possibility, unattainable in the developing embryos, of identifying the nature of the tumoral nerve-growth promoting factor, its even more important contribution was that it provided a fast and most reliable assay to screen normal and neoplastic tissues, as well as organic fluids, as potential sources of this factor. Through the availability of this *in vitro* method, it became possible, in subsequent years, to discover large and small sources of NGF and to ascertain that this protein molecule is synthesized in minute but measurable quantities by a large number of normal and neoplastic cells from different vertebrate species.

The first problem we faced was to ascertain that the *in vitro* effect was called forth by the same agent that was released by the two sarcomas, 180 and 37, in the living embryos. Results reported in my 1958 article,[6] provided irrefutable evidence that the effects described *in vivo* and *in vitro*, though very different from each other, were indeed produced by NGF.

In 1956, chance rather than programmed research, yielded the discovery that the addition of very minute amounts of snake venom to the culture medium elicited the same nerve-growth stimulating properties as the neoplastic NGF. The venom was added to the culture medium to assay it, not as a potential growth agent, but for its high phosphodiesterase content. The experiments performed by Stanley Cohen,

[6]Levi-Montalcini R., Chemical stimulation of nerve growth. In *The Chemical Basis of Development*, W. McElroy and B. Glass eds. (Hopkins Press, Baltimore, 1958) pp. 646–664.

in order to degrade, with this enzyme, the nucleic acids still present in the active tumoral fraction, demonstrated that the venom itself possesses a potent nerve-growth stimulating activity. Cohen succeeded in identifying the active venom agent as a protein particle with a molecular weight in the order of 20,000.

The most purified preparation had a specific activity (on a protein basis) approximately 1,000 times higher than the purest tumoral fraction.[7,8]

The results indicated that in both instances the active material was a protein. A description of this finding is given in paper 9.

While Cohen was elucidating the biochemical nature of the venom NGF, I investigated its biological properties by daily injections of minute amounts of the active fraction into the yolk of developing chick embryos 7 to 9 days of age. The effects called forth by the venom NGF were so similar to those elicited by sarcomas 180 or 37, that it would have been difficult, on the basis of a histological analysis, to decide whether the embryos bore a tumor implant or if, instead, they had been with the purified venom fraction. In both sets of experiments, the effects were identical. Injection of the venom factor also resulted in the penetration of sympathetic nerve fibers into veins of the embryo (paper 9).

The discovery made two years later (1958) that the submaxillary salivary glands of adult male mice synthesize and release into the saliva a protein molecule endowed with the same nerve growth promoting activity as that produced by snake venom,[9] was prompted by the similarity between the functions displayed by these reptilian and mammalian glands. Both organs synthesize enzymes which play an important role in the digestive function as well as in defensive behavior. The rodent submaxillary salivary gland is a sexually dimorphic organ. This dimorphism is restricted to one of the two gland components, which are known as the acinar and the convoluted tubules. The latter have been shown to be endowed with an endocrine function.[10–14] The fact that this protein molecule is synthesized in large amounts by these gland components

[7]Cohen S., A nerve growth factor promoting protein. In *Proceedings of the Symp. on the Chemical Basis of Development*, W. McElroy and B. Glass eds. (Hopkins Press, Baltimore, 1958) pp. 665–679.

[8]Cohen S. and Levi-Montalcini R., A nerve growth-stimulating factor isolated from snake venom, *Proc. Natl. Acad. Sci. USA* **42**, 571–574 (1956).

[9]Cohen S., A nerve growth-promoting protein. In *A Symposium on the Chemical Basis of Development*, W.D. McElroy and B. Glass eds. 665–679 (1958).

[10]Lacassagne A., Dimorphisme sexuel de la glande sous-maillaire chex la Souris, *Compt. Rend. Soc. Biol.* **133**, 180–187 (1940).

[11]Caramia F., Angeletti P.U. and Levi-Montalcini R., Experimental analysis of the mouse submaxillary gland in relationship to its nerve-growth factor content, *Endocrinology* **70**, 915–922 (1962).

[12]Levi-Montalcini R. and Angeletti P.U., Biological aspects of the nerve growth factor. In *CIBA Symposium Growth of the Nervous System*, G.E.W. Wolstenholme and M. O'Connor, eds. (J. & A. Churcill Ltd., London, 1968) pp. 126–142.

[13]Bing J., Poulsen K., Hackenthal E., Rix E. and Taugner R., Renin in the submaxillary gland: a review, *J. Histochem. Cytochem.* **28**, 8, 874–880 (1980).

[14]Murphy R.A., Watson A.Y., Metz J. and Forssmann W.G., The mouse sub-maxillary glands: an exocrine organ for growth factors, *J. Histochem. Cytochem.* **28**, 890–902 (1980).

afforded the possibility of exploring its effects not only in chick embryos, but also in neonatal and adult vertebrates. In view of their small size and ready availability, mice and rats became the object of all subsequent investigations in living organisms and *in vitro*, of the effects elicited by salivary NGF on its target cells.

While we focused our attention on the manifold facets of the growth response, other investigators elucidated the primary and secondary structure of this protein molecule,[15] thus paving the way to the identification, many years later, of the NGF gene.

The main feature of the growth response elicited in its target nerve cells is the massive production of fibrillar material by the cells. In tissue cultures, this effect is already apparent two hours after incubation, and is so extensive as to overshadow other less impressive effects such as the marked increase in the Golgi complex and disarray of the endoplasmic reticulum, as visualized at the electron microscope (papers 12 and 13). Prominent changes are also apparent in the nuclear compartment. The nucleoplasm is more dense than in controls and aggregates of chromatin are irregularly dispersed throughout (paper 14).[16–18]

In vivo and *in vitro* studies demonstrated that the NGF effect on its receptive cells is accompanied by marked enhancement of synthetic and oxidative processes. The presence of a proper energy source is required for the outgrowth of nerve fibers; in its absence the growth of fibers begins but then ceases almost immediately. Experiments with specifically-labeled glucose indicate that this stimulation occurs mainly through a direct oxidative pathway (papers 17, 18 and 19). Lipid synthesis is also stimulated. Experiments carried out with labeled acetate provided evidence that the rate of incorporation into total lipids in sensory and sympathetic ganglia incubated with NGF is increased. Incorporation of labeled amino acids into proteins is enhanced by NGF in sensory and sympathetic ganglia incubated *in vitro* (paper 21). RNA synthesis is also rapidly and markedly stimulated by NGF as indicated by uridine incorporation studies and by comparison of the total RNA content in ganglia cultured with and without NGF.

The effect of NGF on enzymes of catecholamines was recently investigated (paper 20). These findings strongly suggested that the growth effect elicited by NGF specifically affects the synthesis of those enzymes involved in the biosynthesis of the sympathetic neurotransmitter. Three other enzymes studied, namely, dopa-decarboxylase, dopamine-hydroxylase and monoamine oxidase, were also found to

[15]Hogue Angeletti R. and Bradshaw R.A., Nerve growth factor from mouse submaxillary gland amino acid sequence, *Proc. Natl. Acad. Sci. USA* **68**, 2417–2410 (1971).

[16]Levi-Montalcini R., The nerve growth factor: its mode of action on sensory and sympathetic nerve cells. In *The Harvey Lectures*, Academic Press, Series **60**, 217–259 (1965).

[17]Levi-Montalcini R., Developmental neurobiology and the natural history of nerve growth factor, *Ann. Rev. Neurosci.* **5**, 341–362 (1982).

[18]Levi-Montalcini R., Angeletti R. and Angeletti P.U., The nerve growth factor. In *The Structure and Function of Nervous Tissue*, Vol. V, G.H. Bourne, ed. (Academic Press, New York, 1972) pp. 1–38.

increase in NGF-treated ganglia, but in a manner parallel to the increase of total protein, so that their specific activities did not appear significantly changed. These data seem to indicate that the NGF specifically stimulates, in developing sympathicoblasts, the synthesis of tyrosine hydroxylase, a rate-limiting enzyme in the production of norepinephrine.

Dramatic evidence at the histological level (paper 11) and at the ultrastructural level (paper 15) of the physiological role played by NGF in the life of sympathetic nerve cells was provided by the massive destruction of immature sympathetic nerve cells called forth by injection of NGF-specific antibodies in neonatal rodents and other vertebrates.

A recent reinvestigation[19] of such a role, accomplished through the generation of homologous recombinant mice lacking the neurotrophin NGF, fully confirmed the results of the investigation we had performed decades before, with the primitive tools and techniques available at the time. For the first time, it was proved that an endogenously synthesized protein is endowed with a vital role in the life of a neuronal population. The deaths of these cells, following NGF deprivation through the administration of specific antibodies, became known as immunosympathectomy (paper 16).

At the time of its discovery, this phenomenon was reputed to be due to the elimination of target-derived NGF. A massive destruction of immature sympathetic nerve cells also occurs after the administration of pharmacological adrenergic blocking agents. This process became known as chemical sympathectomy (paper 22).

It was demonstrated that these agents prevent, through different mechanisms, the intake of NGF from peripheral target tissues. The hypothesis that death of developing sympathetic and sensory nerve cells deprived of NGF is a passive process has been recently revised. It has been proved, in fact, that it results from the activation of genes regulating a death program. This discovery provided additional evidence of the pivotal role of NGF in the life of its receptive nerve cells.[20–22]

The NGF synthesized and released from target tissues of sympathetic neurons and a proportion of sensory neurons is retrogradely transported through the axons to the cell body of responsive nerve cells. Once the retrograde flow of NGF is established, it must continue for the lifetime of the neuron.[23]

[19] Snider W.D., Functions of the neurotrophins during nervous system development: what the knockouts are teaching us, *Cell* **77**, 627–638 (1994).

[20] Johnson E.M. Jr., Chang J.Y., Koike T. and Martin D.P., Why do neurons die when deprived of trophic factor?, *Neurobiol. of Aging* **10**, 549–552 (1989).

[21] Oppenhein R.W., Cell death during development of the nervous system, *Ann. Rev. Neurosci.* **14**, 453–501 (1991).

[22] Raff M.C., Barres B.A., Burne J.F., Coles H.S., Ishizaki Y. and Jacobson M.D., Programmed cell death and the control of cell survival: Lesson from the nervous system, *Science* **262**, 695–700 (1993).

[23] Hendry I. A., Stockel K., Thoenen H. and Iversen L.L., The retrograde axonal transport of nerve growth factor, *Brain Res.* **68**, 103–121 (1974).

Subsequent studies proved that NGF binds to its receptors on the nerve terminals. The NGF receptor complex internalized in membrane-bound vesicles is delivered intact to the neuronal cell bodies and initiates the signal transductor events that are the key to the mechanism of action of NGF.[24]

[24]Heumann R., Neutrophin signalling, *Current Opinion in Neurobiology* **4**, 668–679 (1994).

Reprinted from THE JOURNAL OF EXPERIMENTAL ZOOLOGY
Vol. 116, No. 2, March 1951

SELECTIVE GROWTH STIMULATING EFFECTS OF MOUSE SARCOMA ON THE SENSORY AND SYMPATHETIC NERVOUS SYSTEM OF THE CHICK EMBRYO [1]

RITA LEVI-MONTALCINI AND VIKTOR HAMBURGER
Washington University

TWENTY-THREE FIGURES

INTRODUCTION

One of the major problems of neuro-embryology concerns the specific affinities of different types of nerve fibers to their respective end-organs. The patterns of selective connections which are the foundations of normal functioning are laid down in early embryonic stages. The effectiveness of the mechanisms which bring about the anchorage of motor fibers in effector organs and of sensory fibers in receptor organs is particularly clearly demonstrated in experiments in which the dorsal or ventral half, respectively, of the spinal cord of Amphibian embryos was removed. The fibers emerging from the isolated ventral half formed typical patterns of motor nerves alone, whereas nerves composed exclusively of sensory fibers formed peripheral patterns of sensory nerves (Harrison, '06; Hamburger, '29; Taylor, '44; Weiss, '47).

The mechanisms by which these specific connections are established have not been elucidated as yet. Their analysis is facilitated by the fact that developing nerve centers are extremely sensitive indicators for the conditions which nerve fibers emerging from them encounter at the periphery. Any change in the milieu in which the outgrowing fibers spread have immediate repercussions on the corresponding nerve

[1] Supported by grant from the Rockefeller Foundation.

centers. For instance, subnormal peripheral conditions in the field of growth of sensory fibers result in an atrophy or degeneration of their pericarya and in the reduction of proliferation and differentiation in adjacent cells within the spinal ganglia. On the other hand, the enlargement of the peripheral field, by limb transplantation, has the opposite effects (Hamburger and Levi-Montalcini, '49). These responses are selective in the sense that only those centers which are actually involved in the innervation of the modified peripheral field show a reaction. Different centers respond independently of each other even if they are adjacent to each other. Once this point is established as a general rule, it can be used as an indicator for affinities between specific nerve fibers and their peripheral fields.

The experiments of limb extirpation and transplantation which have been used in the past to study the effects of peripheral factors on the nerve centers do not permit a penetrating analysis of peripheral affinities. Limbs are heterogeneous structures and their extirpation or transplantation results in a modification of the peripheral fields of *all* components of mixed nerves. We have, therfore, started experiments on chick embryos in which more homogeneous tissues, such as muscle, brain, liver or skin, were implanted in the place of a limb. In this way we hope to create specifically favorable conditions for the growth of one component alone. Our preliminary results are not conclusive. Bueker ('48) following a similar reasoning, made the important discovery that mouse sarcoma 180, when transplanted near the leg of a two to two and one-half-day chick embryo was strongly invaded by sensory fibers, whereas the motor nerves did not enter the tumor. The selective nature of the tumor innervation was deduced from the hyperplasia in the spinal ganglia and the coincident hypoplasia in the motor column of tumor-innervating segments. These results were of particular interest to us in connection with our study of the development of spinal ganglia (Hamburger and Levi-Montalcini, '49). We had observed the existence of two types of cells in the spinal ganglia which can be distinguished by their topographic position, coloration, size and time of differentiation and which respond differentially to limb extirpation and transplantation. The question arose whether the tumor supplies favorable growth conditions for both types or only for one type of sensory neuroblasts. We have, therefore, repeated the experiments with mouse sarcoma 180 (with the consent of Dr. Bueker); in addition we have used mouse sarcoma 37.

We wish to express our appreciation of the competent assistance of Miss Mary Lee Winkler throughout all phases of this investigation. The photographic work was done by Mr. Lothar Pinkers and the drawings by Mr. Paul Shieh.

MATERIAL AND METHODS

The experiments consisted of the transplantation of small pieces of mouse tumor to chick embryos of two and one-half to three days of incubation (Hamburger and Hamilton, '51, stages 20–22). The following tumors were used: sarcoma 180, sarcoma 37 (fibrosarcoma); carcinoma dbrB (mammary gland adenocarcinoma) and carcinoma C3HBA (mammary gland adenocarcinoma). All tumors were obtained from the Jackson Memorial Laboratory, Bar Harbor, Maine.

The transplantations were made as follows: Small pieces of the size of a limb bud of a three-day embryo (approximately 1 mm^3) were excised from a solid part of the tumor. A slit was made at the base of the wing bud or leg bud, respectively, between the limb bud and the somites, and the transplant placed in the slit. It was partly exposed and partly extending into the coelomic cavity, and held in place by the tension of the adjacent tissues. The majority of the transplants were made to the leg level. When it was ascertained that the tumor implants were invaded selectively, the possibility suggested itself that we were dealing with the responses of nerves to specific agents produced by tumors. To check an alternative possibility, namely a reaction to mouse tissue in general, transplantations were made of pieces of mouse placenta taken from mice on the 15th day of pregnancy. The size of these transplants and the transplantation technique were the same as in tumor transplants. Altogether 12 transplants were made; of these 4 were studied on sections.

The embryos were fixed between 4½ and 17 days of incubation. The hosts of placenta transplants were fixed between 5 and 12 days. All embryos were sectioned at 10 to 12 μ. Two staining techniques were used. B. Wenger's modification of Heidenhain's Hematoxylin (B. Wenger, '51) was applied for the study of proliferation and cellular degeneration, in

TABLE 1

Survey of results

TUMOR	NUMBER OF TRANSPLANTS	DIED	FIXED	SECTIONED (DAYS)		
				4½–6	7–10	11–16
Sarcoma 180	138	44	94	17	20	11
Sarcoma 37	92	51	41	11	10	4
Totals	230	95	135	28	30	15

embryos of 4½ to 9 days. De Castro's modification of Cajal's silver impregnation technique was used for the study of nerve cell differentiation and peripheral distribution of nerve fibers in embryos of 6 to 17 days of incubation. Only specimens with optimal and complete silver impregnation were used for the present study. Altogether 73 cases of tumor transplants were used. Table 1 gives a survey of the results.

Area measurements of ganglia were made in the following way: Outline drawings were made of every section in 5- and 6-day embryos and of every other section in older embryos, using the camera lucida. The areas were measured with a planimeter and added. The values are given in arbitrary units.

CHARACTERISTICS OF TUMOR GROWTH

None of the transplants of the two carcinomas developed; sections showed that they were resorbed after transplantation. Their failure to develop under the conditions of our experiment cannot be considered as evidence for incompatibility, since the same carcinomas develop on the chorio-allantoic membrane (Taylor and Carmichael, '49). The two sarcomas showed active and progressive growth. The mortality was within the range of the mortality found in transplantation experiments in general, and there was no indication that either of the two sarcomas had a toxic effect, at least up to 13–14 days of incubation. Many embryos which carried very large tumors at these stages were vigorous and normal. The higher mortality in the sarcoma 37-series was due to an accidental infection in this group.

Embryos sectioned two to three days after operation showed actively proliferating tumor tissue in the body wall. Although the transplants were at first exposed to the surface, the exposed parts were resorbed in many cases and skin covered the site of implantation. In other cases the resorption of the exposed parts was not completed at 5 days, and in these instances necrotic tissue was found on the surface. A tendency to grow inward and toward the coelom was a universal feature in all tumors. Both sarcomas showed a strong affinity for the mesonephros. In every case in which the transplant was placed at or near the mesonephric level, the mesonephric tissue was infiltrated by neoplastic cells. The beginning of its invasion was observed at 5 days. Most tumor transplants which had been originally placed in the wing level anterior to the mesonephros invaded the latter by caudal growth, and transplants in the leg level sent strands of cells craniad into the mesonephric tissue. In all but a few cases, the tumors destroyed eventually more or less extensive areas of the mesonephros; however, they did not interfere with the development of the intact parts of this organ. The metanephros which lies in close vicinity to the mesonephros, and dorsal to it was apparently not attacked. Other viscera, such as liver, lungs and intestine, remained also remarkably free of tumor tissue although the tumor was in many instances favorably located for their invasion.

Whereas both tumors were identical in their consistent invasion of the mesonephric tissue and in their refractoriness to other viscera, their invasive tendencies showed differences in other respects:

Sarcoma 180 forms compact masses which displace other tissues or project as solid spherical or oval-shaped bodies into the coelomic cavity. In some older embryos the right abdominal cavity was completely occupied by a solid tumor mass which compressed the intestine and other viscera (fig. 1). The edge of the tumor was usually well delimited even if it was located inside of the body wall, and only neoplastic tissue invading mesenchyme and the mesonephros was not sharply segregated from host tissue. In a few instances the tumor had replaced parts of limb or pectoral muscles. As a rule the tumors increased in size and became progressively larger in older stages. However, their size varied considerably. Most sarcoma-180 transplants appeared to be in a condition of vigorous growth, and only the very large tumors obtained from embryos of 15–16 days showed necrotic areas in the center. No detailed study was made of the histology of the tumors. A casual inspection showed different types of cells, among them giant multinuclear cells and small cells.

Sarcoma 37 shows a much stronger tendency to infiltrate adjacent tissue by means of long, narrow strands of cells which invade not only the mesonephros but in the same fashion adjacent muscle and connective tissue and creep along nerves. One observes frequently the beginning of the infiltration of muscles by such cell groups which lie between the muscle fibers and eventually replace the latter completely. The strong invasiveness of this tumor is further documented by its tendency to envelop the sympathetic chain ganglia (fig. 16) and even the spinal sensory ganglia. However, neoplastic tissue never penetrated into the nervous structures nor were these structures impaired in their development by the close contact with neoplastic cells. Whereas sarcoma 180 merely attached itself to the inner side of the limb or of the somatopleure, rarely invading the adjacent rib or limb muscles, sarcoma 37 shows no such restraint. It expands in all directions, destroying all musculature and connective tissue which is in

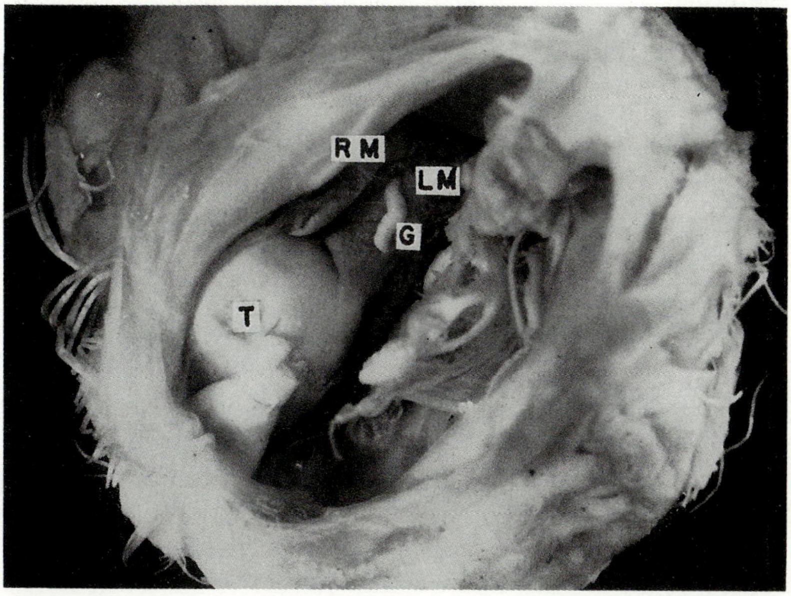

Fig. 1 (180-86; sarcoma 180). Large tumor (T) in right abdominal cavity. g = gonad; lm = left (normal) mesonephros; rm = remnant of right mesonephros.

its path. In contrast to sarcoma 180 the neoplastic tissue of sarcoma 37 has no sharp boundaries except where the bulk of its mass projects into the coelomic cavity. Even in these instances, its edges are usually fringed. The overall growth of sarcoma 37 seems to be more rapid than that of sarcoma 180. Large necrotic areas were found within the tumor as early as at 10 days.

INVASION OF THE TUMORS BY NERVES

In accordance with the observations of Bueker ('48), we have found that the tumors were invaded by nerves. There is no significant difference between the invasion of sarcoma 180 and sarcoma 37, and we shall discuss both together.

The following points concerning the tumor invasion were studied in detail:

1. Nerves and nerve centers in stages preceding the ingrowth of fibers.
2. The process of nerve ingrowth.
3. The source of nerve supply.
4. The endings of nerve fibers in the tumor.
5. The effect of the nerve ingrowth on the nerve center involved; special emphasis was given to the different aspects of their development such as proliferation, cellular differentiation and cellular growth.

In the following presentation it will not always be possible to treat these different components separately since they are intricately interwoven.

I. Period prior to nerve ingrowth

The tumors were implanted at a considerable distance from the central nervous system, namely, ventro-laterally to the somites. Only a few motor fibers have left the cord at the stage of operation. During the first two days after implantation (4 and 5 days of incubation) the nerves follow their normal paths in a ventro-lateral direction. The mixed nerves in these stages are composed exclusively of motor fibers and of the fibers from the early differentiating large ventro-lateral (V-L) cells of the spinal ganglia.

The possibility was considered that agents produced by the tumor might spread, and affect the spinal ganglia and spinal cord in its neighborhood even before nerves reached the tumor. Mitotic counts were made in 5 ganglia which sent fibers towards the tumor and in their contralateral partners (all mitoses from late prophase to early telophase were counted on every section using oil immersion). No significant difference was found in the mitotic activity of the ganglia (see table 2). The differentiation processes in the spinal ganglia and spinal cord were not affected either.

We shall consider next embryos of 5 to 6 days. These are the stages in which nerve fibers reach the level of the tumor. At that stage the tumors have in some cases withdrawn from

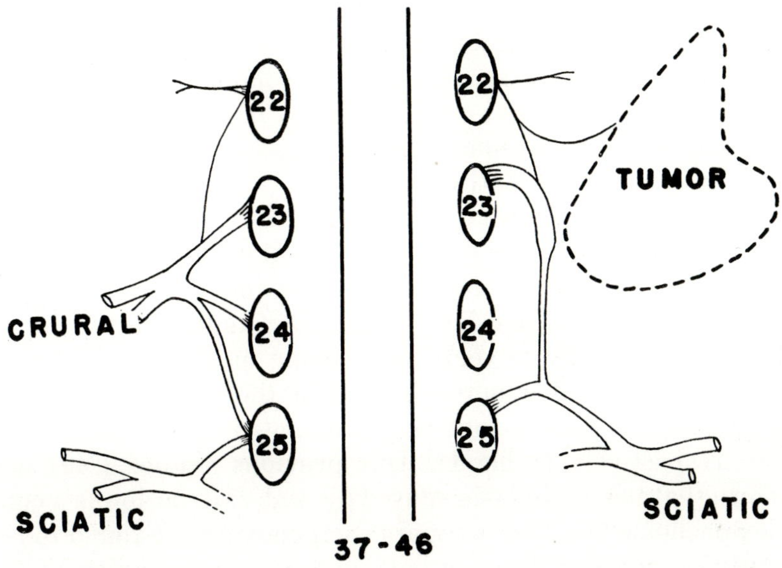

Fig. 2 (37-46; sarcoma 37; $5\frac{1}{2}$-day embryo). Semi-diagrammatic reconstruction of relation of nerves to tumor.

the surface and shifted to a more median position. As a result, limb nerves whose normal pathways in early stages are in the lateral body wall, could by-pass the tumor laterally and enter the limb in the normal fashion. The observation that in these early stages many nerves did establish their normal peripheral connections unhampered by the tumor is important in connection with certain observations made on older embryos (see below). In these instances, no responses were detected in the ganglia or in the central nervous system, even if such nerves had grown along the edge of the tumor. The situation was different in other cases in which the tumor was actually in the pathway of outgrowing limb nerves and formed a barrier at the base of the limb (fig. 2).

In these instances, the nerves were deflected when they reached the tumor. In most cases, they followed the inner edge of the tumor, without sending branches into it. In the case of figure 2 and in another reconstructed case, the crural nerve was completely missing, and residual fiber bundles grew caudad and contributed to the sciatic nerve. In another case, in which the wing nerves were completely blocked, the larger part of the fibers ended abruptly in front of the tumor, and only a few small fiber bundles reached the adjacent body wall along atypical pathways.

In all cases of a substantial block of the ingrowth of nerves into the limb, the motor and sensory centers were hypoplastic. In view of the spectacular hyperplasia found at later stages in ganglia of the same levels, we have studied this point in detail. Area measurements were made of three spinal ganglia which showed hypoplasia, and of the contralateral ganglia. The decrease of total area amounted to 31%, 36%, and 25% respectively (table 4). Our previous studies ('49) had shown that the hypoplasia caused by limb extirpation is a complex phenomenon. The two main components were found to be: reduction of the mitotic activity and cell degeneration of the early differentiating ventro-lateral (V-L) cells. These two points were checked in the present material. Mitotic counts were made on the same three pairs of ganglia which had been used for area measurements, and on two additional pairs. A reduction in the mitotic activity was found ranging from 16% to 37% (table 2). Degenerating cells were found among the V-L cells of hypoplastic ganglia, though their number was smaller than in limb extirpation cases. In the case of limb extirpation the terminations of the blocked nerves at the base of the stump showed neuroma-like conglomerations and coiling of fibers. The same features were observed in nerves blocked by the tumor, though less strikingly. The less extensive reactions in the present material are probably due to the fact that the block was rarely complete and that a number of fibers reached the periphery by atypical pathways.

In no case did nerve bundles enter the tumor at the stages under consideration. It will be recalled that at 5 and 6 days the early differentiating V-L fibers and motor fibers are the only components of the mixed nerves.

TABLE 2

Effect of tumor transplantation on mitotic activity in spinal ganglia (Sarcoma 37 and 180)

CASE NO.	GANGLION NO.	CHRONOLOGICAL AGE IN DAYS	MITOTIC COUNTS		DIFFERENCE %
			Transplant	Control	
180–71	21	5	70	66	+ 6.0
180–32	20	5	124	122	+ 1.6
180–34	20	5	123	131	— 5.3
180–76	17	5½	106	98	+ 8.2
180–72	20	5	140	154	— 9.0
180–33	23	5½	80	127	— 37.0
180–58	15	5½	154	200	— 23.0
37–46	22	5½	90	107	— 15.9
37–46	23	5½	70	106	— 33.9
180–36	23	5½	109	156	— 36.5
37–51	24	6½	110	87	+ 26.4
37–51	25	6½	128	85	+ 50.5
180–20	25	7	70	48	+ 45.9
180–5	17	7	65	38	+ 71.1

To summarize: Up to 6 days, the tumor is not invaded by nerve fibers. Whenever it is in the pathway of nerves, it acts as a barrier for the nerves, and the corresponding nerve centers show signs of impairment. (Motor hypoplasia, degeneration of V-L cells and reduction of the mitotic activity in spinal ganglia.)

Ingrowth and distribution of fibers

From the 7th day on, nerve fibers begin to invade the tumor in large quantities, and by 11 days, the tumors are richly supplied with nerves (figs. 8–10). We shall present first two typical cases, one of 9 days of incubation and one of 15 days.

Sarcoma 180–16. The transplant was placed between the right leg bud and the somites. The embryo was fixed at 9 days. Its right leg was normal.

Tumor. The tumor extends from the level of the 24th to the level of the 30th segment. It is attached to the inner side of the base of the leg. Its bulk forms a solid mass which fills the right half of the coelomic cavity. The mesonephros is largely destroyed.

Leg innervation. The lumbo-sacral plexus is almost normal. It is formed dorsally to the tumor. The crural nerve is almost completely absorbed by the tumor, and only a small branch goes to the periphery. The sciatic nerve by-passes the tumor at its dorso-lateral border and enters the leg in a rather typical fashion. Limb muscles were found to be well innervated.

Nerves in the tumor. The tumor is supplied by the leg nerves 23 to 29 and by the last thoracic nerve, no. 22. The peripheral parts of the tumor and particularly the regions near the entrance points of nerves, are supplied by numerous branches of different diameters which grow in all directions, fuse with each other and subsequently branch again. The central and medio-ventral parts of the tumor are free of nerves. One gets the impression that the rapid growth of the tumor has overtaken that of the fibers. The tumor nerves represent partly side branches of limb nerves and partly independent nerve bundles which grow directly from spinal and sympathetic ganglia into the adjacent neoplastic tissue. Spinal ganglia 22 to 24 which send the bulk of their fibers into the tumor are strikingly hyperplastic (figs. 4 and 5). The hyperplasia is less conspicuous in the posterior ganglia which contribute largely to the sciatic nerve. Sympathetic ganglia innervating the tumor are also hyperplastic.

Spinal cord. The lateral motor column is hypoplastic throughout the leg level. The degree of hypoplasia is higher in the level of atypical nerves (segments 21 to 24) than in the posterior segments. The median part of the dorsal funiculus is reduced in size and the lateral part is enlarged.

Sarcoma 180–86 (fig. 3). Transplant between leg bud and somites. The embryo was fixed at 15 days. The right leg was reduced in size and abnormal in shape.

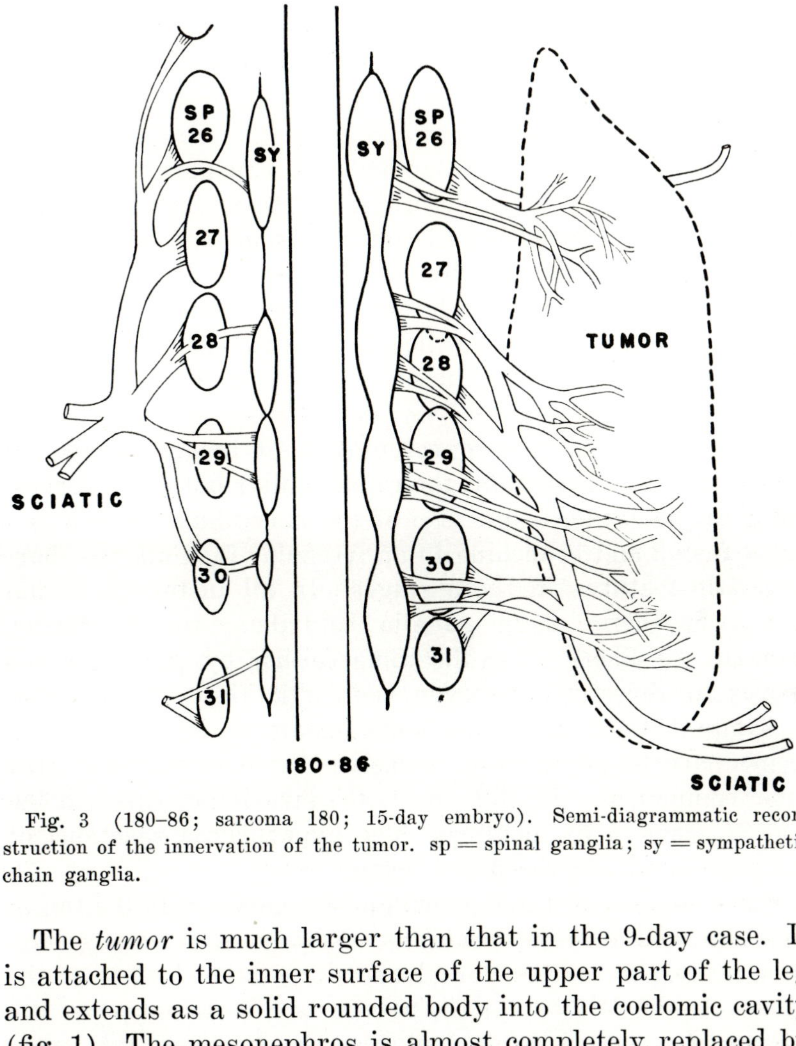

Fig. 3 (180–86; sarcoma 180; 15-day embryo). Semi-diagrammatic reconstruction of the innervation of the tumor. sp = spinal ganglia; sy = sympathetic chain ganglia.

The *tumor* is much larger than that in the 9-day case. It is attached to the inner surface of the upper part of the leg and extends as a solid rounded body into the coelomic cavity (fig. 1). The mesonephros is almost completely replaced by tumor tissue.

Leg innervation. The plexus of the sciatic nerve is rather normal except for nerve 26 which does not join it. The strong sciatic nerve grows across a narrow dorso-lateral portion of the tumor into the limb. Those leg muscles which are present are well innervated.

Nerves in the tumor. The tumor is pervaded by numerous very thick and medium-sized nerve bundles which grow irregularly in all directions, fuse with each other and branch again. Only small areas of the large tumor mass are not reached by fibers. The spinal and sympathetic ganglia which send their fibers into the tumor are considerably hyperplastic. All sympathetic chain ganglia from segments 26 to 31 inclusive are fused into a continuous mass; constrictions indicate the borders of individual ganglia.

Spinal cord. The right lateral motor column is slightly hypoplastic in the anterior leg level. The median part of the dorsal funiculus is reduced in size, and the lateral part is enlarged.

These two cases are representative of stages between 9 and 15 days. In particular, the random distribution of nerves, the great variation in the size of the nerve bundles, the frequent fusion and branching in an irregular fashion, are characteristic features in these stages. In all instances, spinal and sympathetic chain ganglia contribute to the tumor. Sympathetic fibers enter the tumor either by way of mixed nerves or directly, whenever neoplastic tissue is adjacent to sympathetic ganglia. In some instances the contribution of sympathetic fibers is so extensive that it overshadows the other components. In all respects the invasion of sarcoma 180 and 37 seem to be identical. The hyperplastic reactions of the ganglia will be discussed below.

The *beginning* of the ingrowth of the nerves into the tumor was observed in 7-day embryos. Nerves enter in big bundles, preferably at the dorsal and dorso-lateral edge of the tumor. In the 7-day stage the nerves usually do not penetrate deeply. In 8- and 9-day embryos the invasion has made distinct progress. Large nerve bundles penetrate into the deeper parts of the tumor. However, its central and ventral portions are not yet fully supplied by fibers. A complete nerve supply of all parts of the tumor was not observed until 11 days of incubation. From this stage on, medium-sized and small tumors are almost invariably pervaded by nerves throughout their mass. Only in very large tumors is the nerve supply restricted to the more dorsal and marginal areas.

In many instances, nerves were observed to grow across the edge or even across the middle of the tumor to peripheral structures including limbs. These nerves may or may not give off branches into the tumor. We assume that in all cases the connections of nerves with the periphery were established before the tumor reached the nerves.

Termination of nerve fibers (figs. 10 and 11)

It was of particular interest to observe the termination of nerve fibers in neoplastic tissue. As a rule, fibers do not establish contact with individual neoplastic cells. Instead, bundles of different sizes end abruptly without splitting up into individual fibers (fig. 10). One observes occasionally terminal thickenings at the blind ends of fibers, but these cannot be considered as synaptic buttons, since they have no relation to adjacent cells. This type of terminal ending is characteristic of all stages considered in this investigation. In the younger stages this condition may merely indicate that the nerves are still in the process of active growth. Similar growth cones are a characteristic feature of growing nerve fibers. However, the same condition was found at stages in which nerve fibers distributed in normal peripheral fields have already established terminal connections. The abrupt endings of bundles in the tumor in these stages must be considered as atypical.

Two cases in our material deviate from the typical picture. Both are embryos of 13 days of incubation. In both instances parts of the tumor show the typical picture described above; but in other parts the bundles split up into fibers which spread between tumor cells and seem to form a network around individual cells. One might suspect that in other instances similar terminations were present but failed to react to silver. This is not the case; most specimens show an optimal impregnation, identical with that found in the two exceptional cases. At present, we are not in a position to explain the unusual fiber distribution of the latter.

Responses of sympathetic chain ganglia

It was mentioned above that paravertebral sympathetic chain ganglia contribute to the nerve supply of the tumor (figs. 8, 9). Simultaneously with the first entrance and spreading of fibers in the tumors, at 7 days, the sympathetic ganglia which contribute fibers are enlarged, and a progressively increasing hyperplasia was observed in subsequent stages (figs. 6, 13, 16). The increase in the size of ganglia resulted in some instances in the fusion of neighboring ganglia; in such cases, they formed a continuous massive column with constrictions indicating the borders of individual ganglia (see fig. 3). In order to obtain a quantitative estimate of the degree of hyperplasia, area measurements were made for the entire length of the sympathetic chain sending nerves to the tumor. Only the cellular parts of the chain were included. Measurements were made on three embryos on both sides (table 3). The volumes were computed by adding up the figures for areas. The ratios of the values for volumes on the operated and the control side were 6.5:1 for case 180-86 (15 days; fig. 3); 6:1 for a 13-day embryo, and 5.4:1 for an 11½-day embryo.

The enlargement is mainly due to a *cellular hypertrophy*. Figure 6 illustrates the conspicuous increase in cell size which is characteristic of all hyperplastic ganglia.

The question arose whether the enlargement was entirely due to cellular hypertrophy or if the *number of cells* was also increased. Mitotic counts could not give an unequivocal answer in this particular case, for the following reason: The formation of the paravertebral chain is partly brought about by migration of cells from the more ventrally located "pri-

TABLE 3

Paravertebral sympathetic ganglia

CASE NO.	CHRONOLOGICAL AGE IN DAYS	SEGMENTAL LENGTH	VOLUMES [1]			CELL COUNTS [2]			RELATIVE INCREASE IN CELL NUMBERS [3]
			Exp.	Control	Ratio	Exp.	Control	Ratio	
180–103	11½	1668 μ	851.75	157.09	5.4 : 1	4293	13658	1 : 3.2	1.70 ×
180–69	13	3600 μ	2411.0	437.8	6.0 : 1	2702	7525	1 : 2.8	2.10 ×
180–86	15	924 μ	813.78	124.61	6.5 : 1	2680	5678	1 : 2.12	3.07 ×

[1] In arbitrary units.

[2] Three hundred squares covering a total of 2.6 mm^2.

[3] Area/cell numbers.

mary chain'' and only in part by proliferation in loco. Total cell counts would have required the enormous task of counting the entire chain, since ganglia are often fused. In order to obtain an estimate of cell numbers, cells were counted in random samples of equal areas, on both sides. This was done by inserting a micrometer disc (square type) in the ocular and counting the number of cells in 300 squares each, in comparable regions of normal and hyperplastic ganglia. The same three cases were used in which the area measurements were made (table 3). The ratios of cell numbers per unit area were 1:3.2; 1:2.8; and 1:2.12 in favor of the control side. If the enlargement were due exclusively to cellular hypertrophy, this ratio should be close to 1:6, since the average ratio of volumes in the same cases was 6:1. Actually, it was about 1:3. It follows that the number of cells must be approximately twice as large as in the contralateral control ganglia. These figures are based on few cases, and they are merely taken as an indication that the cell numbers in enlarged ganglia are actually increased. In other words, the overall increase in size is due to a greater extent to cellular hypertrophy and to a lesser extent to an increase in proliferation.

The comparison of hyperplastic and normal ganglia shows that the cells in the former are not only larger in size but in more advanced stages of differentiation. Neurofibrils are present as early as at 9 days (figs. 18 and 19), whereas the cells on the control side show no neurofibrils. (The fibers which are regularly observed in the center of normal ganglia can be identified as ingrowing preganglionic fibers.) The advanced stage of differentiation is also attested by the large fiber bundles which emerge from these ganglia and are not found on the control side. This acceleration of differentiation is a response of nerve centers which had never been observed before.

The hyperplastic and hypertrophic effects are observed consistently in sympathetic ganglia which supply nerves to the tumor. A very conspicuous hyperplasia and hypertrophy are found only in cases in which the tumor is closely adjacent to the sympathetic chain, and large fiber bundles enter the tumor directly. It is less pronounced in cases in which sympathetic fibers reach the tumor by way of mixed nerves.

Responses of the spinal ganglia

In previous investigations of the normal development of the spinal ganglia (Levi-Montalcini and Levi, '43; Hamburger and Levi-Montalcini, '49), it was found that the growth and differentiation processes do not occur simultaneously in all sensory cells. Two categories of cells were distinguished on the basis of morphological criteria: early differentiating, ventrolaterally located (V-L) cells and later differentiating mediodorsally located (M-D) cells. The contrast between the two in size and color is particularly striking at 11–13 days of incubation.

The two groups react differently, and independently of each other, in limb extirpation and transplantation experiments. Nevertheless, each of these two groups may be functionally heterogeneous, as we have emphasized before; and, for the time being, we consider the distinction of only two classes as provisional.

1. V-L cells. We have described above the responses of the spinal ganglia to tumors in the early stages in which the mixed nerves reach the level of the tumor (5–6 days of incubation). In these stages, the sensory component of the mixed nerve is exclusively derived from V-L cells, whereas the M-D cells are still undifferentiated. The fibers of the V-L cells do not enter the tumor. Not only do those nerves, which grow along the edge of the tumor, fail to send branches into it, but nerves whose pathways are blocked by the tumor are arrested in their growth. In the latter instances, cell degeneration was found among the V-L cells, and the mitotic activity in the ganglia was reduced. The hypertrophy and accelerated differentiation of the adjacent M-D cells in subsequent stages, which will be discussed below, makes it sometimes difficult to distinguish between the V-L cells and adjacent M-D cells. In some instances, the area occupied by V-L cells was significantly decreased in older embryos; this would indicate a persistence of the early impairment.

2. M-D cells. At 7–8 days of incubation, the picture changes completely. At this critical stage, the nerves begin to send big bundles into the tumor. This stage coincides with a new wave of differentiation in the spinal ganglia, namely that of the M-D cells. At the same time, these cells become larger than those on the contralateral side. They show a strong affinity to silver, and neurofibrils are clearly distinguishable, whereas the corresponding cells of the contralateral side show no sign of differentiation. This point was ascertained both in silver and hematoxylin preparations (figs. 20–23). We are dealing here with the same phenomenon of the acceleration of the differentiation process which was described above for the sympathetic ganglia.

The increase in size becomes more pronounced in the following days. It is very conspicuous at 9 to 13 days, and consistently present in all ganglia whose nerves can be traced into the tumor. This statement is based on observations of 3–5 ganglia of each of over 40 cases. No reaction is found in ganglia which do not contribute nerves to tumors. In order to obtain an estimate of the quantitative aspects of the enlargement of ganglia, area measurements were made in the same way as for sympathetic ganglia in 11 pairs of ganglia, of embryos of 9 to 15 days of incubation. The data are given in table 4. The total increase in size is less pronounced than in sympathetic ganglia. It is again primarily the result of *cellular hypertrophy* which is very conspicuous in every section (figs. 4, 5). A careful observation of the topographic location and of the staining characteristics of the hypertrophic cells shows that they belong to the group of the M-D cells.

It was again of interest to know whether the cellular hypertrophy accounts fully for the increase in area or whether an increase in cell numbers occurs. In the case of spinal ganglia, the condition is more favorable for *mitotic counts* than in sympathetic ganglia, because the ganglia remain separate units even when hyperplastic, and the normal mitotic pattern is well known. Mitotic counts on 4 pairs of ganglia at 7 days of incubation showed a consistent increase in the mitotic activity; this is a reversal of the conditions found a few days earlier in corresponding ganglia (table 2).

The increase in cell numbers was furthermore checked by *cell counts* on 4 pairs of ganglia of embryos between 9 and 15 days of incubation (table 4). Counts were made of all cells, on every other section. No distinction was made between V-L cells and M-D cells. The increase of cell numbers ranges from 20% to 50%. Since all other observations show that only the M-D cells participate in the nerve supply of the tumors, the increase in cell numbers is attributed to this category.

Observations on the spinal cord

The *lateral motor columns* were in no instance hyperplastic. On the contrary, a hypoplasia was observed in many cases. These findings confirm those of Bueker ('48). When present the hypoplasia was limited to a few segments and variable in degree. When the nerves were traced from hypoplastic segments to the periphery, a close correlation was found between location and degree of hypoplasia and the impairment of the limb nerves emerging from these segments. Either the limb innervation had been partially blocked by the tumor, or the limb itself had been impaired by the operation. All instances of motor hypoplasia can be accounted for as the result of these two factors.

The *median motor columns* and the *preganglionic nucleus of Terni* were apparently unaffected by the presence of the tumor. The absence of a response in the latter is surprising in view of the fact that its fibers terminate in sympathetic chain ganglia some of which are tremendously hyperplastic.

The *dorsal funiculus* is formed by afferent dorsal root fibers. Unpublished observations on its normal development (Levi-Montalcini) give evidence that the dorsal root fibers

TABLE 4

Cell counts and area measurements in sensory ganglia

CASE NO.	CHRONOLOGICAL AGE IN DAYS	GANGLION NO.	CELL COUNTS (TOTAL NUMBERS)			AREA [1]			REMARKS
			Exp.	Control	Ratio	Exp.	Control	Ratio	
180–34	5	20				40.29	44.97	0.90 : 1	Preceding first contact of nerves with tumor.
180–59	5	20				38.65	38.21	0.99 : 1	
180–71	5	21				24.08	28.67	0.84 : 1	
180–32	5	20				30.01	31.87	0.94 : 1	
37–36	5	23				45.15	66.45	0.69 : 1	Nerves blocked by tumor.
37–46	5½	23				19.43	30.05	0.64 : 1	
180–58	5½	15				37.77	52.58	0.75 : 1	
180–16	9	23	11913	7916	1.50 : 1	100.42	42.82	2.35 : 1	Spreading of nerves in tumor.
180–53	9	22	9626	7083	1.35 : 1	117.24	36.34	3.20 : 1	
180–53	9	23				118.34	45.94	2.57 : 1	
37–7	9	13				101.85	51.24	1.99 : 1	
37–7	9	14				186.75	99.71	1.97 : 1	
37–7	9	15				288.00	142.35	2.10 : 1	
37–7	9	16				210.00	117.00	1.80 : 1	
37–7	9	17				139.20	90.00	1.55 : 1	
37–9	9	17	8990	7464	1.20 : 1	98.68	36.16	2.70 : 1	
180–103	11½	25				97.92	67.94	1.44 : 1	
180–70	13	20				196.38	151.6	1.40 : 1	
180–86	15	29	5828	4661	1.25 : 1	148.01	60.76	2.40 : 1	
180–99	15 + 18 hr.	15				204.77	196.39	1.10 : 1	
180–102	15 + 18 hr.	25				200.70	168.22	1.20 : 1	

[1] In arbitrary units.

from the V-L cells, which are the first to differentiate, enter the cord laterally, and the incipient dorsal funiculus has a lateral position. This part of the dorsal funiculus is gradually shifted into a more median position, near the dorsal median septum, by the addition of newly entering fibers at its lateral aspect. The progressive increase in the lateral part of the dorsal funiculus coincides with the progressive differentiation of the M-D cells. Therefore, we assume that the lateral part of the dorsal funiculus is derived from M-D cells.

In all 40 cases in which we found a hyperplasia and hypertrophy of the M-D cells in the spinal ganglia, we observed a parallel increase of the lateral part of the dorsal funiculus (fig. 12). The median part was in many, but not in all, cases reduced in size. This reduction was limited to short segments, whereas the size increase of the lateral part extended throughout the level of the tumor. Separate area measurements of the lateral and median parts of the dorsal funiculi were made, using the same technique as was described for the ganglia. The measurements were made throughout the entire length of the sector supplying the tumor, in 4 embryos. Increases in the areas of the lateral parts amounted to 53%, 71%, 78% and 105%, respectively; whereas the median parts were either slightly reduced or unaffected. These observations are consistent with all other data which indicate the independence of the responses of V-L cells and M-D cells to the tumors.

Preliminary data on additional tumor innervation by peripheral sympathetic cells

In addition to the paravertebral sympathetic chain, peripheral sympathetic cell groups are known to be present in a prevertebral position near the dorsal aorta and in the viscera. These scattered ganglia reach their peripheral positions in early stages of embryonic development, and they receive preganglionic fibers through the paravertebral chain. The medullary component of the suprarenal gland is known to have the same origin as the prevertebral ganglia, namely from sympathoblasts which migrated to their terminal position in early phases of development. They differentiate into secretory cells and receive preganglionic fibers. In the chick, the primordia of the medullary adrenal tissue are evident at 6 days of incubation (Brauer, '32).

In all cases of our material in which tumors had reached the location of the suprarenal gland, large groups of sympathetic cells were found in positions where normally only small groups of scattered chromaffine and ganglionic cells are present. In some cases very conspicuous agglomerates of such cells were observed inside of the tumor. These cells have a strong affinity for silver and their nerve fibers are readily traceable in the tumor; they branch there abundantly. The fibers leave the ganglionic masses either in small bundles or in big nerves. The terminations of these fibers in the tumor differ from those derived from other sources. They split up into fine bundles, and individual fibers come in contact with individual neoplastic cells. In many instances, the adrenal medullary primordium of the contralateral side becomes involved; it sends fiber bundles into the tumor across the midline. In cases in which the tumor did not reach the level of the suprarenal primordium, other prevertebral ganglionic masses participated in tumor innervation in a varying degree.

The role of the adrenal primordium and of other prevertebral ganglia in the nerve supply of tumors will be dealt with in more detail in another publication.

The source of tumor nerves

In the following, a brief summary will be given of all data which give evidence that the M-D cells of the spinal ganglia and sympathetic chain ganglia are responsible for the nerve supply of the tumors and that the somatic motor system and the V-L cells do not send fibers into the tumor. (The prevertebral sympathetic system will be disregarded.)

Evidence for participation of sympathetic fibers. Fibers emerging from sympathetic ganglion cells were observed to enter the tumor either directly or by way of mixed nerves (figs. 8, 9). The corresponding ganglia were invariably increased in size.

Evidence for participation of M-D cells. (1) Fibers from spinal ganglia are observed to enter the tumor from 7 days of incubation on (fig. 8), that is synchronously with the observed beginning of neuroblast differentiation and of fiber formation of M-D cells. (2) Spinal ganglia which are connected with tumors are hyperplastic and hypertrophic. (3) Morphological observations show that this enlargement is due to an increase in the size of the M-D cells and not of V-L cells. (4) The lateral part of the dorsal funiculus which is apparently formed by centripetal M-D fibers is conspicuously enlarged in segments supplying nerves to the tumor.

Evidence that V-L cells do not send nerves to the tumor. (1) The fibers of the V-L cells which reach the tumor at 5 to 6 days, either by-pass it without sending branches into it, or, if the tumor blocks their pathway, they are arrested. (2) Degeneration was found in the same stages among the V-L cells. (3) The median part of the dorsal funiculus which is apparently formed by centripetal fibers from V-L cells is either hypoplastic or unchanged but never hyperplastic.

Evidence that the somatic motor system does not send nerves to the tumors. (1) The somatic motor fibers which differentiate very early and combine with the V-L fibers to form mixed nerves, do not enter the tumor and behave in the same way as the V-L fibers at the early stages of 5 and 6 days. (2) A conspicuous hypoplasia of the lateral motor column was found in a large number of cases at the same levels in which ganglia were hypertrophic and hyperplastic. There was no instance of motor hyperplasia. The hypoplasia is correlated with an impaired limb innervation: it is absent when the limb innervation is normal.

DISCUSSION

Bueker ('48) made the important discovery that mouse sarcoma when transplanted to early chick embryos is invaded by sensory fibers whereas motor fibers have no affinity for tumors. These findings were fully confirmed.

The fact that sarcomas of the mouse are heavily invaded by embryonic chick nerves is of considerable interest both from the oncological and from the neurological point of view. In general, very little attention has been paid to the ingrowth of nerves into tumors, and the quantitative aspects of their nerve supply are a matter of controversy (Duncan and Bellegie, '48; Shapiro and Warren, '49). All observations except those of Bueker ('48) were made on adult animals or humans, and no attention was paid to the reaction of the nerve centers involved. In general, the nerve supply seems to be scanty. This is not surprising since most tumors originate long after nerve outgrowth has ceased. The present study which covers the period of nerve outgrowth gives a different picture. Sarcomas 180 and 37, which were implanted prior to nerve outgrowth were supplied with a more dense net of nerve bundles than any normal tissue at any stage of development. The sarcoma tissue provides an extremely favorable medium for the outgrowth of nerve fibers; it stimulates their growth far beyond the normal range. In order to check whether mouse tissue other than sarcoma could stimulate nerve outgrowth, pieces of placenta were implanted as controls (p. 324). They were found to be not innervated. Controls with other tissues are under way.

The selectivity of nerve ingrowth which Bueker described in terms of admission of sensory but not of motor fibers proved to be even more complex and subtle. We found (1) that not all sensory fibers enter the tumor, but only the fibers from the late differentiating medio-dorsal (M-D) cells, (2) that paravertebral sympathetic chain ganglia have a considerable share in the nerve supply, and (3) that the derivatives of the prevertebral sympathetic cell groups, and in particular, the precursors of the suprarenal medulla contribute nerves to the tumor.

As a result of the ingrowth of fibers into the tumor, the corresponding ganglia respond with a conspicuous hyperplasia and hypertrophy which reaches a maximum of 600% in sympathetic ganglia, and of 250% in spinal ganglia (area measurements). It may be mentioned that the highest value of hyperplasia found in spinal ganglia following peripheral overloading by limb transplantation was 140%. We agree with Bueker ('48, p. 383) that the " . . . histochemical properties (of sarcoma 180) . . . favored sensory innervation and hence the enlargement of spinal ganglia."

The responses of nerve centers to changes of the periphery are of a complex nature; they have been discussed recently and will not be recounted here (Hamburger and Levi-Montalcini, '49; '50). It may suffice to state that the effects are not limited to the cells which are directly connected with the periphery, but that indifferent cells adjacent to those cells may become involved; for instance, the mitotic activity and neurodifferentiation of undifferentiated cells may be affected ("field effects"). The hyperplastic reactions in the spinal ganglia and sympathetic ganglia resemble in many respects those observed in cases of peripheral overloading of nerves by transplantation of supernumerary limbs. In particular, the "field effects" were obvious in both instances. Since the tumor-induced increase in size is quantitatively much more pronounced than in the best cases of limb transplantation, the material is unusually favorable for the study of certain details which would have been barely detectable in limb transplantation cases. For instance, the *cellular hypertrophy* and the *acceleration* of the differentiation process of neuroblasts which were distinct both in the spinal and sympathetic ganglia of the present material had not been observed before. The latter finding adds a new and unsuspected facet to the picture of peripheral control of the growth of nerve centers. It becomes more and more obvious that only a dynamic approach can give an adequate understanding of the entire process of nerve cell differentiation, and that observations made on one stage only may be misleading. This is illustrated again in the present instance. The fact that the enlarged cells are at the same time precocious and accelerated in their differentiation complicates the interpretation of the cellular hypertrophy. The hypertrophy is expressed in terms of comparison of cell size in pairs of ganglia, but these cells are not strictly comparable because they are in different stages of differentiation. It is impossible to decide whether the size of hypertrophied cells actually exceeds the normal size characteristic of their stage of differentiation, or whether their normal growth curve is merely accelerated. An excessive growth could be demonstrated only by comparison of complete growth curves of normal and hypertrophied cells which are not available at present.

We have emphasized in previous investigations that the effect of peripheral structures on nerve centers is mediated by nerve fibers and that each center is affected only by its own peripheral field. The present material gives new and definite evidence for this point. The fibers can be traced easily from their source of origin into the tumor (figs. 8, 9), and the study of the early phases of the process of invasion reveals the immediate responses of those centers which supply the nerves. The fact that nerve ingrowth into the tumor and cellular response occur almost simultaneously gives evidence of the extreme sensitivity of neurons to their peripheral milieu. The failure of the fibers to establish normal synaptic connections with neoplastic tissues shows that the fibers can mediate this effect in the absence of specific peripheral terminations.

The most interesting aspect of the invasion of sarcoma by embryonic nerve fibers, namely, the *selective affinity* of some categories of fibers and the refractoriness of others, cannot be explained in terms of the time patterns of the differentiation of nerves. The motor fibers and the sensory V-L fibers, both of which differentiate very early, had an optimal chance to enter the tumor. They did not avail themselves of this opportunity; on the contrary, they showed regressive changes unless they succeeded in by-passing the tumor. One might consider the possibility that in early stages the tumor causes changes in the host tissue immediately surrounding it (necrosis, lysis) which would create a barrier for nerve ingrowth. No such changes were observed. During the days of the establishment of the tumor, neoplastic cells and mesenchymal cells mingled at the edge of the tumor, and nerve fibers were frequently observed in the immediate neighborhood of the tumor without entering it. We conclude that the tumor does not exert a toxic effect; the regressive changes in the nervous system (cell degeneration and reduced mitotic activity) are not different from those observed in limb extirpation experiments; they are explained satisfactorily as effects of the mechanical blocking of normal nerve outgrowth. We did not observe any conspicuous changes in the vascularization of the tumors during the crucial period of the first nerve ingrowth which could account for the different behavior of the early differentiating and late differentiating nerve fibers. All available data indicate that the sarcomas 180 and 37 produce specific growth promoting agents which stimulate selectively the growth of some types of nerve fibers but not of others. The effects are mediated by the nerve fibers to their respective centers.

The question may be raised whether the hyperplasia and hypertrophy is a pathological response or merely the acceleration and quantitative modification of normal developmental processes. Our analysis was limited to morphological features. Neither the cytological structure of the cells nor the mitotic figures showed abnormalities. It appears that the process of differentiation is normal but considerably accelerated.

SUMMARY

Small pieces of mouse sarcomas 180 and 37 were grafted at the base of limb buds of two and one-half- to three-day chick embryos. They grew conspicuously and invaded adjacent tissue, particularly the mesonephros. The following observations were made on the responses of the nervous system to the tumors.

1. The motor fibers and fibers from the early differentiating ventro-lateral (V-L) cells of spinal ganglia which reach

the tumor at 5–6 days fail to enter it; they by-pass it without sending fibers into it or are actually blocked. In the latter instance, regressive changes were observed in the spinal ganglia: degeneration of V-L cells and reduction of the mitotic activity.

2. The picture changes completely at 7 days of incubation. Large nerve bundles are seen to invade the tumors. At 11 days, the density of nerves in the sarcomas exceeds that of any normal tissue at any embryonic stage.

3. The beginning of nerve ingrowth coincides with the beginning of the differentiation and of hyperplastic growth in the paravertebral sympathetic chain ganglia and in the group of late differentiating, medio-dorsal (M-D) cells in the spinal ganglia.

4. Additional evidence is given that the paravertebral sympathetic ganglia and the M-D cells of the spinal ganglia supply nerves to the tumor, whereas the V-L cells of the spinal ganglia and the motor cells are refractory.

5. Prevertebral sympathetic ganglia, and, in particular, sympathetic cells derived from the primordium of the adrenal medulla participate in the nerve supply.

6. The hyperplastic and hypertrophic responses of the sympathetic and spinal ganglia are more striking than those observed under other experimental conditions. They reach maxima of 600% in the former and of 250% in the latter. The increase in size is due largely to cellular hypertrophy, and to a lesser extent, to an increase in cell numbers. An acceleration of the differentiation process was observed in sympathetic ganglia and in the M-D cells of the spinal ganglia.

7. Normal synaptic connections are not a necessary prerequisite for this transmission, since the nerve fibers in the tumors were found to terminate abruptly, without entering into synaptic connections with neoplastic cells.

8. It is concluded that sarcomas 180 and 37 produce specific growth promoting agents which stimulate the growth of sympathetic and of sensory M-D cells and fibers but not of motor and sensory V-L fibers.

LITERATURE CITED

BRAUER, ALFRED 1932 A topographical and cytological study of the sympathetic nervous components of the suprarenal of the chick embryo. J. Morph., *53:* 277–322.

BUEKER, ELMER D. 1948 Implantation of tumors in the hind limb field of the embryonic chick and the developmental response of the lumbo-sacral nervous system. Anat. Rec., *102:* 369–390.

DUNCAN, DONALD AND NICHOLAS J. BELLEGIE 1948 Penetration of an experimental sarcoma by nerve fibers from the spinal cord. Texas Report on Biol. and Med., *6:* 461–469.

HAMBURGER, VIKTOR 1929 Experimentelle Beitraege zur Entwicklungsphysiologie der Nervenbahnen in der Froschextremitaet. Roux' Arch. f. Entwmech. d. Organ., *119:* 47–99.

HAMBURGER, VIKTOR, AND HOWARD L. HAMILTON 1951 A series of normal stages in the development of the chick embryo. J. Morph., *88:* 49–92.

HAMBURGER, VIKTOR AND RITA LEVI-MONTALCINI 1949 Proliferation, differentiation and degeneration in the spinal ganglia of the chick embryo under normal and experimental conditions. J. Exp. Zool., *111:* 457–502.

——— 1950 Some aspects of neuro-embryology. In: *Genetic Neurology*, Univ. of Chicago Press, pp. 128–160.

HARRISON, R. G. 1906 Further experiments on the development of peripheral nerves. Am. J. Anat., *5:* 121–131.

LEVI-MONTALCINI, R., AND G. LEVI 1943 Correlazioni nello sviluppo tra varie parti del sistema nervoso. I. Consequenze della demolizione dell'abbozzo di un arto sui centri nervosi nell'embrione di pollo. Commentationes, Pontif. Acad. Sci., *8:* 527–568.

SHAPIRO, D. M., AND S. WARREN 1949 Cancer innervation. Cancer Research, *9:* 707–711.

TAYLOR, A. CECIL 1944 Selectivity of nerve fibers from the dorsal and ventral roots in the development of the frog limb. J. Exp. Zool., *96:* 159–185.

TAYLOR, A., AND N. CARMICHAEL 1949 The effect on the embryo of continued serial tumor transplantation in the yolk sac. Cancer Research, *9:* 498–503.

WEISS, PAUL 1947 The problem of specificity in growth and development. Yale J. Biol. Med., *19:* 235–278.

WENGER, BYRON 1951 Determination of structural patterns in the spinal cord of the chick embryo studied by transplantations between brachial and adjacent levels. J. Exp. Zool., *116:* 123–163.

PLATE 1

EXPLANATION OF FIGURES

All photographs are of silver-impregnated material (De Castro's modification of Cajal's method); all are unretouched; white demarcation lines are added.

4 and 5 (180-16; sarcoma 180; 9-day embryo). Largest cross sections through hyperplastic spinal ganglion and contralateral control ganglion; both at the same magnification. (Compare size of nuclei in cartilage, C.)
6 and 7 (180-69; sarcoma 180; 13-day embryo). Largest cross sections through hyperplastic sympathetic ganglion and contralateral control ganglion; both at the same magnification. (Compare size of nuclei in cartilage, C.) Strong fiber bundles enter the closely adjacent part of the tumor (T) and branch in it.

EFFECT OF MOUSE TUMOR ON NERVE GROWTH
RITA LEVI-MONTALCINI AND VIKTOR HAMBURGER

PLATE 1

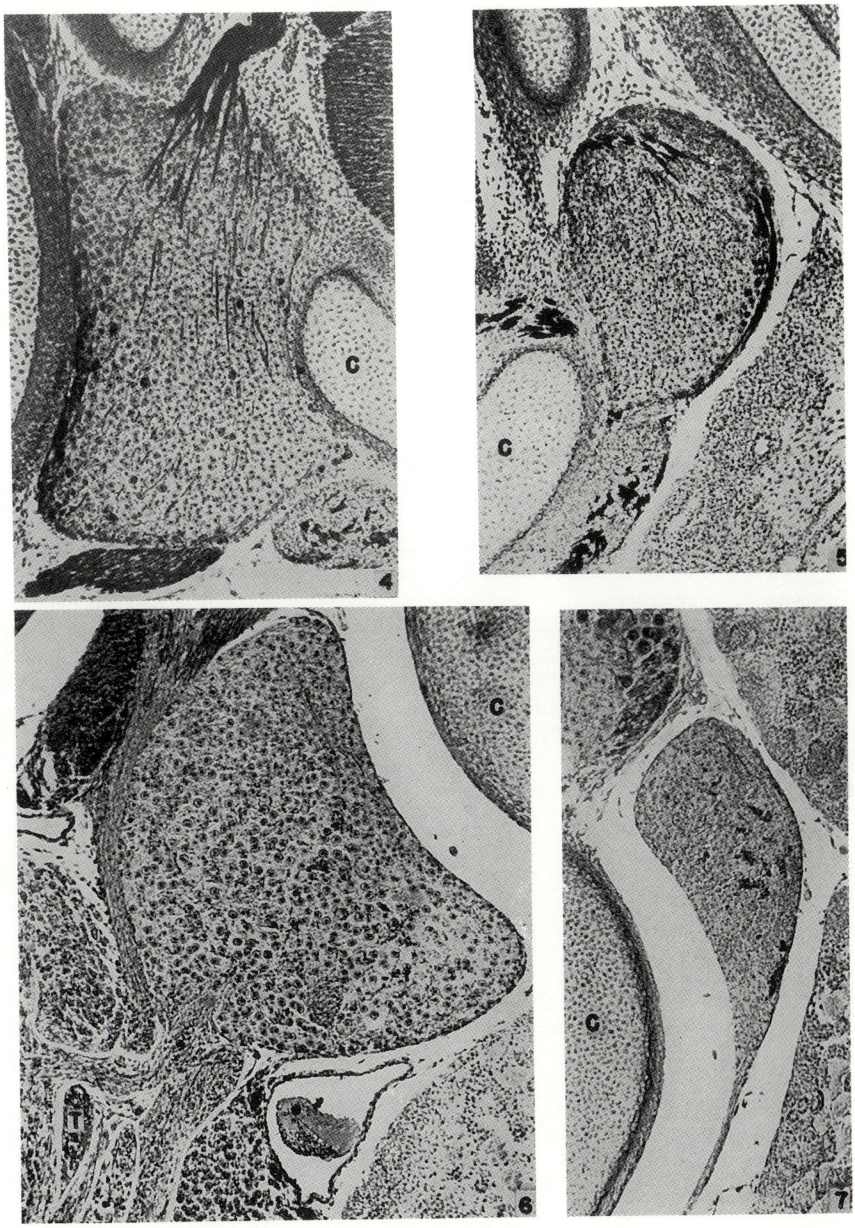

PLATE 2

EXPLANATION OF FIGURES

8 (180-103; sarcoma 180; 11½-day embryo). Sensory and sympathetic fibers combined in a thick bundle enter the tumor (T). sp = spinal ganglion; sy = sympathetic ganglion.

9 (Same case as fig. 8). A nerve bundle, composed only of sympathetic fibers, enters the tumor (T). sp = spinal ganglion; sy = sympathetic ganglion.

10 (180-86; sarcoma 180; 15-day embryo). Thick nerve bundles which do not split up into individual fibers.

11 (Same case as fig. 10.) Terminal part of one bundle at higher magnification.

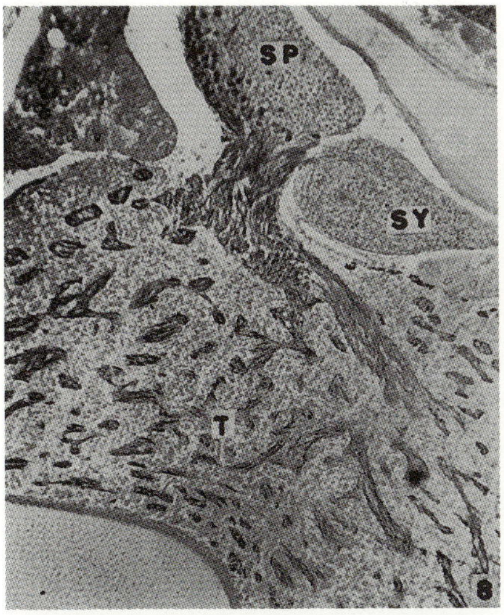
SP
SY
T
8

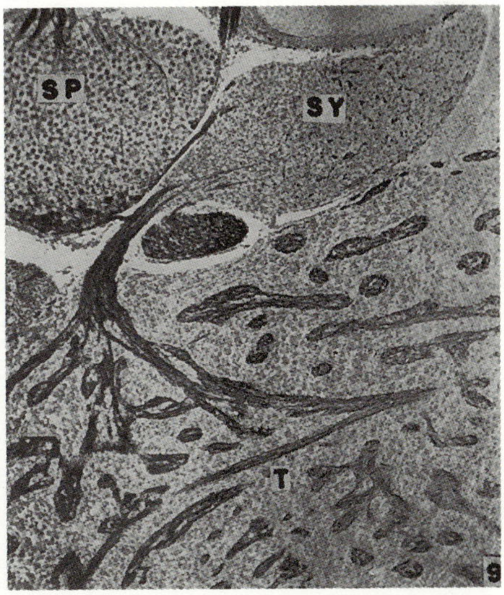
SP
SY
T
9

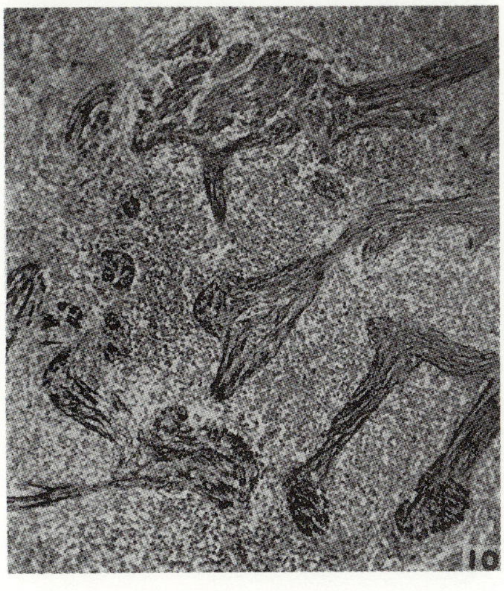
10

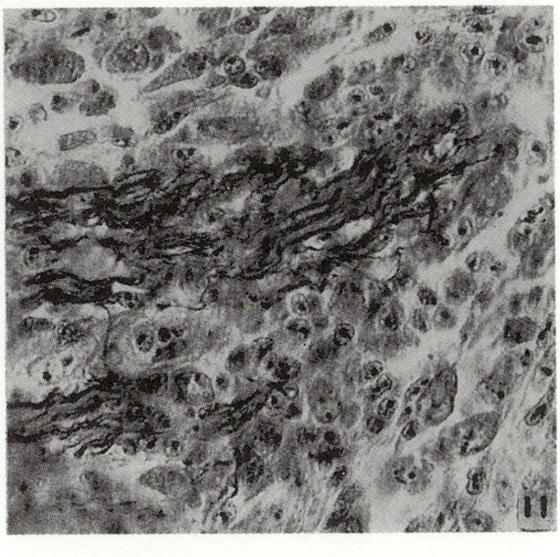
11

PLATE 3

EXPLANATION OF FIGURES

12 (180-16; sarcoma 180; 9-day embryo.) Enlargement of the lateral part of the dorsal funiculus (L), correlated with hyperplasia in M-D cells. The medial part is not affected.

13 (180-104; sarcoma 180; 11½-day embryo.) Hyperplasia of spinal ganglion (Sp) and of sympathetic ganglion (Sy). The decrease in the number of V-L cells (V) is reflected in the decrease in the size of the median part of the dorsal funiculus (M). The lateral part (L) of the dorsal funiculus is enlarged.

14 (180-103; sarcoma 180, 11½-day embryo.) Section through part of tumor (T) to show density and irregular branching of nerve bundles.

Fig. 15 (Same as fig. 14.) Increase in the size of the lateral part (L) and decrease of the median part (M) of the dorsal funiculus. Hypoplasia in the lateral motor column.

PLATE 3

EFFECT OF MOUSE TUMOR ON NERVE GROWTH

RITA LEVI-MONTALCINI AND VIKTOR HAMBURGER

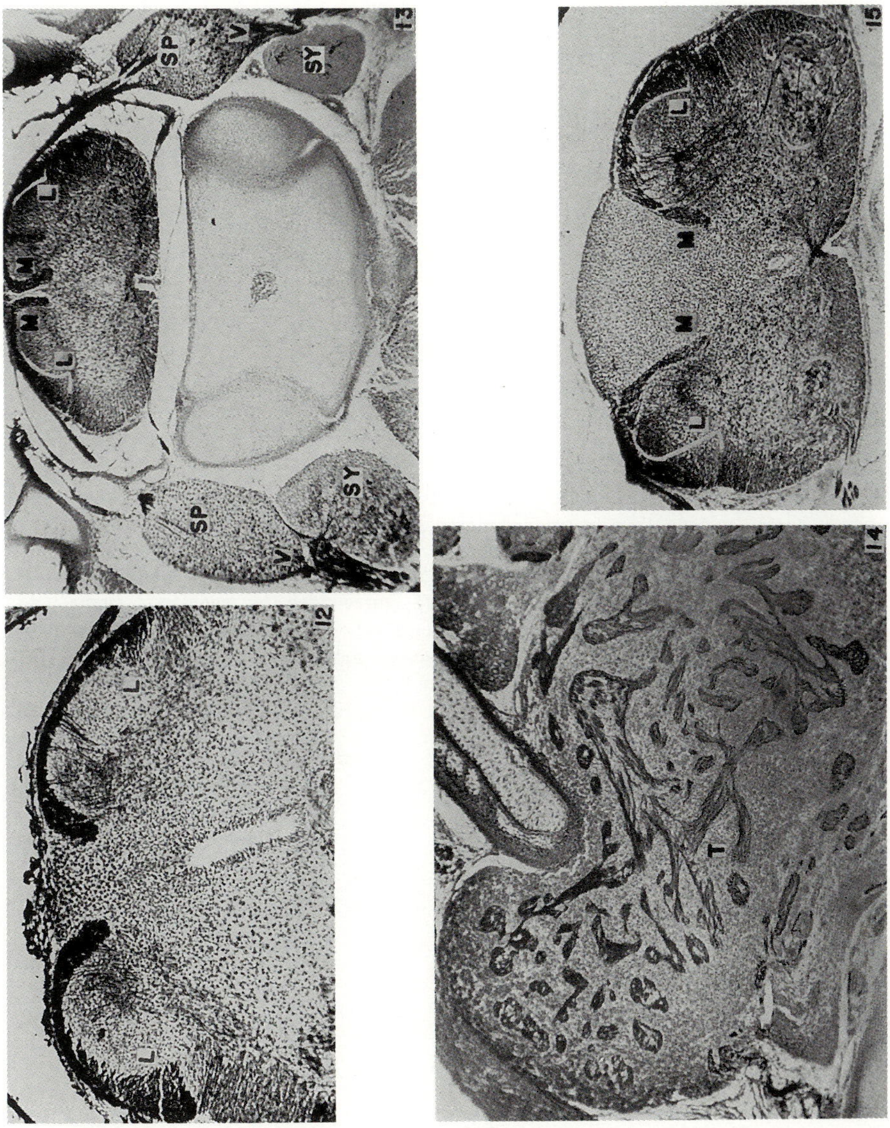

PLATE 4

EXPLANATION OF FIGURES

16 and 17 (37–81; sarcoma 37; 10½-day embryo.) Largest sections through hyperplastic sympathetic ganglion (Sy), surrounded by neoplastic tissue (fig. 16) and control ganglion on contralateral side (sy, fig. 17), both at the same magnification. (Compare size of nuclei of cartilage, C.)

18 and 19 (Same case as figs. 16 and 17.) Parts of sympathetic ganglia of figures 16 and 17, respectively. Notice the advanced stage of differentiation in the hyperplastic ganglion (fig. 18).

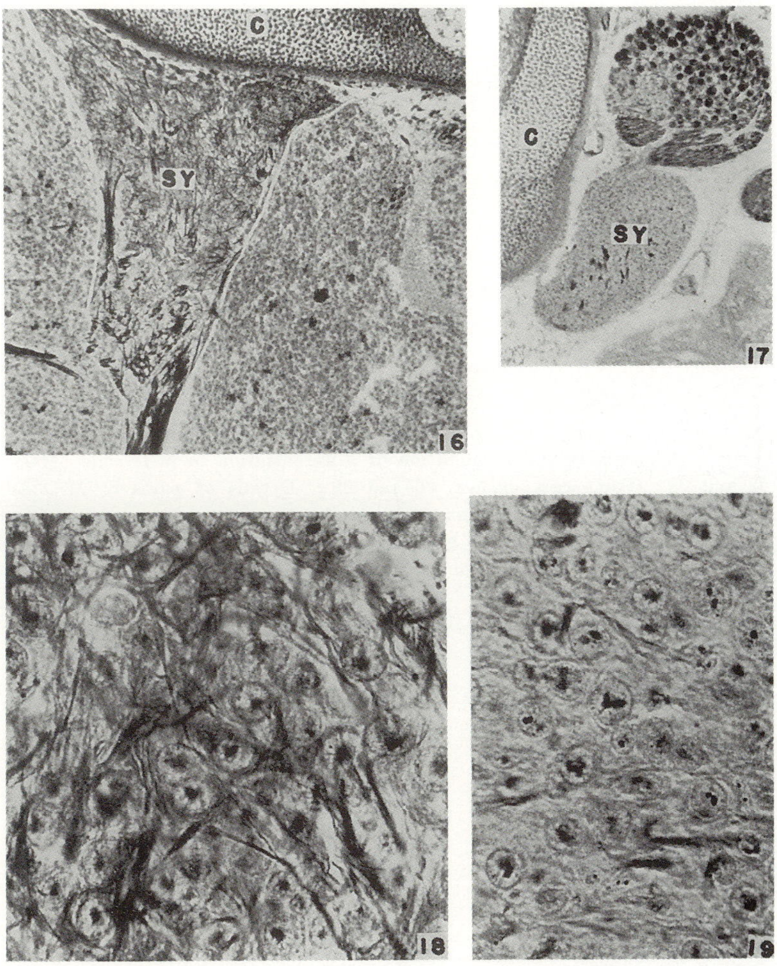
C
SY
16
C
SY
17
18
19

PLATE 5

EXPLANATION OF FIGURES

20 and 21 (180-82; sarcoma 180, 7½-day embryo.) Largest sections through tumor-innervating ganglion (fig. 20) and contralateral ganglion (fig. 21). Beginning of hyperplasia and acceleration of differentiation in M-D cells (MD).

22 and 23 (Same case as figs. 20 and 22.) Details of parts of M-D cells. Advanced stage of differentiation and affinity to silver in hyperplastic cells (fig. 22). The fibers in figure 23 are dorsal root fibers from early differentiating V-L cells which traverse this area.

PLATE 5

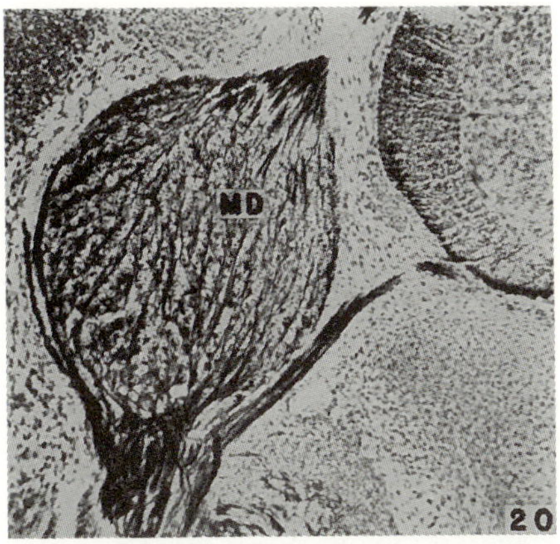

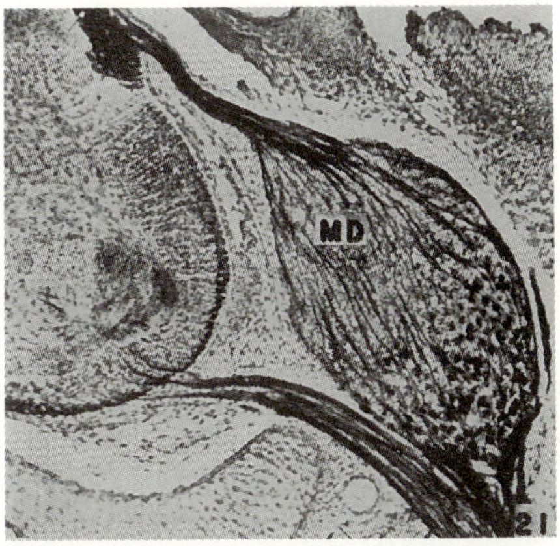

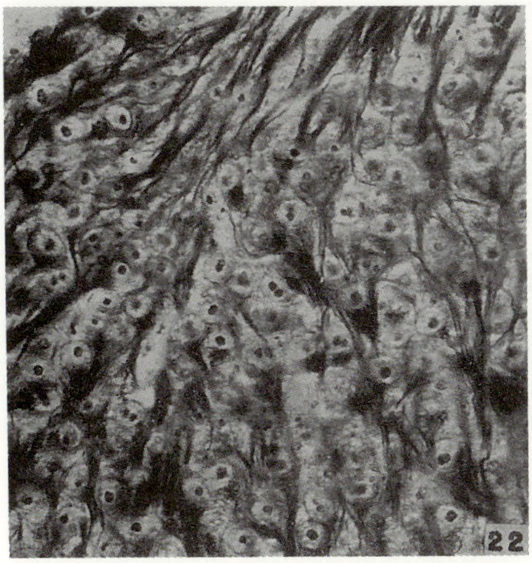

The Chick Embryo in Biological Research, D. A. Karnofsky ed., *Ann. N. Y. Acad. Sci.* **55**, 330–343 (1952).

EFFECTS OF MOUSE TUMOR TRANSPLANTATION ON THE NERVOUS SYSTEM

By Rita Levi-Montalcini
Washington University, St. Louis, Missouri

Mammalian tumors were first grown successfully on the chorio-allantoic membrane of the chick embryo by Murphy.[9] The recent development of this technique has been surveyed by Karnofsky, *et al.*[7] The first intra-embryonic transplants of normal mouse tissue were performed by Rawles.[10] Bueker[2] has combined both techniques: he implanted mouse tumor in the body wall of three-day embryos. This experiment was performed in connection with certain neuro-embryological problems. Previous experiments had shown that the differentiation of primary sensory and motor systems in amphibia and in chick embryos is to a large extent under the control of extrinsic factors.[4] The effect exerted by the periphery is not a species-specific; transplantations in Amphibia (Harrison,[6] Detwiler[3]) had proved the capacity of nerve fibers to invade heteroplastic structures. The same compatibility was demonstrated by Bueker[1] to exist between nerve fibers of the chick embryo and the peripheral field provided by the guinea hen embryo.

By transplanting mouse tumors intra-embryonically, Bueker wanted to test the capacity of nerve fibers to invade tissue of a different class. In addition, it was his intention to confront the nerve fibers with a homogeneous tissue, rather than with a set of heterogeneous and complex structures, as represented by a limb. He performed transplantations of three different tumors: mouse adeno-carcinoma which completely failed to grow, fowl Rous sarcoma which caused extensive hemorrhage and had no effect on the nerve structures, and mouse sarcoma 180, which grew vigorously and was invaded by nerve fibers of the host. The attention was then focused on sarcoma 180.

The characteristics of tumor growth and the response of the spinal nerve centers were investigated in embryos between six and nine days of incubation, that is, between three and six days after the implantation of the tumor. In this short lapse of time, the tumor had infiltrated the somatopleure of the body wall of the embryo and bulged as a conspicuous mass in the coelomic cavity. The mesonephros was partly invaded and destroyed. In eight- to nine-day embryos, nerve bundles were found inside the tumor. The sensory ganglia on the side and at the level of the tumor were considerably enlarged in comparison with the contralateral ganglia. The size increase was 33 per cent on the average. The motor column at the same level was, on the contrary, hypoplastic. The increase in size of the sensory ganglia and the coincident reduction of the motor column were taken as evidence for the sensory nature of the nerve fibers in the tumor. The author concluded that histochemical properties of the tumor favored the branching of sensory nerve fibers, whereas somatic motor fibers were refractory to these agents.

Three important facts were thus established: (1) the sensory ganglia are responsive not only to tissues of the same class, but also to structures of a different class; (2) the effect of sarcoma tissue on nerve fibers is selective, as proved by the admission of sensory fibers and the lack of response of somatic motor fibers; and (3) the nerve fibers act as mediators of this effect on the respective nerve centers, the spinal ganglia.

We have repeated the experiment on a large scale, using mouse sarcomas 180 and 37.[8] The investigation was extended to the entire developmental period, and the silver impregnation technique of Cajal-De Castro was employed. The use of this technique permitted a more penetrating analysis of the relations of nerve fibers and neoplastic cells and it gave more precise information on the pattern of nerve growth inside the tumor at different developmental stages. The results of Bueker were of particular interest to us in connection with our previous study of the development of the spinal ganglia. This investigation had demonstrated in the spinal ganglia two classes of cells, which differ from each other in their developmental pattern, their affinity for silver and their topographic position.[5] It was of interest to determine whether both classes or only one would supply nerves to the tumor.

Our investigation confirmed the findings of Bueker of a selective invasion of the tumor by sensory but not by motor fibers. A careful study of the hyperplastic spinal ganglia led to the conclusion that only the late differentiating medio-dorsal sensory neurons were involved, whereas the early differentiating ventro-lateral cells did not contribute nerves to the tumor. The overall hyperplasia in volume reached a maximum of 250 per cent in the second week of development. Another source of nerve supply to the tumor was identified in the sympathetic ganglia; their contribution of nerves to the tumor seems to be even larger than that of the sensory ganglia if one uses the degree of hyperplasia as a criterion. Sympathetic ganglia adjacent to the tumor underwent a hyperplasia which reached a maximum of 600 per cent. In both sympathetic and spinal ganglia, the overall increase in volume is the result of at least three different factors: increase in cell number, increase in the size of individual cells, and acceleration in the processes of cell differentiation.

The ingrowth of nerve fibers into the tumor begins on the sixth day of development. In the preceding stages (four to six days), nerve fibers grow out from the somato-motor column and from the early differentiating sensory neurons. They fail to invade the tumor, although the latter is readily accessible to both.[18] From the sixth day on, nerves enter the tumor in increasing numbers. Sensory and sympathetic components can be easily traced to their origin (FIGURE 1); in the tumor, the fibers are intermingled and not distinguishable from each other (FIGURE 7). In the majority of the cases, the nerve bundles end abruptly without establishing contact with individual tumor cells. They branch in an irregular fashion and give origin to an extremely dense net of nerve fibers. No embryonic or adult tissue shows a nerve distribution of comparable density.

As a result of this investigation, it was concluded that sarcomas 180 and 37 release a growth promoting agent which stimulates selectively the growth of late-differentiating sensory cells and of sympathetic ganglia. Since the publication of our first results, a large number of additional intra-embryonic transplants were performed, including 95 transplants of sarcoma 180, and 225 of sarcoma 37. An effort was made to raise the embryos to advanced stages. However, only a few embryos older than 14 days were recovered. The high rate of mortality of embryos in later developmental

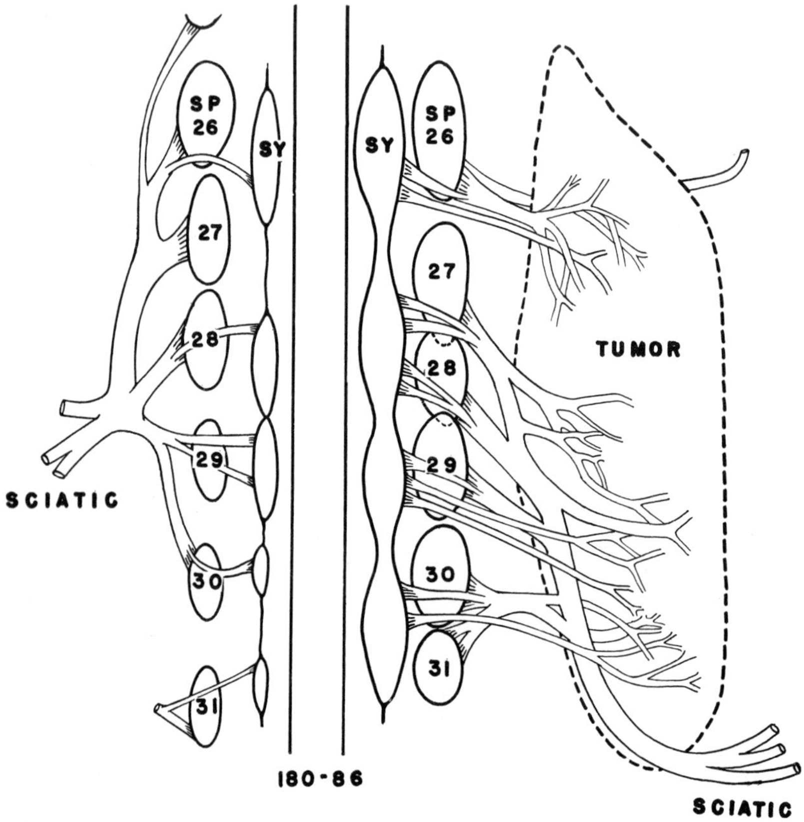

FIGURE 1. Fifteen-day embryo. Semi-diagrammatic reconstruction of the distribution of nerve fibers in the tumor (sarcoma 180). SP, spinal ganglia; SY, sympathetic chain ganglia. (From Levi-Montalcini and Hamburger, 1951, J. Exp. Zool. **116:** FIGURE 3.)

stages seems to be correlated with toxic effects of the tumors, which reached a very large size in the third week of incubation.

It was found that, in addition to nerve centers which are directly connected with the tumor by nerve fibers, paravertebral and prevertebral ganglia which had no fiber connections with the tumor were also hyperplastic. This effect is particularly striking in the thoracic segments of the prevertebral sympathetic chain of 10- to 16-day embryos. Where one finds normally a pair of ganglia at the two sides of the aorta, and two slender strands of nerve fibers connecting these ganglia with the adrenal and hypogastric plexuses, one finds now large ganglionic complexes bulging laterally to the aorta (FIGURES 4, 9, and 10). Thick nerve bundles emerge from their caudal ends and give origin to the prevertebral chains. They are many times larger than those in control embryos. The celiac plexus which, under normal conditions, is rather rudimentary until the tenth day of incubation, is already conspicuous in experimental animals on the tenth day of incubation. It forms a dense meshwork of fibers around the celiac artery, which increases in density during the following days. The celiac plexus is the main source of nerve supply to the abdominal viscera. The pattern of nerve ingrowth into the viscera of normal embryos and embryos carrying tumor transplants will be described below.

In addition to the ganglia adjacent to the aorta, one finds in normal embryos, at the level of the adrenal primordia, groups of nerve cells which are scattered among the strands of cortical cells. Under the impact of the tumor, these clusters of cells are now transformed into large ganglionic masses. An equally striking enlargement was found to occur in the hypogastric ganglia. As a result, the entire prevertebral thoracic and abdominal chain is transformed in some cases into a massive cellular column. Moreover, one occasionally finds ganglionic masses stemming from this column which project into the fringe of the tumor (FIGURE 8); they have no equivalent in normal embryos.

The intrinsic ganglia of the stomach and intestine (Meissner and Auerbach plexuses) were entirely unaffected.

The overgrowth of the ganglia which had no fiber connections with the tumor suggested that this effect on the sympathetic system was mediated through the blood circulation. To test this hypothesis, the tumor was transplanted to the extra-embryonic coelomic cavity, near the presumptive yolk stalk of three-day chick embryos. When it was realized that some sympathetic nerve fibers could still reach the periphery of the tumor, the grafts were made to the allantoic vesicle of four-day chick embryos. A total of 260 transplants were performed: 138 in the yolk stalk and 122 on the allantoic vesicle. The embryos were fixed between seven and 17 days of incubation. They were impregnated with silver and sectioned serially. The tumor grafts were recovered in every case and studied histologically.

Extra-embryonic Transplants

1. *Transplants Near the Presumptive Yolk Stalk.* The tumor became incapsulated in the yolk stalk and bulged laterally in the extra-embryonic coelomic cavity (FIGURE 2). Toward the tenth day of incubation, it had, in some cases, reached the size of a pea. Generally, however, it was of a smaller size. All embryos of this series were fixed between seven and ten days. Sections showed that a few nerves emerging from the extrinsic intestinal ganglia (Remak nerve) had reached the periphery of the tumor and branched in its peripheral parts.

2. *Transplants onto the Allantoic Vesicle.* At four days, the allantois bulges into the extra-embryonic coelomic cavity as a small pear-shaped vesicle. It has not yet come in contact with the chorion membrane. One or two fragments of sarcoma 37 or 180 were grafted onto the allantois

after perforation of the chorion. A considerable number of transplants became established and grew vigorously during the days following the transplantation. At ten days, the size of the grafts was about the same as in the

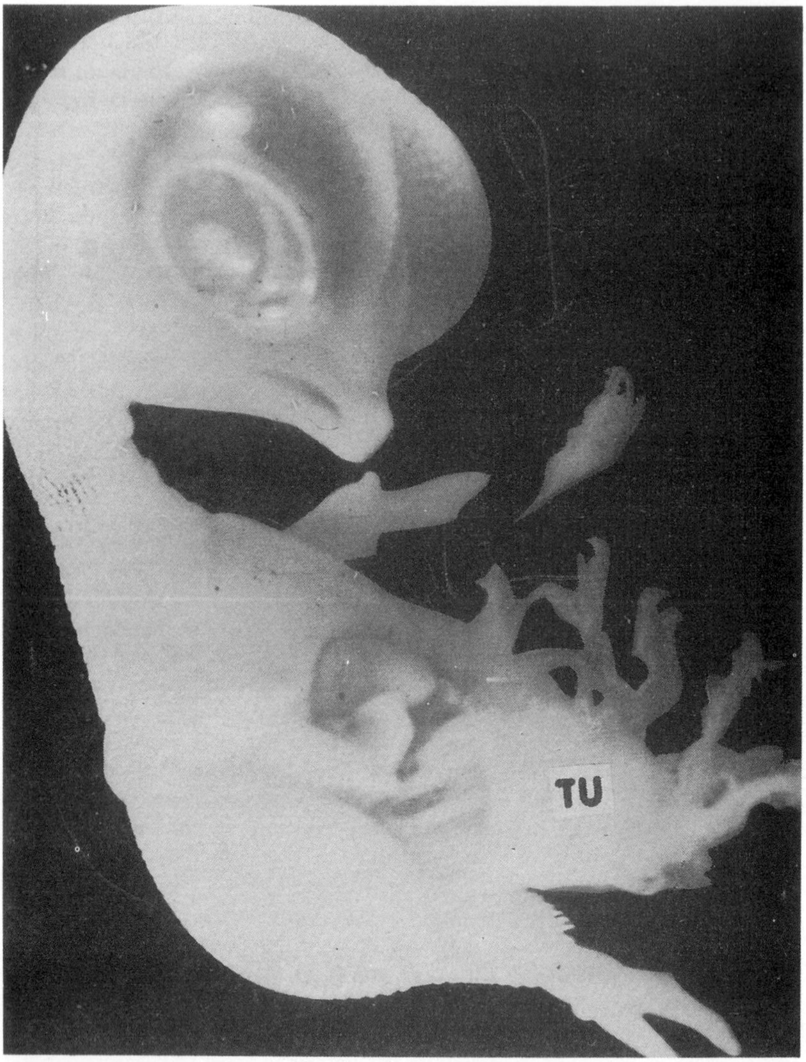

FIGURE 2. Ten-day embryo. Tu, tumor (sarcoma 37) embedded in the yolk stalk.

previous experiments. In the following days, they reached the size of a bean (FIGURE 3) and they were even larger in those embryos which survived until the 16th day of incubation. Up to 12 days, the embryos showed no sign of impairment. From then on, they appeared to be significantly smaller than control embryos: the spleen was enlarged and the liver deeply suffused with bile. Similar and other noxic effects had been described as a result of serial transplantation of mouse adeno-carcinoma to the yolk sac by Taylor and Carmichael.[11] The rate of mortality was highly increased after the 14th

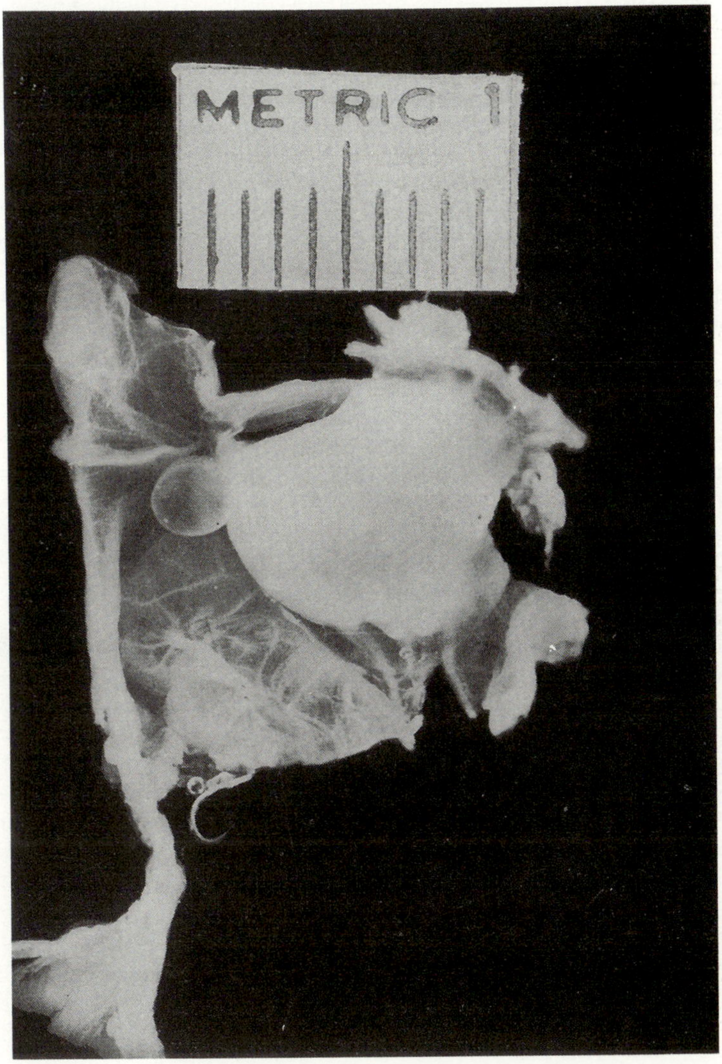

FIGURE 3. Tumor (sarcoma 37) growing on the allantoic membrane of a 14-day embryo.

day, as in the intra-embryonic transplants, and no embryo survived the 17th day of incubation. In many cases, the tumor infiltrated the blood vessels of the chorio-allantoic membrane, resulting in lethal hemorrhages.

In no cases of transplants to the allantoic membrane were nerve fibers found in the tumor. In the majority of extra-embryonic transplants, metastatic tumor cells were found in the host. Small neoplastic nodules were scattered in the mesenteries or became established at the fringe of the thoracic or abdominal organs. The mesonephroi were the preferential site of their formation. In a few cases, metastases completely failed to occur. If the metastatic nodules happened to be within reach of visceral nerve fibers, they were invaded by them. There was no correlation, however, between the hyperplastic condition of the visceral ganglia, as described below, and the nerve connections with tumor metastases. In cases in which metastases did not occur, the effect of the extra-embryonic tumor was the same as in the cases with multiple metastases. Therefore, the occasional ingrowth of nerves into the metastatic nodules was not instrumental in the overall tumor effect. The same reasoning holds for the nerve ingrowth into the peripheral areas of yolk-sac transplants. The yolk transplants and the allantoic transplants, therefore, will be considered together in the following discussion.

The effects of extra-embryonic transplants on the sympathetic paravertebral and prevertebral chain ganglia were qualitatively identical with intra-embryonic transplants of the tumor (FIGURE 4). In all cases, the whole sympathetic system was highly hyperplastic. A total of 35 embryos were investigated in detail. Area measurements were made of the prevertebral sympathetic chain ganglia of the thoracic and abdominal level between the caudal end of the brachial spinal cord and the rostral end of the lumbo-sacral cord. Camera lucida drawings of the contours of the ganglia were made in every other section, and the areas measured with the planimeter; altogether, nine experimental cases and six control embryos were used. The results are represented in FIGURE 5. A hyperplasia of the sympathetic system was already detectable at the end of the seventh day of incubation, whereas the intra-embryonic transplants had no effect in corresponding stages. In the following stages, between eight and 12 days, the effects of intra-embryonic and extra-embryonic transplants are similar in every respect. After the 12th day, the intra-embryonic transplants have a more conspicuous effect than the extra-embryonic transplants. In these late developmental stages, the intra-embryonic tumors have generally reached a larger size than the extra-embryonic transplants, and a relation between the size of the tumor and the amount of the hyperplasia of the sympathetic system suggests itself. No quantitative measurements were made, however.

FIGURE 5 shows that the effect of both intra- and extra-embryonic tumors are progressive and that the effect of the former reaches a maximum of over 800 per cent in 16-day embryos.

Peripheral Distribution of Visceral Nerve Fibers

The next question to be discussed is the distribution of the nerve fibers which emerge from these hyperplastic ganglia.

A complete series of silver-impregnated normal embryos was available for comparison. It may suffice to mention here that, in normal embryos, from the ninth day on, a very small number of visceral nerve fibers can be traced along the main blood vessels into the spleen, ovary, testis, liver, meta-

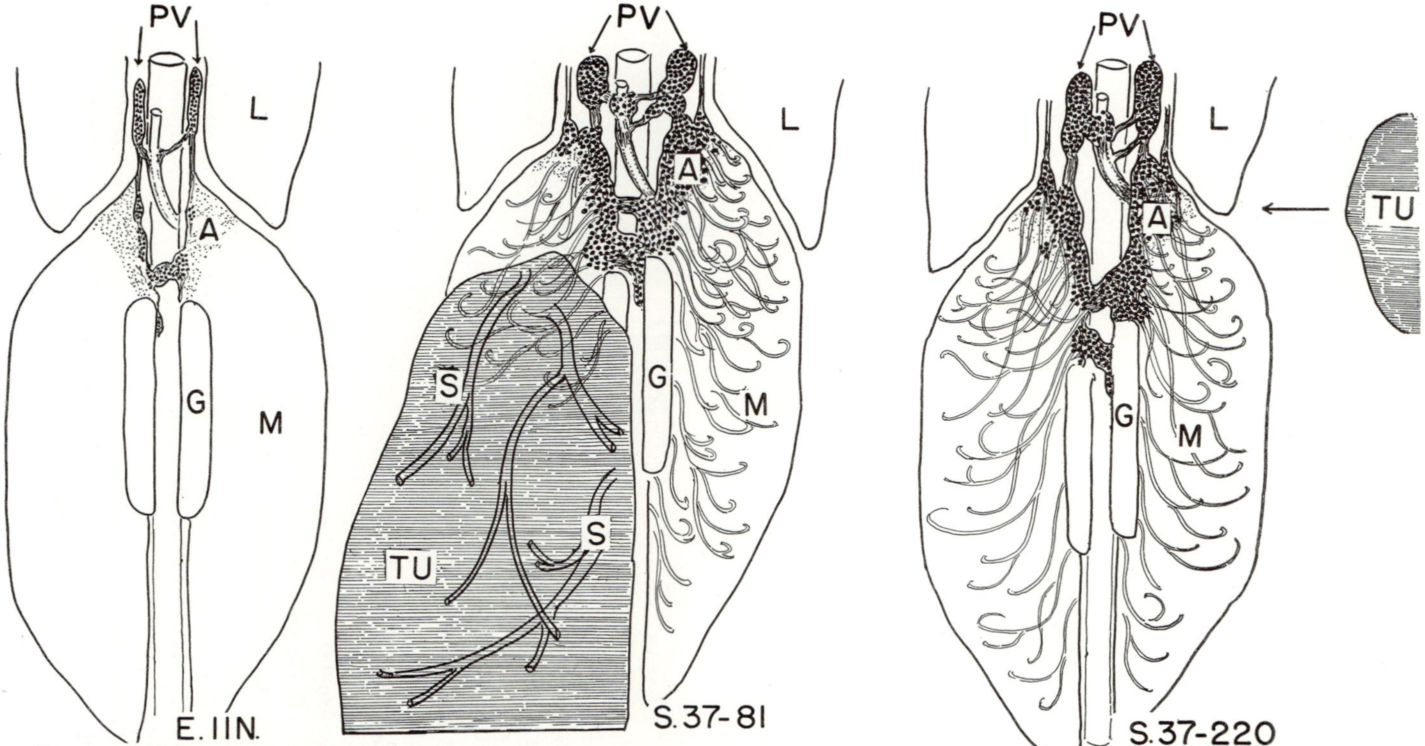

FIGURE 4. Semi-diagrammatic reconstruction of a normal 11-day embryo. (E 11n), of an 11-day embryo carrying an intra-embryonic transplant of mouse sarcoma (S 37–81) and of an 11-day embryo with transplant of sarcoma 37 on the allantoic vesicle (S 37-220). Notice the hyperplastic growth of the prevertebral chain ganglia in embryos carrying tumor transplants. Visceral nerve fibers from these ganglia invade the nearby mesonephroi. A, adrenal; G, gonad; L, lung; M, mesonephros; PV, prevertebral ganglia; S, sensory nerves; Tu, tumor.

nephros, thyroid gland and pancreas. In the mesonephros, a visceral innervation was not observed. In embryos with tumor grafts, all these organs were swamped with nerve fibers from the very beginning of their differentiation. Nerves from the nearby paravertebral and prevertebral ganglia enter the rostral edge of the mesonephros at the end of the seventh day of incubation and grow along the mesonephric tubules, following a rostro-caudal course. In the following days, the number of these nerve bundles

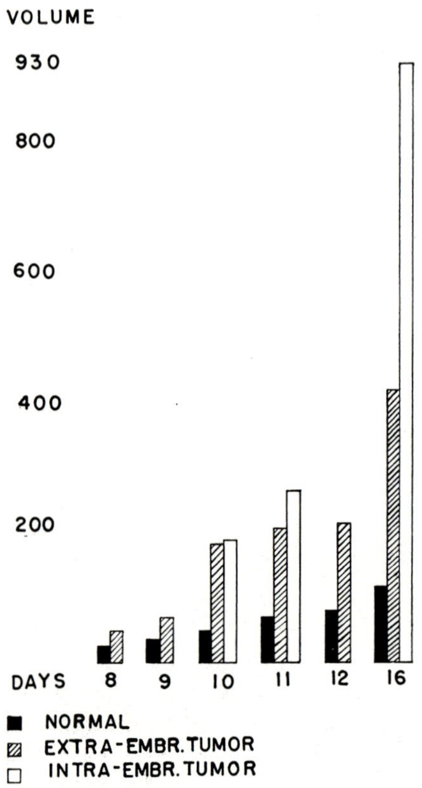

FIGURE 5. Volume increase (in arbitrary units) of the prevertebral thoraco-abdominal chain ganglia in control and experimental embryos, between 8 and 16 days of incubation.

increases rapidly and the entire organ is literally flooded with visceral nerve fibers which can be traced easily to the enlarged ganglia (FIGURE 4). A maximal density of nerve fibers was observed in those cases in which the area measurements had shown a maximal hyperplasia of the sympathetic ganglia. The nerves do not establish connections with the mesonephric tubules; they follow the blood vessels and wrap themselves around them (FIGURES 15, 16).

A similar pattern of nerve distribution was observed in the other organs mentioned above. The spleen, liver, ovary (FIGURES 13, 14), testis, metanephros, pancreas, ultimo-branchial bodies, and thyroid gland (FIGURES 17,

18) are all invaded by nerve fibers as soon as they start to differentiate. In every organ, the nerves enter with the blood vessels. Whereas the pattern of invasion is consistently the same in all cases, the density of nerve fibers

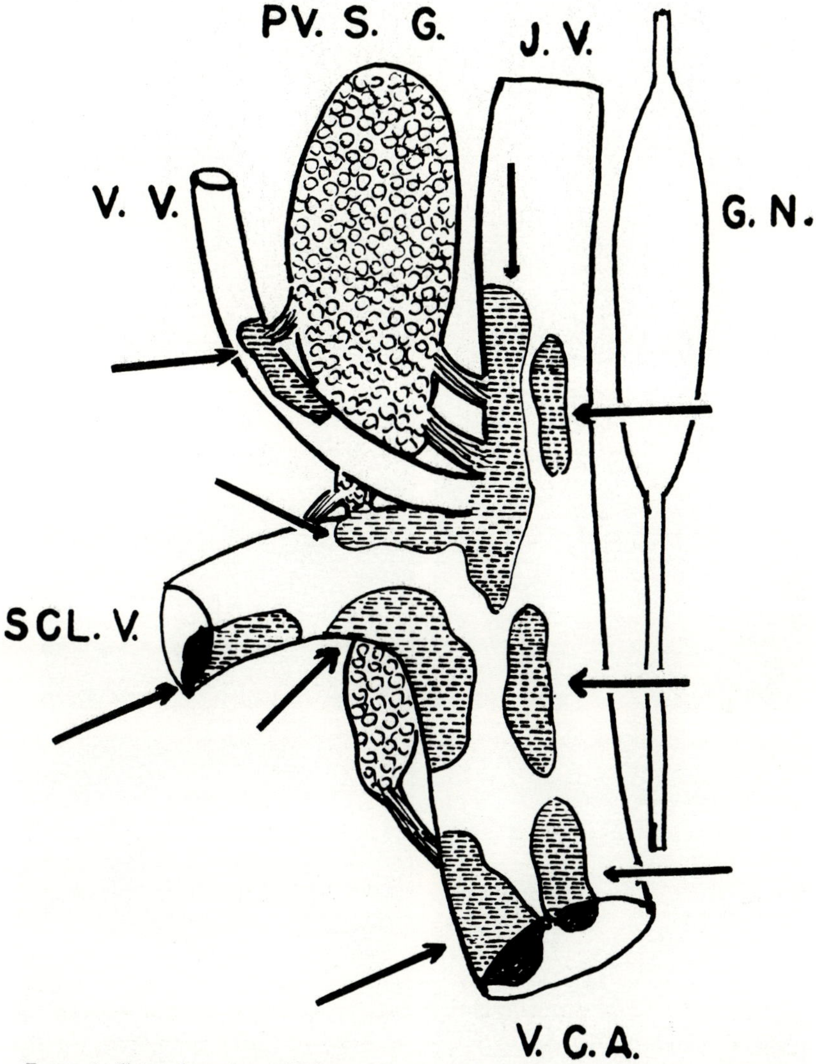

FIGURE 6. Sixteen-day embryo with intra-embryonic tumor (sarcoma 180). Ingrowth of sympathetic nerve fibers into the Jugular, Vertebral, Subclavian, Anterior Caval Veins. GN, Ganglium Nodosum; JV, Jugular Vein; Pv.SG, Paravertebral Sympathetic Ganglion; SCL.V, Subclavian Vein; VCA, Anterior Caval Vein; VV, Vertebral Vein. Arrows point to nerve agglomerations.

shows a high degree of variation in the different embryos. The amount of nerve fibers in the viscera is proportional to the degree of hyperplasia of the visceral ganglia. This, in turn, is dependent on the size of the tumor, as mentioned above.

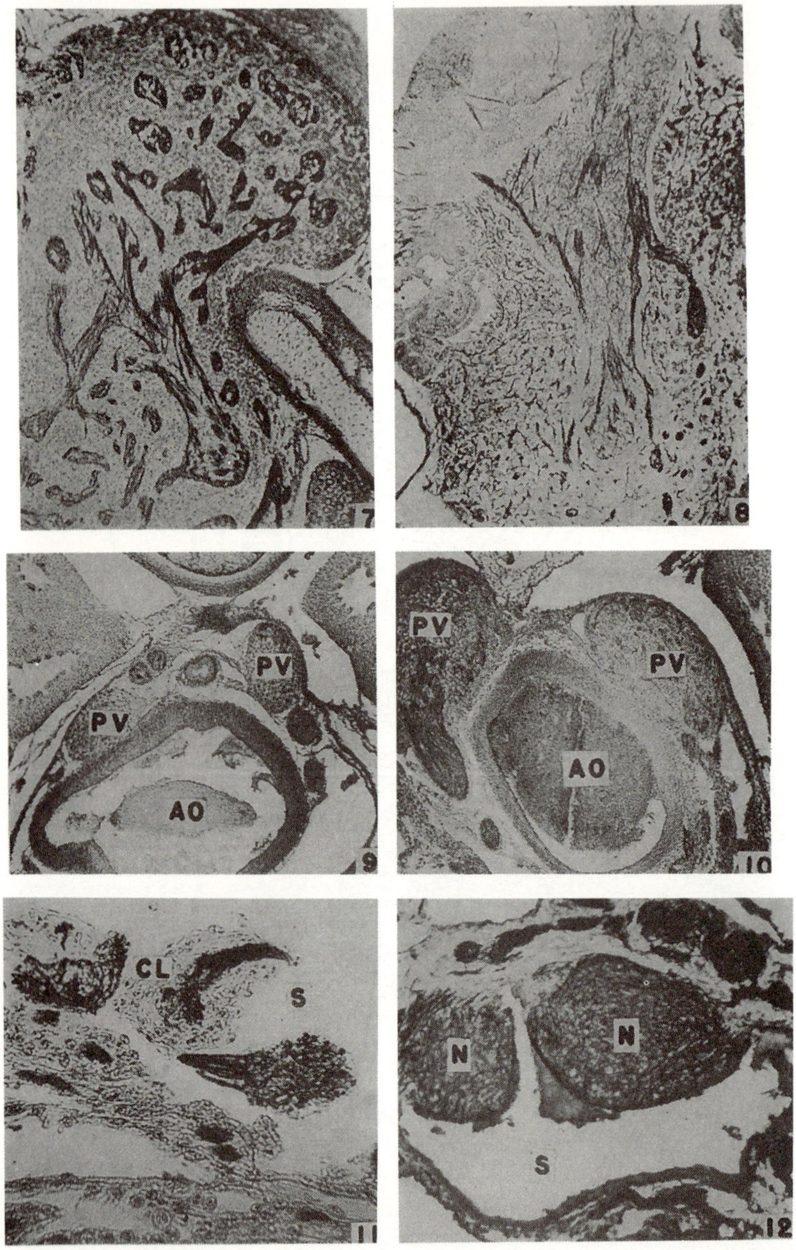

Plate I*

Figure 7. Intra-embryonic tumor (sarcoma 180) in 11½-day embryo. Section through part of the tumor to show density and irregular branching of nerve bundles. (From Levi-Montalcini and Hamburger, 1951, J. Exp. Zool. **116:** figure 14.)

Figure 8. Neoformation of sympathetic ganglion projecting into the fringe of the tumor (sarcoma 180) of an 11-day embryo. Nerve fibers from the ganglionic mass invade the tumor.

Figure 9. Prevertebral sympathetic ganglia (PV) on both sides of the aorta (AO) in a normal 16-day embryo.

Figure 10. The same ganglia in a 16-day embryo with an intra-embryonic tumor (sarcoma 180). AO; Aorta; PV, Prevertebral sympathetic ganglia. Same magnification as figure 9.

Figure 11. Twelve-day embryo with graft of sarcoma 180 on the allantoic membrane. Visceral nerve bundles have perforated the intima and end freely inside the subcardinal vein (S). Blood clots surround the nerve fibers (CL)).

Figure 12. Sixteen-day embryo with intra-embryonic tumor (sarcoma 180). Two large nerve agglomerations of visceral nerve fibers (N) bulge into the subclavian vein (S).

* All photographs are of silver-impregnated material (De Castro's modification of Cajal's method); all are unretouched; white demarcation line added to figure 8.

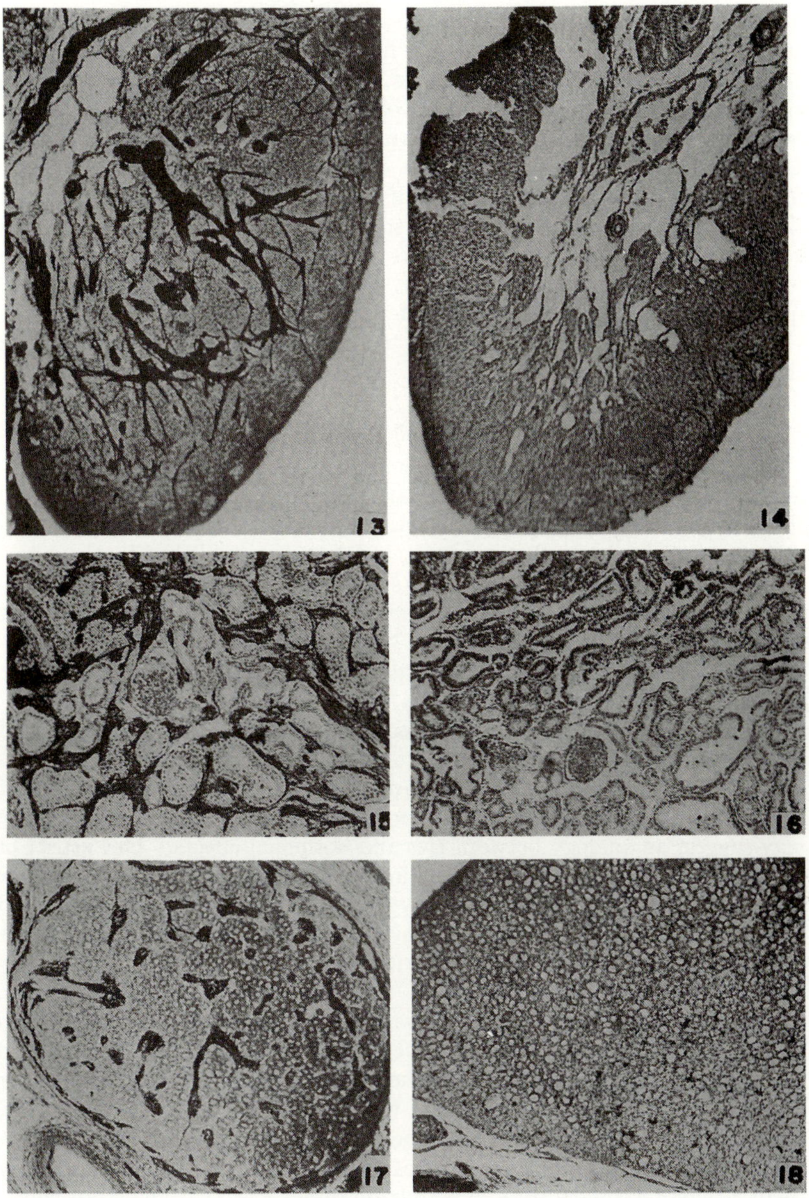

Plate II

Figure 13. Branching of visceral nerve fibers in the ovary of a 16-day embryo with an intra-embryonic tumor (sarcoma 180).
Figure 14. Ovary of a control embryo of 16 days.
Figure 15. Visceral nerve fibers in the mesonephros of a 16-day embryo with an intra-embryonic tumor. The nerve fibers wrap themselves around the blood vessels.
Figure 16. Mesonephros of a control embryo of 16 days.
Figure 17. Thyroid gland of a 16-day embryo with an intra-embryonic tumor (sarcoma 180). The nerves enter the gland with the blood vessels.
Figure 18. Thyroid gland of a control embryo of 16 days.

A very unusual relation was found between sympathetic nerve fibers and the large veins of the thoracic and abdominal level. Whereas one normally finds small numbers, at best, of fine fibers in the media, but none in the intima, the experimental cases show at 11 days the penetration of numerous visceral nerve bundles into the intima of the large veins, such as the jugular, pre-caval, subclavian, and the subcardinal (FIGURES 6, 11, and 12). At 16 days, these nerves have enlarged to a very conspicuous size and bulge into the lumen of these veins in the form of thick nerve agglomerations (FIGURE 12). In some instances, they grow to such a size that the lumen of the blood vessels becomes obliterated. Clusters of blood cells or large blood clots surround the protruding nerve fibers. The arteries and the capillaries are spared from invasion. The death of practically all embryos toward the 17th day of incubation prevented a further analysis of the fate of these aberrant fibers.

Concluding Remarks

The investigation is still in progress and we refrain from a theoretical interpretation of these results. We can state, however, that the humoral transmission of a tumor-produced agent, which affects profoundly the development of the sympathetic system, is definitely established by the extra-embryonic transplants. These results are not in conflict with the previously-advanced hypothesis[2, 18] that nerve fibers mediate the tumor effect whenever they penetrate the tumor. We are not yet in a position to decide if we are dealing with two different modes of action or whether the same mechanism operates in both instances.

We should like to emphasize the exceptional character of the response of the visceral nervous system to the tumor. The results reveal a morphogenetic effect for which there is no parallel. The selective susceptibility of some neurons to the effect of the tumor and the absence of species specificity may suggest an analogy between this and hormonal effects. One aspect of the effect of the tumor does not fit in this picture, however. The tumor promotes an excessive growth of the sympathetic system with total disregard of the requirements of the whole organism. In fact, it upsets its harmonious development to such an extent that it causes profound changes in the sequence of developmental processes. Whereas, under normal conditions, proliferation and differentiation of the sympathetic ganglia are held within limits and kept in pace with the development of other structures, we find that under the impact of the tumor the sympathetic system overtakes other systems. The increase in the size of the sympathetic chain ganglia which, in one particular case, amounts to more than eight times the normal size, is even more impressive if one realizes that the embryos were considerably smaller than normal embryos, due to toxic tumor effects. The latest stages investigated are 16- and 17-day embryos, and there is no indication that the hyperplastic process has come to an end at that time.

The accommodation of an enormous number of nerve fibers in the viscera is equally intriguing. It indicates another severe infraction of the laws which govern the mechanics of development. Under normal conditions, nerve fibers are admitted to organs according to a rigid time schedule, and the quantity of the entering fibers is strictly limited; each organ has a characteristic density, and regulative mechanisms are in operation to prevent hyper-neurotization. In the present instance, all barriers seem to have broken down, and the organs surrender to the invading fibers. It remains to be seen whether some properties of the nerve fibers are changed, such as their invasiveness or growth potential, or whether the organs themselves are directly affected by the tumor. An even more striking instance of the abnormal behavior of visceral nerve fibers is represented by their perforation of the intima of the veins, and the formation of nerve agglomerations in the lumen of these vessels. Nothing can be said concerning the nature of the agent and our research is now heading in this direction.

Two other questions arise in connection with the present investigation:

1. Are the nerve cell bodies or the nerve fibers the immediate target of the tumor agent?

2. What is the source of origin of the greatly enlarged and the newly-formed sympathetic ganglia? In view of their enormous volume and relatively late formation of additional ganglia (after the seventh day), it seems difficult to accept their derivation from the neural crest. In this connection, the contention of Tello[12-14] that the sympathetic cells are mesodermal derivatives and have a common origin with the endothelial cells of thoracic blood vessels deserves serious consideration.

Bibliography

1. Bueker, E. D. 1945. The influence of a growing limb on the differentiation of somatic motor neurons in transplanted avian spinal cord segments. J. Comp. Neur. **82:** 335.
2. Bueker, E. D. 1948. Implantation of tumors in the hind limb field of the embryonic chick and the developmental response of the lumbosacral nervous system. Anat. Rec. **102:** 369.
3. Detwiler, S. R. 1930. Some observation upon the growth, innervation and function of heteroplastic limbs. J. Exp. Zool. **57:** 183.
4. Hamburger, V. 1952. Development of the Central Nervous System. Ann. N. Y. Acad. Sci. **55** (2): 117.
5. Hamburger, V & R. Levi-Montalcini. 1949. Proliferation, differentiation and degeneration in the spinal ganglia of the chick embryo under normal and experimental conditions. J. Exp. Zool. **111:** 457.
6. Harrison, R. G. 1929. Correlation in the Development and growth of the eye studied by means of heteroplastic transplantation. V. Roux'Archiv. **120:** 1.
7. Karnofsky, D. A., L. P. Ridgway, & P. A. Patterson. 1952. Tumor transplantation to the chick embryo. Ann. N. Y. Acad. Sci. **55** (2): 313.
8. Levi-Montalcini, R. & V. Hamburger. 1951. Selective growth stimulating effects of mouse sarcoma on the sensory and sympathetic nervous system of the chick embryo. J. Exp. Zool. **116:** 321.
9. Murphy, J. B. 1912. Transplantability of malignant tumors to the embryo of a foreign species. J. Amer. Med. Assn. **59:** 874.
10. Rawles, M. E. 1940. The development of melanophores from embryonic mouse tissues grown in the coelom of chick embryos. Proc. Nat. Acad. Sci. **26:** 86.
11. Taylor, A. & N. Carmichael. 1949. The effect on the embryo of continued serial tumor transplantation in the yolk sac. Can. Res. **9:** 498.
12. Tello, F. 1925. Sur la formation des chaines primaire et secondaire du grand sympathique dans l'embryon de poulet. Trav. du Lab. de Rech. biol. de l'Univ. de Madrid. **23:** 1.
13. Tello, F. 1945. Algunas observaciones mas sobre las primeras fases del desarollo del simpatico en el pollo. Trab. Inst. Cajal. Inv. Biol. **37:** 103.
14. Tello, F. 1949. Lo evidente y lo duodoso en la genesis del simpatico, con neuvas observaciones. Trab. Inst. Cajal. Inv. Biol. **41:** 1.

Discussion of the Paper

DR. DAVID A. KARNOFSKY (*Sloan-Kettering Institute for Cancer Research, New York, N. Y.*): Dr. Levi-Montalcini has found that the embryos transplanted with Sarcoma 180 die about the 16th day of development, presumably due to the growth of the tumor. We have noted that the chick embryo is relatively resistant to the nitrogen mustards. The injection of a dose of methyl-*bis* (B-chloroethyl) amine hydrochloride, about $\frac{1}{5}$ of that lethal to the chick embryo, is destructive to the sarcoma 180 previously transplanted to the chorio-allantoic membrane. (Ref.: Approaches to Tumor Chemotherapy, A. A. A. S., pp. 293, 1947). After the sarcoma 180 has induced the enlargement of the ganglia in the chick embryo, it may be possible to save the embryo by destroying the tumor with nitrogen mustard, and then see if the overgrowth of nervous tissue persists or regresses.

Reprint from:
Proceedings XIV International Congress of Zoology, Copenhagen 1953. Copenhagen 1956.
(p. 309)

IN VIVO AND IN VITRO EXPERIMENTS ON THE EFFECT OF MOUSE SARCOMA 180 AND 37 ON THE SENSORY AND SYMPATHETIC SYSTEM OF THE CHICK EMBRYO

by RITA LEVI-MONTALCINI (St. Louis)

Small pieces of mouse sarcoma 180 and 37 were grafted at the base of limb buds of two- to three-day chick embryos; they grew conspicuously and became invaded by sensory and sympathetic nerve fibers from the adjacent ganglia. The ganglia supplying nerves to the tumor are highly hyperplastic and hypertrophic. In addition, the sympathetic chain ganglia are enlarged and supernumerary ganglionic masses are formed. Sympathetic nerve fibers invade in large numbers adjacent viscera which normally receive little or no innervation in corresponding developmental stages. Agglomerations of sympathetic fibers protrude into the wall of the veins.

Transplantation of tumors 180 and 37 onto the allantoic membrane of four-day chick embryos produced similar effects on the sympathetic system. It was concluded that sarcomas 180 and 37 produce a diffusible growth stimulating agent which affects sympathetic nerve cells and fibers. The diffusible effect on the sensory ganglia remains to be investigated.

The effect of mouse sarcoma 180 on sensory and sympathetic ganglia of chick embryos was also tested in vitro. Sensory ganglia of seven-day chick embryos were explanted in close proximity to small pieces of sarcoma 180. After 24 hours, the ganglia show a conspicuous outgrowth of nerve fibers whose density is maximal on the side toward the tumor. Control cultures in which the same ganglia were combined with normal embryonic chick tissue (heart myoblasts) show only very few nerve fibers but a large number of migrating fibroblasts. In another control series, in which the same ganglia were confronted with normal embryonic mouse tissue (heart myoblasts) the nerve fiber outgrowth was more pronounced than in the chick tissue controls, but not comparable with the outgrowth called forth by the sarcoma. The effect increases during the following day. Sympathetic ganglia of chick embryos show a similar response to the tumor. Further experiments with these and other tumors are in progress.

BIANCO LUNO A/S, KBH.

Cancer Research **14**, 49–57 (1954).

In Vitro Experiments on the Effects of Mouse Sarcomas 180 and 37 on the Spinal and Sympathetic Ganglia of the Chick Embryo

RITA LEVI-MONTALCINI, HERTHA MEYER, AND VIKTOR HAMBURGER

(Department of Zoology, Washington University, St. Louis, Mo., and the Instituto de Biofisica, Universidade do Brasil, Rio de Janeiro, Brasil)

INTRODUCTION

Previous experiments have given evidence that mouse Sarcomas 37 and 180 produce an agent which promotes the growth of spinal ganglia (1) and of sympathetic ganglia (12) in the chick embryo. This effect was first observed in experiments in which small pieces of tumor were implanted in the body wall of 3-day embryos, where they grew vigorously. Later, the same effects were obtained when the tumors were transplanted extra-embryonically to the allantoic membrane of 4-day embryos (10, 11, 13). The latter result was considered as conclusive evidence that we are dealing with a diffusible agent.

The response of the nervous system was found to be selective and complex. Only the sensory and para- and prevertebral sympathetic ganglia showed a reaction, whereas all centers in the spinal cord were refractory. The hyperplasia of the ganglia was due partly to an increase in cell number and partly to a cellular hypertrophy. The differentiation of nerve fibers was accelerated, and their number was increased to an extraordinary degree. The supernumerary fibers emerging from the hyperplastic ganglia flooded the adjacent viscera: meso- and metanephros, gonads, spleen, adrenal, thyroid, parathyroid glands, and also the tumor, in cases of intra-embryonic transplantations. Normally, these organs receive only a very scant nerve supply or none at all, in corresponding stages of development.

These results confronted us with two major problems: (*a*) the chemical nature of the agent and (*b*) its mode of action. Concerning the latter problem, the previous experiments gave no definite clue whether the agent acts directly on the ganglia or indirectly, by producing complex metabolic changes in the embryo. In this connection, it should be mentioned that the embryos carrying a tumor showed signs of toxic effects (edema, perfusion of the liver by bile, stunted growth) which were eventually fatal to the embryo.

It seemed that the tissue culture method might offer a new approach to the analysis of both problems. This method has several advantages: it permits the direct exposure of the ganglia to the tumor, thus excluding possible influences of the organism; furthermore, extracts of tumors can be easily tested, and other tumors and normal tissues can be screened for possible nerve growth-stimulating effects. Since the behavior of the spinal ganglia of the chick embryo *in vitro*, under normal and experimental conditions, has been studied in great detail (6, 8, 9, 17, 18), one can build on a solid foundation.

MATERIALS AND METHODS

The experiments consisted of the combination of spinal ganglia with fragments of mouse Sarcomas 180, 37, 1, adenocarcinoma dbrB and neuroblastoma C1300. Fragments of embryonic chicken or mouse heart were used as controls. In a limited number of experiments, paravertebral sympathetic ganglia or pieces of spinal cord were exposed to these tumors and to control tissues. Usually, several fragments of the same tumor or of control tissue were placed at a distance of 1–2 mm. from the ganglion; this has been found to be the optimal distance. In no instance were the explants placed in direct contact with each other, but usually the spreading of the tumor resulted in a contact at about 48 hours. In one series, the distance between the ganglion and the tumors was varied, in order to study the range of diffusibility of the tumor agent.

Table 1 gives a summary of all data. It should be pointed out that the number of experiments performed was considerably larger, since, in most instances, several groups of explants were placed

Received for publication August 3, 1953.

in the same hanging drop, at a considerable distance from each other.

The *spinal ganglia* were in most instances lumbo-sacral ganglia of 6- to 7-day embryos. In a few instances, ganglia from 9- and 10-day embryos were used. They were dissected in physiological salt solution under a binocular microscope and explanted *in toto*, that is, surrounded by their capsule.

The isolation of *sympathetic ganglia* is much more laborious than that of spinal ganglia; hence only a limited number of experiments of this type was done. The sympathetic ganglia were obtained from older embryos ranging from 8 to 13 days, since the isolation of younger ganglia is not practicable. Long segments of the paravertebral chain were dissected out and cut into fragments, including one ganglion.

Spinal cord fragments were obtained from 6-day embryos. The lumbo-sacral level was isolated and cut into small square fragments corresponding approximately to one segment. Each fragment was split apart in the median plane, and the lateral halves were explanted separately.

TABLE 1

SURVEY OF EXPERIMENTS
(Numbers of Cases)

Combination of	Spinal ganglion	Sympathetic ganglion	Spinal cord	Chick embryo heart	Chick organs
Sarcoma 180	222	29	6	13	21
Sarcoma 37	45	13	20	6	0
Sarcoma 1	32	2	6	0	0
Adenocarcinoma dbrB	30	0	0	0	0
Neuroblastoma C1300	38	0	0	5	0
Heart tissue (from mouse fetus)	99	9	0	0	10
Heart tissue (from chick embryo)	78	26	14	0	0
Miscellaneous chick organs	16	0	0	0	0
Isolated	108	24	3	7	0
Total	668	103	49	31	31

All *tumors* were obtained from the Jackson Memorial Laboratory at Bar Harbor. In the earlier series, the tumor explants were taken directly from the mouse, but they had a strong inhibitory effect on the ganglia. These unsatisfactory results suggested that an adaptation to the chick embryo might be necessary. For this purpose, the tumors were implanted in the body wall near the hind-limb bud of 2- to 3-day embryos (for technic see [12]) and allowed to grow there for 4–8 days. They were then used for explantation or transferred to other chick embryos for one or more additional passages. Tumors with at least one passage in the chick embryo proved to be optimally effective. Only the "healthy" peripheral parts of the tumors were selected; the central necrotic and hemorrhagic parts were discarded. The pieces were approximately 1 c. mm. in size.

As *controls*, the following *tissues* were used: heart tissue of 7-day (or in a few instances 8- to 10-day) chick embryos, heart tissue of mouse embryos or fetuses, or of newly-born mice.

The *standard culture medium* was chicken plasma and chick embryo extract. The plasma was obtained by the standard technic of bleeding a rooster through the carotid artery. In all experiments, the extract was diluted 1:3 with Earle's solution. The plasma was ordinarily used undiluted; in a few experiments the plasma was diluted 1:1 with Earle's salt solution.

The *hanging-drop technic* was used throughout; Maximow depression slides with two cover glasses were employed in most instances. Most experiments were discontinued after 48 hours. A limited number of cultures which were carried beyond this period were washed every second day with Earle's solution, and nutrient (serum and extract) was added. In all instances, a set of tumor experiments and a set of control experiments were done on the same day, using the same medium, the same technic, and ganglia from the same embryo.

Fixation.—Most cultures were fixed between 24 and 48 hours and impregnated with silver, following Levi's modification of the Cajal-De Castro technic (8). A few slides were fixed in Bouin's fluid and stained with Ehrlich's hematoxylin.

RESULTS

GROWTH OF ISOLATED SPINAL GANGLIA IN VITRO

Ganglia of 6- to 7-day embryos were chosen for our experiments. In these stages, the ganglia are still in a phase of active proliferation, and some of the neuroblasts have not yet sent out nerve fibers (4). In our previous experiments of *in vivo* transplantations of sarcomas, the spinal ganglia had shown the first responses to the tumor agent on the 7th or 8th day of incubation; hence ganglia of 6–7 days were expected to be in the best condition for reaction *in vitro*.

When ganglia of the lumbo-sacral level are cultivated *in vitro* in the absence of other tissues, they show almost no fiber outgrowth during the first 16 hours. The migration of spindle-shaped cells begins at about 10–15 hours. These spindle cells, whose active migration has been observed by all previous workers, represent a heterogeneous population of Schwann cells, satellite cells, and mesenchyme cells of the capsule. At 24 hours, the migration of the spindle cells is well advanced, and a small number of nerve fibers has grown out (Fig. 3).[1] The fibers are distributed irregularly and take a wavy course. Between 24 and 48 hours, the number of nerve fibers increases, but they are still rather sparse (Fig. 7). They have a tendency to fasciculate and to associate with rows and columns of spindle cells. Those fibers which do not join with others follow tortuous routes; occasionally they grow tangentially or in circular paths around the ganglion (8). In accordance with the observations of Levi and Meyer (8), we find that nerve cells do not migrate out of the ganglion, but that the spreading of its surface and the decrease of its density come about by a migration of the spindle cells. Nerve fiber and spindle cell outgrowth continues during the 3d day. After this time, a considerable number of centrally located neurons begin to degenerate. They are disposed of

[1] Since there is no difference between single ganglia and ganglia combined with embryonic chick heart tissue, with respect to nerve fiber growth and spindle cell migration (see p. 53), only the latter experiments were chosen for illustrations.

by macrophages which appear in increasing numbers in older cultures. Our observations on normal ganglia are in agreement with the basic studies of Levi and Meyer (8).

EXPERIMENTS WITH SARCOMA 180

Excellent growth of this tumor was obtained after it had undergone one or more passages in the chick embryo. The migration of cells gets well under way during the first 16 hours. At 24 hours, a uniform margin of typical tumor cells surrounds the explant; the cells are large in size and spindle-shaped: they usually adhere to one another, forming long strands. The growth zone appears to be a homogeneous population of sarcoma cells, not contaminated with small-sized fibroblasts of chick origin (Fig. 14, *s*).

Between 24 and 48 hours, the area of migrating cells is further expanded; the neoplastic cells are healthy. However, the life-span of sarcoma cultures is short, as has been pointed out already by Carrel and Burrows (3). After 48 hours, one finds increasing numbers of degenerating cells and macrophages in the explant and in the growth zone. This occurs even if the cultures are washed and nutritive material is added every other day. The growth conditions could probably be improved by transfer, but in our present investigation, we were particularly interested in the maintenance of the tumor without resection and transfer.

Combination of spinal ganglia with Sarcoma 180.—One or more small pieces of tumor were implanted at distances of 1–2 mm. from a lumbosacral spinal ganglion of a 6- to 7-day embryo. The results obtained in a large number of experiments (Table 1) were entirely consistent and uniform. The nerve fiber outgrowth was definitely precocious in the presence of the sarcoma. At 16 hours, few, if any, fibers are present in control ganglia (see above), whereas in the combination experiments with sarcoma, a rather large number of nerve fibers has grown out at that stage; they are limited to the side facing the tumor. Shortly thereafter, fibers begin to sprout from the entire surface of the ganglion. In contrast to control cultures, no spindle cells have migrated out during this period. At 24 hours, the ganglion represents a remarkable picture (Fig. 1). It is surrounded by a "halo" of nerve fibers. They show maximal density and a very straight course on the side facing the tumor. The tips of the fibers branch profusely and form a brushlike border. This phenomenon has never been observed in normal cultures. On the sides not facing the tumor, the fibers become gradually less dense and longer; the direction of their outgrowth is less straight, and some take more winding routes. However, they never wander in all directions nor do they form bundles, as is characteristic of the control cultures. In all well-growing cultures in which the tumor is rather close to the ganglion, the halo of nerve fibers has a very characteristic contour. Toward the tumor, the border of the fiber tips is sharply demarcated and often ellipsoid in outline. One gets the impression that the fibers face an invisible barrier which none of them trespasses. A close inspection shows that the borderline of the fibers is particularly sharp at the interface between cover glass and plasma clot, whereas the fiber length in deep layers of the same culture is somewhat more uneven. The line of demarcation of the fiber tips becomes less sharp with increasing distance from the tumor, the fibers become gradually longer and more wavy and, at the same time, more variable in length. The migration of the spindle cells is greatly reduced, as compared to the control cultures. They are completely blocked on the side facing the tumor, but a few do grow out on the opposite side. This feature becomes more distinct during the following period.

No significant changes were observed between 24 and 48 hours, but all features described above are accentuated. In many instances, the fibers on the side toward the tumor form a very dense, felt-like matting (Fig. 8). When the tumor is close to the ganglion, neoplastic cells reach the nerve fibers and mingle with them. Very few or no spindle cells are found on the side toward the tumor, but they migrate out in increasing numbers on the other side, where they may occasionally form bundles or ribbons combined with nerve fibers. In cultures older than 48 hours, the results were less consistent than in earlier stages, and a systematic study of this material has not yet been made. In general, we have noticed the disappearance of the sharp contour of the fibers on the side facing the tumor; instead, they show a tendency to form weblike networks superimposed on the tumor cells (Fig. 6). The migration of spindle cells is no longer blocked. A strange phenomenon was observed in older cultures which needs further study: The cellular degeneration which is characteristic of control ganglia of 3–6 days does not seem to occur in the presence of the sarcoma, even though the sarcoma itself shows regressive changes.

Combination of Sarcoma 180 with sympathetic ganglia.—These experiments were done in only a limited number of cases and not continued beyond the 2d day. The growth of sympathetic ganglia of 8- to 13-day chick embryos in tissue culture has been briefly described by Levi and Delorenzi (7). Control ganglia show at 24 hours a rather uniform outgrowth of small spindle cells in all directions. The population of these cells is more homogeneous than in the case of the spinal ganglia, probably because sympathetic ganglia can be isolated free of connective tissue cells. A few very fine nerve fibers are mingled with the spindle cells; they do not form bundles, and they grow in a wavy course. On the following days, the number of nerve fibers increases moderately.

The combination with Sarcoma 180 results in a very precocious and exuberant outgrowth of nerve fibers, closely resembling the pattern found in spinal ganglia (Fig. 14). Again, the fibers facing the tumor are shorter than those on the opposite side, and their contour is sharply delimited. The orientation of the fibers is always radial and very straight, and their density is greater toward the tumor. The fibers are much finer than the sensory fibers, and the halo is formed of a very dense, regular matting.

Combination of Sarcoma 180 with chick embryo heart.—The striking inhibitory effect of Sarcoma 180 on the spindle cells of spinal ganglia raised the question of whether fibroblasts of an entirely different origin might also be affected. Fragments of heart from chick embryos of 7–9 days were used. In most instances, no effect of the sarcoma on the growth pattern of heart fibroblasts was observed, even when the sarcoma was actively growing and close to the heart fragment. In a few instances, the migration of heart fibroblasts seemed to be somewhat impaired. The number of experiments is not sufficiently large to establish this point definitely; however, the effect is not at all comparable to that observed in ganglia.

Experiments with Sarcoma 37

Growth of Sarcoma 37 in vitro.—Consistently good results were obtained with this tumor when it had been grown in the chick embryo for one or more passages. During the first 24 hours, its growth is even more vigorous than that of Sarcoma 180. In distinction to the latter, its cells, which are spherical rather than spindle-shaped, migrate individually and have no tendency to form rows or bands. As a result, the area of expansion of the culture, which increases considerably between 24 and 48 hours, is not as compact as in cultures of Sarcoma 180. After 48 hours, the cultures show signs of deterioration, and an increasingly large number of macrophages is found among the neoplastic cells.

Combination of Sarcoma 37 with spinal ganglia.—The effects on the fiber outgrowth and the spindle cells of spinal ganglia are identical with those observed with Sarcoma 180 (Figs. 4, 10, 12). The density of nerve fiber outgrowth and the demarcation line of the fibers facing the tumor are, if anything, more pronounced than in the case of Sarcoma 180, and the same is true for the inhibitory effect on the spindle cells.

In several series, the distance between the tumor fragments and the ganglion was varied. Maximal effects were obtained when the distance ranged from 1 to 2 mm. At a distance of 3 mm., the effect was somewhat delayed; the fibers were less dense, and their contour toward the tumor was not so sharp as at close distances. However, the general pattern of nerve growth and spindle-cell inhibition was similar to that obtained at shorter distances (Figs. 13, 15). At a distance of 5 mm., a faint effect is still noticeable. On the side toward the tumor, the fibers are distinctly more numerous and longer than on the other sides; however, they are wavy and do not take a straight course.

Combination of Sarcoma 37 with sympathetic ganglia.—This experiment gave exactly the same results as combinations with Sarcoma 180 and therefore requires no separate description.

Explants of spinal cord, alone and combined with Sarcoma 37.—The explants were taken from 6-day embryos. In contrast to cultures of spinal ganglia, no spindle-shaped cells migrate out of the spinal cord (see also [8]). After 24 hours, one finds a limited outgrowth of epithelial sheets which are probably derived from ependymal cells. They are always restricted to some parts of the explant. The nerve fiber outgrowth differs also from that found in sensory ganglia. The fibers are not distributed regularly over the entire surface of the explant, but they are limited to one or a few sites from which they emerge in long strands. Fibers which bridge a localized area of liquefaction show a very straight course, due to passive stretching.

The combination of spinal cord with Sarcoma 37 did not result in an increase of nerve fibers, nor in a change in the general growth pattern. Occasionally, strands of fibers were directed toward the tumor, but the same behavior of fibers was observed in control combinations of spinal cord with embryonic chick heart, and additional fiber groups were always found emerging from other parts of the explant. Altogether, we consider the spinal cord as completely refractory to the tumor. A small number of experiments with Sarcomas 180 and 1 gave the same results.

Experiments with Sarcoma 1

This tumor grows *in vitro* even more actively than do Sarcomas 180 and 37. However, there are some differences in the morphology of the cells and in their distribution pattern. The cells are spindle-shaped and much smaller than those of the other two sarcomas. They migrate individually in a radial direction and distribute themselves very evenly, covering a large area in a short period (Fig. 17). During the 2d day, one finds a considerable number of round cells in the growth zone; to judge from their size and cytological characters, they are transformed neoplastic cells. After 48 hours, the explant is almost invariably surrounded by a liquefied area.

Combination with spinal ganglia.—In all cases, the fiber outgrowth is enhanced, as compared to control cultures, and the density of the fibers is increased on the side facing the tumor. However, the results are not consistent. In the majority of cases, the effects were considerably milder than with other sarcomas. The fibers do not show the growth pattern characteristic of the combinations with Sarcomas 37 and 180. They are much less dense, and they do not grow out in a regular straight radial direction nor do they show a sharp line of demarcation in front of the tumor. During the 2d day, the fibers in the area facing the tumor collect in bundles which penetrate into the growth zone of the tumor. The migration of spindle-shaped cells is apparently not affected. In a smaller number of cases, the effects are stronger and more similar to the effects of the other two sarcomas. A typical halo of nerve fibers is present, and, at the same time, the migration of spindle cells is inhibited on the side facing the tumor. Nevertheless, the fiber density never reaches the same degree as in the other tumors. It is perhaps significant that, in the latter group, the plasma had not been diluted and the medium was therefore more dense; no liquefaction occurred in these cases.

Experiments with Adenocarcinoma dbrB

Typical epithelial growth of this tumor was obtained by placing the tumor fragment on the surface of the medium which had been allowed to begin clotting. Since the tumors had gone through one or several passages in the chick embryo before being used for explantation, they contained chick fibroblasts in their stroma, and these cells migrated out along with the epithelial sheets. As has been observed by others, the epithelial growth of adenocarcinomas *in vitro* is delayed in comparison with the growth of sarcomas.

Combination with spinal ganglia.—In all experiments, several fragments of the tumor were placed near one side of the ganglion. The carcinoma did not stimulate the outgrowth of nerve fibers beyond the normal range of controls, and spindle cells migrated out actively in all directions (Fig. 11). In some instances, they were even more numerous than in control cultures. A striking stimulation of the outgrowth of mouse fibroblasts by carcinomas, *in vitro*, has been described by Ludford and Barlow (15).

Experiments with Neuroblastoma C1300

This tumor is not compact, as are the others, and is rather difficult to grow in tissue culture. However, typical epithelial sheets surrounding the explants were obtained in some cases (Fig. 5, *N*). In addition, a large number of spindle-shaped cells, probably of chick origin, and macrophages were observed in the growth zone.

Combination with spinal ganglia.—In all cases, a number of fragments of different sizes were placed around the ganglion. The results were different from those obtained with other tumors. As a rule, both the fiber outgrowth and the migration of spindle cells were impaired. In a few cases, nerve fibers did grow out, but they were rarely as numerous as in the controls (Fig. 5).

Control Experiments with Embryonic Chicken Heart

The outcome of a large number of combination experiments of spinal ganglia with heart fragments of 7- to 9-day embryos was consistently negative. The presence of the chick tissue did not change the rate, density, and growth pattern of the nerves. The migration of spindle cells from the ganglia was normal in all instances (Figs. 3, 7).

Control Experiments with Heart Tissue of Embryonic, Fetal, or New-Born Mice

This tissue grows fairly well *in vitro* (in a medium of chicken plasma and extract), but its rate of growth is much lower than that of mouse sarcoma cells or of chicken heart fibroblasts.

Combination with spinal ganglia.—Toward the end of the first day, the number of nerve fibers on the side facing the mouse tissue is consistently higher than in isolated control ganglia and in ganglia facing embryonic chicken heart. However, the features which are characteristic of combination cultures with sarcomas are entirely missing (compare Fig. 1 to Fig. 2), and the general appearance resembles that of control ganglia. During the 2d day, the preferential growth of nerve fibers toward the mouse tissue is accentuated, and the over-all density of fibers is greater than in controls, but the difference in the growth pattern, between combinations with sarcoma and combinations with mouse heart, is as distinct as it was before (compare Fig. 9 to Figs. 8, 10).

DISCUSSION

Parallelism of tumor effects in vivo *and* in vitro.—The tissue culture experiments had been undertaken with the expectation that it might be possible to duplicate *in vitro* some of the remarkable effects on ganglia which had been observed *in vivo*. The results obtained in the two sets of experiments show striking similarities, but also differences in some essential points. In both instances, the nerve fiber outgrowth is far beyond the normal range, and the fibers begin to grow out at earlier stages than they do normally. In both experiments, the agent acts at a distance, and the tumor does not require contact with the ganglion or with nerve fibers to exert its influence. Furthermore, the specificity of the target, which was one of the essential features of the sarcoma effect *in vivo*, is also characteristic of the *in vitro* experiments; only spinal and sympathetic ganglia show a response, whereas the cells of the spinal cord are refractory. The parallelism between the two phenomena extends to the quantitative aspects of the response. The two sarcomas, 180 and 37, show very strong effects both *in vivo* and *in vitro*. A third sarcoma, 1, which was tested by Bueker and Hilderman (2) *in vivo*, was found to have only a mild effect on adjacent ganglia and none on remote ganglia. *In vitro*, it was likewise much less effective than the other two sarcomas.

The question arises whether we are dealing with a general tumor effect or with a specific sarcoma effect. The results of both *in vivo* and *in vitro* experiments were clearly in favor of the second alternative. Two epithelial mouse tumors were found to be entirely negative. Neuroblastoma C1300 grew intra-embryonically to a considerable size, but it was not invaded by nerve fibers and did not call forth a hyperplasia of ganglia (2). *In vitro*, this tumor grew in an epithelial fashion, though its growth was not very extensive. It did not stimulate nerve growth, but, on the contrary, had an inhibitory effect on both nerve fibers and spindle cells. Mammary adenocarcinoma dbrB, which was first grown successfully in the yolk sac (16), attained in coelomic transplantations a very conspicuous size, exceeding even Sarcomas 180 and 37 (unpublished observations of Levi-Montalcini). Nevertheless, it had no effect on ganglia, and it was not invaded by nerve fibers. This tumor grew well *in vitro;* it formed epithelial sheets mixed with stroma cells, but it was again entirely ineffective as far as nerve fiber stimulation is concerned.

At present, it seems that the agent which promotes nerve fiber growth is restricted to some sarcomas. The parallelism between the tumor effects *in vivo* and *in vitro* is very striking and suggests strongly that we are dealing in both instances with the same agent.[2]

On the other hand, we wish to emphasize some important differences in the mode of response of the ganglia *in vivo* and *in vitro*. In tissue culture, we have found an outburst of fiber growth which begins as early as 12 hours after explantation, reaches a peak between 24 and 48 hours, and regresses during the following days. In intra- and extra-embryonic transplants, the first responses were observed about 3½–4 days after implantation. Following these initial responses, the effects increased steadily until the death of the embryo. It seems that these differences can be accounted for by differences in the experimental set-up. The embryonic transplantations are done when the embryos are 2–4 days old, at which time the ganglia are in early stages of differentiation. The small tumor fragment undergoes an initial regression (5) and does not reach an appreciable size until the embryo is 6–7 days old. After reaching this stage, the tumor enlarges progressively and increases its activity proportionally. On the other hand, the explanted tumor fragment does not undergo regression. On the contrary, cell migration begins a few hours after transplantation, and mitotic figures are present at that time. These features attest to the strong vitality of the tumor explant from the first hours on. The tumor fragment is placed closely adjacent to a ganglion which is apparently in an optimal stage for reaction; therefore, all conditions are given for an immediate response. The rapid decline of fiber growth after 3 days is paralleled by a concomitant regression of the tumor because of the depletion of the medium which was not renewed.

Two other sarcoma effects on ganglia had been observed in embryonic transplantation: an increase in mitotic activity and a cellular hypertrophy of neuroblasts. These two aspects were not studied in tissue culture.

Among the many problems raised—but not resolved—by our previous transplantation experiments was that concerning the *immediate target* of the tumor agent. Two alternative hypotheses were advanced (11, 13): Either the agent acts directly on the potential neuroblasts or indirectly by breaking down the normal resistance of viscera against hyperneurotization, thus inviting an abnormal inflow of pathfinders into the viscera. According to this second hypothesis, the hyperplasia of the ganglia would be the end result of a complex chain of reactions beginning with a change at the periphery. The nerve fibers would mediate the effect from the periphery to the centers.

[2] In our previous extra-embryonic transplantation experiments (13) only the responses of the sympathetic ganglia were described in detail. However, responses of the spinal ganglia were clearly manifest both in these experiments (unpublished observations) and in our previous intra-embryonic transplanation experiments (12).

If one admits a basic identity of the tumor effects *in vivo* and *in vitro*, the present experiments can be taken as a strong argument in favor of the first hypothesis cited above. In our experiments *in vitro*, we have observed a precocious outburst of fiber formation almost simultaneously in all parts of the ganglion. The fibers grow straight radially, they do not converge toward the tumor, and, what is most important, they do not establish contact with the tumor until after the peak of the reaction has passed.

The growth pattern of nerve fibers in relation to the culture medium and to other conditions.—All arguments presented so far support the assumption that the main features observed *in vitro*, namely, the conspicuous increase in nerve fiber density and the precociousness of fiber outgrowth, are due to a diffusible agent released by sarcomas. However, in tissue culture experiments, the physical condition of the culture medium cannot be ignored, and the question arises whether some of the strikingly regular features of the growth pattern may be due to this factor. This problem is particularly pertinent in experiments with nerve fibers. Weiss (17) has shown that in many instances where the directional outgrowth of nerve fibres, *in vitro* and *in vivo*, had been attributed to a chemical action at a distance ("chemotropism," "neurotropism"), the "guidance" of nerve fibers was actually achieved by the structural organization of the microscopic or micellar constituents of the ground substance; the fibers do not follow diffusion gradients, but specific pathways or track systems established in the substrate on which they grow ("contact guidance"). The density of fiber outgrowth can also be determined in this way, by the channeling of nutrient supply along preferential lines laid down in the matrix (17, p. 437).

Can any of the sarcoma effects on ganglia be attributed to a structural organization called forth by the tumor in the culture medium? In this connection, it should be pointed out that in our experiments the fiber outgrowth is not directed toward the tumor, but straight radially in all directions (Figs. 10, 13, 14). However, the greater density of the fibers on the side facing the tumor (Figs. 1, 4, 8) could be interpreted in this way. This feature is somewhat reminiscent of the "bridge" phenomenon described by Weiss (17). If two spinal ganglia are explanted adjacent to each other, a large tract of nerve fibers and spindle cells eventually connects the two, whereas fiber outgrowth in other directions remains sparse. It is assumed that the two ganglia call forth a dehydration and thus create lines of tension in the micellar components of the medium. These lines would be mutually reinforced in the area between the ganglia and become a preferential pathway for cells and fibers. A similar explanation may hold in our case; however, the gradual decrease of fiber density with increasing distance from the tumor speaks more in favor of a diffusion gradient set up by the tumor.

The very straight, radial course of the nerve fibers in the presence of sarcomas is in contrast to the somewhat wavy and bent course which is characteristic of fibers emerging from isolated ganglia (compare Fig. 7 to Figs. 8, 10, 12). It is conceivable that, under the impact of the very actively growing sarcoma, the matrix surrounding the ganglion undergoes a particularly regular micellar orientation in a radial direction which would be reflected in the straight and almost geometric pattern of nerve growth. A liquefaction of the medium which is sometimes responsible for a straight fiber course plays no role in our experiments.

Physical factors could also be responsible for the sharp and very regular line of termination of the fibers in front of the tumor (Figs. 10, 12, 14). This feature suggests the presence of an invisible barrier in the culture medium. No evidence for an interface in the plasma clot along this line was observed, but the possibility remains that such an interface is created by the apposition of sarcoma and ganglion.

The seemingly paradoxical phenomenon that the nerve fibers are shorter on the side facing the tumor than on the other sides (Figs. 10, 13, 14) could be ascribed to the same factor or to a threshold of tolerance for the sarcoma agent.

An analysis of the role of the culture medium in the nerve growth pattern is in progress.

An inhibitory effect of sarcoma on spindle cells.—Sarcomas 180 and 37 had an inhibitory effect on the migration of spindle cells. The inhibition was always complete on the side facing the tumor (Figs. 10, 12, 14). It decreased with the distance from the tumor, and, in cases of mild tumor effects on the nerves, it was barely noticeable on the side opposite to the tumor. A close correlation between the stimulatory effects on nerve fibers and the blocking effect on spindle cells was a consistent feature of all experiments with these sarcomas.

Temperature experiments with spinal ganglia showed a similar relationship (8). When tissue cultures of spinal ganglia of 11-day embryos were reared at low temperatures (32° C.), the spindle-cell migration was inhibited, but the branching of nerve fibers was enhanced.

Does a causal relation exist between these two phenomena? And, if so, which is the primary response? Our present material does not permit us to give a conclusive answer to these questions, but a few pertinent observations may be mentioned. It was shown above (p. 51) that the nerve fiber outgrowth in sarcoma experiments is very precocious; in fact, it begins a few hours earlier than the spindle-cell migration in control cultures. This indicates that the response of the nerve cells is independent of the spindle cells. Furthermore, a correlation between spindle-cell inhibition and nerve fiber stimulation does not exist in combination experiments with neuroblastoma. This tumor inhibits both nerve fibers and spindle cells. The relation between these two phenomena deserves further study, both *in vivo* and *in vitro*, since it touches upon the basic problem of the interaction of tumors with host tissues.

Ludford (14) and Ludford and Barlow (15) have also observed slight inhibitory effects of mouse sarcoma on mouse fibroblasts *in vitro*. The same authors found a very strong stimulatory effect of carcinomas on the outgrowth of mouse fibroblasts. Our own carcinoma experiments gave only a slight and not consistent effect of this type.

The effects of normal mouse tissues.—Whereas the combination of chick tissue (from embryonic heart) with spinal ganglia gave entirely negative results, the combination of mouse tissue (from embryonic or fetal heart) with chick ganglia resulted in a consistent increase in the number of nerve fibers on the side facing the mouse tissue (Fig. 2). However, as was pointed out above (p. 53), the growth pattern of the nerves did not show any of the characteristics of the sarcoma effects, but was merely an accentuation of the normal pattern (compare Fig. 9 to Fig. 10). This result suggests that the sarcoma agent may be present in low concentration in normal mouse tissue. However, mouse adenocarcinoma and neuroblastoma were negative, and the question of the distribution of the agent in mouse tissue requires further studies. In this connection, some data of Ludford and Barlow (15) are of interest. These authors reported that kidney tissue of mouse embryos has a moderate stimulating effect on mouse fibroblasts which is comparable to that of some mouse carcinomas, but much milder in degree.

SUMMARY

Small fragments of mouse Sarcomas 180 and 37 were placed at a distance of 1–2 mm. from spinal or sympathetic ganglia of a chick embryo in a hanging-drop tissue culture. Under this condition the ganglion produces precociously, within 24 hours, an excessive number of nerve fibers which grow very straight radially in all directions, forming a dense "halo" around the ganglion. Their density decreases, and their length increases, with increasing distance from the sarcoma. In addition, the migration of spindle cells from the ganglia is inhibited by the sarcomas. Mouse Sarcoma 1 has a similar, but milder, effect. Mouse adenocarcinoma dbrB and mouse neuroblastoma C1300 do not stimulate nerve growth. Control experiments with heart tissue from chick embryos were entirely negative, but heart tissue of fetal mice was found to have a mild stimulating effect. However, in the latter instance, the growth pattern is very different from that found in the presence of sarcomas and very similar to that found in normal, isolated ganglia. Sarcomas have no effect on spinal cord fibers.

It is concluded that the mouse sarcomas tested produce a diffusible agent which strongly promotes the nerve fiber outgrowth of ganglia. The results obtained *in vitro* are compared to previous results obtained by intra-embryonic transplantation of the same sarcomas, and the conclusion is reached that the *in vitro* and the *in vivo* effects on the spinal and sympathetic ganglia are due to the same agent.

Magnification of all figures ×65.

FIG. 1.—Lumbar ganglion of 7-day embryo, combined with Sarcoma 180 (*S*). 24 hrs.

FIG. 2.—Lumbar ganglion of 7-day embryo, combined with heart of mouse fetus (to the left of ganglion). 24 hrs. Note fiber outgrowth to the left.

FIG. 3.—Lumbar ganglion of 7-day embryo, combined with heart of chick embryo (*C*). 24 hrs.

FIG. 4.—Lumbar ganglion of 7-day embryo, combined with 2 fragments of Sarcoma 37. 24 hrs.

FIG. 5.—Lumbar ganglion of 7-day embryo (*G*), combined with 2 fragments of neuroblastoma C1300 (*N*). 34 hrs.

FIG. 6.—Lumbar ganglion of 6-day embryo, combined with Sarcoma 180 (*S*). 72 hrs.

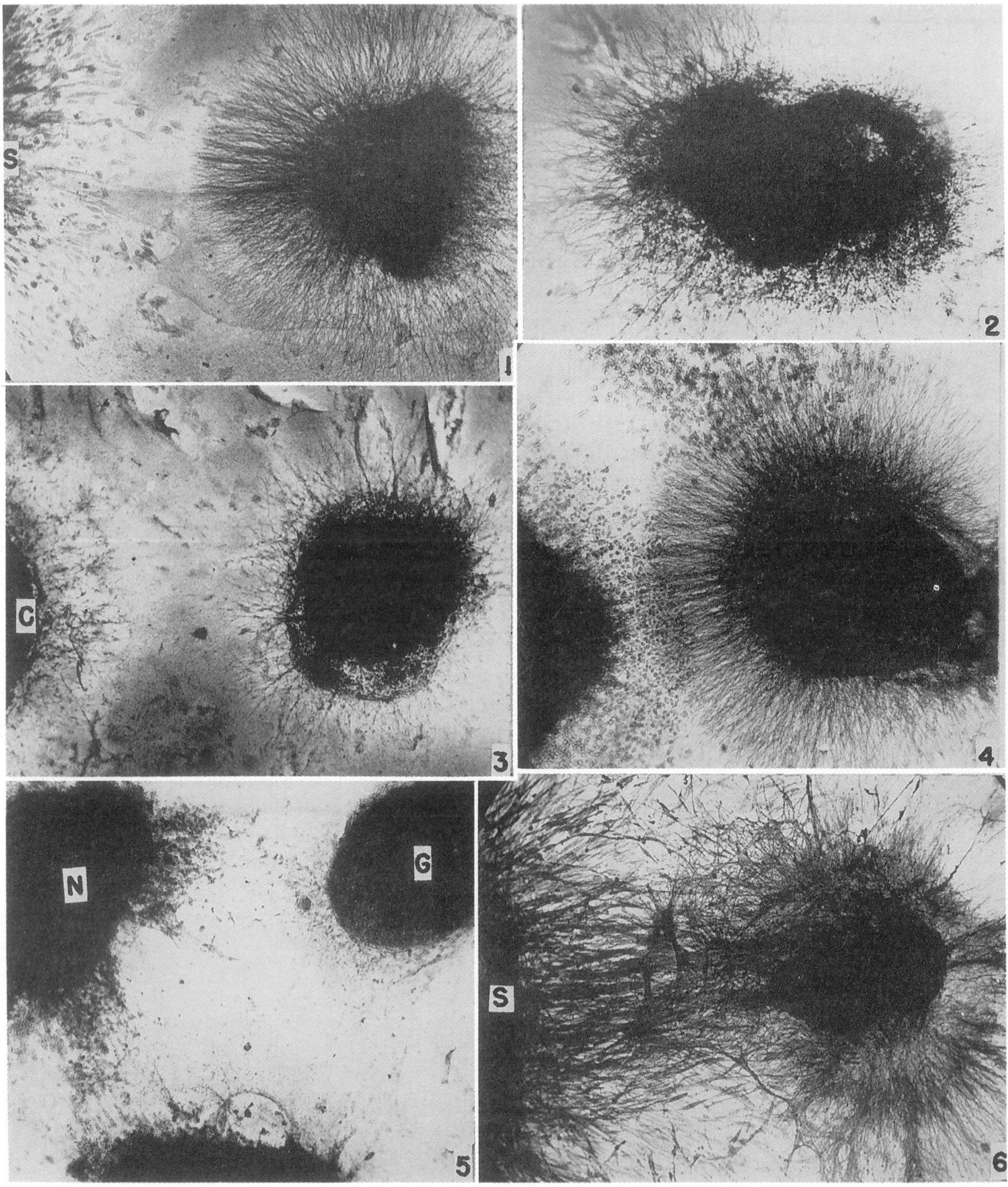
S
1
2
C
3
4
N
G
5
S
6

Magnification of all figures $\times 65$.

FIG. 7.—Lumbar ganglion of 6-day embryo, combined with heart of chick embryo (to the left). 48 hrs.

FIG. 8.—Lumbar ganglion of 7-day embryo, combined with Sarcoma 180 (to the left). 48 hrs.

FIG. 9.—Lumbar ganglion of 7-day embryo, combined with heart of mouse fetus (to the right). 48 hrs.

FIG. 10.—Lumbar ganglion of 7-day embryo, combined with Sarcoma 37 (*S*). 48 hrs.

FIG. 11.—Thoracic ganglion of 7-day embryo, combined with adenocarcinoma dbrB (*A*). 30 hrs.

FIG. 12.—Lumbar ganglion of 7-day embryo, combined with Sarcoma 37. 48 hrs.

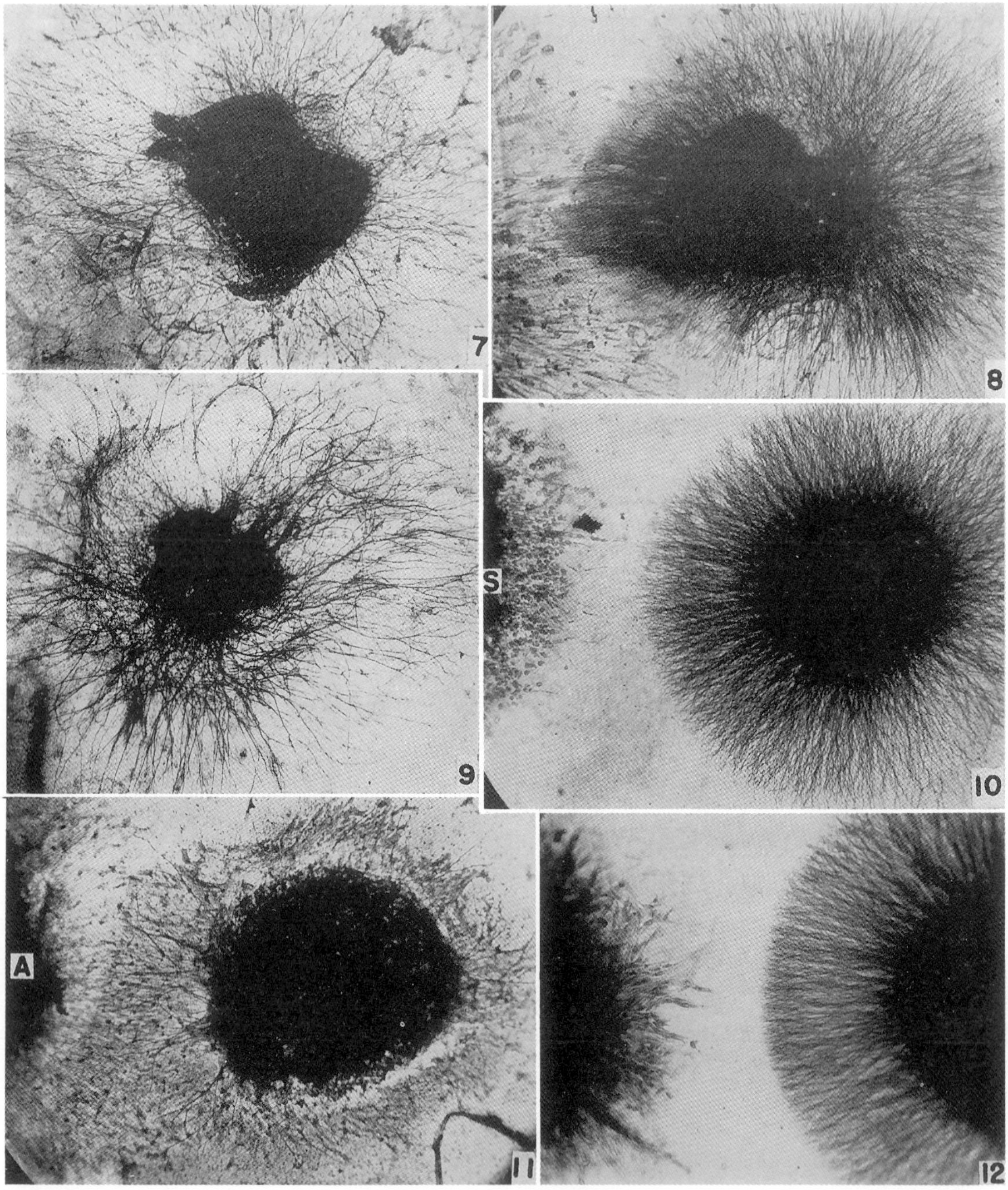
7
8
9
S
10
A
11
12

Magnification of all figures $\times 65$.

FIG. 13.—Lumbar ganglion of 7-day embryo, combined with Sarcoma 37 (to the left). Actual distance between ganglion and sarcoma = 1 mm. (compare with Fig. 15). 48 hrs.

FIG. 14.—Paravertebral sympathetic ganglion of 13-day embryo, combined with Sarcoma 180 (*S*). 24 hrs.

FIG. 15.—Lumbar ganglion of 7-day embryo, combined with Sarcoma 37 (located at some distance from the left lower corner). Actual distance between ganglion and sarcoma = $3\frac{1}{2}$ mm. 48 hrs.

FIG. 16.—Paravertebral sympathetic ganglion of 13-day embryo, combined with Sarcoma 37 (*S*). 44 hrs.

FIG. 17.—Lumbar ganglion of 7-day embryo, combined with Sarcoma 1 (*S*). 48 hrs.

FIG. 18.—Paravertebral sympathetic ganglion of 13-day embryo, combined with heart of chick embryo (*H*). 36 hrs.

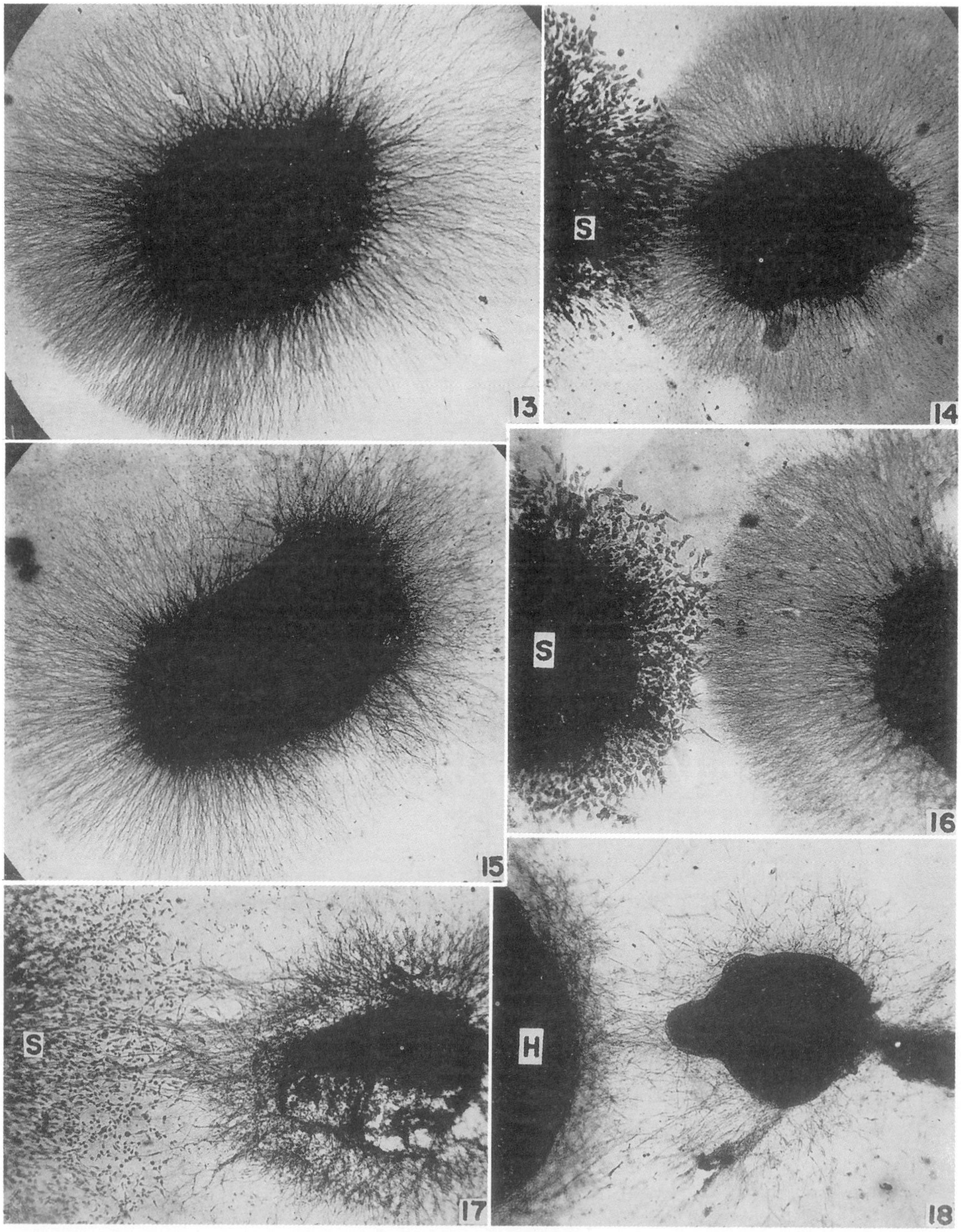
13
S
14
15
S
16
S
17
H
18

ACKNOWLEDGMENTS

The senior author wishes to express her appreciation for a travel grant from the Rockefeller Foundation and for the generous hospitality offered by Dr. C. Chagas, Director of the Institute of Biophysics, University of Rio de Janeiro, Brazil, where the first series of experiments were performed, in cooperation with Hertha Meyer. The experimental work done in St. Louis was supported by grants from the Rockefeller Foundation and the National Institutes of Health.

We wish to express our appreciation of the very able assistance of Mrs. Doris Spruss Fridley and Miss Nancy Starzl. The photographic work was done by Mr. Karl Jacob, Jr., and Mr. A. C. Schewe.

REFERENCES

1. BUEKER, E. D. Implantation of Tumors in the Hind Limb Field of the Embryonic Chick and the Developmental Response of the Lumbo-sacral Nervous System. Anat. Rec., **102**:369–90, 1948.
2. BUEKER, E. D., and HILDERMAN, H. L. Growth-stimulating Effects of Mouse Sarcomas 1, 37, and 180 on Spinal and Sympathetic Ganglia of Chick Embryos as Contrasted with Effects of Other Tumors. Cancer, **6**:397–415, 1953.
3. CARREL, A., and BURROWS, M. T. Cultivation *in Vitro* of Malignant Tumors. J. Exper. Med., **13**:571–75, 1911.
4. HAMBURGER, V., and LEVI-MONTALCINI, R. Proliferation, Differentiation and Degeneration in the Spinal Ganglia of the Chick Embryo under Normal and Experimental Conditions. J. Exper. Zool., **111**:457–502, 1949.
5. KAUTZ, J. Differential Invasion of Embryonic Chick Tissues by Mouse Sarcomas 180 and 37. Cancer Research, **12**:180–87, 1952.
6. LEVI, G. Explantation, besonders die Struktur und die biologischen Eigenschaften der *in vitro* gezüchteten Zellen und Gewebe. Ergebn. Anat. und Entw., **31**:125–707, 1934.
7. LEVI, G., and DELORENZI, E. Trasformazione degli elementi dei gangli spinali e simpatici coltivati *in vitro*. Arch. It. Anat., **33**:443–517, 1935.
8. LEVI, G., and MEYER, H. Nouvelles recherches sur le tissu nerveux cultivé *in vitro:* Morphologie, croissance et relations réciproques des neurons. Arch. de Biol., **52**:133–278, 1941.
9. ———. Reactive, Regressive, and Regenerative Processes of Neurons Cultivated *in Vitro* and Injured with Micromanipulator. J. Exper. Zool., **99**:141–81, 1945.
10. LEVI-MONTALCINI, R. Growth-stimulating Effects of Mouse Sarcoma on the Sensory and Sympathetic Nervous System of the Chick Embryo. Anat. Rec., **109**:59, 1951.
11. ———. Effects of Mouse Tumor Transplantation on the Nervous System. Ann. N.Y. Acad. Sc., **55**:330–43, 1952.
12. LEVI-MONTALCINI, R., and HAMBURGER, V. Selective Growth-stimulating Effects of Mouse Sarcoma on the Sensory and Sympathetic Nervous System of the Chick Embryo. J. Exper. Zool., **116**:321–62, 1951.
13. ———. A Diffusible Agent of Mouse Sarcoma, Producing Hyperplasia of Sympathetic Ganglia and Hyperneurotization of Viscera in the Chick Embryo. J. Exper. Zool., **123**:233–88, 1953.
14. LUDFORD, R. J. The Interaction *in Vitro* of Fibroblasts and Sarcoma Cells with Leucocytes and Macrophages. Brit. M. J., **1**:201–5, 1940.
15. LUDFORD, R. J., and BARLOW, H. The Influence of Malignant Cells upon the Growth of Fibroblasts *in Vitro*. Cancer Research, **4**:694–703, 1944.
16. TAYLOR, A., and CARMICHAEL, N. The Effect on the Embryo of Continued Serial Tumor Transplantation in the Yolk Sac. Cancer Research, **9**:498–583, 1949.
17. WEISS, P. *In Vitro* Experiments on the Factors Determining the Course of the Outgrowing Nerve Fiber. J. Exper. Zool., **68**:393–448, 1934.
18. ———. Experiments of Cell and Axon Orientation *in Vitro:* The Role of Colloidal Exudates in Tissue Organization. *Ibid.*, **100**:353–86, 1945.

Reprinted from the Proceedings of the NATIONAL ACADEMY OF SCIENCES,
Vol. 40, No. 10, pp. 1014–1018. October, 1954

*A NERVE GROWTH-STIMULATING FACTOR ISOLATED FROM SARCOMAS 37 AND 180**

BY STANLEY COHEN, RITA LEVI-MONTALCINI, AND VIKTOR HAMBURGER

DEPARTMENT OF ZOÖLOGY, WASHINGTON UNIVERSITY

Communicated June 29, 1954

A growth-stimulating effect of mouse sarcomas 37 and 180 on the sensory and sympathetic ganglia of chick embryos has been reported.[1] The response of the ganglia to the transplantation of the tumor into 3- to 4-day chick embryos is characterized by a numerical hyperplasia, a cellular hypertrophy, acceleration of differentiation of the ganglia, and an atypical distribution of nerve fibers. The transplantation of the tumor onto the allantoic membrane[2] gave evidence for the humoral nature of the agent.

Explantation of the tumor in vitro in close proximity to sensory or sympathetic ganglia resulted in an exceptional outgrowth of nerve fibers.[3] The parallelism between the tumor effects in vivo and in vitro suggested that we are dealing in both instances with the same agent.

We have now found that cell-free homogenates of the tumors can duplicate in tissue culture the effect of the actively growing tumor. For the assay procedure, hanging-drop preparations were made containing $^1/_3$ plasma (rooster), $^1/_3$ chick extract (5 per cent extract of 10-day chick embryos), and $^1/_3$ of the material to be tested. Saline was used in the controls. Each culture contained a sympathetic ganglion isolated from a 10-day chick embryo. The cultures were observed after 18 hours of incubation at 37° C., and the growth of the fibers was semiquantitatively recorded from 1+ to 4+ (Pl. I, Figs. 1–4). The assay was sensitive to twofold changes in concentration of the active material; smaller changes were not detectable by gross observation of the ganglia.

Preliminary experiments had shown that extracts of sarcomas 37 and 180, which had been grown in the mouse, were almost completely inactive in stimulating the growth of nerve fibers, as were extracts of mouse liver and muscle. However,

PLATE I

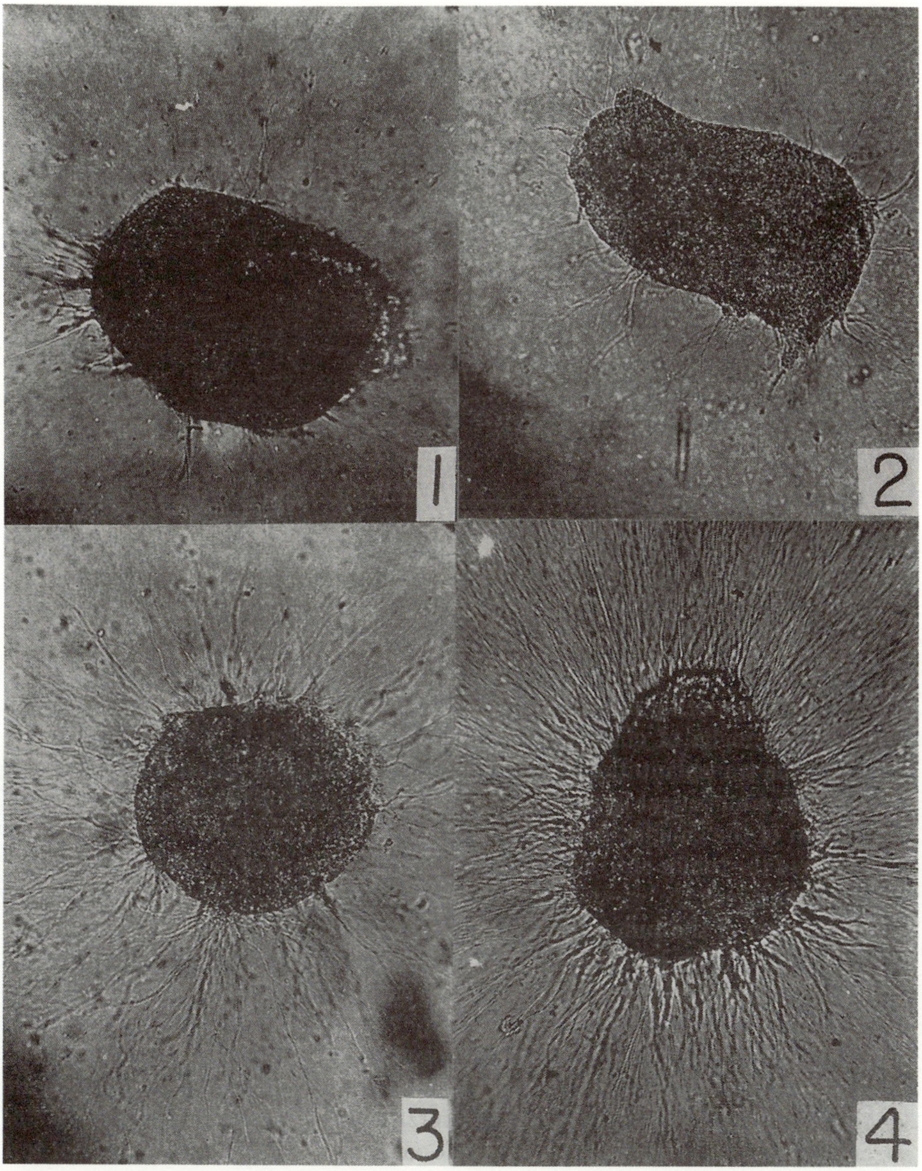

Figs. 1–4.—Microphotographs of living sympathetic ganglia after 18 hours of incubation, showing the nerve growth–promoting effect of increasing concentrations of the tumor extract. The chloroform-treated nucleoprotein fraction was used. Fig. 1, control; Fig. 2, 0.2 mg/ml (recorded as 1 plus); Fig. 3, 0.4 mg/ml (recorded as 2 plus); Fig. 4, 0.8 mg/ml (recorded as 4 plus).

after passage of the sarcomas through the chick embryo, extracts were invariably effective. The tumor was then routinely obtained by transplantation into the body wall of the 3-day chick embryo. The tumors were allowed to grow for 5–7 days before they were harvested.

The intracellular localization of the growth-promoting material was then investigated. Ten per cent homogenates of the tumor were prepared in isotonic sucrose (0.25 *M*, pH 7.4), according to the method of Schneider.[4] The procedure for the isolation of nuclear, mitochondrial, microsomal, and supernatant fractions by differential centrifugation was followed, with two modifications. The nuclear and mitochondrial fractions were washed four times with isotonic sucrose, and the microsomal fraction was isolated by centrifugation at 100,000 × g. Each fraction was then made up to the original volume of the homogenate and dialyzed against saline before assaying for its activity. The results are shown in Table 1. Practically all

TABLE 1

INTRACELLULAR LOCALIZATION OF NERVE GROWTH–PROMOTING FACTOR FROM SARCOMA 37 AFTER PASSAGE THROUGH CHICK

Cell Fraction Added	Growth of Nerve Fibers	Cell Fraction Added	Growth of Nerve Fibers
None	−	Mitochondria	±
Whole homogenate	++++	Microsomes	++++
Nuclei	−	Supernatant	−

the activity resided in the microsomal fraction, which contained approximately 16 per cent of the dry weight of the original tumor. In some instances the microsomal fraction showed a somewhat greater activity than the original homogenate.

The clear reddish pellet of microsomal material may be dispersed in distilled. water and the activity completely sedimented by centrifugation at 100,000 × g for 1 hour. However, in slightly alkaline solutions (pH 9–10) some of the activity remained in the supernatant fluid after centrifugation. At pH 5.6 all the activity in the microsome fraction was precipitated; the precipitate could be redispersed at pH 7.4.

The suspension of microsomes in distilled water could be further fractionated by the addition of streptomycin, which precipitates highly polymerized nucleic acids and nucleoproteins.[5] Streptomycin sulfate (from stock 0.2 *M* solution, pH 7.2) was added to a final concentration of 0.02 *M*. The mixture was allowed to stand for two hours in the cold and then centrifuged for 5 minutes at 8,500 × g. The clear reddish supernatant (containing home-proteins) showed no absorption peak at 260 mμ and no nerve growth–promoting activity. The precipitated streptomycin-nucleoprotein complex was dispersed in a solution containing 0.2 *M* sodium bicarbonate and 0.2 *M* sodium chloride. The streptomycin was then removed by dialysis for 24 hours against 0.2 *M* sodium bicarbonate and, finally, 24 hours against distilled water. This fraction, containing all the microsomal nucleic acid, possessed practically all the activity of the whole homogenate. The ratio of the absorption peak at 260 mμ to that at 280 mμ was 1.61.

PLATE II

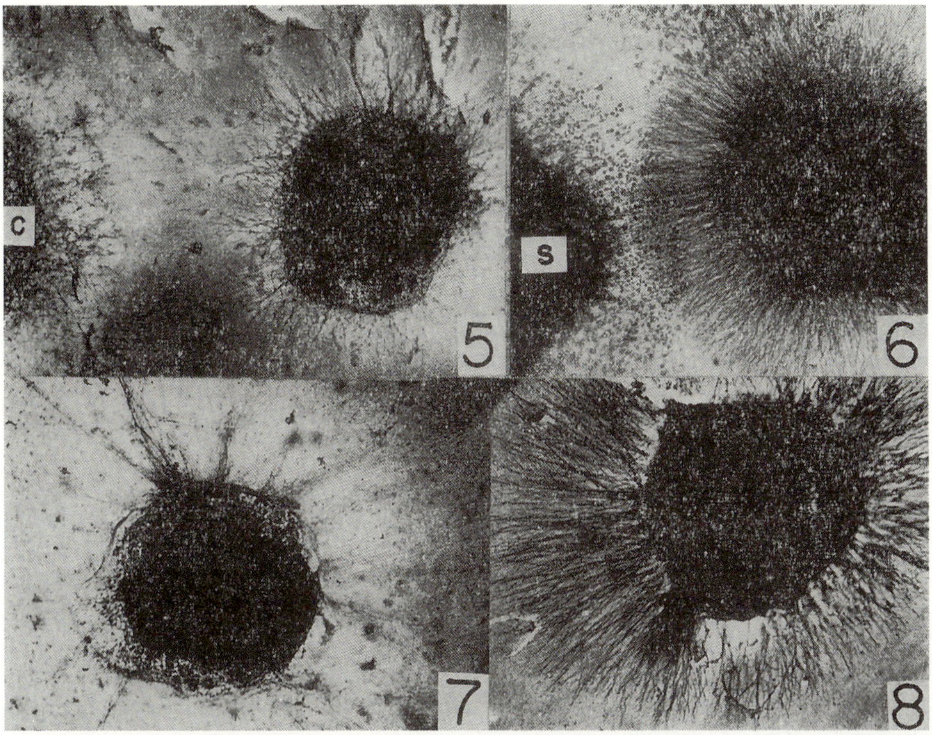

FIGS. 5–8.—Microphotographs of silver-impregnated sensory ganglia, comparing the effect of the intact tumor with the growth-stimulating effect of the cell-free extract of the same time. Fig. 5, control lumbar ganglion of 7-day embryo combined with heart of check embryo (*C*); Fig. 6, ganglion combined with two fragments of sarcoma 37 (*S*); Fig. 7, control ganglion; Fig. 8, ganglion growing in a medium to which the cell-free extract of the tumor was added.

One volume of chloroform was added to 10 volumes of the nucleoprotein solution (in distilled water). After gentle mixing by inversion for 10 minutes, the resulting milky suspension was centrifuged at 8500 × g for 5 minutes, and the clear, somewhat opalescent, supernatant fluid was found to contain from 50 to 100 per cent of the original activity. The solution showed a typical nucleoprotein absorption curve, with a peak at 260 mμ and a 260 to 280 mμ absorption ratio of 1.78. The active material was heat-labile; the activity was completely destroyed by heating for 5 minutes at 80° C. The material was nondialyzable. When adjusted to pH 9–10, the activity no longer was sedimented in the ultracentrifuge (1 hour at 100,000 × g). This soluble active material could again be precipitated with streptomycin. Our purest preparation contained approximately 3 per cent of the dry weight of the original tumor. It was found to contain 66 per cent protein (as determined by the procedure of Lowry *et al.*,[6] using bovine albumin as a standard), 26 per cent ribosenucleic acid (determined by the oricinol procedure,[7] using adenosine as a standard) and less than 0.3 per cent desoxyribosenucleic acid (determined by the diphenylamine reaction,[7] using desoxyribose as a standard).

The nerve growth–stimulating effects of the intact tumor, when growing in close proximity to a sensory ganglion, are compared in Plate II, Figures 5–8, with the growth-stimulating effects of extracts obtained from these tumors.

Our investigations are now directed toward (*a*) the further elucidation of the nature of the active material, (*b*) the duplication of the effect of the growing tumor in the living embryo with the active material isolated from the tumors, and (*c*) an examination of the metabolic response of the nerve cells under the influence of the growth-promoting agent.

* This work has been supported in part by a grant from the National Institutes of Health of the United States Public Health Service. A preliminary report was presented at the forty-fifth annual meeting of the American Association for Cancer Research at Atlantic City, April, 1954.

[1] E. D. Bueker, *Anat. Record,* **102,** 369–390, 1948; R. Levi-Montalcini and V. Hamburger, *J Exptl. Zool.,* **116,** 321–362, 1951; R. Levi-Montalcini, *Ann. N.Y. Acad. Sci.,* **55,** 330–343, 1952.

[2] R. Levi-Montalcini and V. Hamburger, *J. Exptl. Zool.,* **123,** 233–288, 1953.

[3] R. Levi-Montalcini, H. Meyer, and V. Hamburger, *Cancer Research,* **14,** 49–57, 1954.

[4] W. C. Schneider, *J. Biol. Chem.,* **176,** 259–266, 1948.

[5] S. S. Cohen, *J. Biol. Chem.,* **168,** 511–526, 1947; H. Von Euler and L. Heller, *Arkiv Kemi* **26A,** No. 10, 1–16, 1948; R. J. Kutsky, *Proc. Soc. Exptl. Biol. Med.,* **83,** 390–395, 1948.

[6] O. H. Lowry, N. J. Rosebrough, A. L. Farr, and R. J. Randall, *J. Biol. Chem.,* **193,** 265–275 1951.

[7] W. C. Schneider, *J. Biol. Chem.,* **161,** 293–303, 1945.

Reprinted from the Proceedings of the NATIONAL ACADEMY OF SCIENCES,
Vol. 42, No. 9, pp. 695–699. September, 1956.

*IN VITRO AND IN VIVO EFFECTS OF A NERVE GROWTH-STIMULATING AGENT ISOLATED FROM SNAKE VENOM**

BY RITA LEVI-MONTALCINI AND STANLEY COHEN

DEPARTMENT OF ZOÖLOGY, WASHINGTON UNIVERSITY, ST. LOUIS, MISSOURI

Communicated by V. Hamburger, July 19, 1956

INTRODUCTION

Previous reports have dealt with a nerve growth-stimulating effect of mouse sarcomas 180 and 37 on sensory[1] and sympathetic ganglia of the chick embryo.[2] The effect consisted in a striking increase in size and cell number of sensory and sympathetic ganglia adjacent to the tumor and of an extensive invasion of their fibers into the tumor. Transplantation of the tumor onto the allantoic membrane had a growth-promoting effect similar to the intra-embryonic transplantations.[3, 4] In the allantoic transplants the tumor was not accessible to nerve fibers; the effect was therefore mediated through the circulation which the tumor shared with its host. Embryos bearing mouse tumors in the abdominal wall or on the allantoic membrane differed from controls also in another respect: the viscera were overloaded with sympathetic nerve fibers, whereas control embryos showed practically no visceral innervation at corresponding stages of development. Large nerve bundles were also found in blood vessels, where they formed agglomerates projecting into the lumen of the vessel. The agent released by sarcomas 180 and 37 has therefore a dual effect: it stimulates the overgrowth of sensory and sympathetic ganglia and is responsible for the aberrant distribution of sympathetic nerves in the viscera and blood vessels. Two alternative hypotheses were offered: either the agent affects primarily the growth and differentiation processes of nerve cells, or changes of the periphery are the primary effect and the overgrowth of the ganglia is an indirect effect.[4] The following experimental result favored the first hypothesis.

Sensory and sympathetic ganglia were explanted in vitro in proximity to a fragment of mouse tumor. Under these conditions the ganglia are directly exposed to the tumor, and influences of the organism are excluded. Ganglia explanted together with normal embryonic tissues of chick or mouse served as controls. Whereas very few nerve fibers emerged from the control ganglia in the first 24 hours, a dense fibrillar "halo" surrounded the ganglia exposed to the tumor at the end of the first day.[5] These experiments indicated a direct action of the neoplastic agent on nerve cells. The tissue-culture method was then used to test the effect of a cell-free homogenate of the tumor. When it was found to possess the same activity as the growing tumor in vitro, the investigation was directed toward the purification and identification of the agent.[6, 7]

In the course of this work another growth-stimulating agent was discovered in snake venoms. The properties of this agent will be dealt with in another communication.[8] It is the object of the present report to describe its biological effects on sensory and sympathetic ganglia in vitro and in vivo and to compare them with those of the tumor factor.

EXPERIMENTS IN VITRO

The investigation was centered on sensory ganglia of 7-day chick embryos. The standard hanging-drop technique was used. The medium consisted of one part plasma, one part synthetic medium, and one part of the snake venom to be assayed. We used crude venom from *Crotalus adamanteus* (rattlesnake) and *Agkistrodon piscivorus* (moccasin) and a protein fraction (DEAE fraction) derived from the moccasin venom.[8] In each culture three to four ganglia were explanted. A total of several hundred ganglia were tested. A series of controls was run in which one part of an isotonic saline solution was added to the medium instead of the snake-venom fraction. The cultures were incubated at 37° and examined after 6, 12, and 24 hours.

Results.—The addition to the medium of 7 μg/ml of the crude snake venom or of 1 μg/ml of the DEAE fraction had identical effects. The following description therefore applies to the experiments performed with both. The results were in fact remarkably consistent, and only minor fluctuations were observed.

Six hours after the beginning of the experiment, a dense "halo" of short nerve fibers surrounded the ganglia. The fibers increased rapidly in length and density in the following hours. At 24 hours the fibrillar halo had reached its maximal density (Fig. 1). The characteristics of this most unusual fiber outgrowth were identical with those described in previous investigations dealing with mouse tumor;[5, 6] they will therefore not be recounted here (Figs. 3 and 4). It may only be mentioned that the cultures differed from controls also in another respect. Whereas in control cultures the fiber outgrowth was very scanty in the first 24 hours and the nerve fibers grew in a wavy, irregular fashion (Fig. 2), in the present experiments the fibers grew straight, radially, and showed extensive branching, forming a brushlike border at their tips. In most cases, a second, shorter halo formed in the clot between the twelfth and the twenty-fourth hour. As in the previous experiments with tumor, very few or no spindle cells were found among the nerve fibers in the first 24 hours. In the following days, however, a large number of these cells migrated from the explant between the fibers. In the present experiments a considerable degree of liquefaction of the plasma occurred. As a result, a disintegration of the fibrillar halo was observed in many of the cultures after 48–72 hours. The cultures were then discontinued.

A more limited number of experiments were performed with sympathetic ganglia explanted from 8- to 9-day chick embryos. The results were in all respect similar to the ones described above.

EXPERIMENTS IN VIVO

It had been suggested previously that the in vitro and in vivo effects of tumors were produced by the same agent. However, preliminary experiments of injections of active tumor fractions into the yolk of 7-day embryos were negative. We attribute this failure to the fact that the pure fraction was available only in very small amounts and that probably the quantities injected were considerably smaller than those produced continually by living tumor. The striking similarity between the in vitro effects of the tumor and of the snake venom, and the fact that the activity of the purest fraction of snake venom is 1,000 times that of the purest tumor fraction (on a dry-weight basis), suggested a repetition of the injection experiments with snake venom.

The effect of the venom was assayed by injecting it into the yolk of embryos ranging between 6 and 8 days of development. A small window was opened in the shell, and the fraction DEAE of the venom (containing 150 μg protein/ml together with 100 μg/ml of terramycin) was injected into the yolk with a 1-ml. syringe.

In these preliminary experiments the total amount of venom injected, the stage of the embryos, and the length of time of the experiments were varied, in order to determine (*a*) the tolerance of the embryo, (*b*) the stages of development which are sensitive to the effect of the venom, and (*c*) the length of time required to secure an effect.

The injections were repeated two or three times a day. A total of 120 embryos were injected; 17 were recovered and impregnated with silver following the Cajal–De Castro technique. They were sectioned serially and used for the present investigation.

Results.—The toxicity of the venom was reflected in the severe mortality and in the diffuse edema which affected the surviving embryos. The results proved beyond doubt that the venom possesses a growth-stimulating effect on sensory and sympathetic ganglia of the embryo similar in every respect to the effect of mouse sarcomas. The increase in size of the sympathetic system and, in particular, of its prevertebral chain was so striking that it overshadowed the reaction of the sensory ganglia. The same results were obtained in embryos bearing mouse sarcomas. The analysis was then focused on the sympathetic system.

Of the 17 embryos, 11 had been injected over a period of between 36 hours and 4 days, and 6 over a period of only 26 hours. Ten of the 11 embryos injected for a longer period showed a considerable increase in size of the sympathetic ganglia and an atypical distribution of nerve fibers in the viscera. The intensity of the effect varied in different embryos: it increased parallel to the length of the time the embryos had been exposed to the venom. A definite increase in the prevertebral sympathetic complex was observed in one embryo injected over a period of 36 hours, from 6 to $7^1/_2$ days, with a total amount of 0.5 ml. Nerve bundles emerging from the prevertebral complex entered the adjacent mesonephroi (Figs. 9 and 10). No nerves are found in the mesonephroi at corresponding stages in control embryos. The nine other embryos which showed a positive effect were fixed 2, 3, and 4 days after the beginning of the experiment. The total amounts of venom injected were: 0.6 ml. in one, 1.3 ml. in four, 1.8 ml. in two, and 3.0 ml. in two cases. A maximal effect was observed in the two embryos injected over a period of 4 days with 3.0 ml. and recovered at 10 days. The embryos injected for only 26 hours showed no effect.

In one of the embryos injected over a period of 4 days, one ganglion of the prevertebral sympathetic complex at the level of the adrenal was measured. Each section was projected with the help of the camera lucida, and the contour of the ganglion was drawn on cardboard. The total number of sections was then weighed, and the weight was compared to that of the same prevertebral ganglion of a control embryo.

The results gave an increase in weight of 3.2 times the control. Since the weight is proportional to the volume, the figure indicates a similar increase in volume. Estimates of cell numbers were obtained by inserting a micrometer disk in the ocular and counting the number of cells in 200 squares each in comparable regions of ganglia from normal and injected embryos, using oil immersion. The results are

1,066 cells in 200 squares of the injected embryo and 2,158 in the control. The ratio of cell number per unit area was therefore 1:2. The volume ratios were then divided by the cell-count ratios to obtain an estimate of the relative increase in cell numbers. The latter amounts to 1.6. Hence the over-all increase in size is due to a greater extent to cellular hypertrophy and to a lesser extent to an increase in cell numbers. Comparable results were obtained in embryos bearing tumors.[2]

A comparison of hyperplastic and normal ganglia (Figs. 5 and 6) shows that the cells of the injected embryos are not only larger than the controls but also in a more advanced stage of differentiation. This is attested by the presence of neurofibrils in the cytoplasm of almost all the sympathetic neurons in the experimental cases, but only in very few of the controls (Figs. 7 and 8). Furthermore, the nerve bundles which emerge from the hyperplastic and hypertrophic ganglia are much larger than those in the controls.

COMPARISON OF THE IN VIVO EFFECTS OF MOUSE SARCOMAS AND SNAKE VENOM

The results of the preliminary experiments with snake venoms reported here point to a striking similarity between the two phenomena.

In both groups of experiments not only the mesonephroi (Figs. 9 and 10) but also other viscera—thyroid, metanephroi, spleen, and liver—are invaded by sympathetic fibers. In the experiments with snake venom as in those with tumor, nerve bundles perforated the wall of thoracic veins and projected inside the blood vessels (Figs. 11 and 12). Since the experiments with venom were not performed in as late developmental stages as those with tumor, the neuromas were of small size and comparable only to neuromas found in 9–10-day embryos bearing a mouse tumor. The quantitative data reported above are also in line with those given for tumor effects.[2, 4]

In both groups of experiments no morphological changes were detected in any part of the nervous system other than spinal and sympathetic ganglia.

CONCLUDING REMARKS AND SUMMARY

A nerve growth-promoting agent was discovered in snake venoms, and its in vitro and n vivo effects were investigated.

In both sets of experiments the nerve fiber outgrowth from sensory and sympathetic ganglia was found to be far beyond the normal range. Furthermore, the experiments in vivo gave evidence of a remarkable increase in cell size and cell numbers in sympathetic ganglia and of an acceleration of their differentiative processes. The nerve fibers produced in excess entered into the adjacent viscera.

These observations show that the snake venoms duplicate the effects of a "protein fraction" of mouse sarcomas in vitro and the effect of living sarcomas in vivo. The similarity of the two agents is further emphasized by the data which were obtained in attempts to purify them. In both instances, we are apparently dealing with a protein or a substance bound to protein.[7, 8] The purest fraction of snake venom obtained so far is approximately 1,000 times as effective as our purest fraction of mouse tumor (on a dry-weight basis).[8] This fact as well as the ready availability of different snake venoms will facilitate the further analysis of the phenomena reported above.

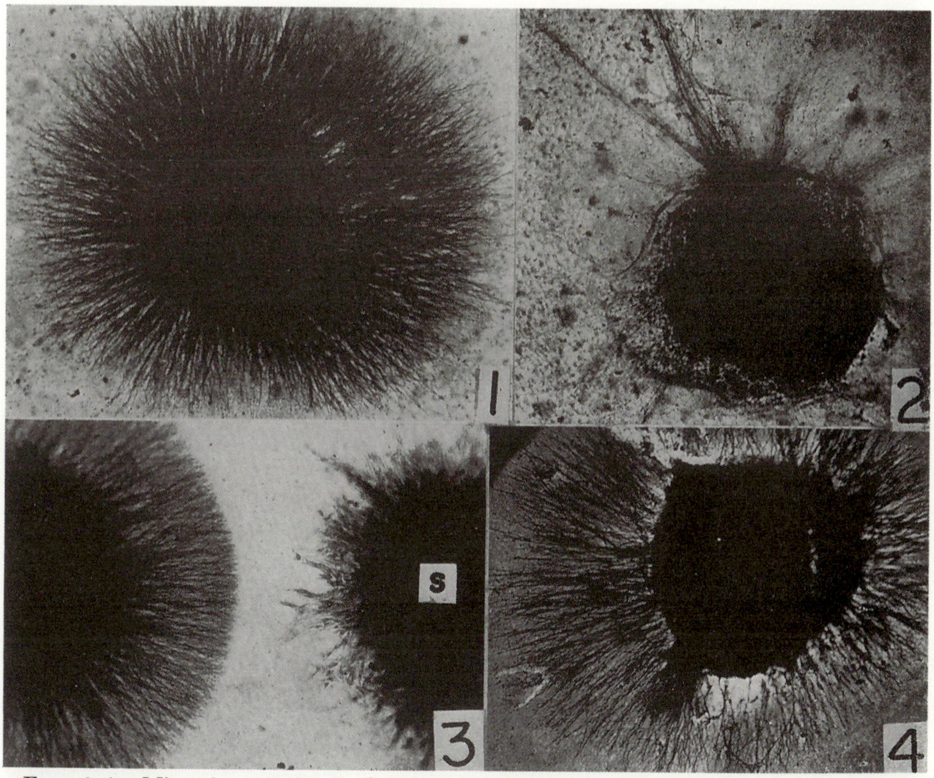

FIGS. 1–4.—Microphotographs of silver-impregnated sensory ganglia, comparing the effect of the snake venom, of the intact tumor, of the cell-free extract of the same tumor, and a control ganglion. All ganglia were isolated from 7-day embryos and cultured for 24 hours. Fig. 1, sensory ganglion growing in a medium to which the fraction DEAE of moccasin venom had been added. Fig. 2, control ganglion. Fig. 3, ganglion combined with a fragment of sarcoma 37 (*S*). Fig. 4, ganglion growing in a medium to which cell-free extract of the tumor has been added.

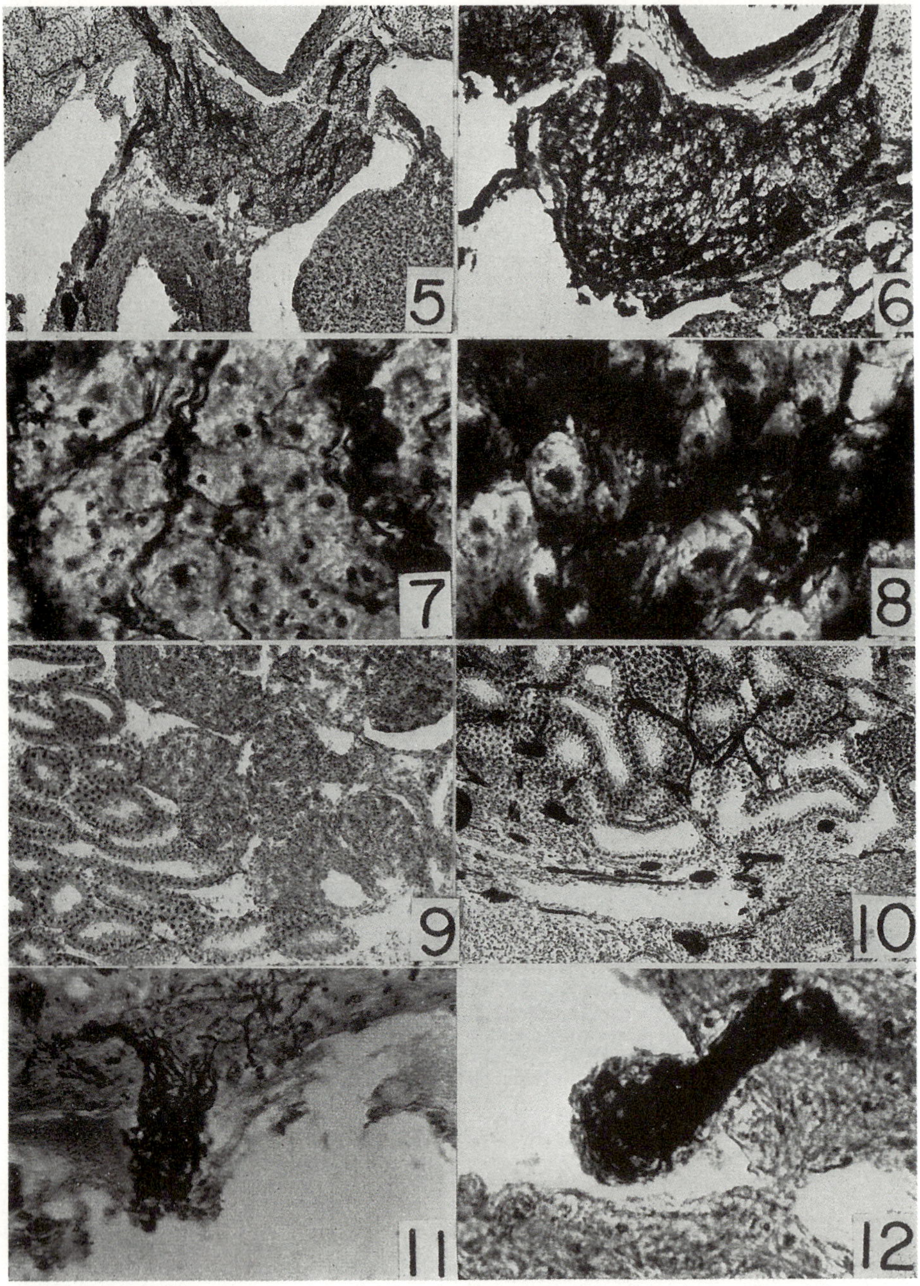

FIGS. 5–12.—Microphotographs of silver-impregnated control and experimental embryos. Fig. 5, prevertebral sympathetic ganglion of a 10-day embryo. Fig. 6, prevertebral sympathetic ganglion of a 10-day embryo injected with moccasin snake venom (fraction DEAE) for 4 days. Figs. 7 and 8, enlarged sectors of Figs. 5 and 6, respectively. Figs. 9 and 10, mesonephros of a control and of an injected 10-day embryo. Notice the sympathetic nerve fibers between the mesonephric tubules in the injected embryo. Fig. 11, neuroma projecting into the lumen of a vein of a 9-day embryo with an extra-embryonic sarcoma 37. Fig. 12, same, in a 10-day embryo, injected with moccasin venom, fraction DEAE (see text).

Our failure, so far, to obtain positive effects after in vivo injection of mouse tumor fractions was attributed to the relative weakness of the tumor fraction rather than to the presence of two different agents in the tumor—one that would be responsible for the in vivo effects and one responsible for the in vitro effects. The present finding that the much stronger venom factor reproduces both effects in every detail gives further strong support to our contention.

* This work has been supported by Grant B-463 from the National Institute of Neurological Diseases and Blindness of the National Institutes of Health, Public Health Service, by a grant of the American Cancer Society, recommended by the Committee on Growth, National Research Council, and by a contribution from an institutional grant of the American Cancer Society to Washington University. The able assistance of Miss Barbara Bankhead is gratefully acknowledged.

[1] E. D. Bueker, *Anat. Record*, **102**, 369–390, 1948.

[2] R. Levi-Montalcini and V. Hamburger, *J. Exptl. Zool.*, **116**, 321–362, 1951.

[3] R. Levi-Montalcini, *Ann. N.Y. Acad. Sci.*, **55**, 330–343, 1952.

[4] R. Levi-Montalcini and V. Hamburger, *J. Exptl. Zool.*, **123**, 233–288, 1953.

[5] R. Levi-Montalcini, H. Meyer, and V. Hamburger, *Cancer Research*, **14**, 49–57, 1954.

[6] S. Cohen, R. Levi-Montalcini, and V. Hamburger, these PROCEEDINGS, **40**, 1014–1018, 1954.

[7] S. Cohen and R. Levi-Montalcini, *Cancer Research* (in press).

[8] S. Cohen and R. Levi-Montalcini, these PROCEEDINGS, **42**, 571–574, 1956.

Reprinted from the Proceedings of the NATIONAL ACADEMY OF SCIENCES
Vol. 46, No. 3, pp. 373–384. March, 1960.

EXCESSIVE GROWTH OF THE SYMPATHETIC GANGLIA EVOKED BY A PROTEIN ISOLATED FROM MOUSE SALIVARY GLANDS*

BY RITA LEVI-MONTALCINI AND BARBARA BOOKER

DEPARTMENT OF ZOÖLOGY, WASHINGTON UNIVERSITY, ST. LOUIS

Communicated by Viktor Hamburger, January 4, 1960

The discovery that mouse sarcomas produce a diffusible agent which selectively promotes the growth of the sensory and sympathetic ganglia in the chick embryo,[1, 2] confronted us with three major problems: (*a*) the characterization of the growth agent, (*b*) the analysis of its mode of action on the embryonic nerve cells which were increased in number and in size under the impact of the agent to an extent unmatched by any previous experimental device, and (*c*) the question of the significance and distribution of nerve-growth agents in tumors and other tissues.

These problems were investigated using the method of explanting embryonic ganglia *in vitro* together with the tumor or other tissues to be tested. The finding that fragments of tumor explanted in proximity to the ganglia elicited exuberant nerve fiber outgrowth, and that this effect is apparent within 5–6 hours, greatly speeded up and simplified the bio-assaying procedures, thus making possible the biochemical approach to our problem.[3] The isolation of a nerve-growth promoting protein fraction from the neoplastic tissue was the first step in this direction.[4] In experiments aimed at the further characterization of the nature of the agent, Cohen made use of snake venom as a source of proteolytic enzymes. This led to the discovery that the snake venom also contains a nerve-growth promoting factor which replicates in all respects the effects of mouse sarcomas on the sensory and sympathetic ganglia of the chick embryo *in vitro* and *in vivo*.[5, 6] Since the snake venom is produced in a modified salivary gland, the mouse salivary glands were tested as another possible source of the nerve-growth promoting factor. The extract of the mouse submaxillary glands proved to contain a nerve-growth promoting agent even more potent than the two mentioned above.[7, 8] It was found that this agent likewise promotes the growth of the sensory and sympathetic embryonic ganglia of the chick embryo, leaving other systems unaffected. The similarity extends also to other facets of the phenomenon such as the massive neurotization of the viscera of the chick embryo. The potency of this factor, its ready availability, its presence in various tissues of different vertebrates, suggested an investigation of its effects on a wider scale. In all previous experiments the chick embryo was the test object. We have now investigated the effects of the nerve-growth factor in mammals. The results of *in vitro* experiments on mouse and rat sensory and sympathetic ganglia were reported.[7, 8] We will report in this paper the results of *in vitro* effects on human fetal ganglia and of *in vivo* experiments on the sympathetic ganglia of newborn, young, and adult mice. We will also report on experiments designed to establish the relationship between salivary glands and nerve-growth agents. In a second paper, we will present evidence of the near-total disintegration of the sympathetic ganglia in mammals as a result of daily injections of the antiserum to the nerve-growth factor. The agents used in all these experiments were prepared by Dr. S. Cohen and they are described by him in another paper in this issue.

In vitro Effects on Sensory and Sympathetic Ganglia of Human Fetuses.—Four human fetuses, three about two-and-a-half months old and one about three-and-a-half months old, were obtained from the Medical School of Washington University. They resulted from hysterotomy for therapeutic abortion and were made available as soon as delivered.† The sensory and sympathetic ganglia of these fetuses were explanted in tissue culture: the same hanging drop technique was used as in all previous experiments.[3, 5, 7] The medium consisted of one part of chicken plasma, one part of synthetic medium, and one part of the purified salivary fraction. Control cultures were prepared using one drop of physiological solution instead of the salivary fraction. A total of forty experimental and forty control cultures were prepared; in each culture a large number of ganglia were explanted. The effects of the salivary extract on ganglia of the two-and-a-half month fetuses were similar in all respects to the effects elicited by mouse sarcomas, snake venom, and the mouse salivary extract on ganglia of chick embryos (Fig. 3), mouse, and rat embryos. A dense halo of nerve fibers surrounded the ganglia in the experimental cultures; it was already apparent 12 hours after the beginning of the experiment and increased in density and in size at the end of the first day (Fig. 5). A comparable, but somewhat slower, reaction was observed in ganglia explanted from the older fetus. A considerable degree of liquefaction of the cultural medium took place on the second and third day of culture; the experiments were then discontinued. Control ganglia showed few or no nerve fibers in the area surrounding the explant 24 hours after the preparation of the culture (Fig. 2).

In vivo Effects on the Sympathetic Ganglia of Newborn and Adult Mice.—(*a*)*Material and methods:* The effect of the salivary extract was tested on newborn and adult mice. Solutions of the active extract at different concentrations and different degree of purification were at first assayed in three groups of newborn mice and then on adult mice.

Group 1: Ten newborn mice were injected daily with the salivary extract referred to in Cohen's paper as fraction CM-1 (these PROCEEDINGS, this issue). This solution was injected subcutaneously in the amount of 0.05 ml per gm of body weight at a concentration of 1,500 units of biological activity per ml. One unit of biological activity is defined as that amount per ml required to elicit a 3+ response

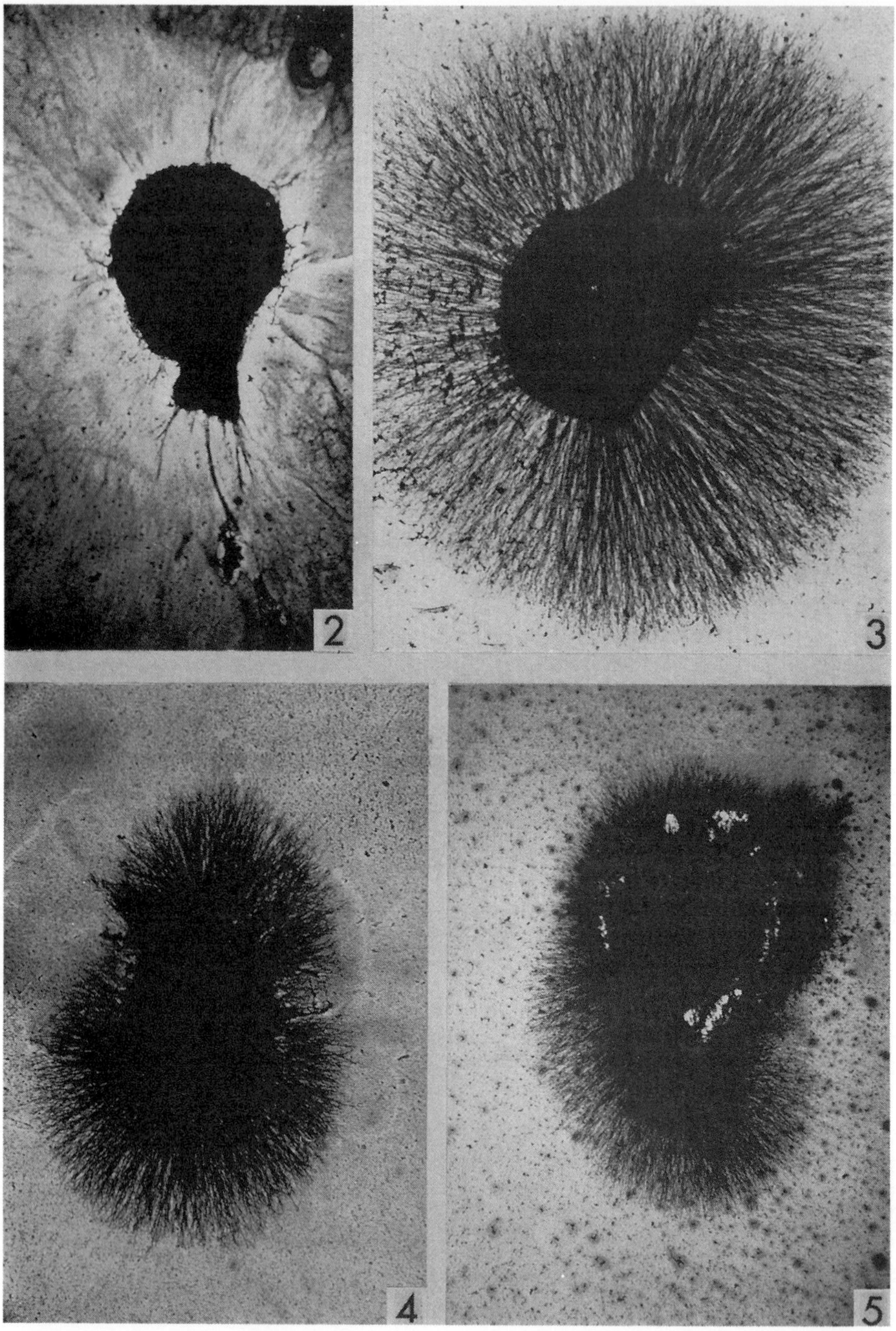

PLATE I.—Microphotographs of sensory ganglia after 24 hours *in vitro*. Silver impregnation. FIG. 2.—Ganglion of a $2^1/_2$ months human fetus in the standard control medium. FIGS. 3, 5.—Ganglia of a 7-day chick embryo (Fig. 3) and of a $2^1/_2$-month human fetus (Fig. 5) in a medium containing the purified salivary protein at a concentration of 1:18,000. FIG. 4.—Ganglion of a 7-day chick embryo in a medium containing mouse serum at a concentration of 1:30. Short and dense halo in Figs. 4 and 5 characteristic of strong and slightly inhibitory effects of the growth agent at high concentration.

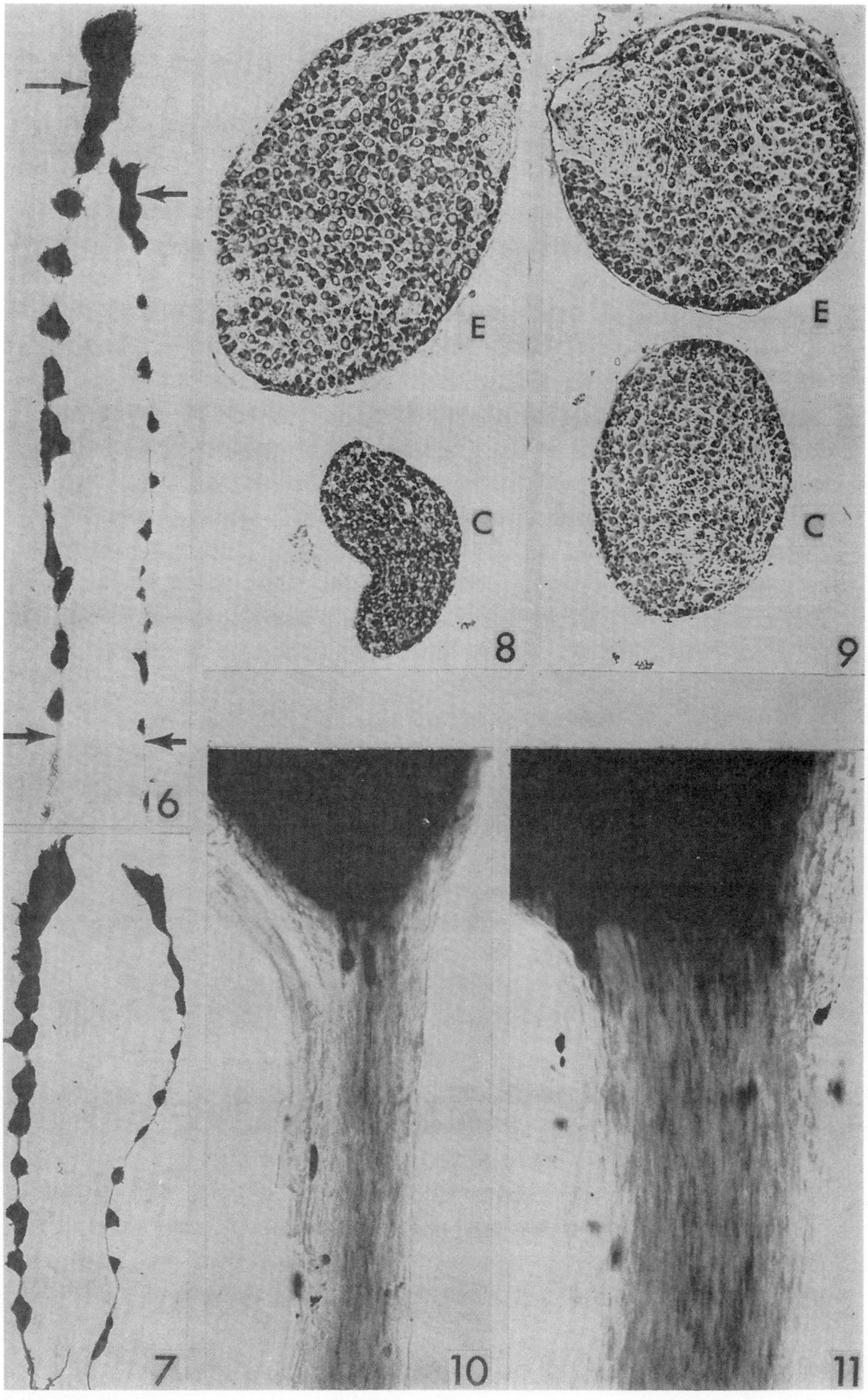

PLATE II.

FIGS 6, 7.—Whole mounts of the sympathetic thoracic chain ganglia of experimental (E) and control (C) mice 19 and 12-day old respectively. Experimental mice injected with the CM-3 salivary fraction since birth. FIG. 8.—Transverse section of stellate ganglia in experimental (E) and control (C) ganglia of Fig. 6. Sections through levels indicated by upper arrows in both chains of Fig. 6. FIG. 9.—Transverse section of superior cervical ganglia of two adult mice. E, mouse injected for 1 week with the CM-1 salivary fraction. C, control ganglion. FIGS. 10, 11.—Sympathetic nerve trunk of control and experimental chain ganglia of Fig. 6e at higher magnification. Low arrows in Fig. 6 indicate areas enlaged in Figs 10, 11.

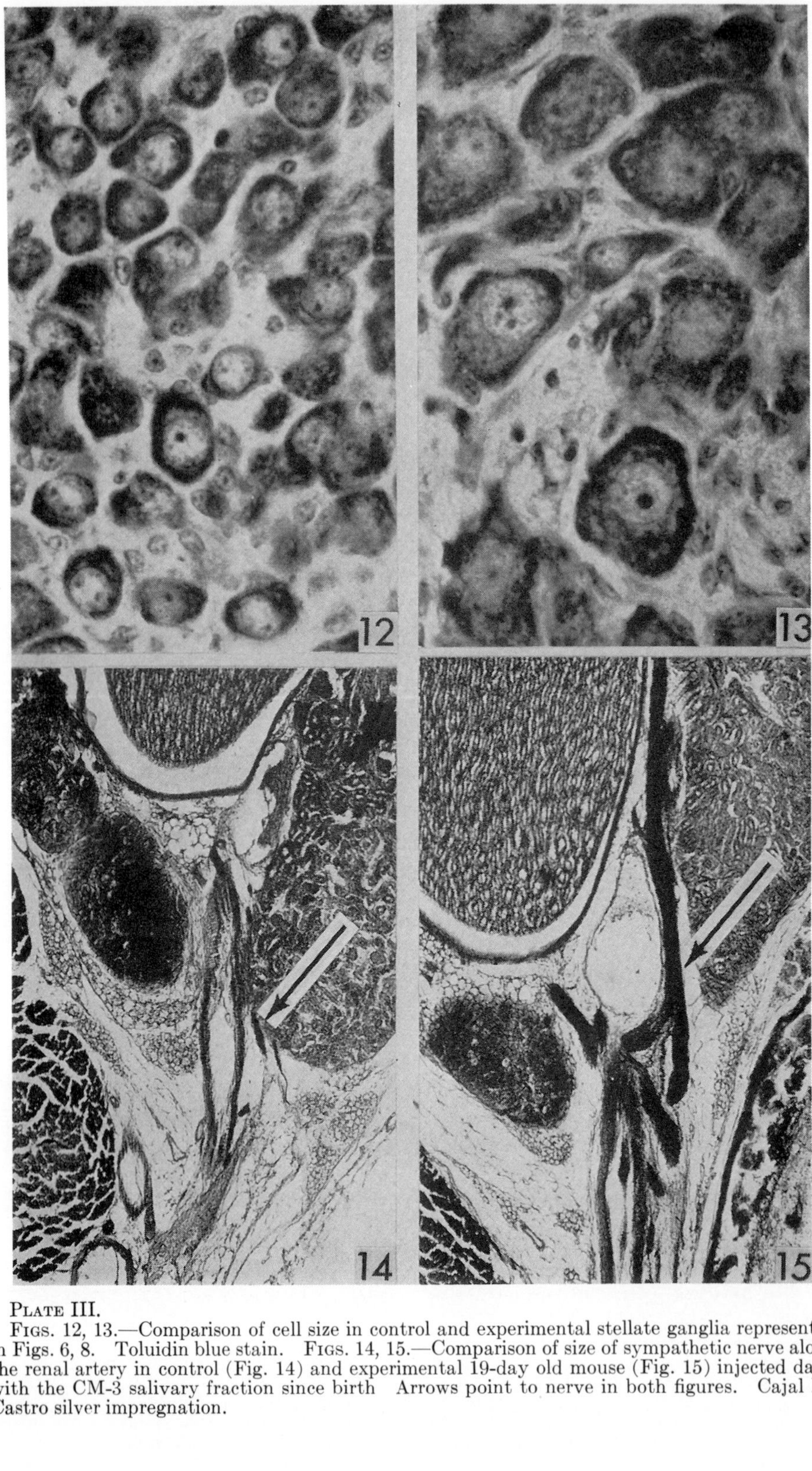

PLATE III.

FIGS. 12, 13.—Comparison of cell size in control and experimental stellate ganglia represented in Figs. 6, 8. Toluidin blue stain. FIGS. 14, 15.—Comparison of size of sympathetic nerve along the renal artery in control (Fig. 14) and experimental 19-day old mouse (Fig. 15) injected daily with the CM-3 salivary fraction since birth Arrows point to nerve in both figures. Cajal De Castro silver impregnation.

in tissue culture. The injected and control mice of the same litters were sacrificed between the 7th and the 14th day and examined for effects on the nervous system and other systems. When it was found that such treatment had consistently elicited a significant increase in the sympathetic ganglia, a second series of experiments was performed with a more concentrated solution.

Group 2: Thirty newborn mice were injected with the CM-1 fraction at a concentration of 6,000 units per ml. The solution was injected daily, in the same amount as used in previous experiments. The injected and untreated mice of the same litters were sacrificed every day or every other day between the second and the thirtieth day.

Group 3: Ten newborn mice were injected daily, from birth, with a more highly purified fraction referred to as CM-3 fraction in the paper by S. Cohen. The solution was injected in the same amount and in the same concentration as in the experiments of Group 2. The injected and control mice were sacrificed between 1 and 9 days and at 12 and 19 days respectively.

Group 4: Fifty adult mice were injected with the CM-1 fraction at the same concentration and in proportional doses per body weight as in groups 2 and 3. The injected and control mice were sacrificed in groups of 3 to 4 between the end of the first week and the fourth week and compared with controls.

The injected and control mice of all 4 groups were dissected and used for investigation of the sympathetic ganglia; other components of the nervous system were also examined. The sympathetic ganglia were either dissected out and stained with hematoxylin and toluidin blue for counts of mitotic figures and of nerve cells, or they were left *in situ* and studied after silver impregnation and sectioning of the whole organism. This material was used for the study of the neurofibrillar differentiation of nerve cells and of the peripheral distribution of sympathetic nerve fibers in injected and control animals.

Counts of mitotic figures were made in sympathetic ganglia of mice of the third group between 12 hours and 9 days. The mitoses were counted in each section of the superior or stellate ganglia in control and treated mice.

Area measurements were made in ganglia dissected from mice 5 to 27 day old, of groups 2 and 3. Each section was projected with the help of the camera lucida and the contour of the section was drawn on cardboard. The total number of sections was then weighed and the weight compared with that of the same control ganglion. Since the weight is proportional to the volume, the figures indicate a similar ratio between the volume of experimental and control ganglia. A total of 18 control and 18 experimental ganglia were measured and compared. The same technique was used on a small number of ganglia of the fourth group.

Cell counts were performed in sympathetic ganglia of groups 3 and 4, by inserting a micrometer disk in the ocular and counting all nerve cells in every other section of the experimental and control ganglia.

Two additional series of experiments were performed. In the first series we tested the serum of adult and weanling mice for the presence of the nerve-growth factor. A total of 150 adult mice of both sexes and of 30 weanling mice were tested. The blood was either collected from the blood vessels immediately after decapitation of mice in light chloroform anesthesia, or it was drained directly from the aorta and the heart in mice anesthetized with nembutal. The blood was allowed to clot at room temperature and then stored for half an hour in the refrigerator. The serum was then collected in separated vials and each specimen tested on sensory ganglia of 8-day chick embryos explanted *in vitro* with the usual hanging drop technique. All sera were also tested at dilutions of 0.1, 0.01, and 0.001.

In the second series of experiments we extirpated the submaxillary and sublingual salivary glands in 25 adult and 5 weanling mice, and we inspected the sympathetic ganglia of the operated and control mice between 2 weeks and 6 months after the operations.‡

(*b*) *Effects of the purified salivary extract on newborn mice:* The newborn mice injected with the CM-1 fraction in the weak and even more in the strong concentration, exhibited side effects which will be mentioned but not described in detail. From the third day on, the growth rate dropped sharply and at the end of the first month, the mice were barely larger than at the end of the first week. The hair growth was severely impaired, the lids opened 6 to 7 days earlier than in controls and the cutting of the inferior and superior incisors and their calcification took place 5 days earlier than in controls. The animals recovered if the injections were discontinued at the end of the first month; dwarf mice resumed growth and three months later they did not differ from untreated controls at a macroscopical inspection. The effects mentioned above did not affect the vitality of the mice in the doses used. None of the forth injected mice died as a result of the treatment. No side effects were observed in newborn mice injected with the more purified fraction CM-3. They were healthy and vigorous as controls. Obviously, we are dealing with 2 factors, one of which was removed in the process of purification.

Both fractions CM-1 and CM-3 evoked a marked overgrowth of the sympathetic ganglia (Figs. 6, 7, 8, 9). In the following we will consider only the results of the injections of fractions CM-1 and CM-3 in the concentration of 6,000 units per ml (groups 2 and 3).

The average *volume increase* of the superior cervical ganglia in group 2, as determined by comparison of 13 experimental and 13 control ganglia was 3 : 1. The corresponding figures for the same ganglia of mice injected with fraction CM-3 and controls (group 3) were higher: measurements of 3 experimental and 3 control ganglia at 12 days gave a ratio of 4.1 : 1 and on 2 experimental and 2 control ganglia at 19 days gave 6.4 : 1. More long range experiments are planned to establish the ceiling of this effect.

The results of *cell counts* in 12 control and 12 experimental ganglia of mice injected since birth with the CM-3 fraction are given in Table 1. The increase in cell number in the injected mice over the controls averages 2.5 and 2.18 respectively in one 12- and one 19-day old mouse. The results of cell counts in adult mice, as given in the same Table for 5 controls and 2 experimental ganglia, show no increase in cell number in the injected adult mice over the controls. It is of interest to note that the total cell population in the normal stellate ganglia is 13,000 whereas the same ganglion in an experimental mouse has a population of over 30,000. The mechanism of this cell increase will be dealt with below.

The concomitant *increase in cell size* in injected animals is shown in Figures 8 and 9 for the stellate ganglia of the 19-day old mice and adult mice respectively. No size measurements were made. Since the volume increase in the ganglia was 4 to 6 times (see above) whereas the increase in cell number in the same ganglia was about twofold, we conclude that cellular hypertrophy has a greater share in the end effect than the increase in cell number.

Cytological examination showed that the hypertrophic neurons differ from controls also in the more intense basophilia and in the size of the nucleoi which are much larger than in controls (Figs. 12, 13). A parallel increase in neurofibrillar material in the hypertrophic nerve cells is apparent in the silver-stained ganglia. Observations of ganglia dissected from three-day old mice injected with the CM-1 or CM-3 fractions since birth, indicate that the size increase is already evident at that time.

It was of interest to decide whether the increase in cell number in the injected mice is due to an increase in *mitotic activity* or to other mechanisms such as the production of a larger number of sympathetic nerve cells at the expense of germinal or pluripotential cells present in the ganglia. The presence in the ganglia of a fairly

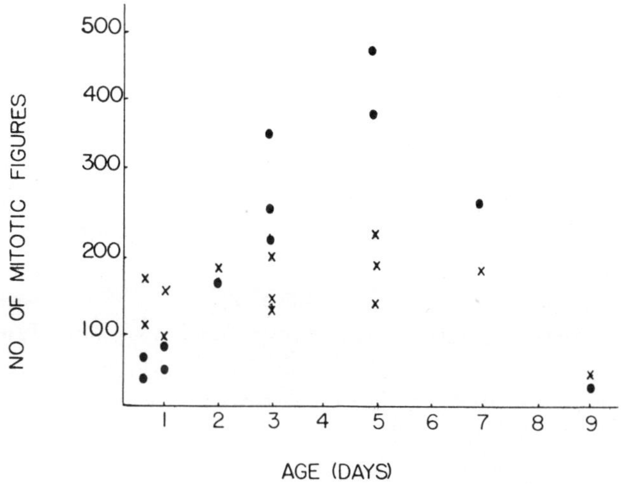

FIG. 1.—Effect of the nerve-growth factor on mitotic activity. Crosses and dots indicate number or mitoses per ganglion (superior cervical ganglion) in control and experimental mice respectively.

large number of small-sized cells, beside the satellite cells, could suggest this possibility. On the other hand, since the mitotic activity is still high at birth and comes to an end at 9 days, the injection of the nerve-growth factor in newborn mice could also affect this process. Counts of mitotic figures in control and experimental ganglia between 1 and 9 days showed a sharp increase in the mitotic activity of the experimental ganglia between 3 and 7 days with a peak at 5 days. These results, presented in Figure 1, favor the hypothesis that the increase in nerve cells in the injected mice is due to an increase in mitotic activity. Although there is no way of deciding how many of the dividing cells are neuroblasts, the results correlate well with the finding of an increase in cell number in later stages; they are also in agreement with observations on the effects of nerve-growth agents on mitotic activity in the spinal ganglia of the chick embryo.[9]

(*c*) *Peripheral distribution of sympathetic nerve fibers:* One would expect that the increase in cell number and cell size in sympathetic ganglia of the injected mice would result in a parallel increase in the size of nerves emerging from the enlarged ganglia. This increase was in fact observed and it is documented in Figures 10 and 11. A comparison of the cephalic, thoraic, and abdominal regions of experimental and control mice shows an increased density and thickness of the nerve plexuses around the blood vessels of the injected animals. In the kidney, sympathetic nerve bundles were traced along the intrarenal blood vessels and also among the renal tubules, in much larger number in the experimental than in control animals. Figures 14 and 15 show the increase in thickness of the sympathetic nerve along the renal artery in a 19-day injected mouse compared to a control of the same age.

These effects, the increase in number and size of sympathetic nerve cells and the hyperneurotization of the viscera, are very similar to the effects called forth by mouse sarcomas, snake venom, and the salivary extract in the chick embryo.[1, 2, 5, 7] The two sets of experiments differ in two respects: (1) While in chick embryos, the sympathetic nerve fibers produced in excess follow in many instances anomalous routes and even force their way into the lumen of blood vessels, no such deviations were observed in mice injected with the salivary gland fractions. Thus the quantitative rather than the qualitative aspect of the distribution of sympathetic nerves seems to be affected in mammals in post-natal periods. (2) Whereas these agents evoke in the chick embryo a striking response also in the sensory ganglia,[2, 9] this effect is barely noticeable in newborn mice. Measurements of a number of sensory ganglia showed only a slight increase in the experimental material.

TABLE 1

Total Nerve Cell Numbers in Sympathetic Ganglia of Mice Injected with Salivary Gland Fractions

Fraction	Age	Ganglia	Control	Experimental	Ratio Exp./Cont.
CM-3	12	Thoracic N. 6	1,365	3,212	...
		Thoracic N. 7	1,165	2,731	...
		Thoracic N. 8	1,155	4,422	...
		Thoracic N. 9	1,026	2,849	...
		Thoracic N. 10	1,465	2,567	...
			Total 6,176	Total 15,781	2.5
CM-3	19	Stellate			
		Thoracic N. 2	14,400	30,800	...
		Thoracic N. 3	1,644	3,684	...
		Thoracic N. 4	1,770	2,133	...
		Thoracic N. 5	1,094	3,144	...
		Thoracic N. 6	820	2,328	...
		Thoracic N. 7	1,362	2,758	...
		Thoracic N. 8	1,392	4,162	...
			Total 22,482	Total 49,009	2.18
CM-1	Adult	Stellate	14,000	13,689	...
		Stellate	12,000	...	...
			Average 13,000	13,689	1.05
	Adult	Sup. Cerv.	13,300	15,770	...
		Sup. Cerv.	16,415	...	...
		Sup. Cerv.	15,652	...	...
			Average 15,122	15,770	1.03

(*d*) *Effects of fraction CM-1 on adult mice:* The daily injection of 0.05 ml per body weight of fraction CM-1 (6,000 units per ml) was well tolerated and the injected mice did not show any adverse effects. The sympathetic ganglia were considerably larger than in controls: the difference was evident at a macroscopic inspection of the 50 injected and 50 control mice. Area measurements of the ganglia in a few of cases gave a twofold enlargement, compared with controls. The size increase was due to cellular hypertrophy of the individual neurons (Fig. 9). The cell number was apparently not increased (see Table 1).

Nerve-Growth Effects of Blood Serum of Adult and Young Mice.—Previous experiments *in vitro* gave evidence of a mild nerve-growth effect elicited by embryonic mouse heart on sensory and sympathetic ganglia of chick embryo.[3] Traces of the same activity were then found in homogenates of striated muscle of adult mice. Bueker[10] detected activity in partially purified preparations of thymus, kidney, and muscle. We found in some instances evidence of activity in the urine of adult mice. The same activity was detected in mouse saliva. We then proceeded to test the serum of adult and weanling mice.

Observations to be reported in detail elsewhere disclosed a considerable sexual dimorphism in the sympathetic nerve cells of adult mice. In male mice, the sympathetic neurons are considereably and consistently larger and more intensely stained with basic dyes than in female mice of the same size. These results suggested tests of the blood serum of mice of the two sexes separately. The results are summarized in Table 2. A maximal effect was obtained from the serum of 15 male

TABLE 2

NERVE-GROWTH EFFECTS OF MOUSE SERUM *in vitro*

Age	Concentration	Response*			
		++++	++	+	±
Adult males	$^1/_{10}$	15			
	1	10	30	20	25
Adult females	1	1	7	10	32
Weanling	1		4		26

* ++++ to ± give degrees of effects from maximal to barely detectable.

mice and one female mouse. In 15 cases with a 4+ effect at a dilution of 1 : 10 (Fig. 4) an effect was still detectable at a dilution of 1 : 1000. The nerve-growth effect of the serum collected from adult females or from weanling mice of both sexes was consistently milder than that of the serum of adult males. These sex and age differences have a parallelism in the finding of Cohen (these PROCEEDINGS, this issue) of a higher specific activity in the submaxillary salivary gland of male than of female mice and of much lower activity in the same gland of weanling mice. These results indicate that the nerve-growth factor is present in higher concentration in the blood of adult male than female mice and that it is in even lower concentration in the blood of weanling mice. They also show a considerable variation among mice of the same group. Experiments in progress are expected to answer the question of whether such variations are correlated with physiological differences in the tested animals and if the stress resulting from the administration of the anesthetic may account for such variations.

Effects of the Extirpation of the Submaxillary and Sublingual Glands in Adult and Weanling Mice.—The operated mice were sacrificed between two weeks and six months after the operations and were compared with controls of the same size.

When mice in the weanling stage were operated, the controls were selected from the same litters.

In the first days after the operation, the animals showed signs of discomfort and the fur became ruffled. In the following days, the mice recovered, but in most instances the fur remained deranged. The operated adult males differed from controls also in another respect. They became much more tame than controls and easier to handle. The correlation between submaxillary salivary glands and other endocrine glands, in particular sex glands, have been investigated by many authors[11, 12] and will not be discussed here. They are mentioned, however, since they may have a bearing on the present results.

In all the experimental and control mice, the sympathetic chain ganglia were dissected out, stained with toluidin blue, and sectioned at 10 microns. Since the superior cervical ganglion might have been indirectly affected by the extirpation of part of its peripheral field of innervation, the salivary complex, we used instead the stellate ganglion which does not contribute to the innervation of the salivary glands.

Results.—Twenty operated adult males and twenty controls were examined between three weeks and two months after the operation. In five mice deprived of the salivary glands a slight decrease in size of the sympathetic nerve cells was detected. The nerve cells also stained less intensely than controls with toluidin blue and appeared similar to sympathetic nerve cells of female mice. No differences were noticeable in the other 15 operated mice. Five adult female mice were operated and compared with controls three weeks after the operation. No size differences were detected between these and control nerve cells. Equally negative were the results in 5 mice operated during the weanling stage. One of these was compared with a control of the same litter, six months after the operation, the other mice one month after the operation. In all instances, the sympathetic chain ganglia appeared of the same size as the controls and the histological examination of the stellate ganglia revealed no changes in cell size.

The serum of mice deprived of the salivary glands was also tested for the nerve-growth factor. In one mouse deprived of the glands two months earlier the serum evoked a 4+ effect *in vitro* as the serum of control mice. In the other specimen the effect varied from a 2+ to a barely detectable effect. These results indicate that the nerve-growth agent is present, even in the absence of the salivary glands.

Summary.—Previous work has provided evidence for the presence of nerve-growth promoting agents in a variety of biological materials: mouse sarcomas, snake venoms, and mouse submaxillary salivary glands. In the present investigation the effect of the active fraction isolated from the mouse salivary glands was tested *in vitro* on ganglia of human fetuses; it was found that it elicits the same effects as on ganglia of other species. The active fraction was then injected in new-born and in adult mice. In all instances the injection resulted in a marked increase of the sympathetic ganglia; the response varied with the age of the animal, the amount, and the purity of the fraction injected. In some instances a sixfold increase in size was observed. Slight nerve-growth promoting effects of different mouse tissues had been observed *in vitro* in previous experiments. We have now found evidence for the presence of the nerve-growth factor in the serum of adult and weanling mice. Maximal effects were obtained from the serum of adult male mice; the serum of female mice is less effective. This sex difference is paralleled by a sex difference in the size of adult sympathetic nerve cells.

* This work has been supported by a grant from the National Science Foundation and by a contribution from an Institutional grant of the American Cancer Society to Washington University.

† We wish to express our appreciation to Dr. Willard Allen of the Department of Gynecology and Obstetrics of the Washington School of Medicine who made this material available to us.

‡ The microphotographs in this and in the following paper were made by Mr. Cramer Lewis of the Department of Illustration of the Washington University School of Medicine.

[1] Levi-Montalcini, R., *Ann. N. Y. Acad. Sci.*, 330–343 (1952).

[2] Levi-Montalcini, R., and V. Hamburger, *J. Exptl. Socl.*, **123**, 233–288 (1953).

[3] Levi-Montalcini, R., H. Meyer, and V. Hamburger, *Cancer Research*, **14**, 49–57 (1954).

[4] Cohen, S., R. Levi-Montalcini, and V. Hamburger, these Proceedings, **40**, 1014 (1954).

[5] Levi-Montalcini, R., and S. Cohen, these Proceedings, **42**, 571–574 (1956)

[6] Cohen, S , and R. Levi-Montalcini, these Proceedings, **42**, 695–699 (1956).

[7] Levi-Montalcini, R., in *The Chemical Basis of Development*, (Baltimore: The John Hopkins Press, 1958), pp. 645–664.

[8] Cohen, S., in *The Chemical Basis of Development*, (Baltimore: The John Hopkins Press, 1958), pp. 665–676.

[9] Levi-Montalcini, R., and V. Hamburger, *J. Expl. Zool.*, **116**, 321–362 (1951).

[10] Bueker, E. D., I. Schenklein, and J. Bane, *Anat. Record*, **133**, 256 (1959).

[11] Lacassagne, A., *Compt. Rend. Soc. Biol.*, **133**, 280–281 (1940).

[12] Junqueira, L. C. U., A. Faier, M. Rabinovitch, and L. Frankenthal, *J. Cell. and Comp. Physiol.*, **34**, 129–158 (1949).

Reprinted from the Proceedings of the National Academy of Sciences
Vol. 46, No. 3, pp. 384–391. March, 1960.

*DESTRUCTION OF THE SYMPATHETIC GANGLIA IN MAMMALS BY AN ANTISERUM TO A NERVE-GROWTH PROTEIN**

By Rita Levi-Montalcini and Barbara Booker

Department of Zoölogy, Washington University, St. Louis

Communicated by Viktor Hamburger, January 4, 1960

The observation by S. Cohen that a rabbit antiserum against a protein fraction of the salivary gland inactivates the *in vitro* nerve-growth effects of this protein (pp. 302–311, these Proceedings) suggested to him to test the effects of the antiserum on newborn mice. The finding of a remarkable decrease in size of the sympathetic ganglia of the injected mice prompted an extensive investigation of the effects of the antiserum on the sympathetic ganglia of mice and other mammals. The results of this study are reported in the following pages.

Materials and Methods.—Newborn mice were injected daily with 0.05 ml of the rabbit antiserum per 1.5 gm of body weight.† Controls of the same litters were either injected with serum of a normal rabbit, or they were not treated at all. Twenty experimental and twenty control mice were sacrificed between the 12th hour after the first injection and the 25th day. Two groups of newborn mice, injected for 8 and 20 days respectively after birth, were sacrificed three and four months after the termination of the treatment. An equal number of untreated mice of the same litters were available for control. The same techniques as used in the previous experiments (pages 373–384) were used for area measurements, cell and mitotic counts, and for histological examination.

The effect of the rabbit antiserum was then tested on newborn rats, rabbits, and one pair of kittens. The amount injected was in the same proportion to the body weight as in the experiments with newborn mice. A limited number of experiments were also performed on adult mice and on one pair of young adult squirrel monkeys.

Effects of the Antiserum on Newborn Mice.—The injection of the rabbit antiserum in newborn mice did not affect their development in any way. Mice injected daily for a period of 25 days did not differ from mice injected with normal rabbit antiserum or from untreated controls. Since mice injected with the normal serum were identical in all respects with untreated controls, including the development of the sympathetic ganglia, only untreated animals were used for comparison in most experiments.

The effects of the antiserum on the development and growth of the sympathetic ganglia were dramatic in their sudden outburst, in their severity, and in the end effects. An inspection of sympathetic ganglia at different trunk and head regions in the same animals showed that all ganglia were affected in the same way; therefore, the data presented here on the superior cervical ganglia can be considered as representative of the degree of reduction which took place in all sympathetic ganglia.

A decrease in mitotic figures is already noticeable 12 hours after the first injection; the decrease becomes more pronounced at the end of the first day. Between the second and the third day, the mitotic activity is sharply reduced and the number of degenerating cells is greatly increased. The neuroblasts which are present are considerably smaller than the control cells. As a result, the volume of the ganglia is reduced to approximately $^1/_6$ of the controls on the fourth day. No cell counts were performed at this stage since most of the neuroblasts are still small and not easily distinguishable from satellite cells.

Between the fourth and the ninth day, the process of cell atrophy and death continues at a somewhat slower pace; the disintegrative processes of preceding stages have in fact so much reduced the population of nerve cells as to set limits to the process itself.

The effects of the daily injections of antiserum in a 9-day mouse are illustrated in Figure 5. The products of disintegrating cells which in earlier stages encumbered the ganglia have now disappeared. A few nerve cells, smaller than controls but otherwise in apparently normal condition, are scattered among a uniform population of satellite cells. The latter do not seem to be affected by the antiserum at this stage. A reduction in their number takes place in more advanced stages when only a few satellites are still present in each section of the diminutive ganglia (Fig. 8).

At 20 days, counts of nerve cells in the superior cervical ganglia of experimental and control mice show that the sympathetic nerve cells are reduced in one instance to less than 1 per cent of the control (Fig. 1, Table 1). The reduction of volume of the same ganglion was to 4.3 per cent of the control. The process seems to have reached its peak. Counts of cells in another pair of experimental and control ganglia of a 25-day-old mouse give slightly higher figures: 1.7 per cent of the cell population of the control are present in the experimental ganglion. Such differences may well be accounted for as individual variations. It should be noted that a considerable variation in the cell population was also observed in ganglia of untreated mice (see Table 1). The results, seem, however, to indicate that the process has not further advanced, and that the few cells which escaped death in earlier stages may now be able to survive.

It was of interest, in this connection, to examine the injected mice some months after the termination of the treatment. Seven newborn mice of two litters were injected daily with the antiserum until the eighth and the twentieth day respectively. Experimental and untreated mice of the same litters were examined three to four months later. The treated mice did not differ from controls in their general appearance. The lack of adverse effects on the visceral functions of the injected mice was illustrated by the fact that one of these mice delivered a normal litter.

Two experimental and two control mice of each group were sacrificed at the end of the third month and the sympathetic chain ganglia were dissected out and compared. Two more pairs of mice injected for 8 days were sacrificed at the end of the fourth month (Figs. 2, 8). The results are given in Table 1. The reduction in cell number in the superior cervical ganglia fluctuates between 2.56 per cent and 0.84

TABLE 1

EFFECT OF ANTISERUM INJECTIONS ON THE VOLUME AND CELL NUMBER OF THE SYMPATHETIC GANGLIA IN MAMMALS*

Mammal	Age, Days	No. of Inject.†	Ganglion	Volume Ratio, Exp./Cont.	Number of Cells Cont.	Number of Cells Exp.	Cell Ratio, Exp./Cont.
Mouse	2	2	Sup. Cerv.	0.37:1			
Mouse	4	4	Sup. Cerv.	0.16:1			
Mouse	9	9	Stellate	0.13:1			
Mouse	20	20	Sup. Cerv.	0.04:1	13,300	91	0.70:100
Mouse	25	25	Sup. Cerv.	..	16,415	279	1.70:100
Mouse	90	8	Sup. Cerv.	..	16,447	421	2.56:100
Mouse	120	8	Sup. Cerv.	..	14,800	140	0.94:100
Mouse	120	20	Sup. Cerv.	..	13,000	110	0.84:100
Mouse	Adult	20	Sup. Cerv.	..	15,652	5,328	34.00:100
Rat	4	4	Sup. Cerv.	6.64			
Rat	7	7	Sup. Cerv.	10.54	32,000	2,310	7.00:100
Rabbit	$3^1/_2$	3	Sup. Cerv.		66,300	6,200	9.00:100
Rabbit	5	5	Sup. Cerv.		67,000	9,300	14.00:100
Rabbit	7	7	Sup. Cerv.		68,000	11,050	16,00:100
Cat	7	7	Sup. Cerv.		114,000	8,600	7.70:100
Monkey	7	7	Sup. Cerv.	1.65			

* Details of experiments in text.
† Beginning at birth, except for adult mice.

per cent of the corresponding control ganglia. In all instances, the residual cells were much smaller and did not stain with basic dyes as deeply as the controls. The function of the surviving cells will be tested in other mice of the same litters which are still alive. The 20-day treatment, as well as the 8-day treatment, is therefore sufficient to destroy 97 per cent to 99 per cent of the sympathetic nerve cells. Since the mice were examined some months after the end of the treatment, it is proved that the damage inflicted to nerve cells is irreversible. Experiments with higher amounts of antiserum and possibly an even more potent antiserum will show whether it may be possible to attain a total extermination of the sympathetic nerve cells.

Effects of the Antiserum in Other Newborn Mammals.—The few experiments performed up to now on newborn rats, rabbits, and kittens are not sufficient to give more than a very incomplete idea of the effects of the antiserum in these species.

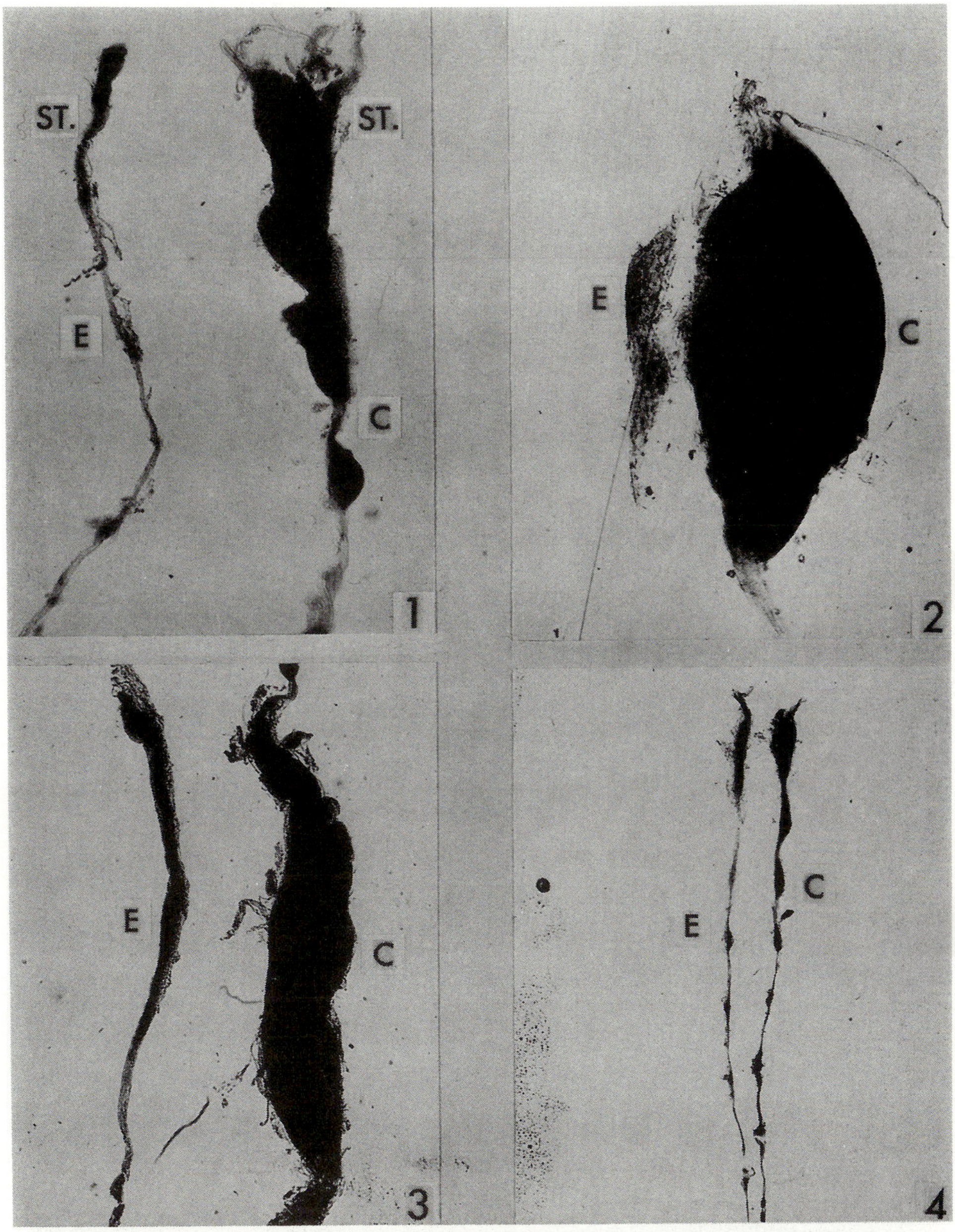

PLATE I.—Effects of rabbit antiserum against the purified protein of mouse salivary gland.

FIG. 1.—Stellate (St.) and first ganglia of sympathetic thoracic chains in control (C) and experimental (E) 20-day old mice. Experimental mouse injected daily since birth with the antiserum to the mouse salivary protein. 30×. FIG. 2.—Superior cervical ganglia in control (C) and experimental (E) 4-month old mice. Experimental mouse injected daily from birth to the 8th day with the antiserum to the salivary protein. 33×. FIG. 3, 4.—Effects of the antiserum in a 7-day old rat (FIG. 3) and in a $3^1/_2$-day old rabbit (FIG. 4); both injected daily since birth. FIG. 3, stellate ganglia in control (C) and experimental rat (E). 30×. FIG. 4, stellate and sympathetic thoracic chains in control (C) and experimental (E) rabbit. 4×.

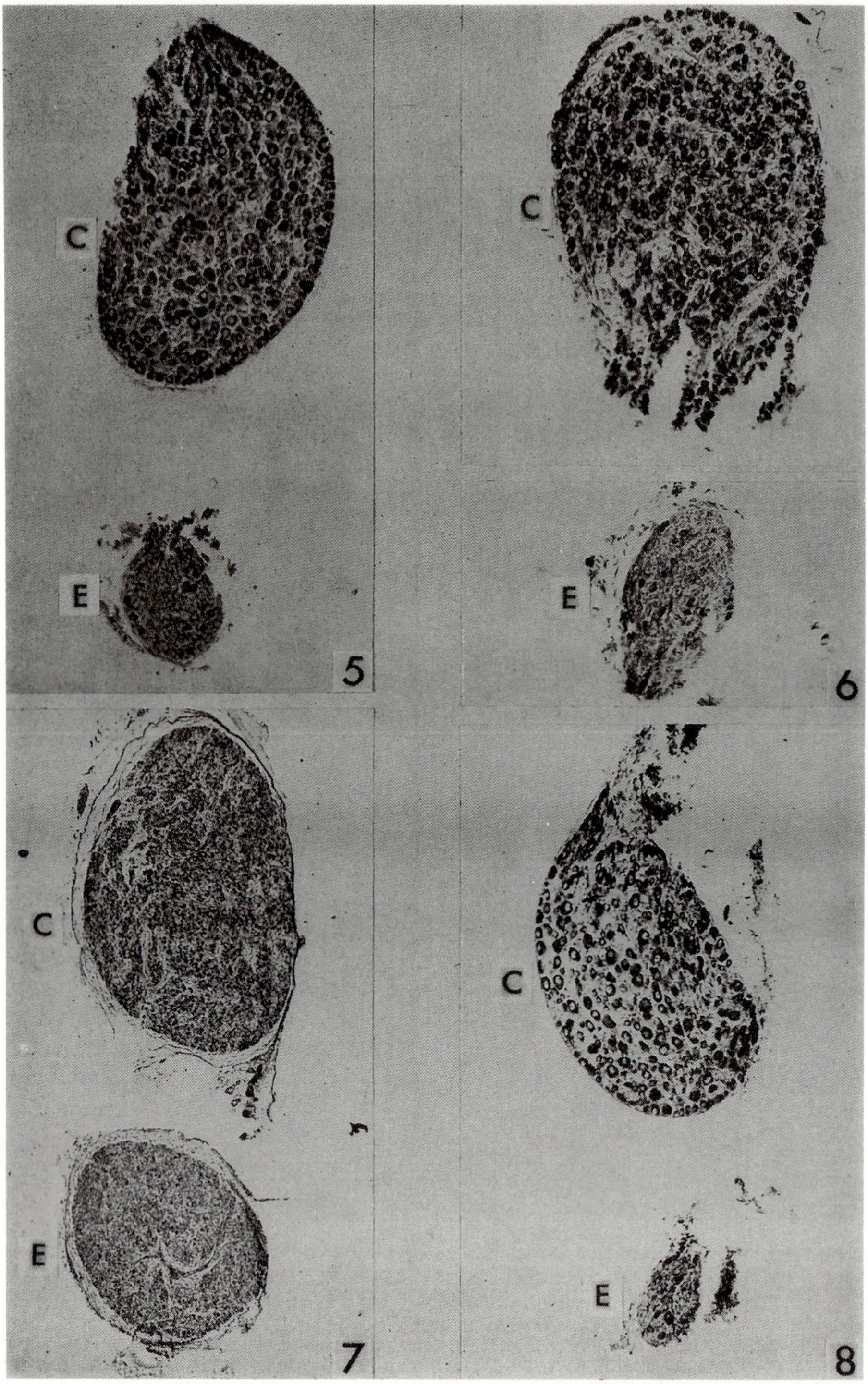

PLATE II.—Transverse sections of superior cervical ganglia in control (C) and experimental (E) animals, injected with rabbit antiserum to the salivary protein.

FIG. 5.—9-day old mice; E, injected daily since birth. 90 ×. FIG. 6.—7-day old rats; E, injected daily since birth. 90×. FIG. 7.—7-day old kittens; E, injected daily since birth; the ganglion consists of satellite cells and few nerve cells. 30 ×. FIG. 8.—4-month old mice; E, injected from birth to the 8th day. 60×.

The results obtained are, however, so impressive as to justify the contention that the antiserum affects the sympathetic ganglia of these mammals in a similar way as described above for newborn mice (Figs. 3, 4, 6, 7).

Two newborn rats were injected with the antiserum (0.05 ml per 1.5 gm of body weight). The experimental and control animals were sacrificed respectively at 5 and 7 days. The results observed are so similar to the ones described in mice at the same developmental stages as to make a description unnecessary. The number of cells in the superior cervical ganglion of one injected rat at 7 days was reduced to 7 per cent of the control (Table 1). The surviving cells are reduced in size when compared to controls but otherwise apparently in good condition (Figs. 3, 6).

Three newborn rabbits were injected with the rabbit antiserum and sacrificed at $3^1/_2$, 5, and 7 days. Since the ganglia are rather large in this species and the nerve cells easily distinguishable from satellites even in earlier stages, cell counts were feasible in three-and-a-half day old ganglia. A high reduction in the number of nerve cells was observed: only 9 per cent of the control population is present in the injected rabbit at this stage (Fig. 4). The two rabbits of the same litter, injected and sacrificed two and four days later respectively, showed a somewhat less severe atrophy of 14 per cent and 16 per cent respectively. Furthermore, the surviving cells were not reduced in size as in the younger animal. Investigation of animals injected for a longer period will answer the question whether rabbits develop a resistance to the antiserum which could account for such results.

The effects of the antiserum in one adult mouse, one young squirrel monkey, and a newborn kitten are shown in Table 1.

Discussion.—The results reported in this and in the previous paper bring to light the remarkable capacity of sympathetic nerve cells of newly born mammals to increase in size and number when injected with a mouse salivary gland protein, as well as their vulnerability to an antiserum to this protein.

The possibility that the sympathetic ganglia in man may be likewise affected by the nerve-growth factor is suggested by the *in vitro* experiments reported in the preceding paper.

The salient aspects of these phenomena are: their universality, their target specificity, and the unsuspected growth potentialities of young and adult nerve cells which materialize under the impact of nerve-growth promoting agents.

Of particular significance is the new finding that the antiserum against the salivary gland protein not only inactivates this protein, but, if injected alone into a newborn mammal, destroys its sympathetic cells almost totally in a remarkably short time. No other organs and tissues appear to be affected. On the other hand, normal serum leaves the sympathetic system as well as all other systems intact.

The question arises whether the near-total destruction of the sympathetic nerve cells is a direct or an indirect effect of the antiserum. It is conceivable that in the normal animal a factor circulates which is necessary for growth and maintenance of the sympathetic system and that the antiserum inactivates this factor. This hypothesis is supported by the detection of the nerve-growth agent in the blood of young and adult mice. Alternatively, the antiserum may exert a cytotoxic affect by combining with the antigen in or on the surface of the nerve cells. The same questions are raised in other investigations of the cytotoxic effects of the antisera.[1] Immunochemical techniques, such as the detection of antibodies labeled with a radioactive isotope or with a fluorescent dye, may show if there is localization of the antiserum on the sympathetic nerve cells and if so, provide evidence in favor of the second alternative. These experiments will be performed in the near future.

The results reported in these and in previous papers[2–5] give evidence for the existence of closely related nerve-growth promoting proteins in two seemingly unrelated sources: mouse sarcomas and salivary glands. One may wonder whether both structures produce these proteins. We have positive evidence that the mouse sarcomas manufacture the nerve-growth agent even when they grow in the chick embryo.[6,7] Corresponding experiments in which the salivary glands were transplanted in the chick embryo were negative so far. Furthermore the nerve-growth agent was detected in serum of adult mice which had been deprived of their salivary glands two months earlier. The sympathetic ganglia of these mice did not show appreciable differences from control mice. Since the salivary glands and their homologue, the snake venom glands, contain the agent in high concentration, the only alternative would be that these glands store, but do not produce, the growth agent. If this suggestion would prove to be correct, we would still be faced with the problem of identifying the source of the nerve-growth agent in the organism. The possibility ought to be considered that the nerve-growth agent is not manufactured in any particular organ or organs, but in the mesenchymal tissue. The only evidence in favor of this hypothesis comes from the *in vivo* and *in vitro* experiments with mouse sarcomas and from *in vitro* experiments with explants of embryonic mouse heart.[8] Epithelial structures like carcinomas proved to be completely deprived of the property of stimulating nerve growth *in vivo*[9] as well as *in vitro*.[8] Research in progress is now aimed at the investigation of this problem.

Finally, we should like to call attention to other aspects of these experiments. They gave evidence of the high tolerance of the organism for such deviations from normality as a sixfold increase in volume of the sympathetic ganglia, or their near-total extinction. The injected animals apparently did not differ from controls when raised and observed in the sheltered conditions of the laboratory. Physiological tests are expected to reveal these differences and will therefore be applied to the treated and control animals.

Summary.—We have reported the effects of an antiserum to the nerve-growth factor which selectively destroys most of the sympathetic nerve cells in a variety of newborn mammals. Daily injections for a period of 8 days resulted in the disappearance of up to 99 per cent of the sympathetic nerve cells. This process is irreversible. It is suggested that the salivary glands do not produce the nerve-growth agent but merely accumulate and store it. The mode of action of the antiserum and the possible role of the nerve-growth agent as a maintenance factor for sympathetic neurons are under investigation.

* This work has been supported by a grant from the National Science Foundation and by a contribution from an institutional grant of the American Cancer Society to Washington University.

† For preparation of the antiserum see paper by Cohen, S., 302–311, these PROCEEDINGS.

[1] Edds, M. V. (Ed.), *Conference on Immunology and Development* (Chicago: The University of Chicago Press, 1956).

[2] Cohen, S., and R. Levi-Montalcini, these PROCEEDINGS, **42**, 695–699 (1956).

[3] Levi-Montalcini, R., and S. Cohen, these PROCEEDINGS, 571–574 (1956).

[4] Levi-Montalcini, R., in *The Chemical Basis of Development* (Baltimore: The John Hopkins Press, 1958), pp. 645–664.

[5] Cohen, S., in *The Chemical Basis of Development* (Baltimore: The John Jopkins Press, 1948) pp. 665–676.

[6] Levi-Montalcini, R., *Ann. N. Y. Acad. Sci.*, 330–343 (1952).

[7] Levi-Montalcini R., and V. Hamburger, *J. Exptl. Zool.*, **123**, 233–288 (1953).

[8] Levi-Montalcini, R., H. Meyer, and V. Hamburger, *Cancer Research*, **14**, 49–57 (1954).

[9] Bueker, E. D., *Cancer Research*, **17**, 190–199 (1957).

Reprinted from DEVELOPMENTAL BIOLOGY, Volume 7, March 1963

DEVELOPMENTAL BIOLOGY, 7, 653–659 (1963)

Essential Role of the Nerve Growth Factor in the Survival and Maintenance of Dissociated Sensory and Sympathetic Embryonic Nerve Cells *in Vitro*[1]

RITA LEVI-MONTALCINI[2] AND PIETRO U. ANGELETTI

Istituto Superiore di Sanità Department of Biological Chemistry, Rome, and Washington University, Department of Zoology, St. Louis, Missouri

Accepted November 26, 1962

INTRODUCTION

Evidence was presented in previous work (Levi-Montalcini and Cohen, 1960; Levi-Montalcini and Angeletti, 1960; Cohen, 1960) that the growth and maintenance of sympathetic nerve cells is under control of a specific protein factor referred to as nerve growth factor (NGF). The same factor has a potent growth-promoting effect on sensory embryonic nerve cells (Levi-Montalcini, 1958), but it does not affect the sensory nerve cells once they are fully differentiated. The NGF is normally present in the receptive nerve cells and is detectable in the blood and body fluids of mammals (Levi-Montalcini and Angeletti, 1961) and in embryonic material (Bueker *et al.*, 1960). The above findings are based on experiments in intact animals and on ganglia dissected out from the embryos and cultured *in vitro*. In the present paper we shall report on the effect of the same factor on dissociated sensory and sympathetic embryonic nerve cells explanted *in vitro* in chemically defined media.

MATERIAL AND METHODS

The NGF was extracted from the mouse submaxillary salivary glands and purified by the procedure described by Cohen (1960). Chromatography on a carboxymethyl cellulose column with a gradient salt

[1] This work was supported by a NIH Grant B.3777.

[2] Visiting Professor at the Department of Biological Chemistry, Istituto Superiore di Sanità, Rome.

elution was used in the final step. The purest NGF preparation had a specific activity of 0.01 μg per milliliter when tested in tissue culture by the method previously described (Levi-Montalcini, 1958). Sterilization of the NGF was obtained by ultrafiltration under pressure. Eagle's medium from Microbiological Associates was used either alone or with the addition of 10% horse serum. In most of the experiments, antibiotics were added in the usual doses to the medium. All cultures were carried out in small plastic dishes (Falcon disposable dishes), in a thermostat at 37° flushed with 4% CO_2 regulated with a flowmeter. In some experiments coverslips previously prepared with reconstituted rat collagen (Bornstein, 1958) were placed on the bottom of the dishes. Sensory ganglia were dissected out from 7- to 10-day chick embryos and collected in physiological solution. They were then transferred in 0.5% trypsin solution and incubated at 37°C for 20 minutes. About 50 ganglia were used in each experiment. Upon careful washing in several changes of physiological solution, the ganglia were gently dissociated by suction with a micropipette and dispersed in 0.5 ml of Eagle's medium. Thus a homogeneous suspension of nerve and supporting cells was prepared. Four drops of the above suspension were placed in each of 8 plastic dishes prepared with the culture medium prewarmed and equilibrated with 4% CO_2. Sympathetic ganglia dissected out from 9- to 14-day chick embryos were prepared as above. The only difference consisted in a longer exposure to the action of trypsin, which was prolonged up to 35 minutes. The volume of the medium in each dish was 1.5 ml. In the experimental cultures, the NGF was added in a final concentration of 0.05 μg/ml every day. In the control cultures the NGF, inactivated by heating 5 minutes at 80°C, was added in the same amount. The culture medium was renewed every 3 days. At 24-hour intervals, the cultures were fixed and stained with toluidine blue and the Cajal-DeCastro technique for silver impregnation.

RESULTS

The observations to be reported are based on the study of 200 cultures of sensory nerve cells and 50 cultures of sympathetic nerve cells. In a first series of experiments, the culture medium (Eagle) was supplemented with 10% horse serum. In a second group, Eagle's medium alone was used. The response of the sensory and sympathetic nerve cells to the NGF was essentially the same with some differences to be briefly considered below.

Sensory Nerve Cells

Eagle's medium supplemented with 10% horse serum proved to be inadequate for the maintenance of the nerve cells. Twelve hours after the beginning of the incubation period, a rather large number of nerve cells in the control cultures succeeded in establishing contact with the plastic or with the collagen-coated glass surface. In contrast with the supporting cells (fibroblasts and satellites) which spread on the bottom of the dishes, the nerve cells appeared round and did not show any fiber outgrowth. At the end of the first day, most of these cells showed signs of deterioration such as diffuse cytoplasmic granulation and pycnosis of the nuclei. At the end of the second day, all the nerve cells had undergone massive degeneration. Cellular debris and dead nerve cells floated freely in the medium. As a result, the cultures consisted only of fibroblasts and satellite cells (Fig. 4).

A strikingly different picture was presented by the experimental cultures supplemented with the active NGF. Twelve hours after the beginning of the incubation period, a large number of nerve cells had established firm contact with the dish surface. They showed two oppositopolar processes growing out from the ovoidal cells. These cells were intermingled with a large number of supporting cells. Twenty-four hours later, the nerve fibers had considerably increased in length and had built a fibrillar net which covered part of the bottom of the dishes (Fig. 5). At the nodal points of this network, the nerve cells stood out in small groups of 5 or 6 units or as isolated cells on the background of the supporting cells. In the living material as well as after toluidine or silver stains, the cells appeared to be in excellent condition with large nuclei and nucleoli (Fig. 3). The inspection of the same cultures in the following days up to the end of the first week, did not reveal substantial differences from the above picture. The nerve cells underwent only a moderate increase in size, but no otherwise appreciable morphological changes. In order to see whether the horse serum was a necessary ingredient, a number of cultures were prepared in 100% Eagle's medium alone. Control and experimental cultures did not substantially differ from the ones described above (Figs. 1 and 2).

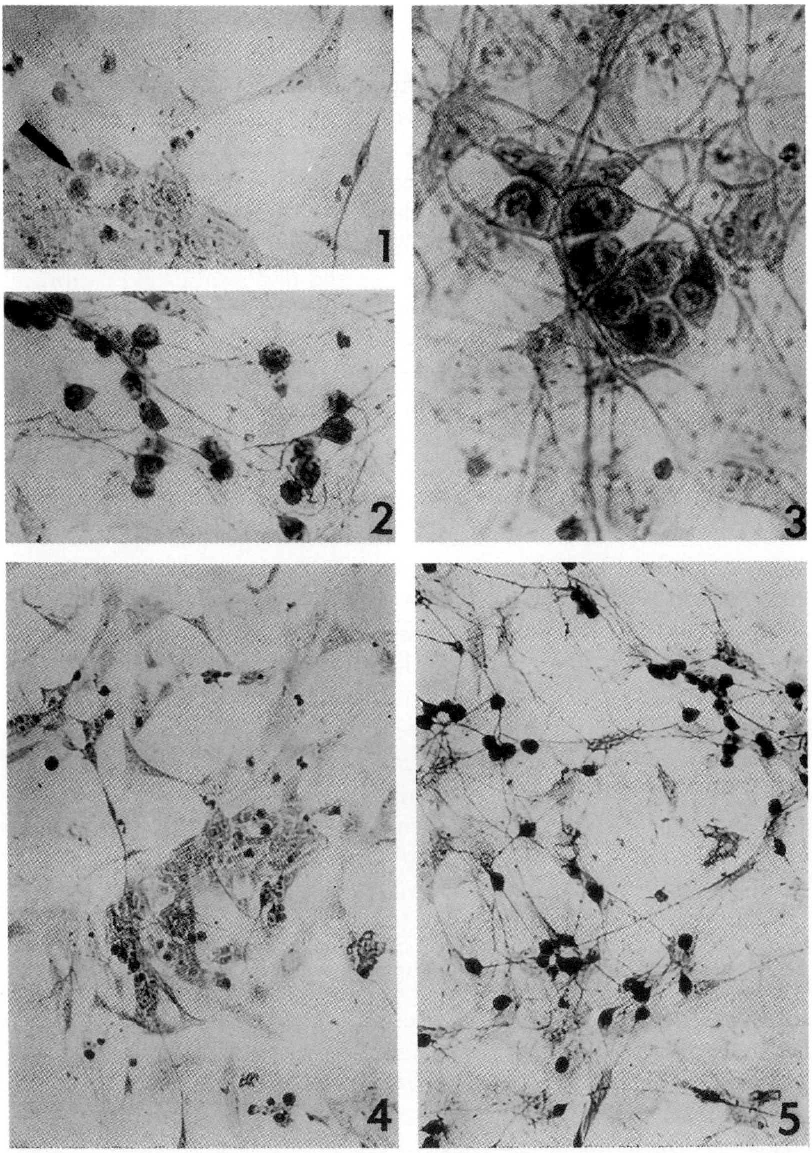

FIGS. 1 and 2. Sensory nerve cells explanted from 9-day chick embryos and cultured for 48 hours in Eagle's medium alone (Fig. 1) and in Eagle's medium supplemented with the NGF (Fig. 2). Arrow in Fig. 1 points to necrotic nerve cells.

Sympathetic Nerve Cells

Since the results conform in general to the results obtained in the cultures of dissociated sensory cells, we will only briefly mention the differences between the two series of experiments.

The addition of the NGF to the culture medium resulted in the survival of a large number of the dissociated sympathetic cells. At the end of the first day of culture, the sympathetic cells had spread on a large area of the bottom of the dishes and had established reciprocal contact through the long axonal processes. A singular observation was the scarcity of satellites or other cell types, which contrasted with the picture offered by the dissociated sensory ganglia. Since satellites are indeed present in fairly large number in the intact sympathetic ganglia, one wonders why only few of them should be present in the cultures. In the following days, the sympathetic nerve cells which appeared at first as isolated units or small cell groups, showed a tendency to adhere to each other and to give origin to large cell agglomerates. These agglomerates in turn lost contact with the plastic or glass surface, detached in part from these surfaces, and eventually became entirely free and floated in the medium. It is still to be ascertained whether the formation of these large cell agglomerates is due to mechanical causes such as traction exerted by the long axons which pull the cells from the dish surface and brings two or more cells in reciprocal contact, or is due in part to weak adhesiveness of these cell types with the plastic or collagen-coated cover slides.

In the control cultures, a few sympathetic nerve cells were present at the end of the first day of incubation. The neurons progressively decreased in number in the following days, and practically no nerve cells were present after the third day of culture.

DISCUSSION

During the past decade, tissue culture has provided a most useful tool to investigate the nutritional requirements of different cell types. The final goal of such investigations was to determine a well-defined

---

FIG. 3. Sensory nerve cells cultured for 7 days in Eagle's medium + 10% horse serum + NGF.

FIGS. 4 and 5. Two-day cultures of sensory nerve cells from 10-day chick embryos in Eagle's medium + 10% horse serum (Fig. 4) and in Eagle's medium + 10% horse serum + NGF (Fig. 5).

chemical medium that would provide the necessary ingredients for the growth and maintenance of given cell types. As pointed out by Parker, however, "although certain cell lines have become adapted to continuous multiplication in chemically defined media, it is still necessary in most undertakings to depend on naturally occurring substances derived from the organism" (Parker, 1961). This is particularly true for cultures of highly specialized cells such as nerve cells. From the classic investigations by Levi and co-workers (Levi and Meyer, 1941) to the recent work by Nakai (1956) and Peterson and Murray (1960), sensory nerve cells have proved to grow well *in vitro* provided that some naturally occurring ingredients such as plasma, serum, and embryonic extract be added to the medium. Our experiments performed on isolated sensory and sympathetic embryonic nerve cells proved that neither the simple Eagle's solution nor Eagle's solution supplemented with 10% horse serum is adequate to support the survival of these cells. As reported, the addition of a specific protein (NGF) proved to be essential to their maintenance. The cell requirement for this factor appears to be extremely low. Denaturation of the protein by heating abolishes its effect. Both observations rule out the hypothesis that the NGF merely serves as a source of cell proteins and suggest instead that it acts in a catalytic fashion. Its normal occurrence in body fluids and in the embryonic extract could well explain the requirement for these natural ingredients in cultures of dorsal root ganglia as indicated by the quoted authors.

The reported effect of the NGF on isolated sensory and sympathetic nerve cells are well in line with the results of our previous experiments in the living organism. On the basis of all the experimental evidence, one may conclude that the NGF plays a role in the life of the receptive cells which has no parallel in other well-known secondary growth regulators, such as nutrients, vitamins, and hormones. This agent acts at an extremely low concentration, is highly specific, is normally present in the reacting cells, and moreover proves to be essential to their subsistence. It is conceivable, as suggested by P. Weiss that "each specific cell type reproduces its protoplasm by a mechanism in which key compounds (templates) characteristic of the individual cell type act as catalysts" (Weiss and Kavanau, 1957).

It is tempting to assign to the NGF such a fundamental role.

SUMMARY

Sensory and sympathetic nerve cells were dissociated by trypsinization and cultured in CO_2-conditioned liquid media. Eagle's medium alone and Eagle's medium supplemented with 10% horse serum proved to be inadequate for the survival and maintenance of these cells. Addition of a specific protein factor (nerve growth factor) at a concentration of 0.05 μg per milliliter of the culture medium, resulted in the survival and maintenance of the sensory and sympathetic nerve cells. These results add new evidence in support of the concept that the NGF plays an essential role in the life of the above cells.

REFERENCES

BORNSTEIN, M. B. (1958). Reconstituted rat-tail collagen used as substrate for tissue cultures on cover-slips. *Lab. Invest.* **7,** **134–140.**

BUEKER, E. D., SCHENKEN, I., and BANE, J. L. (1960). The problem of distribution of a nerve growth factor specific for spinal and sympathetic ganglia. *Cancer Research* **20,** 1220–1228.

COHEN, S. (1960). Purification of a nerve-growth promoting protein from the mouse salivary gland and its neuro-cytotoxic antiserum. *Proc. Natl. Acad. Sci. U. S.* **46,** 302–311.

LEVI, G., and MEYER, H. (1941). Nouvelles recherches sur le tissu nerveux cultivé in vitro. Morphologie, croissance et relations réciproques des neurones. *Arch. biol. (Liége)* **52,** 133–278.

LEVI-MONTALCINI, R. (1958). Chemical stimulation of nerve growth. *Symposium on Chem. Basis Develop., Johns Hopkins Univ. 1958, McCollum-Pratt Inst. Contrib. No.* **234,** 646–664.

LEVI-MONTALCINI, R., and ANGELETTI, P. U. (1960). Biological properties of a nerve-growth promoting protein and its antiserum. *In* "Regional Neurochemistry" (S. S. Kety and J. Elkes, eds.), pp. 362–377. Pergamon, New York.

LEVI-MONTALCINI, R., and ANGELETTI, P. U. (1961). Growth control of the sympathetic system by a specific protein factor. *Quart. Rev. Biol.* **36,** 99–108.

LEVI-MONTALCINI, R., and COHEN, S. (1960). Effects of the extract of the mouse submaxillary salivary glands on the sympathetic system of mammals. *Ann. N. Y. Acad. Sci.* **85,** 324–341.

NAKAI, J. (1956). Dissociated dorsal root ganglia in tissue culture. *Am. J. Anat.* **99,** 81–129.

PARKER, R. C. (1961). Methods of tissue culture. Hoeber (Harper), New York.

PETERSON, E. R., and MURRAY, M. A. (1960). Modification of development in isolated dorsal root ganglia by nutritional and physical factors. *Develop. Biol.* **2,** 461–476.

WEISS, P., and KAVANAU, J. L. (1957). A model of growth and growth control in mathematical terms. *J. Gen. Physiol.* **41,** 1–47.

Reprinted from Science, January 10, 1964, Vol. 143, No. 3602, pages 105–110

Growth Control of Nerve Cells by a Protein Factor and Its Antiserum

Discovery of this factor may provide new leads to understanding of some neurogenetic processes.

Rita Levi-Montalcini

Long before the possibility of experimentally modifying the developing nervous system was conceived, it was realized that end organs play an all-important role on the associated nerve structures. For it is the very nature of nerve cells to establish contact with other cells such as muscle fibers, glandular tissue or other nerve cells. Such close morphological and functional connections could hardly be conceived without a bondage which would interlock the life of the two partners: both would be severely affected if such links were broken. While the mutual dependency of nerve cells and associated structures was clearly documented in the adult organism, it remained for the study of the embryo to bring into focus, in a deceptively clear and persuasive way, the role of such end organs in the growth, differentiation, and maintenance of nerve cells.

The great experimental embryologist, R. G. Harrison, first realized the advantages offered by the embryo as the object of study of the differentiating nerve centers, and it was the nervous system of the amphibian larva which was first challenged to solve problems it had never faced, such as providing for the innervation of organs and limb rudiments borrowed from larvae belonging to species of different size. In thrusting his glass needle into the soft body of the larva and in performing heteroplastic transplantations, the embryologist wanted to test the capacity of the developing nervous system to establish functional connections with foreign tissues. Even more, he wanted to see how the nerve centers and the growing nerve fibers would adjust to the different geometrical dimensions and structural configuration of the foreign organs. The results indicated a remarkable flexibility on the part of the developing nervous system which readily adapts to a new situation even to the point of accelerating the growth rate of its nerve fibers to keep pace with the fast-growing limb rudiments of a smaller species (*1*).

These results also proved that the embryonic nervous system is highly receptive to influences exerted by tissues which supply the peripheral field to the outgrowing nerve fibers. That this influence is not species-specific is documented by the growth response called forth by organs or limb rudiments of a different species. Such experiments and others dealing with heterotopic transplantations proved that peripheral fields are interchangeable, provided that some of the basic rules are respected: motor and sensory nerve fibers have their own fields of innervation; deprivation of one given sector of its motor or sensory nerve contingent does not favor invasion by the other. The same holds true with respect to somato-motor and somato-visceral fibers. The somato-motor fibers end in the voluntary striated muscles while the somato-visceral make synaptic contact with nonstriated muscles, heart, and glandular tissues. Hence embryonic nerve fibers, as well as mature nerve fibers, have no access to peripheral end organs which belong to a category different from their own. In one instance only were embryonic nerve fibers of a given type seen to make functional connections with tissues of an entirely different type from that with which they normally connect. This was the case when intrinsic nerve fibers from the exposed stump of a sectioned neural tube wandered into the surrounding tissues and established functional connections with muscles (*2*).

When Harrison presented in the Harvey lecture the results of the experimental analysis of the developing nervous system in the amphibian larvae, the report, although impressive for the ingenuity of the experiments and the talent displayed in the analysis of the response of nerve centers to all possible changes inflicted to their peripheral fields of innervation, may have raised some doubt as to the information one could gain in this way on the factors which control growth and differentiation in the nervous system. The concluding sentence in the lecture pointed to a sideline of these investigations which seemed to Harrison even more promising than the main object of the research. "By ablation and transplantation of parts of the embryo it is possible to fashion almost any kind of nervous system desired and subsequently to study its function without the disturbing effects of trauma. This is one of the most promising lines of investigation leading off from the field covered in the present lecture" (*1*).

Almost 30 years later, one must admit that these predictions did not materialize, while instead some valuable information was gained by pursuing the main line of research opened by Harrison and his students, and by carrying further the inquiry into the control mechanisms of the developing nervous system.

Considerable progress was achieved by replacing the amphibian larva with the chick embryo. It was V. Hamburger who conceived and contributed most of the work on the effects of the

The author is professor of zoology at Washington University in St. Louis, Missouri.

peripheral field of innervation on sensory and motor spinal systems in the chick embryo (*3–5*). Even though these same nerve structures were also the object of experimental analysis in the amphibian larvae, a considerable amount of work had centered on the causal analysis of brain structures and associated sense organs; for the vertebrate brain, with all its structural and functional complexity, holds such a fascination for man that it is difficult for him to resist the temptation of exploring it in the hope of learning more about his own brain.

Compared to the brain, the spinal cord appears to be of almost diagrammatic simplicity. How deceptive in fact this is, was to be learned through long and laborious attempts to explore the mechanisms which control the growth and differentiation of the spinal cord and of the sensory ganglia. The chick embryo proved to be particularly suited for this analysis. Its nervous system is more complex, but it lends itself better than the amphibian nervous system to the analysis, for its nerve centers are more clearly segregated and defined and their strong affinity for silver permits a visualization of the nerve structures far superior to that in amphibians. But it was a new approach to the problem which brought a significant contribution to our admittedly still very rudimentary knowledge of neurogenesis. Most of the previous work had focused on the study of the end effects of early surgical interventions on the developing nervous system. The results of heteroplastic and heterotopic organ transplantations were in fact explored in the fully developed larvae, long after the problem of adjusting to the new situation had been solved by the nerve centers and contact had been firmly established between the growing nerve fibers and the new periphery. It now became clear that this static approach would leave unanswered the main question: how does the nervous system solve the problem with which it is confronted? It is, in fact, the mechanics of developmental processes rather than their final completion that is of primary interest to the student of neurogenesis. This realization prompted a detailed analysis of the nervous system of the chick embryo from the stage of 38 hours of incubation, when the first wave of differentiation starts in the rostral part of the neural tube, to the time of hatching. The study of normal embryos was rewarding. It made clear the composite nature of apparently simple neurogenetic processes and, in so doing, brought into focus some basic problems which lent themselves to the experimental approach.

In this article I will briefly outline some of the developmental neurogenetic processes which were the objects of investigation in recent years.

At the time of its early formation, that is during the first week of incubation, the nervous system of the chick embryo (and the same is true for all vertebrates) is the stage for massive cell movements, segregation of cell groups, and degeneration of other cell populations. Migration of neuroblasts, that is, nerve cells at an early stage of their differentiation, occurs throughout the entire length of the neural tube from its apical to its caudal end. Migrating nerve cells move in the same orderly fashion as ant or termite armies and, like them, they keep in close reciprocal contact during the long journey which takes place in the dense matrix of the brain vesicles and in the spinal cord. Closely timed inspection of the developing nervous system, hour after hour and day after day, revealed the complexity and extension of these active cell displacements which may take several days to be completed (*6*, *7*). At the same time as these migratory movements occur, other nerve cell populations undergo disintegration and death and are wiped out, in a matter of a few hours, by intervening macrophages. Finally other populations form and achieve their differentiation without undergoing either degenerative or migratory processes. Thus the final product, the differentiated nervous system with its motor, sensory, and associative centers well segregated from each other and interconnected through hundreds of fiber tracts, is greatly different from the embryonic nervous system. The developmental history of each nerve center should therefore be traced back to its early inception in order to know how such a center attains its final size and position. Differentiation itself, in the central nervous system, does not seem to obey the rules which operate in other systems. Some nerve cells show all marks of morphological and functional differentiation at an early developmental stage, while others retain the characteristics of immature cells till the end of the incubation period and then suddenly transform into fully differentiated neurons. Differences in apparently homogeneous cell populations become manifest as differentiation progresses and hitherto similar cells diverge from each other in their growth rate, structure, and end connections. These differences become even more apparent under experimental conditions. Of the experimental work performed on the developing nervous system of the chick embryo, I will briefly mention here only some of the results of experiments aimed at the analysis of spinal ganglia confronted with a larger peripheral field or, conversely, deprived of their field of innervation. The main contribution of these experiments was not so much to stress the all-important role of the periphery on associated sensory centers, as to reveal the extent of these effects on the different developmental aspects which we are used to refer to with the rather general and vague term of "differentiation." The closely timed study of the response of these ganglia to the extirpation of the limb bud in two-day chick embryos, gave evidence of a sharp decrease in the mitotic activity of ganglia destined to innervate the amputated limb rudiment. Since dividing cells lack any differentiative mark and have no fiber connection with the periphery, these effects were designated with the noncommittal term of "field effects" or remote rather than direct effects (*5*). If the adverse effects on the mitotic activity already reduce the size of the sensory cell population, the same populations are even more severely affected by a destructive process which attacks cells in early differentiative stages in the same ganglia. About half of these cells undergo sudden death between the fourth and the fifth day of incubation. Dead cells are removed in a few hours by intervening macrophages, while the surviving cells undergo slow regressive changes but survive till the end of the incubation period. The striking similarity between this massive cell disintegration and the physiological disintegrative processes already mentioned, raises the question of whether in both instances, death might not follow the depletion of some agent essential to growth and differentiation of these nerve cells. Other instances of massive cell degeneration will be considered subsequently.

Thus the severe atrophy of ganglia deprived of their peripheral field of innervation is the end result of a number of detrimental effects. Why some cells should undergo sudden death while others, also deprived of their end organs, should suffer only regressive changes compatible with life is one of the many questions still unanswered (*5*). Equally extensive and complex is the growth response of the same ganglia confronted with the task of providing the innervation of an additional limb bud. The mitotic activity sharply increases in these ganglia, and all subsequent developmental steps from early to advanced differentiative processes are likewise enhanced. While suggesting that the periphery releases some agents which control th mitotic activity as well as differentiation in the sensory nerve cells associated with these end organs, the results gave no clue concerning the nature of these agents and we would today have progressed no further than stating the problem, were it not for a fortuitous discovery which led to the identification and isolation of one of such factors.

Growth Response of Nerve Cells to a Tumor Factor

When in 1947 E. Bueker implanted a fragment of mouse sarcoma 180 into the body wall of a 2-day chick embryo, he had in mind to test the capacity of sensory and motor nerve fibers of the host to innervate a fast-growing tissue. Since the tumor, at variance with a limb bud or other embryonic organs, is homogeneous in structure, the experiment was also expected to answer the question of whether the tumor would admit all or only a given type of nerve fibers. In his 1948 paper (*8*) Bueker reported that nerve fibers from the adjacent sensory ganglia had gained access to the tumor and that these ganglia underwent what seemed to be a moderate increase in size of the same order of magnitude as that resulting from the implantation of an additional limb bud. The author saw in these results a confirmation of the hypothesis that the final size attained by the sensory ganglia is dependent upon the extension of their peripheral field of innervation. The fact that sensory but not motor fibers branched into the tumor confirmed the hypothesis of a selective affinity of nerve fibers for peripheral end organs. Finally, the observation that only sensory ganglia connected through nerve fibers with the tumor increased in size seemed to emphasize the similarity between these effects and the effects elicited by an additional limb rudiment and to give support to the hypothesis that nerve fibers are instrumental in channeling the effects of the periphery on the associated nerve centers. Viewed in this light, the results of the tumor implantation conformed to the results of limb or organ transplantation. Fifteen years had to elapse before the impact of this ingenious experiment on the classic and somewhat ill-defined concepts of "peripheral field effects" was fully realized.

Of the extensive work performed during the last twelve years, I shall consider here only that not fitting into previous schemes, thus paving the way to new experiments and to the revision of old concepts. A reinvestigation of the tumor effects (*9*) led to the discovery of new facets of the phenomenon. It was found that the growth response of the sensory ganglia contributing fibers to the tumor far exceeds the response called forth by an additional limb bud implanted in the embryo at the same developmental stage, that is, at 3 days of incubation. Whereas sensory ganglia supplying nerve fibers to an additional limb undergo a size increase ranging between 20 and 40 percent in excess of the controls, the tumor evokes an increase in volume of the same ganglia two to three times their normal size. Nerve fibers branching into the tumor make no attempt to establish synaptic contact with the neoplastic cells, but wander around in tortuous circuits, and build a dense fibrillar net around and between the cells. If the size increase of the sensory ganglia and the peripheral branching of their fibers departed already to a considerable extent from the effects evoked by an additional limb, the growth response of the sympathetic ganglia adjacent to the tumor had no precedent in any previously observed effects. These ganglia appeared in fact to be five to six times larger than controls. In addition, atypical accessory ganglionic agglomerates were partly imbedded in the tumor and partly irregularly scattered on the fringe of the neoplastic mass. Normal and atypical sympathetic ganglia sent large fiber bundles into the tumor where they intermingled with the sensory fibers and, together with them, took possession of that bizarre "peripheral field" which mimicked like a caricature the effects of normal peripheral end organs on associated nerve centers. The viscera of the embryo, which normally lack a sympathetic innervation up to the end of the third week of incubation, were now flooded with sympathetic nerve fibers from the end of the first week of incubation.

In some organs like the mesonephros, which during its short life cycle is deprived of sympathetic innervation, the density of nerve fibers was such as to force the tubules apart. All the available space in this and other organs like the ovary, the spleen, and the thyroid was filled with large nerve bundles of such size and density that the characteristic structures of the invaded organs were overshadowed.

These most unusual findings led me to propose a different interpretation of this extraordinary growth response. The observed effects could in fact be better explained by assuming that the tumor harbors a growth factor which selectively enhances the nerve fiber outgrowth from the sensory and sympathetic ganglia of the host (*10*). This hypothesis received full confirmation from experiments of extra-embryonic tumor transplantation. Fragments of sarcoma 180 or sarcoma 37 (both elicit the same effects) were grafted onto the chorio-allantoic membrane of 4- to 6-day embryos. In such a position the tumors and the embryos share the circulation, but no direct contact is established between the embryonic and the neoplastic tissues. The effects compared in all respects with the effects of intra-embryonic tumor transplants, thus giving decisive evidence for the release into the blood stream of a "nerve growth factor" (NGF) produced by the neoplastic cells. In intra- as well as in extra-embryonic tumor transplantations, thick nerve bundles were also found in the lumen of large and small veins. Some vessels were in fact filled with dense nerve agglomerates to the point that the blood circulation was greatly obstructed (*10, 11*).

These discoveries, while giving new support to the hypothesis that the tumor harbors a nerve growth factor, also suggested a different experimental approach to the problem. In order to isolate and identify the active agent, a less complex system than the developing embryo was needed. The tissue-culture technique seemed to offer considerable advantages over the embryo by making it possible to test the effects of the tumor directly on isolated sensory and sympathetic ganglia. These experiments were first performed at the Biophysics Institute of Rio de Janeiro in association with H. Meyer. Sensory and sympathetic ganglia of 7- to 9-day chick embryos were confronted in vitro with fragments of sarcoma 180 or sarcoma 37.

In control cultures the same ganglia

were combined with tissue fragments explanted from chick or mouse embryos. The culture medium consisted of a drop of chicken plasma and a drop of embryonic extract, later replaced with a similar amount of amino acid solution. The cultures were incubated at 37°C and examined at 6-hour intervals. The results confirmed the expectations. As early as 5 hours after the beginning of incubation, ganglia combined with a fragment of sarcoma 180 or sarcoma 37 showed striking differences from controls. While these controls produced only few and sparse nerve fibers, ganglia adjacent to a tumor fragment were surrounded by a dense halo of nerve filters which increased in density and length during the following 24 hours (*12*). Once decisive evidence was obtained that this in vitro effect was evoked by the same neoplastic factor which enhances sensory and sympathetic nerve fiber outgrowth in vivo, the tissue-culture method replaced almost entirely the laborious and time-consuming technique of implanting the tumor in the chick embryo and analyzing the effects 2 to 3 weeks later. These results marked the turning point of the investigation. As a result, some of the picturesque and esthetically attractive features of the embryological research were lost, but the investigation itself gained in precision and depth, since for the first time the chemical approach to the problem was made possible.

While biochemists consulted earlier had refrained from the too difficult task of identifying the tumor factor which evokes the growth effects in the embryo, the same problem presented in the simplified version of identifying the factor which promotes the nerve fiber outgrowth in vitro now seemed accessible to analysis. The biochemist, S. Cohen, agreed to join our group and undertake the task of identifying the agent. Ever since this in vitro technique was first devised in 1953, it has become the method of choice for the test of nerve growth factors in our laboratory as well as in other laboratories which are also engaged in the study of the same problem (*13*). It was only after the nerve growth factor became available in a purified form and in large quantities, that it was possible to assay its effects in the living organism. The embryos and the newborn animals then resumed their roles as the major objects of this investigation.

The Nerve Growth Factor: Its Nature, Biological Sources, and Effects

It is the rule rather than the exception in research that new leads should come from accidental findings and that these leads, when followed, would channel the main investigation into a new direction. Of all sciences, biological sciences still remain the most dependent upon such fortuitous leads to unravel some of the intricate mechanisms of life. The discovery of a potent nerve growth factor in snake venom and in the mouse salivary glands which occurred shortly after the discovery of the tumor nerve growth factor, and the recognition that both agents have many properties in common with the tumor factor, can well be listed among such fortuitous events. In 1954, Cohen had succeeded in isolating a nucleoprotein from sarcomas 180 and 37, endowed with the growth-promoting properties of the tumors and of their extracts (*14*). It was now of interest to establish whether the activity resided in the intact nucleoprotein fraction or if the nucleic acids or the proteinic component of such fraction elicited the growth effects. In this connection, snake venom was used by Cohen because of its phosphodiesterase content. Its addition to the culture medium in minute amounts was aimed at the degradation of the nucleic acid component of the active fraction. The unexpected and startling outcome of these experiments was the finding that the snake venom itself harbors a nerve growth factor. In fact ganglia cultured in a medium containing the purified venom at a concentration of 0.05 μg per ml developed an exceedingly dense halo of nerve fibers. This observation prompted a search for the venom effects in the living embryo. The results gave additional evidence for the striking similarity between the tumor and the venom effects. Daily injections of 0.5 μg of the purified venom into the yolk of embryos between the 6th and the 9th day of incubation, evoked the same nerve growth effects as intra-embryonic or extra-embryonic tumor transplantation (*15*). It now became clear why the injection of the tumor extract had failed to produce any growth effect. The specific activity of the venom is in fact about 1000 times higher than the same activity in the tumor homogenate. The intact tumor evokes the growth effects, in spite of its relatively low content of the NGF, through continuous release of this factor by living and actively dividing cells. Both the tumor and the snake venom factors are heat labile, nondialyzable, destroyed by acid (0.1*N* HCl), stable to alkali (0.1*N* NaOH), and to 6*N* urea. The chemical analysis of the venom factor was pursued farther than had been possible with the tumor factor. Upon acid hydrolysis and two-dimensional paper chromatography, the amino acid pattern was qualitatively identical to the chromatogram of crystalline bovine albumin. The activity is destroyed upon incubation with proteolytic enzymes and upon incubation with antiserum to the snake venom. The estimated molecular weight as determined with the Spinco analytical ultracentrifuge is of the order of 20,000 (*16*).

While the discovery of the growth-promoting factor in snake venom could not possibly have been anticipated, the finding of a similar factor in the submaxillary salivary glands of the mouse was the result of a planned search. The occurrence of two agents with nearly identical biological properties in mouse tumors and in the snake venom suggested the existence of other possible sources of this agent. The mouse salivary glands, being in many respects homologous to the snake venom glands, were thought of by Cohen as another possible source of a NGF factor. Experiments in vitro fully confirmed the guess (*17*). The submaxillary salivary glands of the adult male mouse harbor a factor endowed with the same nerve growth promoting activity as the sarcomas and the snake venom (Figs. 1 and 2). Its specific activity is of the order of 6000 to 10,000 times higher than the specific activity of tumors and about ten times higher than that of the snake venom. Its effects in vitro and in vivo on sensory and sympathetic ganglia of the chick embryo are qualitatively identical with the effects of the two other factors. The active agent was identified by Cohen in a protein particle that is heat labile, destroyed by acid, resistant to alkali, and nondialyzable. Like the NGF isolated from the snake venom, it is destroyed by proteolytic enzymes. Its molecular weight is of the order of 44,000 (*18*).

The high specific activity of the salivary NGF which in vitro evokes the halo effect at a molar concentration of 1×10^{-9}, suggested a test of its effects in mammals. Daily injections of the NGF into newborn mice in the amount

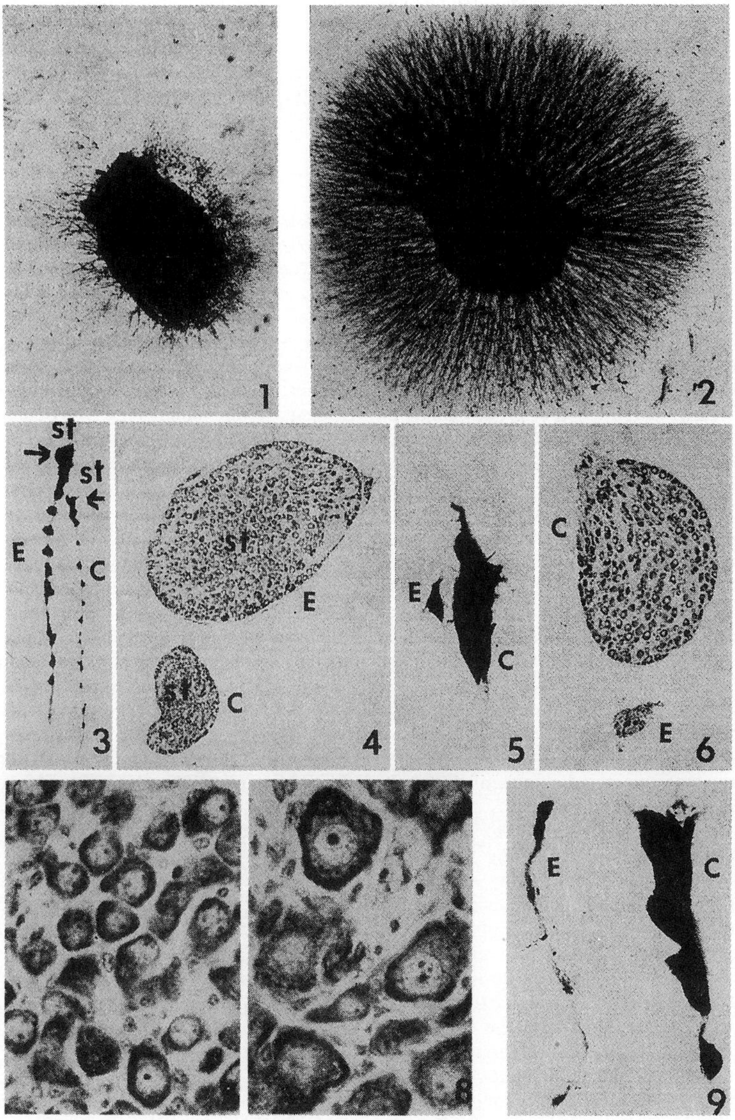

Figs. 1 and 2. Photomicrographs of 7-day sensory ganglia of chick embryo after 24 hours of culture in vitro. Fig. 1. Ganglion in a control medium. Fig. 2. Ganglion in a medium supplemented with the salivary NGF at a concentration of 0.01 micrograms per ml. Silver impregnation. Fig. 3. Whole mounts of the sympathetic thoracic chain ganglia of experimental (*E*) and control (*C*) mice 19 days old. Experimental mouse injected daily with salivary NGF from birth. *ST*, stellate ganglia. Fig. 4. Transverse sections of stellate ganglia (*ST*) in experimental (*E*) and control (*C*) ganglia of Fig. 3. Sections through levels indicated by arrows in both chains in Fig. 3. Fig. 5. Whole mounts of control (*C*) and experimental (*E*) superior cervical ganglia of 3-month-old mice injected for a week after birth with the antiserum to the salivary NGF. Fig. 6. Transverse sections through the superior cervical ganglia of control (*C*) and experimental (*E*) 9-month-old mice. Experimental mouse injected for three consecutive days after birth with the antiserum to the salivary NGF. Figs. 7 and 8. Comparison of cell size in control (Fig. 7) and experimental (Fig. 8) stellate ganglia represented in Figs. 3 and 4. Fig. 9. Whole mounts of stellate ganglia of control (*C*) and experimental (*E*) 9-month-old mice. Experimental mouse injected for three consecutive days with the antiserum to the salivary NGF.

of 0.5 μg per gram of body weight, evoked a striking increase of the sympathetic ganglia, already well apparent 48 hours after the first two injections. In 2- to 3-week old mice injected daily, beginning at birth, the same ganglia are four to six times larger than controls, while no changes are apparent in the parasympathetic and sensory ganglia nor in any other nerve cell population (Figs. 3 and 4). Cell measurements and cell counts show that the NGF calls forth an increase in cell number as well as increase in size of individual nerve cells (Figs. 7 and 8). When the mitotic activity comes to an end at 9 days after birth, the NGF evokes cell hypertrophy but no increase in cell number. In adult mice, sympathetic neurons attain a size two times larger than controls (*17*). The universality of the NGF effects was verified by in vitro experiments on ganglia of several species, man included. Ganglia explanted from human fetuses which became available from surgical abortions, show the same growth response to the NGF as ganglia of birds and rodents. The essential role of this protein in the life of the target nerve cells is further indicated by in vitro experiments on sensory and sympathetic nerve cells dissociated through trypsinization and explanted in liquid media. Nearly all of these cells disintegrate in the first 24 hours in a medium consisting of the Eagle solution alone or supplemented with 10 percent horse serum. The same nerve cells survive up to the third week in this medium when the NGF is added at a concentration of 0.05 μg per milliliter of solution (*19*).

Antiserum to the Nerve Growth Factor

An antiserum to the NGF was prepared by S. Cohen by injecting the purified NGF protein into the pads of rabbits together with Freund adjuvant. The antibody titer was first assayed in vitro. When it was found that its addition to the NGF solution at progressively higher dilutions inactivated the growth effects of the NGF in vitro, the antiserum was injected into newborn mice and the injection was repeated daily for the first week after birth. Three weeks later, treated and control mice of the same litter were killed. On macroscopic inspection, the sympathetic chain ganglia of the injected mice were virtually absent. Upon microscopic analysis, the diminutive ganglia consisted of satellites and a few neurons amounting to 3 to 5 percent of the normal cell population (Figs. 5, 6, and 9). Similar results were obtained with newborn rats, rabbits, and kittens. The injection of the antiserum in adult animals produces less drastic but still severe detrimental effects, while no adverse effects are apparent in other nerve cell populations or in other cell types (*20*).

Ever since these results were first reported in 1960, hundreds of mice and rats were injected with the antiserum in this and other laboratories. The nearly total atrophy of the sympathetic chain ganglia was in all instances ascertained by microscopic inspection up to two years after the treatment. Of interest is the observation that the lack of any sympathetic control is compatible with a normal life, at least under the sheltered conditions of the laboratory. The treated animals are comparable to controls in growth and in other respects. They are now the object of extensive physiological, pharmacological, and behavioral investigations.

Reevaluation of Control Mechanisms of Nerve Cells by Extrinsic Agents

One may ask whether the identification and isolation of specific nerve growth factors in some mouse sarcomas, in the snake venom, and in the mouse salivary glands, has any bearing upon the problem which promoted this search, namely the control mechanism of the end organs on the associated nerve structures. At the time of the discovery of the growth effects elicited by chorioallantoic tumor transplantations on the sympathetic system of the embryo, this response appeared quite different from the growth response evoked by additional limb buds or sense organs on the nervous system of the host. It differed from it not only in the magnitude of the effects but also in many other respects. The massive invasion of the embryonic viscera by large sympathetic nerve bundles and the penetration of these nerve fibers into the lumen of blood vessels indicated such a profound deviation from normality as to raise the question of whether the sympathetic nerve cells had not undergone a radical transformation into malignant cells. Such a hypothesis was readily dismissed but still we were not prepared to see any similarity between the effects elicited by the tumor and the effects evoked by normal embryonic tissues. The discovery of the nerve growth factor in the snake venom and in the mouse salivary glands proved that the production of the NGF is not the prerogative of neoplastic cells. A number of experiments devised and performed together with P. U. Angeletti proved that granuloma tissue, experimentally produced in a variety of mammals, releases the NGF (*21*). The same factor was detected in the sympathetic ganglia and in the serum of mammals, man included (*22*). Its presence was likewise ascertained in embryonic tissues (*13*). Hence we came to the conclusion that the nerve growth promoting protein is a normal constituent of the sympathetic cells and is normally present in the blood and body fluids of birds and mammals. Leaving open the long debated and still unanswered question of its main source of production in the organism, all evidence seems to us to favor the hypothesis that this protein which we designated as the "nerve growth factor" plays a most important role in the life cycle of the sympathetic nerve cells. As for the sensory cells, we have evidence that the same protein is required during their early developmental stages. Why the mature sensory neurons should become refractory to the NGF is one of the many unsolved aspects of this problem.

The identification of the NGF in a protein which is normally present in the developing and in the mature organism, raises the question of whether the end organs might not affect the growth and differentiative processes of associated nerve centers through the release of specific growth factors. In this way one could explain the mitotic effects and the excessive branching and altered peripheral distribution of nerve fibers subsequent to the implantation of additional organs or limb rudiments, which could not be otherwise explained. In contrast to the tumor and the purified nerve growth factors, end organs have a very restricted field of action. End organs affect only the growth of nerve centers which provide their innervation. Such differences could be of a quantitative rather than a qualitative order and could be correlated with differences in the production and release of growth factors in the two sets of experiments. Production and discharge may in fact be very limited in cases of implantation of limbs or additional organs.

In suggesting that peripheral structures and the NGF might in the final analysis operate in a similar way, we do not imply that the released growth factors should be the same. On the contrary, there is reason to believe that each nerve cell type might be receptive to only one specific factor.

Summary

Evidence is presented that sympathetic nerve cells of birds and mammals are receptive to the growth promoting effects of a protein (NGF) which was isolated from some mouse sarcomas, snake venom and mouse salivary glands. This same protein is a normal constituent of the sympathetic cells and is present in the blood and body fluids of a variety of vertebrates, man included. An antiserum to the NGF selectively destroys the sympathetic nerve cells of newborn animals without affecting other nerve cells or organs. The "immunosympathectomized" animals are comparable to controls in growth and viability.

These results give evidence for the essential role of this particular protein in the growth, differentiation, and maintenance of sympathetic nerve cells. They also suggest that other nerve cells might also depend upon specific factors for their differentiation and growth. These results are discussed in the general framework of neurogenetic problems.

References and Notes

1. R. G. Harrison, *Proc. Roy. Soc. London Ser. B* **118**, 155 (1935).
2. P. Weiss, *J. Exptl. Zool.* **113**, 397 (1950).
3. V. Hamburger, *ibid.* **68**, 449 (1934).
4. ———, *Physiol. Zool.* **12**, 268 (1939).
5. ——— and R. Levi-Montalcini, *J. Exptl. Zool.* **111**, 457 (1949).
6. R. Levi-Montalcini, *J. Morphol.* **86**, 253 (1950).
7. ———, in *Nature of Biological Diversity*, J. Allen, Ed. (McGraw-Hill, New York, 1963), pp. 261–295.
8. E. D. Bueker, *Anat. Record* **102**, 369 (1948).
9. R. Levi-Montalcini and V. Hamburger, *J. Exptl. Zool.* **116**, 321 (1951).
10. R. Levi-Montalcini, *Ann. N.Y. Acad. Sci.* **55**, 330 (1952).
11. ——— and V. Hamburger, *J. Exptl. Zool.* **123**, 233 (1953).
12. R. Levi-Montalcini, R. Meyer, V. Hamburger, *Cancer Res.* **14**, 49 (1954).
13. E. D. Bueker, I. Schenkein, J. L. Bane, *ibid.* **20**, 1220 (1960).
14. S. Cohen, R. Levi-Montalcini, V. Hamburger, *Proc. Natl. Acad. Sci. U.S.* **40**, 1014 (1954).
15. R. Levi-Montalcini and S. Cohen, *ibid.* **42**, 695 (1956).
16. S. Cohen and R. Levi-Montalcini, *ibid.*, p. 571.
17. R. Levi-Montalcini and B. Booker, *ibid.* **46**, 373 (1960).
18. S. Cohen, *ibid.*, p. 302.
19. R. Levi-Montalcini and P. U. Angeletti, *Develop. Biol.* **7**, 653 (1963).
20. R. Levi-Montalcini and S. Cohen, *Ann. N.Y. Acad. Sci.* **85**, 324 (1960).
21. R. Levi-Montalcini and P. U. Angeletti, in *Regional Neurochemistry*, S. S. Kety and J. Elkes, Eds. (Pergamon, New York, 1960), p. 362.
22. ———, *Quart. Rev. Biol.* **36**, 99 (1961).

24 J. ULTRASTRUCTURE RESEARCH 36, 24–36 (1971)

Ultrastructural Changes in Sympathetic Neurons of Newborn and Adult Mice Treated with Nerve Growth Factor[1]

P. U. ANGELETTI, R. LEVI-MONTALCINI, AND F. CARAMIA

Laboratorio Biologia Cellulare del CNR, Rome and Department of Biology, Washington University, St. Louis, Missouri 63130

Received July 10, 1970

Purified nerve growth factor (NGF) induces a marked hypertrophy of the sympathetic chain ganglia when injected into newborn and adult mice. A 5-day treatment of newborn or adult mice with daily injections of NGF (20 μg/g body weight) results in both instances in a striking increase in wet weight of sympathetic ganglia. When examined at the electron microscope, the NGF-treated neurons appear very enlarged: the endoplasmic reticulum and free ribosomes are very abundant. The cytoplasmic area is filled with the interlacing bundles and threads of neurofilaments. This extraordinary production of neurofilaments and neurotubuli appears to be the most striking feature of the NGF-treated neurons.

The treatment with a specific nerve growth factor (NGF) of newborn and adult mice results in a marked volume increase of the sympathetic ganglia of the experimental animals. Extensive studies performed in our laboratory in these years showed that this effect is due to increase in cell number and cell size in newborn animals, while only the latter effect obtains in adult specimens (*4*, *7*).

In vitro the NGF elicits the production of a dense fibrillar halo by sensory and sympathetic ganglia dissected out from 8-day chick embryos and cultured in a semi-solid medium enriched with this factor. The effect, which materializes in a few hours, brings in sharp relief the most outstanding feature of this growth factor, which is to greatly enhance nerve fiber production by its target cells (*3*, *6*).

The life-long responsiveness of the sympathetic neurons to the NGF raised the question whether the growth effects obtained *in vivo* compare at the ultrastructural level to those observed *in vitro* in embryonic sensory and sympathetic nerve cells (*8*).

The results reported in this article show that the *in vivo* growth response also consists of a marked hypertrophy of neurons and is characterized by a massive production

[1] Reprint requests should be addressed to Dr. Rita Levi-Montalcini, Department of Biology, Washington University, St. Louis, Missouri 63 130.

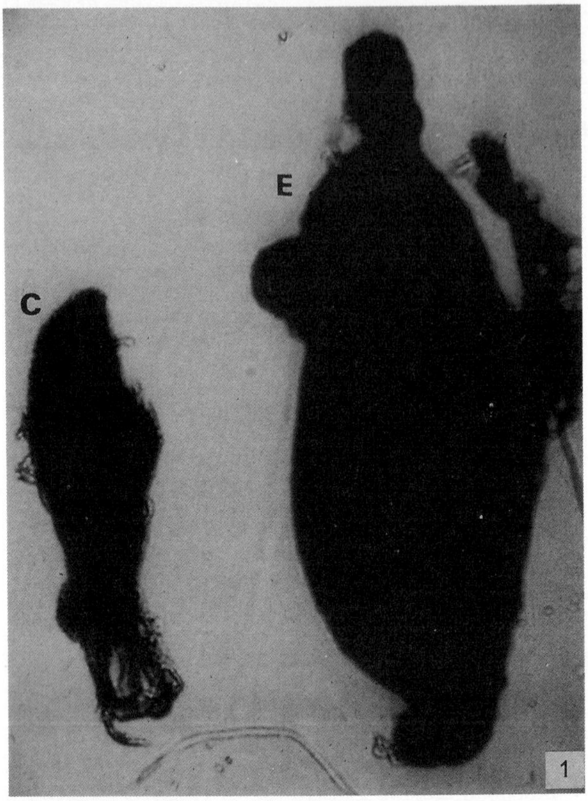

FIG. 1. Whole mounts of superior cervical ganglia of control (C) and NGF–treated (*E*) 5-day-old mice.

of neurofilaments and neurotubules in the cytoplasm of these cells. This effect is obtained in newborn and in adult animals.

MATERIALS AND METHODS

The nerve growth factor was purified from mouse salivary glands as previously described (*1*, *2*). The purified NGF is homogeneous by several physicochemical and immunological criteria and has a molecular weight of 30 000 (*10*). Its specific activity is of the order of 0.01 μg protein per milliliter. In a first group of experiments, newborn mice were injected daily for 5 days with NGF (20 μg/g body weight); control animals of the same litter received saline injections.

In a second group of experiments, adult female mice (2 months old) were injected intravenously for 5 days with the NGF in the amount of 20 μg/g body weight.

Control animals of the same litter were injected with the physiological solution. At the end of the treatment, the animals were sacrificed and the superior cervical ganglia were dissected out. Some ganglia of both control and treated animals were weighed with a torsion balance. Others were immediately fixed in cold 3% glutaraldehyde in 0.1 *M* phosphate buffer at pH 7.3 for 3 hours. They were then washed in two changes of the same cold buffer and postfixed for 2 hours in 2% osmium tetroxide. The ganglia were then dehydrated in ethanol solution, placed in toluene and embedded in Epon. Sections were cut with a Porter Blum MTI microtome, stained with uranyl acetate and lead citrate, and examined with a Philips 300 electron microscope. Some sections were stained with toluidine blue and studied at the light microscope.

OBSERVATIONS

The 5-day treatment of newborn mice with NGF results in a 4- to 5-fold volume increase of sympathetic chain ganglia (Fig. 1). At the light microscope, cross sections of superior cervical ganglia show a marked size increase of neurons in NGF-treated mice. The number of nerve cells is more than doubled as compared to controls. In the majority of nerve cells the cytoplasmic area is enlarged and intensely basophilic, particularly in the cell outer layer. Nuclei and nucleoli are markedly larger than in untreated neurons.

When examined at low magnification in the electron microscope, sections of superior cervical ganglia from NGF-treated mice showed obliteration of intercellular spaces between neurons, in contrast to the always well-defined intercellular spaces seen in control sections (Figs. 2 and 3). The endoplasmic reticulum of NGF treated neurons appears highly developed throughout the cytoplasm and, in particular, in the cell outer layer. It consists of numerous but short profiles randomly distributed, which in some instances exhibit a parallel array of cisternae. The free ribosomes are very abundant throughout the cytoplasm. The mitochondria do not apparently differ from those of controls.

The Golgi apparatus has undergone a remarkable increase and shows numerous dilated vesicles (Fig. 4). While in control neurons it is preferentially supranuclear or lateral to the nucleus, in NGF-treated neurons the Golgi apparatus extends also into other areas.

FIG. 2. Electron micrograph of sympathetic neurons from superior cervical ganglion of control 5-day-old mouse. *N*, nucleus; *Nu*, nucleolus. ×6 500.

FIG. 3. Electron micrograph of sympathetic neurons from superior cervical ganglia of 5-day-old mouse treated since birth with NGF. *N*, nucleus; *Nu*, nucleolus. Notice the abundance of the neurofibrillar material (arrows). ×6 500.

FIGS. 4 and 5. Electron micrograph of neurons from superior cervical ganglia of 5-day-old NGF-treated mouse, at higher magnification.

FIG. 4. Dilation of the Golgi apparatus (G), longitudinal section of a bundle of neurofilaments (*f*). ×32 500.

FIG. 5. Cross section of a large bundle of neurofilaments (*f*) in an area rich in free ribosomes. ×47 000.

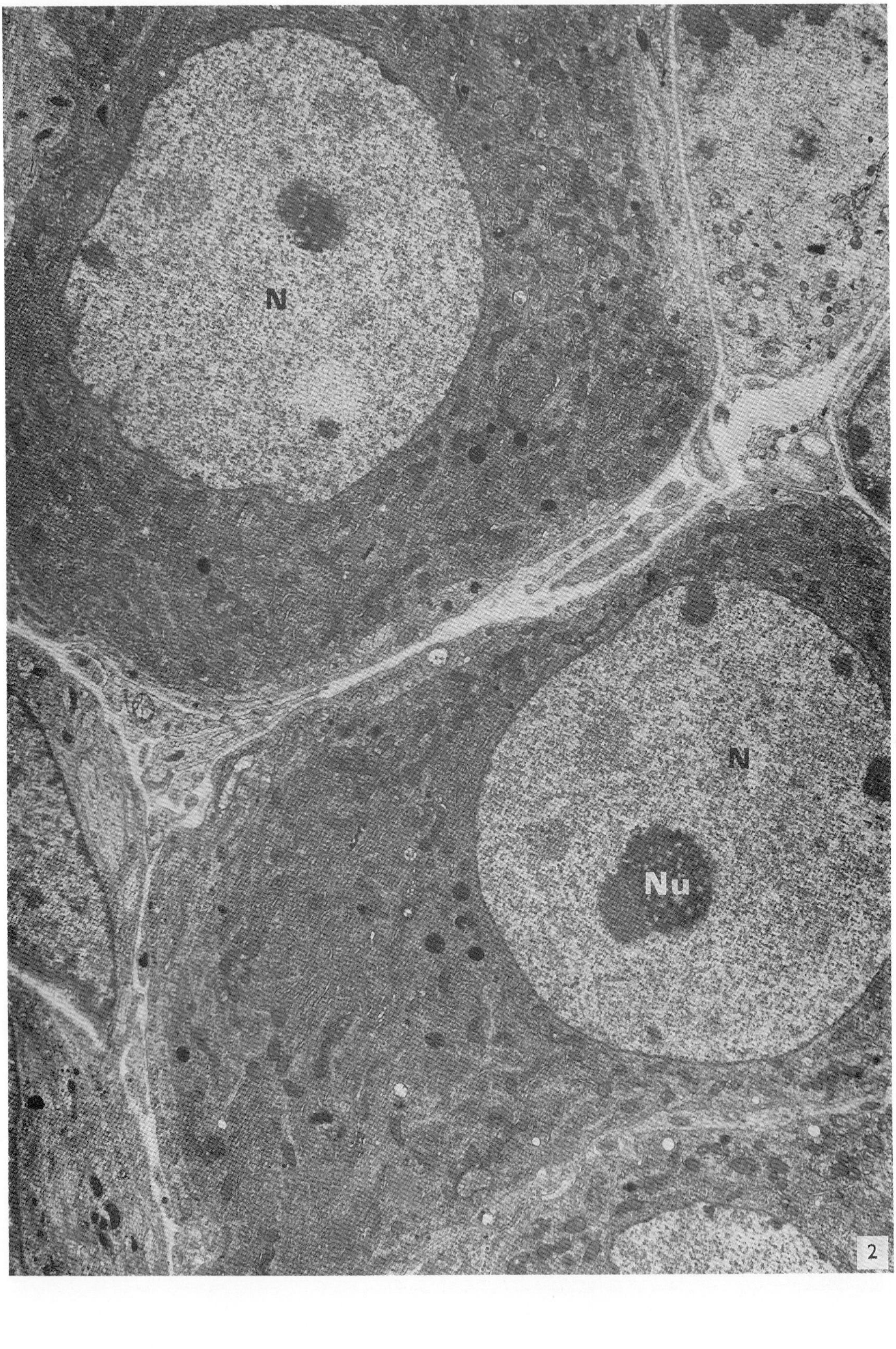
N
N
Nu
2

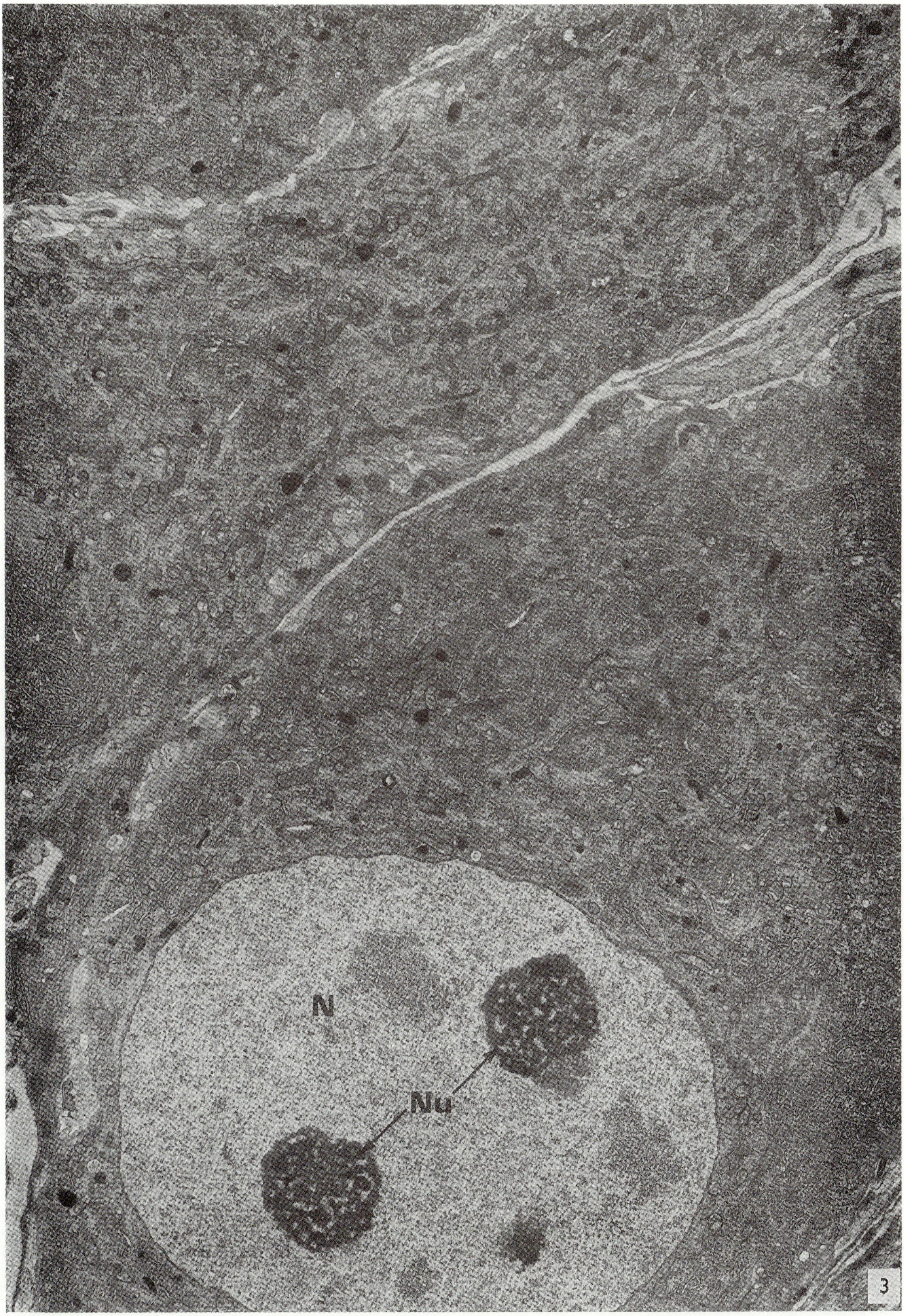
N
Nu
3

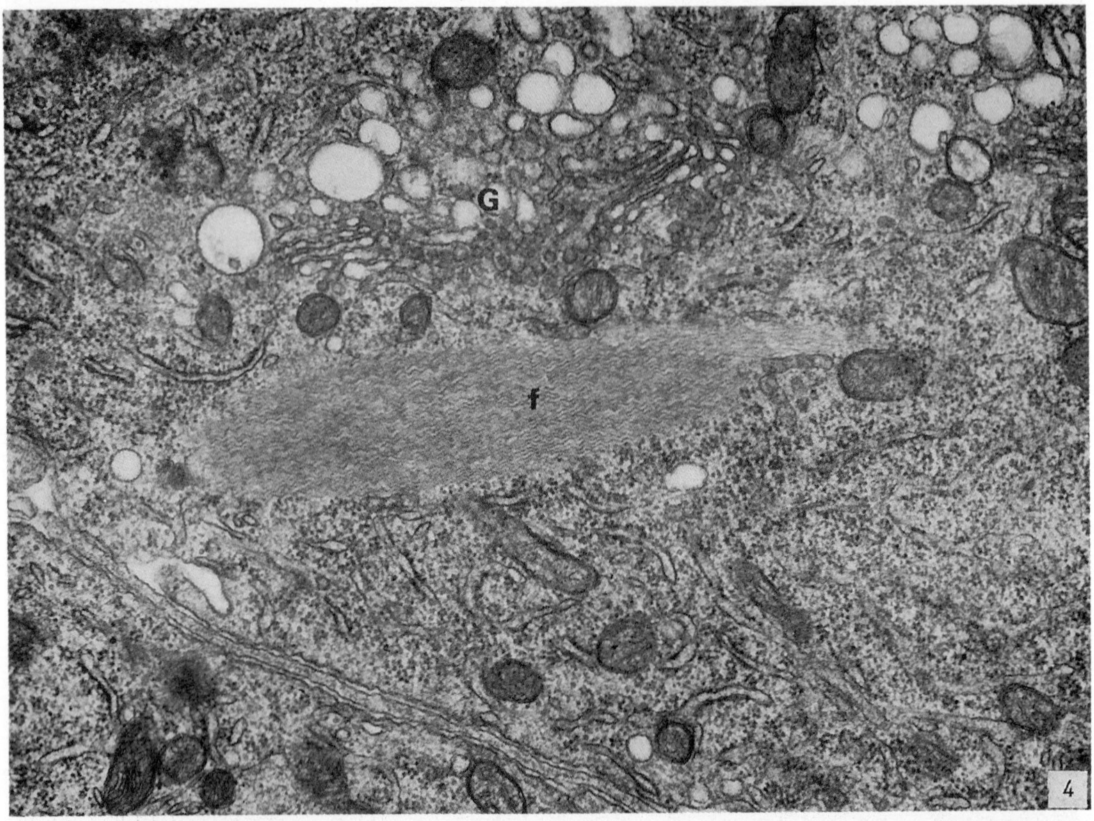
G
f
4

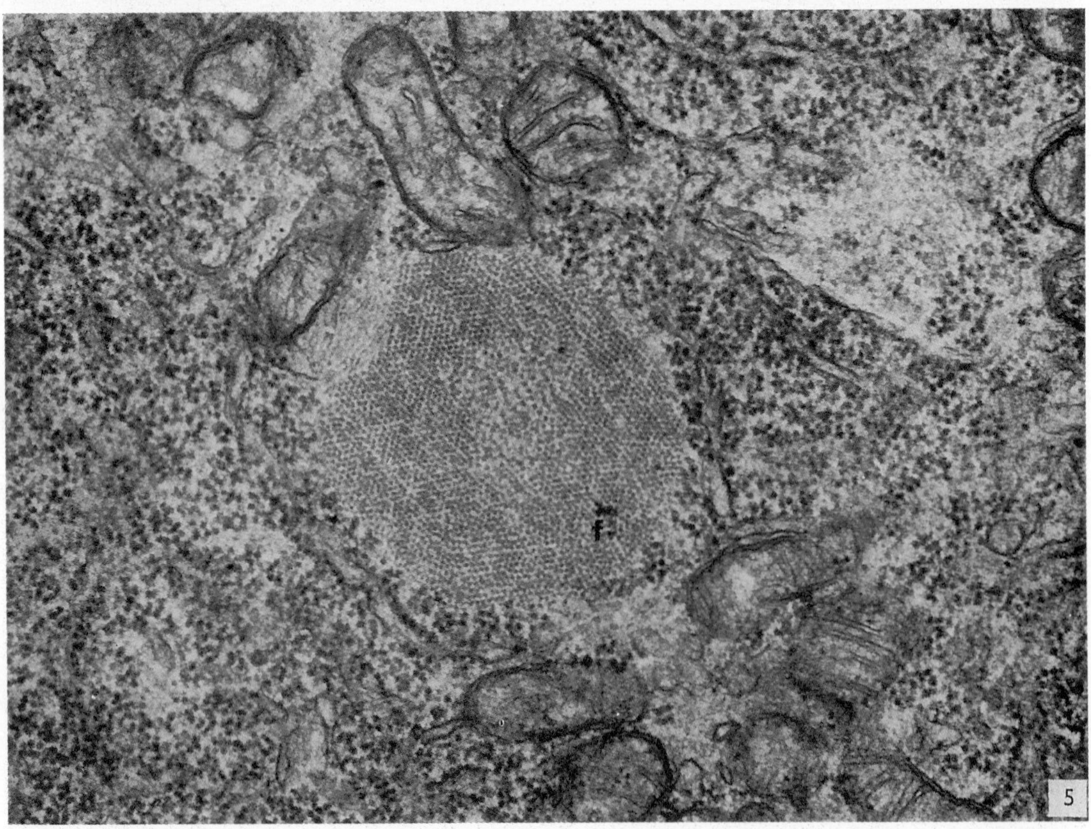
f
5

The most prominent feature of NGF-treated neurons is the massive increase of neurofibrillar material. At low magnification numerous bundles of low density crisscross the cytoplasm in all directions. They consist of neurofilaments and neurotubules, as is evident at higher magnification (Figs. 4 and 5). In some instances, formations resembling the microfibrillar paracrystalline inclusions described by Seite (*9*) are seen (Fig. 5).

The 5-day NGF treatment of adult mice results in a 2-fold increase in wet weight of sympathetic ganglia. The treated animals were in excellent condition and their body weight was the same as that of controls.

Cross sections of superior cervical ganglia of control and experimental ganglia examined at the light microscope show a striking hypertrophy of the NGF-treaed neurons (Figs. 6 and 7). The size increase of the cells is to a large extent due to an increment of the cytoplasmic area, which shows a texture quite different from that of control neurons. The contour of the nucleus, which is sharply defined in these cells under normal conditions, is much less evident in experimental neurons. Nucleoli are prominent and much larger than in controls. Glial and satellite cells do not differ from those of untreated ganglia.

As described in ganglia of newborn mice, nerve cells come in contact with each other, due to the conspicuous enlargement of the cytoplasm; the cellular contour which is ovoidal in controls has now assumed a polyhedric shape (Figs. 8 and 9). Satellite cells appear as squeezed in the tight space between neurons.

In the cytoplasm, the endoplasmic reticulum is abundant and represented by short profiles with associated ribosomes (Fig. 10). The free ribosomes are more numerous than in untreated neurons, and they are grouped in rosettes or small clusters densely packed in compact areas. These zones of rich endoplasmic reticulum and free ribosomes are crossed in all directions by interlacing strands of fibrillar material (Fig. 10). The presence of these extraordinarily abundant bundles and threads of neurofilaments is indeed the most striking feature in the NGF-treated neurons. They appear more numerous around the nucleus, where they form a low density ring encircling the

FIGS. 6 and 7. Cross section of control (Fig. 6) and NGF–treated (Fig. 7) superior cervical ganglia of adult mice (same magnification). The experimental mice were injected with NGF for 5 days. Notice the conspicuous enlargement of the neurons, while glial cells are of the same size in control and experimental section. Light micrograph, toluidine blue stain.

FIG. 8. Electron micrograph of sympathetic nerve cells from the superior cervical ganglion of control adult mouse. Two elongated neurons, well separated from each other by intercellular space and glial cells. In the cytoplasm are numerous mitochondria. The endoplasmic reticulum is formed by lamellae in parallel arrays. Few lipoid droplets are seen. ×6 500.

FIG. 9. Electron micrograph of sympathetic nerve cells from superior cervical ganglion of adult mouse injected for 5 days with NGF. Conspicuous enlargement of the neurons results in their reciprocal contact and obliteration of intercellular space. In the cytoplasm the low density areas due to interlacing strands of neurofilaments are very abundant and are particularly evident around the nucleus. ×6 500.

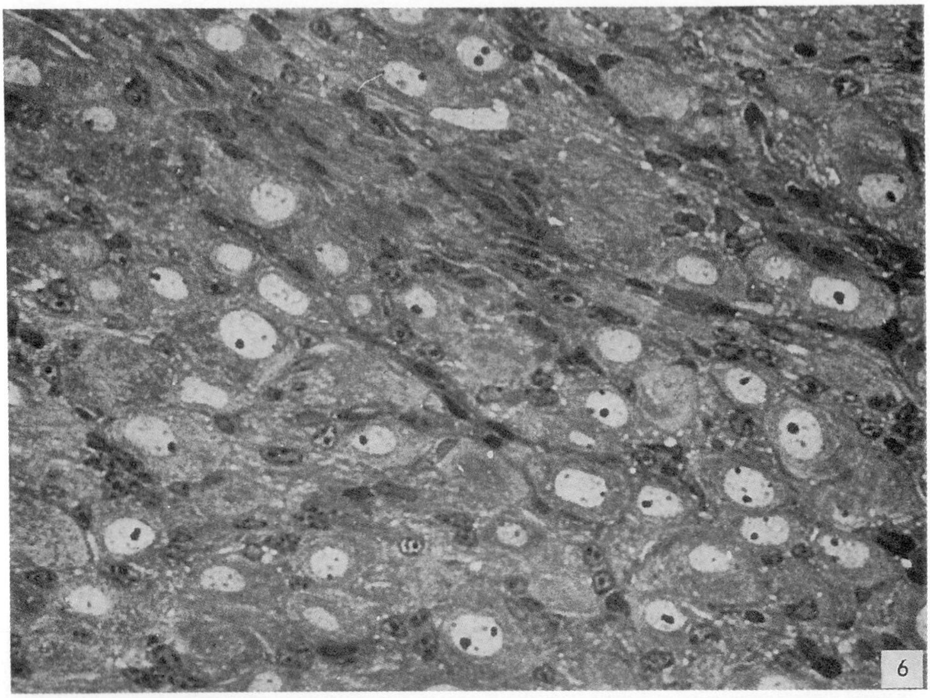

6

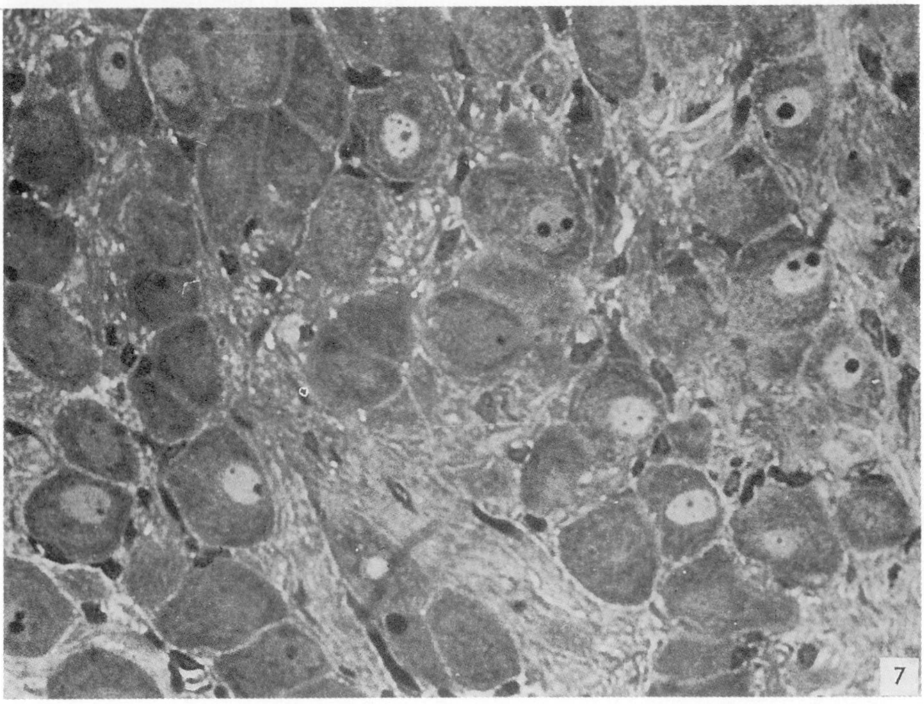

7

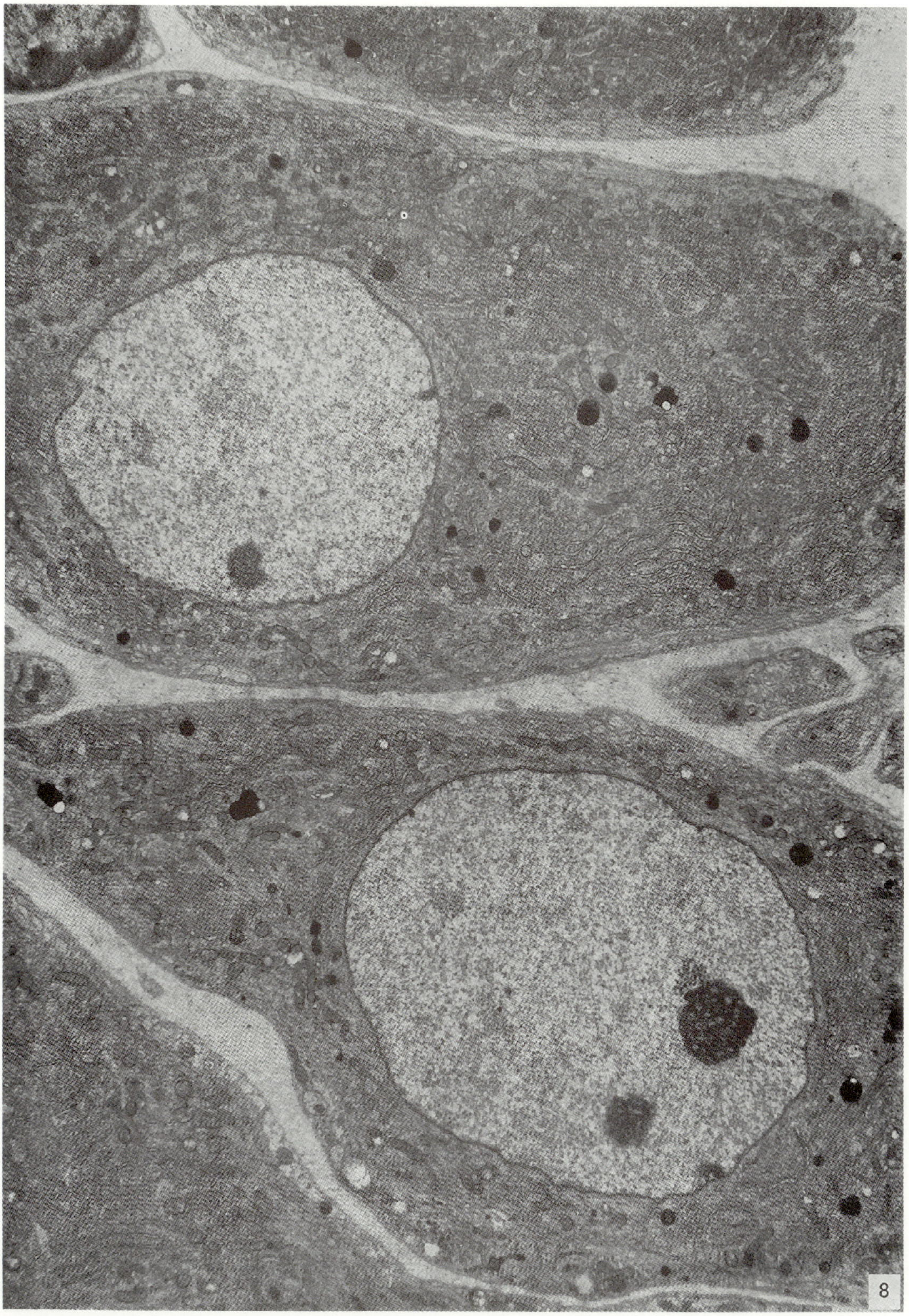
8

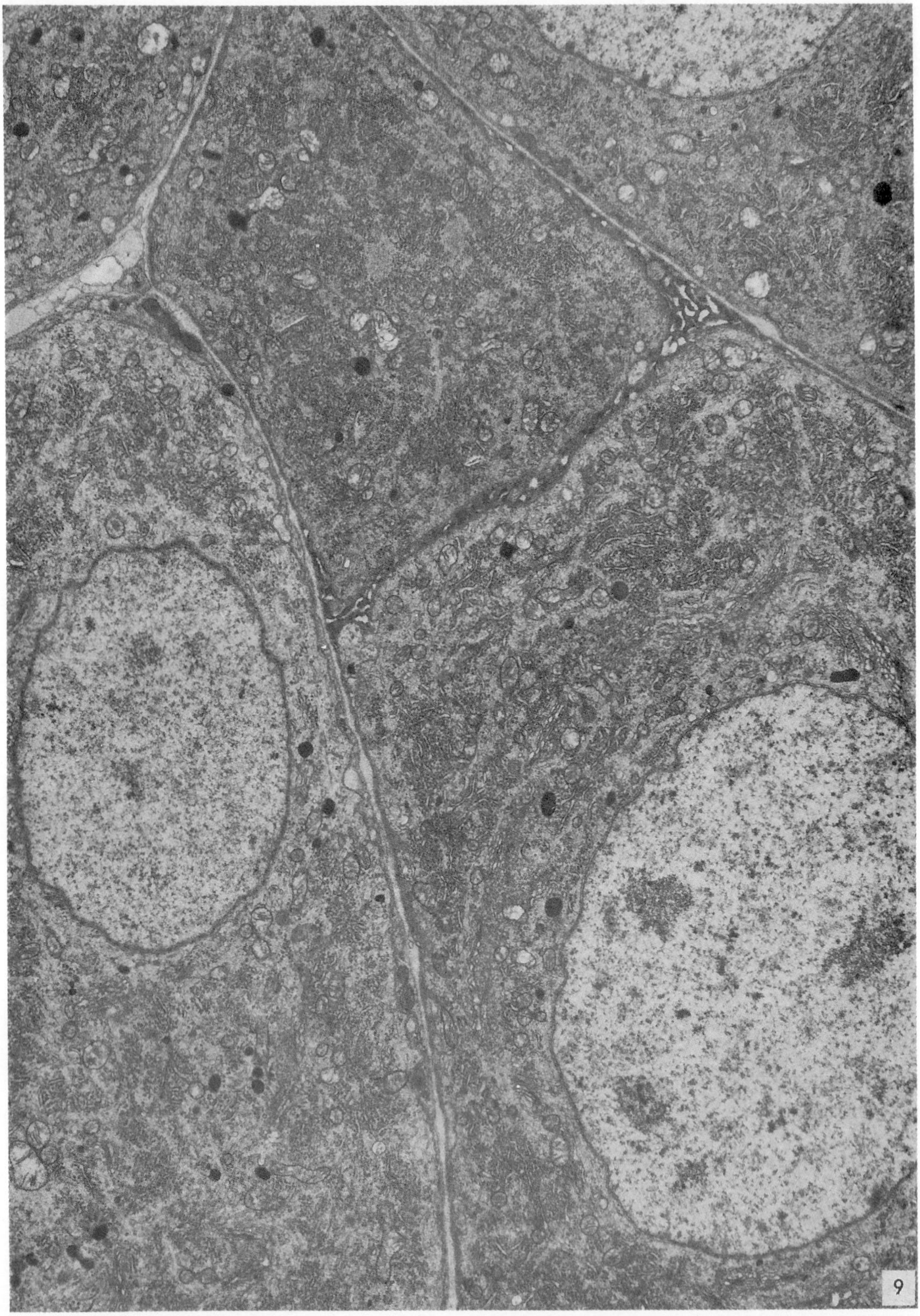
9

nuclear envelope (Fig. 7). Because of this fibrillar network the endoplasmic reticulum appears to be confined in small distinct areas segregated from each other. At high magnification it appears formed by short dilated cisternae which are no longer oriented in parallel arrays (Fig. 10).

The Golgi apparatus, which in control neurons is located around the nucleus, in experimental neurons is found distributed in various areas all over the cytoplasm. The mitochondria do not substantially differ from those of controls.

DISCUSSION

It was the object of the studies reported above to explore at the ultrastructural level the changes called forth by the nerve growth factor in sympathetic nerve cells of newborn and adult mice and to compare these changes to those elicited by the same factor in embryonic sensory nerve cells of 8-day chick embryos incubated *in vitro* for 4–12 hours in a medium enriched with the NGF.

As reported in a previous article (*8*), the main changes called forth by the NGF on embryonic sensory nerve cells are a dilation of the Golgi apparatus, a marked increase in ribosomes and in the number of membrane elements of the endoplasmic reticulum, and, more prominent than all other changes, a massive increase of neurofilaments, which fill practically all cytoplasm of these immature nerve cells.

The results obtained in the present experiments compare to a remarkable extent to those obtained in embryonic sensory nerve cells incubated for short periods *in vitro*, in spite of the fact that the experimental conditions, as well as the target cells, differ from those of the previous experiments (*3*, *5*, *7*).

Also in the adult neurons the outstanding feature of the NGF effect is the massive increase in neurofibrillar material which fills all cytoplasmic areas of these neurons and accounts for their marked enlargement.

Since neurofibrils are a unique differentiative mark of nerve cells, their increase represents the end result of the growth processes stimulated by the NGF in its target cells. It remains to be explained why other nerve cells do not respond in a similar way to this factor. Conceivably, the primary site of action of the NGF is not at the level of the neurofibrillar formation. Their striking increase may well result from the stimulation of some more basic and highly specific processes in the sequence of events which result in growth and differentiation of the receptive nerve cells. The massive production of neurofibrils should therefore be regarded as a consequence of the enhancement

FIG. 10. Electron micrograph of part of the cytoplasm of a neuron from the superior cervical ganglion of an adult mouse treated for 5 days with NGF. Bundles of neurotubuli (*t*) cross the cytoplasm in all directions, delimiting compact areas of endoplasmic reticulum (*ER*), rich in short lamellae and clusters of free ribosomes. *m*, mitochondria. ×18 500.

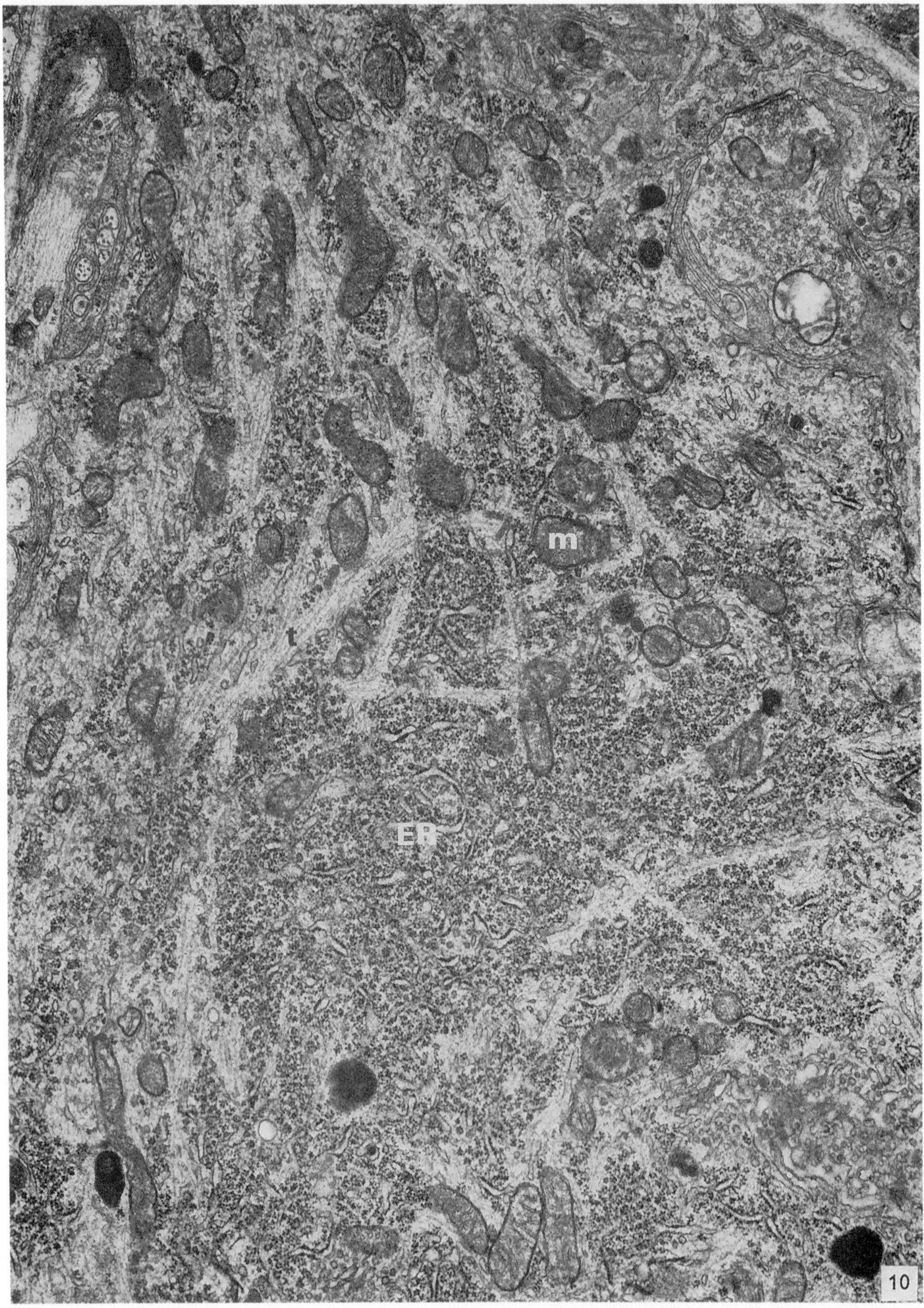
m
t
ER
10

of all anabolic processes stimulated to their maximal rate in the cells triggered by the NGF. In line with this concept is the observation that all cell organelles known to be involved in the cell synthetic processes, such as the nucleoli, the Golgi complex, the endoplasmic reticulum, and the free ribosomes, are increased in sympathetic nerve cells exposed to the action of the NGF.

This work was supported in part by grants from the National Institutes of Health (NS-03777), from the National Science Foundation (GB-1633OX) and from the Air Force Office of Scientific Research (69-1824).

REFERENCES

1. Angeletti, P. U., Levi-Montalcini, R. and Calissano, P., *Advan Enzymol.* **31**, 51 (1968).
2. Bocchini, V. and Angeletti, P. U., *Proc. Nat. Acad. Sci. U. S.* **64**, 787 (1969).
3. Levi-Montalcini, R., *Harvey Lect.* **60**, 217 (1966).
4. Levi-Montalcini, R. and Angeletti, P. U., *Quart. Rev. Biol.* **36**, 99 (1961).
5. Levi-Montalcini, R. and Angeletti, P. U., *in* Wolstenhoime, C. E. W. and O'Connor, M. (Eds.), *Growth Nervous System, Ciba Found. Symp.* p. 126 (1968).
6. Levi-Montalcini, R. and Angeletti, P. U., *Physiol. Rev.* **48**, 534 (1968).
7. Levi-Montalcini, R. and Booker, B., *Proc. Nat. Acad. Sci. U. S.* **42**, 373 (1960).
8. Levi-Montalcini, R., Caramia, F., Luse, S. A. and Angeletti, P. U., *Brain Res.* **8**, 347 (1968).
9. Seite, R., *Z. Zellforsch. Mikrosk. Anat.* **101**, 621 (1969).
10. Zanini, A., Angeletti, P. and Levi-Montalcini, R., *Proc. Nat. Acad. Sci. U. S.* **61**, 835 (1968).

Reprinted from
Brain Research
Elsevier Publishing Company, Amsterdam
Printed in the Netherlands

ALTERATIONS IN THE FINE STRUCTURE OF NUCLEOLI IN SYMPATHETIC NEURONS FOLLOWING NGF-ANTISERUM TREATMENT

RITA LEVI-MONTALCINI, FELICE CARAMIA* AND PIETRO U. ANGELETTI

Department of Biology, Washington University, St. Louis, Mo. (U.S.A.) and Istituto Superiore di Sanità, Rome (Italy)

(Accepted July 24th, 1968)

INTRODUCTION

The term 'immunosympathectomy' was coined to describe the destruction of sympathetic chain ganglia of newborn mammals following the injection of an antiserum to the nerve growth factor[2,4–6]. Ever since this dramatic effect was discovered, the question arose as to the mechanism and site of action of the antiserum in the target sympathetic nerve cells. Previous studies at the ultrastructural level by Sabatini and co-workers[10] indicate the early appearance of degenerative changes in the nuclear compartment of sympathetic cells exposed *in vivo* or *in vitro* to the action of the antiserum. The observations to be reported below describe alterations in the fine structure of the nucleoli and suggest these as the possible primary lesions inflicted by the antiserum.

MATERIAL AND METHODS

The nerve growth factor (NGF) was isolated and purified from male mouse submaxillary glands according to the procedure adopted in our laboratory[1,2]. Briefly, crude extracts of the glands in Tris-buffer, pH 7.2, were fractionated through a Sephadex G-100. The active fractions were pooled and chromatographed on DEAE-cellulose at pH 7.2. The unadsorbed protein peak containing most of the biological activity was dialyzed against acetate buffer 50 m*M*, pH 5, and then fractionated through a CM-cellulose column with a gradient salt elution. The last protein peak containing all the NGF activity was pooled, dialyzed and rechromatographed twice under the same conditions. The final preparation was chromatographically pure and showed one single cathodic band on acrylamide electrophoresis at pH 4. Two milligrams of the NGF preparation were used to immunize each rabbit with Freund adjuvant. A week later the rabbits received a second dose of 1 mg NGF + incomplete

* Permanent address: Istituto di Patologia Generale, Università di Roma, Rome, Italy.

Freund. After 10 days, the rabbits received a series of 2 i.v. injections of NGF (100 μg per rabbit). The first bleeding was done a week after the last i.v. injection. Quantitative precipitin reaction, complement fixation, immunoelectrophoresis and *in vitro* inhibition of NGF activity were carried out with the antisera.

The *in vivo* effect was checked by injecting the antisera into newborn mice (0.5 ml/g body weight) for various periods of time. Superior cervical ganglia from antiserum treated and from control mice were removed under a dissecting microscope and fixed for 2 h in 3% glutaraldehyde in phosphate buffer 0.1 *M*, pH 7.4, with added $CaCl_2$. The specimens were washed for 2 h in the same buffer (4 changes) and postfixed in osmic hydroxide 1.33% for 2 h. After dehydration in ethanol, the tissue was immersed for 1 h in toluene and then embedded in Epon 8/12. Sections were cut with a Potter–Blum–Servall MT_2 microtome with a diamond knife and then mounted on copper-grid 200 mesh without any supporting film. The specimens were stained with uranyl acetate in 50% ethanol or in lead hydroxide, according to Millonig[7]. They were examined in an RCA 3 H electron microscope at 8000 up to 50,000 magnification. [^{14}C]Uridine (New England Nuclear Corp.) was used as precursor in incorporation studies. The isotope was injected i.p. (5 μC/g) into experimental and control mice 3 h before sacrifice. The tissue specimens were homogenized in cold trichloroacetic acid (TCA) 5%, the precipitate was washed several times in TCA and finally dissolved in formamide. Radioactive samples were then counted in a liquid scintillation system (Packard Tri-Carb).

RESULTS

The nerve growth factor used to immunize rabbits had a specific activity of about 0.002 μg/ml; thus 1 biological unit (B.U. = the minimum amount in protein required in tissue culture to give a 3+ response) was of the order of 0.1 ng (ref. 3). In the inhibition test, 1 ml of antiserum completely inhibited 150,000 B.U. On immunoelectrophoresis, a single precipitin band was observed over a wide range of Atg-antibody concentration.

Daily injection of this antiserum into newborn mice for a 5-day period brings about a 95% reduction of the superior cervical ganglion as judged by cell count. Morphological changes of the As-treated ganglia are already apparent at the light microscope after 24 h and become more and more dramatic in the following days. The time-course sequence of these lesions has been carefully investigated and reported in previous papers[3–5]. On the basis of these observations, we centered our attention on the early phases of the As treatment, namely at 12 and 24 h after the injection. At later stages a large variety of advanced and diffuse degenerative lesions are observed in the majority of the neurons.

Control ganglia

Large and small neurons are readily distinguishable by electron microscopy in sections of superior cervical ganglion of 1-day-old mouse. The large neurons have a

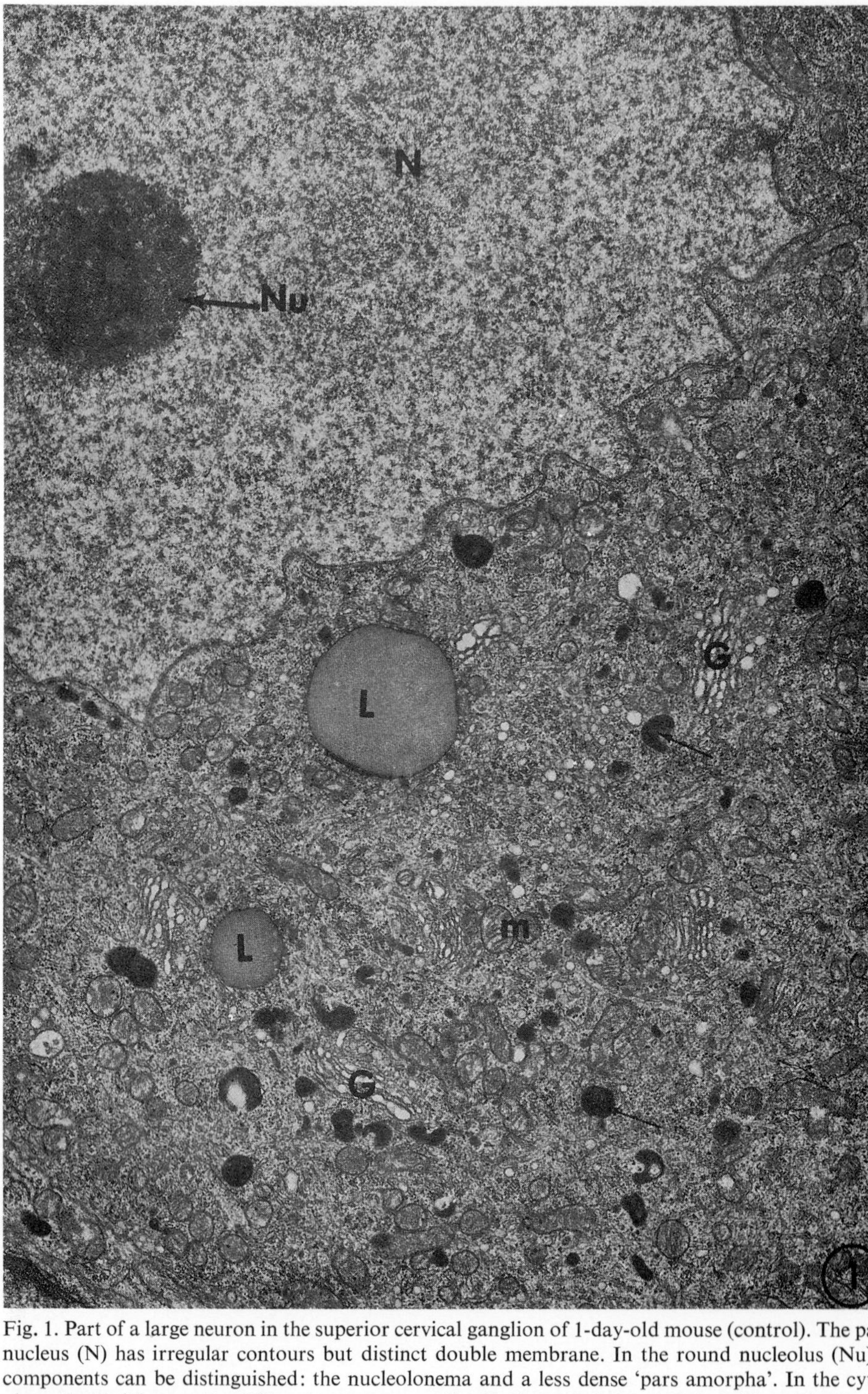

Fig. 1. Part of a large neuron in the superior cervical ganglion of 1-day-old mouse (control). The pale nucleus (N) has irregular contours but distinct double membrane. In the round nucleolus (Nu) 2 components can be distinguished: the nucleolonema and a less dense 'pars amorpha'. In the cytoplasm the Golgi apparatus (G) and numerous small mitochondria (m) are present. The endoplasmic reticulum is formed by short cisternae and clusters of free ribosomes. Small and very dense bodies, membrane bound (arrows), are seen. Few lipid droplets (L).

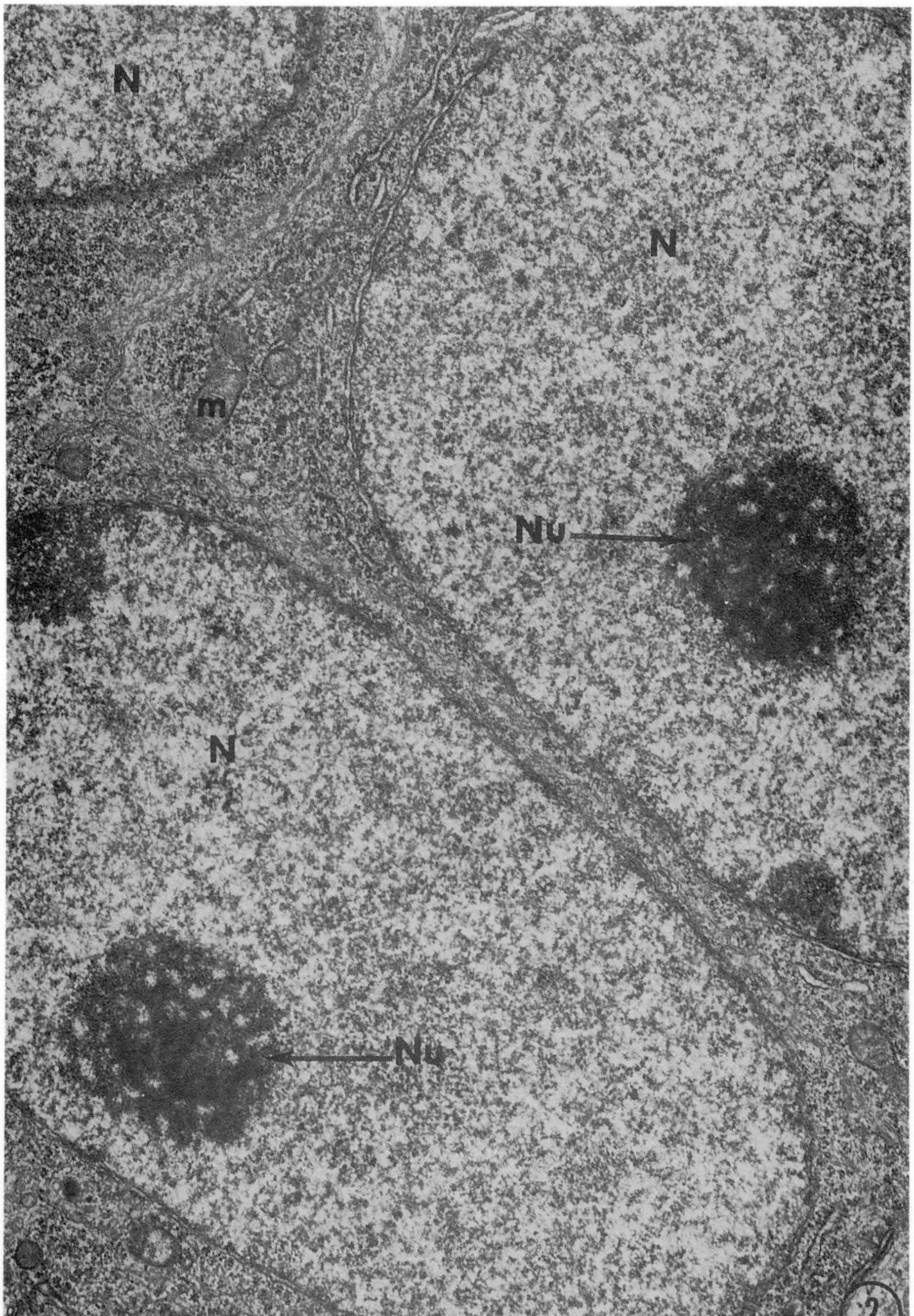

Fig. 2. Superior cervical ganglion of 1-day-old mouse (control). Two small neurons showing the typical organization of the nucleolus. Note that free ribosomes are always organized in rosettes.

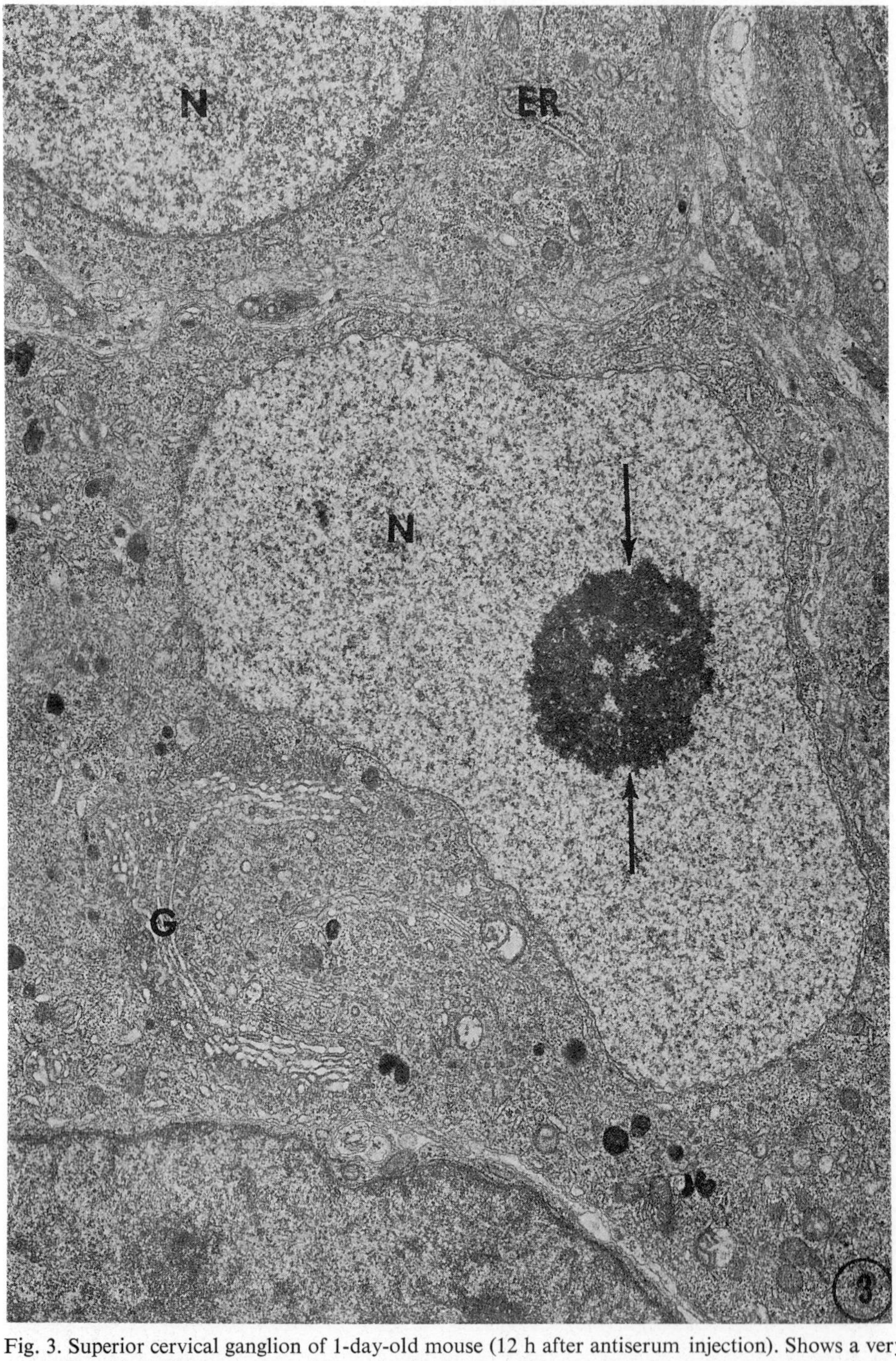

Fig. 3. Superior cervical ganglion of 1-day-old mouse (12 h after antiserum injection). Shows a very early change in the fine structure of the nucleolus (arrows). This appears denser than normal while the separation between the nucleolonema and the pars amorpha is less distinct. Within the nucleolar mass one sees a small and very dense round formation. In the cytoplasm the ribosomes are no longer grouped in clusters and rosettes. N, nucleus; G, Golgi apparatus; ER, endoplasmic reticulum.

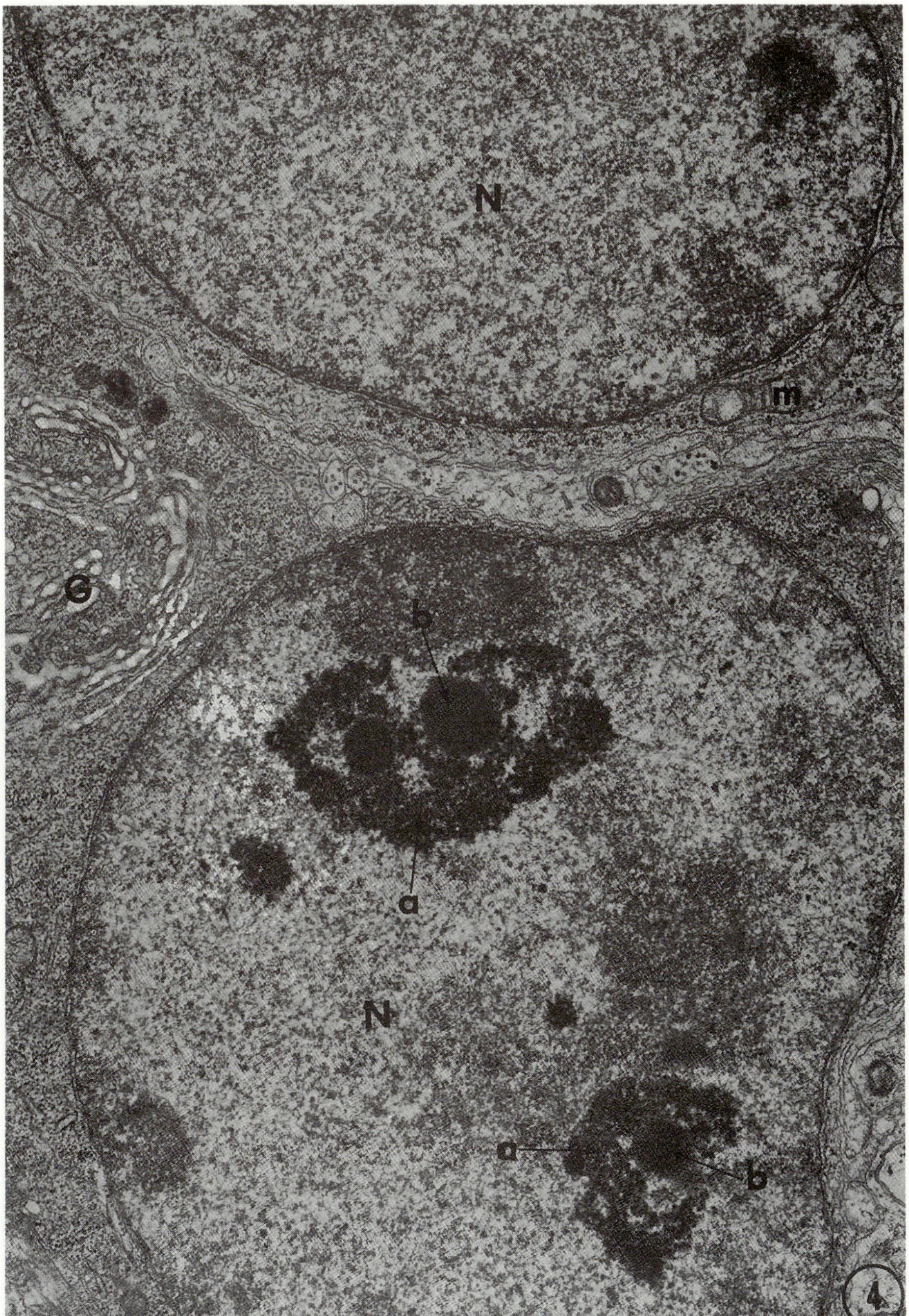

Fig. 4. Superior cervical ganglion of 1-day-old mouse (12 h after the As injection). The nucleus (N) of the small neuron (bottom) shows 2 very dense and irregular formations which present an external dense granular zone (a) surrounding 1 or 2 very homogeneous dense round bodies (b). Adjacent to these altered nucleoli a rather large granular mass can be seen, frequently terminating in the nuclear envelope.

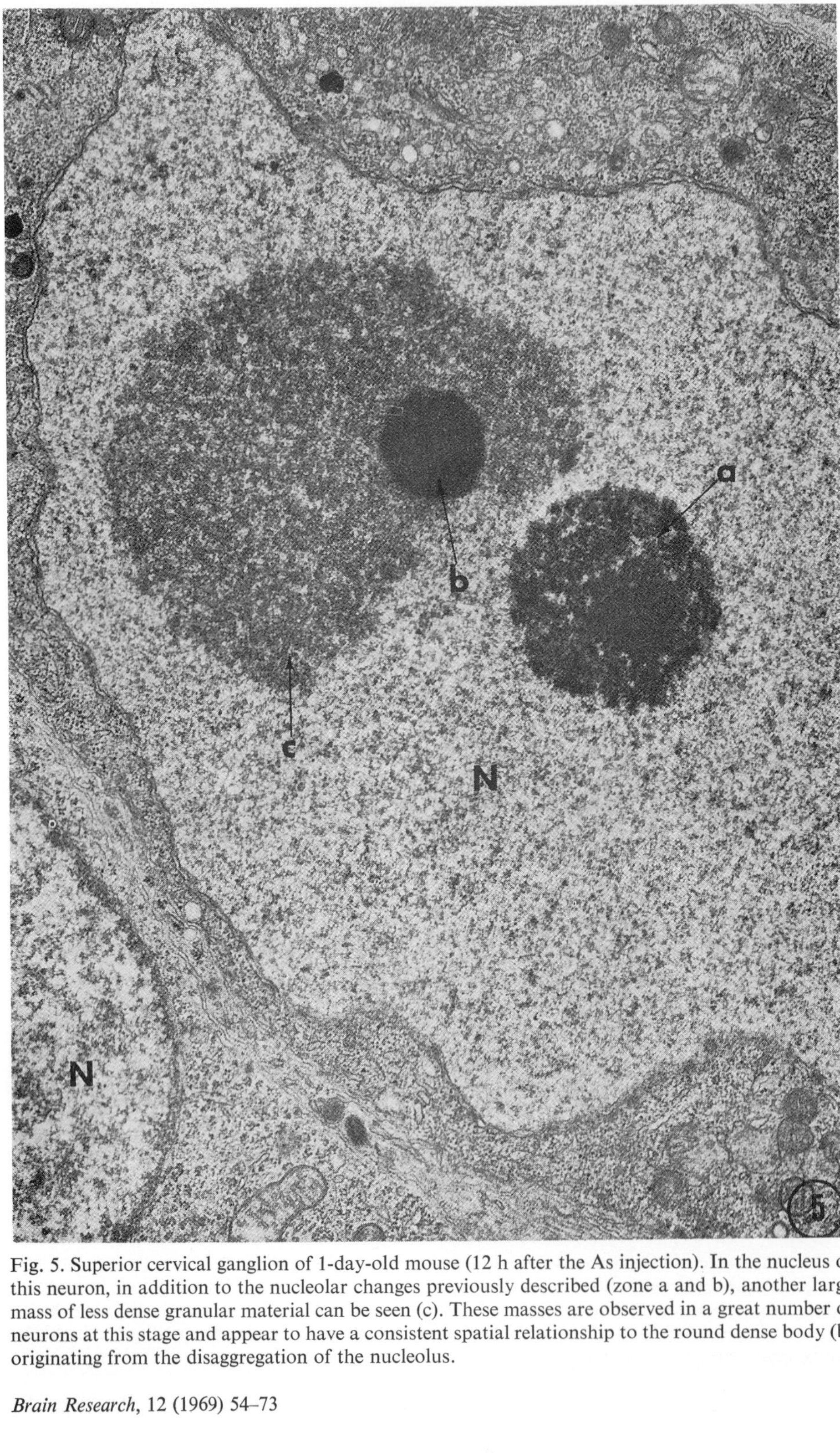

Fig. 5. Superior cervical ganglion of 1-day-old mouse (12 h after the As injection). In the nucleus of this neuron, in addition to the nucleolar changes previously described (zone a and b), another large mass of less dense granular material can be seen (c). These masses are observed in a great number of neurons at this stage and appear to have a consistent spatial relationship to the round dense body (b) originating from the disaggregation of the nucleolus.

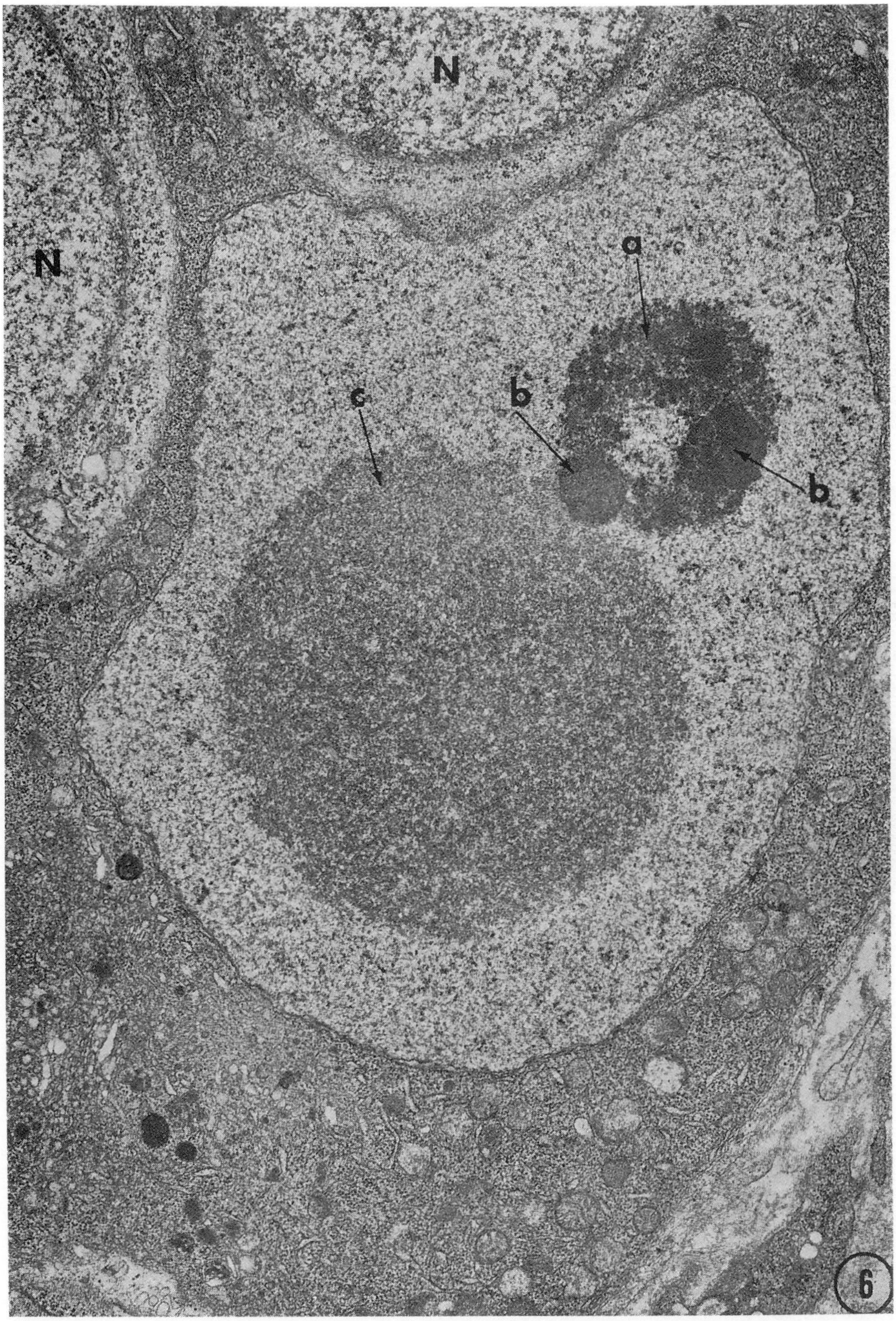

Fig. 6 Superior cervical ganglion (12 h after the As injection). Progressive enlargement of the granular mass which fills a large part of the nucleus. Again this altered nuclear structure is in relation to the dense round body (b). Alterations of the mitochondria are visible in the cytoplasm. Note the different structure and density of the nucleus of this neuron as compared to the nuclei of the two small neurons in the upper part of the picture.

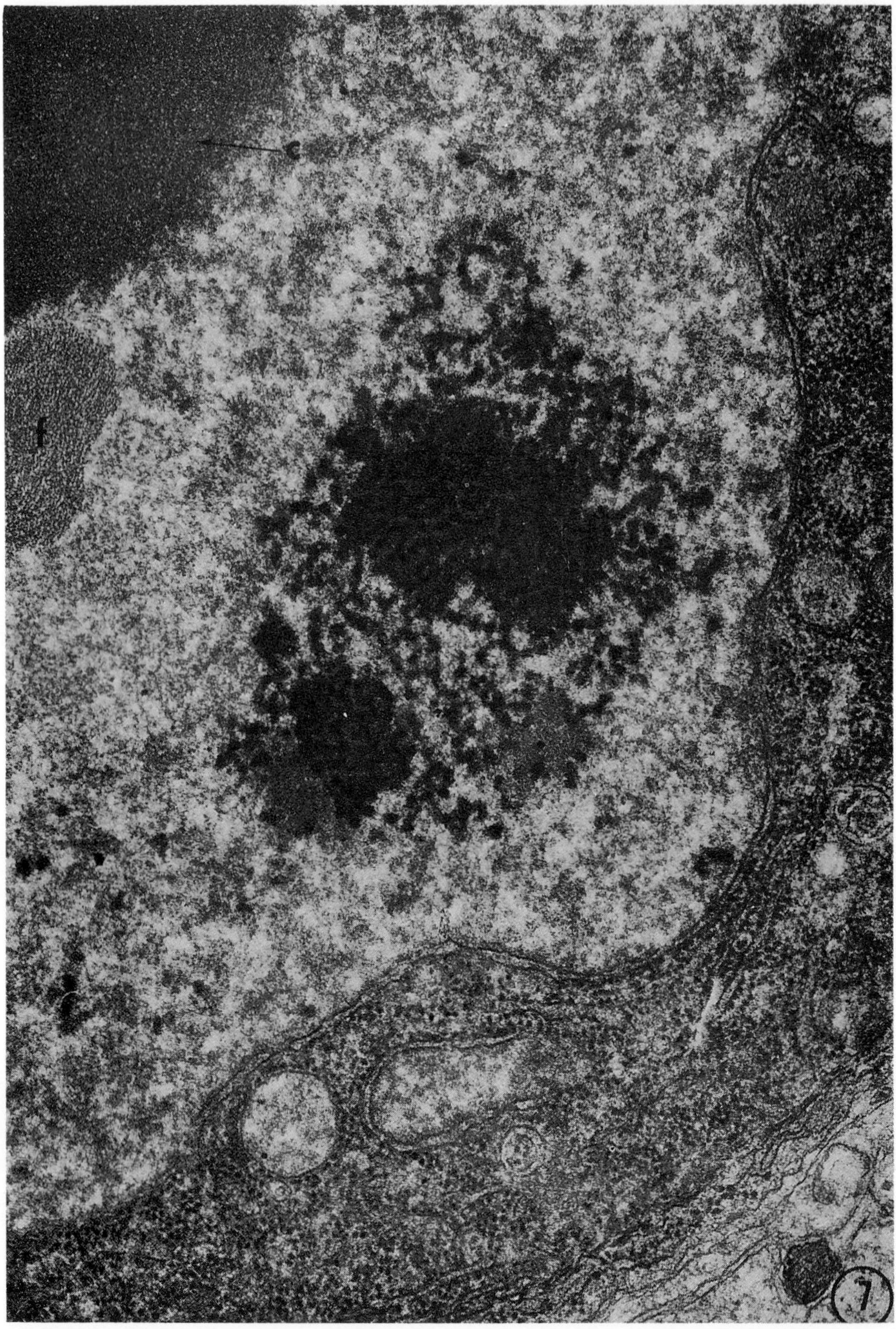

Fig. 7. Superior cervical ganglion of 1-day-old mouse (12 h after the As injection). Disaggregation of the nucleolar structure seen in some neurons, appearing as close network of dense tubules. At left of the picture, a large mass of filaments can be seen (f) and part of large zone c.

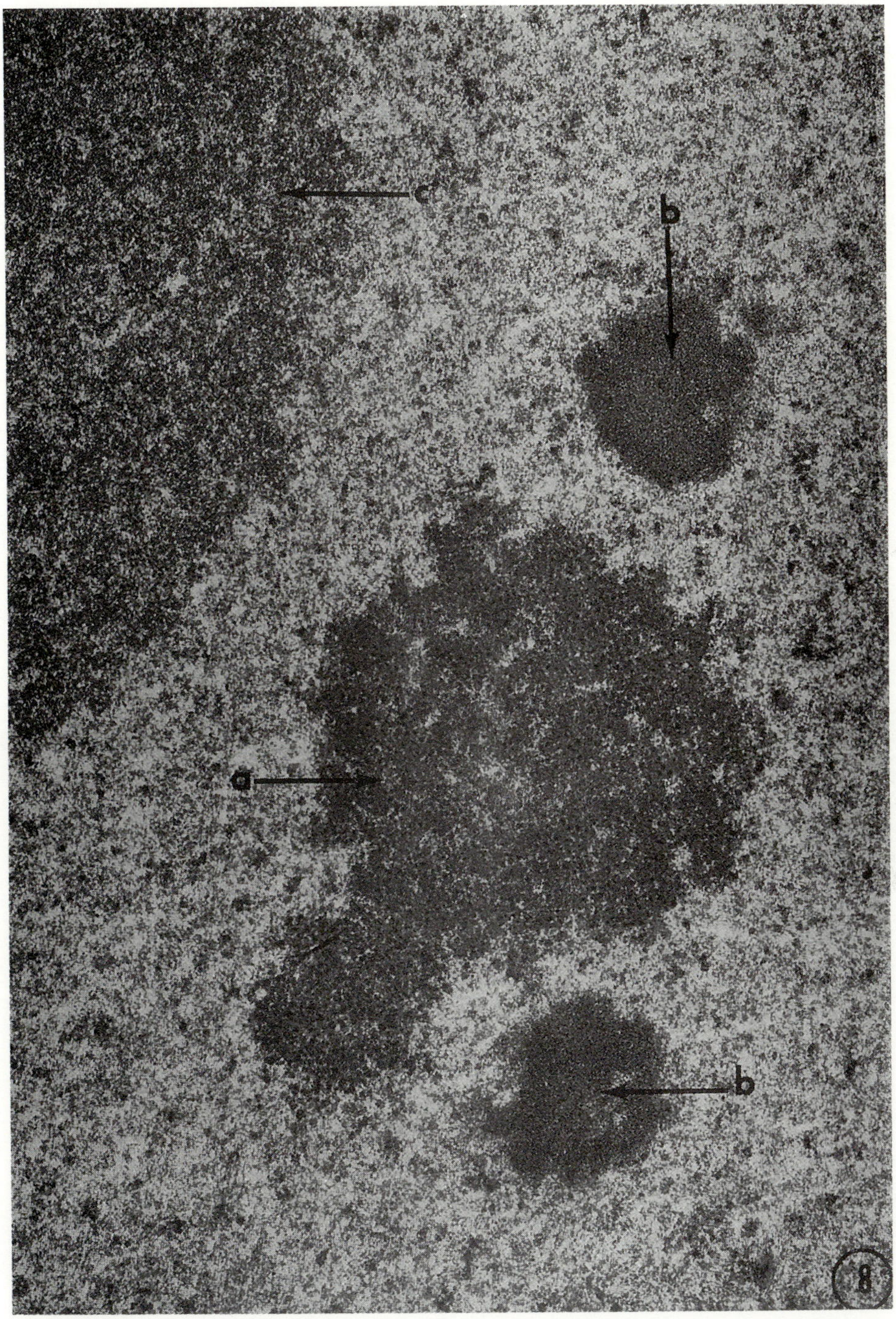

Fig. 8. Different granular material which forms the abnormal structures previously described within the nucleus, shown at higher magnification.

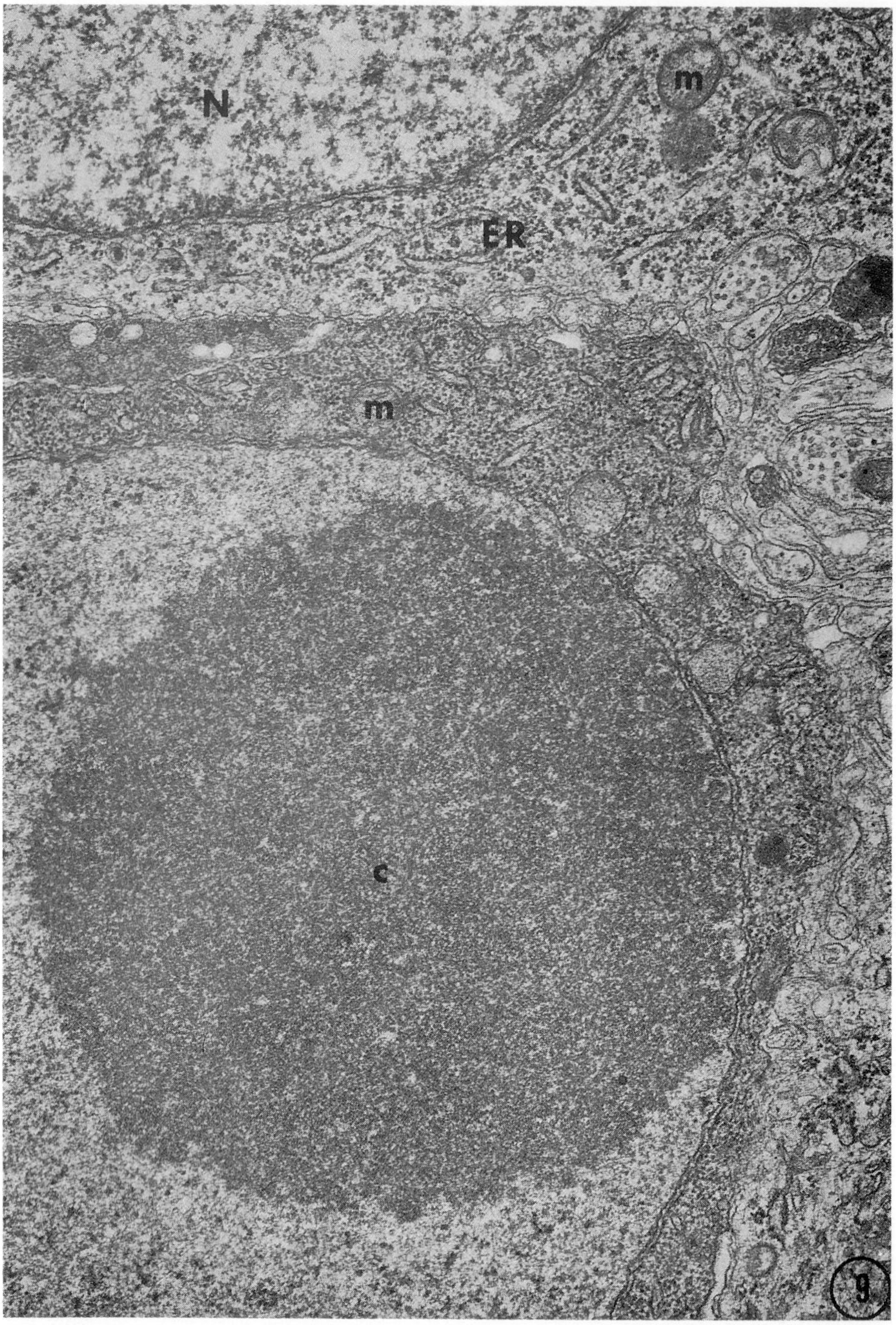

Fig. 9. Superior cervical ganglion of 1-day-old mouse (12 h after As injection), showing a neuron with more advanced nuclear lesions. The large granular mass (c) appears now somewhat denser and compact. The surrounding nuclear chromatin is also denser than normal. Condensed cytoplasm with free ribosomes diffuse throughout. Note for comparison another neuron (upper part) which apparently has not yet been affected by the As.

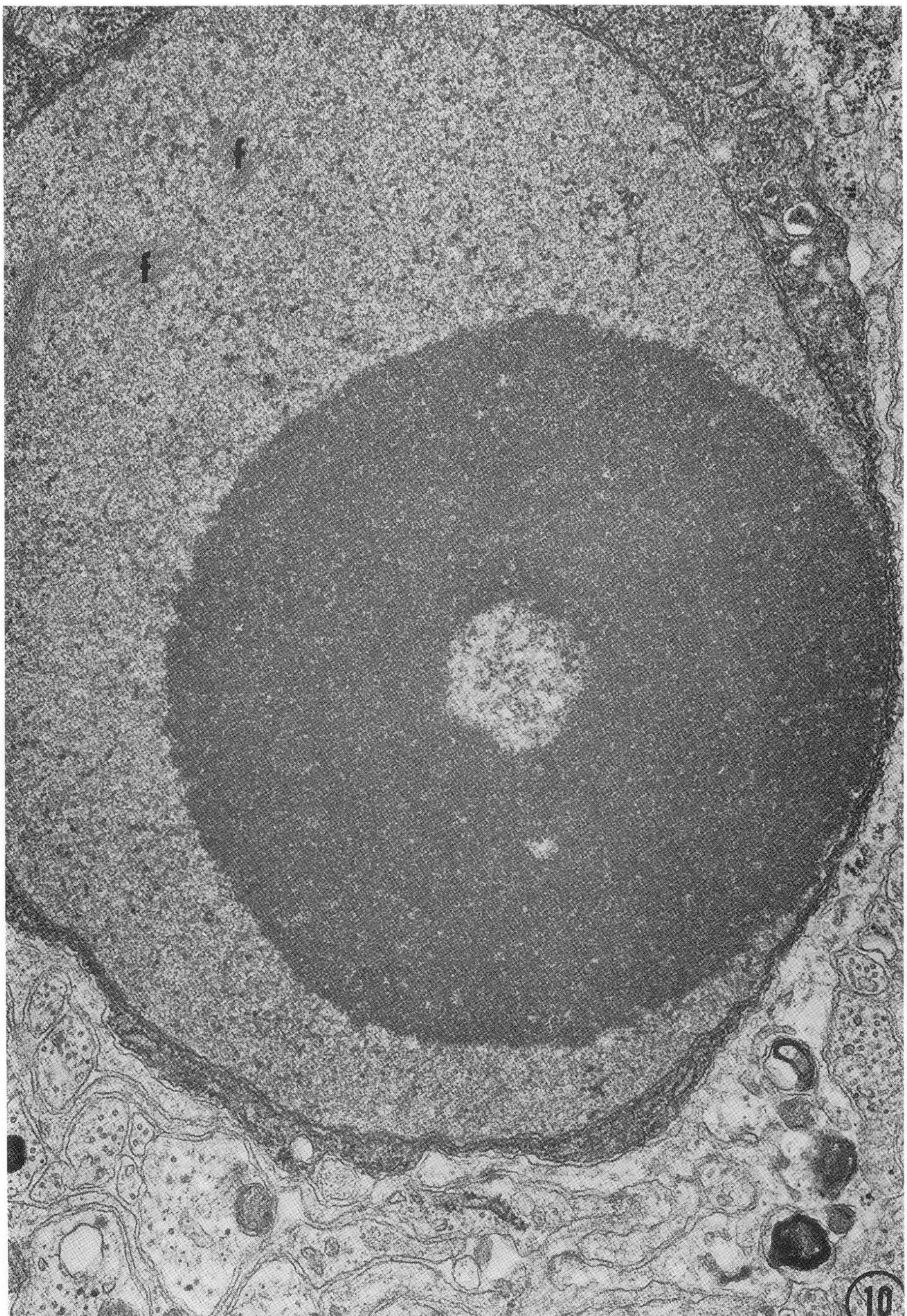

Fig. 10. Superior cervical ganglion of 1-day-old mouse (24 h after the As injection). Advanced lesion of the neuron by the As; the cytoplasm is reduced to a very thin sheet surrounding a large nucleus with altered chromatin filled with the granular mass (c). Presence of filaments (f).

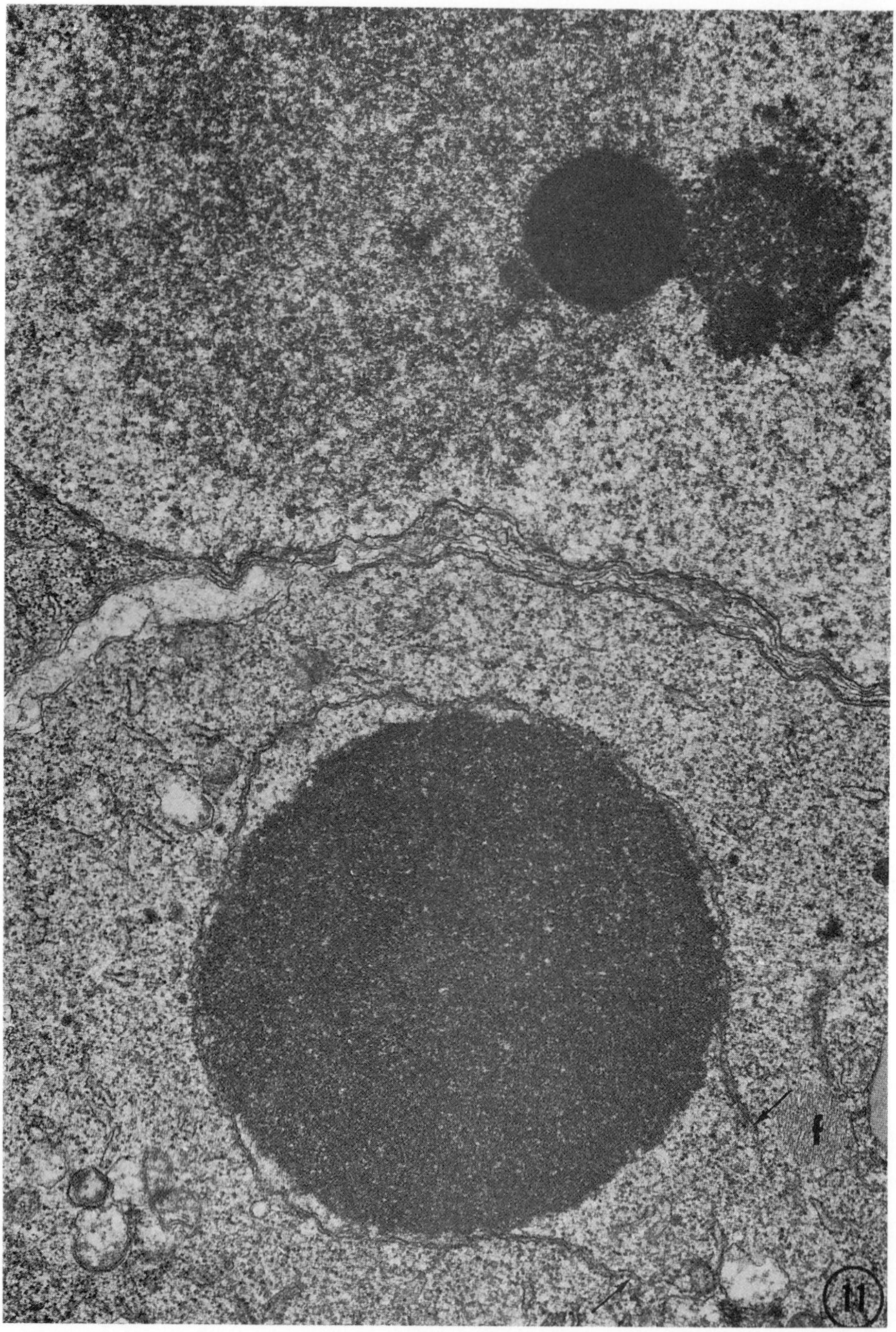

Fig. 11. Two stages of the nuclear lesion. In the upper part a neuron with the three zones: a, b, c. In the lower part the nucleus is now occupied by a large condensed c zone surrounded by the nuclear membrane open at the point indicated by the arrows.

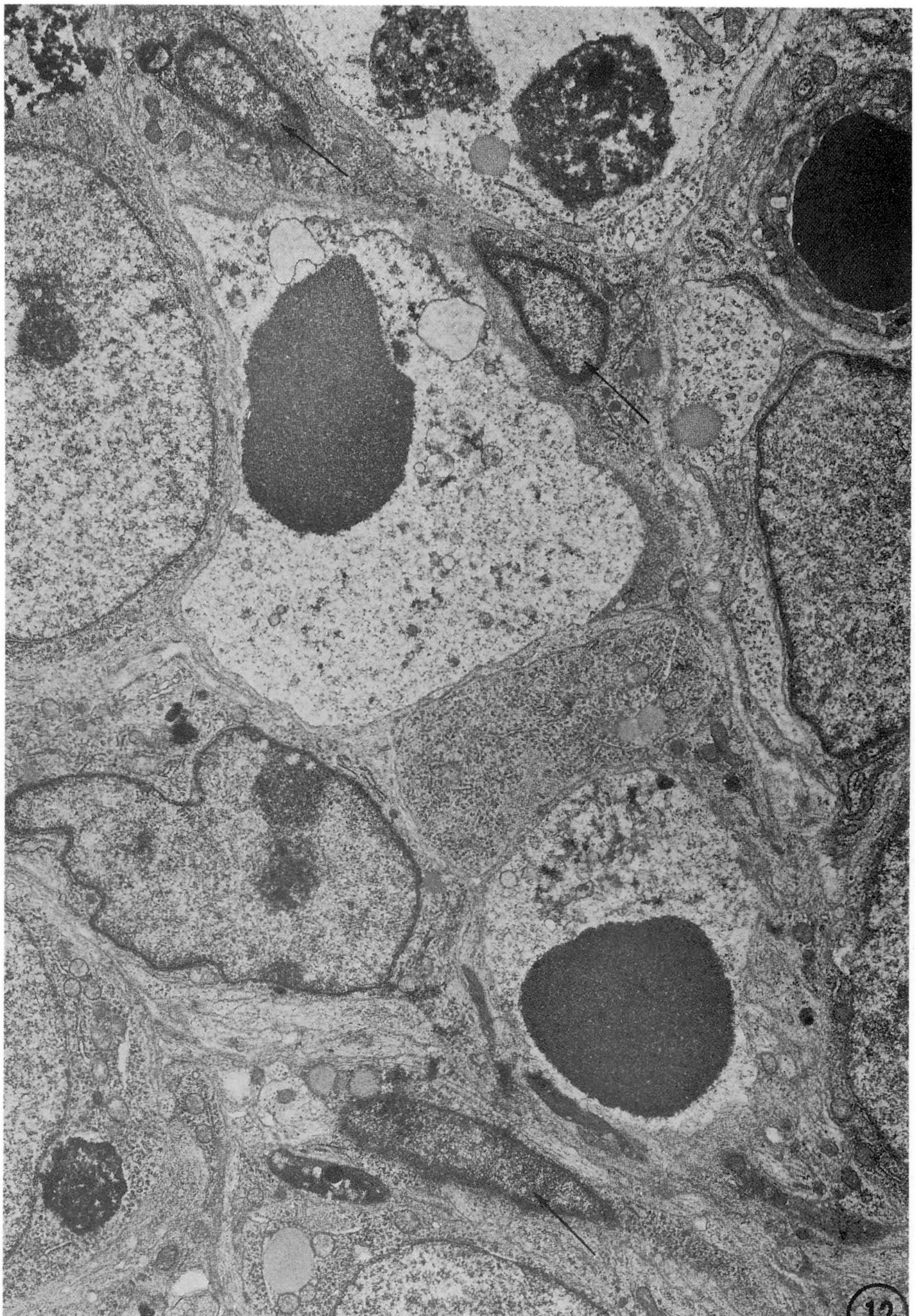

Fig. 12. Superior cervical ganglion of 1-day-old mouse (24 h after As injection). Advanced lesions in several neurons shown at low magnification. Note in contrast the preservation of satellite cells (arrows).

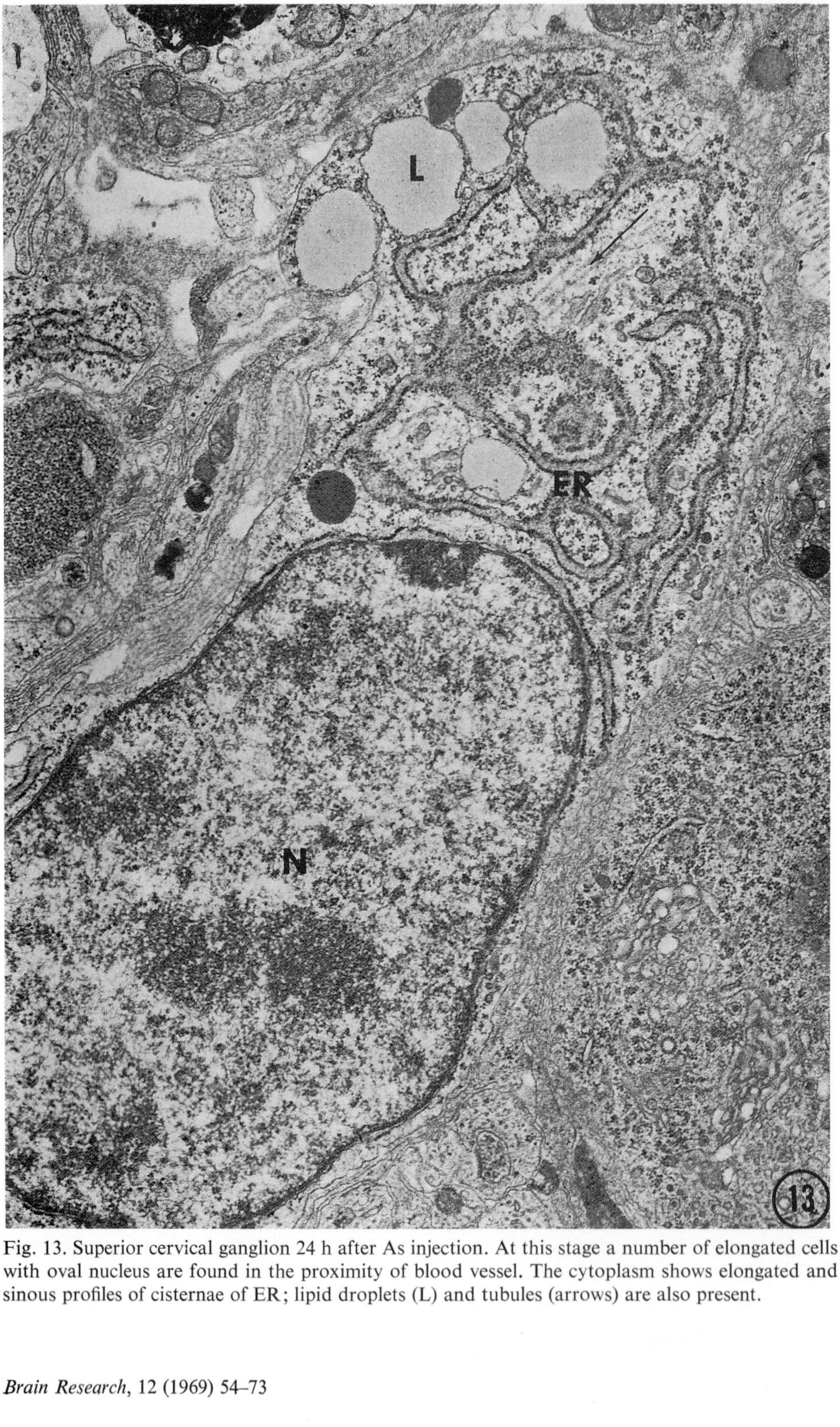

Fig. 13. Superior cervical ganglion 24 h after As injection. At this stage a number of elongated cells with oval nucleus are found in the proximity of blood vessel. The cytoplasm shows elongated and sinous profiles of cisternae of ER; lipid droplets (L) and tubules (arrows) are also present.

very pale nucleus, round or with irregular contour, limited by a distinct double membrane.

The nucleolus is oval or round in shape and presents a typical structure consisting of nucleolonema and a less dense pars amorpha. The cytoplasm is dense and very rich in organelles; numerous mitochondria, round or elongated, are seen; the Golgi apparatus appears as a unique large complex or distributed in multiple small areas (Fig. 1) formed by vacuoles and interconnected tubules.

Large droplets of lipids are also found in proximity to the nucleus. The endoplasmic reticulum is represented by short and disorganized cisternae with a few ribosomes attached. The free ribosomes are always arranged in clusters and rosettes. Where the cytoplasm is less dense, neurofilaments can be seen running in all directions, sometimes sparse and sometimes in bundles. The smaller neurons have a round or oval nucleus, always with regular contours. The nucleolus is similar in size and structure to that seen in larger neurons (Fig. 2). The cytoplasm, less abundant, sometimes only a thin sheet around the nucleus, rarely contains lipid droplets or dense bodies. The free ribosomes are very abundant and typically arranged in rosettes. The profiles of cisternae of the endoplasmic reticulum are elongated and parallel to the nuclear envelope.

Experimental ganglia: early lesions

The most common and impressive alteration is observed in the fine structure of the nucleolus. The nucleolar changes to be described below are seen in the majority of the neurons, even when other nuclear or cytoplasmic alterations are still not appreciable. The nucleolus appears denser and internally disorganized; it is difficult now to distinguish the nucleolonema from the pars amorpha in view of the presence of clumping, irregular aggregations (Fig. 3) of very coarse, granular material. In some cells this initial alteration in the fine structure of the nucleolus is accompanied by a disorganization of the free ribosomes in the cytoplasm, no longer grouped in rosettes but sparse and uniformly distributed through the cytoplasm (Figs. 3, 4). In other neurons, also 12 h after the As injection, the nucleolar lesion is more advanced; the typical organization of the nucleolar material is completely changed and now abnormal structures are seen within the nucleus. Some of the nucleolar material is condensed to form one or two very dense round bodies (zone b, Fig. 4) which are surrounded by coarse and dense granular material (zone a). The nucleus gives the impression of being denser than normal and the chromatin seems to clump in large areas, more pronounced in the vicinity of the altered nucleolus. At this stage, the nuclear envelope appears still intact and the plasma membranes are also apparently normal. The only cytoplasmic alteration is the above mentioned disorganization of ribosomes.

Advanced lesions

Twenty-four hours after the As injection, the nucleoli of virtually all neurons are abnormal and present various degrees of disorganization. Three distinct formations

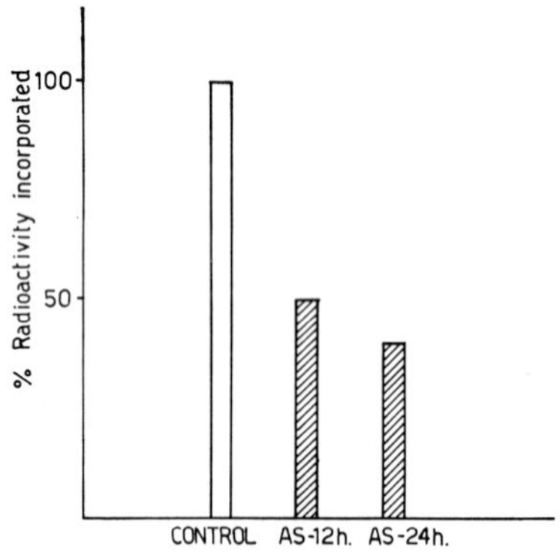

Fig. 14. *In vivo* incorporation of [^{14}C]uridine into superior cervical ganglia of 1-day-old mice. The isotope was injected 3 h before sacrifice. The NGF-antiserum was injected 12 and 24 h before sacrifice in the experimental animals. Normal rabbit serum was injected in controls. The radioactivity incorporated per mg protein in the experimental ganglia is expressed as per cent of the respective controls.

are now observed in most of the nuclei: the 2 zones a and b, originated from separation and condensation of the nucleolar material, and a third large area of less dense granular material (zone c). The prominent aggregate of granules in the nucleoplasm appears to have a spatial relationship with the nucleolar body (zone b) which is frequently included in the c zone or located at its periphery (Figs. 5, 6, 11).

Around this progressively larger granular aggregate (c) the nuclear chromatin appears somewhat denser. Inspection of these formations at high magnification (Figs. 7, 8) shows that the 3 zones (a, b and c) have a different organization. At subsequent stages, zone c tends to expand progressively and finally occupies most of the nuclear space; at the same time it becomes denser (Figs. 9, 10), while the nuclear envelope breaks in one or more points and the cytoplasmic and nuclear material become intermixed. The initial cytoplasmic alterations are now more pronounced; the free ribosomes are completely disorganized and the mitochondria begin to show signs of degeneration. Vacuolization, pyknosis, rupture of the plasma membranes and other signs of cytolysis are seen in many neurons. At this stage (24 h) numerous macrophages and histiocytes are seen in the ganglia (Figs. 12, 13).

Impaired RNA synthesis

The very early and diffuse changes of the fine nucleolar structure following As-administration correlate well with the severe impairment of RNA synthesis in the ganglia under the same experimental conditions. Control and experimental mice were given labeled uridine (5 μC/g i.p.) 3 h before sacrifice. Superior cervical ganglia were dissected out, weighed and homogenized in cold TCA. The TCA precipitate was repeatedly washed, finally dissolved in formamide and counted in a liquid scintillation system. As a control, pieces of liver from normal and As-treated animals were similarly processed. Uridine incorporation into TCA-insoluble material was practically identical in the liver of normal and experimental animals; in the superior cervical ganglia, however, already 12 h after the As injection there was a marked and constant reduction of uridine incorporation. As shown in Fig. 14, the radioactivity incorporated (per unit fresh weight) in the experimental ganglia was about 50% of the respective control values after 12 h and still reduced to 40% after 24 h.

DISCUSSION

Since the early discovery of the striking cytotoxic effect produced in the sympathetic chain ganglia of newborn mammals by an antiserum to the nerve growth factor, the rapid appearance of morphological lesion of the treated neurons was stressed. Twelve hours after the first subcutaneous injection of the antiserum a sharp decrease in mitotic activity of sympathetic neuroblasts was observed; the number of mitotic figures still decreases in the following 24 h and is practically nil in the following days[5,6]. At the same time, differentiative processes, which are very active in control ganglia, cease in the experimental animals and progressively larger numbers of neuroblasts undergo disintegration and are removed by macrophages which surround the necrotic cells. Satellite cells are spared such effects during the early phases and only in subsequent days undergo atrophy. The rapidity of the cytotoxic effects of the As on sympathetic neuroblasts *in vivo* as well as *in vitro* was also well documented by Sabatini *et al.* in an electromicroscopic investigation. Already a few hours after the antiserum injections, alterations in the fine structure of some neurons were observed[10]. These changes were predominantly seen in the nuclear compartment and consisted in folding of the nuclear envelope and condensation of chromatin into large and small masses. Comparatively less striking were the cytoplasmic alterations at these early stages.

In a subsequent investigation it was observed that these early ultrastructural changes following the As treatment are accompanied by a marked impairment of RNA synthesis in treated ganglia[11].

The results of the present investigation focus attention on the marked and very early changes in the fine structure of the nucleoli. Twelve hours after the As injection, in most neurons the nucleolus shows remarkable morphological alterations which become more prominent at later stages. The results reported above seem to indicate that the alteration of the structural configuration of the nucleoli is the first observable cytotoxic effect of the antiserum. The various nucleolar components, which in the normal functional state are related to each other in a well-defined macromolecular organization, undergo a process of separation and coalescence into masses of different density and structure. Around these abnormal nucleolar formations the nuclear chromatin seems to undergo a process of condensation and alteration which increases progressively with time.

The results of incorporation experiments add further evidence for the severe and early nucleolar damage which follows the As treatment. At 12 h, uridine incorporation is reduced to about 50% of the control values. Taking into consideration that satellite cells are not affected by the antiserum, the sharp decrease in uridine incorporation suggests that RNA synthesis is severely impaired in most of the nerve cells. While there is no doubt that the cytotoxic effect elicited by the NGF-antiserum is due to an antigen–antibody reaction taking place in the target nerve cells, the sequence of the structural and functional alterations differs from that normally observed in other immunological reactions. In our experimental system it appears to be an initial localization of ultrastructural changes in the nuclear compartment and particularly in the nucleolar structure. Lesions of the nucleoli of mammalian cells, comparable to those described above, were observed with actinomycin D and with some other toxic compounds[8,9,12], known to inhibit nuclear RNA synthesis. Although the mechanism of action of the NGF-antiserum must obviously be different from those of these simple chemicals, even in this case the very early loss of structural integrity of the nucleolus may be related to a derangement of RNA synthesis in the neurons affected by the As. This may simply mean that the nucleolar structure reflects an active functional state of this organelle and when its synthetic processes are blocked, its structural organization collapses. At present we have no definite evidence as to the site of action of the NGF in the receptive nerve cells. Although the cell membrane would be the favored candidate, the possibility that the antibodies may get inside the cell, through pinocytosis or other processes, and reach their specific antigen in some cell compartment cannot be excluded. Experiments now in progress with ferritin-labeled As are expected to shed some light on this problem.

SUMMARY

Antiserum to the nerve growth factor (NGF) produces a severe cytotoxic effect on sympathetic nerve cells when injected into newborn mice. The early cytological lesions were studied by electron microscopy in superior cervical ganglia of 1-day-old mice 12 and 24 h after antiserum treatment. In line with previous observations, early and marked ultrastructural changes were observed in the nuclear compartment. Alterations in the fine structure of nucleoli of the nerve cells are very prominent by 12 h after the As injection. The early alterations consist of condensation and disarrangement of the nucleolar components, which become progressively more severe in the following hours. Distinct dense bodies form from the separation and condensation of the nucleolar material and eventually the nuclear envelope breaks down and cytoplasmic and nuclear material mix. In later stages, marked disorganization of ribosomes and mitochondria is apparent in the cell cytoplasm. Incorporation experiments show that RNA synthesis is markedly reduced in the experimental ganglia already 12 h after the As injection.

ACKNOWLEDGEMENTS

Research sponsored by the Air Force Office of Scientific Research, Office of Aerospace Research, U.S. Air Force, under AFOSR Grant Nr. 1182-67, and by a grant from the Consiglio Nazionale delle Ricerche, Rome.

REFERENCES

1 ANGELETTI, P. U., LEVI-MONTALCINI, R., AND CALISSANO, P., The nerve growth factor (NGF): chemical properties and metabolic effects, *Advanc. Enzymol.*, 31 (1968) 51–75.
2 COHEN, S., Purification of a nerve growth promoting protein from the mouse salivary gland and its neurocytotoxic antiserum, *Proc. nat. Acad. Sci. (Wash.)*, 46 (1960) 302–311.
3 LEVI-MONTALCINI, R., The nerve growth factor: its mode of action on sensory and sympathetic nerve cells, *Harvey Lect.*, 60 (1966) 217–259.
4 LEVI-MONTALCINI, R., AND ANGELETTI, P. U., Immunosympathectomy, *Pharmacol. Rev.*, 18 (1966) 619–628.
5 LEVI-MONTALCINI, R., AND BOOKER, B., Destruction of the sympathetic ganglia in mammals by an antiserum to the nerve-growth promoting factor, *Proc. nat. Acad. Sci. (Wash.)*, 42 (1960) 384–391.
6 LEVI-MONTALCINI, R., AND COHEN, S., Effects of the extract of the mouse salivary glands on the sympathetic system of mammals, *Ann. N.Y. Acad. Sci.*, 85 (1960) 324–341.
7 MILLONIG, G., A modified procedure for lead staining of thin sections, *J. biophys. biochem. Cytol.*, 11 (1961) 736–739.
8 REDDY, J., HARRIS, C., AND SVOBODA, D., Inhibition by lasiocarpine of RNA synthesis, RNA polymerase and induction of tryptophan pyrrolase activity, *Nature (Lond.)*, 217 (1968) 659–661.
9 REYNOLDS, R. C., MONTGOMERY, P. O'B., AND KARNEY, D. H., Nucleolar 'caps' — a morphologic entity produced by the carcinogen 4-nitroquinoline *N*-oxide, *Cancer Res.*, 23 (1963) 535–538.
10 SABATINI, M. T., DE IRALDI, A. P., DE ROBERTIS, E., Early effects of the antiserum (AS) against the nerve growth factor (NGF) on the structure of sympathetic neurons, *J. exp. Neurol.*, 12 (1965) 370–383.
11 SABATINI, M. T., LEVI-MONTALCINI, R., E ANGELETTI, P. U., Effetti precoci del siero anti-Nerve Growth Factor (anti-NGF) sulla cellula nervosa, *Ann. Ist. sup. Sanità*, 2 (1966) 349–355.
12 SCHOEFL, G. I., The effect of actinomycin D on the fine structure of the nucleolus, *J. Ultrastruct. Res.*, 10 (1964) 224–243.

PHARMACOLOGICAL REVIEWS
VOL. 18, NO. 1, PART I

H. IMMUNOSYMPATHECTOMY[1]

RITA LEVI-MONTALCINI AND PIETRO U. ANGELETTI

Department of Zoology, Washington University, St. Louis and Istituto Superiore di Sanità, Roma

"The sympathetic division of the autonomic system, despite its extensive range of action, is not essential for existence. The two ganglionic chains, as well as the prevertebral ganglia have been removed from animals and yet, within the confines of the laboratory, such altered animals seem not to differ from the normal. When they are exposed to heat or cold, however, or when subjected to lowering of blood through hemorrage, they show definite deficiencies. These deficiencies more clearly define the statement, previously made, that the autonomic system is concerned with preserving steady states in the fluid matrix of the organism; it is especially the sympathetic division which has the general function of maintaining homeostasis" (4).

This paragraph in the first chapter of Cannon and Rosenbleuth's classical monograph on autonomic neuroeffector systems, summarizes the results of a long series of experiments initiated by Cannon and his associates in 1929 and pursued in the subsequent years. The experiments consisted in the surgical removal of the sympathetic ganglia in adult cats. They were performed step-wise and brought to completion in a 6-month period. Practically all of the paravertebral and prevertebral ganglia were removed, and the animal was then tested under various conditions and compared to intact animals. Thirty years later, in 1959, a much simpler technique became available to remove the sympathetic system, or rather to prevent its differentiation in newborn animals, without otherwise interfering with the differentiation and growth of other structures or organs (6, 15). This technique consists in the injection into newborn animals of an antiserum to a protein which proved to be essential for the life of the sympathetic nerve cells (13). The almost total and permanent destruction of the para- and prevertebral sympathetic ganglia ensues. No noxious effects or other effects are apparent in other sectors of the nervous system or in any other organ or system. From 1959 to the present time, thousands of mice and rats, deprived in this way of the sympathetic division of the autonomic system, have been raised in our laboratories and elsewhere and used for structural, physiologic, and pharmacologic investigations. Although at present most of these studies are still in progress and far from being completed, they have already brought some significant contributions to our knowledge of the structural and physiologic organization of the sympathetic system. In the following pages we will consider at first the morphologic changes produced by the antiserum in the sympathetic ganglia at different time intervals after the treatment. Then we will consider the results of physiologic and pharmacologic studies performed in the immunosympathectomized animal.

[1] The work performed in our laboratory was supported by grants from the National Institute of Neurological Diseases and Blindness (NIM) and from the National Science Foundation.

PRODUCTION AND ASSAY OF THE ANTISERUM TO THE NERVE GROWTH FACTOR

A specific antiserum to the nerve growth factor (NGF) was first prepared by Cohen (6). For routine preparation of the antiserum 4 mg of the purified NGF (active *in vitro* at a concentration of 0.05 μg/ml) are dissolved in 1 ml of complete Freund adjuvant. The thick emulsion is then injected into the foot pads of a rabbit. Three weeks later, the rabbit is bled and the serum tested in tissue culture with the assay devised by R. Levi-Montalcini, currently used in all studies on the NGF (12). The titer of the antiserum is defined as the reciprocal of the maximal dilution of the serum which completely inactivates 1 biologic unit of the NGF. Hyperimmunization of the animal is achieved by two or three subcutaneous injections of the NGF into the rabbit 3 weeks after the first injection. As a rule, the titer of a good antiserum ranges between $\frac{1}{1000}$ and $\frac{1}{2000}$. The antiserum is then injected subcutaneously in newborn mice or other newborn mammals in the amount of 0.05 ml per g of body weight. Daily injections of the antiserum for 2 to 3 days, result in the permanent destruction of almost all of the sympathetic ganglia in mice. This was ascertained by extensive histologic studies performed in treated animals at long intervals after the treatment. Two-year-old mice, injected at birth with the antiserum, are deprived of practically the entire sympathetic para- and prevertebral ganglia. The persistence of peripherally located adrenergic neurons, will be considered below.

The effects of the antiserum were also explored in other newborn mammals. In rats, the species most extensively investigated besides mice, the effects are similar, although not as extensive as in mice. Rabbits and kittens were injected at birth and explored a week later. At this time, the sympathetic cell population of the superior cervical ganglia was reduced to less than 10% of the controls. No long-range experiments were performed to ascertain whether this reduction of the nerve cell population persists as such in the adult animal.

A much less drastic effect is obtained in the adult mouse. The repeated injection of the antiserum for up to a month results in a moderate atrophy of the sympathetic ganglia. Volume determinations indicate that the atrophy is of the order of 40% (15). Since the treated animals were sacrificed soon after the injections, it remains to be established whether the atrophic cells undergo any recovery in the following days. Lower vertebrates such as amphibia, reptiles, and birds do not react to the antiserum.

MORPHOLOGIC CHANGES INDUCED BY THE ANTISERUM IN THE SUPERIOR CERVICAL GANGLION AND IN THE THORACIC CHAIN GANGLIA

From the beginning of these investigations, all morphologic studies centered on the superior cervical, stellate, and thoracic paravertebral ganglia in mice. On first inspection, it appeared that the effects of the antiserum were equally extensive in all para- and prevertebral ganglia. More recent morphologic and pharmacologic studies indicate differences which will be considered in detail in a following section. The results to be reported here apply therefore to these ganglia in mice. Similar results were obtained in rats, rabbits, and kittens.

Twelve hours after the first subcutaneous injection of the antiserum, a decrease in mitotic activity becomes evident. Between the second and the third day, the mitotic activity still decreases, and it is practically nil in the subsequent 2 days. At the same time, the processes of differentiation, which are very active in control ganglia, cease in the experimental animals, and a progressively larger number of neuroblasts undergo disintegration and are removed by macrophages which surround the necrotic cells. Between the fourth and the ninth day, the processes of cell atrophy and death continue at a somewhat slower pace because of the drastic reduction in the cell population produced in the first days of the treatment. Whereas the nerve cells show all these signs of disintegration, satellite cells are spared such effects during the early phases of the treatment, but in subsequent stages they also undergo atrophy and the ganglia are reduced to exceedingly small and sclerotic nodules (figs. 1–4). On histologic analysis, the ganglia are seen to consist of a small population of glial cells and few sparse nerve cells of a size considerably smaller than in controls.

Electronmicroscopic analysis of the same ganglia, performed by Sabatini *et al.* (17) gave evidence of the extraordinary rapid effects of the antiserum *in vivo* as well as *in vitro*. The injection of the antiserum in the infant mouse results 2 hr later in "a striking folding of the nuclear envelope and the condensation of chromatin into large and small masses .... In the cytoplasm there is dilatation of the cisternae of the vacuolar system including the perinuclear ones forming the nuclear envelope. The plasma membrane is corrugated and in some neurons it may show an intense process of vesciculation with formation of pynocitic vesicles" (17). Between 24 and 30 hr, most of the nerve cells are in an advanced state of cytolysis while the Schwann cells, which are not altered, show masses of engulfed material and cytolysosome particles. The effects of the antiserum are even more impressive *in vitro* and occur earlier. In the presence of the antiserum, extensive cytolytic lesions occur in ganglia cultured 4 hr *in vitro*. The effects *in vivo* and *in vitro* are suggestive of an antigen-antibody reaction.

DIFFERENTIAL RESPONSE OF THE SYMPATHETIC GANGLIA OF NEWBORN MICE AND RATS TO THE ANTISERUM

In most instances, the cell population of the superior cervical ganglion in mice is reduced to 2 to 5% of controls, while in rats the residual cell population amounts to 10 to 15%. Similar mass atrophy occurs in the stellate as well as in all other paravertebral ganglia. Equally affected is the prevertebral complex, which in rodents consists of an irregularly shaped ganglionic mass in close apposition to the celiac artery (fig. 4). From this large ganglion, two thin fibrous nerve strands emerge and adhere to the aorta. At the bifurcation of the aorta, the two strands normally join and a nodule of variable size but usually barely detectable is present. Small cell agglomerates of the same or even larger size than the above nodule, are irregularly scattered along the fibrous strands. In rodents therefore, most if not all the innervation of the abdominal viscera is from the celiac ganglionic complex, which as reported above, is severely affected by the antiserum to the antiserum to the NGF. The near-total atrophy of all para- and prevertebral

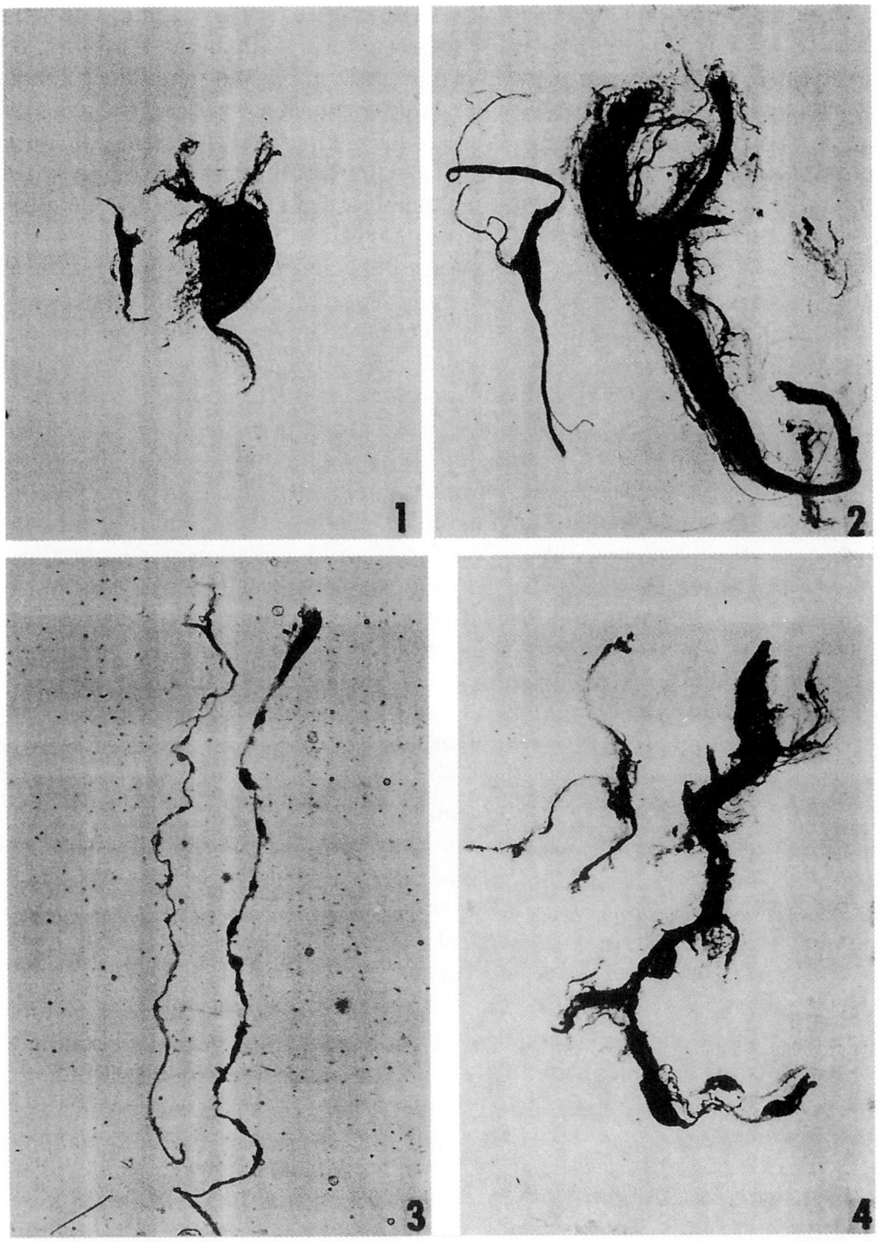

FIGS. 1–4. Whole mounts of superior cervical ganglia of control mice and mice treated with the antiserum to nerve growth factor (fig. 1); of superior cervical ganglia of control and treated rats (fig. 2); of stellate and thoracic chain ganglia of control and treated mice (fig. 3); and of celiac ganglia of control and treated mice (fig. 4). In all figures, the atrophic ganglia of the immunosympathectomized animals are to the left and the control ganglia to the right.

ganglia (from an embryologic viewpoint the superior cervical ganglia belong to the prevertebral chain ganglia) is obtained only if the antiserum is administered immediately after birth. Less complete and variable results are obtained when the treatment starts a few days later. In these cases, with increasing age the residual cell population becomes progressively larger, particularly in the celiac ganglionic complex, where nerve cells develop an antiserum resistance sooner than in other ganglia. We believe that these differences merely reflect a different degree of maturation of the sympathetic neurons at the time of birth in different areas.

An interesting instance of antiserum resistance is observed in the peripherally located adrenergic ganglia which provide for the innervation of the vas deferens or the uterus. A histochemical investigation by Hamberger *et al.* (10) showed the persistence of the fluorescent adrenergic nerve fibers around the vas deferens of rats treated with antiserum. The site of origin of these fibers was traced to a large ganglionic complex in close apposition to the caudal portion of the vas deferens. This ganglion is only slightly reduced in size in the experimental animals and the individual neurons are apparently normal. Hamberger *et al.* concluded that these adrenergic neurons are structurally and functionally different from other adrenergic neurons. Such findings, as they point out, "call for caution in interpretation of pharmacological data obtained with this test organ" (10). Experiments to be reported in the following section further stress the marked differences between this peripheral component of the sympathetic nervous system and the para- and prevertebral chain ganglia.

CONTENT, UPTAKE, AND METABOLISM OF NOREPINEPHRINE (NE) IN PERIPHERAL ORGANS OF IMMUNOSYMPATHECTOMIZED ANIMALS

A marked reduction in the NE content in the heart, spleen, and lungs in immunosympathectomized mice and rats was first reported by us and confirmed in other laboratories (13, 22). No changes were detected in the NE content in the central nervous system of the same animals. These results are in agreement with our previous findings that no morphologic lesions are apparent in the central nervous system of animals treated with the antiserum (15).

It is generally agreed that the NE content of various organs depends chiefly on the supply of adrenergic fibers (7). Degeneration of the adrenergic nerves to an organ results in the disappearance of NE, while the adrenaline (E) content of the same organ, which as a rule amounts to 2 to 15% of the total catecholamine content, is not changed (8). Also it is generally accepted that the neurotransmitter is accumulated in the peripheral part of the axons and is stored in a particulate form in the nerve axons. Recently, the elegant fluorescence technique devised by Falck *et al.* (9) provided a most valuable tool for visualizing the localization of the neurohormone in peripheral organs and tissues. A systematic study by Hamberger *et al.* (10) of immunosympathectomized animals gave evidence for the total absence of adrenergic fluorescent fibers in the iris (figs. 5, 6), in the submaxillary salivary glands, and the heart and skeletal muscle vessels, whereas a normal contingent of fluorescent fibers was found, as already mentioned, around the vas deferens, in the hypothalamus, and in the medulla. Also the system of

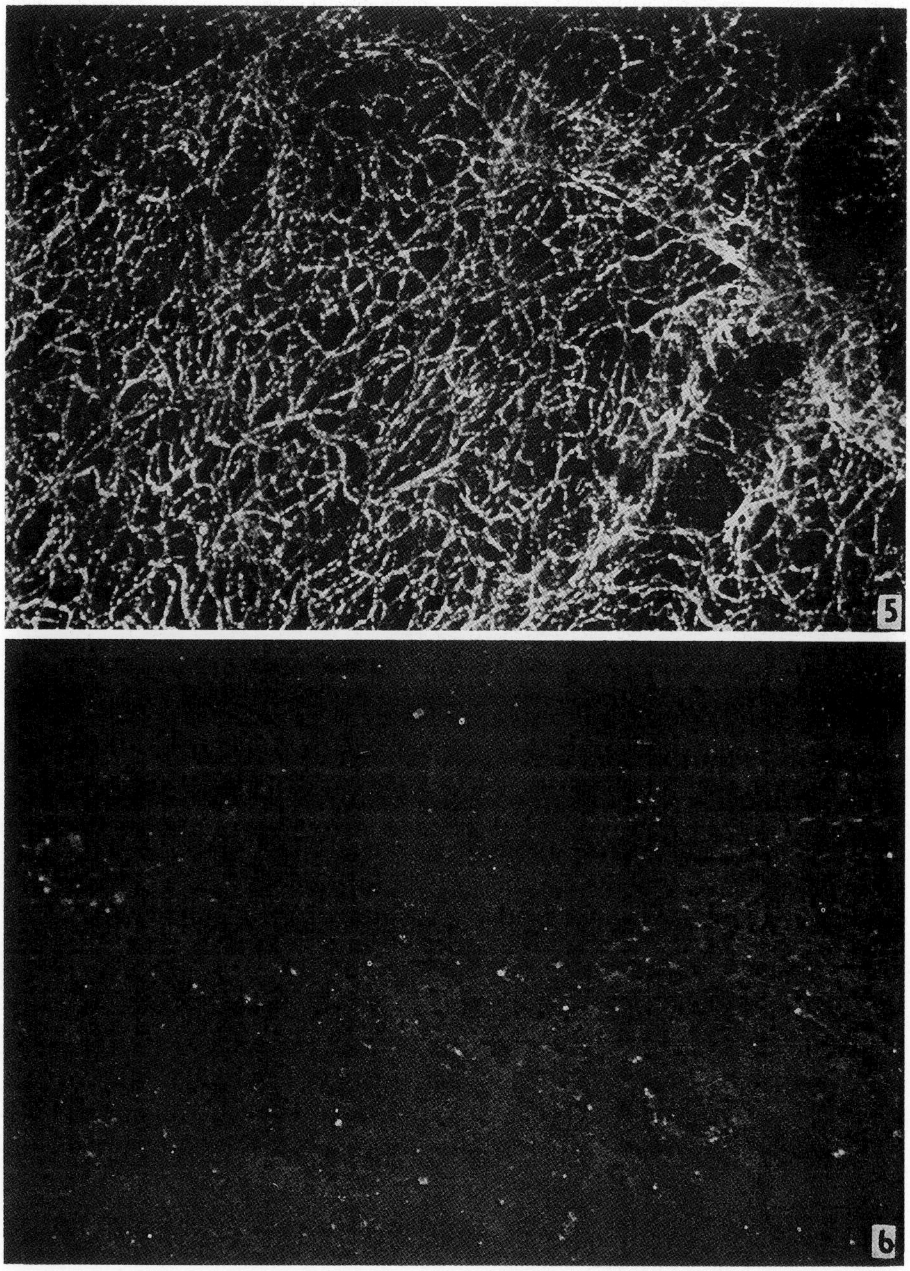

FIGS. 5 AND 6. Stretch preparation of iris mounted whole. The striking network of fluorescent terminals in the normal rat (fig. 5) is completely lacking after immunosympathectomy (fig. 6). (From: Hamberger, B., Levi-Montalcini, R., Norberg, K. A. and Sjöqvist, F.: Monoamines in immunosympathectomized rats. Int. J. Neuropharmacol. **4**: 91–95, 1965.)

fluorescent adrenergic nerve terminals around the intramural parasympathetic nerve cells first described by Norberg (16) seems intact or at least partly intact (the technique does not allow for a rigorous quantitative determination).

A further confirmation of the depletion of NE from storage sites in the peripheral organs of immunosympathectomized rats came from electron microscopic studies of particulate material collected by differential centrifugation through a sucrose density gradient from the homogenate of heart on normal and treated rats. Dense-core granular vesicles are present in large number in controls, whereas they occur extremely rarely if at all in the heart of rats treated with the antiserum. The same vesicles are entirely absent from the submaxillary fraction of immunosympathectomized mice and rats but are intact in the vas deferens (19). Disappearance of the storage sites from peripheral organs of rats treated with the antiserum was further demonstrated by Shore *et al.* (18) in studies of binding and release of metaraminol.

The uptake of labeled NE in peripheral organs of experimental and control rats has been studied in many laboratories. All authors (11, 19, 22) report a marked reduction (up to 98%) in the H^3-NE uptake by organs such as the heart, spleen, and salivary glands in the animals treated with the antiserum, while the uptake is normally reduced to 50 to 60% of controls in the intestine and is normal or even increased in the vas deferens and uterus.

Of considerable interest was a comparison by Carpi and Oliverio (5) of the excretion of NE in immunosympathectomized and control rats, under basal and emergency conditions and after adrenal demedullation. The elimination of NE in the experimental animal under basal conditions is reduced below the level of sensitivity of the method but reaches measurable level after injection of amphetamine. Adrenal demedullation calls forth a stable elimination of NE, while the excretion of E falls below the limits of sensitivity of the method in both treated and control rats. Urinary E reappears in both groups after injection of amphetamine. A paradoxical response to amphetamine injection took place in the immunosympathectomized and demedullated rats. The urinary excretion of NE rose in fact to higher levels in the treated animals than in controls. The authors write that "these data indicate that in the immunosympathectomized rats there exists a noradrenergic system, the activity of which, latent in basal conditions, becomes evident in emergency conditions." They suggest the extra-adrenal chromaffin system as the possible vicarious system. The alternative hypothesis (20) that the urinary NE in the experimental rats could originate from the prevertebral ganglia, seems difficult to accept in view of the findings of the almost total lack of NE excretion in these rats under basal conditions. Also it would be difficult to explain how the residual population of small cells in the prevertebral ganglia could bring the NE urinary excretion above normal levels in the rats treated with the antiserum after demedullation and injection of amphetamine.

PHYSIOPHARMACOLOGIC CHANGES IN THE IMMUNOSYMPATHECTOMIZED ANIMAL

Observations of large colonies of immunosympathectomized animals over a period of 5 years fully confirm the original findings by Cannon that elimination of sympathetic nerve function does not affect the vitality of animals kept under the sheltered conditions of the laboratory. We may add, on the basis of our experience, that the lack of the sympathetic system since birth does not interfere in any appreciable way with the normal development of the animals. Clear-cut differences, however, can be brought to light in conditions aimed to challenge the sympathetic function. At present only few preliminary studies have been performed along this line. They deal mainly with the study of the reactivity of the vascular system to electrical or chemical stimulation.

Intra-arterial injections of two doses of NE resulted in a considerably higher pressor response in treated than in control rats, while no differences were observed to injections of E (3).

Zaimis *et al.* (21, 22) conducted a series of investigations on immunosympathectomized rats. On gross inspection, enophthalmos was the only feature which distinguished the treated rats from controls. Under anesthesia, the resting heart rate of the immunosympathectomized rats was normal but the average resting blood pressure was 70 mm Hg while in controls was 100 mm Hg. The pressor as well as the positive inotropic responses to injected E and NE were significantly greater in treated than in control rats.

Peripheral responses were studied by Brody (3), using a technique devised by the author for perfusion of the hindquarters. The intravenous injection of the ganglionic stimulant dimethylphenyl piperazine did not alter the perfusion pressure in rats treated with the antiserum, while it increased the blood pressure to the same extent as in controls. This latter effect was abolished in the experimental rats by adrenal demedullation, thus indicating that the effect on blood pressure was obtained through release of catecholamines from the adrenal medulla rather than through stimulation of adrenergic nerve fibers. The electrical stimulation of the lumbar sympathetic chain evoked a rather large pressor response in controls but no response in the immunosympathectomized rats. The magnitude of the potentials recorded from the lumbar sympathetic nerves of the immunized animals was considerably smaller than the magnitude of the discharge in control animals (2). This low activity which failed to produce peripheral vasoconstriction upon nerve stimulation is likely to result from the small nerve cell population resistant to the antiserum treatment. As an alternative hypothesis, the author considered the possibility that low activity might result from fibers which "subserve a function other than the release of norepinephrine." Under normal conditions, the low potentials of these fibers may be masked by the potentials of the larger contingent of adrenergic nerve fibers (2).

The heart rate, the electrocardiographic pattern, and the body temperature of experimental and control rats were recorded under normal conditions or when the animals were exposed to a cold environment (1). No significant differences were found between the two groups. A dramatic difference in the electrocardiographic pattern and cold resistance became apparent following adrenal demedullation. Under these conditions, the treated animals cannot maintain body temperature, which falls rapidly and already 2 hr after exposure to cold 4°C fluctuates between 30° and 24°, whereas it remains normal in controls. After 8 hr, the temperature reaches a minimum of 14°, whereas in controls it is still 36°. In these conditions the electrocardiograms have an agonal pattern. The authors concluded that in the immunosympathectomized animals the adrenal medulla displays a role as a catecholamine-releasing organ.

CONCLUDING REMARKS

It may be at present premature to attempt an evaluation of the services of the immunosympathectomized animal as a tool to explore the structure and function of the sympathetic nervous system. The investigations, listed rather than critically evaluated in the previous pages, indicate the wide range of areas which can benefit from the availability of these animals. Their advantages over animals surgically deprived of the sympathetic ganglia are too obvious to need comments. It may instead be of interest to consider one aspect of immunosympathectomy which appears at first sight to limit the usefulness of this tool of investigation. A small nerve cell population ranging in size from 2 to 10 % of the controls persists in para- and prevertebral sympathetic ganglia of treated mice and a somewhat larger cell population persists in the same ganglia of the treated rats. Pharmacologic and physiologic tests suggest that this residual cell population has little if any functional activity. A quite different situation occurs in the peripherally located adrenergic neurons which provide to the innervations of the vas deferens and the uterus. These large cell agglomerates are only slightly reduced in the immunosympathectomized animals and the individual neurons are comparable in size and morphologic features to controls. Histochemical and pharmacologic studies also showed that this nerve cell population is not affected by the antiserum treatment. These results indicate a clear-cut difference between nerve cells hitherto considered similar and labeled as "adrenergic neurons." It remains to be seen whether the small residual cell population of adrenergic neurons in the para- and prevertebral ganglia belongs to this same category as the adrenergic neurons which provide to the innervation of the vas deferens or the uterus. At present we would favor another alternative, namely that these para- and prevertebral sympathetic neurons owe their resistance to the antiserum to the fact that they have already attained a considerable degree of differentiation when the treatment was first initiated. The size of this population varies considerably if the antiserum is injected in the first day after birth or some time later. In the first instance the residual adrenergic neurons are in number of 2 to 3 % but their number may increase up to 30 % if the treatment starts 7 to 10 days after birth. This seems not to be the case for the ganglia which innervate the vas deferens and the uterus.

It can be anticipated that the immunosympathectomized animals as well as animals with greatly enlarged sympathetic ganglia (treated with the nerve growth factor) will provide a most valuable material for the exploration of all aspects of sympathetic nervous function. It is fortunate that colonies of animals practically deprived of the sympathetic system or conversely with hypertrophic ganglia became available at a moment when the interest in the structure and function of this system is again very vivid and the study is made possible by the elaboration of new and more precise techniques of investigation.

REFERENCES

1. Berti, F., Lentati, R. and Usardi, M. M.: Effects of cold exposure on heart function in immunosympathectomized rats. Med. Pharmac. Exper., in press.
2. Brody, M. J.: Electrical activity in sympathetic nerves of immunologically sympathectomized rats. Proc. Soc. Exp. Biol., N.Y. **114:** 565–567, 1963.
3. Brody, M. J.: Cardiovascular responses following immunological sympathectomy. Circulation Res. **15:** 161–167, 1964.
4. Cannon, W. B. and Rosenblueth, T.: Autonomic Neuro-effector Systems, XIV, 229 pp., Macmillan Co., New York, 1937.
5. Carpi, A. and Oliverio, A.: Urinary excretion of catechol amines in the immunosympathectomized rat. Balance phenomena between the adrenergic and the noradrenergic system. Int. J. Neuropharmacol. **3:** 427–431, 1964.
6. Cohen, S.: Purification of a nerve growth promoting protein from the mouse salivary gland and its cytotoxic antiserum. Proc. Nat. Acad. Sci., Wash., **46:** 302–311, 1960.
7. Euler, U. S. v.: Noradrenaline. Charles C Thomas, Springfield, Ill., XXI-382 pp., 1956.
8. Euler, U. S. v.: Neurotransmission in the adrenergic nervous system. Harvey Lectures Series **55:** 43–65, 1961.
9. Falck, B., Hillarp, N.-Å., Theme, G. and Torp, A.: Fluorescence of catecholamines and related compounds condensed with formaldehyde. J. Histochem. Cytochem. **10:** 348–354, 1962.
10. Hamberger, B. R., Levi-Montalcini, R., Norberg, A. and Sjökvist, F.: Monoamines in immunosympathectomized rats. Int. J. Neuropharmacol. **4:** 91–95, 1965.
11. Iversen, L. L., Glowinski, J. and Axelrod, J.: Reduced uptake of tritiated noradrenaline in tissues of immunosympathectomized animals. Nature, Lond. **206:** 1222–1223, 1965.
12. Levi-Montalcini, R.: Growth control of nerve cells by a protein factor and its antiserum. Science **143:** 1–6, 1964.
13. Levi-Montalcini, R. and Angeletti, P. U.: Noradrenaline and monoaminoxidase content in immunosympathectomized animals. Int. J. Neuropharmacol. **1:** 161–164, 1962.
14. Levi-Montalcini, R. and Angeletti, P. U.: Essential role of the nerve growth factor in the survival and maintenance of dissociated sensory and sympathetic nerve cells *in vitro*. Develop. Biol. **7:** 653–659, 1963.
15. Levi-Montalcini, R. and Booker, B.: Destruction of the sympathetic ganglia by an antiserum to a nerve-growth protein. Proc. Nat. Acad. Sci., Wash. **46:** 324–331, 1960.
16. Norberg, K. A.: Adrenergic innervation of the intestinal wall studied by fluorescence microscopy. Int. J. Neuropharmacol. **3:** 379–382, 1964.
17. Sabatini, M. T., Pellegrino de Iraldi, A. and De Robertis, E.: Early effects of the antiserum (AS) against the nerve growth factor (NGF) on the fine structure of sympathetic neurons. J. exp. Neurol., in press.
18. Shore, P. A., Busfield, D. and Alpers, H. S.: Binding and release of metaraminol: Mechanism of norepinephrine depletion by a-methyl-m-tyrosine and related agents. J. Pharmacol. **146:** 194–199, 1964.
19. Sjökvist, F., Titus, I., Michelson, A., Taylor, F. Jr. and Richardson, K. C.: Uptake and metabolism of *d,l*-norepinephrine 7 H^3 in tissues of immunosympathectomized mice and rats. Life Sci. **4:** 1125–1133, 1965.
20. Vogt, M.: Sources of noradrenaline in the "immunosympathectomized" rat. Nature, Lond. **204:** 1315–1316, 1964.
21. Zaimis, E.: The immunosympathectomized animal: a valuable tool in physiological and pharmacological research J. Physiol. **177:** 35–36, 1964.
22. Zaimis, E., Berk, L. and Callingham, B. A.: Biochemical and functional changes in the sympathetic nervous system of rats treated with nerve-growth factor-antiserum. Nature, Lond. **206:** 1220–1222, 1965.

Eur. J. Clin. Biol. Res. **16**, 866–874 (1971).

GROWTH REGULATION OF THE SYMPATHETIC NERVOUS SYSTEM : IMMUNOSYMPATHECTOMY AND CHEMICAL SYMPATHECTOMY

by P. U. ANGELETTI and R. LEVI-MONTALCINI

(Lab. Biologia Cellulare, CNR, Istituto Superiore di Sanita, Roma)*

SUMMARY

Two different methods for a selective and complete destruction of the peripheral sympathetic nervous system are briefly presented. By using a specific antiserum to a Nerve Growth Factor, an « Immunosympathectomy » is obtained in newborn mammals. A « Chemical sympathectomy » can be induced by administration of a dopamine-analogue, 6-hydroxy-dopamine.

RÉSUMÉ

Deux méthodes différentes pour une destruction sélective et complète du système nerveux sympathique périphérique sont brièvement décrites.

INTRODUCTION

Among other sections of the nervous system, the sympathetic division of the autonomic system, is perhaps one of the most widely investigated in terms of its structural and functional specificity. The work done in the past two decades, has shown that the growth and differentiation of sympathetic neuroblasts can be specifically altered in at least three different experimental conditions. The first condition occurs when a specific Nerve Growth Factor (NGF) is given to newborn animals during development of the sympathetic system. As a result of this treatment a dramatic hyperplasia and hypertrophy of all peripheral sympathetic ganglia ensues (6, 8). The second condition occurs when an antiserum to the NGF is administered to newborn animals; in this instance a specific cytotoxic effect leads to the death and disappearance of practically all peripheral sympathetic neurons (7, 9). Under a third experimental condition, an analogue of dopamine, 6-hydroxidopamine, given during development of the sympathetic system, causes a selective and complete destruction of all paravertebral and prevertebral ganglia (2, 3).

The « Immunosympathectomy » induced by NGF antiserum and the « chemical sympathectomy » by 6-hydroxydopamine will be reviewed in this article.

* Viale Regina Elena, 299, 00161 - Rome.

Immunosympathectomy

In 1959, the antiserum to the salivary NGF was produced by. Cohen by injecting the purified NGF protein into rabbits with Freund's adjuvant (4). The antibody titer was first assayed « in vitro ». The assay technique consists of the addition of 0.2 ml of the antiserum at progressively higher dilutions to 0.02 μg of the purified NGF in 0.2 ml of isotonic saline. The mixture is incubated at 20° C for one hour and then added to the culture medium in the proportion of one third of the total volume of the medium. Readings are made after twelve and twenty-four hours of incubation at 37 °C. The titer of the antiserum is expressed as the reciprocal of the highest dilution that results in complete inactivation of the NGF. Under the best conditions, the titer reaches the value of 2400 or even higher. Aliquots of this antiserum are then injected subcutaneously into newborn mice, rats or other mammals in the amount of 0.05 ml/g of body weight. Daily injections of the antiserum in newborn mice for five days result in the permanent destruction of 92-95 % of the sympathetic nerve cell population of the paravertebral sympathetic chain ganglia (9). A somewhat higher percentage of sympathetic nerve cells persists in the prevertebral ganglionic complex ; this varies according to the potency of the antiserum. As a rule, 10 % of these cells persists in the celiac ganglionic complex, the most extensively investigated of all prevertebral ganglia. The persistence of adrenergic ganglionic complexes that innervate the male and female sex organs is the only instance of adrenergic neurons resistant to the antiserum. These same cells fails to respond to the NGF.

A detailed histological study of the superior cervical ganglion and of other ganglia in mice treated with the antiserum and sacrified between twelve hours and ten days after the beginning of the treatment showed the rapid sequence of degenerative effects called for by the antibodies to the NGF. Twelve hours after the first injection of the antiserum, a decrease in mitotic activity becomes evident. At the same time, the process of differentiation which is very active in control ganglia, ceases in experimental animals and a progressively larger number of neuroblasts undergo disintegration and are removed by macrophages surrounding these necrotic cells. Between the fourth and the ninth days, the process of cell atrophy and death continues at a slower pace because of the drastic reduction in the cell population produced in the first days of the treatment. While the nerve cells show all these signs of disintegration, satellite cells are spared such effect during the early phases of the treatment but in a subsequent stage they also undergo atrophy and the ganglia are reduced to exceedingly small sclerotic nodules. Electron microscopy in experimental ganglia examined 12 hours after the injection of the antiserum, shows cells with incipient regressive changes and others in an advanced state of deterioration. This differential response to the antiserum, already seen in our earlier studies using light microscopy, happens because immature nerve cells and cells in a more advanced stage of differentiation are present in the same ganglion ; the former are far more vulnerable than the latter to the action of the antibodies to the NGF.

On electron microscopy the first detrimental changes induced by the antiserum are observed in the nucleoli, which appear denser than controls and disorganized; the nucleolonema is no longer distinguishable from the pars amorpha in view of the presence of clumping irregular aggregations of coarse, granular material. More advanced nucleolar lesions are seen in other neurons also 12 days after the antiserum injection; the normal organization of the nucleolar material is completely changed and abnormal structures are apparent within the nucleus. Some of the nucleolar material is condensed to form one or two very dense round bodies which are surrounded by coarse and dense granular material. In the vicinity of the altered nucleolus the chromatin is denser than normal and forms large clumps. In the cytoplasm the ribosomes are no longer grouped in rosettes but are sparse and uniformly distributed through the cytoplasm. Twenty-four hours after the antiserum action, all nucleoli of nerve cells present various degrees of disorganization. At later stages the nuclear envelope breaks in one or more places and the cytoplasmic and nuclear material become intermixed : the free ribosomes are completely disorganized and the mitochondria show signs of degeneration. Vacuolization, pyknosis, rupture of the plasma membrane and other signs of cytolysis are seen in many neurons (10).

Chemical sympathectomy

More than seven years ago the dopamine analogue, 2, 4, 5. Trihydroxyphenilethylamine or 6-hydroxydopamine was introduced in experimental pharmacology and found to produce a marked depletion of norepinephrine stores in sympathetically innervated organs (11).

At variance with other depleting drugs, it was soon observed that the effect of 6-OHDA on both storage and release of catecholamines in peripheral organs was extraordinarily long lasting (5, 11). Irreversible damage of noradrenaline (NA), stores or replacement of the physiological transmitter by 6-OHDA or its metabolites, were postulated as possible mechanism of this depletion. The former hypothesis was later confirmed in an elegant series of ultrastructural and biochemical investigations by Thoenen and Tranzer, which demonstrated that after 2-3 days treatment with 6-OHDA the adrenergic nerve endings were in process of degeneration and two weeks later virtually no adrenergic nerve terminals could be detected (12, 13). In adult animals the selective degeneration of adrenergic nerves is restricted or, at least, largely predominant in the distal part of the axon and the perikaryon is spared. For this reason, 3-4 months after treatment nerve terminals with dense core granules had regenerated in all the organs studied (12). Parallel to the regeneration of the nerve endings there was a restoration of the NA content in the same organs.

The knowledge that sympathetic nerve cells respond in a markedly different way during early or late stages in their life cycle, suggested the possibility that these cells may be more vulnerable to the action of 6-hydroxydopamine during early periods of their development. The experimental findings, to be reviewed below, fully confirmed this assumption : administration of 6-OHDA in newborn animals, at doses which do not result in any detrimental effect on body growth of treated animals, causes the selective and permanent destruction of all sympathetic ganglia (2, 3). In a first series of experiments, 6-OHDA was injected into newborn mice at a dosage of 50 μg/g body weight, daily, for seven days.

One experimental and one control animal of the same litter were sacrified every other day during the first week and then at the end of the first and second week after discontinuation of treatment. On inspection treated and control mice did not differ from each other in body weight somatic development or vitality (fig. 1). The superior cervical ganglia, stellate thoracic paravertebral chain ganglia, celiac ganglion, and the ganglionic complex around the vas deferens were dissected out from both groups and used for whole mounts, histological and ultrastructural studies. Brain, spinal cord, spinal ganglia, adrenal glands, thymus, spleens and testis were also examined. No differences were observed at the macro- and microscopic examination, in brain, spinal cord, spinal ganglia or any of the above mentioned organs between control and experimental animals fixed few days after the beginning of treatment. A marked size decrease was instead found in sympathetic ganglia of experimental animals. These differences between control and experimental animals became more pronounced in subsequent days and at the end of treatment, in ten-day old mice, the sympa-

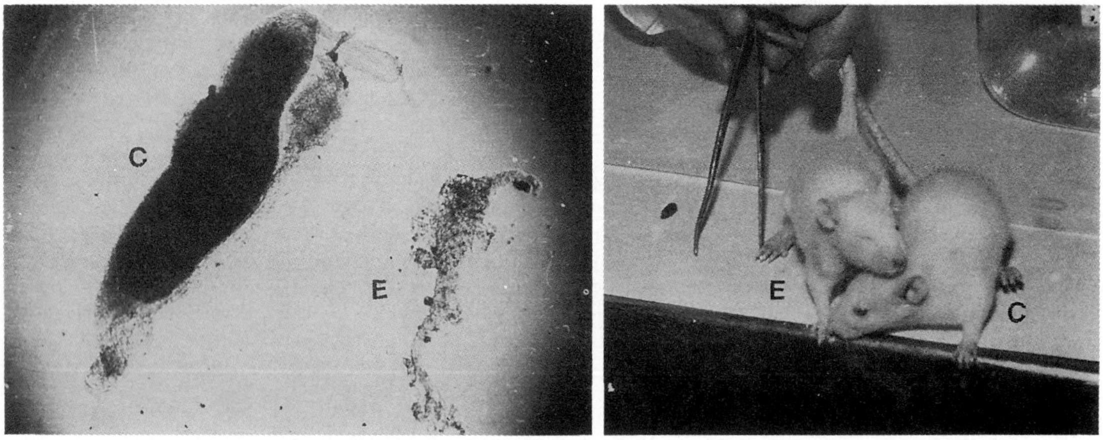

FIG. 1.

Young rat treated at birth with 6-OHDA for a week period (E) with a litter-mate control (C). Note the ptosis in the sympathectomized animal. Control and experimental superior cervical ganglia on the left.

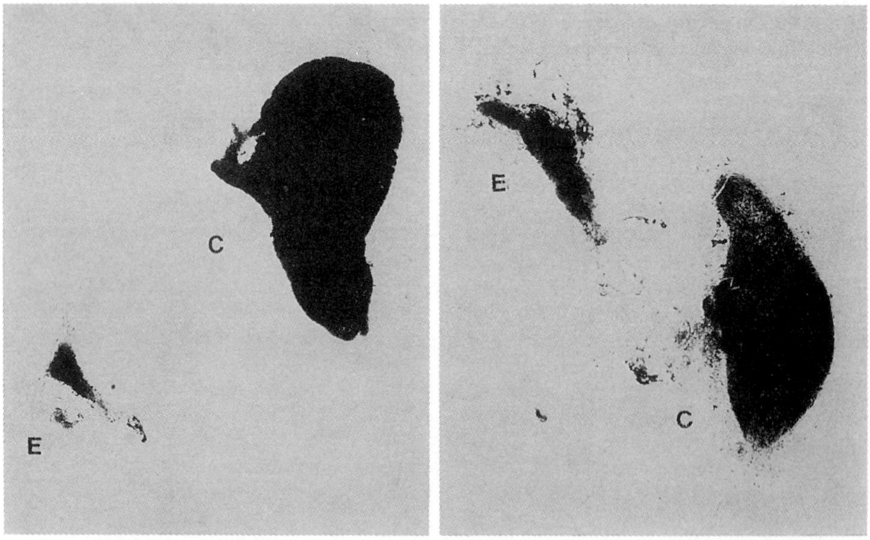

FIG. 2.

Superior cervical ganglia from mice (left) and rats (right). C = control ; E = 6-OHDA treated.

thetic cervical ganglia of 6-OHDA treated animals are reduced to small sclerotic nodules barely recognizable under the stereomicroscope (fig. 2). The histological studies of ganglia examined at dif ferent time intervals show the dramatic sequence of degenerative changes in sympathetic neurons from the third to the seventh day when all the nerve cells are destroyed. Glial and other satellite cells do not appears directly affected at these stages. After three days of treatment, most of the sympathetic nerve cells show degenerative marks such as vacuolization, enlarged and pale nuclei, rupture of cell membranes. Some neurons are pyknotic and only few cells still appear normal at this early stage; they are usually the smaller and less

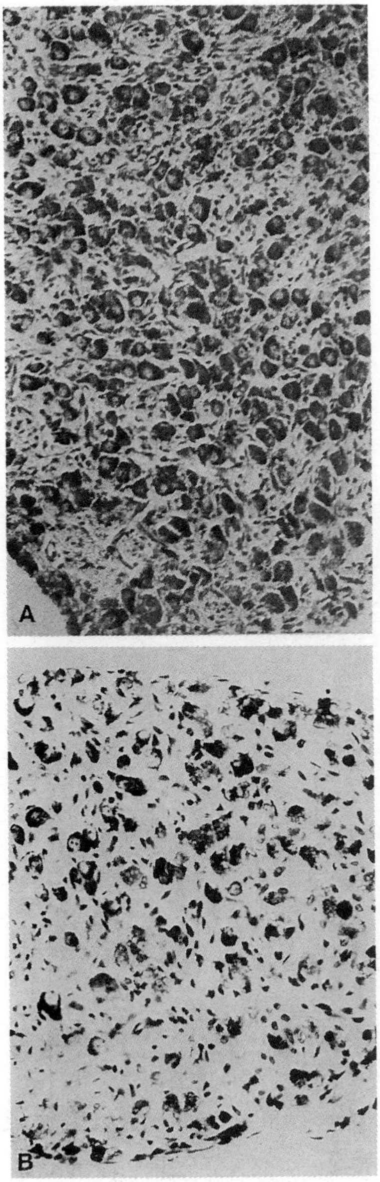

FIG. 3.

Cross section from superior cervical ganglia of mice after 3 days (A) and 5 days (B) treatment with 6-OHDA. (× 380.)

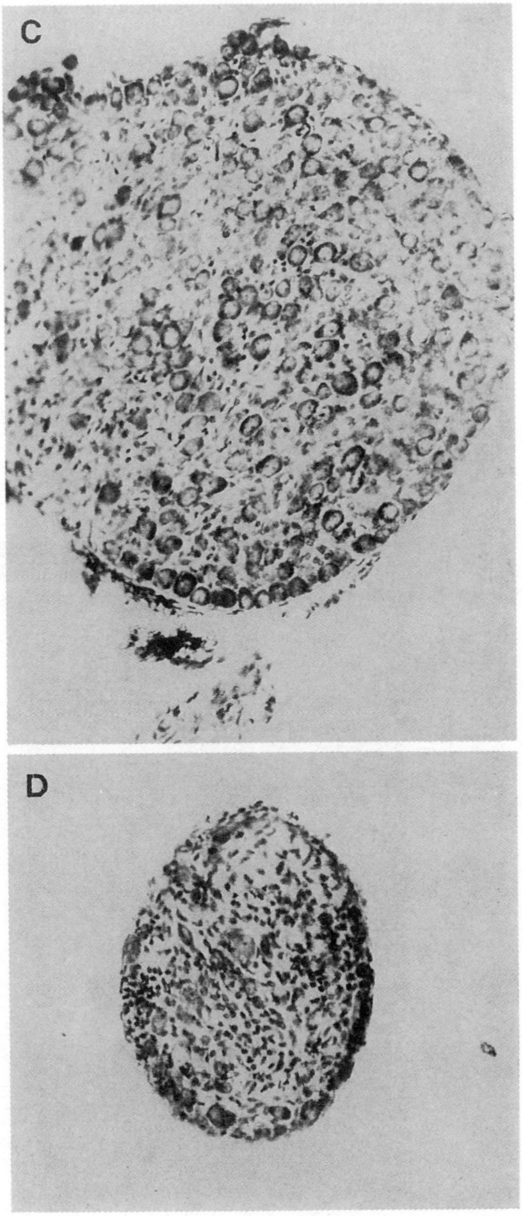

FIG. 4.

Cross sections from superior cervical ganglia of mice after 7 days (C) and 10 days (D) treatment with 6-OHDA. (× 160).

differentiated neuroblasts. In subsequent days all neurons are severely affected and no nerve cells appear intact (fig. 3, 4). In mice three weeks old, sacrified two weeks after discontinuance of 6-OHDA treatment all the sympathetic chain ganglia are reduced to minute sclerotic nodules where no nerve cells are anymore detectable in cross section. No microscopic alterations were observed in other tissues and organs; in the adrenal glands where no degeneration was ever found during the treatment with 6-OHDA, the adrenal medulla showed, at this stage, clear signs of hypertrophy. Examination of sympathetic ganglia during 6-OHDA treatment, at the electron microscope revealed an impressive sequence of degenerative changes in the cell ultrastructure. The alteration are very preminent in the cytoplasmic area and consist of shrinkage of mitochondria, disorganization of

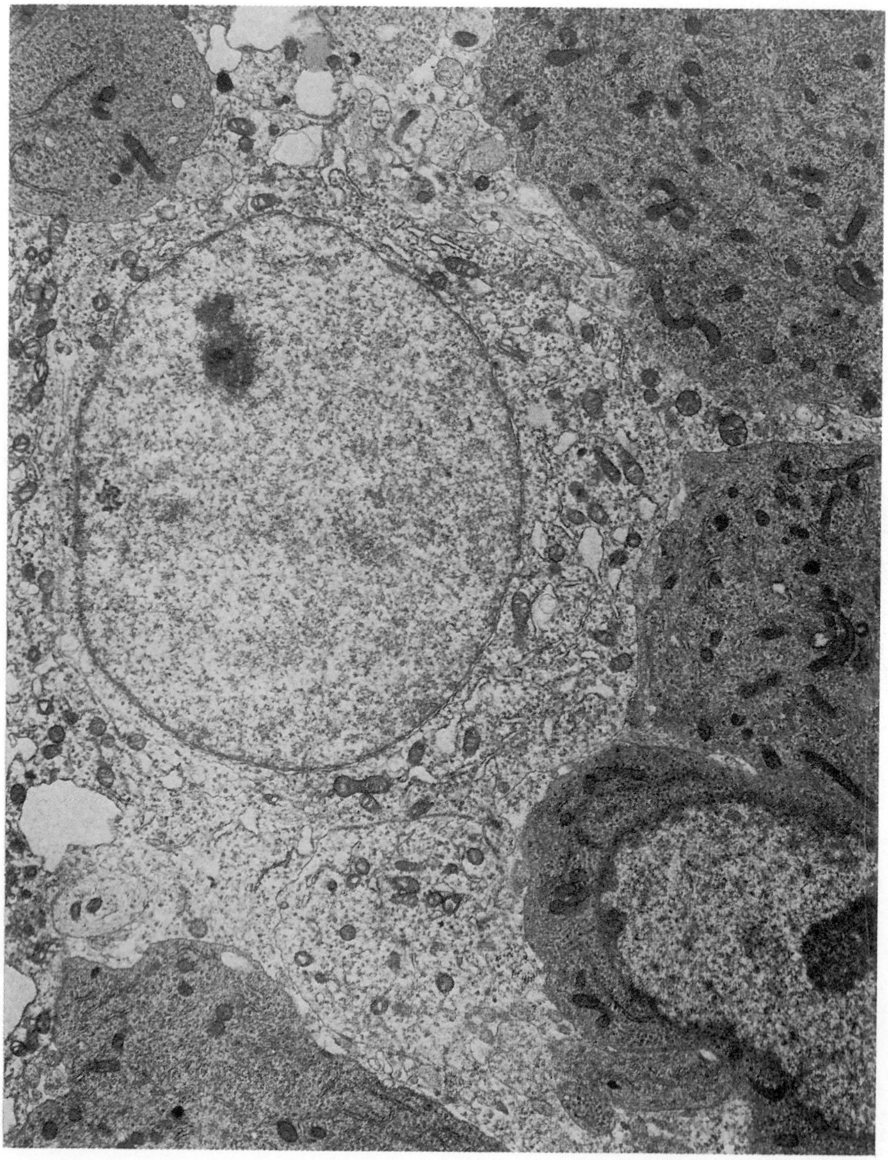

FIG. 5.

Electronmicrophotograph of sympathetic neurons from superior cervical ganglion of a 4-day old mouse sacrified 8 hours after injection of 6-OHDA. Notice the widespread cytoplasmic lesions, shrinking of mitochondria and extensive vacuolization. Some small neuroblasts are not yet affected. (× 18 000).

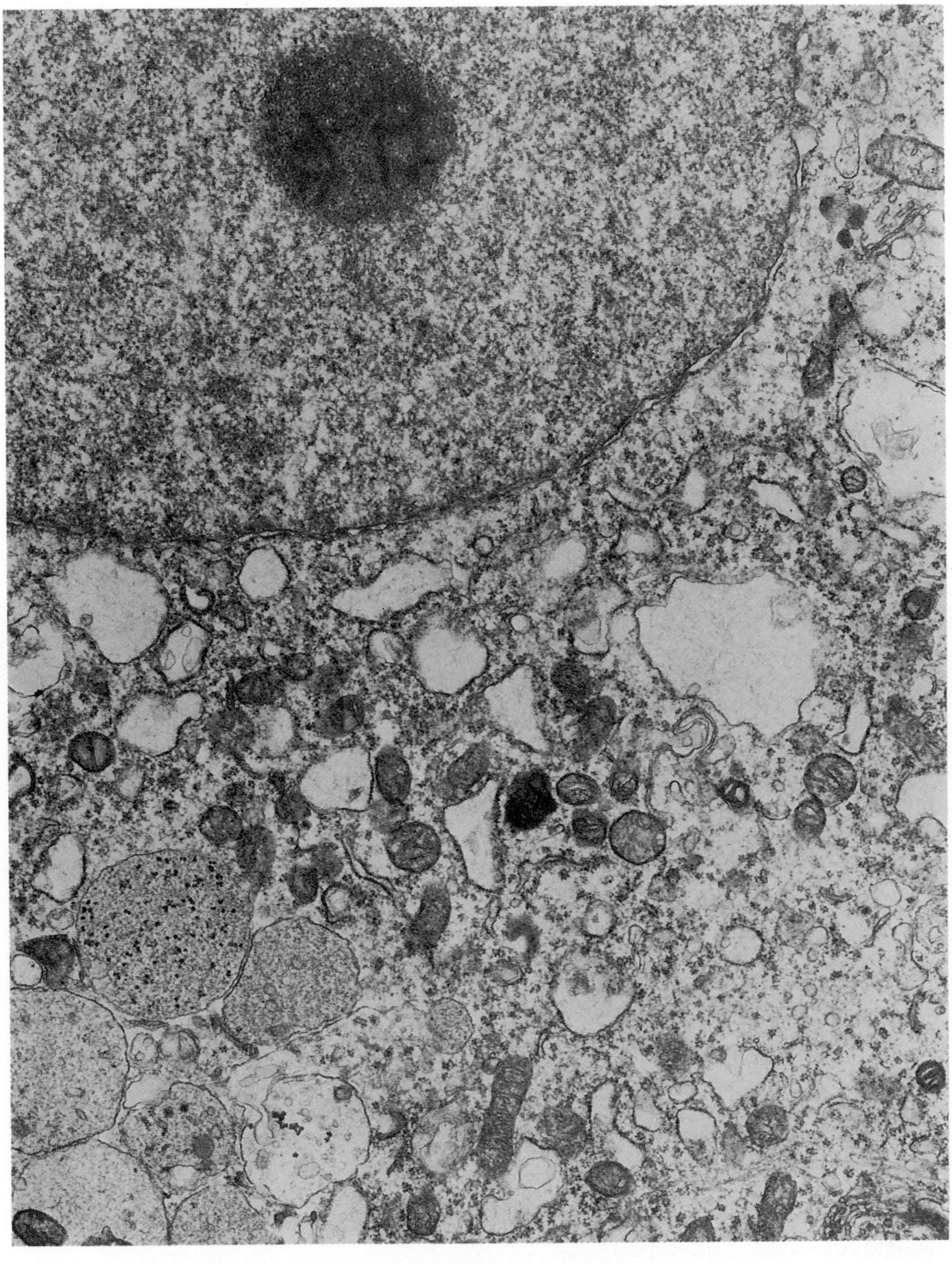

FIG. 6.

Electronmicrophotograph of sympathetic neuron from superior cervical ganglion of a 4-day-old mouse 8 hours after 6-OHDA injection. Note the widespread cytoplasmic alterations. (× 28 000).

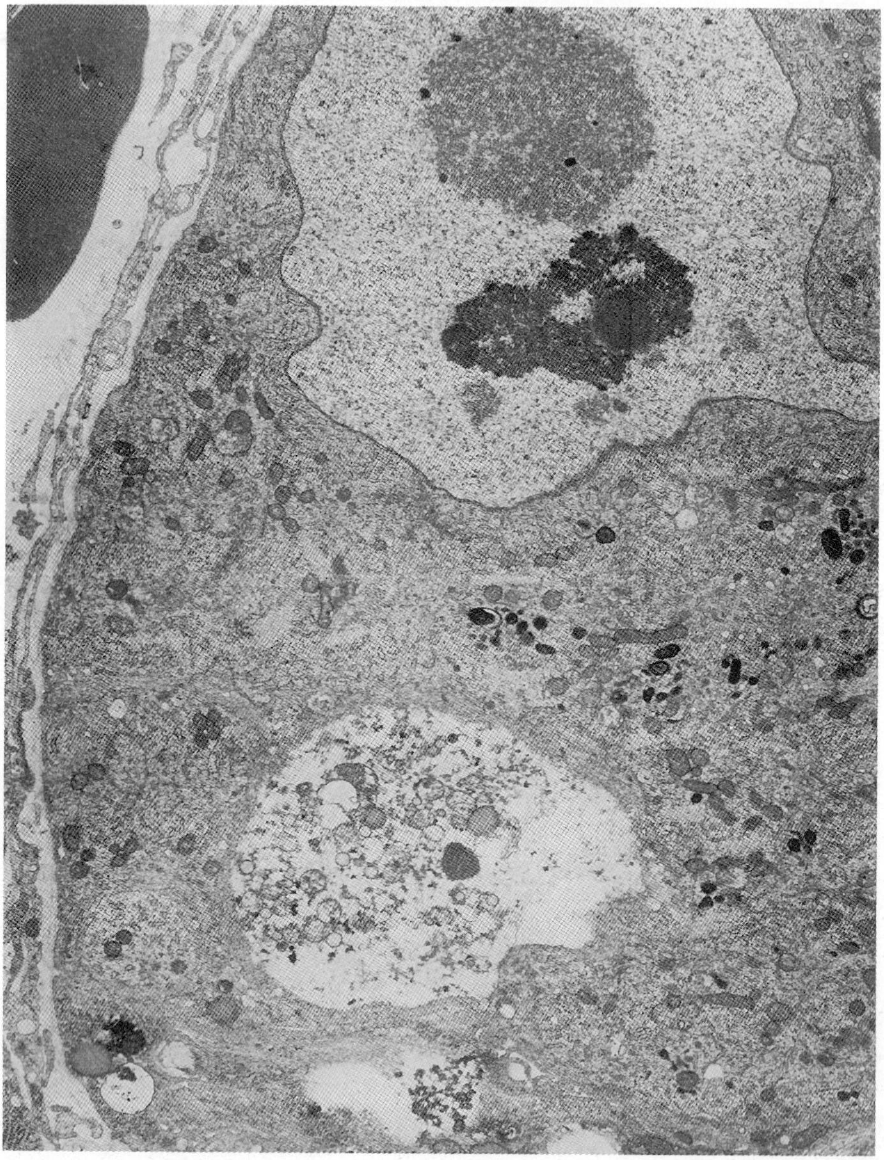

FIG. 7.

Electronmicrophotograph of sympathetic neuron from superior cervical ganglion of a 4-day-old mouse 24 hours after 6-OHDA injection.
A large cytoplasmic vacuolization. Alteration of the nuclear structure. (× 18 000).

the endoplasmic reticulum, extensive vacuolization, rupture of cell membranes. At later stage also the nuclear compartment shovs marked alterations, with condensation of chromatin, disorganization of the nucleoli and rupture of the nuclear membrane (fig. 6, 7). Parallel to the morphological alteration of the ganglia, a drastic and permanent reduction of the sympathetic function takes place in mice and rats after the 6-OHDA treatment (1). Thus the uptake and accumulation of H^3 norepinephrine (NE) in peripheral organs was drastically reduced. In brain, the accumulation of labeled NE is very low compared to other tissues; however, even in the brain the radioactivity recovered from 6-OHDA-treated animals was significantly reduced (from 30 to 60 % of control values). The impairment of the NE-uptake in peripheral organs of 6-OHDA treated animals remained unchanged with time and essentially the same results were obtained in mice and rats of 1, 3 and 6 months of age. The NE content of some tissues from normal and treated mice and rats was determined by enzymatic methods. As shown in tables I and II,

TABLE I

Norepinephrine content in organs
of 6 months old mice treated at birth with 6-OHDA

Organ	Animals	
	Controls	Experimental
Heart	0.85 ± 0.05*	0.07 ± 0.03
Spleen	0.56 ± 0.03	0.05 ± 0.02
Pancreas	0.65 ± 0.03	0.05 ± 0.03
Brain	0.48 ± 0.02	0.15 ± 0.05

* Figures = mean μg/gr and S.E.

TABLE II

Norepinephrine content in organs
of 6 months old rats treated at birth with 6-OHDA

Organ	Animals	
	Controls	Experimental
Heart	0.92 ± 0.04*	0.06 ± 0.01
Spleen	0.65 ± 0.03	0.08 ± 0.03
Pancreas	0.42 ± 0.04	0.05 ± 0.02
Brain	0.55 ± 0.03	0.19 ± 0.06

* μg/gr ± S.E.

the NE content in heart, spleen and pancreas was reduced to less than 10 % of that found in normal controls. In brain the NE content was significantly reduced, although to much less extent than in other organs.

The investigations reported above have clearly shown that a selective destruction of developing sympathetic systems in mammals can be achieved in two entirely different ways. In the « Immunosympathectomy » specific antibodies to the nerve growth factor selectively induce a cytotoxic effect on the sympathetic neurons which results in their death and disappearance, while other nerve cells and tissue are completely spared. The mechanism by which the NGF antiserum exerts such a specific cytotoxic effect is not known although clearly it results from the reaction of antibodies with the nerve growth factor somewhere in the cells compartments. A full explanation of the process which leads to the immunosypathectomy must await a complete knowledge of the localization and mode of action of NGF in the sympathetic nerve cells. Here it may be of interest to consider one aspect of immunosympathetcomy, related to the differential response of various sympathetic ganglia. A small nerve-cell population ranging from 2 to 10 % of the controls, persists in para-and prevertebral sympathetic ganglia. These cells appear to be resistent to the action of the antiserum.

Furthermore, peripherally located adrenergic neurons which provide the innervation of the vas deferens and uterus are not affected at all by the antiserum treatment. These results indicate a clear-cut difference between nerve cells hitherto considered similar and labeled as « adrenergic neurons ». Preliminary observations in mice treated with 6-OHDA indicate that also under these experimental conditions the sympathetic neurons innervating the sex organs are resistant to the toxic action of the dopamine analogue. As for the mechanism of action of 6-OHDA in developing sympathetic neuroblasts, there is no reason to believe that it is different from that demonstrated by Thoenen and Tranzer (12). Owing to its ability to take up and concentrate catecholamines from the environment, the cell picks up the dopamine analogues which then exerts its toxic effect. The major difference between the adult sympathetic cells and the young sympathicoblast is that in the former case the lesions are mainly confined to the end-terminal, while the perikaryon is spared. In the latter case, instead, not only the end terminals but the cell body are affected by the toxic action of 6-OHDA. The different response is obviously related to the different stage of differentiation of the cell and to its degree of functional maturity.

REFERENCES

1. ANGELETTI P. U. Chemical sympathectomy in newborn animals. Neuropharmacology, 1971, *10*, 55.

2. ANGELETTI P. U. and LEVI-MONTALCINI R. Specific cytotoxic effect of 6-hydroxydopamine on sympathetic neuroblasts. Arch. ital. Biol., 1970, *108*, 213.

3. ANGELETTI P. U. and LEVI-MONTALCINI R. Sympathetic nerve cell destruction in newborn mammals by 6-hydroxydopamine. Proc. nat. Acad. Sci., U.S., 1970, *65*, 114.

4. COHEN S. Purification of a nerve growth promoting protein from the mouse salivary gland and its neurocytotoxic antiserum. Proc. nat. Acad. Sci., U. S., 1960, *46*, 302.

5. LAVERTY R., SHARMAN D. F. and VOGT M. Action of 2, 4, 5 trihydroxyphenylethylamine on the storage and release of noradrenaline. Brit. J. Pharmacol., 1965, *24*, 549.

6. LEVI-MONTALCINI R. The nerve growth factor: its mode of action on sensory and sympathetic nerve cells. Harvey Lect., 1966, *60*, 217.

7. LEVI-MONTALCINI R. and ANGELETTI P. U. Immunosympathectomy. Pharmacol. Rev., 1966, *18*, 619.

8. LEVI-MONTALCINI R. and ANGELETTI P. U. Nerve growth factor. Physiol. Rev., 1968, *48*, 534.

9. LEVI-MONTALCINI R. and BOOKER B. Destruction of the sympathetic ganglia evoked by an antiserum to protein isolated from mouse salivary glands. Proc. nat. Acad. Sci., U.S., 1960, *42*, 378.

10. LEVI-MONTALCINI R., CARAMIA F. and ANGELETTI P. U. Alterations in the fine structure of nucleoli in sympathetic neurons following NGF-antiserum treatment. Brain Res., 1969, *12*, 54.

11. PORTER C. C., TOTARO J. A. and STONE C. A. Effect of 6-hydroxydopamine and some other compounds on the concentration of NE in the hearts of mice. J. Pharmacol. exp. Ther., 1963, *140*, 308.

12. THOENEN H. and TRANZER J. P. Chemical sympathectomy by selective destruction of adrenergic nerve endings with 6-hydroxydopamine. Naunyn-Schmiedeberg's Arch. exp. Path. Pharmak., 1968, *261*, 271.

13. TRANZER J. P. and THOENEN H. An electron microscope study of selective, acute degeneration of sympathetic nerve terminals after administration of 6-hydroxydopamine. Experientia, 1968, *24*, 155.

R. L. Dehaan and H. Ursprung eds., *Organogenesis* 187–198 (1965).

7

The Action of Nerve Growth Factor on Sensory and Sympathetic Cells

RITA LEVI-MONTALCINI AND PIETRO U. ANGELETTI*

Department of Zoology
Washington University
St. Louis, Missouri

The early discovery of a nerve growth-promoting agent released by fragments of some mouse tumors in tissue culture (Levi-Montalcini, 15; Levi-Montalcini et al., 25) led to the isolation and purification of the nerve growth factor (NGF). The biological properties of this factor have been extensively studied and described in recent years (Levi-Montalcini and Cohen, 23; Levi-Montalcini and Angeletti, 20; Levi-Montalcini, 18). The NGF acts directly on the receptive nerve cells, namely the sensory cells, during an early phase of their life cycle, and on the sympathetic cells throughout all their life. The in vitro effect consists of the production of a dense halo of nerve fibers by explanted ganglia; the halo is apparent after only 6 hours of incubation. The growth effect is elicited by NGF at a concentration of 10^{-9} *M*. Evidence has been given that NGF plays an essential role in the growth and maintenance of the receptive cells. When dissociated sensory or sympathetic neuroblasts are cultured in minimum essential medium, their survival is strictly conditioned to the presence of NGF (Levi-Montalcini and Angeletti, 21). The in vivo effect of this growth factor was investigated in the chick embryo as well as in newborn and adult mammals. In the chick embryo it calls forth a striking hypertrophic and hyperplastic response in sensory and sympathetic ganglia and hyperinnervation of the viscera (Levi-Montalcini and Hamburger, 24). In newborn and adult mammals the effect is restricted to the sympathetic ganglia. Newborn mice injected daily with NGF (10 μg/g body weight) show at the end of the first

* The work described in this paper was supported by grants from the National Institute of Neurological Diseases and Blindness, Public Health Service (NB-01602, NB-03777), the National Science Foundation (GB-1862), and Merck, Sharp and Dohme Research Laboratories.

week of treatment a volume increase of ten times that of controls. The over-all effect is due to increase in cell number and cell size. Adult mammals respond to NGF with marked hypertrophy of the same cells (Levi-Montalcini and Booker, 22).

The nerve growth factor is a protein. Its molecular weight has been estimated to be of the order of 44,000. Its biological activity is heat labile, is destroyed by proteolytic enzymes, and is unstable to acid pH but resistant to alkaline pH. It is endowed with antigenic properties and when injected into rabbits with Freund's adjuvant, induces the production of specific antibodies (AS) (Cohen, 6). Injection of AS into newborn mammals results in the almost total atrophy of the sympathetic ganglia. The destruction of the sympathetic cells is permanent. The effects cited have been described in detail in previous publications (Levi-Montalcini and Angeletti, 19; Levi-Montalcini, 17).

The growth effect elicited by NGF on responsive nerve cells has now been investigated at the metabolic level. The results of these studies will be dealt with in the following sections.

NGF Effect on Glucose Metabolism

It was demonstrated that NGF, to be effective, needs a source of energy which may be supplied by glucose or mannose. Total deprivation of carbohydrates from the medium results in a lack of fiber outgrowth from explanted ganglia. When labeled glucose is added to the culture medium, the growth effect is accompanied by increased oxidation of glucose to CO_2. This increased glucose oxidation is not prevented by glycolytic inhibitors such as fluoride. These results suggested the existence of a direct oxidative pathway in the responsive nerve cells (Cohen, 5). Experiments were performed to study the effect of NGF on the utilization of glucose by nerve cells. The relative yields of $C^{14}O_2$ from glucose-1-C^{14} and glucose-6-C^{14} were investigated in sensory and sympathetic ganglia incubated in Warburg flasks with and without NGF (Angeletti et al., 2). In all these experiments, carried out for a 3-hour period, there was a marked increase in the oxidation of 1-C-glucose and only slight increase in the yield of CO_2 from 6-C-glucose. These experiments, performed at first with intact ganglia, were then repeated with a suspension of cells obtained after trypsinization of the ganglia (Table 7.1). In a second set of experiments the in vivo effect of NGF on glucose utilization was studied. The NGF was injected in newborn mice 12 to 24 hours before sacrifice. The ganglia were then dissected out and incubated in a Warburg flask with labeled glucose. Here again, the oxidation of glucose-1-C^{14} was significantly increased, thus suggesting an influence of NGF on the hexosemonophosphate pathway. Control experiments with different nervous tissue known not to be responsive to NGF did not show any change in glucose metabolism. Since it is generally believed that the oxidation of 6-C-glucose reflects mainly the activity of the glycolytic pathway, while the oxidation of 1-C-glucose is due to the activity of both this pathway and of the hexosemonophosphate shunt (Horecker, 12), we tested in our experimental material the effect of glycolytic inhibitors on glucose oxidation. Addition of NaF to the medium at a concentration of 10^{-2} M causes a strong inhibition of the utilization of 6-C-glucose, but only a slight inhibition of 1-C-glucose (Fig. 7.1). These findings are in line with previous observations on sensory ganglia explanted in semisolid medium, where the addition of fluoride, while inhibiting the growth of fibroblasts, would enhance the nerve fiber outgrowth elicited by NGF (Cohen, 5).

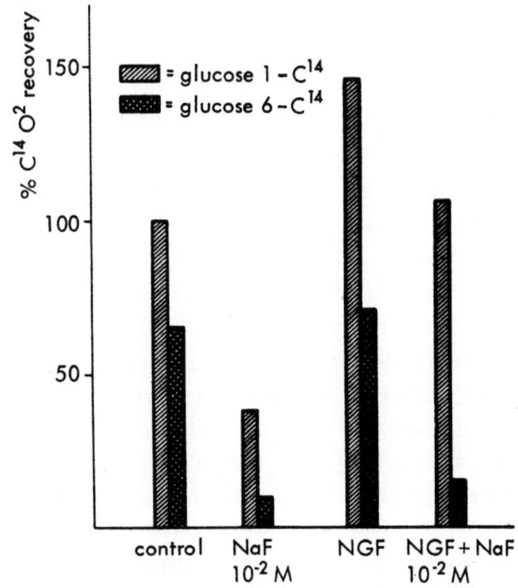

Fig. 7.1 Effect of fluoride on glucose oxidation to CO_2 by sensory ganglia incubated with and without NGF. (Reproduced from Angeletti et al., *Biochim. Biophys. Acta* 90:445–450, 1964.)

NGF Effect on Lipid Biosynthesis in Sensory and Sympathetic Ganglia

The effect of the NGF on lipid synthesis was investigated by studying the incorporation of labeled acetate into lipids by sympathetic and sensory ganglia explanted in vitro and incubated in the presence or absence of NGF (Angeletti et al., 1). Pools of contralateral sensory ganglia from 8-day chick embryos, or pools of six superior cervical ganglia from newborn mice were incubated in Maximow dishes containing 1 ml of Eagle medium added with 1-C^{14}-acetate (1 μc/ml). The incubation was carried out for various lengths of time in a CO_2-conditioned

TABLE 7.1 Metabolism of 1-C^{14}- and 6-C^{14}-glucose to CO_2 by Cell Suspensions from Sympathetic Ganglia, Sensory Ganglia, Brain, and Medulla of Chick Embryos[a]

SUSPENSION	MEDIUM	$C^{14}O_2$ FROM 1-C^{14}-GLUCOSE	$C^{14}O_2$ FROM 6-C^{14}-GLUCOSE	1-C^{14}-GLUCOSE/ 6-C^{14}-GLUCOSE
Sympathetic cells	Control	1464	1008	1.4
	+ NGF	1870	1050	1.8
	Control	882	624	1.4
	+ NGF	1140	700	1.6
	Control	720	460	1.5
	+ NGF	1098	510	2.0
Sensory cells	Control	3120	2350	1.4
	+ NGF	3870	2170	1.8
	Control	2640	1594	1.6
	+ NGF	3300	1440	2.3
	Control	1945	1660	1.2
	+ NGF	2625	1840	1.5
Medulla cells	Control	1134	794	1.4
	+ NGF	1082	720	1.5
Brain cells	Control	1236	770	2.1
	+ NGF	1128	680	1.95

[a] Incubation 3 hours, temperature 37° C.

incubator at 37° C. The NGF extracted and purified with the usual procedure was added at concentration of 0.05 μg/ml. At the end of the incubation, the ganglia were removed and washed several times with cold acetate, then homogenized in 0.5 ml of a chloroform–ethanol mixture (2 : 1) for the extraction of the total lipids (Folch et al., 8). The homogenates were washed three times with saline and each time the ethanol–chloroform extract was separated by centrifugation. Finally, an aliquot of the radioactive extract was mixed with a liquid scintillator and measured with a liquid scintillation counter. In both control and experimental ganglia, the acetate-C^{14} was actively incorporated into total lipid and the rates of incorporation were found to be linear within the first hours. The addition of NGF to the medium greatly increased the rate of acetate incorporation in both sympathetic and sensory ganglia. It was significantly increased after 2 hours of incubation, and at the fourth hour it was more than 50 percent above the control values (Fig. 7.2, Table 7.2). In all these experiments the NGF concentration in the medium was 0.05 μg/ml. In tissue culture of explanted ganglia this dose calls forth a dense halo of nerve fiber outgrowth within 6–8 hours after the beginning of incubation. As reported elsewhere, with higher concentrations of NGF (10–100 times higher) the nerve fiber outgrowth does not take place (Levi-Montalcini, 16); the reason for this lack of effect is still unexplained. Experiments on acetate incorporation performed with different doses of NGF show that the rate of incorporation of labeled acetate into mixed lipids increases only at an optimal concentration of NGF (0.1–0.05 μg/ml). Lower or higher doses have no effect. The mechanism whereby the stimulation of lipid biosynthesis in ganglia treated with NGF takes place

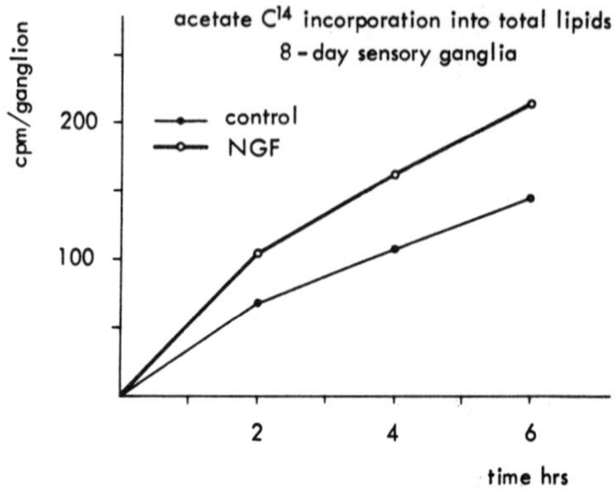

Fig. 7.2 NGF effect on acetate-C^{14} incorporation into total lipids of 8-day sensory ganglia. Experimental conditions as described in the text. (Reproduced from Angeletti et al., *Biochim. Biophys. Acta* 84:778–781, 1964.)

TABLE 7.2 Incorporation of Acetate-C^{14} into Lipids by Superior Cervical Ganglia of Newborn Mice (2-day-old). Pool of 4 Ganglia Incubated in Eagle Medium plus Acetate-C^{14} (1 μc/ml)[a]

Experiment	Time (hrs.)	Radioactivity into Lipid Extract CPM
Control	4	1440
Exper.		2460
Control	4	1920
Exper.		2670
Control	6	2520
Exper.		3900

[a] Reproduced from Angeletti et al., *Biochem. Biophys. Acta* 90:445, 1964.

is a matter for further investigation. In this connection, it is of interest that NGF increases glucose utilization through a direct oxidative pathway, which according to some reports is activated in certain tissues during lipogenesis (Glock, McLean, and Whitehead, 10). Furthermore, it has been demonstrated that TPNH is a necessary co-factor in the reductive synthesis of fatty acids (Lynen, 26). It is attractive, in view of these findings, to consider the two metabolic changes induced by NGF in nerve cells as reciprocally correlated. To investigate this point further, the effect of a strong oxidizing agent such as methylene blue on both acetate incorporation and glucose metabolism was investigated in our experimental system. It was possible to show that methylene blue increases glucose metabolism through the hexose shunt by serving as hydrogen acceptor oxidizing TPNH formed by triphosphopyridine nucleotide (TPN). If TPNH is required for fatty acid synthesis, then the oxidation of TPNH by methylene blue should decrease the rate of lipid synthesis. In some preliminary experiments, in fact, addition of methylene blue (10^{-4} *M*) to the incubation medium effectively reduces acetate incorporation in control and experimental ganglia, while the oxidation of glucose-C^{14} is enhanced.

NGF Effect on Protein Synthesis

This effect was investigated by studying the incorporation rate of labeled amino acids into protein by sympathetic and embryonic sensory ganglia (Angeletti et al., 3). The ganglia were incubated in basal Eagle medium supplemented with L-leucine-C^{14} and L-threonine-C^{14}. The NGF was added to the medium at an effective concentra-

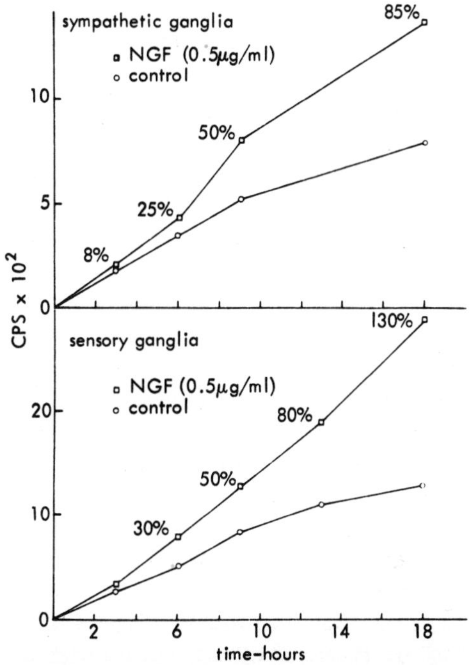

Fig. 7.3 NGF effect on amino acid incorporation (L-leucine-C^{14} and L-threonine-C^{14}) by sensory and sympathetic ganglia incubated in basal Eagle medium. (Reproduced from Angeletti et al., *Biochim. Biophys. Acta* 95:111–120, 1965.)

tion of 0.05 μg/ml. The rate of incorporation was found to be linear during the first 6–8 hours. The addition of NGF stimulated an incorporation of labeled amino acids into protein which was more than 30 percent above control values after 6 hours of incubation. The results are shown in Figure 7.3. In some experiments, the NGF was injected into newborn mice (30 μg/g of body weight) 12 hours before sacrifice. The superior cervical ganglia were then rapidly removed and incubated in the radioactive medium. Also in these circumstances there was marked stimulation of isotope incorporation into ganglia from NGF-treated mice. Control experiments carried out with other tissues (brain, spinal cord, sensory ganglia from 14-day chick embryos) showed no effect of the nerve growth factor on incorporation of amino acid.

The possibility that the increased rate of amino acid incorporation into protein might be due to the effect of the NGF on amino acid transport across the membrane was next explored. There is, in fact, evidence that some hormones (growth hormone and insulin in muscle, estrogen in uterus) produce a parallel increase of intracellular amino acid concentration and of amino acid incorporation into protein (Manchester and Young, 27; Mueller et al., 28). In our system the problem was investigated by studying the effect of NGF on the accumulation of a non-metabolized amino acid (α-aminoisobutyric acid) in sensory and sympathetic ganglia in vitro. The ganglia were incubated in Krebs Ringer buffer and in Eagle medium containing α-aminoisobutyric acid-1-C^{14}; the NGF was added at concentrations ranging from 0.5 to 0.05 μg/ml. Incubation was carried out for various lengths of time, then the ganglia were quickly removed, washed three times in saline solution, and finally homogenized in cold trichloroacetic acid. The radioactivity was then counted on trichloroacetic acid extracts and in samples of the incubation media. Parallel experiments were performed with leucine-C^{14} to check the NGF effect on amino acid incorporation into protein. As can be seen from Figure 7.4, the NGF had no effect on accumulation of α-aminoisobutyric acid within the ganglia, whereas it effectively increased the incorporation of leucine into protein. At variance with other hormonal effects, then, NGF does not seem to influence primarily the mechanism of membrane transport. Therefore, the stimulation of protein synthesis in the ganglia cannot be attributed to an increased availability of intracellular substrate.

The effect of NGF on intracellular protein turnover was investigated by measuring the rate of disappearance of

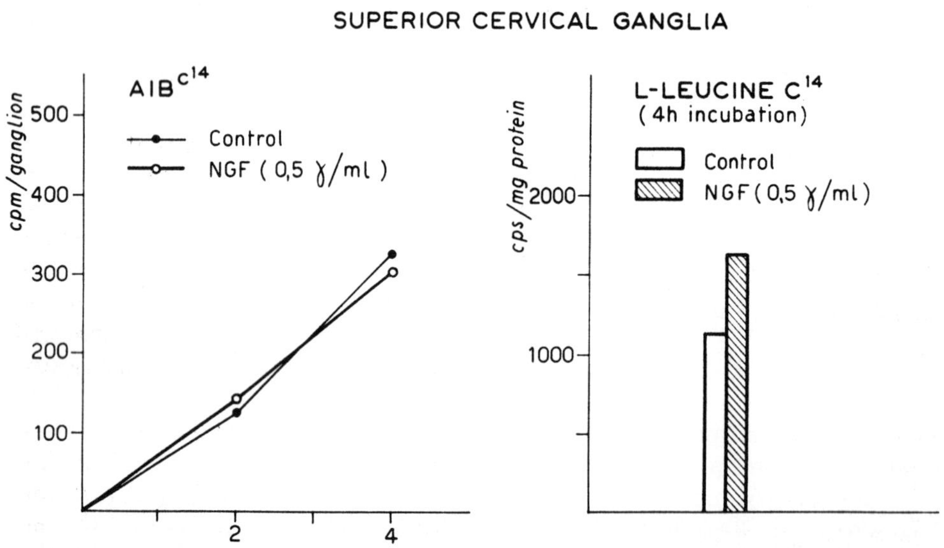

Fig. 7.4 NGF effect on α-aminoisobutyric acid-C^{14} accumulation into superior cervical ganglia and on L-leucine-C^{14} incorporation into proteins of the same ganglia. Experimental conditions as described in the text.

radioactivity in ganglia preincubated in hot medium for a short time and transferred to a cold medium with or without addition of a chaser. It appears that increase in protein synthesis under NGF stimulation is accompanied by an increase in protein turnover.

Effect of NGF on RNA Synthesis

When sensory ganglia are incubated in Eagle medium containing uridine-H^3, the NGF results in a marked increase of uridine incorporation (Toschi et al., 30). Five hours later, the incorporation of uridine into the treated ganglia exceeds that of control ganglia by more than 100 percent. Figure 7.5 shows the relative amount of protein and RNA labeled in sensory ganglia incubated with or without NGF. The rise in uridine-H^3 labeling of RNA in the treated ganglia precedes the increase in amino acid incorporation.

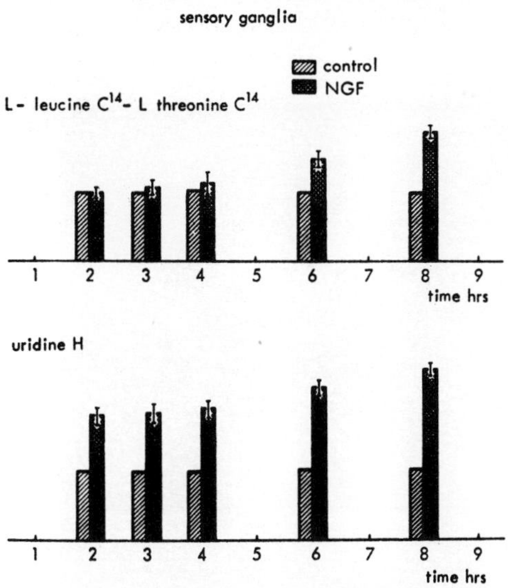

Fig. 7.5 Time sequences of NGF-stimulated acceleration of protein (top) and RNA (bottom) synthesis in 8-day chick embryo sensory ganglia. Incubation in Eagle basal medium added with L-leucine-C^{14}, L-threonine-C^{14} (2 μc/ml), or with uridine-H^3 (10 μc/ml). Results expressed as percent of the control values. (Reproduced from Angeletti et al., *Biochim. Biophys. Acta* 95:111–120, 1965.)

In order to gain more information on stimulation by the NGF on RNA synthesis, pulse-labeling experiments were performed for various lengths of time. In a typical experiment, sensory ganglia were incubated for 4 hours with NGF and then exposed to uridine-H^3 for 60 minutes. The RNA was then extracted and analyzed by sucrose gradient. Preliminary data indicate that rapidly labeled RNA is rather heterogenous with a predominance of heavy components. Radioautographic controls demonstrate that the nuclei of nerve cells treated with NGF are much more heavily labeled than nuclei of control cells (Fig. 7.6).

Effects of Specific Metabolic Inhibitors

The possibility that NGF stimulates protein synthesis by regulating the rate of an RNA functionally similar to the messenger-RNA described in bacterial systems was explored by using some specific metabolic inhibitors. Puromycin, known to inhibit protein synthesis at the ribosomal level, added to the culture medium at a concentration of 20 μg/ml, suppressed amino acid incorporation in both control and experimental ganglia. Uridine incorporation, however, was only slightly reduced (to 70 percent of normal values). Under these conditions, the NGF stimulation of RNA synthesis was still clearly evident. After 6 hours of incubation in presence of the inhibitor, the ratio of radioactivity incorporated into treated ganglia and controls was 2:1.

In order to test whether or not complete suppression of protein synthesis would prevent the NGF effect on RNA synthesis, puromycin (40 μg/ml) was added to the medium 30 minutes before the addition of NGF; the ganglia were incubated for 4 hours and then exposed for 60 minutes to labeled uridine. The results of these experiments are shown in Figure 7.7 which indicates that: (1) Inhibition of protein synthesis at the ribosomal level does not prevent the NGF effect on the synthesis of RNA. (2) This effect is not mediated by *de novo* synthesis of some protein molecules in the cell.

Actinomycin-D, a powerful inhibitor of DNA-primed RNA synthesis, at a concentration of 1 μg/ml in our system, was found to reduce uridine incorporation to 10–15 percent of normal values in both control and experimental ganglia. Protein synthesis was only slightly inhibited in both control and experimental ganglia but the NGF stimulation of amino acid incorporation was completely abolished (Fig. 7.8). These findings give support to the hypothesis that the primary action of NGF is at the nuclear level, at a step which involves new synthesis of DNA-primed RNA.

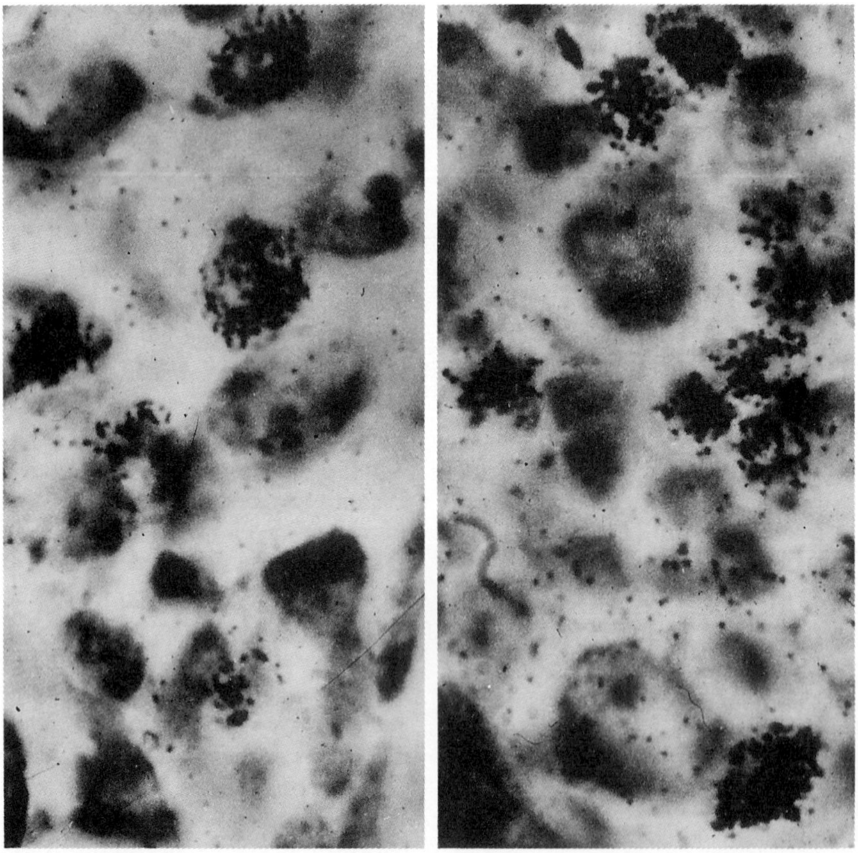

Fig. 7.6 Radioautography of 8-day sensory ganglia incubated for 2 hours in Eagle medium added with uridine-H^3 (1 μc/ml) and NGF (0.05 μg/ml).

Additional evidence was gained in experiments where sensory ganglia were incubated with or without NGF for a 4-hour period and then transferred to a medium with radioactive amino acids containing actinomycin-D for 4 more hours. Under these conditions, there was a clear-cut stimulation of amino acid incorporation into treated ganglia. A conceivable explanation of these results is that preincubation with NGF would call forth an increased synthesis of RNA which, in turn, is reflected in a stimulation of protein synthesis.

Discussion

An attempt is made in the present paper to correlate the nerve growth effect, investigated in the past mainly at

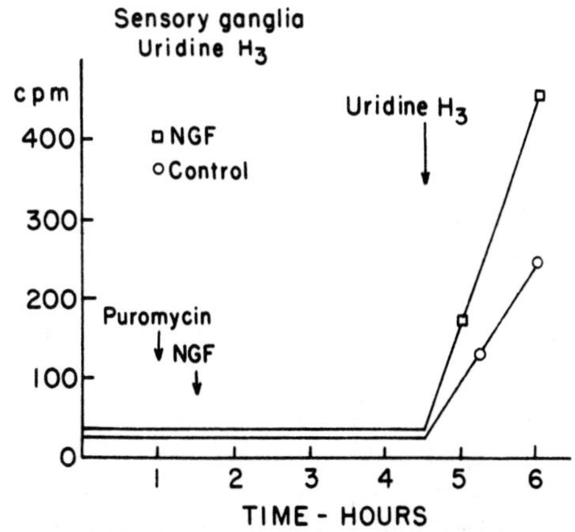

Fig. 7.7 Effect of puromycin (40 μg/ml) on the NGF stimulation of uridine incorporation in sensory ganglia. The inhibitor was added to the medium 30 minutes before the addition of NGF (0.05 μg/ml). Uridine-H^3 added at arrow. (Reproduced from Angeletti et al., *Biochim. Biophys. Acta* 95:111–120, 1965.)

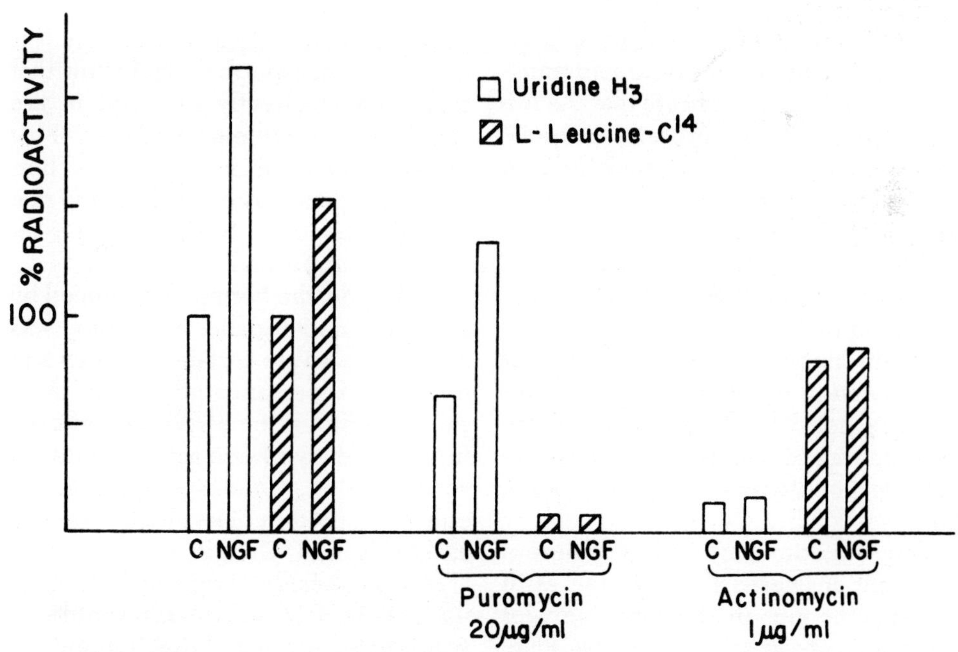

Fig. 7.8 Effect of puromycin and of actinomycin on uridine-H^3 and on L-leucine-C^{14} incorporation by 8-day sensory ganglia incubated in Eagle medium with and without NGF. Results expressed as percent of control values. (Reproduced from Angeletti et al., *Biochim. Biophys. Acta* 95:111–120, 1965.)

a morphological level, with measurable effects on various metabolic pathways. The analysis of these effects and of their temporal sequence is a prerequisite to an understanding of its mechanism of action. The data gathered so far, although not complete, allow some considerations of the metabolic changes induced by the NGF, and suggest some working hypotheses amenable to experimental testing.

We shall first consider the NGF effect on protein metabolism in receptive nerve cells, since the problem of protein synthesis holds a key position in any attempt to explain growth and its regulation.

Evidence has been presented that the addition of very minute amounts of NGF to sensory or sympathetic ganglia cultured in liquid media increases the incorporation of amino acids into protein and apparently also increases the rate of protein turnover. The experiments on α-aminoisobutyric acid transport in our system indicate that the initial step in protein synthesis, the transfer of amino acids from extracellular to intracellular compartments, is not affected by the NGF, thus ruling out a primary NGF effect on amino acid transport across the cell membrane. Pertinent in this connection are the recent observations that the kinetics of incorporation of amino acids into protein are not dependent upon the rate at which amino acids enter the intracellular pool (Kipnis et al., 14; Segal, 29).

The experiments reported herein have not answered the question as to the specific site in the machinery of protein synthesis at which the NGF action is exerted. However, if one compares the lag preceding the increase in the rate of amino acid incorporation to the immediate stimulation in RNA synthesis, it would appear that NGF stimulates the production of an essential component of the cell synthetic system (m-RNA) rather than activating pre-existing cell components. Experiments with specific inhibitors lend support to this hypothesis. Actinomycin-D, in fact known to inhibit DNA-primed RNA synthesis, entirely abolishes the NGF effect on amino acid incorporation. Puromycin on the other hand, although suppressing protein synthesis, does not prevent the initial NGF stimulation of RNA synthesis.

The above considerations are obviously in the frame of currently accepted models suggested for the genetic control of protein synthesis in bacteria (Jacob and Monod, 13). Although definite evidence is still lacking that the rapidly synthesized RNA in the nucleus of mammalian cells plays the same role as in bacteria, it is difficult to escape the conclusion that this RNA is directly involved in the regulation of protein synthesis. That growth factors as well as some hormones might act by controlling the rate of synthesis of m-RNA has been repeatedly proposed.

Ecdysone, the hormone produced by the prothoracic gland in insects, was shown to activate certain gene loci and elicit the appearance of puffs in large salivary chromosomes. These puffs exhibit a high synthetic activity with respect to RNA, which seems related to enhanced protein biosynthesis (Clever and Karlson, 4; Gall, 9).

Insulin, besides other effects, stimulates both RNA and protein synthesis. It has been suggested that it might act by controlling the synthesis of some messenger-RNA (Wool and Munro, 31). Recently, however, it has been shown that the insulin effect on protein synthesis can be dissociated from its effect on RNA synthesis, since actinomycin-D abolishes the latter, leaving the former unaffected (Eboué-Bouis et al., 7).

The sequence of RNA and protein synthesis during early estrogen action has been the object of several investigations. It has been shown that many of the responses of the uterus to estrogen, including the striking increase of RNA synthesis, are blocked by prior puromycin treatment (Mueller et al., 28; Hamilton, 11). The hypothesis has been advanced that the primary effect of estrogen might be an increased synthesis of RNA polymerase (28). At variance with these findings, the stimulation of RNA synthesis by NGF in the responsive nerve cells is apparently not dependent upon a previous stimulation of protein synthesis.

In suggesting the stimulation of nuclear RNA synthesis as the site of action of NGF, we leave unanswered the question of how this stimulation takes place. Since the NGF is normally present in responsive nerve cells and has proved to be essential to their survival, it can be inferred that it is part of a basic growth control mechanism of these cells.

When, under experimental conditions, larger concentrations of the NGF are suddenly made available to these cells, they are brought to an excited state in which all anabolic processes are stimulated to their maximal rate. The increased rate of protein and lipid synthesis, as well as the enhanced carbohydrate utilization, reflects this shift in cell homeostasis, which ultimately results in the remarkable cellular overgrowth which we have described.

References

1. Angeletti, P. U., A. Liuzzi, and R. Levi-Montalcini, "Stimulation of Lipid Synthesis in Sensory and Sympathetic Ganglia by a Specific Nerve Growth Factor," *Biochim. Biophys. Acta* 84:778–781 (1964).
2. ——, ——, ——, and D. Gandini-Attardi, "Effect of a Nerve Growth Factor on Glucose Metabolism by Sympathetic and Sensory Nerve Cells," *Biochim. Biophys. Acta* 90:445–450 (1964).
3. ——, D. Gandini-Attardi, G. Toschi, M. L. Salvi, and R. Levi-Montalcini, "Metabolic Aspects of the NGF Effect on Sympathetic and Sensory Ganglia: Protein and RNA Synthesis," *Biochim. Biophys. Acta* 95:111–120 (1965).
4. Clever, U., and P. Karlson, "Induktion von Puff-veränderungen in den Speicheldrüsenchromosomen von Chironomus tentans durch Ecdyson," *Exptl. Cell. Res.* 20:623–626 (1960).
5. Cohen, S., "Purification and Metabolic Effects of a Nerve-growth-promoting Protein from Snake Venom," *J. Biol. Chem.* 234:1129–1137 (1959).
6. ——, "Purification of a Nerve-growth-promoting Protein from the Mouse Salivary Gland and Its Neurocytotoxic Antiserum," *Proc. Natl. Acad. Sci. U. S.* 46: 302–311 (1960).
7. Eboué-Bouis, D., A. M. Chambaut, P. Volfin, and H. Clauser, "Action of Insulin on the Isolated Rat Diaphragm in the Presence of Actinomycin-D and Puromycin," *Nature* 199:1183–1184 (1963).
8. Folch, J., M. Lees, and G. H. Sloane Stanley, "A Simple Method for the Isolation and Purification of Total Lipides from Animal Tissues," *J. Biol. Chem.* 226:497–509 (1957).
9. Gall, J. G., "Chromosomes and Cytodifferentiation," *Cytodifferentiation and Macromolecular Synthesis*, M. Locke, ed. New York: Academic Press, pp. 119–143 (1963).
10. Glock, G. E., P. McLean, and J. K. Whitehead, "Pathways of Glucose Utilization in the Mammary Gland of the Rat," *Biochim. Biophys. Acta* 19:546–547 (1956).
11. Hamilton, T. H., "Sequences of RNA

and Protein Synthesis during Early Estrogen Action," *Proc. Natl. Acad. Sci. U. S.* 51:83–89 (1964).

12. Horecker, B. L., "Interdependent Pathways of Carbohydrate Metabolism," *The Harvey Lectures*. New York: Academic Press, vol. 57, pp. 35–61 (1962).
13. Jacob, F., and J. Monod, "Genetic Regulatory Mechanisms in the Synthesis of Proteins," *J. Mol. Biol.* 3:318–356 (1961).
14. Kipnis, D. M., E. Reiss, and E. Helmreich, "Functional Heterogeneity of the Intracellular Amino Acid Pool in Mammalian Cells," *Biochim. Biophys. Acta* 51:519–524 (1961).
15. Levi-Montalcini, R., "Effects of Mouse Tumor Transplantation on the Nervous System," *Ann. N. Y. Acad. Sci.* 55:330–343 (1952).
16. ——, "Chemical Stimulation of Nerve Growth," *A Symposium on the Chemical Basis of Development*, W. D. McElroy and B. Glass, eds. Baltimore: The Johns Hopkins Press, pp. 645–665 (1958).
17. ——, "Analysis of a Specific Nerve Growth Factor and of Its Antiserum," *Sci. Rept. Ist. Super. Sanita* 2:345–368 (1962).
18. ——, "Growth Control of Nerve Cells by a Protein Factor and Its Antiserum," *Science* 143:105–110 (1964).
19. ——, and P. U. Angeletti, "Biological Properties of a Nerve Growth Promoting Protein and Its Antiserum," *Regional Neurochemistry*. London: Pergamon Press, pp. 362–377 (1960).
20. ——, and ——, "Growth Control of the Sympathetic System by a Specific Protein Factor," *Quart. Rev. Biol.* 36:99–108 (1961).
21. ——, and ——, "Essential Role of the Nerve Growth Factor in the Survival and Maintenance of Dissociated Sensory and Sympathetic Embryonic Nerve Cells *in vitro*," *Develop. Biol.* 7:653–659 (1963).
22. ——, and B. Booker, "Excessive Growth of the Sympathetic Ganglia Evoked by a Protein Isolated from Mouse Salivary Glands," *Proc. Natl. Acad. Sci. U. S.* 46:373–384 (1960).
23. ——, and S. Cohen, "Effects of the Extract of the Mouse Submaxillary Salivary Glands on the Sympathetic System of Mammals," *Ann. N. Y. Acad. Sci.* 85: 324–341 (1960).
24. ——, and V. Hamburger, "A Diffusible Agent of Mouse Sarcoma, Producing Hyperplasia of Sympathetic Ganglia and Hyperneurotization of Viscera in the Chick Embryo," *J. Exptl. Zool.* 123:233–288 (1953).
25. ——, H. Meyer, and V. Hamburger, "*In vitro* Experiments on the Effect of Mouse Sarcomas 180 and 37 on the Spinal and Sympathetic Ganglia of the Chick Embryo," *Cancer Res.* 14:49–57 (1954).
26. Lynen, F., "Participation of Acyl-CoA in Carbon Chain Biosynthesis," *J. Cellular Comp. Physiol.* 54:33–43 (1959).
27. Manchester, K. L., and F. G. Young, "The Effect of Insulin on Incorporation of Amino Acids into Protein of Normal Rat Diaphragm *in vitro*," *Biochem. J.* 70: 353–358 (1958).
28. Mueller, G. C., J. Gorski, and Y. Aizawa, "The Role of Protein Synthesis in Early Estrogen Action," *Proc. Natl. Acad. Sci. U. S.* 47:164–169 (1961).
29. Segal, S., "Hormones, Amino Acid Transport and Protein Synthesis," *Nature* 203:17–19 (1964).
30. Toschi, G., D. Attardi-Gandini, and P. U. Angeletti, "Effect of a Specific Neuronal Growth Factor on RNA Metabolism by Sensory Ganglia from Chick Embryo," *Biochem. Biophys. Res. Commun.* 16:111–115 (1964).
31. Wool, J. G., and A. I. Munro, "An Influence of Insulin on the Synthesis of a Rapidly Labeled RNA by Isolated Rat Diaphragm," *Proc. Natl. Acad. Sci.* 50: 918–923 (1963).

Reprinted from *Proceedings of the Royal Society of Medicine*, May 1965, Vol. 58, No. 5, pp. 357–360
(Section of Experimental Medicine and Therapeutics, pp. 1–4).

Biological Aspects of Specific Growth Promoting Factors

Professor Rita Levi-Montalcini
(*Department of Zoology, Washington University, St Louis, Missouri, USA*)

The recognition of the growth promoting activity elicited by a variety of agents belonging to the class of hormones as well as to the less well defined category of 'growth factors' dates back to the second half of the past century. However, only much more recently have the growth effects elicited by various agents been submitted to a rigorous analysis and attempts made to elucidate how they act on the target cells. All these investigations centred mainly if not exclusively on the study of hormones: the ill-defined class of 'growth factors' was only occasionally the object of analysis. In these few instances the study was restricted to the exploration of the effects elicited by given growth agents on cell populations cultured in liquid or semi-solid media.

This paper discusses two growth factors which were recently characterized and assayed *in vivo* and *in vitro*. Both factors markedly enhance the growth of given cell types leaving all other cells unaffected; both factors play an outstanding role in the life of the target cells. Similar control mechanisms may exist for all cell types and this possibility raises the question of the biological significance of 'specific growth factors' and of their relationship to hormones.

The Nerve Growth Factor

In 1952 we reported on the discovery of a diffusible nerve growth factor released by mouse sarcomas 180 and 37 (Levi-Montalcini 1952). This report was based on a long series of studies of the effects elicited by these tumours in the chick embryo (Levi-Montalcini & Hamburger 1951). When implanted into the body wall of 3-day chick embryos, or on to the chorioallantoic membrane of 4–6-day chick embryos, both tumours cause striking developmental changes in the sympathetic para- and pre-vertebral ganglia. The ganglia markedly increase in size, due to an increase in cell number and cell size of the sympathetic neurones. The ganglia reach a volume four to six times larger than control ganglia and produce an excessive number of nerve fibres which enter in large bundles into most of the embryonic viscera. They also force their way into the lumen of large and small blood vessels. The density of these fibres in some organs like the mesonephros (which normally is not innervated) is such as to disrupt the structural and functional organization of the organ.

A less striking but marked increase in size is also apparent in sensory ganglia of embryos bearing intracœlomatic transplants of the same tumours. Unlike the sympathetic ganglia, the sensory ganglia are receptive to the growth promoting effects of mouse sarcomas only during a restricted period of their early differentiation. In chick embryos this period occurs between the sixth and the twelfth day of incubation (Levi-Montalcini 1963).

The effect of the tumour growth factor was then investigated in sensory and sympathetic ganglia cultured *in vitro* by the hanging drop technique. It was shown that in proximity to a fragment of the neoplastic tissue both types of ganglia produce a dense halo of nerve fibres in eight to twelve hours, while no fibres grow from control ganglia in the same period (Levi-Montalcini *et al.* 1954). The growth factor present in these mouse sarcomas was identified in a heat-labile protein particle (Cohen *et al.* 1954).

While these studies were in progress, two much more potent sources of the same or of closely related nerve growth factors were discovered in snake venom (Cohen & Levi-Montalcini 1956) and in mouse salivary glands (Levi-Montalcini & Cohen 1960). The assumption that these newly discovered nerve growth factors are very similar if not identical with the neoplastic nerve growth factors was based on extensive biochemical investigations. When injected into the yolk of developing chick embryos, both factors elicit the same growth response from sympathetic ganglia as the implanted mouse sarcomas. Their *in vitro* effects on sensory and sympathetic ganglia are identical with the effects elicited by a fragment of mouse sarcoma or cell-free extract of the mouse tumours (Levi-Montalcini 1958). The biochemical analysis of the venom and salivary factors indicated that in both instances the active agent was a protein (Cohen 1958).

The nerve growth factor (NGF) was further purified from mouse salivary glands, where it is present in a concentration about six thousand times higher than in mouse tumours, and was then used in all subsequent investigations.

The purest preparation of the salivary NGF, which was identified as a single protein component with a sedimentation constant of 4·3S. (Cohen 1960) was then assayed in newborn mice. Daily injections of this fraction for nine days result in the production of para- and pre-vertebral sympathetic ganglia ten to twelve times larger than control ganglia. This remarkable growth effect is due to increased mitotic activity as well as to an increase in the size of individual neurones. These cells grow much larger than fully differentiated neurones of adult mice. In the adult, the NGF elicits increase in size of sympathetic nerve cells but no increase in their number (Levi-Montalcini & Booker 1960*a*). The marked hypertrophy and hyperplasia of sympathetic ganglia of treated animals result in turn in the hyperinnervation of the viscera which does not appear to interfere with the normal function of these organs. The treated animals in fact develop as well as controls.

While these results gave evidence of the striking growth response called forth by the NGF in sympathetic nerve cells of embryonic, newborn and adult organisms, the following observations lead us to consider the NGF as essential to the very life of the responsive nerve cells.

In 1959 Cohen produced an antiserum to the NGF by injecting the purified growth factor with Freund adjuvant into the pads of rabbits (Cohen 1960). The injection of minute amounts of this antiserum in newborn mammals immediately blocks the growth and differentiation of developing sympathetic nerve cells and kills already differentiated neurones. As a result, the sympathetic ganglia undergo almost total atrophy (Levi-Montalcini & Booker 1960*b*). The disappearance of 95–98% of the whole sympathetic cell population of para- and pre-vertebral ganglia was verified by histological analysis of the treated animals one to three years after the injection of the antiserum. No adverse effects are apparent in other tissues or organs. The dramatic and almost instantaneous destruction of the sympathetic cell population upon injection of the antiserum to the NGF indicates that these cells cannot survive in the absence of this growth factor. Additional evidence in support of the hypothesis that the NGF plays a fundamental role in the life of sympathetic nerve cells came from experiments performed *in vitro* on ganglia treated with trypsin. Under these conditions, individual nerve cells survive and develop in liquid synthetic media enriched with NGF at a molar concentration of 10^{-6}. In its absence all nerve cells rapidly disintegrate and die (Levi-Montalcini & Angeletti 1963).

Thus we came to the conclusion that the NGF plays a role in the life of the sympathetic nerve cells which has no parallel in other well known growth regulators such as nutrients, vitamins and hormones. The NGF acts in extremely low concentration, is highly selective and specific in its effects, is normally present in the receptive nerve cells (Levi-Montalcini & Angeletti 1961) and moreover proves to be essential to their subsistence, at least during an early phase of their differentiation.

The mechanism whereby the NGF calls forth this remarkable response from the receptive nerve cells was recently the object of a series of metabolic studies (Toschi *et al.* 1964, Angeletti *et al.* 1964, Levi-Montalcini & Angeletti 1965). Evidence was presented that the NGF markedly stimulates the rate of amino-acid incorporation and turnover in embryonic sensory nerve cells *in vitro* and in embryonic and mature sympathetic nerve cells *in vivo* and *in vitro*. Under these conditions, the NGF calls forth an even higher increase in RNA synthesis as proved by experiments with labelled uridine. Experiments with the inhibitors puromycin and actinomycin-D suggest that the stimulation of protein synthesis by NGF depends on a primary effect on nuclear RNA synthesis.

The Epidermal Growth Factor

While the analysis of the NGF was in progress and attempts were being made to elucidate its mechanism of action, another equally potent and highly specific growth factor was discovered in the same mouse submaxillary salivary glands by Stanley Cohen. This factor, designated from its activity as the epidermal growth factor (EGF), is present in high concentration in the tubular portion of the submaxillary salivary glands of the adult male mouse (Cohen 1962). It is in this part of the gland that the NGF was also detected. The two growth factors are in fact so closely linked to each other that they can be separated only by extensive purification procedures. It was the presence of some side-effects on the skin elicited by crude NGF which gave Stanley Cohen the lead for the discovery and further purification of EGF.

EGF is a protein. Cohen has calculated that its molecular weight is of the order of 14,000. It is heat-stable, not dialysable, destroyed by incubation with proteolytic enzymes and inactivated by a specific antiserum. Two amino acids, phenylalanine and lysine are lacking. Their absence in the purified fraction is taken as the criterion of purity since 'it might be expected that contaminating polypeptides would contain these amino acids' (Cohen 1964).

EGF has a remarkable growth effect on epithelial structures. When daily injections of 0·5 μg per 1·5 g of body weight are given to newborn mice for six days, a marked increase occurs in thickness and precocious keratinization of the skin as well as an increase in height of the epithelium lining the oral cavity, the œsophagus and the stomach mucosa. *In vitro*, EGF elicits a striking growth effect on fragments of embryonic chick skin. While control epithelial tissues, cultured in liquid synthetic media, remain unchanged for the first forty-eight hours of culture, comparable skin fragments cultured in the presence of EGF at the concentration of 0·1 μg/ml proliferate markedly. Over longer periods of culture, EGF stimulates the subsequent keratinization of the epidermal cells.

Production and Biological Significance of the Growth Factors

As mentioned above, two highly specific and potent growth factors were isolated from the tubular portion of the adult mouse submaxillary salivary glands, and their properties were assayed in living organisms as well as *in vitro*. At present, experiments are in progress to search for other possible growth promoting activities of the salivary gland extract on other cell types beside the nerve cells and the epithelial cells.

As for the production of the two growth factors isolated in large quantities from the above source, two alternative possibilities were considered: either the two factors are synthesized and released from the salivary gland in the same way as hormones are produced by endocrine glands, or they are produced elsewhere in the organism and are released through the salivary gland excretory channels. The second alternative seems likely (Levi-Montalcini & Angeletti 1960, Levi-Montalcini 1962). The finding that NGF is present also in mouse sarcomas and in granuloma tissue (Levi-Montalcini & Angeletti 1960) suggested the possibility that this growth promoting factor might be produced in the organism everywhere by cells of mesenchymal type and be utilized by nerve cells of the type described above. If this hypothesis turned out to be correct, one would still be faced with many unsolved problems such as its presence in large quantity in the mouse salivary gland and in the snake venom gland and its absence from the salivary glands of other vertebrates. Also it remains to be seen whether other specific growth factors such as those described above exist and, if so, whether they bear any relationship with the hormones.

This last question is not only a semantic one. Indeed our hesitation in defining the category of the growth factors reflects our ignorance of the precise role of these agents which seem to share some properties with hormones but are unlike them in many important respects. A hormone, as defined by the Shorter Oxford dictionary, is a 'substance formed in an organ and serving to excite some vital process'. These substances, as is well known, are the product of specialized endocrine glands.

The two specific growth factors, NGF and EGF do not fit this concept of a hormone. There is no indication that they are produced in a well-defined organ or organs. Their primary effect is to stimulate the growth of the responsive cells rather than to stimulate their functional activity. It was the excessive proliferation and overgrowth of the sympathetic nerve cells which impressed us from the beginning of this investigation. The massive invasion of embryonic organs by sympathetic nerve fibres as well as the penetration of these fibres into the blood vessels, while stressing the magnitude of the growth response, also indicated that this excessive growth of the sympathetic system was detrimental rather than beneficial to the organism as a whole.

That NGF plays a most important, and we should say essential, role in the life of the target cells is shown by the dramatic effects elicited by the antiserum to NGF and by the demonstration that isolated nerve cells can only survive and grow in the presence of NGF. This cannot be said of hormones which stimulate the receptive organs or tissues but are not indispensable for their survival and growth.

A last but not less important difference between hormones and our growth factors is temporal in nature. Hormones display their function rather late in life. Although hormonal effects are already apparent during fœtal life, it is in the post-natal life and in the fully grown organisms that the role of hormones becomes prominent. Our growth factors on the contrary are most important during early growth and differentiation of the target cells. Indeed some of these cells, such as the sensory nerve cells, are receptive to the growth effect of the NGF only during a very restricted and early period of their growth. Even the sympathetic nerve cells, which respond to this agent throughout life, show a maximal growth response during the early phase of their differentiation. The same is true for the epidermal growth factor.

It is tempting to suggest that specific growth factors such as those described might be regarded as a sort of more primitive and fundamental integrative system than hormones. They are possibly metabolites released by cells still not organized in well defined organs and utilized by other cells as growth factors. Since their main function is indeed to promote growth in the responsive cells, the non-committal term of 'growth factors' seems to be appropriate at present, though we should be ready to replace it with a more precise term as our knowledge of these remarkable biological agents gains in precision and depth.

Acknowledgments: The work described in this paper was supported by grants from the National Institute of Neurological Diseases and Blindness, Public Health Service, Bethesda, Maryland, from the National Science Foundation, Washington DC, from the American Cancer Society, and from the United Cerebral Palsy Foundation.

REFERENCES

Angeletti P U, Liuzzi A, Levi-Montalcini R & Attardi D G
(1964) *Biochem. biophys. Acta* **90**, 445
Cohen S
(1958) In: Chemical Basis of Development. Ed. W D McElroy & B Glass. Baltimore; p 665
(1960) *Proc. nat. Acad. Sci., Wash.* **46**, 302
(1962) *J. biol. Chem.* **237**, 1555
(1964) *Nat. Cancer Inst. Monogr.* No. 13, p 13
Cohen S & Levi-Montalcini R
(1956) *Proc. nat. Acad. Sci., Wash.* **42**, 571
Cohen S, Levi-Montalcini R & Hamburger V
(1954) *Proc. nat. Acad. Sci., Wash.* **40**, 1014
Levi-Montalcini R
(1952) *Ann. N.Y. Acad. Sci.* **55**, 330
(1958) In: Chemical Basis of Development. Ed. W D McElroy & B Glass. Baltimore; p 646
(1962) *Sci. Rep. Super. Sanità* **2**, 345
(1963) In: The Nature of Biological Diversity. Ed. J Allen. New York; p 261
Levi-Montalcini R & Angeletti P U
(1960) In: Regional Neurochemistry. Ed. S S Kety & J Elkes New York; p 362
(1961) *Quart. Rev. Biol.* **36**, 99
(1963) *Develop. Biol.* **7**, 653
(1965) In: Symposium on Organogenesis. Eds. R L. DeHaan & H Ursprung. Baltimore (in press)
Levi-Montalcini R & Booker B
(1960*a*) *Proc. nat. Acad. Sci., Wash.* **46**, 373
(1960*b*) *Proc. nat. Acad. Sci., Wash.* **46**, 384
Levi-Montalcini R & Cohen S
(1960) *Ann. N.Y. Acad. Sci.* **85**, 324
Levi-Montalcini R & Hamburger V
(1951) *J. exp. Zool.* **116**, 321
Levi-Montalcini R, Meyer H & Hamburger V
(1954) *Cancer Res.* **14**, 49
Toschi G, Attardi D G & Angeletti P U
(1964) *Biochem. biophys. Res. Common.* **16**, 111

Printed in England by Staples Printers Limited, at their Tabernacle Street, London, E.C.2, establishment

PHYSIOLOGICAL REVIEWS
Vol. 48, No. 3, July 1968
Printed in U.S.A.

Nerve Growth Factor

RITA LEVI-MONTALCINI AND PIETRO U. ANGELETTI

Department of Biology, Washington University, St. Louis, Missouri; and Istituto Superiore di Sanita, Rome, Italy

THE TERM "NERVE GROWTH-PROMOTING FACTOR" was introduced in 1954 to designate a nucleoprotein particle isolated from mouse sarcomas 180 and 37 and endowed with the property of enhancing growth and differentiation in sensory and sympathetic nerve cells (31). Two additional and much more potent sources of this factor were discovered in 1956 and in 1958 in snake venom and in mouse submaxillary salivary glands (22, 30, 59, 74). The availability of large quantities of the growth-promoting protein extracted from these two sources made possible a better characterization of this factor, so that since 1956 (30) it has been identified as a protein rather than as a nucleoprotein molecule. The biological properties of this agent, which became known as the "nerve growth factor" (NGF), its range of action on the target cells, and its sources, metabolic effects, mechanism of action, and biological significance are the subject of this review.

EARLY LITERATURE

In 1948 Bueker reported on the profuse sensory innervation of fragments of mouse sarcoma 180 implanted into the body wall of 3-day-old chick embryos (12). The experiments were part of a more general plan of investigation of the effects elicited by the peripheral end organs on associated nerve structures. The finding that the sensory ganglia providing fibers to the neoplastic tissue were enlarged seemed to fit well in the accepted concept that the size of a given nerve center is dictated by the size of its peripheral field of innervation.

An extensive and detailed inspection of embryos bearing transplants of mouse sarcomas 180 or 37 revealed new facets of this phenomenon and suggested an entirely different interpretation of the effects evoked by the neoplastic tissue. It was found that sympathetic ganglia of embryos bearing transplants of sarcomas 180 or 37 underwent a far more impressive size increase than sensory ganglia. This effect was not restricted to ganglia adjacent to the tumor but was also very evident in sympathetic ganglia remote from the implant and not connected by nerve fibers with the neoplastic tissue (76). The hypertrophic and hyperplastic sympathetic ganglia produced an extraordinary number of nerve fibers that massively invaded the embryonic viscera and even penetrated inside the blood vessels (58). The suggestion was made that both sarcomas 180 and 37 released a diffusible growth factor into the host circulation that was responsible for the excessive overgrowth of the ganglia and the anomalous distribution of sympathetic nerve fibers. This hypothesis received strong support from experiments of extraembryonic tumor implantation. Fragments of mouse sarcomas 180 or 37 were grafted onto the chorioallantoic membrane of 4- to 6-day chick embryos. In these experiments the tumors and the embryo shared the circulation but no direct contact was established between the embryonic and the neoplastic tissues. The effects compared in all respects with the effect of intraembryonic tumor transplants, thus giving additional evidence for the release into the blood stream of a nerve growth factor produced by neoplastic cells (58, 77).

In 1953 a simple in vitro technique was devised to explore the effects elicited by the two mouse sarcomas 180 and 37 on sensory and sympathetic ganglia. Fragments of these tumors were cultured in a semisolid medium consisting of chicken plasma and embryonic extract. Sensory and sympathetic ganglia dissected out from 7- to 9-day chick embryos were incubated in this medium close to the neoplastic tissues. Experimental and control cultures consisting of ganglia combined with sarcomas 180 or 37, or ganglia combined with embryonic mouse or chick tissue, were inspected 10–24 hr later. The production of a dense, fibrillar halo of nerve fibers by ganglia adjacent to the neoplastic tissues, but not in controls, was taken as evidence for the in vitro release of a growth-promoting agent by these tumors (78). No comparable effect was obtained with other tumors. Since this first experiment was performed, the in vitro production of a fibrillar halo by explanted sensory and sympathetic ganglia has become the most reliable index of the presence of the nerve growth-promoting factor in the culture medium and has been used in all tests performed to detect the presence of this factor in other presumptive sources.

The finding that an extract of mouse sarcomas 180 or 37 retains the growth-promoting activity of intact tumors made possible a chemical study of the growth factor and recognition of its presence in subcellular components. The factor was identified in heat-labile, nondialyzable nucleoprotein particles present in the microsomal fraction of neoplastic cells (31). In order to purify further the growth factor, Cohen made use of snake venom as a source of phosphodiesterase. His experiments, aimed at the enzymatic inactivation of the nucleic acids present in the active fraction, gave an unexpected and unforeseeable result: the snake venom itself was found to possess a nerve growth-promoting activity far more pronounced than that of the tumors under investigation (23).

In 1958 a third and even more potent source of the NGF was discovered in mouse submaxillary salivary glands (24, 59). In vivo and in vitro experiments established the remarkable similarity between the growth effects elicited by this, by the mouse sarcomas, and by snake venom factors.

In 1959 a new facet of the phenomenon came to light with the discovery that a specific antiserum to the salivary NGF destroys the sympathetic ganglia of newborn mammals. The term "immunosympathectomy" was coined to designate this procedure by which animals could be deprived of their sympathetic functions.

CHARACTERISTICS OF THE GROWTH RESPONSE OF SENSORY AND SYMPATHETIC NERVE CELLS IN THE LIVING ORGANISM

Sensory Ganglia

Early studies of the effects elicited by mouse sarcomas 180 and 37 on sensory ganglia of chick embryos were at first interpreted to mean that the growth response is restricted to the ganglia contributing to the innervation of the implant (12, 76). A size increase in these ganglia is apparent on the 6th day of incubation, shortly after their nerve fibers have established contact with the neoplastic cells. From the 6th to the 11th day of incubation, sensory nerve fibers grow vigorously and branch profusely into the rapidly expanding neoplastic tissue. Toward the end of the 11th day of incubation, those sensory ganglia that send nerve fibers into the tumor are three times larger than contralateral ganglia that have no access to the implant. The size increase is due to an intense mitotic activity much more pronounced than in contralateral ganglia, to acceleration of differentiation of early neuroblasts, and to hypertrophy of differentiated nerve cells (76). After the 7th day, a marked difference becomes noticeable between two types of embryonic sensory nerve cells. These two cell types were the object of previous extensive investigations. It was found that they differ from each other in size, developmental pattern, and peripheral field of distribution. Since they form two groups well segregated from each other, they were designated, on the basis of their position in the sensory ganglia of the chick embryo, as the ventrolateral (V-L) and the mediodorsal (M-D) nerve cells (48, 77).

It was now found that the M-D, but not the V-L, cells are receptive to the growth-stimulating effect elicited by the tumors. The growth response elicited by mouse sarcomas 180 or 37 is maximal between the 7th and the 9th day of incubation. It decreases thereafter and is apparent no more after the 15th day of incubation. An extensive study of the distribution of sensory nerve fibers in the tumor between the 7th and the 13th day of incubation showed that nerve fibers branch profusely into the tumor but do not establish contact with the neoplastic cells. In this respect they differ markedly from embryonic sensory nerve fibers, which soon after they reach their end organs, muscles, or subcutaneous tissue establish provisional contact in the form of end boutons. In the tumor, sensory nerve fibers branch in all directions, building a fiber net far more dense than in any embryonic structures (76, 77).

Soon after the two more potent sources of the nerve growth factor were discovered in snake venom and in the mouse submaxillary salivary glands (22, 30, 59), they were assayed for their growth-promoting activity in chick embryos. At first the purified venom and then the purified salivary factor, when this became available 2 years later in 1958, were injected daily into the yolk sac of chick embryos between the 7th and the 10th day of incubation, in the amount of 50 μg of the purified protein per dose. In both experimental series the M-D sensory nerve cells exhibited a marked increase in number and in size. These growth effects differ in only one respect from those elicited by implanted fragments of mouse sarcomas. In the former, but not in the latter (sarcoma experiments), all sensory spinal and most of the cephalic ganglia of the injected embryos are hyperplastic and hypertrophic; in other words the growth effects are generalized and not, as in experiments of tumor transplantation, restricted to ganglia adjacent to and supplying fibers to the tumor (59, 62). It is likely, though not proved, that quantitative rather than qualitative differences in the NGF from these several sources account for the above results. As is reported in a later section, the NGF is present at a concentration respectively 1000 and 6000 times lower in mouse sarcomas than in snake venom and mouse salivary glands (22, 62).

Injections of a tumor extract in the chick embryos do not elicit any growth effect on sensory ganglia. The hyperplastic and hypertrophic effects evoked by the grafted tumor are conceivably due to the continuous release of minute quantities of NGF from the actively growing neoplastic cells, which would affect only closely adjacent ganglia and result in a local rather than a generalized effect. Further evidence of a gradient in the growth response elicited by mouse sarcomas came from studies on sympathetic nerve cells (reported below). Sensory nerve fibers, produced in excess by hypertrophic and hyperplastic sensory ganglia, collect in thick nerve bundles that branch under the skin of the embryo and even perforate it and appear at the body surface (59). No increase is seen in the enteroceptive and proprioceptive innervation of the embryos.

Sympathetic Ganglia

The response of sympathetic ganglia to the NGF from tumors, venom, or submaxillary glands has been the object of more extensive investigation than the response of sensory ganglia to the same factors. Chick embryos were used at first, but when NGF from mouse submaxillary glands became available, the response was studied in newborn and adult mice. (The results are reported later.)

Growth response of sympathetic ganglia in the chick embryo. Implantation of fragments of sarcomas 180 or 37 into the body wall of 3-day chick embryos results in a growth response of sympathetic ganglia of the host that far exceeds in magnitude that of sensory ganglia.

The growth effects become apparent in sympathetic ganglia of chick embryos toward the 6th day of incubation and consist of an increase in mitotic activity, acceleration of maturation processes, and hypertrophy of differentiated nerve cells. In these respects, sympathetic nerve cells behave in the same way as the M-D sensory nerve cells. They differ in the following aspects: *1)* The whole sympathetic nerve cell population, and not only a percentage of these cells, is receptive to the action of the growth factor released by neoplastic cells. *2)* The growth response persists also in sympathetic nerve cells at advanced differentiative stages, until the end of the incubation period. *3)* The magnitude of this response far exceeds that of sensory ganglia. The volume of the sympathetic ganglia in an embryo bearing implants of mouse sarcomas 180 or 37 is of the order of six times that of controls at the 11th day of incubation. New sympathetic ganglionic complexes are formed near the implanted tumor and contribute with large nerve bundles to its innervation. *4)* Nerve fibers produced in excess by sympathetic ganglia remote from the tumor invade most of the embryonic viscera, building a very dense nerve net in the parenchyma of the mesonephros, metanephros, ovary, testis, thyroid, spleen, liver, pancreas, and feather bulbs. A few organs and embryonic structures are spared such nerve invasion; these are the heart, muscles, and respiratory and digestive tracts (58, 77). No satisfactory explanation is offered for such a differential hyperneurotization of embryonic organs and structures under the action of the tumoral NGF. Two alternative hypotheses have been tentatively suggested: *1)* The NGF released by the tumors collects in pools in some but not other organs and structures and exerts a neurotactic effect on sympathetic nerve fibers. In favor of this hypothesis is the exceedingly dense nerve network that forms in the mesonephros between the 6th and the 11th day and in the metanephros only in a subsequent developmental period, when this latter organ takes over the embryonic excretory functions. The formation of large neuromas of sympathetic nerve fibers inside the veins of embryos bearing implants of sarcomas is also suggestive of a similar effect. *2)* As an alternative hypothesis, one can conceive of a sort of a built-in barrier against sympathetic nerve fibers in some organs that from early embryonic stages receive a large quota of parasympathetic nerve fibers. These are the heart and the respiratory and digestive tracts (58, 77).

The finding of an exceedingly large number of sympathetic nerve fibers in the organs and in veins of embryos bearing transplants of mouse sarcomas 180 or 37 first suggested that the tumors release into the circulation of the host a diffusible growth factor responsible for the observed effects. The hypothesis was tested by implanting fragments of sarcomas 180 or 37 onto the chorioallantoic membrane of 4- to 6-day chick embryos. In this position the tumor and the embryo share the circulation but no direct contact is established between the neoplastic and the embryonic tissues. Yet in this experimental series also the sympathetic ganglia underwent marked hypertrophic and hyperplastic changes, and sympathetic nerve fibers produced in excess by these ganglia invaded embryonic viscera and blood vessels (62, 65). These results gave definite evidence in favor of the hypothesis that a diffusible growth-promoting agent is released by these tumors (58).

A systematic comparative study of the volume increase of sympathetic ganglia of embryos bearing intraembryonic or extraembryonic tumor transplants showed that growth effects are more marked in the former than in the latter experimental groups. Furthermore, ganglia adjacent to the tumor in intraembryonic transplants are larger than ganglia remote from the graft. These results, described further in later sections, indicate that the magnitude of the growth effects elicited by the tumoral NGF correlates with the amount of this factor present in the circulation as well as in end organs. Differential NGF release by sarcomas 180 and 37 from different sources and by other mouse sarcomas is considered on p. 545, 546.

Daily injections of the NGF extracted from snake venom and from mouse submaxillary salivary glands into the yolk of developing chick embryos from the 7th to the 10th day of incubation (50 μg of the purified protein per dose) result in a marked increase of sympathetic ganglia of the host, hyperinnervation of the viscera, and penetration of sympathetic nerve fibers into the veins of the host. The biological effects are therefore in all respects similar to the effects elicited by implanted fragments of mouse sarcomas 180 or 37 (74, 75).

NGF effects in newborn and adult mice. Although the implantation of fragments of mouse sarcomas 180 or 37 in newborn and adult mice did not evoke any growth effect in sensory or sympathetic ganglia of the host (Levi-Montalcini, unpublished results), the injection of the purified salivary nerve growth factor in newborn and adult mice called forth a massive enlargement of sympathetic ganglia of treated animals. Lack of any appreciable effect in sensory ganglia of these mice is in line with the results obtained in the chick embryo and shows once more that these cells are receptive to the growth-promoting activity of the NGF only during a restricted period of their early embryonic life (71).

The purified salivary NGF was injected subcutaneously in newborn mice in the amount of 0.05 ml/g of body weight, the injections being repeated daily for periods ranging from 3 to 27 days. The NGF protein is present in this preparation at the concentration of about 300 μg/ml. Since newborn mice weigh at birth between 0.5 and 0.7 g, they received therefore 15–20 μg per dose of the NGF protein. In biological units this would correspond to 1500–2000 units per dose.

The injected and control mice at the end of the experimental period were sacrificed, and the sympathetic para- and prevertebral chain ganglia were dissected out and compared. The superior cervical ganglia and in some instances also the stellate and thoracic ganglia were sectioned serially at 10 μ, stained, and used for volume determinations, mitotic counts, and counts of the whole nerve cell populations. The mitotic activity as determined between the 3rd and the 9th day after birth is much higher, and the total nerve cell population is between 2 and 3 times larger in NGF-treated than in control mice. At 9 days mitotic activity comes to an end both in normal and NGF-treated animals. In the subsequent developmental periods, the injections of the NGF call forth hypertrophic but no hyperplastic effects (71). The overall size increase of para- and prevertebral sympathetic ganglia varies from one specimen to another and depends on the strength and purity of the injected NGF preparation. In earlier experiments the superior cervical ganglia of mice injected from the day of birth to the 27th day were found to be 6 times larger than the same ganglia of littermates (71). In subsequent experiments performed with an NGF deprived of toxic side effects and further purified, higher doses of the active protein could be injected. The superior cervical ganglia of mice treated with this preparation from birth to the 9th day reached a volume about 12 times as large as those of controls (62, 63). These results bring to light one of most interesting and still unexplained aspects of the phenomenon, namely, the almost unlimited growth potentialities of sympathetic nerve cells of infant mice treated with the NGF.

Newborn mice injected with a non-highly purified NGF exhibited side effects such as stunted growth, lack of hair, opening of the eyelids 6–7 days earlier than littermates, and eruption and calcification of the inferior and superior incisors 3–4 days earlier than untreated mice (64, 71). Subsequent studies confirmed these results (15). Stunted mice recovered when the NGF treatment was discontinued; they resumed growth and 3 months later did not differ noticeably from littermates. Since these prominent side effects did not occur in mice injected with a more highly purified NGF, the conclusion was reached that "obviously we are dealing with two factors, one of which was removed in the process of purification" (71, p. 379). The validity of this hypothesis was proved 2 years later, when Cohen succeeded in isolating the factor responsible for the precocious opening of the eyelids and incisor eruption in another protein molecule that became known as the "epidermal growth facter" (26). Since the chemical nature and the biological properties of this new specific factor resemble in many respects those of the NGF they are briefly considered later.

Extensive cytological examinations showed that the hypertrophic neurons differ from controls not only in their larger size but also in the more intense basophilia and marked increase in size and number of the nucleoli and increase in neurofibrils (71). Determinations showed that the norepinephrine content in the NGF-treated ganglia increases proportionally to the cell volume increase (3–4 times in the cases examined), whereas epinephrine is barely detectable and compares with that of controls (33).

Peripheral distribution of sympathetic nerve fibers in NGF-treated mice. Histological analysis of experimental and control mice stained with specific silver techniques and sectioned serially showed a marked hyperinnervation of the viscera, the external tunica of blood vessels, and the hair bulbs in NGF-treated mice. A detailed analysis of the peripheral distribution of adrenergic nerve fibers in control and NGF-treated mice, performed with the histochemical technique devised by Falck and others (36), to visualize the adrenergic transmitter in sympathetic post-ganglionic neurons and in their fibers (87), gave an impressive demonstration of the increased density of the adrenergic ground plexus in the iris, submaxillary and parotid glands, and intramural ganglionic plexuses of the intestinal tract in the NGF-treated mice. Furthermore, varicose adrenergic terminals were also detected in areas that are normally devoid of adrenergic innervation. For example, the normally negligible amount of parenchymal adrenergic fibers in the sublingual gland was increased to a considerable number of terminals surrounding the acinar portion of the gland. Similar numerical increase in the adrenergic nerves was found in the extraorbital lacrimal glands, the pancreatic insulae, and the adrenal cortex. One singular exception to the rule that sympathetic adrenergic ganglia and their peripheral fibers undergo size and numerical increase in NGF-treated animals was found in the ganglionic complexes that innervate the vas deferens and the uterus. Cell counts and volume determination of these ganglia showed no appreciable increase in experimental mice (Levi-Montalcini, unpublished observations). The fluorescent histochemical technique utilized by Olson likewise showed no increase in the adrenergic ground plexus in the vas deferens (87). Increased catecholamine content in ganglia and several peripheral organs of mice, rats, and kittens injected with the NGF has been demonstrated with quantitative fluorometric techniques by other authors (34).

CHARACTERISTICS OF THE IN VITRO GROWTH RESPONSE OF SENSORY AND SYMPATHETIC GANGLIA

Fibrillar Halo

As briefly mentioned previously, the in vitro assay of the NGF was first devised to test the effect of mouse sarcomas on sensory and sympathetic ganglia dissected out from 8-day chick embryos and cultured in a semisolid medium near the neoplastic tissue (78). The same test was then used to assay the biological activity of purified NGF added at different concentrations to the culture medium, or of fluids and organ extracts suspected of harboring the NGF. The production of the fibrillar halo is in fact such a consistent and stereotyped phenomenon that the "halo effect" became the unequivocal sign for the presence of the NGF in biological fluids or tissue extracts, even when present only in trace amounts, from the time this test was first devised 15 years ago to the present.

Since the size of the halo and the density of nerve fibers that form it vary with the amount of NGF present in the culture medium (although the variations are not proportional to the NGF protein content in the culture medium), it became necessary to grade the response and to establish the biological unit of the NGF activity (60, 62). This is defined as the amount of purified NGF required to produce in a 12- to 18-hr period a fibrillar halo of uniform density around the ganglion, where each ray is represented by a nerve fiber ending abruptly in the medium and forming with the other fibers a circular or ellipsoidal, perfectly geometrical ring around the explant. The amount of NGF protein required to elicit this effect corresponds to about 1 ng (10^{-9}) of the NGF purified according to Cohen's procedure (24). There is, however, a certain complexity in the apparently simple problem of establishing a dose-effect relation, as well as much confusion in the literature in this respect, as a result of the fact that with super- or submaximal doses of the NGF these quantitative relationships became obscured. It is therefore necessary to consider the whole spectrum of the NGF's effects from the highest to the lowest concentration of the extract to be assayed, if one wishes to establish the amount of NGF present in the preparation. Unfortunately this criterion has been seldom used and, as a result, many data in the literature on the NGF specific activity in solutions or organ extracts under investigation are of doubtful significance.

An additional reason for conflicting results in the literature may be due to the way the in vitro tests for NGF activity are run in different laboratories. This is quite an important point in evaluating the results of these tests. When we first devised the tissue culture technique and then used it to assay the NGF from snake venom or mouse salivary gland, we found that we could evoke the fibrillar halo characteristic of one biological unit with dilutions as high as 10^{-10} or even 10^{-11} μg NGF/ml of medium if the same pipette was used for subsequent dilutions of the NGF preparation. We then found that if instead the pipette was changed for every successive dilution, only traces of NGF activity were found in dilutions higher than 10^{-8} μg NGF/ml of medium. These results can be explained by the assumption that a few NGF molecules may adhere to the glass pipette and in this way are carried from one dilution to the other, thus giving a very inaccurate evaluation of the activity of the NGF. An extraordinarily high NGF activity found by some laboratories in the crude extract or in the further purified extract from mouse salivary glands may well be due to neglect of this rule of procedure.

Next we briefly consider the characteristics of the in vitro growth response to higher or lower NGF doses.

A two- to fivefold increase in the NGF concentration that corresponds to 1 biological unit, or in other words an increase from 1 up to 2–5 NGF biological units in the culture medium, results in the production of a fibrillar halo much denser and shorter than that elicited by 1 biological unit. A 10- to 100-fold increase in the concentration of the NGF (10–100 biological NGF units in the medium) calls forth, respectively, an exceedingly short fibrillar halo or no fiber outgrowth at all. This effect examined in ganglia in vivo, or stained and studied in toto in the whole mount, was at first taken as evidence of an inhibitory effect elicited by high doses of NGF (60, 64). Closer inspection of hundreds of ganglia cultured for 12–24 hr in the presence of 10–100 NGF biological units, then fixed and stained with specific silver techniques and sectioned at 10 μ, offered a different explanation of the lack of fiber outgrowth from the explanted ganglia. In the presence of a high NGF quantity the nerve fibers, instead of growing in a radial direction around the explant, build a dense fibrillar capsule around the ganglion. The thickness of the capsule increases in width with increased amounts of NGF; the production of neurofibrillar material, therefore, not only is not inhibited but is further enhanced by the addition to the culture medium of supermaximal NGF doses. The study of sensory ganglia cultured in the presence of one or more NGF biological units and in control media shows an additional feature of the NGF effects. Whereas in media deprived of the NGF the nerve cells undergo an almost total disintegration after 18–24 hr of culture, a high percentage of sensory nerve cells is present in cultures enriched with 1 biological unit of the NGF. Where 100 or 1000 NGF biological units are present in the medium, all sensory nerve cells are well preserved at the end of the 1st day of culture (62, 70). Although it remains to be explained why supermaximal doses of the NGF elicit the formation of a fibrillar capsule rather than a fibrillar halo, these results are against the assumption of inhibitor or toxic effects by high NGF doses. Submaximal NGF quantities, added to the culture medium in amounts corresponding to 0.5, 0.1, or 0.05 NGF biological units, result in the formation of a fibrillar halo of increased length and decreased fiber density. Eventually only a few irregular nerve fibers branch around the explant. Since ganglia in control media may occasionally also produce some nerve fibers in the first 24 hr of culture, it is difficult to evaluate the quantity of NGF present in the medium at these very low concentrations. Reliable data can be obtained, therefore, only by adding the tissue extract or the body fluid to the culture medium at progressively higher dilutions and by then inspecting the entire experimental series. If the undiluted extract or fluid contains less than 0.5 NGF biological units, the results should be considered as negative. The possibility should always be kept in mind, however, that the lack of the fibrillar halo may result from a supermaximal rather than submaximal NGF quantity in the extract to be assayed.

Short- and Long-Term Cultures of Sympathetic Ganglia in Semisolid Media

Short-term cultures of sympathetic ganglia, dissected out from chick embryos between the 7th and the 9th day of incubation and cultured in control media or in media enriched with the NGF, have given results so similar and in fact so nearly identical to the results described for sensory ganglia that further description is unnecessary. It is of interest to consider instead the NGF effects on sympathetic ganglia from newborn mice or chick embryos maintained in vitro for long periods of time in a medium containing chick embryo extract, human placental extract, ox serum ultrafiltrate, and a balanced salt solution. In the experimental cultures the NGF was added in the amount respectively of 1, 10, 100, and 400 biological units (32). The NGF-rich and control cultures were examined every few days up to a month. The best effects were obtained when the NGF was added to the culture medium at the supermaximal doses of 100–400 biological units/ml. Cultures examined in vivo or after fixation and toluidine stain consisted of a sympathetic nerve cell population much larger than in control cultures. The cytoplasm of experimental sympathetic cells showed denser Nissl bodies and nuclei and nucleoli twice as large as in controls. Likewise neurofibrils were markedly increased as shown by treatment with specific silver preparations. The potent stimulating effect of the NGF in these long-term cultures was also shown in electrophysiological experiments on sympathetic chain ganglia from newborn mice. The nerve impulse propagated for much longer distances in the treated ganglion chains than in controls. These results suggest a more precocious maturation of sympathetic nerve cells in the presence of the NGF than in media deprived of this factor (32).

NGF Effect on Dissociated Sensory and Sympathetic Nerve Cells in Liquid Media

Additional evidence for the vital role played by the NGF in the survival and growth of embryonic sensory and sympathetic nerve cells came from experiments on nerve cells dissociated with trypsin and cultured in liquid media consisting of Eagle amino acid solution or Eagle medium supplemented with 10% horse serum. In the experimental cultures the NGF was added in the amount of 5–100 biological units/ml. Neither the Eagle solution alone nor this solution supplemented with 10% horse serum is adequate for survival and further growth of embryonic sensory and sympathetic nerve cells (67). At the end of the first 24 hr of culture most of the dissociated nerve cells have undergone disintegration, and 48 hr later the control cultures consist of fibroblasts and satellite cells with few or no nerve cells. In the presence of NGF, nerve cells survive in excellent condition and undergo vigorous growth, as proved by the very dense nerve net that covers the dish bottom. These cultures can be kept alive for as long as a month. These experiments prove that the requirement of nerve cells for this protein is exceedingly low. Denaturation of the protein by heating abolishes the effect. Both observations rule out the hypothesis that the NGF merely serves as a source of cell proteins and suggest instead that it acts in a catalytic fashion.

NGF Effects on Sensory Nerve Cells at Ultrastructural Level

The electron-microscopic analysis by Crain et al. (32) of sensory ganglia cultured in semisolid media for periods ranging from 30 min to 48 hr, showed highly ordered mosaics of cytoplasmic granules in the NGF-treated cells, whereas the same structures were seen only seldom in control sensory cells. The hypothesis was proposed that these paired rows of granules might represent an early stage in the formation of additional ER cisternae or, alternatively, that they might be related to the early formation of neurotubules. Conclusions as to their possible significance, however, should wait for confirmation of these findings.

A more recent study of embryonic sensory cells, cultured in semisolid media for periods ranging from 4 to 12 hr in the presence of 10–100 biological units of the NGF and in control media, showed striking changes in the ultrastructure of the nerve cells. The nuclei developed an irregular contour and a more dense and clumped chromatin in comparison with the controls. In the cytoplasm the Golgi apparatus was dilated; in some cells there was a massive increase in ribosomes associated with an increased number of membranous elements of endoplasmic reticulum. Neurofilaments and neurotubules, sparse in control cultures, were already prominent after 4 hr of incubation in the presence of the NGF. After 12 hr the cytoplasm of most ganglionic cells was packed with large masses of neurofilaments (73).

BIOLOGICAL SOURCES OF NGF IN THE ORGANISM

It is apparent from the brief historical survey already given that a number of tissues, organs, and body fluids share the property of releasing the NGF. A systematic screening of potential sources of this factor, performed during the years from the beginning of this investigation to the present day, has revealed such a large number of sources that one is led to speculate as to the significance of its widespread occurrence and also about the possibility of detecting the site or sites of its production in the organism. [A similar situation obtains for two other specific growth factors, viz. the "erythropoietin-stimulating factor" and the "epidermal growth factor" considered later. More than half a century after the discovery of the erythropoietin-stimulating factor (21), its source of origin in the organism is still a matter of discussion. Likewise the origin of the epidermal growth factor remains to be identified.] Here we list and briefly consider the tissues, organs, secretions, and excretions endowed with a nerve growth-promoting activity, and then consider the problem of its source of origin in the organism.

Mouse Sarcomas: Characteristics of NGF-Releasing Neoplastic Cells

Mouse sarcomas 180 and 37 share a common laboratory origin from mouse salivary carcinomas that, through serial transplantations in mice, underwent progressive structural and cytological changes until they had lost all characteristics of their tissue or origin (66). Both tumors consist of a polymorphic cell population with small and large cells closely intermingled; one or the other cell type may prevail in tissue samples dissected out from one or the other mouse sarcoma. In vivo and in vitro experiments indicate that a much higher NGF activity is present in the neoplastic large-cell type (77). Tumors consisting only or mainly of the large cells are highly invasive on implantation into chick embryos. The release of large quantities of the NGF from these transplants is indicated by the massive invasion of the tumor by sensory and sympathetic nerve fibers and by marked hyperplastic and hypertrophic effects on sympathetic ganglia. These changes are seen in ganglia adjacent to the tumor and contributing to its innervation, as well as in those remote from the implant and sending their nerve fibers into the embryonic viscera, veins, and other structures. Implanted fragments of the same tumors consisting only or mainly of small cells evoke a milder growth effect. Here only sensory and sympathetic ganglia innervating the tumor undergo size increase; the effect in these instances is strictly local (76).

Both tumors, sarcomas 180 and 37, retain their nerve growth-promoting activity indefinitely on serial passages in vitro as cell monolayers (66). Differential centrifugation of the tumor homogenate has given evidence that the NGF protein is present in the microsomal fraction of neoplastic cells (31).

NGF Activity in Other Tumors, Tissues, and Body Fluids

The finding of the nerve growth-promoting activity in mouse sarcomas 180 and 37 raised the question of whether the observed effects are due to some specific growth properties of neoplastic cells. A systematic screening of a large number of tumors ruled out this hypothesis. A mild growth effect on sensory and sympathetic ganglia of chick embryos is elicited by mouse sarcoma 1 in experiments either in vitro or in vivo (13, 14, 78), whereas other equally rapidly growing tumors such as a variety of mouse carcinomas, melanomas, and neuroblastomas of mouse or rat origin (14) do not evoke any growth effect. Human neuroblastomas behave in a somewhat different way. Although the tumor's extract has no nerve growth promoting activity, the sera of children bearing the tumor have consistently higher NGF titers than sera of healthy children or children affected with other tumors. A somewhat milder nerve growth effect was obtained from sera of children bearing fibrosarcoma (18). The authors submit the hypothesis that neuroblastoma may elaborate but not store the NGF.

The high NGF activity released by mouse sarcomas 180 and 37 suggested the possibility that any mesenchymal tissue in a phase of active growth might also release this factor. The hypothesis was tested by producing granuloma tissue, explanting it in vitro, and assaying its effects on adjacent sensory and sympathetic ganglia of chick embryos. [A description of the technique used to produce granulomas is given elsewhere (64).] The experiments showed that granulomas from mice, rats, guinea pigs, rabbits, or monkeys elicit in vitro the same growth effects as fragments of mouse sarcomas 180 and 37. Granuloma extracts therefore possess a nerve growth-promoting activity of the same potency as mouse sarcomas.

Screening of normal embryonic and postembryonic tissues and of organs of chicks, mice, rats, and other species revealed the presence of the NGF in trace amounts in the chick embryo (16) and in a number of organs and tissue extracts of mice and rats (64). The NGF is also present in urine and saliva from mice (71). Of particular interest is the discovery of this nerve growth-promoting protein in the serum of all vertebrates assayed for this purpose, man included. Likewise of considerable interest is the finding that the NGF is a normal constituent of embryonic sensory and of embryonic and mature sympathetic nerve cells (64). The rather high NGF content of sympathetic ganglia is revealed by comparing the amounts of extracts of various tissues (measured on the basis of their protein content) that upon addition to 1 ml of the culture medium result in the formation of the fibrillar halo. This amount corresponds (see p. 541) to 1 biological unit of the NGF. Similar fibrillar halos are produced by the addition to the culture medium of 15,000 μg of sarcoma extract or 200 μg of the extract of sympathetic ganglia (22). The much higher NGF activity of snake venom and of mouse salivary glands is considered later.

NGF Activity in Snake Venom

The chemical properties of this potent nerve growth-promoting factor are considered in a later section. A study of the venom gland in vipers is in progress (Levi-

Montalcini and Angeletti, unpublished observations). Here we only mention that this gland has a structural configuration similar to that of mammalian parotid glands. One biological unit of the NGF is present in 3–6 μg of venom when added to 1 ml of tissue culture.

Mouse Submaxillary Salivary Glands and Their Role in NGF Production

One biological unit of NGF is present in 1.5 μg of protein of the crude extract of the submaxillary salivary glands of adult male mice (24). The discovery in 1958 of this new, most potent source of the NGF and its easy availability have channeled almost all subsequent work into the purification of this factor, its chemical characterization, and study of its biological properties. These topics are considered in other sections of this review; here we consider the morphological characteristics of this gland and the role of sex hormones in NGF production.

The mouse submaxillary salivary glands consist of two morphologically and functionally distinct components: the acini and the convoluted tubules. The former contain mucopolysaccharides; their high amylase content is regarded as the result of the in situ production of this enzyme. The convoluted tubules, in contrast, are characterized by their high content in proteolytic and hydrolytic enzymes and their sexual dimorphism (56). The male gland differs from the female in the structural and biochemical configuration of these tubules. In males they undergo a marked size increase at puberty; proteolytic and hydrolytic enzymes begin to be synthesized in these tubules at that time (52). Additional evidence for the role of sex hormones in calling forth these structural and biochemical changes has come from other experimental studies. Testosterone injections in female mice result in a sudden increase in the size of the tubules and in a parallel increment in their protease content, whereas castration sharply decreases the size of the tubules and the enzyme content in the male gland (52). There is extensive direct and indirect evidence for the localization of the NGF protein in the convoluted tubules of these glands (24, 64). The NGF is absent from the glands of newborn mice at a time when the tubules have not yet undergone differentiation; it appears at puberty and increases in subsequent developmental stages to reach a plateau in the adult male. Its concentration is 10 times higher in male than in female glands. Injections of testosterone result in a marked increase in the NGF content in the female gland, whereas castration sharply decreases the content of the same protein in males (20, 68). Further evidence for the localization of the NGF in the gland's tubular components is obtained by treating frozen sections of the gland of adult males with specific fluorescent antiserum to the NGF. The tubular but not the gland's acinar portion shows in these preparations an intense and localized fluorescence (43, 65). Injections of the antiserum to the NGF in the puberal mice result in marked regressive changes in the epithelial cells lining the tubules, and this provides additional evidence for the localization of the NGF in these cells (20). Other similar investigations indicate that the salivary glands play an active role in maintaining the NGF level in the blood of mice (103); but, paradoxically, no proof was obtained in favor of the hypothesis that the salivary glands are essential or even necessary for the NGF production. Against this assumption are a number of observations that are listed here only briefly. *1*) Extirpation of the submaxillary salvary glands does not result in any adverse effect in sympathetic nerve cells, such as one would expect if the specific growth factor were produced by these glands as a hormone to be supplied to its target cells (74). *2*) Sympathetic nerve cells of male and female mice are the same size, and yet the male salivary glands have a much higher NGF content than the female glands. *3*) The NGF has not been detected in the salivary glands of other mammals even though it is present in trace amounts in the serum of all mammals tested.

The question has been raised, but still is not answered (70), whether this protein present in such a large quantity in male mouse salivary glands might not serve other purposes connected with digestion or perhaps with the gland's other function, namely the production and release of toxic substances in the saliva. The fact that this protein is present in snake venom makes this hypothesis plausible.

Whether the submaxillary glands synthesize the NGF protein in the gland's tubular portion or whether this protein is produced elsewhere, then exported and stored in the tubules, has been answered recently in favor of the first alternative. In vitro and in vivo experiments performed in three laboratories (19, 70, 103) gave decisive evidence for the in situ production of the NGF. In view of the considerable interest of these findings, we will briefly mention some of the experimental work that established this point.

The in vitro experiments consisted of the incubation of slices of salivary glands from adult male mice in Eagle medium supplemented with radioactive amino acids. Incubations were carried out for 60 min and 2 hr. The incorporation of labeled amino acids into protein was found to increase linearly with time. Likewise the NGF underwent a similar linear increase (70).

The in vivo experiments consisted of the injection with a microsyringe of radioactive amino acid in one of the two lobes of the salivary gland in adult male mice. The animals were sacrificed after 2 hr; the two lobes were separately homogenized, and the experimental and control lobes were pooled in two groups and then centrifuged. The results clearly showed a differential labeling of the NGF from the two lobes. In the lobe injected with the radioactive amino acids the labeled NGF was markedly higher than in the other lobe. These results give additional evidence for the in situ production of the NGF present in the salivary glands. In fact one would expect the NGF of the two lobes to have the same, or nearly the same, specific radioactivity if this protein were synthesized not in the salivary glands but elsewhere and then exported to the salivary glands and stored there (70).

Convincing indirect evidence for the in situ production of NGF was obtained by Burdman and Goldstein (19). These authors injected intravenously the specific antiserum to the NGF in mice. It was assumed that the antiserum present in the circulation of the treated animals would interfere with the accumulation of large quantities of the NGF in the salivary gland if this protein were produced elsewhere and then transferred through the circulation into the salivary gland. Experiments performed in many different conditions and in animals at different developmental stages consistently showed no effect by the antiserum on the content of the NGF in the submaxillary salivary gland. Also it was shown by the same authors that intravenous injection of large quantities of NGF in mice could not be localized in the submandibular gland cells of the same animals as determined by fluorescent-antibody technique. They therefore came to the conclusion that the NGF is synthesized and stored in cells of the tubular portion of this gland.

GROWTH INHIBITION OF SYMPATHETIC NERVE CELLS BY ANTIBODIES TO NGF

The selective and permanent destruction of 90–95% of sympathetic nerve cells in newborn mice and other newborn mammals injected with a specific antiserum to the NGF was first reported in 1960 (72, 75). The antiserum was produced by Cohen by injecting the purified NGF protein into rabbits with Freund adjuvant, using techniques reported in the original articles (24, 65). Here we shall only briefly consider the in vivo and in vitro effects of the antiserum, since they bring once more into focus the key role of the NGF in the life of sympathetic nerve cells.

The antibody titer is first assayed in vitro. The assay technique consists of the addition of 0.2 ml of the antiserum at progressively higher dilutions to 0.2 μg of the purified NGF in 0.2 ml of isotonic sodium chloride (65). The mixture is incubated at 26 C for 1 hr and then added to the culture medium in the proportion of one-third of the total volume of the medium. Readings are made after 12 and 24 hr of incubation at 37 C. The titer of the antiserum is expressed as the reciprocal of the highest dilution that results in complete inactivation of the nerve growth factor. Under the best conditions the titer reaches the value of 2400 or even higher. Aliquots of this antiserum are then injected subcutaneously into newborn mice, rats, or other mammals in the amount of 0.05 ml/g of body weight. Daily injections of the antiserum in newborn mice for 3–5 days result in the permanent destruction of 95–98% of the sympathetic nerve cell population of the paravertebral sympathetic chain ganglia (72). A somewhat higher percentage of sympathetic nerve cells persists in the prevertebral ganglionic complex; this varies according to the potency of the antiserum. As a rule some 15–20% of these cells persist in the celiac ganglion, the most extensively investigated of all prevertebral ganglia. The persistence of adrenergic ganglionic complexes that innervate the male and female sex organs is the only instance of adrenergic sympathetic cells resistant to the antiserum (69). These ganglia also fail to respond to the nerve growth factor (62, 63).

Histological, physiological, and pharmacological studies on immunosympathectomized animals are reported in detail in a number of publications (53, 69, 104, 105). For the purpose of this review, it is of interest to consider the earliest effects elicited by the antiserum on its target cells, as revealed by examinations using both optic and electron microscopes. For reasons of convenience the superior cervical ganglia of newborn mice injected in vivo with the antiserum, or dissected out and incubated for short time periods in presence of the antiserum, were the object of choice in most of these investigations.

Regressive changes in sympathetic nerve cells are already visible by optic microscopy 12 hr after the first injection of the antiserum. Whereas in control ganglia mitotic activity in the superior cervical and other sympathetic ganglia is very high at birth and continues to the 9th day, in ganglia of antiserum-treated mice mitotic figures are rarely observed and most of these appear atypical. A number of pyknotic cells are scattered throughout the ganglia; the number of dead and degenerating cells increases rapidly between the end of the 1st and the 2nd subsequent days. By the end of the 2nd and 3rd day of incubation, all mitotic activity has ceased. Neuroblasts are much smaller than controls and the cytoplasm is almost deprived of ribonucleic acid, as shown by a failure to stain with basic dyes and by the fact that nucleoli are barely visible (72). Dead cells and cell debris are scattered in large numbers among the neuroblasts and satellite cells that are present in larger proportion than in controls. In the following days satellite cells also undergo regression and the ganglia are reduced to diminutive sclerotic nodules barely detectable with the aid of a dissecting microscope.

Electron-microscopic studies of the same ganglia give evidence of the extraordinarily rapid effects elicited by the antiserum in vivo and in vitro. Two hours after the injection of the antiserum the sympathetic nerve cells show folding of the nuclear envelope, condensation of chromatin into large and small masses, dilation of the vacuolar system, and mitochondrial alterations. The in vitro effects of the antiserum are even more marked; extensive cytolytic lesions are already apparent after 4 hr of incubation (91). A comparison of these changes with those seen in electron-microscopic studies of tumor cells cultured in the presence of specific antisera shows that there are substantial differences between the latter and the effect mentioned above. Nuclear damage seems to be practically absent in the initial phase of the reaction of neoplastic cells to antisera (42, 47, 57). On the other hand, these data and other studies now in progress (unpublished observations from our laboratory) suggest that the nucleus of sympathetic nerve cells is the primary target of the antiserum to the NGF.

Further evidence for the precocious and lethal effects caused by the antiserum to the NGF comes from studies of the incorporation of uridine into RNA in sympathetic nerve cells. Four hours after the injection of the antiserum the incorporation of labeled uridine in the superior cervical ganglia of newborn mice is reduced to half of the control values (92).

CHEMICAL PROPERTIES OF NGF

NGF From Snake Venom

After the first discovery of NGF activity in snake venom, various samples of dried venoms from species of the three families of poisonous snakes were tested for their biological activity. All venoms examined showed similar growth-promoting properties. The crude venoms of *Elapsidae* and *Viperidae* were found to be 2–4 times more potent than *Crotalidae* (23). The venom of the moccasin was used by Cohen for purification of the NGF (22); approximately 25% of the activity present in 1 g of crude venom was recovered in 5.2 mg of protein, a purification of about 40-fold.

On a Spinco analytic ultracentrifuge, only a single component was detectable, with an S_{20} of 2.2 S; the molecular weight was estimated to be about 20,000. The 280/260 absorption ratio was found to be 1.3; on acid hydrolysis and two-dimensional paper chromatography, the amino acid pattern was quantitatively identical to similar chromatograms prepared with bovine albumin. By the orcinol procedure, 1.6% of hexose was found to be present. Two additional lines of evidence support the view that the biological activity is associated with a protein: *1*) the biological activity was completely destroyed on incubation with proteolytic enzymes (trypsin, pepsin, chymotrypsin); *2*) the biological activity was lost on incubation with antiserum to snake venom. The possibility that the growth-promoting activity of the venom was due to one of the enzymes known to be present in the venom was carefully investigated by Cohen. Activity of a number of enzymes was assayed in snake venom and in the purified NGF but none was found to be associated with the nerve growth factor (22, 23).

In the original purification procedure devised by Cohen, which yields a pure and active NGF protein of about 20,000 mol. wt., a treatment with 6 M urea is used in the first step in order to avoid loss of biological activity. When milder fractionation procedures are used, the NGF activity is found associated with proteins in various molecular aggregates (2). Thus, when crude venom is fractionated on Sephadex G-100, the biological activity appears in a broad area corresponding to molecular weights around 40,000. Upon chromatography on a DEAE-cellulose column at pH 7.4, all the biological activity is recovered in the unabsorbed peak; when this component is analyzed by sucrose density-gradient centrifugation, the NGF activity appears to be associated with proteins of molecular weight in the range of 20,000. By chromatography of this fraction on a CM-cellulose column at pH 5, two active components are obtained. Both bind strongly to the column and are eluted with high salt concentration. On disc electrophoresis, on acrylamide gel at pH 6.6, the first component gives two closely spaced bands and the second component a single sharp band moving fast to the negative pole. In the ultracentrifuge these forms of NGF appear to have sedimentation coefficients where S_{20} is 2.5 and 1.5, respectively. Molecular weights of about 22,000 and 12,000 were calculated by gel-filtration experiments (2). The specific activity of the larger NGF molecule is 2 times higher than that of the smaller NGF species. Complete amino acid analysis performed on these two forms of NGF from *Crotalus adamanteus* resulted in an almost identical amino acid composition for both of them. This finding suggests that they have the same primary structure and that the major differences between them must reside in their molecular size and/or conformation. Immunochemical analysis, using a microcomplement fixation, appears to confirm this assumption, since the C′-fixation curves of the various NGF samples are completely superimposable.

On the basis of their amino acid composition, a minimal molecular weight of 5000 has been calculated for the NGF from *Crotalus adamanteus*. This may actually represent the molecular weight of the NGF monomer, although a biologically active NGF of this size has not yet been isolated. It is suggested that the snake venom NGF is a protein that exists in multiple aggregate states of similar subunits. Definite correlation between the observed biological activity and a given form of the NGF must await further characterization of the basic monomer and of each pure aggregate.

NGF From Mouse Submaxillary Salivary Glands

As already reported the submaxillary glands of adult male mice are by far the richest source of NGF. Crude extracts are active in tissue culture at concentrations as low as 1–2 μg/ml. Cohen devised a procedure that gives a 100-fold purification of the active agent with an overall yield of about 20% (24). The purified NGF appears to be a protein with an S_{20} of 4.33 S and a molecular weight estimated at about 44,000. Biological activity is destroyed on incubation with proteolytic enzymes and is inhibited by specific antibodies. When tested in tissue culture with embryonic sensory ganglia, the most purified preparation evoked the characteristic fibrillar halo at the concentration of 0.015 μg/ml of medium. By immunological tests (in vitro inhibition of biological activity) a low degree of cross reaction was observed between the salivary and the venom NGF (24). Further studies are now in progress in our laboratory on the molecular properties of the NGF present in snake venom and in salivary glands and on the evolutionary changes that this protein exhibits from reptiles to mammals.

In crude extracts of submaxillary glands the NGF activity appears associated with a large molecular aggregate of over 120,000 mol. wt. This aggregate, on gel-filtration and ion-exchange chromatography, shows one single component when analyzed at neutral pH on starch-gel electrophoresis (79, 93); when analyzed instead by electrophoresis on acrylamide gel at pH 8.3, the same high-molecular-weight NGF preparation shows a number of components of different mobility. On elution of the gel sections, part of the applied activity can be recovered, and this is all present in a fine sharp band near the origin; the same active component migrates to the negative pole at pH 6.6. Dissociation of the large molecular aggregate also occurs on dialysis against acetate buffer (0.01 M, pH 5); chromatography of this material on a CM-Sephadex column at the same pH gives three main components: a pregradient peak comprising a group of rather acidic proteins, a second peak emerging between 0.1–0.2 M NaCl, and a third peak eluted at 0.5 M NaCl. The first and second peaks are completely inactive, whereas the applied NGF activity is recovered in the last peak. When analyzed by disc electrophoresis, this fraction shows a single basic component, which in the ultracentrifuge has a sedimentation coefficient of 2.3 S_{20}.

Our viewpoint is that the other protein components separated by CM-chromatography do not contribute to the biological activity of the NGF. This belief is supported by the following arguments.

1) The CM fraction obtained according to Cohen or with slight modification of the original procedure (24) when assayed in vitro shows a specific activity higher than each of the fractions obtained during the purification steps from the crude extract to this final NGF preparation. This result, first reported by Cohen in 1960 (24), has been consistently confirmed in all subsequent studies performed in our laboratory.

2) This same NGF form is fully active when injected in vivo. The impressive growth effects elicited in newborn mice were, in fact, obtained by injecting this NGF preparation.

3) The antiserum to the NGF was obtained by immunizing rabbits by injection of this NGF molecule. Thus, antibodies specific against this single protein call forth the destructive effects reported in this article. Since a full biological activity is present also in NGF preparations with higher molecular weight, the hypothesis is proposed that, as in the case of the snake venom NGF, the salivary NGF might also aggregate in multiples of the same NGF molecule. The two sedimentation coefficients of 2.3 and 4.4, alternatively obtained in various preparations of the CM fraction, suggest that the latter may represent a dimer of the former.

Other Models Proposed for NGF Molecule

Studies on the chemical structure of NGF isolated from the mouse submaxillary glands were performed by Schenkein and Bueker (94, 95), and more recently by Varon et al. (100, 101). Both models will be briefly considered below.

The purification procedure followed by Schenkein and Bueker consisted first of a precipitation of the gland homogenate with streptomycin sulfate, followed by alcohol precipitation of the supernatants; the ethanol precipitation was dissolved and salted out with ammonium sulfate at 40% and 90% saturation. This last precipitate, containing the biological activity, was then fractionated on a CM-column washed with alkali and then with distilled water until neutrality. Under these conditions the active material did not bind to the column and was eluted with the first protein peak. Further examination of this fraction by paper electrophoresis at pH 7.4 showed three bands that were labeled A, B, and C. In vitro bioassay of the individual bands showed no significant biological activity when tested separately, but when combined bands A and C gave a positive tissue culture test. According to the authors, the two components A and C were then obtained on a preparative scale by using electrophoresis on a starch column. Again, in vitro bioassay showed that fractions A and C singly possessed only a weak biological activity, whereas on recombination of the two fractions a restoration of the full biological activity followed. Dialysis after lyophilization of fractions A and C showed that C but not A was dialyzable. Quantitative amino acid analysis of fraction A showed the presence of all essential amino acids with the exception of methionine; a minimum molecular weight of 8600 was calculated. In the ultracentrifuge this fraction showed two peaks with S_{20} of 2.41 and 4.24 respectively. Data on a quantitative amino acid analysis of fraction C allowed the calculation of a minimum molecular weight of about 3500.

In comparing these data with those originally reported by Cohen, Schenkein and Bueker suggested the possibility that the species with an S_{20} of 4.24 might represent a dimerized form of A that in combination with C (mol. wt. 3500) would give the fully active NGF form of about 42,000 mol. wt. In summary, according to these results the active salivary NGF appears to be a protein, which under specific experimental conditions splits into two inactive components, one of which (C) seems to be a rather small polypeptide; recombination of the two parts restores the biological activity. Confirmation of these results from other laboratories is greatly to be desired.

A second model for the NGF molecule has been proposed by Varon et al. on the basis of association and dissociation experiments (100, 101). These authors used a three-step procedure to isolate the NGF from crude extracts of submaxillary glands; they used gel filtration on Sephadex G-100, chromatography at neutral pH on DEAE-cellulose, and a final gel filtration on Sephadex G-150. The final product accounted for 2% of the protein in the gland and, according to the authors, for 80% of the activity. The minimum amount in protein content required for giving a 3+ response in tissue culture (corresponding to our definition of 1 biological unit of the NGF) would be of the order of 13 ng; this specific activity is in the same range as that obtained in our laboratory according to Cohen's procedure.

This fraction migrated on electrophoresis at neutral pH as a single component, which thus represented the NGF. Sedimentation analysis showed a major component with an $S_{20,w}$ of 7.1; the molecular weight of this NGF species was calculated to be approximately 140,000. This form of NGF appeared to be stable only within a limited range of pH (5–8); outside these limits the large aggregate dissociated into a number of subunits whose sedimentation coefficients were sufficiently close to each other as to sediment in a single boundary with an S_{20} of 2.6, indicating a molecular weight in the range of 30,000. The dissociation of the NGF appeared to be reversible within given pH limits; thus over 80% of the high molecular weight was recovered if the completely dissociated system at pH 3.8 was dialyzed at neutral pH. The same dissociation was found on exposure to pH 9.7, but after exposure to pH 10.3 the reversibility was less complete; exposure to pH higher than 13.3 resulted in no recombination when the pH was adjusted to neutrality. According to Varon et al. the biological activity of the large aggregate decreased to about 25% of its value on dissociation of the large aggregate and was restored after dialysis at pH values under conditions where physical measurements showed recovery of the high-molecular-weight NGF form. On acrylamide gel electrophoresis at alkaline pH, after irreversible dissociation of NGF at high pH, three groups of proteins were observed: *1)* a group of three acidic subunits, *2)* a second group of three components of intermediate charge, and *3)* a third group consisting apparently of a single basic subunit. Activity measurements on gel sections indicated that only the basic subunit had NGF activity. By chromatography on CM-cellulose at acid pH the three groups could be separated from each other, and again the most basic component was found to possess NGF activity, though considerably lower than that of the original NGF aggregate. From the various components separated on CM column, the original high-molecular-weight NGF species could be regenerated on dialysis at neutral pH, and apparently the activity level was increased more than twofold as a result of this recombination.

In summary, according to this model three types of different subunits account for the NGF molecule in its native state. Only one of these subunits, however, carries the specific biological activity; the others simply act as regulatory subunits "which permit control of biological activity through allosteric effects" (100, p.

1788). Varon et al. speculate about the possibility that interaction of the subunits may be of relevance in the extraganglionic environment or that they may serve to stabilize the activity of the active subunit or to maintain a constant concentration of this subunit through a dissociation equilibrium. The validity of this attractive model rests on the possibility of obtaining more convincing evidence in its favor. Perhaps the weakest point, as we see it, is that the NGF biological activity was judged only on the basis of in vitro experiments and never assayed in vivo. The same criticism applies to the model proposed by Schenkein and Bueker. As discussed previously, the in vitro response has only a limited value, and a twofold or even fivefold increase or decrease of biological activity may be very difficult to disclose for the reasons given on p. 542. Until more rigorous quantitative assays of the NGF become available using in vitro techniques, it seems highly desirable to grade this activity with parallel in vivo tests of each NGF preparation.

METABOLIC EFFECTS OF NGF

The growth effect elicited by NGF in the receptive nerve cells has been investigated biochemically; the main results of these studies are dealt with in the following sections.

Effect on Glucose Metabolism

Cohen first investigated the effect of NGF purified from snake venom on glucose oxidation by embryonic sensory ganglia cultured in vitro (23). He observed that the presence of glucose or mannose is required for the outgrowth of nerve fibers. If no energy source was added the outgrowth of fibers was initiated but ceased almost immediately. Glucose could not be replaced by D-fructose, L- or D-arabinose, D-ribose, D-galactose, glucuronic acid, gluconic acid, malic acid, L-ketoglutaric acid, succinic acid, or fumaric acid. Lactate and pyruvate could partially replace glucose. Using ^{14}C-labeled glucose he observed that the NGF increased the oxidation of carbon 1 of glucose by 41–54%; less of the C-6 of glucose was oxidized by the ganglia and the stimulation brought by the NGF was 12–26%. The presence of fluoride or cyanide did not prevent the nerve fiber outgrowth nor the increase of glucose oxidation. These results were confirmed and extended with NGF purified from the mouse submaxillary gland. The relative yields of $^{14}CO_2$ from glucose-1-^{14}C were investigated in sensory and sympathetic ganglia incubated in Warburg flasks with and without NGF (7). In all these experiments, carried out for a 3-hr period, there was a marked increase in the oxidation of glucose-1-^{14}C and only a slight increase in the yield of CO_2 from 6-^{14}C-glucose (6). Similar results were obtained when the NGF was injected in vivo in newborn animals and glucose oxidation was studied in the superior cervical ganglia. Control experiments with different nervous tissues known to be unresponsive to NGF did not show any change in glucose metabolism. The addition of fluoride to the medium at a concentration 10^{-2} M did not prevent the stimulation of oxidation of 1-C-glucose (6, 25). These results show that the NGF stimulates glucose utilization by the receptive nerve cells and that such stimulation appears to be mainly on a direct oxidative pathway. A number of enzyme activities representative of carbohydrate metabolism were also investigated in sensory and sympathetic ganglia incubated for different lengths of time with or without NGF. No evidence was obtained for some specific enzyme induction under the NGF stimulation (7); the common finding was instead that the NGF maintained all the enzyme activities tested at levels higher than controls. The effect was particularly evident in long-term experiments (6–24 hr) where most of the enzyme activities in the control ganglia underwent a progressive decline, whereas they were maintained at a steady-state level in the presence of NGF (4, 81, 84).

Effect on Lipid Metabolism

The growth response evoked by the NGF in vitro as well as in vivo is accompanied by an overall increase of synthetic processes in the receptive neurons. Experiments carried out with labeled acetate gave evidence that the rate of lipid synthesis is markedly stimulated in embryonic sensory and sympathetic ganglia incubated with NGF (5, 80). Within a 4-hr period of incubation the radioactivity incorporated into total lipids was, on the average, 60–90% greater in NGF-treated ganglia than in controls. Increased lipid synthesis was also observed in superior cervical ganglia from mice explanted in vitro and cultured for various periods of time (80). In experiments with embryonic sensory ganglia the lipid extracts were analyzed by thin-layer chromatography, followed by autoradiography. No significant differences in the relative intensity of labeling of the various components were detectable. The increase in the rate of lipid formation by the NGF-stimulated cells could be correlated with an increased availability of $NADPH_2$ (a necessary cofactor in the reductive synthesis of fatty acids) and/or with the induction of new enzyme molecules. The results of experiments with specifically labeled glucose indicate that the hexose monophosphate pathway is indeed activated in sensory or sympathetic ganglia under the influence of the nerve growth factor; the two major enzymes of the pentose pathway are specific for NADP, which is reduced to $NADPH_2$, and this may well serve as a reservoir of reducing potential for the synthesis of fatty acids. It is of interest to note that the stimulatory effect of NGF on acetate incorporation was completely abolished by actinomycin-D at a dosage that did not depress significantly the basal rate of incorporation in control ganglia (80). The significance of this finding is discussed below.

More recently the effect of NGF on lipid metabolism in embryonic sensory ganglia was investigated by using DL-2-^{14}C-mevalonic acid as a precursor (82, 83). Contrary to the results with ^{14}C-acetate, the presence of NGF in the culture medium resulted in a 25% decrease of incorporation of labeled mevalonate after 4, 6, and 8 hr of incubation. Moreover, almost all the mevalonic acid incorporated was present in a sterol-like compound that was isolated and partially characterized. This steroid appears to be present in embryonic ganglia in fairly large amount (5% of dry weight) up to certain developmental stages (13–14 days), and it decreases thereafter while cholesterol concentration gradually increases. Since NGF significantly decreased the synthesis of this steroid in embryonic ganglia explanted in vitro, the possibility that this metabolic effect might be related to some structural changes taking place in the stimulated nerve cells appears worth investigating.

Effect on Protein and RNA Synthesis

The growth effect elicited by the NGF on receptive nerve cells is accompanied by a rapid increase in net protein synthesis. Cohen first examined the effect of the growth factor on the incorporation of ^{14}C-lysine into protein of embryonic sensory ganglia cultured in vitro (23). Under the influence of the venom NGF, the incorporation was increased 58–72% within a 20-hr period. The addition to the culture medium of the amino acid analog *p*-fluorophenylalanine (0.2 mM) almost completely inhibited the NGF effect.

More extensive studies of the NGF effect on protein synthesis were carried out with chick embryonic sensory and sympathetic ganglia and with superior cervical ganglia from newborn mice. The incorporation of a number of radioactive amino acid was consistently enhanced by the presence of the nerve growth factor in the medium even after only 3 hr of incubation. Other experiments indicated that the intracellular protein turnover is also stimulated by the growth factor (3).

A preliminary analysis of soluble proteins from embryonic sensory ganglia incubated for various periods of time in the presence of radioactive precursors revealed that the NGF selectively stimulates the rate of synthesis of some classes of proteins. When total soluble proteins were analyzed by sucrose density-gradient centrifugation, the radioactive profile of the NGF-treated samples consistently showed a relative increase in the labeling of heavier proteins (prealbumin zone) as compared with the controls. By column chromatography on DEAE-cellulose, soluble proteins from ganglia incubated 3–8 hr with and without NGF were separated in a discrete number of components; here the experimental pattern showed a marked increase in the labeling of the more acidic proteins as compared with controls. Selective stimulation of more negatively charged proteins was also observed when the protein extracts were fractionated by disc electrophoresis on acrylamide gel (39). These data indicate that the NGF modifies to variable degrees the rate of synthesis of different classes of proteins The addition of puromycin to the medium suppressed amino acid incorporation in both control and NGF-treated ganglia and inhibited the outgrowth of nerve fibers. Under these conditions, however, the stimulatory effect of NGF on RNA synthesis was not suppressed.

The stimulation of RNA synthesis, as indicated by incorporation of radioactive precursors, appears to be one of the earliest metabolic effects elicited by the NGF on explanted ganglia cultured in vitro. In early experiments carried out with snake venom NGF, the incorporation of ^{14}C-adenine into RNA increased from 40 to 70% in a 20-hr period (23). The stimulatory effect of NGF on the incorporation rate of labeled precursors into RNA is even more striking in shorter periods of incubation. When embryonic sensory ganglia were incubated in a minimum essential medium containing uridine ^{3}H, the incorporation rate into NGF-treated ganglia exceeded that of control ganglia by more than 100% as early as 2 hr of incubation (3, 98). Time-sequence studies indicated that the NGF stimulation of RNA synthesis preceded that of protein synthesis (3, 62). Furthermore, ganglia preincubated with puromycin were still responsive to the NGF stimulation of RNA synthesis, thus suggesting that this effect is not mediated by de novo synthesis of some protein within the cells. Actinomycin D, at concentrations of 0.5 and 1 μg/ml, was found to cut down uridine incorporation to 15% in both control and experimental ganglia; at this concentration protein synthesis was only slightly inhibited for the first 3–4 hr of incubation but NGF stimulation of amino acid incorporation was completely abolished (3). The above findings seem to indicate that one of the primary effects elicited by the growth factor is at a step that involves new syntheses of DNA-primed RNA, which in turn is responsible for the enhancement of protein synthesis. Density-gradient profiles of ^{3}H-uridine-labeled RNA extracted from ganglia incubated with and without NGF did not show any significant qualitative difference (98). The effect of NGF on the total RNA content and base ratios was also investigated in isolated neuroblasts from embryonic sensory ganglia cultured in vitro (17). A significant increase was found in the amount of RNA per cell after 17 hr of incubation only in the presence of NGF. No differences, however, were observed in base ratios between NGF-treated and control cell RNA's. In a more recent study Toschi et al. (99) analyzed the RNA extracted from ganglia incubated 5 hr with and without NGF in a medium containing ^{32}P-orthophosphate. The RNA's were chromatographed on a methylated albumin column and the various components were analyzed for their nucleotide composition. The main labeled component outlined by chromatography appeared to be a heavy, nonribosomal RNA, with a predominant nucleotide composition of a DNA-like type. These results indicate that during the first 5 hr of incubation, there is only limited synthesis of R-RNA by sensory ganglia with or without NGF; during this time the cells produce high-molecular-weight RNA species with DNA-like base composition whose synthesis is greatly stimulated by the growth factor.

NGF AND OTHER SPECIFIC GROWTH FACTORS

The discovery of the epidermal growth factor (EGF) follows so closely that of the NGF, and its early history is so much interwoven with that of the NGF, that a brief report on its chemical properties, biological activity, and mechanism of production seems justified in a review devoted to the NGF. Moreover, both of these factors and also the erythropoietin-stimulating factor (ESF) show striking similarities in their biological activity as well as in other characteristics. Future studies on the NGF and on any of the other two agents should benefit therefore from a comparative analysis and a joint effort to elucidate the biological significance not only of one but all of them.

Epidermal Growth Factor

The epidermal growth factor (EGF) was discovered by Cohen in 1962 (25). Its action was first noted when the nerve growth factor was injected in newborn mice, but at that time it was not recognized as a separate factor (71). The treated mice differed from untreated littermates not only in the marked volume increase of the sympathetic ganglia, but also in their stunted growth, failure of development of hair, precocious opening of the eyelids, precocious eruption of the upper and lower incisors, and calcification of these incisors before those of controls. Although some effects such as dwarfism or failure of hair growth could be supposed to be the result of some toxic effect of the injected, not highly purified preparation, the eyelid and incisor effects could hardly be explained that way. Using a biological assay based on the precocious opening of the eyelids, which occurred 6 days before controls in infant mice, Cohen was able to purify and identify this factor. He found the activity present in a heat-stable, nondialyzable protein whose more distinctive chemical characteristic was the absence of phenylalanine and lysine. Its molecular weight, estimated from the number of amino acid residues, is approximately 15,000. The factor is antigenic but no biological activity has been detected of injection of the antiserum into newborn animals (25). Additional evidence for the protein nature of this factor was obtained from studies of its chemical composition, electrophoretic and immunophoretic analysis, and evidence of its inactivation by proteolytic enzymes. The purified EGF was injected daily into newborn mice and rats in the amount of 0.5–1 $\mu g/g$ of body weight for periods of time ranging from 8 to 12 days. Histological evidence was obtained that the precocious eyelid opening is a reflection of enhanced epidermal keratinization and an increase in the overall thickness of the epidermis (26–29). Similar epithelial changes take place on the whole body surface, as well as in the epithelium lining the oral cavity, esophagus, and other sectors of the digestive tract. Newborn rabbits and dogs are likewise receptive to the EGF. In rodents 12–20 days of age, the injection of the EGF for a 3- to 4-week period results in a visible increase in the diameter of the tail and a thickening of the epithelium covering the plantar surface of the feet, while the fur-covered skin is not affected by the EGF. In vitro studies on epithelial tissues dissected out from different sources disclosed a very potent growth-stimulating effect that materializes in short-term experiments (9, 27). These studies gave evidence for a direct stimulatory action of the EGF on the target cells and at the same time offered an opportunity for analyzing the metabolic effects called forth by the EGF and for exploring its mechanism of action at the cellular and subcellular levels. Of these more recent investigations and those still in progress, one needs to mention only that the EGF markedly stimulates the net accumulation of protein and RNA in the target cells. Cell-free extracts from epidermis cultured in the presence of the EGF are more than twice as active as extracts from control cultures in incorporating labeled amino acids (49). This difference has been accounted for in the ribosomal fraction of the cells (50). A primary action of the EGF on nuclear RNA is considered unlikely on the basis of previous experiments with actinomycin D and puromycin: the latter but not the former prevents the EGF effect (49). In this respect the NGF and the EGF effects seem to be at variance (see p. 558).

Other features of the EGF are worth mentioning since they are similar, and in some respects are in fact identical, to those of the NGF. The EGF is present in large quantity in the submaxillary glands of the adult male mouse but it is absent in the same gland of animals before puberty and could not be detected in the salivary glands of female mice or of other mammals. Extirpation of the mouse salivary glands performed in studies of the NGF (62) did not result in any appreciable "deprivation effects" in epidermal cell systems. No evidence was obtained from in vivo experiments as to whether the EGF is synthesized in the salivary glands or somewhere else and merely stored in these organs. Extrapolation from results obtained with the NGF (70) suggests the first alternative as the more likely. Tissue culture experiments indicate that serum from different sources stimulates epidermal growth (28). Argyris and Argyris (10, 11) and Patterson (88) observed an in vivo stimulation elicited by some tumors on adjacent epidermal tissue. It remains to be seen whether the EGF or some different growth-stimulating factor is responsible for these latter effects.

Erythropoietin-Stimulating Factor

An erythropoietin-stimulating activity was discovered at the beginning of the century by two French authors, Carnot and Déflandre, and named after its action, hemopoietine (21). In the 6 decades that followed, this agent was referred to as hemopoietine, erythropoietin-stimulating factor (ESF), or simply as erythropoeitin. It has been the object of investigations that have become particularly intensive in this last decade, as exemplified by the review article of Gordon in 1959 (46) and by the symposium on erythropoiesis devoted to the analysis of this factor that appeared in 1962 (51). The chemical nature of ESF and the main features of this growth response are considered briefly.

The ESF has been assumed by most investigators to be a glycoprotein of fairly large molecular size. Although at present, as pointed out by Goldwasser, all data are still too inaccurate to yield more than a rough estimate (44) the molecular weight appears to be of the order of 60,000–70,000. The active principle is said by most authors to be a non-species-specific, heat-stable, nondialyzable substance, destroyed on incubation with proteolytic enzymes, and endowed with antigenic properties (45). This last property was studied particularly by Garcia and Schooley (41), who, on extraction and purification of the ESF from the urine of severely anemic patients, succeeded in preparing a rabbit antiserum and in proving its in vivo anti-ESF activity. The assumption of a specific anti-ESF activity is based on the observation of a marked reduction in radioiron incorporation in normal mice receiving injections of this antiserum. According to the authors, "These results support the concept that erythropoietin is involved in the normal homeostatic control of erythropoiesis" (40, p. 57). This important finding is quoted because of the similarity to the effects elicited by a specific antiserum to the NGF (see p. 549, 550).

The ESF is normally detectable in the plasma and urine of animals rendered anemic by cobalt treatment, exposure to lowered oxygen pressure, or treatment with phenylhydrazine (45, 85). Its presence in the urine of patients suffering from hypoplastic anemia (40, 85) and in human blood (55) has also been reported. Both the plasma and urine of animals made anemic with various treatments are routinely used for the extraction and further purification of the ESF. The site of formation of the ESF is far from settled (45, 51). At present the kidney is reputed to be the major site of production of this factor, in spite of the fact that the ESF is found only in minimal quantity in this organ (38, 44, 45, 90). It is generally conceded that other tissues and also tumors (102) release ESF. Significant information about the mechanism of its action and the structural and biochemical events called forth by the ESF in the target cells was obtained from in vitro studies of the effect of this factor on bone marrow populations (1, 35, 54, 86, 89). Bone marrow stem cells respond to the ESF in vitro by increased stroma and RNA synthesis (35, 37). The stimulated RNA synthesis is abolished in the presence of actinomycin. These and other data reported by Goldwasser (44) suggest that the ESF exerts its action by regulating gene activity in erythroid cells. Tissue culture and perfusion techniques indicate that ESF increases the rate of red cell production and maturation as well as the release of erythrocytes into the circulation. A basic question remains unanswered (44): whether the ESF is a single substance with more than one site of action or whether the purest preparation yet available consists of more than one substance, each with its site of action. Goldwasser has recently proposed interesting models to account for the plural structural and biochemical events elicited by this factor (45).

Granulocytosis-Inducing Factor

A systematic screening of the submaxillary salivary gland extract for other potential growth-promoting activities resulted in the discovery of a protein fraction endowed with the property of greatly increasing the number of granulocytes in the circulation of newborn mice and of all other mammals so far investigated (8). Preliminary in vivo and in vitro experiments (70) suggest that a direct stimulatory action is elicited by this protein on bone marrow stem cells, which are thus channeled in the granulocytic differentiative pathways. Studies in progress are directed to the further purification of this protein fraction and to determine if it is another specific growth factor.

One may state that the growth response systems considered above share several common major features and differ from each other in only a few aspects. To summarize these features, the similarities and differences between the NGF, EGF, and ESF are listed.

1) All three factors were identified in protein molecules. *2*) They are normally present at a very low concentration in the blood, organ extracts, and urine (NGF and ESF) of individuals belonging to different species, man included. *3*) Two of these factors (NGF and EGF) are found in large quantities in the male mouse submaxillary salivary glands, which provide the source for the extraction of these factors. The kidney is reputed to play an important role in ESF production in spite of the fact that the ESF is present only in trace amounts in this organ. Considerations previously presented oppose the concept that the salivary glands produce the NGF and the ESF in an endocrine-gland fashion and supply these factors to their target cells. Likewise the kidney does not seem to fit in the role of an endocrine organ with respect to the ESF. The problem of the physiological source of these three factors remains unsolved. For each factor the possibility is suggested that more than one tissue or organ is engaged in its production and that the target cells receive it from the other cells that may be assembled in organs or disseminated everywhere in the organism. *4*) All three factors are endowed with antigenic properties, though only in two systems (NGF and ESF) does the antigen-antibody reaction in the living organism result in destructive effects on the target cells. *5*) The growth response is restricted to one cell type for the EGF and the ESF, whereas for the NGF two closely related nerve cell types respond to its action during embryonic life; only one of the two types of nerve cells, those in sympathetic ganglia, retains this capacity beyond the embryonic stages of life. *6*) The growth factors exert their action directly on the target cells as proved by in vitro studies with the NGF, EGF, and ESF. *7*) In all instances the effect materializes soon after the beginning of the incubation period and is characterized by clear-cut morphological and biochemical events in the target cells. *8*) The same mechanism of action has been suggested for the NGF and ESF. In both instances studies with specific inhibitors of RNA and protein synthesis suggest that the primary site of action is at the transcription level, whereas similar experiments suggest that the EGF elicits a primary change in the ribosomes of its target cells. *9*) A striking parallel is found in the extent of the growth response of the two cell types, epithelial and nerve cells, respectively, to the EGF and the NGF. In both instances the growth response is particularly impressive in cells in their early stages of differentiation and consists of an increase in the mitotic activity and acceleration of maturation processes. A size increase in fully differentiated sympathetic nerve cells finds no parallel in epithelial cells. This difference, however, is readily explained on the basis of intrinsic differences in the two cell types; nerve cells but not epithelial cells retain throughout life the capacity of increasing in size. The main feature of the growth response to the ESF is a numerical increase in immature red cells or reticulocytes. Since this factor exerts its action mainly on stem cells in bone marrow it has been referred to as the "primary inducer of erythroid differentiation" (44). This definition should be avoided, however, since the ESF differs in many respects from classical inducers (see following section), and also because this expression does not take into consideration other ESF effects, such as the early release of immature newly formed red cells into the circulation, the increase of several enzymatic activities in immature red cells (44), or the facilitation of the entry of iron into some cells of the marrow, which seems to indicate an influence on iron transport due to membrane changes (45). Specific growth factors and their possible significance are considered next.

BIOLOGICAL SIGNIFICANCE OF NGF

Before attempting an evaluation of the physiological significance of the NGF we shall briefly consider the biochemical properties of this factor and the main features of the growth response elicited in the target cells.

At the time of its discovery in 1954, the NGF was identified by Cohen et al. in a nucleoprotein molecule (31) and later in a protein molecule (30). Extensive analyses performed in several laboratories in subsequent years led to different models of the NGF molecule, but substantially confirmed the earlier findings on the protein nature of this factor. Though complete agreement has yet to be reached on its ultimate size, and though its molecular structure is still the object of intensive investigation, it may be stated that the biological activity resides in certain parts of the polypeptide chain. Hence the NGF can be definitely set apart from the large and heterogenous class of substances endowed with some growth-promoting activity such as vitamins, oligominerals, auxins, and nonprotein hormones. It belongs instead to another equally large and no less heterogenous class of substances endowed with biological activity that share with the NGF the property of having a protein structure. We shall briefly consider three large groups of these substances, namely the enzymes, inducers, and protein hormones.

The possibility that the NGF owes its biological action to an enzymatic property was not readily discarded (24). It had in its favor the finding that the NGF is found in large quantity in snake venom and in mouse salivary glands, which are characterized by a wealth of hydrolytic enzymes. Furthermore, the observation that the NGF level in the mouse salivary glands under normal and experimental conditions closely follows the level of esterase and protease seemed to make this hypothesis even more plausible. However, no catalytic activity has been detected as yet in the most purified NGF preparation, in spite of extensive studies directed to uncovering this presumptive function. In the absence of experimental evidence, it seems at present unjustified to speculate further on the enzymatic nature of the NGF.

We shall now consider similarities and differences between the NGF, inducers, and hormones.

Inducers and the NGF share some properties in common. In both instances the activity is bound to protein molecules released by some cells and interacting with other cells. In induction these molecules play a key role in the differentiation of the reacting cells. Undifferentiated cells (if one deals with induction of primary embryonic systems), or cells in early differentiative stages (if secondary embryonic systems are concerned), are channeled into one of several alternative differentiative pathways on their reception of the protein molecules that transmit the inducer. The reaction is time-bound; it takes place only if the target cells are in the responsive stage (and this varies for each system) and ceases soon after the same cells have acquired biochemical and structural differentiative marks as a result of the induction. Induction, therefore, is regarded as the switching on of dormant genes in the target tissue, which triggers a series of processes leading to the synthesis of specific differentiative products. Inducer agents have been recently isolated in highly purified form [we refer the reader to the excellent recent review articles by Tiedemann (96, 97)] and are known to retain their activity when tested on in vivo and in vitro systems. Under normal conditions, however, there is evidence that induction occurs only through a cell-to-cell contact, or at best at very short distances of the order of a few microns.

None of the above conditions applies to the NGF. Although it is mainly effective on the embryonic target cells, these cells must already have gone through the first differentiative phases before they acquire the capacity of responding to the NGF. One of the two cell types retains throughout all its life cycle the capacity of responding to the NGF. The response is stimulus-bound but not time-bound. It consists of a marked increment in the cells' growth potentialities as well as acceleration of early differentiative phases. In another way the NGF differs from inducers. The NGF molecule is present in the living organism in the body fluids and organ extracts. On injection, it circulates in the blood and reaches its target even if remote from the injection point. Although the differences listed above are contrary to the concept that the NGF belongs to the class of inducers, it should be stressed that in spite of these obvious differences they share an important aspect in common. In both instances a protein molecule released by some tissues plays a conspicuous role in the fate of other cell types. One could hardly conceive of the harmonious development of highly complex systems such as those of pluricellular organisms if interaction among tissues and dependent differentiation did not occur early in ontogenesis.

A thinner demarcation line can be drawn between the NGF and protein hormones (61). We are dealing in both instances with protein molecules released in the circulation and acting on target cells, whether assembled in organs or disseminated in the organism, or even dissociated in vitro. Both the NGF and protein hormones share the property of activating anabolic processes and, as a result, of enhancing growth in the target organs or cells. The NGF, as well as many protein hormones, exerts its influence on immature and mature receptive cells, and this influence is stimulus-bound. On withdrawal of the agent the effect ceases. The NGF, however, differs from "conventional hormones" in many other ways.

A basic tenet of classic endocrinology is the production of the hormone by organs or tissues easily identifiable on the basis of histological and biochemical data. A time-honored procedure is the extirpation of the organ or tissue that is the presumptive hormone producer and the subsequent finding that the hormone disappears from the circulation and that deprivation effects can be recognized in the target tissues. Extirpation of the mouse salivary glands in young and adult specimens fails to produce such effects. Though decisive evidence has been obtained for the in situ synthesis of the NGF, the exclusion of the mouse salivary glands does not result in any appreciable changes in its target cells, the sympathetic neurons. Further evidence contrary to the assumption that the NGF synthesized by the mouse salivary glands is utilized in hormone fashion by the sympathetic nerve cells has come from a number of well-ascertained facts reported earlier in this review.

In two other respects the NGF differs from "conventional" hormones: the magnitude of the effects and the strict specificity of the growth response. Both are unmatched by any of the known protein and nonprotein hormones.

The last and perhaps the most significant feature of the NGF system is represented by the dramatic effects elicited by the specific antiserum to this growth-promoting protein. The sudden disintegration and death of immature sympathetic nerve cells of newborn mice injected with this antiserum find no parallel in any antiserum effects to hormones. There is also another way to demonstrate the key role of the NGF in the life of the receptive cells. As reported above, embryonic sensory and sympathetic nerve cells survive and grow in a minimum liquid medium only if the medium is enriched with the NGF at a concentration as low as 10^{-7} g/ml.

Although the NGF differs in many significant ways from protein hormones it presents striking features in common with other specific growth factors. The two that come close to the NGF are, as reported, the EGF and the ESF.

The above considerations suggest that it is appropriate to list the NGF and other specific growth factors in a class distinct from those of inducers and hormones. Since the NGF and the other growth factors share many properties in common with these other substances, it is tempting to consider them as a class fitting between the inducers and the hormones. If this suggestion is valid it may be possible to discover other specific growth factors that will prove to play a key role on other cellular types in the same specific fashion as the NGF. Pending this evidence we shall, for the time being, consider the NGF as a protein endowed with a major function in the development and growth of its target cells.

SUMMARY AND CONCLUDING REMARKS

Although the primary site and the mechanism of action of the NGF on the target cells remain at present a matter of conjecture, and although it seems advisable to wait for more information before attempting to identify the former and explain the latter, one may with profit summarize the main features of the growth response elicited by the NGF on the receptive nerve cells and also outline a program for future work.

Among growth control mechanisms that operate in embryonic and post-embryonic life, the control exerted by the NGF on sensory and sympathetic nerve cells stands out by virtue of the magnitude of its effects, its target specificity, and the plurality of its actions.

The target cells become receptive to the NGF only after they have acquired unmistakable differentiative marks characteristic of one or the other of the two cell types. The NGF molecule is a normal constituent of the embryonic sensory and the embryonic and mature sympathetic nerve cells, and it is also present in many organs and tissue extracts; it is found in trace amounts in the serum of all mammals, man included. These findings lend support to the hypothesis that this protein is endowed with physiological significance. The same protein elicits striking growth effects on sensory and sympathetic nerve cells, but its range of action differs in magnitude and in time extension on the two cell types: only sympathetic nerve cells remain receptive to its action throughout their life cycle. The key role played by the NGF in the economy of these cells is further documented by the cytotoxic lethal effects elicited on immature sympathetic nerve cells by a specific antiserum to this molecule.

Since the magnitude of the growth response keeps pace with the amount of NGF supplied to the living organism, and the effect fades away soon after discontinuation of the treatment, a feedback mechanism must, under physiological conditions, control the production and release of this molecule. Neither its source of origin nor these hypothetical regulatory mechanisms have been identified. Equally unexplained at present is the role, if any, played by the salivary glands and the venom glands in supplying the receptive nerve cells with this factor.

Similar and equally challenging yet unsolved questions are raised by the discovery of two other specific growth factors, the epidermal growth factor and the erythropoietin-stimulating factor.

Progress in identifying the origin and in understanding the mechanism of action of any one of these three factors will advance our knowledge of the other two and possibly of still other unknown but similar factors. Future research should therefore aim not only at a more exhaustive and penetrating analysis of the NGF molecule and of its effects at a biochemical and ultrastructural level, but also at the investigation of the more general features of these growth regulatory factors that do not seem to fit into any of the well-known categories of agents that regulate growth and differentiative mechanisms in animal cells.

The investigations from the writers' laboratories have been supported in part by grants from the Public Health Service (NB-01602, NB-03777), the National Science Foundation (GB-1862), the Air Force Office of Scientific Research (AF-AFOSR-1182-67), The John A. Hartford Foundation (SD-333), and the Consiglio Nazionale delle Ricerche.

REFERENCES

1. ALPEN, E. L. Actions of erythropoietin in artifically-maintained bone marrow. In: *Erythropoiesis*, edited by L. O. Jacobson and M. Doyle. New York: Grune and Stratton, 1962, p. 304–309.
2. ANGELETTI, P. U., P. CALISSANO, J. S. CHEN, AND R. LEVI-MONTALCINI. Multiple molecular forms of the nerve growth factor. *Biochim. Biophys. Acta* 147: 180–182, 1967.
3. ANGELETTI, P. U., D. GANDINI-ATTARDI, G. TOSCHI, M. L. SALVI, AND R. LEVI-MONTALCINI. Metabolic aspects of the effect of nerve growth factor on sympathetic and sensory ganglia: protein and ribonucleic acid synthesis. *Biochim. Biophys. Acta* 95: 111–120, 1965.
4. ANGELETTI, P. U., R. LEVI-MONTALCINI, AND P. CALISSANO. The nerve growth factor: chemical properties and metabolic effects. *Advan. Enzymol.* 31: 51–75, 1968.
5. ANGELETTI, P. U., A. LIUZZI, AND R. LEVI-MONTALCINI. Stimulation of lipid biosynthesis in sympathetic and sensory ganglia by a specific nerve growth factor. *Biochim. Biophys. Acta* 84: 778–781, 1964.
6. ANGELETTI, P. U., A. LIUZZI, R. LEVI. MONTALCINI, AND D. G. GANDINI-ATTARDI. Effects of a nerve growth factor on glucose metabolism by sympathetic and sensory nerve cells. *Biochim. Biophys. Acta* 90: 445–450, 1964.
7. ANGELETTI, P. U., A. LIUZZI, AND F. POCCHIARI. Carbohydrate metabolism in embryonic sensory ganglia: effect of a specific protein nerve growth factor (NGF). In: *1st Intern. Meeting Neurochemistry (Strasbourg). Abstr. of Communications, 1967.*
8. ANGELETTI, P. U., M. L. SALVI, F. CAPANI, AND L. FRATI. Granulocytosis-inducing factor from the mouse submaxillary gland. *Biochim. Biophys. Acta* 111: 344–346, 1965.
9. ANGELETTI, P. U., M. L. SALVI, R. L. CHESANOW, AND S. COHEN. Azione dell' "Epidermal Growth Factor" sulla sintesi di acidi nucleici e proteine dell' epitelio cutaneo. *Experientia* 20: 146–148, 1964.
10. ARGYRIS, T. S. The growth promoting effects of tumors on tissues. In: *Advances in Biology of Skin. Carcinogenesis*. New York: Pergamon, 1966, vol. V, p. 55–73.
11. ARGYRIS, T. S., AND B. F. ARGYRIS. Differential response of skin epithelium to growth-promoting effects of subcutaneously transplanted tumor. *Cancer Res.* 22: 73–77, 1962.
12. BUEKER, E. D. Implantation of tumors in the hind limb field of the embryonic chick and developmental response of the lumbosacral nervous system. *Anat. Record* 102: 369–390, 1948.
13. BUEKER, E. D. Hypertrophy and hyperplasia of sympathetic and spinal ganglia in the chick embryo induced by sarcoma 1 (abstr.). *Anat. Record* 112: 317, 1952.
14. BUEKER, E., AND H. L. HILDERMAN. Growth-stimulating effects of mouse sarcomas 1, 37 and 180 on spinal and sympathetic ganglia of chick embryos as contrasted with effects of other tumors. *Cancer* 6: 397–415, 1953.
15. BUEKER, E. D., AND I. SCHENKEIN. Effects of daily subcutaneous injections of nerve growth stimulating protein fractions on mice during post natal to adult stage. *Ann. N. Y. Acad. Sci.* 118: 147–232, 1964.
16. BUEKER, E. D., I. SCHENKEIN, AND J. L. BANE. The problem of a nerve growth factor specific for spinal and sympathetic ganglia. *Cancer Res.* 20: 1220–1228, 1960.
17. BURDMAN, J. Early effects of a nerve growth factor on RNA content and base ratios of isolated chick embryo sensory ganglia neuroblasts in tissue culture. *J. Neurochem.* 14: 367–371, 1967.
18. BURDMAN, J. A., AND M. N. GOLDSTEIN. Long-term tissue culture of neuroblastomas. III. In-vitro studies of a nerve growth stimulating factor in sera of children with neuroblastoma. *J. Cancer Inst.* 33: 123–133, 1964.
19. BURDMAN, J. A., AND M. N. GOLDSTEIN. Synthesis and storage of a nerve growth protein in mouse submandibular glands. *J. Exptl. Zool.* 160: 183–188, 1965.

20. CARAMIA, F., P. U. ANGELETTI, AND R. LEVI-MONTALCINI. Experimental analysis of the mouse submaxillary salivary gland in relationship to its nerve-growth factor content. *Endocrinology* 70: 915–922, 1962.
21. CARNOT, P., AND C. DÉFLANDRE. Sur l'activité hémopoïetique des différents organes au cours de la régénération du sang. *Compt. Rend. Acad. Sci.* 143: 432–435, 1906.
22. COHEN, S. A nerve growth-promoting protein. In: *Chemical Basis of Development*, edited by W. D. McElroy and S. Glass. Baltimore: Johns Hopkins Press, 1958, p. 665–667.
23. COHEN, Purification and metabolic effects of a nerve growth-promoting protein from snake venom. *J. Biol. Chem.* 234: 1129–1137, 1959.
24. COHEN, S. Purification of a nerve growth promoting protein from the mouse salivary gland and its neurocytotoxic antiserum. *Proc. Natl. Acad. Sci. U.S.* 46: 302–311, 1960.
25. COHEN, S. Isolation of a mouse submaxillary gland protein accelerating incisor eruption and eyelid opening in the new-born animal. *J. Biol. Chem.* 237: 1555–1562, 1962.
26. COHEN, S. Isolation and biological effects of an epidermal growth-stimulating protein. In: *Metabolic Control Mechanisms in Animal Cells*, edited by W. J. Rutter. Washington, D.C.: Natl. Cancer Inst. Monograph 13, 1964, p. 14–37.
27. COHEN, S. The stimulation of epidermal proliferation by a specific protein (EGF). *Develop. Biol.* 12: 394–407, 1965.
28. COHEN, S. Growth factors and morphogenetic induction. In: *Developmental and Metabolic Control Mechanisms and Neoplasia. A Collection of Papers Presented at the Nineteenth Annual Symposium on Fundamental Cancer Research, 1965.* Baltimore: Williams & Wilkins, 1965, p. 251–272.
29. COHEN, S., AND C. A. ELLIOT. The stimulation of epidermal keratinization by a protein isolated from the submaxillary salivary gland of the mouse. *J. Invest. Dermatol.* 40: 1–5, 1963.
30. COHEN, S., AND R. LEVI-MONTALCINI. A nerve growth stimulating factor isolated from snake venom. *Proc. Natl. Acad. Sci. U.S.* 42: 571–574, 1956.
31. COHEN, S., R. LEVI-MONTALCINI, AND V. HAMBURGER. A nerve growth stimulating factor isolated from sarcomas 37 and 180. *Proc. Natl. Acad. Sci. U.S.* 40: 1014–1018, 1954.
32. CRAIN, S. M., H. BENITEZ, AND A. E. VATTER. Some cytological effects of salivary nerve growth factor on tissue cultures of peripheral ganglia. *Ann. N. Y. Acad. Sci.* 118: 206–231, 1964.
33. CRAIN, S. M., AND R. G. WIEGAND. Catecholamine levels of mouse sympathetic ganglia following hypertrophy produced by salivary nerve-growth factor. *Proc. Soc. Exptl. Biol. Med.* 107: 663–665, 1961.
34. EDWARDS, D. C., E. L. FENTON, S. KAKARI, B. J. LARGE, L. PAPADAKI, AND E. ZAIMIS. Effects of nerve growth factor in new-born mice, rats and kittens. *J. Physiol. (London)* 186: 10–12P, 1966.
35. ERSLEV, A. J. Erythropoietin in vitro. In: *Erythropoiesis*, edited by L. O. Jacobson and M. Doyle. New York: Grune and Stratton, 1962, p. 275–285.
36. FALCK, B., N.-A. HILLARP, G. THIEME, AND A. TORP. Fluorescence of catecholamines and related compounds condensed with formaldehyde. *J. Histochem. Cytochem.* 10: 348–354, 1962.
37. FISCHER, S. Studies on the mechanism and site of action of erythropoietin. In: *Erythropoiesis*, edited by L. O. Jacobson and M. Doyle. New York: Grune and Stratton, 1962, p. 204–215.
38. FISHER, J. W., N. P. SANZARI, B. J. BIRDWELL, AND J. J. CROOK. The role of the kidney in erythropoietin production. In: *Erythropoiesis*, edited by L. O. Jacobson and M. Doyle. New York: Grune and Stratton, 1962, p. 78–86.
39. GANDINI-ATTARDI, D., P. CALISSANO, AND P. U. ANGELETTI. Protein synthesis in embryonic sensory ganglia: effect of the nerve growth factor on soluble proteins. *Brain Res.* 6: 367–370, 1967.
40. GARCIA, J. F., AND J. C. SCHOOLEY. An immunological study of human urinary erythropoietin. In: *Erythropoiesis*, edited by L. O. Jacobson and M. Doyle. New York: Grune and Stratton, 1962, p. 56–57.
41. GARCIA, J. F., AND J. C. SCHOOLEY. Immunological neutralization of various erythropoietins. *Proc. Soc. Exptl. Biol. Med.* 112: 712–714, 1963.
42. GOLDBERG, B., AND H. GREEN. The cytotoxic action of immune gamma globulin and complement Krebs ascites tumor cells. I. Ultrastructural studies. *J. Exptl. Med.* 109: 505–510, 1959.
43. GOLDSTEIN, M. N., AND J. A. BURDMAN. Studies of the nerve growth factor in submandibular glands of female mice treated with testosterone. *Anat. Record* 151: 199–208, 1965.
44. GOLDWASSER, E. Biochemical control of erythroid cell development. In: *Current Topics in Developmental Biology*, edited by A. Monroy and A. A. Moscona. New York: Academic, 1966, vol. 1, p. 173–209.
45. GOLDWASSER, E. Erythropoietin induction of red cell differentiation. In: *Experimental Biology and Medicine. Morphological and Biochemical Aspects of Cytodifferentiation*, edited by E. Hagen, W. Wechsler, and P. Zilliken. Basel: Karger, 1967, vol. 1, p. 234-243.
46. GORDON, A. S. Hemopoietine. *Physiol. Rev.* 39: 1–40, 1959.
47. GREEN, H. R., A. FLEISCHER, P. BARROW, AND B. GOLDBERG. The cytotoxic action of immune gamma globulin and complement on Krebs ascites tumor cells. II. Chemical studies. *J. Exptl. Med.* 109: 511–521, 1959.
48. HAMBURGER, V., AND R. LEVI-MONTALCINI. Proliferation, differentiation and degeneration in the spinal ganglia of the chick embryo under normal and experimental conditions. *J. Exptl. Zool.* 111: 457–502, 1949.
49. HOOBER, J. K., AND S. COHEN. Epidermal growth factor. I. The stimulation of protein and ribonucleic acid synthesis in chick embryo epidermis. *Biochim. Biophys. Acta* 138: 347–356, 1967.
50. HOOBER, J. K., AND S. COHEN. Epidermal growth factor. II. Increased activity of ribosomes from chick embryo epidermis for cell-free protein synthesis. *Biochim. Biophys. Acta* 138: 357–368, 1967.
51. JACOBSON, L. O. Sites of formation of erythropoietin. In: *Erythropoiesis*, edited by L. L. Jacobson and M. Doyle. New York: Grune and Stratton, 1962, p. 69–70.
52. JUNQUEIRA, L. C. U., A. FAJER, M. RABINOVITCH, AND L. FRANKENTHAL. Biochemical and histochemical observations on the sexual dimorphism of mice submaxillary glands. *J. Cellular Comp. Physiol.* 34: 129–158, 1949.

53. KLINGMAN, G. I. In utero immunosympathectomy of mice. *Intern. J. Neuropharmacol.* 5: 163–170, 1966.

54. KORST, D. R., E. P. FRENKEL, AND J. E. WILHELM. Studies using short-term bone marrow culture. I. Ferrokinetics and erythropoietin. In: *Erythropoiesis*, edited by L. O. Jacobson and M. Doyle. New York: Grune and Stratton, 1962, p. 310–320.

55. KURATOWSKA, Z., E. KOWALSKI, B. LIPINSKI, AND E. MICHALAK. Human blood as a source for preparation of erythropoietin. In: *Erythropoiesis*, edited by L. O. Jacobson and M. Doyle. New York: Grune and Stratton, 1962, p. 58–63.

56. LACASSAGNE, A. Dimorphisme sexuel de la glande sousmaxillaire chez la souris. *Compt. Rend. Soc. Biol.* 133: 180–181, 1940.

57. LATTA, H. Some cytological effects of antibodies. In: *Mechanism of Hypersensitivity*, edited by J. H. Shaffer, G. A. Logrippo, and M. W. Chase. New York: Little, Brown, 1959, p. 123–141.

58. LEVI-MONTALCINI, R. Effects of mouse tumor transplantation on the nervous system. *Ann. N. Y. Acad. Sci.* 55: 330–343, 1952.

59. LEVI-MONTALCINI, R. Chemical stimulation of nerve growth. In: *Chemical Basis of Development*, edited by W. D. McElroy and B. Glass. Baltimore: Johns Hopkins Press, 1958, p. 646–664.

60. LEVI-MONTALCINI, R. The nerve growth factor. *Ann. N. Y. Acad. Sci.* 118: 149–168, 1964.

61. LEVI-MONTALCINI, R. Biological aspects of specific growth promoting factors. *Proc. Roy. Soc. Med.* 58: 357–360, 1965.

62. LEVI-MONTALCINI, R. The nerve growth factor: its mode of action on sensory and sympathetic nerve cells. *Harvey Lectures Ser.* 60: 217–259, 1966.

63. LEVI-MONTALCINI, R. Differentiation and growth control mechanisms in the nervous systems. In: *Experimental Biology and Medicine. Morphological and Biochemical Aspects of Cytodifferentiation*, edited by E. Hagen, W. Wechsler, and P. Zilliken. Basel: Karger, 1967, vol. 1, p. 170–182.

64. LEVI-MONTALCINI, R., AND P. U. ANGELETTI. Biological properties of a nerve-growth promoting protein and its antiserum. In: *Regional Neurochemistry, Proc. 4th Intern. Neurochemical Symp.*, edited by S. S. Kety and J. Elkes. New York: Pergamon, 1961, p. 362–376.

65. LEVI-MONTALCINI, R., AND P. U. ANGELETTI. Growth control of the sympathetic system by a specific protein factor. *Quart. Rev. Biol.* 36: 99–108, 1961.

66. LEVI-MONTALCINI, R., AND P. U. ANGELETTI. Considerations on a nerve growth promoting activity of some tumors. In: *Biological Interactions in Normal and Neoplastic Growth*, edited by M. J. Brennen and W. L. Simpson. Boston: Little, Brown, 1962, p. 225–237.

67. LEVI-MONTALCINI, R., AND P. U. ANGELETTI. Essential role of the nerve growth factor in the survival and maintenance of dissociated sensory and sympathetic nerve cells in vitro. *Develop. Biol.* 7: 653–659, 1963.

68. LEVI-MONTALCINI, R., AND P. U. ANGELETTI. Hormonal control of the NGF content in the submaxillary salivary glands of mouse. In: *Salivary Glands and their Secretions*, edited by L. M. Sreebny and J. Meyer. Oxford: Pergamon, 1964, p. 129–141.

69. LEVI-MONTALCINI, R., AND P. U. ANGELETTI. Immunosympathectomy. *Pharmacol. Rev.* 18: 619–628, 1966.

70. LEVI-MONTALCINI, R., AND P. U. ANGELETTI. Biological aspects of the nerve growth factor. In: *Ciba Found. Symp. Growth of the Nervous System*, edited by G. E. W. Wolstenholme and M. O'Connor. London: Churchill, 1968, p. 126–147.

71. LEVI-MONTALCINI, R., AND B. BOOKER. Excessive growth of the sympathetic ganglia evoked by a protein isolated from mouse salivary glands. *Proc. Natl. Acad. Sci. U.S.* 42: 373–384, 1960.

72. LEVI-MONTALCINI, R., AND B. BOOKER. Destruction of the sympathetic ganglia in mammals by an antiserum to the nerve-growth promoting factor. *Proc. Natl. Acad. Sci. U.S.* 42: 384–391, 1960.

73. LEVI-MONTALCINI, R., F. CARAMIA, S. A. LUSE, AND P. U. ANGELETTI. In vitro effects of the nerve growth factor on the fine structure of the sensory nerve cells. *Brain Res.* 8: 347–362, 1968.

74. LEVI-MONTALCINI, R., AND S. COHEN. In-vitro and in-vivo effects of a nerve growth stimulating agent isolated from snake venom. *Proc. Natl. Acad. Sci. U.S.* 42: 695–699, 1956.

75. LEVI-MONTALCINI, R., AND S. COHEN. Effects of the extract of the mouse salivary glands on the sympathetic system of mammals. *Ann. N. Y. Acad. Sci.* 85: 324–341, 1960.

76. LEVI-MONTALCINI, R., AND V. HAMBURGER. Selective growth-stimulating effects of mouse sarcoma on the sensory and sympathetic nervous system of the chick embryo. *J. Exptl. Zool.* 116: 321–362, 1951.

77. LEVI-MONTALCINI, R., AND V. HAMBURGER. A diffusible agent of mouse sarcoma producing hyperplasia of sympathetic ganglia and hyperneurotization of the chick embryo. *J. Exptl. Zool.* 123: 233–388, 1953.

78. LEVI-MONTALCINI, R., H. MEYER, AND V. HAMBURGER. In-vitro experiments on the effects of mouse sarcoma 180 and 37 on the spinal and sympathetic ganglia of the chick embryo. *Cancer Res.* 14: 49–57, 1954.

79. LEVI-MONTALCINI, R., L. TENTORI, G. VIVALDI, P. U. ANGELETTI, AND G. B. MARINI-BETTOLO. Composizione in ammino acidi del fattore di crescita del sistema nervoso (NGF). *Gazz. Chim. Ital.* 95: 333–337, 1965.

80. LIUZZI, A., P. U. ANGELETTI, AND R. LEVI-MONTALCINI. Metabolic effects of a specific nerve growth factor (NGF) on sensory and sympathetic ganglia. Enhancement of lipid biosynthesis. *J. Neurochem.* 12: 705–708, 1965.

81. LIUZZI, A., P. CALISSANO, A. GRASSO, P. U. ANGELETTI, AND R. LEVI-MONTALCINI. Effect of a nerve growth factor (NGF) on some enzyme of glucose metabolism in the responsive nerve cells. In: *Proc. 3rd Meeting Fed. Europ. Biochem. Soc.* Oxford: Pergamon, 1967, p. 334.

82. LIUZZI, A., AND F. H. FOPPEN. Content of sterol-like compound 121 in various animal tissues. *Ann. Ist. Sup. Sanita.* In press, 1968.

83. LIUZZI, A., AND F. H. FOPPEN. Sterol-like compound from sensory ganglia: effect of nerve growth factor (NGF) and insulin on its biosynthesis. *Biochem. J.* 107: 191–196, 1968.

84. LIUZZI, A., F. POCCHIARI, AND P. U. ANGELETTI. Glucose metabolism in embryonic ganglia: effect of nerve growth factor (NGF) and insulin. *Brain Res.* 7: 452–454, 1968.

85. LOWY, P. H., AND H. BORSOOK. Preparation and properties of erythropoietin concentrates from rabbit plasma and human urine. In: *Erythropoiesis*, edited by L. O. Jacobson and M. Doyle. New York: Grune and Stratton, 1962, p. 33–42.

86. MATOTH, Y. In vitro methods in the study of erythropoietin. In: *Erythropoiesis*, edited by L. O. Jacobson and M. Doyle. New York: Grune and Stratton, 1962, p. 299–303.

87. OLSON, L. Outgrowth of sympathetic adrenergic neurons in mice treated with a nerve growth factor. *Z. Zellforsch.* 81: 155–173, 1967.

88. PATTERSON, W. B. Induction of normal cell proliferation by tumor cells. In: *Biological Interactions in Normal and Neoplastic Growth*, edited by M. J. Brennan and W. L. Simpson. Boston: Little, Brown, 1962, p. 247–252.

89. POWSNER, E. R., AND L. BERMAN. The effect of erythropoietin on erythroblast maturation in bone marrow cultures. In: *Erythropoiesis*, edited by L. O. Jacobson and M. Doyle. New York: Grune and Stratton, 1962, p. 286–298.

90. REISSMANN, K. R., AND T. NOMURA. Erythropoietin formation in isolated kidneys and liver. In: *Erythropoiesis*, edited by L. O. Jacobson and M. Doyle. New York: Grune and Stratton, 1962, p. 71–77.

91. SABATINI, M. T., A. P. DE IRALDI, AND E. DE ROBERTIS. Early effects of the antiserum (AS) against the nerve growth factor (NGF) on the structure of sympathetic neurons. *J. Exptl. Neurol.* 12: 370–383, 1965.

92. SABATINI, M. T., R. LEVI-MONTALCINI, AND P. U. ANGELETTI. Effetti precoci del siero antinerve growth factor (anti-NGF) sulla cellula nervosa. *Ann. Ist. Sup. Sanita.* 2: 349–355, 1966.

93. SALVI, M. L., P. U. ANGELETTI, AND L. FRATI. Frazionamento delle proteine solubili della ghiandola sottomascellare del topo: localizzazione di alcune componenti biologicamente attive. *Il Farmaco* 20: 12–21, 1965.

94. SCHENKEIN, I., AND E. D. BUEKER. Dialyzable cofactor in nerve growth promoting protein from mouse salivary glands. *Science* 137: 433–434, 1962.

95. SCHENKEIN, I., AND E. D. BUEKER. The nerve growth factor as two essential components. *Ann. N. Y. Acad. Sci.* 118: 171–182, 1964.

96. TIEDEMANN, H. The molecular basis of differentiation in early development of amphibian embryos. In: *Current Topics in Developmental Biology*, edited by A. Monroy and A. A. Moscona. New York: Academic, 1966, vol. 1, p. 85–112.

97. TIEDEMANN, H. Inducers and inhibitors of embryonic differentiation: their chemical nature and mechanism of action. In: *Experimental Biology and Medicine. Morphological and Biochemical Aspects of Cytodifferentiation*, edited by E. Hagen, W. Wechsler, and P. Zilliken. Basel: Karger, 1967, vol. 1, p. 8–21.

98. TOSCHI, G., E. DORE, P. U. ANGELETTI, R. LEVI-MONTALCINI, AND C. H. DE HAËN. Characteristics of labelled RNA from spinal ganglia of chick embryo and the action of a specific growth factor. *J. Neurochem.* 13: 539–544, 1965.

99. TOSCHI, G., E. DORE, AND O. SAPERA. Studies on RNA synthesized in vitro by sensory ganglia of chick embryo. In: *1st Intern. Meeting Neurochemistry (Strasbourg). Abstr. of Communications, 1967.*

100. VARON, S., J. NOMURA, AND E. M. SHOOTER. Subunit structure of a high-molecular weight form of the nerve growth factor from mouse submaxillary gland. *Proc. Natl. Acad. Sci. U.S.* 57: 1782–1789, 1967.

101. VARON, S., J. NOMURA, AND E. M. SHOOTER. The isolation of the mouse nerve growth factor protein in a high molecular weight form. *Biochemistry* 6: 2202–2209, 1967.

102. WALDMANN, T. A., AND W. F. ROSSE. Sites of formation of erythropoietin. In: *Erythropoiesis*, edited by L. O. Jacobson and M. Doyle. New York: Grune and Stratton, 1962, p. 87–92.

103. WEIS, P., AND E. D. BUEKER. Ligation of mouse submandibular glands and its effect on components of nerve growth factor. *Proc. Soc. Exptl. Biol. Med.* 121: 1135–1140, 1966.

104. ZAIMIS, E. The immunosympathectomized animal: a valuable tool in physiological and pharmacological research. *J. Physiol. (London)* 177: 35–36, 1964.

105. ZAIMIS, E., L. BERK, AND B. A. CALLINGHAM. Biochemical and functional changes in the sympathetic nervous system of rats treated with nerve-growth factor-antiserum. *Nature* 206: 1220–1222, 1965.

Reprinted from
Proc. Nat. Acad. Sci. USA
Vol. 68, No. 7, pp. 1598–1602, July 1971

Selective Induction by Nerve Growth Factor of Tyrosine Hydroxylase and Dopamine-β-Hydroxylase in the Rat Superior Cervical Ganglia

(dopa decarboxylase/monoamine oxidase)

H. THOENEN*, P. U. ANGELETTI†, R. LEVI-MONTALCINI†, AND R. KETTLER*

*Department of Experimental Pharmacology, Hoffman-La Roche, Basel, Switzerland; and
†Laboratorio di Biologia Cellulare, C.N.R., Via Romagnosi 18A, 00196 Rome, Italy

Contributed by R. Levi-Montalcini, May 6, 1971

ABSTRACT **Treatment of newborn rats with 10 μg/g of nerve growth factor for 10 days enhanced not only the growth but also the differentiation of neuroblasts in superior cervical ganglia. These morphological changes were accompanied by selective induction of tyrosine hydroxylase and dopamine β-hydroxylase (EC 1.14.2.1), whereas the total and specific activities of other enzymes involved in biosynthesis or metabolic degradation of norepinephrine rose only in proportion to the increase in volume of the sympathetic ganglia. There are remarkable similarities between this effect of nerve growth factor and the induction of trans-synaptic enzymes by increased activity of the sympathetic nervous system.**

Ever since the discovery that a specific nerve growth factor (NGF) enhances growth and differentiative processes of sympathetic neurons (1, 2), the question has been raised whether this factor selectively stimulates metabolic pathways characteristic of this particular nerve cell type. Sympathetic neurons offer an almost ideal system to test this hypothesis, since the biochemical correlates of their function are rather well known. However, it was not until recently that sufficiently sensitive methods became available to measure the activity of enzymes involved in the synthesis and metabolic degradation of the adrenergic neurotransmitter in small tissue samples (3–6). It thus became possible to study whether the hyperplastic and hypertrophic ganglia of newborn rats treated with NGF also differ from controls in the specific activity of these enzymes.

The results to be reported here show that NGF, besides its characteristic morphological effects, also produces a selective induction of tyrosine hydroxylase and dopamine β-hydroxylase (EC 1.14.2.1), enzymes located in adrenergic neurons (7). In contrast, the activities of Dopa decarboxylase (EC 4.1.1.26) and monoamine oxidase (EC 1.4.3.4), enzymes which are present also in other cell types (7), increase only in proportion to the rise in volume of the adrenergic neurons in the superior cervical ganglia.

METHODS

Newborn rats (Wistar descent) of either sex were injected subcutaneously daily for 5 and 10 days with 10 μg/g of NGF dissolved in 0.9% NaCl. The NGF was prepared according to the method of Bocchini and Angeletti (8). The animals were killed by transection of the thorax above the heart. The superior cervical ganglia were removed under the dissecting microscope, fixed in Bouin's solution, and stained with toluidine blue. Ganglia of 5- and 10-day-old experimental and control littermates were then sectioned serially at 10 μm. Cell counts were performed by inserting a micrometer disk into the ocular and counting all nerve cells in every other section of the experimental and control ganglia. For ultrastructural studies, the ganglia were fixed for 2 hr in 3% glutaraldehyde in 0.1 M phosphate buffer (pH 7.4) with added $CaCl_2$; the specimens were washed for 2 hr in the same buffer and postfixed in osmic hydroxide (1.33%) for 2 hr. Sections were cut with a Porter–Blum–Sorvall MT-2 microtome, stained with uranyl acetate in 50% ethanol or in lead hydroxide, and examined in a Philips 300 electron microscope. For the determination of tyrosine hydroxylase activity, each pair of ganglia was homogenized in 0.5 ml of ice-cold 0.25 M sucrose; for dopamine β-hydroxylase, in 0.5 ml of 0.005 M phosphate buffer (pH 7.5) containing 0.1% Triton X-100; for dopa decarboxylase, in 0.5 ml of 0.1 M phosphate buffer (pH 7.0); and for monoamine oxidase in 0.1 M phosphate buffer pH 7.2.

The activity of tyrosine hydroxylase was determined by the method of Levitt *et al.* (3), dopamine β-hydroxylase according to Duch *et al.* (4), dopa decarboxylase according to Håkanson and Owman (5), with modifications described in detail by Thoenen *et al.* (9), and monoamine oxidase according to Wurtman and Axelrod (6).

Enzyme activities are all expressed both in terms of product formed per hr per pair of ganglia (total activity) and in terms of product formed per hr per mg protein. The proteins were determined by the method of Lowry *et al.* (10).

Enzyme kinetic parameters were determined by the method of Wilkinson (11) with a digital computer according to a program by Cleland (12).

RESULTS

Morphological effects of NGF on sympathetic ganglia in newborn rats

The effects of daily injections of NGF in newborn rats for periods of 5 and 10 days are illustrated in Figs. 1–7. The

TABLE 1. *Cell number in superior cervical ganglia of 5-day-old NGF-treated and control rats*

	Number of sections	Cell number
Control	141	29,800
NGF-treated	216	74,000

Abbreviation: NGF, nerve growth factor.

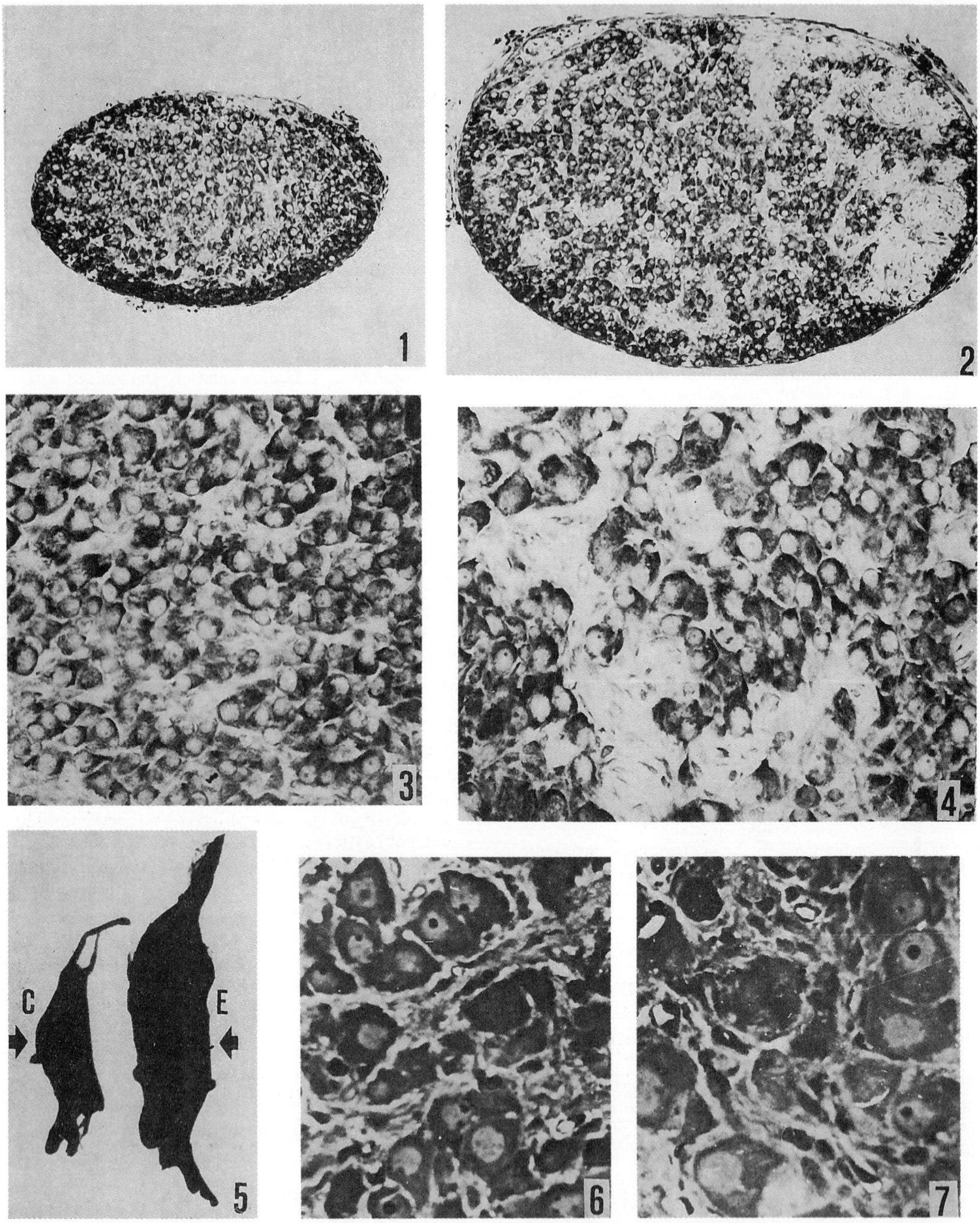

Figs. 1–7. Transverse sections through superior cervical ganglia of control (Fig. 1, ×114) and experimental (Fig. 2, ×114) animals (5-day-old rats), shown in whole mounts in Fig. 5 (×13). Arrows in Fig. 5 indicate the level of sections. Figs. 3 and 4 (×399) show size difference in sympathetic nerve cells of control (Fig. 3) and experimental (Fig. 4) ganglia reproduced in Figs. 1, 2, and 5. Note large fiber bundles among cells in NGF-treated ganglion. Figs. 6 and 7 (×855) show difference in size of sympathetic nerve cells of superior cervical ganglia of 10-day-old control (Fig. 6) and NGF-treated (Fig. 7) rats.

volume difference of superior cervical ganglia of NGF-treated and control rats sacrificed at 5 days is shown in Fig. 5. Transverse sections of the same ganglia at low (Figs. 1 and 2) and high (Figs. 3 and 4) magnification give evidence for the size increase of sympathetic neurons in the ganglion of the experimental rat and for a marked difference in the texture of

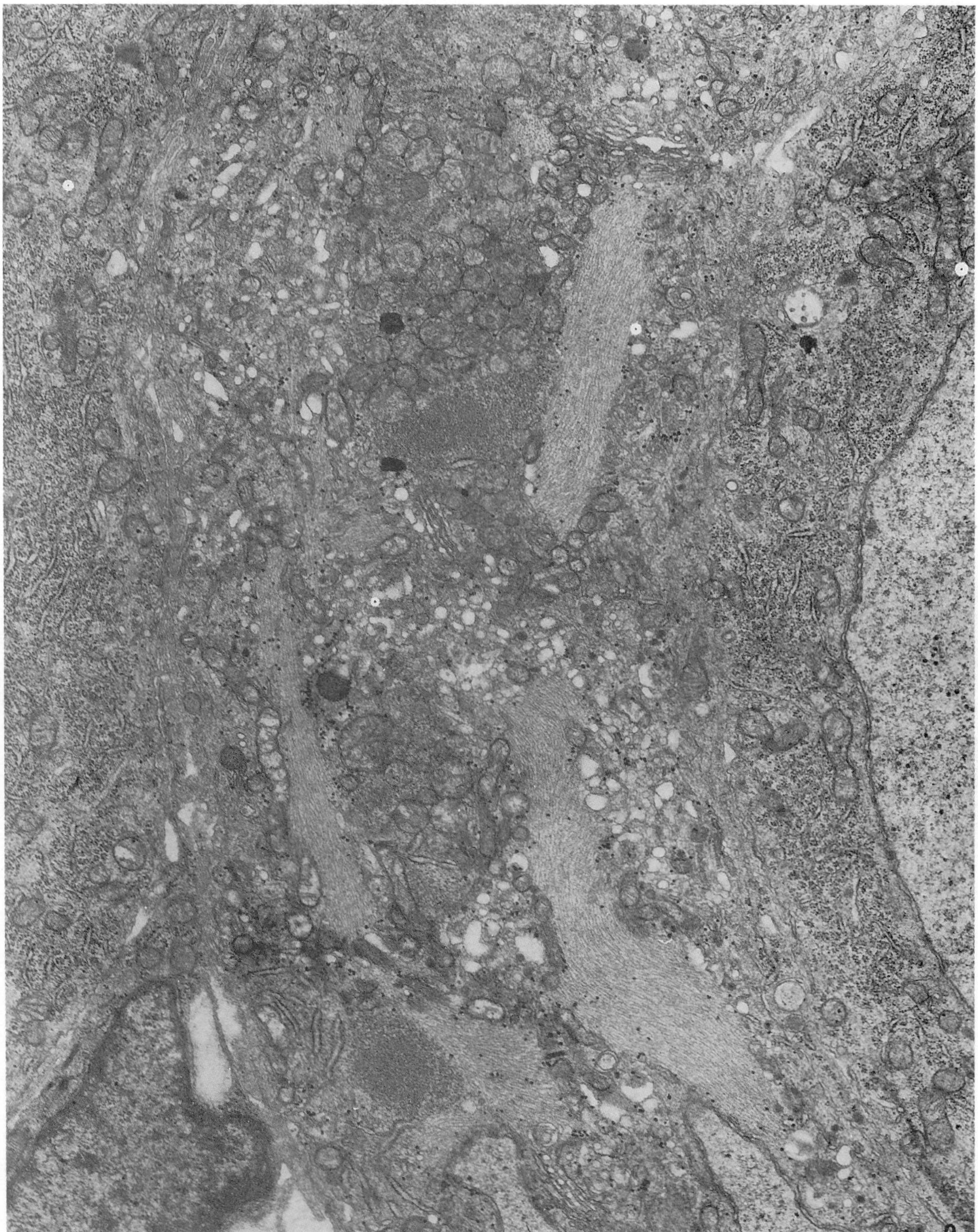

FIG. 8. Electron micrograph of cytoplasmic area of sympathetic neuron of 10-day-old rat treated since birth with NGF. Large neurofilament bundles fill part of the field. (×13,940)

the ganglia. While in controls (Figs. 1 and 3) the nerve cell population is compact and evenly distributed, large fiber bundles segregate the population in the experimental ganglion into cell groups (Figs. 2 and 4). The total numbers of sections and of nerve cells in these ganglia are given in Table 1. Figs. 6 and 7 show transverse sections of superior cervical ganglia of 10-day-old NGF-treated and control rats. The size increase of individual neurons is even more marked than in younger

TABLE 2. *Effect of NGF on specific and total activities of enzymes in superior cervical ganglia of newborn rats*

	Controls		NGF-treated	
	Specific	Total	Specific	Total
Tyrosine hydroxylase*	1.9 ± 0.06	0.10 ± 0.004	8.4 ± 0.9	1.9 ± 0.02
Dopamine β-hydroxylase†	11.8 ± 0.1	1.1 ± 0.04	42.0 ± 2.6	13.0 ± 0.6
Dopa decarboxylase‡	0.32 ± 0.03	0.02 ± 0.001	0.51 ± 0.03	0.13 ± 0.07
Monoamine oxidase§	160 ± 8	23.5 ± 2.2	189 ± 12	57.8 ± 8.5

Activities (mean ± SE, n = 6–8) are amounts of product formed per hr per mg protein (specific) and amounts of product formed per hr per pair of ganglia (total): * nanomoles of dopa, † picomoles of octopamine, ‡ micromoles of dopamine, § nanomoles of indoleacetic acid.

TABLE 3. *Kinetic characterization of tyrosine hydroxylase and dopamine β-hydroxylase in superior cervical ganglia of NGF-treated rats and controls*

	Tyrosine hydroxylase K_m (μM tyrosine)	Dopamine β-hydroxylase K_m (μM tyramine)
Controls	20.5 ± 3	277 ± 34
NGF-treated	23.0 ± 2	256 ± 32

Newborn animals were treated for 10 days with 10 μg/g of NGF daily.

specimens. The results of these studies, in agreement with those reported in mice (1, 2), indicate that the NGF calls forth hyperplastic and hypertrophic effects. Volume measurement of the superior cervical ganglia of 10-day-old experimental and control rats showed that the former is five times larger than the latter. On a protein basis, the average protein content per pair of ganglia amounts to 225 ± 10 μg in the NGF-treated rat as compared to 65 ± 2 μg in controls.

Fig. 8 shows an electron micrograph of sympathetic nerve cells of the superior cervical ganglion of a rat treated with NGF from birth to the tenth day of age. The most impressive effect is the massive production of neurotubules and neurofilaments.

This effect is in all respects similar to those elicited by NGF on sensory neurons in chick embryos *in vitro* and on sympathetic neurons in newborn mice (13, 14).

Effect of NGF on total and specific enzyme activity in superior cervical ganglia

Fig. 9 and Table 2 show that treatment of newborn rats for 10 days with NGF produced an 18-fold increase in total activity of tyrosine hydroxylase (product formed per hr per pair of ganglia) and a 13-fold increase in that of dopamine β-hydroxylase. The activities of dopa decarboxylase and monoamine oxidase rose only 4.5-fold and 2.5-fold, respectively. The

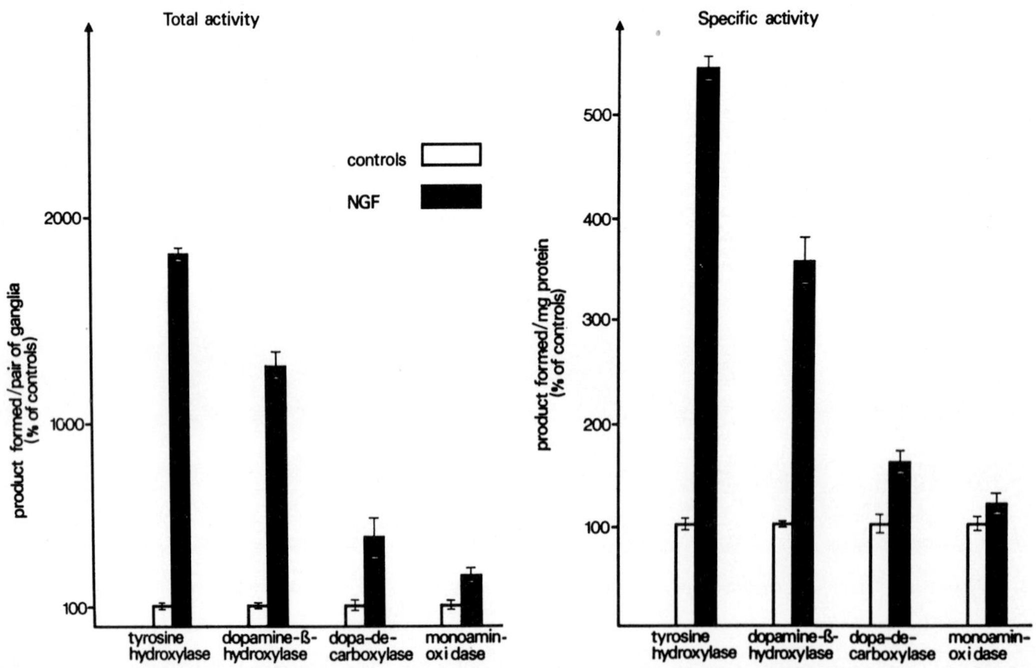

FIG. 9. Effect of NGF on enzymes involved in synthesis and metabolic degradation of norepinephrine. Newborn rats were treated with 10 μg/g of NGF for 10 days. The activity of all enzymes studied in the superior cervical ganglia is expressed in % of controls both for total (product formed per hr per pair of ganglia) and specific activity (product formed per hr per mg protein).

selectivity of the increase in enzyme activity becomes even clearer if one compares specific enzyme activity (product formed per hr per mg protein) of NGF-treated animals with that of untreated controls. The ratio between the specific activity of NGF-treated animals and controls amounts to 5.4 for tyrosine hydroxylase, 3.6 for dopamine β-hydroxylase, 1.5 for dopa decarboxylase, and 1.2 for monoamine oxidase (Fig. 4). The small increase in specific activity of the last two enzymes could be explained by a relative increase in the volume of neuronal versus satellite cells. In superior cervical ganglia of controls, the neuronal cells represent about 0.75 of the total volume.

Mechanism of increased tyrosine hydroxylase and dopamine β-hydroxylase activity

It could be assumed that the increased activity of these two enzymes results from a direct activation by NGF. To examine this possibility, we added 1–30 μg of NGF to enzyme preparations (100 μl) used for the determination of tyrosine hydroxylase and dopamine β-hydroxylase activity. In both cases the enzyme activity remained unchanged after the *in vitro* addition of NGF. To determine whether the NGF promotes the formation of an activator, inhibits the production of an inhibitor, or enhances selectively the synthesis of tyrosine hydroxylase and dopamine β-hydroxylase, we combined enzyme solutions from controls and NGF-treated animals in several proportions. Both for tyrosine hydroxylase and dopamine β-hydroxylase, the activities of these combinations were always additive and therefore did not provide evidence for the loss of an inhibitor or appearance of an activator in the enzyme preparations of animals treated with NGF. Also, the fact that no significant change in K_m values (Table 3) occurred speaks in favor of an increase in the amount of enzyme protein.

DISCUSSION

Evidence is here presented that NGF produces a selective 15- to 20-fold increase in the activity of tyrosine hydroxylase and dopamine β-hydroxylase, enzymes that are exclusively located in adrenergic neurons (7). In contrast, the activity of dopa decarboxylase and monoamine oxidase rose only about in proportion to the increase in volume of sympathetic ganglia, corresponding to a 4-fold increase in their protein content.

The considerable rise in the activity of tyrosine hydroxylase and of dopamine β-hydroxylase does not result from the formation of activators or disappearance of inhibitors, but rather from an increased synthesis of new enzyme protein. We are therefore dealing with the selective induction of two enzymes that are exclusively located in adrenergic neurons by a specific growth factor.

Enhanced activity of the peripheral sympathetic nervous system resulting from cold exposure, or from depletion of the adrenergic transmitter stores by reserpine produces, similar changes in the pattern of enzymes (15–17) involved in the synthesis of norepinephrine as does administration of NGF. This finding raises the question whether similar basic mechanisms are responsible in the two instances. It remains also to be elucidated whether this specific growth factor and the functional activity of neurons play similar roles in the growth regulation of the target cells during ontogenesis and in their adaptation to changed functional stages.

1. Levi-Montalcini, R., *Harvey Lect.*, **60**, 217 (1966).
2. Levi-Montalcini, R., and P. U. Angeletti, *Physiol. Rev.*, **48**, 534 (1968).
3. Levitt, M., J. W. Gibb, J. W. Daly, M. Lipton, and S. Udenfriend, *Biochem. Pharmacol.*, **16**, 1313 (1967).
4. Duch, D. S., O. H. Viveros, and N. Kirshner, *Biochem. Pharmacol.*, **17**, 255 (1968).
5. Håkanson, R., and C. Owman, *J. Neurochem.*, **13**, 597 (1966).
6. Wurtman, R., and J. Axelrod, *Biochem. Pharmacol.*, **12**, 1433 (1964).
7. Thoenen, H., "Catecholamines", in *Handbook of Experimental Pharmacology* (Springer, 1971), in press.
8. Bocchini, V., and P. U. Angeletti, *Proc. Nat. Acad. Sci. USA*, **64**, 787 (1969).
9. Thoenen, H., R. Kettler, W. Burkard, and A. Saner, *Naunyn-Schmiedebergs Arch. Pharmakol.*, in press.
10. Lowry, O. H., N. J. Rosebrough, A. L. Farr, and R. J. Randall, *J. Biol. Chem.*, **193**, 265 (1951).
11. Wilkinson, G. N., *Biochem. J.*, **80**, 324 (1961).
12. Cleland, W. W., *Nature*, **198**, 463 (1963).
13. Levi-Montalcini, R., F. Caramia, S. A. Luse, and P. U. Angeletti, *Brain Res.*, **8**, 347 (1968).
14. Angeletti, P. U., R. Levi-Montalcini, and F. Caramia, *J. Ultrastruct. Res.*, 1971, in press.
15. Mueller, R. A., H. Thoenen, and J. Axelrod, *J. Pharmacol. Exp. Ther.* **169**, 74 (1969).
16. Mueller, R. A., H. Thoenen, and J. Axelrod, *Mol. Pharmacol.*, **5**, 463 (1969).
17. Molinoff, P. B., S. Brimijoin, R. Weinshilboum, and J. Axelrod, *Proc. Nat. Acad. Sci. USA*, **66**, 453 (1970).

Reprinted from
Brain Research
Elsevier Publishing Company, Amsterdam
Printed in the Netherlands

Selective accumulation of [^{125}I] labelled nerve growth factor in sympathetic ganglia

Nerve growth factor (NGF) enhances the growth of sympathetic nerve cells in both newborn and adult animals[8]. The stimulation of sympathetic ganglia during NGF treatment is characterized by an overall increase of metabolic processes[9] and, furthermore, by a selective and marked induction of some key enzymes of the catecholamine biosynthetic pathway[11]. Recently, the primary and secondary structure of the NGF polypeptide moiety has been elucidated[2], thus facilitating studies of the structural–functional relationships of this unique protein. Since there are two tyrosyl residues present per polypeptide chain, an attempt was made to label NGF with ^{125}I for use as a tool in the investigation of its distribution and turnover in various tissues *in vivo*. Using the high specific activity NGF thus prepared, it has been shown that there is a selective accumulation of NGF into sympathetic ganglia.

The NGF used was isolated from male mouse submaxillary glands as described by Bocchini and Angeletti[4], except that the final step of the purification procedure (CM-cellulose chromatography) was repeated twice in order to obtain a preparation free from contamination. Radioactive iodine (^{125}I, carrier free) was purchased from the Radiochemical Centre (Amersham). Iodination of NGF was carried out at pH 8 in the presence of Chloramine T (ref. 6). In a typical experiment, 400 μl of ^{125}I (2 mCi) and 100 μl of Chloramine T (1 mg/ml) in 0.15 *M* Tris–HCl pH 8 were added to 500 μg of NGF in 200 μl of the same buffer. The reaction was stopped after 30 min by the addition of 100 μl of sodium metabisulfite (0.75 mg/ml) in Tris buffer. The labelled NGF was separated from unreacted iodine by use of a Sephadex G-25 column (Pharmacia), 0.5 cm $\times$ 20 cm, equilibrated with phosphate buffered physiological saline. The biological activity of both cold and radioactively labelled NGF samples was assayed in tissue culture[8]. All samples were counted in a Wallac gamma spectrometer.

Newborn and adult Swiss mice were used in all experiments. Labelled samples were injected subcutaneously or intravenously as indicated. Superior cervical ganglia were dissected out using a binocular microscope and weighed on a torsion balance. Samples of other tissues were also collected and weighed. No differences in results were obtained if the tissue was counted in large or small pieces.

A typical sample of [^{125}I]NGF prepared by the above procedure had a specific activity of 1.85 $\times$ 10^6 counts/min/μg protein. A control sample of bovine serum albumen (Sigma) possessed a specific activity of 2.6 $\times$ 10^6 counts/min/μg protein. The labelled NGF retained full biological activity. Although this method of labelling does not quantitatively modify all tyrosyl residues, NGF has been found to be completely active when modified at both tyrosyl residues by other reagents, indicating that this amino acid is not a part of the active site[3].

In a first series of experiments, the plasma half-life of the radioactive NGF was determined. Adult male mice weighing 30 g were injected via the tail vein with 2 μg [^{125}I]NGF (3.6 $\times$ 10^6 counts/min). Such doses are lower than those normally used to produce the classical 10- to 12-fold ganglion size increase (5–10 μg/g body weight/day), but do produce biological and metabolic effects within a few h[9]. Five μl of blood were collected from each mouse within 10 min of the injection, and at various time intervals thereafter. The samples were diluted in physiological saline before counting. The plasma half-life of NGF under these experimental conditions was calculated to be 1 h 50 min.

The rapid disappearance of NGF from the plasma is accompanied by a differential accumulation of this factor in various organs and tissues. Five-day-old mice were injected subcutaneously with [^{125}I]NGF (10^6 counts/min). Groups of three animals were sacrificed after 12, 24 and 48 h. The superior cervical ganglia, submaxillary glands, heart, liver, kidney, and brain were dissected out, weighed and counted. As shown in Table I, the specific radioactivity was markedly higher in the superior cervical ganglia than in all other tissues examined. The decay was also slower in sympathetic ganglia than in other structures.

Fig. 1 shows the percent of radioactivity found in various tissues relative to skeletal muscle of newborn and adult male mice injected with [^{125}I]NGF 24 h before sacrifice. As indicated, the specific activity in the sympathetic ganglia is about 12 times higher than in muscle. Comparatively higher values are obtained in kidney, submaxillary glands, and adrenals than in brain or skeletal muscle. When [^{125}I]albumen was injected into newborn mice, there was no evidence of selective accumulation in sympathetic ganglia either 12 or 24 h after injection. On the contrary, the specific radioactivity of superior cervical ganglia was lower than that found in other tissues such as liver or kidney.

The possibility that the selective accumulation of NGF in the sympathetic ganglia might depend on the functional and morphological integrity of the end terminals was explored by comparing the concentration of labelled NGF in ganglia of control animals and of mice treated with 6-hydroxydopamine (50 μg/g body weight). Newborn mice were injected subcutaneously, while adult mice were injected intraperitoneally. Experimental animals were treated with 6-hydroxydopamine or saline

TABLE I

In vivo DEGRADATION OF [^{125}I]NGF INCORPORATED IN VARIOUS TISSUES

5-day old mice were injected subcutaneously with 10^6 counts/min ^{125}I-NGF (specific activity 1.4 $\times$ 10^6 counts/min/μg) in phosphate buffered saline and sacrificed 12, 24 and 48 h later. Each value is the average ($\pm$ S.E.) of 3 experiments.

Tissue	*Counts/min/mg tissue*		
	12 h	*24 h*	*48 h*
Superior cervical ganglia	1400 $\pm$ 250	865 $\pm$ 160	125 $\pm$ 32
Submaxillary glands	422 $\pm$ 65	160 $\pm$ 45	84 $\pm$ 25
Heart	250 $\pm$ 36	69 $\pm$ 12	24 $\pm$ 10
Liver	260 $\pm$ 41	71 $\pm$ 22	29 $\pm$ 11
Kidney	570 $\pm$ 86	139 $\pm$ 32	55 $\pm$ 18
Brain	73 $\pm$ 15	24 $\pm$ 8	4

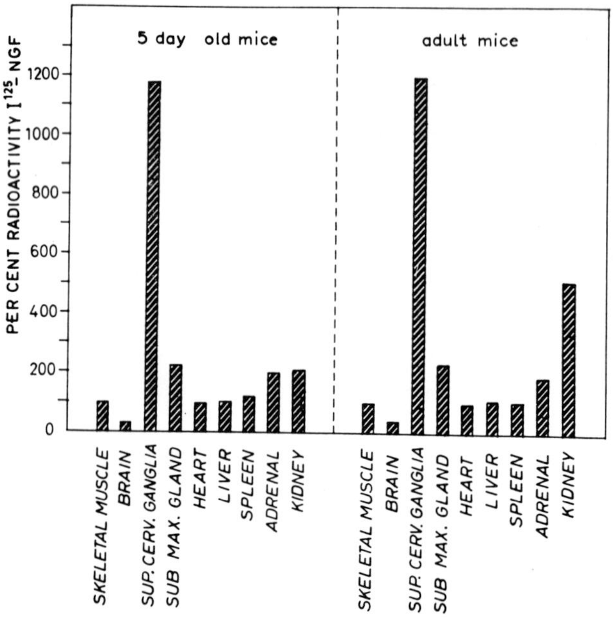

Fig. 1. Distribution of labelled NGF in various tissues of newborn and adult mice. [^{125}I]NGF (specific activity, 1.4×10^6 counts/min/μg) was injected subcutaneously in newborn mice and intravenously in adult mice. The animals were sacrificed after 16 h and the tissues removed, weighed and counted. The values are expressed as percent of the specific radioactivity (counts/min/mg tissue) found in skeletal muscle.

24 h prior to injection with radioactive NGF. They were sacrificed 12 h thereafter, and the radioactivity was measured in superior cervical ganglia and in the heart. The results given in Fig. 2 show that the NGF accumulation in sympathetic ganglia is markedly reduced in animals treated with 6-hydroxydopamine compared to controls, while the rate of plasma turnover was constant.

The *in vivo* effects of NGF are restricted to the sympathetic ganglia in newborn and adult animals[8,9]. The basis of this specificity is still unknown, yet elucidation of this aspect of the problem would be most important to understanding the mechanism of action of NGF. The experiments reported above were designed togain information about the fate of exogenous NGF in newborn and adult animals. The results indicate a rapid plasma turnover of NGF when introduced intravenously and a differential distribution of the labelled protein in peripheral organs and tissues. The very low radioactivity found in brain demonstrates the inability of the NGF molecule to cross the blood–brain barrier. The relatively higher concentration of labelled NGF in kidney and salivary glands compared to that of organs such as heart, liver, and muscle can be accounted for in part by their excretory function. In the case of the adrenal gland, whose specific radioactivity appears in the same range as in kidney and salivary glands, there may be a selective accumulation of NGF, which is, however, far below that found in sympathetic ganglia. In the latter, in fact, it reaches values as much as 12-fold higher than in skeletal muscle.

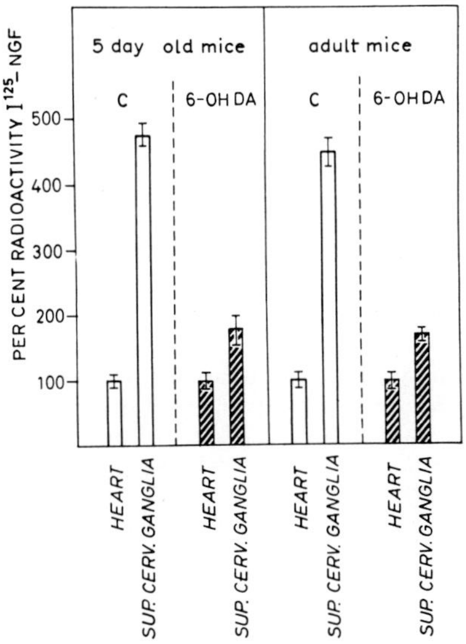

Fig. 2. Effect of 6-hydroxydopamine on accumulation of NGF in heart and superior cervical ganglia of newborn and adult male mice. See text for details.

These findings raise the question of how NGF gains access to its target cells, whether at the level of the nerve terminals or the cell body. The preliminary experiments performed with 6-hydroxydopamine-treated animals would seem to favour the former hypothesis. This drug has been shown to produce selective degeneration of end terminals in adult animals[12], whereas in newborn animals, the peripheral lesions are later followed by retrograde degeneration of the cell bodies[1]. In the experiments reported above, pretreatment with 6-hydroxydopamine virtually eliminated the differential accumulation of NGF by superior cervical ganglia relative to other tissues. It is important to distinguish whether this uptake occurs via the end terminals with subsequent retrograde transport to the cell body[7,10], at the perikaryon itself, or indeed, whether NGF acts wholly at the level of the plasma membrane as demonstrated for insulin by Cuatrecasas[5]. Experiments towards this end are currently in progress.

The authors would like to acknowledge the invaluable advice of Drs. R. and S. Tosi during the course of these studies.

Laboratory of Cell Biology,
C.N.R., Rome (Italy)

R. H. ANGELETTI
P. U. ANGELETTI
RITA LEVI-MONTALCINI

1 ANGELETTI, P. U., AND LEVI-MONTALCINI, R., Sympathetic nerve cell destruction in newborn mammals by 6-hydroxydopamine, *Proc. nat. Acad. Sci. (Wash.)*, 65 (1970) 114–121.

2 ANGELETTI, R. H., AND BRADSHAW, R. A., The amino acid sequence of 2.5S mouse submaxillary gland nerve growth factor, *Proc. nat. Acad. Sci. (Wash.)*, 68 (1971) 2417–2420.

3 ANGELETTI, R. H., FRAZIER, W. A., AND BRADSHAW, R. A., Topography of nerve growth factor: reactivity of tyrosine and tryptophan, *Abst. VIIIth FEBS Mtg.*, (1972) 462.

4 BOCCHINI, V., AND ANGELETTI, P. U., The nerve growth factor: purification as a 30,000 molecular weight protein, *Proc. nat. Acad. Sci. (Wash.)*, 64 (1969) 787–794.
5 CUATRECASAS, P., Insulin-receptor interaction in adipose tissue cells: direct measurement and properties. *Proc. nat. Acad. Sci. (Wash.)*, 68 (1971) 1264–1268.
6 HUNTER, W. M., AND GREENWOOD, F. C., Preparation of iodine-131 labelled human growth hormone of high specific activity. *Nature (Lond.)*, 194 (1962) 495.
7 KRISTENSSON, K., AND OLSSON, Y., Retrograde axonal transport of protein. *Brain Research*, 29 (1971) 363–365.
8 LEVI-MONTALCINI, R., The nerve growth factor: its mode of action on sensory and sympathetic nerve cells, *Harvey Lect.*, 60 (1966) 217–259.
9 LEVI-MONTALCINI, R., AND ANGELETTI, P. U., The nerve growth factor. *Physiol. Rev.*, 48 (1968) 534–569.
10 SJÖSTRAND, J., KRISTENSSON, K., AND OLSSON, Y., Retrograde axonal transport of exogenous proteins in the hypoglossal nerve, *IIIrd Int. Congress Neurochem.*, Budapest, (1971) 156.
11 THOENEN, H., ANGELETTI, P. U., LEVI-MONTALCINI, R., AND KETTLER, R., Selective induction of tyrosine hydroxylase and dopamine-hydroxylase in the rat superior cervical ganglia by nerve growth factor. *Proc. nat. Acad. Sci. (Wash.)*, 68 (1971) 1598–1602.
12 THOENEN, H., AND TRANWER, J. P., Chemical sympathectomy by selective destruction of adrenergic nerve endings with 6-hydroxydopamine, *Naunyn.-Schmiedeberg's Arch. exp. Path. Pharmak.*, 261 (1968) 271–288.

(Accepted August 14th, 1972)

PART III

WIDENING OF THE NGF SPECTRUM OF ACTION

A complete inventory of all NGF biological effects discovered in this second period of NGF studies would convey the impression that this most versatile molecule is endowed with almost unlimited capacity to exert its control on the target cells by resorting to different strategies according to the cell types and their pre- or post-differentiative stages.

An alternative and more plausible hypothesis[1] is that the broad spectrum of action of NGF derives from the remarkable flexibility of cells to respond to the same signal(s) by activating or repressing one of the vast repertory of intracellular mechanisms characteristic of each cell type (paper 24).

In Part III of this volume, I wish to call attention to some fundamental similarities in the response of NGF target cells. In all these cells, NGF dependence is maximal during early developmental stages. It declines thereafter, but still persists in fully differentiated cells. The similarity in the response elicited in homologous cells in animal species belonging to different vertebrate phyla indicates that the NGF receptor and the cells' intracellular mechanisms have not undergone substantial changes throughout the long evolutionary process from fishes to mammals. Their preservation supports the concept that this polypetide is endowed with an all-important biological role.

Evidence that the NGF spectrum of action is much wider than had at first been conceived also came from a new interpretation of findings which had remained unexplained ever since the earliest studies on the effect produced in chick embryos and neonatal rodents by NGF extracted from different sources. I had, in fact, noticed small and larger sympathetic nerve cell aggregates in sites where they were not observable in control specimens. The correct interpretation of these findings was suggested by the realization that sympathetic and chromaffin cells, not only originate from a common neural crest precursor, but are also committed to a similar differentiative program. Experiments performed by other investigators[2] demonstrated that an *in vitro* cultured clonal cell line, derived from a pheochromocytoma rat tumor (PC12 line), acquires several morphological and biochemical markers of sympathetic cells cultured in media supplemented with minute amounts of NGF. Upon NGF withdrawal, cells revert to their original neoplastic phenotype. These findings introduced

[1]Calissano P., Cattaneo A., Aloe L. and Levi-Montalcini R., The nerve growth factor (NGF). In *Hormonal Proteins and Peptides* (Academic Press, 1984) Vol. **XII**, pp. 1–56.

[2]Greene L.A. and Tischler A.S., Establishment of a noradrenergic clonal line of rat adrenal pheochromocytoma cells which respond to nerve growth factor, *Proc. Natl. Acad. Sci. USA* **73**, 2424–2428 (1976).

a new and most valuable model for the exploration of the mechanism of action of NGF at the morphological, physiological, biochemical and molecular levels. Shortly thereafter, it was also found that chromaffin cells dissociated from rodent adrenal medulla and cultured in NGF-containing media undergo morphological differentiation into sympathetic-like nerve cells.[3] It then occurred to me that the previously found ectopic sympathetic cell aggregates might derive from the property of NGF to alter the program in stem cells (not yet committed) to become sympathetic rather than chromaffin cells.

In order to explore this phenomenon in living organisms, we administered NGF to chick embryos and to rodent fetuses. The selection of the phenotypic chromaffin cell type in chick embryos is accompanied by impressive morphogenetic changes, such as the formation of large ganglionic complexes of sympathetic nerve cells positioned inside a diminutive adrenal gland consisting of cortical cell strands and entirely lacking the gland medullary component (papers 23, 26). In rat fetuses, NGF was administered at later developmental stages when chromaffin cell precursors had already settled inside the adrenal gland; the switch from glandular to sympathetic nerve cells took place *in situ*, resulting in the replacement of the medullary gland component with large aggregates of sympathetic cells. The phenotypic modulation produced by NGF is not restricted to precursors of adrenal medullary cells, but occurs also in other chromaffin stem cell derivatives that give rise to abdominal paraganglia, and carotid and aortic bodies (paper 23).

The widening spectrum of NGF action was brought to our attention by the discovery of:

1. its action on the neuroendocrine axis;
2. its activity in neuronal cell populations of the CNS of higher and lower vertebrates;
3. its effects on cells belonging to the immune system;
4. its role in aggressive and defensive behavior.

The most relevant findings reported in this section, are briefly outlined below.

In 1977, we reported that NGF injections in neonatal rodents result in a striking numerical and size increase in mast cells which play the main role in allergic reactions through the release of numerous biologically active compounds. These cells derived from the hematopoietic system are present in all peripheral tissues in close proximity to innervating sensory and sympathetic nerve fibers; they are also found in the CNS.

As described and illustrated in paper 25, it is unlikely that this NGF effect on mast cells is due to its antigenic properties, since other biologically active polypeptides and proteins do not elicit this effect. In support of this hypothesis is also the

[3]Unsiker K., Krisch B., Otten U. and Thoenen H., Nerve growth factor-induced fiber outgrowth from isolated adrenal chromaffin cells: impairment by gluco-corticoids, *Proc. Natl. Acad. Sci. USA* **75**, 3498–3502 (1978).

finding that NGF administration to neonatal rats results in an approximately 10-fold numerical increase of these cells. No increase is apparent upon NGF injections in adult rats, as would be expected from a NGF-induced immunological response. In subsequent years it was demonstrated that NGF stimulates B and T lymphocyte proliferation, and selected antibody production. NGF also promotes hematopoietic cell growth and proliferation, and helps in maintaining the viability and functional properties of polymorphonuclear neutrophils. These NGF effects are consistent with the presence of NGF receptors on monocytes and lymphocytes. Lymphocytes produce NGF, suggesting an autocrine or paracrine action for NGF in the development and regulation of immune cell responses.[4–6]

The discovery of a severe neuroendocrine syndrome in prenatal, NGF antibody-treated rats revealed the key role played by NGF in the development of the Hypothalamo-Pituitary Adrenal (HPA) axis. This syndrome was produced by injection of purified NGF antibodies in rat fetuses (paper 27). In addition to the sensory deficit caused by marked atrophy of sensory dorsal root and cephalic ganglia, the prenatal NGF antibody-treated rats exhibit a severe neuroendocrine deficit syndrome characterized by progressive loss of weight, and apathetic, sluggish behavior. About 60% of the treated pups die before the end of the third week; those that survive show stunted growth, marked atrophy of the thyroid gland and impairment of other components of the endocrine system.

Another landmark in the life history of NGF was marked in 1984 with the discovery that the spectrum of action of this growth factor is not restricted to two cell lines of the Peripheral Nervous System (PNS), the sympathetic adrenergic neurons and a portion of sensory nerve cells, but also extends to some CNS neuronal cell populations.[7] Their identification was achieved by resorting to the specific retrograde axonal transport technique of radio labelled NGF, mentioned in Part II of this volume.

After some unsuccessful attempts based on the expectation of finding these neurons in the two aminergic CNS nerve cell populations (locus ceruleus and substantia nigra), Seiler and Schwab found them instead in unexpected CNS regions. "Surprisingly", they reported, "NGF was found to be retrogradely transported from the hippocampus to large cholinergic neurons" localized in some septal forebrain nuclei. The same technique was used to identify, in the neocortical frontal and occipital regions, the site of synthesis of NGF and its receptive neurons in other cholinergic nerve

[4]Brodie C. and Gelfand E.W., Functional nerve growth factor receptors on human B lymphocytes, Interaction with IL-2, *J. Immunol.* **148**, 3492–3497 (1992).
[5]Otten U., Ehrard P. and Peck R., Nerve growth factor induces growth and differentiation of human B lymphocytes, *Proc. Natl. Acad. Sci. USA* **86**, 10059–10063 (1989).
[6]Thorpe L. W., Werrbach-Perez K. and Perez-Polo J. R., Effects of nerve growth factor expression on interleukin-2 receptors on cultured human lymphocytes, *Ann. NY Acad. Sci.* **496**, 310–311 (1987).
[7]Seiler M. and Schwab M., Specific retrograde transport of nerve growth factor (NGF) from neocortex to nucleus basalis in the rat, *Research* **300**, 33–39 (1984).

cells in the forebrain basal centers and the striatum. Evidence for the role played by NGF in these neurons was found by other investigators, who demonstrated that local injection of exogenous NGF increases the activity of the enzyme involved in the synthesis of the cholinergic neurotransmitter acetylcholine.[8,9]

Subsequent studies provided even more compelling evidence in favor of the role attributed to endogenous NGF in the functional activity, and even in the survival, of cholinergic basal forebrain and striatum neurons.[10,11] Elimination of the target regions or surgical sectioning of their nerve fiber connections with cholinergic neurons result in severe shrinkage of these nerve cells and in a decrease of their functional activity. Local supply of exogenous NGF leads to the reestablishment of the normal function of these cholinergic nerve cells.[12]

The main contribution of these studies has been to reveal the NGF activity in cholinergic neurons in the CNS which play a key role in cognitive and higher brain functions. These findings prospect the possibility that an exogenous NGF supply might mitigate memory and learning deficiencies which are consequent to the impairment of these neurons.

Evidence in support of this hypothesis has been obtained in animal models, primates included.[13–15]

Studies that were performed in our laboratory in 1985 demonstrated that NGF enhances the differentiation, not only of cholinergic brain neurons, but also of other noncholinergic CNS nerve cell nuclei in lower vertebrates: the amphibian Xenopus

[8]Gnahn H., Hefti F., Heuman R., Schwab M.E. and Thoenen H., NGF-mediated increase of choline acetyltransferase (ChAT) in the neonatal rat forebrain: evidence for a physiological role of NGF in the brain?, *Dev. Brain Res.* **9**, 45–52 (1983).

[9]Mobley C.W., Rutkowsky J.L., Tennekoon G.I., Buchanan K. and Johnson M.V., Choline acetyltransferase activity in striatum in neonatal rats is increased by nerve growth factor, *Science* **229**, 284–287 (1985).

[10]Cuello C., Trophic responses of forebrain cholinergic neurons: a discussion, *Prog. Brain Res.* **98**, 265–277 (1993).

[11]Olson L., Wetmore C., Stromberg I. and Ebendal T., Endogenous and exogenous nerve growth factor in the central nervous system. In *Transmission in the Brain: Novel Mechanism of Neural Transmission*, Fuxe K. and Agnati L.F. eds. 455–462 (1991).

[12]Fischer W. and Björklund A., Loss of AChE- and NGFr-labelling precedes neuronal death of axotomized septal-diagonal band neurons: reversal by intraventricular NGF-infusion. In Nerve growth factor: amelioration of degenerative changes in cholinergic neurons of the brain, W. Fischer ed., *Exp. Neurol.* 195–219 (University of Lund, Sweden, 1991).

[13]Hellweg R., Fischer W., Hock C., Gage F.H., Biörklund A. and Thoenen H., Nerve growth factor levels and choline acetyltransferase activity in the brain of aged rats with spatial memory impairments, *Brain Res.* **537**, 123–130. In Nerve growth factor: amelioration of degenerative changes in cholinergic neurons of the brain, W. Fischer ed. (University of Lund, Sweden, 1991) pp. 101–108.

[14]Fischer W., Björklund A., Chen K. and Gage F.H., NGF improves spatial memory in aged rodents as a function of age, *J. Neurosci.* 229–261 (1991).

[15]Tuszynski M.H., Sang U.H., Amaral D.G. and Gage F.H., Nerve growth factor infusion in the primate brain reduces lesion-induced cholinergic neuronal degeneration, *J. Neurosci.* **10**, 3604–3614 (1990).

tadpoles (paper 28). More recently, a function for NGF in activating non-cholinergic CNS brain populations in higher vertebrates has been reported.[16,17]

Demonstration of the role played by NGF in cholinergic and non-cholinergic CNS populations further stimulated interest in the mechanism of action of this protein molecule at the molecular level. This interest, in turn, resulted in the identification of the human and rodent NGF gene[18,19] and in the elucidation of the processes involved in reception and transduction of the NGF signal.

The biological actions of NGF are mediated by specific cell surface receptors. In neurons, for example, NGF receptors are present on the surface membrane of undifferentiated cells and on the nerve endings of axons and dendrites of the same fully differentiated cells. Binding of NGF to its receptors is followed by internalization of the NGF-receptor complex and distribution into different cellular compartments, including lysosomes which eventually destroy the factor.

There are two structurally unrelated receptors for the NGF family.[20] The neurotrophins bind with low affinity ($K_d = 1nM$) to the so-called p75 NGF receptor. The second NGF binding protein is a member of the Trk family of tyrosine kinase receptors, gp 140^{TrkA}. The receptor gp 140^{TrkA} has many of the properties characteristic of a high-affinity receptor class ($K_d = pM$), and its interaction with NGF appears to be responsible for mediating neuron survival and differentiation.[21,22]

The NGF signaling process is initiated by ligand-induced gp 140^{TrkA} dimerization which mediates phosphorylation of several key enzymes involved in distinct signal transduction cascades, and includes second messenger production and activation of genetic transcriptional programs.[23]

While these studies were in progress, another novel approach was adopted in our laboratory to understand the significance of the endogenous production of NGF in mouse hypothalamus following intermale aggressive behavior.

Papers 29 and 30 prospect a solution to the long unsolved problem, namely, why the male mouse submaxillary salivary gland should produce such a large quantity

[16]Carmignoto G., Maffei L., Candeo P., Canella R. and Comelli C., Effect of NGF on the survival of rat retinal ganglion cells following optic nerve section, *J. Neurosci.* **9**, 1263–1272 (1989).
[17]Chen S. and Bentivoglio M., Nerve growth factor receptor-containing cholinergic neurons of the basal forebrain project to the thalamic reticular nucleus in the rat, *Brain Res.* **606**, 207–212 (1993).
[18]Frankee U., De Martinville B., Coussens L. and Ullrich A., The human gene for the ß subunit of the nerve growth factor is located on the proximal short arm of chromosome 1, *Science* **222**, 1248–1250 (1983).
[19]Scott J., Selby M., Urdea M., Quiroga M., Bell G.I. and Rutter W.J., Isolation and nucleotide sequence of a cDNA encoding the precursor of mouse nerve growth factor, *Nature* **302**, 538–540 (1983).
[20]Meakin S. and Shooter E.M., The nerve growth family of receptors, *TINS* **15**, 9, 323–331 (1992).
[21]Barbacid M., Nerve growth factor: a tale of two receptors, *Oncogene* **8**, 2033–2042 (1993).
[22]Chao M.V., Neurotrophin receptors: a window into neuronal differentiation, *Neuron* **9**, 583–593 (1992).
[23]Heumann R., Neurotrophin signalling, *Neurobiology* **4**, 668–679 (1994).

of NGF, a protein molecule with apparently no connection with the digestive function. These considerations suggested to me that NGF synthesized in the convoluted tubules of these glands could be endowed with a modulatory role in the intraspecific aggressive behavior typical of these rodents. Experiments reported in paper 31 provided unequivocal evidence that, during experimentally induced aggressive behavior between adult male mice, the basal NGF level in the bloodstream is markedly higher than under normal conditions. Paper 30 demonstrated that, in the hypothalamus, NGF messenger RNA and the NGF molecule increase during aggression and anxiety.

The sudden increase in brain NGF level which follows a psychosocial stressful event could serve some still unknown CNS process leading to a renewal of brain plasticity at the adult stage. In fact, it appears that stress can rapidly change behavioral response and promote, for example, imprinting-like phenomena in adults, in whom some social stimuli suddenly become relevant and produce long-lasting behavioral alterations. Endogenous NGF has been suspected of being involved in adult CNS plasticity, since the normally ongoing release of NGF evokes collateral sprouting from the terminal field of adult neurons.[24]

The 1986 review paper (paper 31) presents an overview of the main findings on NGF, its mechanism of action and the growth response elicited in receptive neuronal and nonneuronal nerve cells.

The 1987 Nobel Lecture is presented in paper 32.

[24]Alleva E. and Aloe L., Physiological roles of nerve growth factor in adult rodents: a biobehavioral perspective, *Int. J. Comp. Psychol.* **2**, 4, 213–230 (1989).

Histochemistry and Cell Biology of Autonomic Neurons, SIF Cells, and Paraneurons,
edited by O. Eränkö et al.
Raven Press, New York © 1980

Tropic, Trophic, and Transforming Effects of Nerve Growth Factor

Rita Levi-Montalcini and Luigi Aloe

Laboratorio di Biologia Cellulare CNR, Rome, Italy

> Experiments are the only means of knowledge at our disposal. The rest is poetry, imagination.
>
> *Max Planck*

It is the rule rather than the exception in experimental sciences that the results seldom if ever measure up to the expectation. It is likewise a widespread habit to ignore those findings that do not fit in preconceived schemes. The history of the nerve growth factor (NGF) from its rather remote beginnings to the present day exemplifies such constant refusal to conform to predictions based on previous findings and on firmly established dogmas in developmental neurobiology. In the case of NGF, however, the discrepancies between anticipation and results followed a reverse trend from that mentioned above: The findings almost invariably went beyond the expectation, uncovering new facets of the phenomenon under investigation and forcing us to extend these studies and at the same time to focus our attention on new features of this remarkable stimulus-response system.

Although it is not the purpose of this presentation to review the main steps of this never ending investigation, we shall, in a way of introduction to the report of recent developments, briefly mention the most striking deviations of the earlier findings from those that had been anticipated.

It was the excessive production and atypical distribution of sympathetic nerve fibers into the viscera and inside the vascular bed of avian embryos bearing grafts of mouse sarcomas 180 or 37 (27) that called for a revision of the hypothesis that implantation and vigorous growth of these tumors elicits a growth response from sensory (10) and sympathetic (39) ganglia innervating these tumors which is similar even if more impressive than that elicited by implantation of additional limb buds borrowed from other embryos. The finding that extraembryonic transplants of these two tumors elicited the same effects as intraembryonic transplants (27,40) strongly supported the hypothesis that these mouse sarcomas produce a growth effect on sensory and sympathetic embryonic ganglia by releasing a humoral agent which reaches the nerve cells through vascular channels rather than through, or concomitance with, the conventional generally accepted mediation through nerve fibers.

The tissue culture hanging-drop technique provided an ideal system to explore the morphological effects called forth by (at that time) still unknown tumoral factor (28,41) and to obtain preliminary information on its chemical nature, tentatively identified in a nucleoprotein moiety. The same *in vitro* technique made it possible to assay other tissues and humoral organic agents as potential sources of the nerve growth tumoral factor which ever since 1954 became known as the nerve growth factor (18). It was this technique which revealed the existence of two much more potent NGF sources than mouse sarcomas in snake venom (17) and in mouse submaxillary salivary glands (29). These discoveries not only invalidated the hypothesis that the nerve growth-promoting effect elicited by the two mouse sarcomas could reside in a specific property of neoplastic cells (possibly a virus particle) but made available much larger quantities of NGF, which afforded the possibility of much more extensive and rigorous studies of its biochemical properties and of the growth response elicited in its target cells. In 1956 NGF was identified by Cohen in a protein molecule isolated from snake venom (15), and in 1958 an almost identical molecule endowed with the same biological properties was extracted and purified from mouse salivary glands (16,29). With the elucidation in 1971 of the primary amino acid sequence of the 2.5S salivary NGF achieved by Angeletti and Bradshaw (6), this protein molecule became the first growth specific factor to be characterized at the chemical level.

Studies during this last decade centered on the exploration of the binding properties of NGF to its membrane receptors (8,23,42), the evidence of its retrograde axonal transport (22) from target tissues, and the molecular basis of its mechanism of action (12,13). Here we first consider some of the recent developments which brought to our attention new features of the NGF trophic effects and uncovered two other NGF properties: (a) its neurotropic effects on developing sympathetic fibers, and (b) its transforming effects of neoplastic or immature chromaffin normal cells in sympathetic nerve cells.

RECENT STUDIES ON NGF TROPHIC AND PROTECTIVE EFFECTS ON IMMATURE SYMPATHETIC NERVE CELLS

Previous studies (9,30,36,37,45) gave unequivocal evidence for the magnitude of the nerve growth-promoting effect elicited by the NGF purified from snake venom and mouse salivary glands, and at the same time proved that NGF displays a much more essential role in the life of target cells than was first conceived. It was in fact shown that a specific antiserum to NGF injected in neonatal rodents and other neonatal mammals produces massive destruction of sympathetic para- and prevertebral ganglia (38). This dramatic effect, which deprives the animal of the sympathetic function for its entire life and which became known as immunosympathectomy (35), gave evidence for the essential role played by NGF on the sympathetic cells during their early phases of growth and differentiation. *In vitro* experiments provided additional evidence in favor of a NGF unique role. Sensory and sympathetic nerve cells dissociated from ganglia of 8-day chick embryos do not survive when cultured in minimal essential media deprived of serum. They survive instead indefinitely and build a dense fibrillar network which covers the entire surface of the culture dish when minute quantities (10 ng/ml culture) of NGF are added daily to these otherwise inadequate media (34). The much more marked vulnerability of immature sympathetic nerve cells to the effects of a specific antiserum to NGF than that of differentiated noradrenergic neurons suggested that these studies be extended to the effects of pharmacological compounds which in one or another way interfere with the transmission of the nerve impulse from the end terminals of sympathetic cells to their target tissues in fully developed organisms. The first of these compounds to be tested was a dopamine derivative known as 6-hydroxydopamine (6-OHDA), which blocks the transmission of the nerve impulse in sympathetic nerve fibers for a 6- to 8-week period (46). The cause of this effect is the accumulation of this compound in the adrenergic nerve end terminals where it selectively accumulates, producing degenerative reversible lesions of these endings (47). Studies performed in our laboratory in 1969 showed that 6-OHDA injections in neonatal rodents result in the massive death of para- and prevertebral sympathetic ganglia comparable to that produced by a specific antiserum to NGF. This effect became known as chemical sympathectomy (4). Subsequent studies provided strong evidence in favor of the hypothesis that nerve cell death is due to the block of retrograde axonal transport of the NGF produced by peripheral tissues. This process, which is essential for growth and differentiation of the immature nerve cells, is impeded by the accumulation of 6-OHDA in the growing adrenergic endings. During the following years it was discovered that two other compounds, guanethidine and vinblastine, also produce, although through entirely different mechanisms (5,20,44), the destruction of para- and prevertebral sympathetic ganglia when injected in neonatal rodents. These findings raised the question of whether an exogenous supply of NGF would counteract the degenerative effects produced by these three agents (6-OHDA, guanethidine, and vinblastine) in immature sympathetic nerve cells. As reported in previous articles (25,33), the results fully confirmed our expectation of an NGF protective effect on these cells. Of particular interest and worth some comment was the finding that, in the case of a combined 6-OHDA and NGF treatment, the NGF not only prevented the death of the immature sympathetic nerve cells but produced an extraordinary volume increase of sympathetic para- and prevertebral sympathetic ganglia, which over a 3-week period of daily treatment reached a volume 30 times larger than that of the same ganglia in control littermates and about three times larger than that of ganglia of littermates injected with NGF alone (3). Structural, ultrastructural, fluorescence, and biochemical studies provided an explanation of the cause of this paradoxical growth effect. Even in the presence of NGF, 6-OHDA accumulates in the noradrenergic nerve endings and blocks the transmission of the nerve impulse and the retrograde axonal transport of NGF from the endogenous peripheral NGF sources. The simultaneous supply of NGF far in excess of that which normally reaches the cells from peripheral tissues produces an increase in size and number of nerve cells comparable to that elicited by injections of the same NGF doses in intact neurons. In addition, and at variance with the latter effects, the chemically axotomized axons produce an enormous amount of neurofibrillar material which is routed in collaterals sprouting from the proximal segment of the chemically transected axons. A similar effect obtains in surgically axotomized and NGF-treated ganglia (2). The formation of this extraordinary number of collaterals, which force the cells apart and form a dense fibrillar capsule around the ganglia, is due to the lack of negative feedback control mechanisms which under normal conditions prevent extra production of nerve fibers, as shown in ingenious and rigorous experiments performed by Diamond and co-workers (19). Neither guanethidine nor vinblastine interfere with the two-way traffic at the terminal endings of axons with their target tissues; the combined treatment of NGF with either one or the other of these two compounds results therefore in an effect comparable to, but not exceeding, that elicited by NGF alone (25,44).

NGF NEUROTROPIC EFFECT

One of the most debated and controversial problems in developmental neurobiology is defining the role of endogenous and exogenous factors in guiding nerve fibers toward their target organs. Three basic mechanisms have been proposed over the years to explain the formation of specific neuronal circuitry: (a) an elaborate predetermined program encoded genetically in each neuron that unfolds according to rigid and unmodifiable rules; (b) a random process of trial and error in which growing nerve fibers that make the right connections are consolidated and those that fail are reabsorbed; and (c) a general program of circuit formation that is brought to completion by an interplay between genetic and extrinsic factors. The first mechanism can be discounted since it would require very large amounts of genetic information—much more than could be encoded in the entire complement of DNA in the cell nucleus of each neuron. The second mechanism is also most unlikely because such a random process of circuitry formation would be extremely time-consuming and wasteful of energy and resources. The third mechanism therefore seems the most probable: The circuitry between neuronal cell population in the central nervous system and between motor, sensory, and autonomic neurons and their peripheral effectors or receptors is established through a combination of genetic and extrinsic factors. The existence of such extrinsic factors was first suggested by Cajal, who visualized them as chemical signals issued from peripheral tissues that would direct the growing nerve fibers toward their matching target cells (11). At that time and during subsequent years, however, the methodology for detecting these hypothetical chemical factors in the developing embryos was not yet available, and the neurotropism theory (the name suggested by Cajal to designate this effect) was dismissed in favor of an even more putative role played by mechanical "contact guidance factors," or adhesion of the fibers to a solid substrate in the growth of nerve fibers and formation of correct end connections. The discovery of NGF, however, offered the opportunity of reconsidering the concept of neurotropism under far more favorable conditions than had been previously available. Here we only briefly mention experiments performed in our laboratory and described in detail in previous publications (32,43) which gave unequivocable evidence in favor of an NGF neurotropic role on growing sympathetic nerve fibers in neonatal rodents. NGF was injected with the aid of a microcapillary glass pipette into the brain of neonatal mice and rats in the proximity of the two nuclei cerulei. The experiments were aimed at exploring whether intracerebral monoaminergic nerve cells, and in particular those of the locus ceruleus, would undergo a size increase in the presence of NGF. The results did not confirm this hypothesis but brought to light a different and unforeseeable effect. Extensive studies performed with the fluorescence microscope on the serially sectioned cerebrospinal axis processed according to the Falck-Hillarp technique showed in fact that sympathetic nerve fibers emerging from the greatly enlarged paravertebral sympathetic ganglia gained access inside the spinal cord and the brainstem through the dorsal roots of spinal nerves and the sensory-motor roots of the lowest cephalic nerves, and settled, respectively, in the dorsal funiculi of the spinal cord and ventrolateral cordons of the brainstem. Collaterals emerging from these ectopic fiber bundles entered the white and gray matter in close association with blood vessels. Discontinuation of the NGF intracerebral injections resulted in the fading and final disappearance of these parasitic systems. The hypothesis that the entrance of sympathetic nerve fibers into the central nervous system of intracerebrally NGF-injected rodents is produced by an NGF diffusion gradient from the site of injection into the floor of the fourth ventricle to the spinal cord realized through its transport along sensory and motor roots to the adjacent ganglia is supported by the marked volume increase of these ganglia and the demonstration of a high NGF level in the brain and spinal cord of the injected pups 24 to 30 hr after the last injection. More recent ingenious *in vitro* experiments performed in two other laboratories provided strong additional support in favor of a clear-cut neurotropic NGF effect on sympathetic sensory and nerve fibers growing out of dissociated nerve cells of both types (14,26).

NGF IN VITRO TRANSFORMING EFFECT OF PHEOCHROMOBLASTS IN SYMPATHETIC NERVE CELLS

Greene and Tischler discovered that cells of the PC 12 line isolated from a rat pheochromocytoma tumor respond to NGF by acquiring properties characteristic of sympathetic neurons upon addition of NGF to the culture medium. This in turn brought to light a new feature of the biological activity of this protein molecule (21). Subsequent studies by Unsicker et al. (48) provided evidence for a similar *in vitro* effect elicited by NGF on explants of rat's adrenal medulla and on the chromaffin cells dissociated from this gland. These studies proved the property of NGF to divert neoplastic and normal chromaffin cells cultured *in vitro* in sympathetic nerve cells. On the basis of these findings, Unsicker and co-workers submitted the hypothesis that these *in vitro* effects were at least in part attributed to the fact that chromaffin and sympathetic cells, which have their origin in a common precursor, under normal conditions undergo transformation in glandular cells under the influence of glucocorticoid hormones released from the adjacent cortical cells of the adrenal gland. Although this is a plausible hypothesis, experiments performed *in vivo* in our laboratory showed that NGF injections in intact rat fetuses and chick embryos also elicit this effect in an even more massive and impressive way than on dissociated neoplastic or normal chromaffin cells *in vitro.*

TRANSFORMATION OF IMMATURE CHROMAFFIN CELLS IN SYMPATHETIC NERVE CELLS IN RAT PUPS AFTER PRE- AND POSTNATAL NGF INJECTIONS

As reported in a previous article (1), NGF injections in 16 to 17-day-old rat fetuses administered through the intact tubes, and resumed immediately after birth on a daily basis from the first day to days 10 to 20, result in massive transformation of the immature chromaffin cells in sympathetic nerve cells. In the experiments briefly considered here and reported more extensively elsewhere, we increased the prenatal treatment to two NGF injections rather than one, given at 15.5 and 17 days; we also extended the studies of the effects of this pre- and postnatal treatment to the chromaffin cells of the abdominal paraganglia and the carotid bodies. The effects of this more intensive NGF prenatal treatment on the adrenal medulla is illustrated in Fig. 1a,b, which compares the two largest cross sections of the adrenal gland in a control and an experimental 2-week-old pup. The experimental pup received two intrauterine injections between the end of the 15th and 17th gestational days and then daily postnatal injections of NGF (10 μg/g body weight) until the day of sacrifice. It is also evident from comparison of the two microphotos that the intensive NGF treatment resulted not only in massive transformation of chromaffin tissue into sympathetic nerve cells but also in the production of an enormous number of nerve fibers, which encapsulated the transformed nerve cells and enlarged the medullary area at the expense of the surrounding cortical zones. The latter are infiltrated with large nerve fiber bundles, which further reduces the space available for cortical cells of the three lateral, medial, and internal zones. Figure 1c,d compares the two largest sections of the adrenal glands of a control and an experimental rat of the same age as those portrayed in Fig. 1a,b. The glands were processed according to the Falck–Hillarp technique, and the sections were photographed with the fluorescence microscope. Note the enormous enlargement of the fluorescent medullary area in Fig. 1d, as compared to Fig. 1c and the presence of intensely fluorescent fiber bundles in the cortical zones of the experimental but not the control gland. The fluorescence of the control adrenal medulla was intensely yellow, whereas

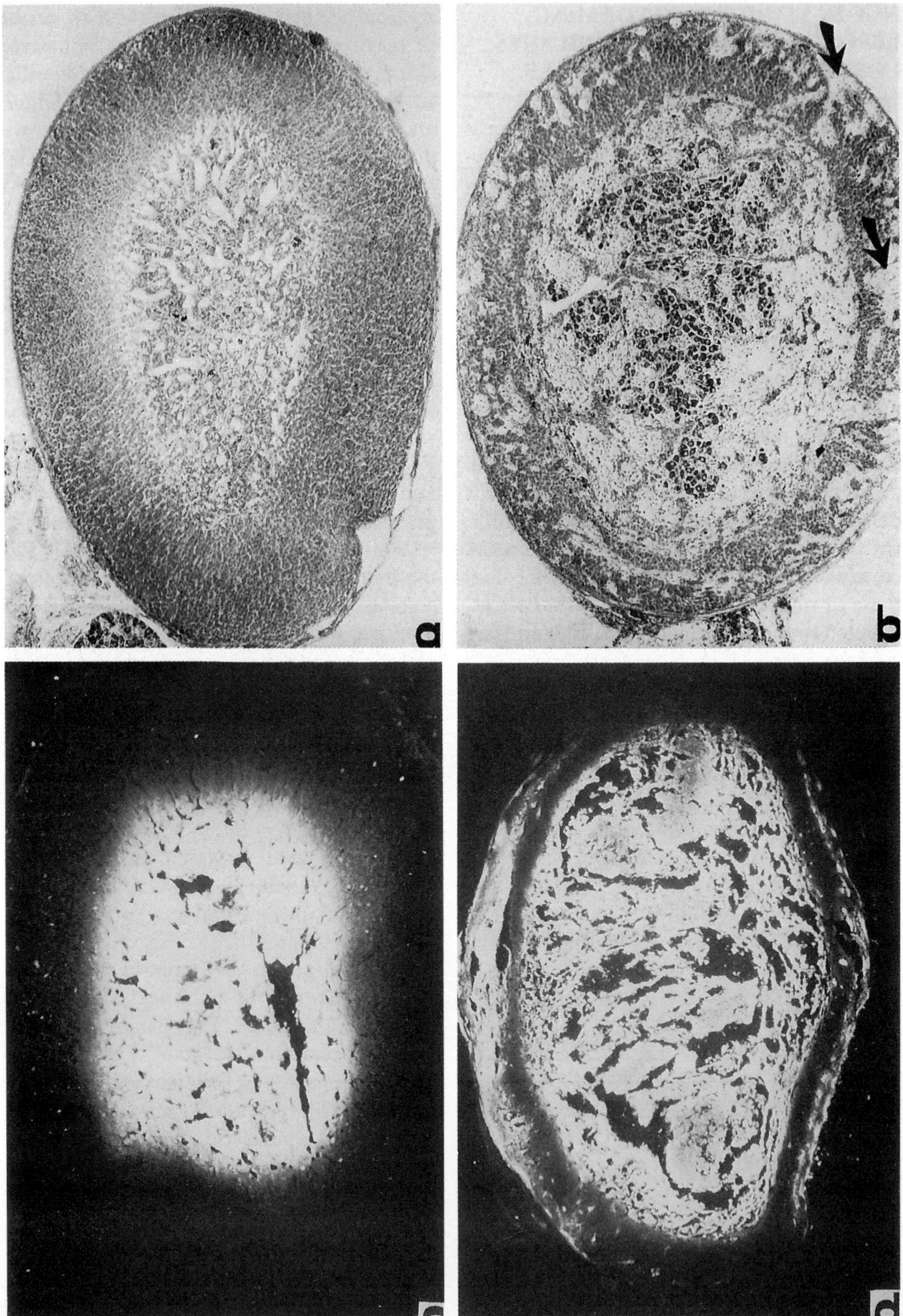
a
b
c
d

the experimental section exhibited the typical green color of noradrenaline-containing cells and fiber bundles. Figure 2 compares at low and high magnification the carotid bodies and surrounding tissues in a control and an experimental rat pup of the same series as those illustrated in Fig. 1. The small spindle-shaped glomus cells of the control body intermingled with interstitial cells of similar size and appearance are, in the experimental organ, transformed into a homogenous population of much larger cells indistinguishable from sympathetic neurons of the adjacent superior cervical ganglion apparent in Fig. 1a,b. The marked volume increase of the experimental carotid body, obvious from a comparison of Fig. 2a and b, is due to the increased size of individual cells, but foremost to the production of a large number of nerve fibers which join the postganglionic nerves of the adjacent superior cervical ganglion.

Studies still in progress on the abdominal paraganglia showed the identical pattern of transformation of glandular cells into nerve cells as depicted in the adrenal medulla and the carotid bodies.

TRANSFORMATION OF IMMATURE CHROMAFFIN CELLS IN NERVE CELLS IN CHICK EMBRYOS

The technique devised by Auerbach and co-workers (7) to cultivate chick embryos in Petri dishes from the 3rd to the 19th day of incubation provided an invaluable tool to inspect the effects of intraembryonic NGF injections from the 4th to the 17th day of embryonic development. These studies could not be performed *in ovo* because of the movements of the embryo, its envelopment in the chorioallantoic membranes, and the hemorrhage that follows to the blind insertion of the injection needle into the embryos. Under excellent conditions of visibility it was possible to inject microgram quantities of NGF directly into the body of the embryos with a microcapillary needle. We report on the preliminary results of these studies.

Figure 3 illustrates two transverse sections of 5-day-old chick embryos processed according to the Falck technique and photographed in the fluorescence microscope. On the fourth day the control embryo received an injection of physiological solution and the experimental embryo received 5 μg NGF. The marked increase in fluorescent cells migrating from the neural crest is apparent in the experimental embryo as compared to the control. Of particular interest is the formation of two intensely fluorescent cell aggregates in close proximity to the medial aspect of the primordium of the adrenal gland in the experimental embryo, whereas these cells are entirely absent in the control.

Older control and experimental embryos sacrificed daily between the 7th and the 18th days were fixed according to the Cajal De Catro silver technique and sectioned serially. Figure 4a,b illustrates two transverse sections of 11-day-old control and experimental embryos at the level of the adrenal medulla and mesonephros, and Fig. 4c,d shows sections at the same level in two control and experimental embryos sacrificed on the 18th day of incubation. Before commenting on the results, it is worth noting that the adrenal gland in the avian species differs markedly from that of mammals. Chromaffin cells in long thin strands or islands are intermingled with cords of cortical cells. The two cell types are

◀
FIG. 1. a and **b**: Transverse sections of adrenal glands of 3-week-old rat pups. **a**: Control. **b**: Pup injected during fetal and postnatal life with NGF as explained in the text. Arrows in **b** point to fiber bundles produced by the transformed sympathetic cells in the adrenal medulla, infiltrating the three zones of the adrenal cortical gland. Further explanation in text. Toluidine blue. ×50. **c** and **d**: Histofluorescence preparations of transverse sections of control (**c**) and NGF-injected (**d**) pups of the same age as the specimen in **a** and **b**. Further explanation in text. ×50.

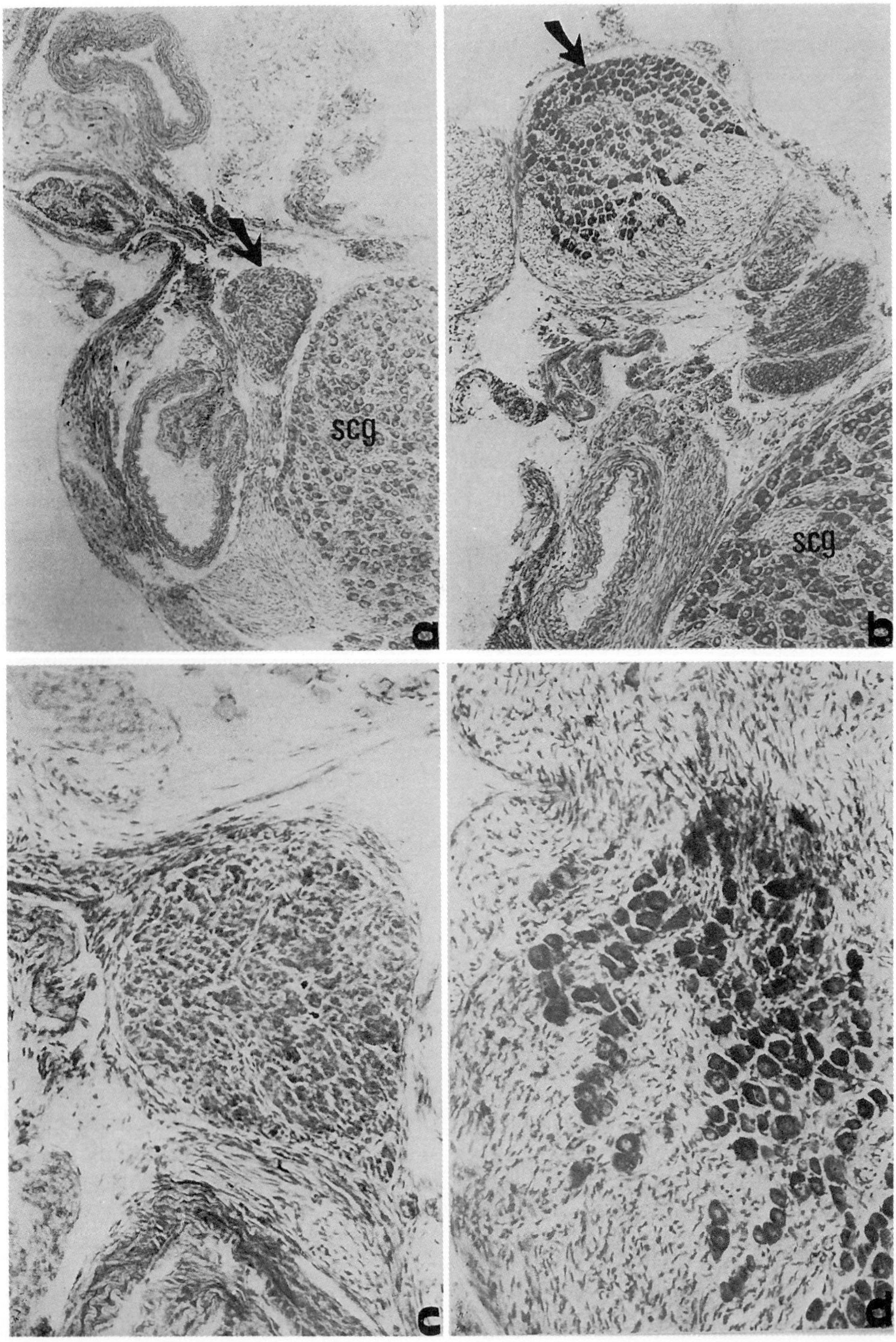

FIG. 2. a and **b**: Arrows point to carotid bodies in control (**a**) and NGF-injected (**b**) 3-week-old rat pups. (scg) Superior cervical ganglion. Further explanation in text. Toluidine blue. ×60. **c** and **d**: Carotid bodies of control (**c**) and NGF-injected (**d**) pups shown at higher magnification than in **a** and **b**. ×200.

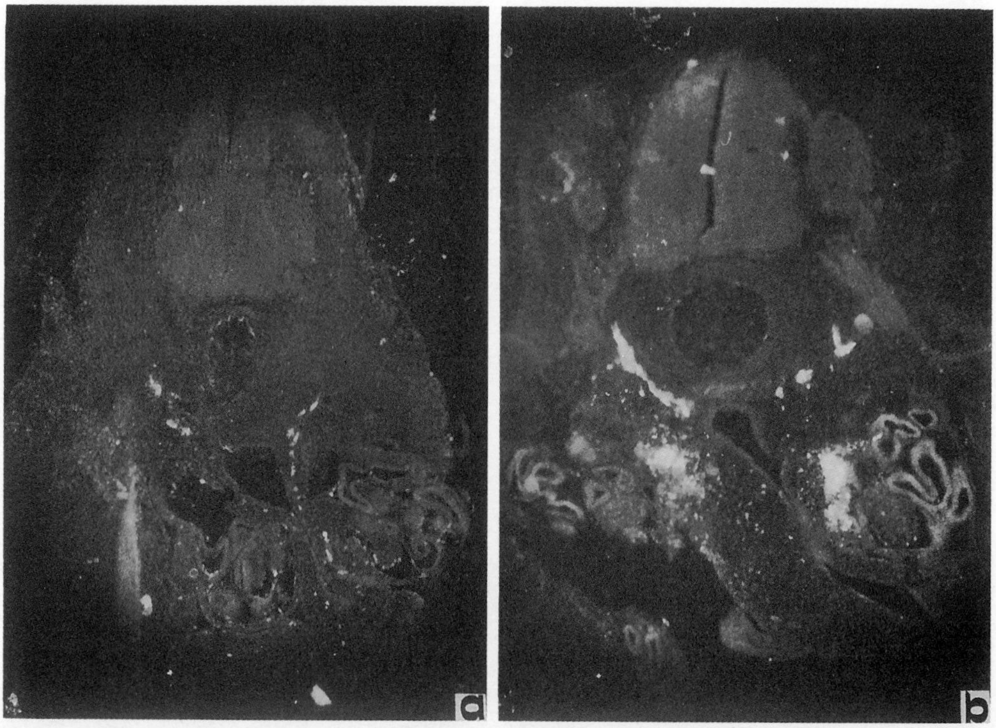

FIG. 3. Histofluorescence preparations of cross sections of 5-day-old chick embryos: (**a**) control and (**b**) embryo injected with NGF as explained in text. Notice marked increased number of fluorescent cells in **b** and the formation of two fluorescent cell aggregates in proximity to the medial aspect of the primordium of the unstained adrenal gland. Further explanation in text. ×36.

distinguishable from each other by the different shades of color in routine histological preparations and silver-stained sections. When the specific chromaffin staining technique is used, the chromaffin cells exhibit a yellow-green color on the light background of the unstained cortical cells.

The still not fully differentiated adrenal glands in 11-day embryos differ markedly in the control and experimental specimens. In the former (Fig. 4a), the glands appear as two large ovoidal bodies in close apposition to the adjacent mesonephros. In the experimental embryo (Fig. 4b), the two glands at this low magnification appear to consist of two components: a lateral irregularly shaped body (marked with a continuous solid line) and a medial intensely stained body (marked with a white dotted line). Inspection at higher magnification showed that the latter darker areas consist of dense aggregates of nerve cells similar in all respects to the cells of para- and prevertebral sympathetic ganglia, apparent in the same microphoto. Figure 4c,d depicts the adrenal glands in a control and an experimental 18-day embryo. The dotted line encircles the intensely darkly stained agglomerates of sympathetic nerve cells which surround, as large irregularly shaped cell aggregates, the greatly reduced adrenal gland. Note that, besides this marked volume reduction, the experimental gland consists of a uniform and apparently homogenous cell population, whereas the larger control gland consists of dark and light colored cords and cell islands. The former are cortical cells, and the latter (also examined in chromaffin-stained sections) are glandular

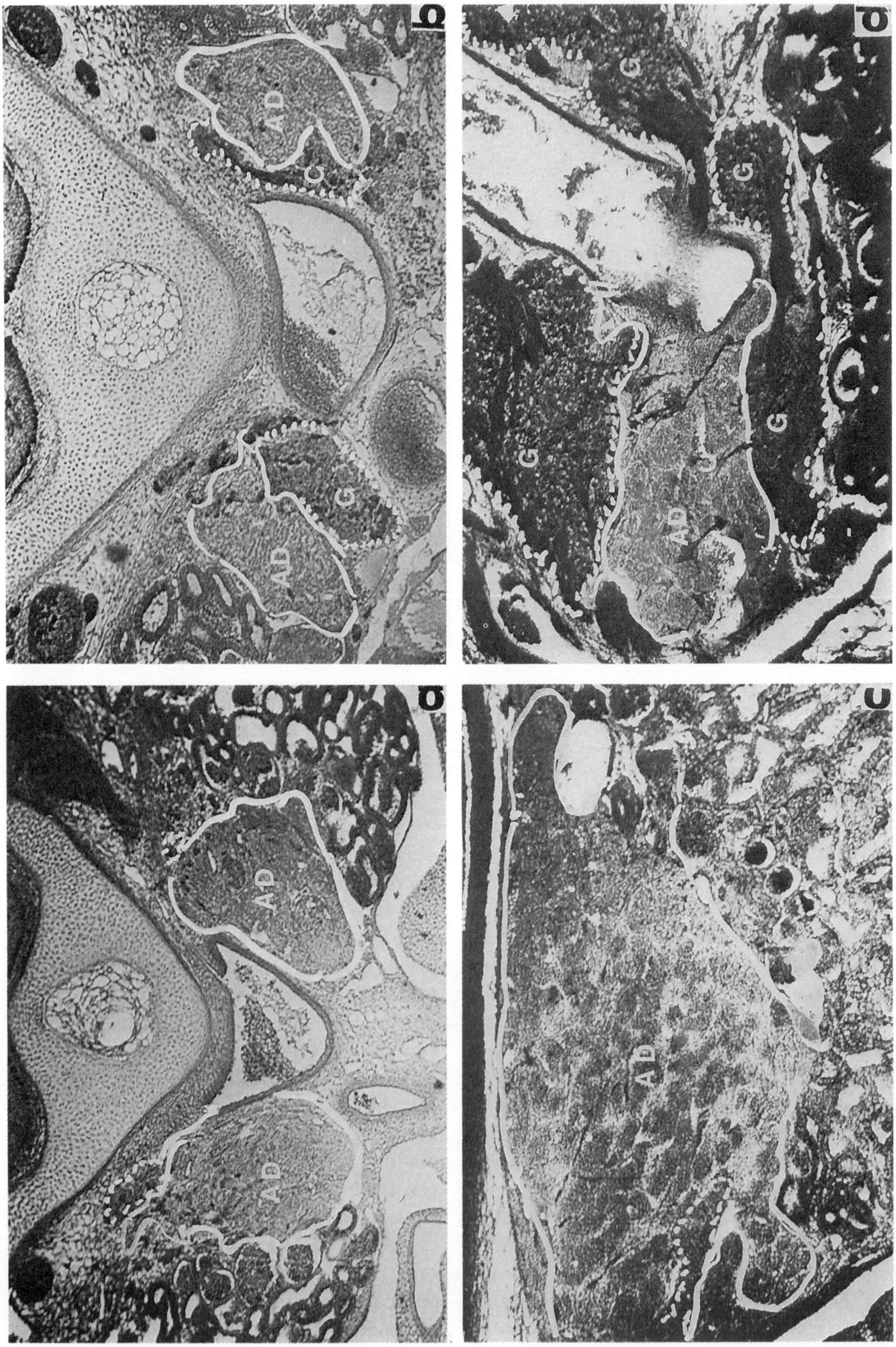
AD
G

medullary cells. This cell population is absent in the experimental gland. In the control gland a diminutive dark cell aggregate consists of sympathetic nerve cells. In the experimental specimen fiber bundles assembled in large fascicles direct their course toward adjacent blood vessels, which are wrapped by these newly formed nerve fibers and invade the adjacent mesonephros in large number.

CONCLUDING REMARKS

From the beginning of these studies we realized that the effects elicited by a humoral factor released by two mouse sarcomas did not fit in any of the preconceived schemes of stimulus-response systems described by previous investigators who had watched the growth response of spinal ganglia elicited in amphibian larvae and avian embryos after implantation of additional organs or other tissues. We could not anticipate at that time, however, to what extent this newly discovered phenomenon departed from other known effects, nor could we foresee the magnitude and plurality of the responses elicited by this molecule which is synthesized in large amount in some glands and released in minute quantities by a large variety of vertebrate cell lines. Perhaps the most challenging aspect of this never-ending investigation is to have shed light not on one but on an ever-increasing list of events which materialize during the early phases of growth and differentiation of sympathetic nerve cells and cells stemming from the same neural crest precursors. Studies now in progress on the effects of a monospecific antiserum to NGF prospect a new and even broader field of action of NGF than that outlined in this chapter.

As stated by Kuhn "In Science . . . novelty emerges with difficulty, manifested by resistance, against a background provided by expectation" (24). In the case of NGF-target cell interaction, 30 years of investigations have seen this barrier broken down and, again to use an expression of Kuhn "The initially anomalous has become the anticipated." It is in this spirit of expectation of other novelties that we conclude this report of the recent developments in the "NGF uncharted route" (31).

REFERENCES

1. Aloe, L., and Levi-Montalcini, R. (1979): Nerve growth factor in-vivo induced transformation of immature chromaffin cells in sympathetic neurons: Effect of antiserum to the nerve growth factor. *Proc. Natl. Acad. Sci. USA,* 76:1246-1250.
2. Aloe, L., and Levi-Montalcini, R. (1979): Nerve growth factor induced overgrowth of axotomized superior cervical ganglia in neonatal rats: Similarity and differences with NGF effects in chemically axotomized sympathetic ganglia. *Arch. Ital. Biol. (in press).*
3. Aloe, L., Mugnaini, E., and Levi-Montalcini, R. (1975): Light and electron microscope studies on the excessive growth of sympathetic ganglia in rats injected daily from birth with 6-OHDA and NGF. *Arch. Ital. Biol.,* 113:326-353.
4. Angeletti, P.U., and Levi-Montalcini, R. (1970): Sympathetic nerve cell destruction in newborn mammals by 6-hydroxydopamine. *Proc. Natl. Acad. Sci. USA,* 65:114-121.
5. Angeletti, P.U., and Levi-Montalcini, R. (1972): Growth inhibition of sympathetic cells by some adrenergic blocking agent. *Proc. Natl. Acad. Sci. USA,* 69:86-88.
6. Angeletti, R.H., and Bradshaw, R.A. (1971): Nerve growth factor from mouse submaxillary gland: Amino acid sequence. *Proc. Natl. Acad. Sci. USA,* 68:2417-2420.
7. Auerbach, R., Kubai, L., Knighton, D., and Falkman, J. (1974): A simple procedure for long-term cultivation of chick embryos. *Dev. Biol.,* 41:391-394.

◀
FIG. 4. **a** and **b**: Transverse sections of control (**a**) and NGF-injected (**b**) 11-day-old chick embryos at the level of the adrenal glands. The contour of the adrenal gland (AD) in the control and the experimental case is shown with the solid white line. The dotted line in **b** encircles large sympathetic cell aggregates (G) lacking in the control case. Further explanation in text. Silver technique. ×80. **c** and **d**: Adrenal glands of control (**c**) and NGF-injected (**d**) chick embryos sacrificed at 18 days. The contour of the adrenal gland (AD) is shown in the control and the experimental case with a solid white line. The dotted line encircles the large sympathetic nerve cell (G) aggregates in **d** and the diminutive sympathetic nerve cell cluster in **c**. Silver technique. ×60.

8. Banerjee, S.P., Snyder, S.H., Cuatrecasas, P., and Greene, L.A. (1973): Binding of nerve growth factor receptors in sympathetic ganglia. *Proc. Natl. Acad. Sci. USA,* 70:2519-2523.
9. Boyd, F.L., Bradshaw, A.R., Frazier, A.W., Hogue-Angeletti, A.R., Jeng, I., Pulliam, M.W., and Szutowicz, A. (1975): Nerve growth factor. In: *Minireviews of Neurosciences: Life Sciences,* p. 29. Pergamon Press, New York.
10. Bucker, E.D. (1948): Implantation of tumors in the hind limb field of the embryonic chick and developmental response of the lumbosacral nervous system. *Anat. Rec.,* 102:369-390.
11. Cajal y Ramon (1913): *Degeneration and Regeneration of the Nervous System,* translated by R. May. Reprinted 1968 by MIT Press, Cambridge, Mass. (original Spanish edition, 1913).
12. Calissano, P., Levi, A., Alemà, S., Chen, J.S., and Levi-Montalcini, R. (1975): Studies on the interaction of the nerve growth factor with tubulin and actin. In: *26 Colloquium Mosbach, Molecular Basis of Motility,* edited by L. Heilmeyer, J.C. Rüegg, and Th. Wieland. Springer Verlag, New York.
13. Calissano, P., Monaco, G., Castellani, L., Mercanti, D., and Levi, A. (1978): Nerve growth factor potentiates actomyosin adenosine triphosphatase. *Proc. Natl. Acad. Sci. USA,* 75:2210-2214.
14. Campenot, R. (1977): Local control of neurite development by nerve growth factor. *Proc. Natl. Acad. Sci. USA,* 74:4516-4519.
15. Cohen, S. (1958): A nerve growth promoting protein. In: *Chemical Basis of Development,* edited by W.D. McElroy and S. Glass, pp. 665-667. Johns Hopkins Press, Baltimore.
16. Cohen, S. (1960): Purification of a nerve growth promoting protein from the mouse salivary gland and its neurotoxic antiserum. *Proc. Natl. Acad. Sci. USA,* 46:302-311.
17. Cohen, S., and Levi-Montalcini, R. (1956): A nerve growth stimulating factor isolated from snake venom. *Proc. Natl. Acad. Sci. USA,* 42:571-574.
18. Cohen, S., Levi-Montalcini, R., and Hamburger, V. (1954): A nerve growth stimulating factor isolated from sarcoma 37 and 180. *Proc. Natl. Acad. Sci. USA,* 40:1014-1018.
19. Diamond, J., Cooper, E., Turner, C., and MacIntyre, L. (1976): Trophic regulation of nerve sprouting. *Science,* 193:371-377.
20. Eränkö, O., and Eränkö, L. (1971): Histochemical evidence of chemical sympathectomy of guanethidine in newborn rats. *Histochem. J.,* 3:451-456.
21. Greene, L.A., and Tischler, A.S. (1976): Establishment of a noradrenergic clonal line of rat adrenal pheochromocytoma cells which respond to nerve growth factor. *Proc. Natl. Acad. Sci. USA,* 73:2424-2428.
22. Hendry, I.A., Stoeckel, K., Thoenen, H., and Iversen, L.L. (1974): Retrograde axonal transport of the nerve growth factor. *Brain Res.,* 68:103-121.
23. Herrup, K., and Shooter, E.M. (1973): Properties of nerve growth factor receptors of avian dorsal root ganglia. *Proc. Natl. Acad. Sci. USA,* 70:3884-3888.
24. Kuhn, T.S. (1975): *The Structure of Scientific Revolution.* International Encyclopedia of University Sciences. University of Chicago Press, Chicago, Illinois. Second edition.
25. Johnson, E.M., and Aloe, L. (1974): Suppression of the in vitro and in vivo cytotoxic effects of guanethidine in sympathetic neurons by nerve growth factor. *Brain Res.,* 81:519-538.
26. Letourneau, P.C. (1978): Chemotactic response of nerve fiber elongation to nerve growth factor. *Dev. Biol.,* 66:183-196.
27. Levi-Montalcini, R. (1952): Effects of mouse tumor transplantation on the nervous system. *Ann. NY Acad. Sci.,* 55:330-343.
28. Levi-Montalcini, R. (1953): In vivo and in vitro experiments on the effect of mouse sarcoma 180 and 137 on the sensory and sympathetic system of the chick embryo. In: *Proceedings XIV International Congress of Zoology,* Copenhagen, p. 309.
29. Levi-Montalcini, R. (1958): Chemical stimulation of nerve growth. In: *Chemical Basis of Development,* edited by W.D. McElroy and S. Glass, pp. 646-664. Johns Hopkins Press, Baltimore.
30. Levi-Montalcini, R. (1966): The nerve growth factor: Its mode of action on sensory and sympathetic nerve cells. *Harvey Lect.,* 60:217-259.
31. Levi-Montalcini, R. (1975): An uncharted route. In: *Neuroscience: Paths of Discovery,* pp. 245-255. MIT Press, Cambridge, Mass.
32. Levi-Montalcini, R. (1976): The nerve growth factor: Its role in growth, differentiation and function of the sympathetic adrenergic neuron. *Prog. Brain Res.,* 45:235-258.
33. Levi-Montalcini, R., Aloe L., Mugnaini, E., Oesch, F., and Thoenen, H. (1975): Nerve growth factor induced volume increase and enhanced tyrosine hydroxylase synthesis in the chemically axotomized sympathetic ganglia of newborn rats. *Proc. Natl. Acad. Sci. USA,* 72:595-599.
34. Levi-Montalcini, R., and Angeletti, P.U. (1963): Essential role of the nerve growth factor in the survival and maintenance of dissociated sensory and sympathetic nerve cells in vitro. *Dev. Biol.,* 7:653-659.
35. Levi-Montalcini, R., and Angeletti, P.U. (1966): Immunosympathectomy. *Pharmacol. Rev.,* 18:619-628.
36. Levi-Montalcini, R., and Angeletti, P.U. (1968): Nerve growth factor. *Physiol. Rev.,* 48:534-565.
37. Levi-Montalcini, R., Angeletti, R.H., and Angeletti, P.U. (1972): The nerve growth factor. *Struct. Function Nerv. Tissue,* 5:1-38.
38. Levi-Montalcini, R., and Booker, B. (1960): Destruction of the sympathetic ganglia in mammals by an antiserum to the nerve growth promoting factor. *Proc. Natl. Acad. Sci. USA,* 42:384-391.
39. Levi-Montalcini, R., and Hamburger, V. (1951): Selective growth-stimulating effects of the mouse sarcoma on the sensory and sympathetic nervous system of the chick embryo. *J. Exp. Zool.,* 118:321-362.

40. Levi-Montalcini, R., and Hamburger, V. (1953): A diffusible agent of mouse sarcoma producing hyperplasia of sympathetic ganglia and hyperneurotization of chick embryo. *J. Exp. Zool.,* 123:233-288.
41. Levi-Montalcini, R., Meyer, H., and Hamburger, V. (1954): In vitro experiments on the effects of mouse sarcoma 180 and 137 on the spinal and sympathetic ganglia of the chick embryo. *Cancer Res.,* 14:49-57.
42. Levi-Montalcini, R., Revoltella, R., and Calissano, P. (1974): Microtubule proteins in the nerve growth factor mediated response. *Recent Prog. Horm. Res.,* 38:635-669.
43. Menesini Chen, M.G., Chen, J.S., and Levi-Montalcini, R. (1978): Sympathetic nerve fibers ingrowth in the central nervous system of neonatal rodents upon intracerebral NGF injections. *Arch. Ital. Biol.,* 116:53-84.
44. Menesini Chen, M.G., Chen, J.S., Calissano, P., and Levi-Montalcini, R. (1977): Nerve growth factor prevents vinblastine destructive effects on sympathetic ganglia in newborn mice. *Proc. Natl. Acad. Sci. USA,* 74:5559-5563.
45. Mobley, W.C., Server, A.C., Ishii, D. N., Riopelle, R.J., and Shooter, E.M. (1977): Nerve growth factor. *N. Engl. J. Med.,* 297:1149-1158.
46. Porter, C.C., Totaro, J.S., and Stone, C.A. (1963): Effect of 6-hydroxydopamine and some other compounds on the concentration of norepinephrine in the hearts of mice. *J. Pharmacol. Exp. Ther.,* 140:308-316.
47. Tranzer, J.P., and Thoenen, H. (1968): An electron microscopic study of selective acute degeneration of sympathetic nerve terminals after administration of 6-hydroxydopamine. *Experientia,* 24:155-156.
48. Unsicker, K., Krisch, B., Otten, U., and Thoenen, H. (1978): Nerve growth factor induced fiber outgrowth from isolated adrenal chromaffin cells: Impairment by glucocorticoids. *Proc. Natl. Acad. Sci. USA,* 75:3498-3502.

Experimental Cell Research 154 (1984) 1–9

SPECIAL ARTICLE

The Nerve Growth Factor

Established Findings and Controversial Aspects

PIETRO CALISSANO, ANTONINO CATTANEO, SILVIA BIOCCA, LUIGI ALOE, DELIO MERCANTI and RITA LEVI-MONTALCINI

Istituto di Biologia Cellulare, CNR, Via Romagnosi 18A, I-00196 Rome, Italy

NGF Exerts a Trophic, Neurotropic and Differentiative Action

The finding that some mouse tumours enhance growth of sensory [1] and sympathetic ganglia of chick embryos [2] was shortly after followed by the discovery that this effect was due to a humoral factor released in the circulation of the host by these grafted tumours [3]. In order to identify the factor a faster and more convenient bioassay was successfully developed resorting to the tissue culture technique [4] which not only confirmed the humoral nature of the tumoral factor, but provided dramatic evidence of its powerful nerve growth-promoting activity, hence the name NGF [5]. The in vitro bioassay currently offers the most impressive and unequivocal test for the identification of NGF in biological samples.

Subsequent experiments performed in living organisms, organ cultures and dissociated nerve cells demonstrated that NGF exerts a triad of actions which can be defined as *trophic*, *tropic* and *differentiative*. Although trophic and differentiative effects are not easily distinguishable in sensory and sympathetic cells where the two events overlap, they became amenable to a rigorous analysis with the discovery that NGF modulates the phenotypic expression of both normal and neoplastic chromaffin cells (see below).

The *trophic*, essential role played by NGF in sensory and sympathetic ganglia during developmental stages is manifested by the absolute requirement of this protein for in vitro survival and regeneration of neurites [6]. Neither the sensory nor the sympathetic ganglia, however, consist of homogeneous populations of cells fully dependent on NGF for survival and differentiation. As far as sensory ganglia are concerned, they consist of two distinct neuronal cell populations—one (~60%) highly receptive to NGF, the other non-responsive [2, 7] or, according to recent studies, weakly responsive [8].

The sympathetic ganglia consist of long and short adrenergic neurons according to their location, distant from or adjacent to, the innervated organs. Only the long adrenergic neurons (in turn consisting of different NGF-responsive popula-

0014-4827/84 $03.00

tions [9], which build up the entire para- and prevertebral chains and constitute the large majority of the sympathetic nervous system) respond to NGF, while there is no evidence that the short adrenergic neurons are receptive [10]. Other factors may act upon these NGF-non-responsive populations as suggested by numerous in vitro studies [11–15]. Injections of NGF cause a marked volume increase of sympathetic para- and pre-vertebral ganglia in chick embryos and in neonatal rodents [16, 17] due to several additive effects: precocious differentiation, increase of nerve cell size, excessive neurite production and survival of neurons otherwise doomed to die. These effects are accompanied by stimulation of the synthesis of their specific neurotransmitters [18].

The vital role played by NGF in the life of sympathetic nerve cells is even more dramatically demonstrated by the finding that removal of circulating NGF in living animals via administration of specific antibodies [19, 20], prevention of its access to the perikaryon with administration of drugs which block its uptake from nerve endings [21], or retrograde axonal transport [22–24], results in death of the vast majority of sympathetic cells. This is also the case when the NGF supply from nerve terminals is surgically interrupted by transection of sympathetic nerve fibres [25]. Degeneration and death caused by these diverse treatments is generally referred to as *immuno*, *chemical*, or *surgical* sympathectomy and is, in all instances, prevented by concomitant exogenous administration of NGF.

In chemical sympathectomy, NGF administration not only counteracts destruction of sympathetic ganglia, but it also induces their excessive volume increase caused by the overproduction of collaterals inside and around the ganglia [26]. This phenomenon is most likely due to interruption of a feedback mechanism which, under normal conditions, sets limits to axonal elongation and branching, once fibres have established connections with end organs.

Further evidence in favour of the physiological role played by NGF in the survival not only of sympathetic but also of sensory cells of mammalian fetuses came from two immunological experimental approaches consisting, respectively, in (*a*) inducing production of NGF autoantibodies in pregnant rats and guinea pigs [27], or (*b*) administering antimouse NGF antibodies to rat fetuses [28]. In both instances, auto- or hetero-NGF antibodies not only cause massive destruction of sympathetic ganglia but also a marked reduction of sensory dorsal root ganglia. Furthermore, at birth the prenatally treated rodents and guinea pigs develop sensory deficits, and neonatal rodents manifest a deficiency syndrome characterized by stunted growth and apathetic behaviour [28]. An attractive possibility is that this syndrome is caused by neutralization by NGF antibodies of a NGF or NGF-like molecule(s) playing an important role in a critical period of development of central neuronal populations.

NGF not only induces outgrowth or regeneration of neurites from target cells, but it also guides growing axons or collaterals along its own concentration gradient (*neurotropism*). This action was first suggested by the finding that in chick embryos bearing NGF-releasing sarcomas, sympathetic nerve fibres grew ectopically, entering into veins which were the most likely channels of NGF [17]. Additional firm evidence for a NGF neurotropic role was obtained in subsequent experiments demonstrating that NGF injections into the floor of the fourth ventricle of neonatal rats caused fibre bundles originating from sympathetic ganglia to penetrate inside the neural tube ending at the level of intracerebral NGF pools [29, 30]. In vitro experimental approaches gave more rigorous proof that neurites of NGF target cells grow along a NGF concentration gradient [31, 32] and deflect their route according to the changed position of the NGF source [33]. While these studies unequivocally established the NGF neurotropic effect as independent from its trophic action, they leave unanswered the question of whether this is exerted through a local control of growth cone motility [34, 35], altered adhesion of this locomotory organelle to the substratum [36], or via other mechanisms [37].

As already mentioned, it is difficult to distinguish between NGF trophic action, manifested as an absolute in vitro and in vivo requirement of this protein by sensory and sympathetic cells, and an effect of NGF on their differentiation into mature nerve cells. Definitive evidence for a NGF action on the phenotypic expression of target cells has been obtained with normal and neoplastic chromaffin cells. In the presence of NGF, a clonal cell line derived from a rat pheochromocytoma (PC12), ceases dividing and undergoes morphological changes which, within 2–3 weeks, result in the formation of an intricate network of electrically excitable neurites [38, 39]. These cells acquire, to a large extent, the phenotype of sympathetic cells, although only some subclones derived from PC12 cells respond with an increase in tyrosine hydroxylase, the rate-limiting enzyme of catecholamine biosynthesis [40, 41]. In recent years, PC12 cells became a model system [42] to study the mechanism of action of NGF, although their neoplastic nature warns against extrapolating all their responses to that of normal target cells. NGF-induced outgrowth of neurites in PC12 cells has been distinguished in *generation* and *regeneration* of neurites, the latter occurring when previously differentiated cells are divested of their processes and replated. Neurite generation occurs slowly and requires RNA transcription (*priming period*), whereas regeneration is much faster and is independent of any macromolecular synthesis [43].

These findings led to the proposal that neurite outgrowth from sensory or sympathetic ganglia explanted in vitro is due to regeneration rather than generation of neurites (also suggested by the speed of this process and its independence from any macromolecular synthesis), while priming of these cells, i.e., induction of neurite growth, occurred earlier in vivo during the first days of development. Priming of PC12 cells may also be responsible for, and connected with, NGF-induced mitotic arrest which is accompanied by marked inhibition of the synthesis of a single-stranded DNA-binding protein, possibly implicated in DNA replication [44].

The process of priming by NGF of PC12 cells can be taken as the first experimental evidence of a crucial action of this factor in the unfolding of their differentiative program, which would depend upon this protein and not on other factors or hormones for the very generation of neurites and dendrites. The property of modulating phenotypic expression of target cells is further demonstrated by the effect exerted by NGF on normal chromaffin cells or on their precursors. Stem cells originating from the neural crest and committed to become chromaffin cells can be induced to differentiate into sympathetic nerve cells in chick embryos injected with large doses of NGF [45, 46]. When NGF is administered to rat fetuses, the switch from glandular to sympathetic nerve cells also takes place, resulting in the replacement of chromaffin cells from adrenal medulla and paraganglia, as well as cells from carotic bodies, with sympathetic neurons.

Analogous effects can be obtained in vitro in chromaffin cells dissociated from rodent adrenal medulla. In the presence of NGF they undergo morphological differentiation into sympathetic-like cells, an effect counteracted by dexamethasone [47].

NGF Action Is Not Confined to Neural Crest Derivatives

Cells which are known to respond to NGF are: (*i*) sympathetic and sensory cells throughout their lives; (*ii*) neoplastic and immature chromaffin cells. Susceptibility to the action of NGF appears, in general, inversely related to the developmental stage of the animal: maximal during early phases of embryonic life and progressively restricting in the adult. Since all these cells originate from the neural crest, it is conceivable that the spectrum of NGF action is broader and includes other, as yet unidentified, cells. A study now in progress in our laboratory on the effect of NGF in tadpoles of the amphibian *Xenopus laevis* (Levi-Montalcini & Aloe), is uncovering new target nerve cells in the central nervous system.

Besides these newly found NGF-receptive cells, another non-neuronal cell which shares many properties with adrenergic sympathetic cells, the mast cell, came to the fore as its number very markedly increases upon NGF administration to neonatal rats [48]. The same cells explanted in vitro release histamine when exposed to this growth factor [49].

The Mechanism of Action of NGF Is Still Largely Unknown

It is now well established that NGF-target cells possess receptors located on the cell soma and at the terminal nerve endings [42, 50]. It is not yet clear, on the contrary, whether there is one or more molecularly distinct type of NGF receptors. Kinetic studies [51–54] have not yet discriminated between the possibility of (*i*) ligand-induced conversion from one type into another; or (*ii*) existence of lower affinity sites with no relevance to transduction of NGF messages. Possibly this problem will soon be solved with their isolation and physicochemical characterization [55, 56]. Moreover, it is not yet clear whether NGF receptors operate as *transducers* or *translocators* of the NGF message. In the first case, one would anticipate that some sort of intracellular second messenger is involved as a transducer and amplifier of NGF interaction with its receptors, but its nature, if any, has still eluded identification, since it does not appear to be any of the cyclic nucleotides so far identified. Other hypotheses involving permeability changes of monovalent cations [57, 58], Ca^{2+} [59] or protons [60] have been postulated, but it remains to be established whether these effects are primary or secondary events of the cell response to NGF.

Alternatively, either the receptor, NGF itself, or a complex of the two, may be responsible for triggering the cell response within the cell compartment. One cannot exclude, of course, that NGF operates through different mechanisms according to the type of evoked biological response. NGF–receptor complexes are internalized, both from nerve endings [61, 15] as well as from cell soma [62–64]. As already mentioned, the pool of NGF internalized from nerve fibre terminals is transported in a retrograde fashion up to the cell perikaryon [15, 61, 65]. The demonstration that interference in this process by different chemical or surgical manipulations leads to death of sympathetic cells would strongly favour the conclusion that NGF internalization is crucial, at least for its trophic action.

Intracellular NGF is found in lysosomes, in discrete dots around the nuclear membrane and in contact with other intracellular organelles [66–68]. It is not yet clear whether at least part of the NGF escapes endocytic vesicles, as currently demonstrated for a large number of peptides, and, in the affirmative case, where these single molecules are located. Potential intracellular targets of this pool of NGF are microtubules [69] and, to a lesser extent, microfilaments [70], as demonstrated by in vitro studies on fixed and permeabilized cells [71].

mRNA for Mouse Salivary NGF Codes for a Much Larger NGF Precursor

The information obtained after isolation of a cDNA encoding a precursor of mouse submaxillary gland NGF [72], soon followed by that of an analogous cDNA encoding a precursor of human NGF [73], together with previous knowledge about the molecular structure of their final product (the biologically active NGF) allows drawing a precise picture about the beginning and the final step of NGF synthesis. The controversial aspect deals with the intermediate step(s), namely where processing occurs and what enzyme(s) is responsible for this event.

Mouse salivary NGF, as sequenced (118 amino-acid residues) [74, 75] or employed for biological studies (two subunits each of 13 250 D), is not synthesized as such but is produced under the form of a polypeptide of 307 amino-acid residues [72]. This proNGF contains not only NGF (amino acids 188–305) but an amino terminal sequence, possibly functioning as a signal for its translocation into the cisternae of endoplasmic reticulum, and other sites postulated to undergo processing by specific esteroproteases. Thus, NGF appears to belong to the class of *polyproteins*, as is the case for most hormones and biologically active peptides, and its processing into distinct molecular entities is most probably mediated by some proteases. Indeed, an arginine-specific protease belonging to the family of serine proteases is found associated with NGF together with another protein of unknown function to form a 7S complex [76, 77]. In this complex, NGF is referred to as β subunit, while the protease(s) and the other protein(s) are defined respectively as γ and α subunits. As recently pointed out [78], however, it is not yet clear whether actual processing of the NGF precursor is carried out by the γ subunit or by other enzymes. Nor is it known whether processing itself also occurs in other NGF-producing cells and, in the affirmative case, whether the final secreted product is identical to submaxillary NGF or if it is of a different size and biological activity. Dissociation of the three subunits forming the 7S complex of the submaxillary gland occurs at very low NGF concentrations or by removal from β NGF of C and N-terminal residues [79]. This yields the so-called 2.5S NGF [80] which is distinct from β NGF for lack of 8 amino acids residues at C-terminal, and an N-terminal arginine, but possesses the same biological activity. Thus, we may roughly divide the formation of biologically active NGF, beginning at its synthesis, into two steps: (*i*) processing from a large precursor and subsequent postulated association to its processor enzyme; (*ii*) dissociation from it, to become the unique molecular entity recognized by specific receptors. As for the sources of NGF, the prevailing concept is that, at variance with classical hormones, synthesis of NGF is *multifocal* and its discharge may occur in proximity of target cells or of their terminal nerve endings, not necessarily through the blood circulation. This hypothesis should be unambiguously verified with the presently available immunological and recombinant cDNA techniques. Sources which produce NGF in exceedingly large amounts (μg/mg total proteins) but which release this protein in exocrine fluids are: glands of several poisonous snakes [81], the submaxillary gland of mice [19] and of *Mastomys nataliensis* [82], the guinea pig prostate [83] and bovine seminal fluid [84]. The functional role of this large NGF pool is still unknown (for a discussion see [85]). The same biologically active molecule is synthesized and secreted in micro quantities by several normal and neoplastic cell lines [86–90], from lower to higher vertebrates, man included. The synthesis of NGF in mouse salivary gland is under control of testosterone [91, 92] and thyroxine [93], suggesting that an analogous hormonal control is also operative in other NGF producing cells.

After three decades of studies since its discovery, NGF is finding its proper place in developmental neurobiology, and its important functional role would suggest the existence of a family of similar molecules acting upon other neuronal target cells. From fish and amphibians up to mammals, the action of NGF—though restricted mostly to certain neuronal populations—seems to be a crucial one. The demonstration that single genes (oncogenes) are responsible for cell transformation and that some of them share sequence homologies with growth factors [94] or their receptors [95] suggests that other protein factors may be operative in the opposite manner by activating differentiative programs of target cells. NGF appears to belong to this second category and offers a model for search of other differentiating gene products. The recent localization of human NGF gene in the short arm of chromosome 1 [96] which, in neuroblastomas is a site of frequent deletions, offers the possibility of further assessing this hypothesis.

This work has been supported by grant no. 1-848 March of Dimes to P. C. and by a grant from CNR, Progetto Finalizzato Malattie del Sistema Nervoso to R. L. M.

REFERENCES

1. Bueker, E D, Anat rec 102 (1948) 369.
2. Levi-Montalcini, R & Hamburger, V, J exp zool 116 (1951) 321.
3. Levi-Montalcini, R, Ann NY acad sci 55 (1952) 330.
4. Levi-Montalcini, R, Mayer, H & Hamburger, V, Cancer res 14 (1954) 49.
5. Cohen, S, Levi-Montalcini, R & Hamburger, V, Proc natl acad sci US 40 (1954) 1014.
6. Levi-Montalcini, R & Angeletti, P U, Dev biol 7 (1963) 653.
7. Levi-Montalcini, R & Hamburger, V, J exp zool 123 (1953) 233.
8. Hamburger, V & Yip, J W, J neurosci 4 (1984) 767.
9. Edgar, D, Barde, Y A & Thoenen, H, Nature 282 (1981) 294.
10. Levi-Montalcini, R, Immunosympathectomy (ed G Steiner & E Schonbaum) p. 55. Elsevier/North-Holland, Amsterdam (1972).
11. Ebendal, T, Dev biol 72 (1979) 276.
12. Richardson, P M & Ebendal, T, Brain res 246 (1982) 57.
13. Ebendal, T, Norrgren, G & Hedlund, K O, Med biol 61 (1983) 65.
14. Varon, S & Adler, R, Curr top dev biol 16 (1980) 207.
15. Thoenen, H & Barde, Y A, Physiol rev 60 (1980) 1284.
16. Levi-Montalcini, R & Booker, B, Proc natl acad sci US 46 (1960) 373.
17. Levi-Montalcini, R, Harvey lecture 60 (1966) 217.
18. Thoenen, H, Angeletti, P U, Levi-Montalcini, R & Kettler, R, Proc natl acad sci US 68 (1971) 1598.
19. Cohen, S, Proc natl acad sci US 46 (1960) 302.
20. Levi-Montalcini, R & Booker, B, Proc natl acad sci US 46 (1960) 384.
21. Angeletti, P U & Levi-Montalcini, R, Proc natl acad sci US 65 (1970) 114.
22. Calissano, P, Levi, A, Alemà, S, Chen, J S & Levi-Montalcini, R, Molecular basis of motility (ed L M G Heimeyer, J C Ruegg & T H Wieland) p. 186. Springer Verlag, Berlin (1976).
23. Menesini-Chen, M L, Chen, J S, Calissano, P & Levi-Montalcini, R, Proc natl acad sci US 74 (1977) 5559.
24. Johnson, E M, Brain res 141 (1978) 105.
25. Hendry, I A, Brain res 90 (1975) 235.
26. Levi-Montalcini, R, Aloe, L, Mugnaini, E, Oesch, F & Thoenen, H, Proc natl acad sci US 72 (1975) 595.
27. Johnson, E M, Gorin, P D, Brandeis, L D & Pearson, J, Science 210 (1980) 916.
28. Aloe, L, Cozzari, C, Calissano, P & Levi-Montalcini, R, Nature 291 (1981) 413.
29. Levi-Montalcini, R, Progr brain res 45 (1976) 235.
30. Menesini-Chen, M G, Chen, J S & Levi-Montalcini, R, Arch ital biol 116 (1978) 53.
31. Charlwood, K A, Lamont, D M & Banks, B E C, Nerve growth factor and its antiserum (ed E Zaimis) p. 102. The Athlone Press, New York (1972).
32. Letourneau, P C, Dev biol 66 (1978) 183.
33. Gundersen, R W & Barrett, J N, Science 206 (1979) 1079.
34. Campenot, R B, Proc natl acad sci US 74 (1977) 4516.
35. Seeley, P J & Greene, L A, Proc natl acad sci US 80 (1983) 2789.
36. Gundersen, R W & Barrett, J N, J cell biol 87 (1980) 546.
37. Pfenninger, K H & Johnson, M P, Proc natl acad sci US 78 (1981) 7797.
38. Greene, L A & Tischler, A S, Proc natl acad sci US 73 (1976) 2424.
39. Dichter, M A, Tischler, A S & Greene, L A, Nature 268 (1977) 501.

40. Hatanaka, H, Brain res 222 (1981) 225.
41. Goodman, R & Herschman, H, Proc natl acad sci US 75 (1978) 4587.
42. Greene, L A & Shooter, E M, Ann rev neurosci 3 (1980) 353.
43. Burstein, D E & Greene, L A, Proc natl acad sci US 75 (1978) 6059.
44. Biocca, S, Cattaneo, A & Calissano, P, Proc natl acad sci US 81 (1984) 2080.
45. Aloe, L & Levi-Montalcini, R, Proc natl acad sci US 76 (1979) 1246.
46. Levi-Montalcini, R & Aloe, L, Histochemistry and cell biology of autonomic neurons, SIF cells and paraneurons (ed O Erankö, S Soinila & H Paivarinta) vol. 25, p. 3. Raven Press, New York (1982).
47. Unsicker, K, Krisch, B, Otten, U & Thoenen, H, Proc natl acad sci US 75 (1978) 3498.
48. Aloe, L & Levi-Montalcini, R, Brain res 133 (1977) 358.
49. Bruni, A, Bigon, E, Boarato, E, Leon, A & Toffano, G, Febs lett 138 (1982) 190.
50. Mobley, W C, Server, A C, Ishi, D N, Riopelle, R J & Shooter, E M, New engl j med 297 (1977) 1096.
51. Frazier, W A, Boyd, L F & Bradshaw, R A, J biol chem 249 (1974) 5513.
52. Sutter, A, Riopelle, R J, Harris-Warrick, R M & Shooter, E M, J biol chem 254 (1979) 5972.
53. Tait, J F, Weinman, S A & Bradshaw, R A, J biol chem 256 (1981) 11086.
54. Cattaneo, A, Biocca, S, Nasi, S & Calissano, P, Eur j biochem 135 (1983) 285.
55. Massague, J, Guillette, B J, Czech, M P, Morgan, C J & Bradshaw, R A, J biol chem 256 (1981) 9419.
56. Puma, P, Buxser, S E, Watson, L, Kelleher, D J & Johnson, G L, J biol chem 258 (1983) 3370.
57. Varon, S & Skaper, S D, Somatic and autonomic nerve–muscle interactions (ed G Burnstock, G Urbova & R O'Brien). Elsevier/North-Holland, Amsterdam (1983).
58. Skaper, S D & Varon, S, Brain res 197 (1980) 379.
59. Schubert, D, La Corbière, M, Whitlock, C & Stallcup, W, Nature 273 (1978) 718.
60. Boonstra, J, Van der Saag, P T, Moolenaar, W H & De Laat, S W, Exp cell res 131 (1981) 452.
61. Hendry, I A, Stockel, K, Thoenen, H & Iversen, L L, Brain res 68 (1974) 103.
62. Calissano, P & Shelanski, M L, Neuroscience 5 (1980) 1033.
63. Biocca, S, Levi, A & Calissano, P, J receptor res 1 (1980) 373.
64. Levi, A, Schechter, Y, Neufeld, E J & Schlessinger, J, Proc natl acad sci US 77 (1980) 3463.
65. Stockel, K, Schwab, M & Thoenen, H, Brain res 99 (1975) 1.
66. Hogue-Angeletti, R, Sticker, A & Ganatos, N K, Brain res 241 (1982) 145.
67. Marchisio, P C, Naldini, L & Calissano, P, Proc natl acad sci US 77 (1980) 1656.
68. Rohrer, H, Schafer, T, Korsching, S & Thoenen, H, J neurosci 2 (1982) 687.
69. Calissano, P & Cozzari, C, Proc natl acad sci US 71 (1974) 2131.
70. Calissano, P, Monaco, G, Castellani, L, Mercanti, D & Levi, A, Proc natl acad sci US 75 (1978) 2210.
71. Nasi, S, Cirillo, D, Naldini, L, Marchisio, P C & Calissano, P, Proc natl acad sci US 79 (1982) 820.
72. Scott, J, Selby, M, Urdea, M, Quiroga, M, Bell, G & Rutter, W J, Nature 302 (1983) 538.
73. Ullrich, A, Gray, A, Berman, C & Dull, T J, Nature 303 (1983) 821.
74. Hogue-Angeletti, R & Bradshaw, R A, Proc natl acad sci US 68 (1971) 2417.
75. Hogue-Angeletti, R, Mercanti, D & Bradshaw, R A, Biochemistry 12 (1973) 90.
76. Varon, S, Nomura, J & Shooter, E M, Biochemistry 6 (1967) 2202.
77. — Ibid 7 (1968) 1296.
78. Bradshaw, R A, Nature 303 (1983) 751.
79. Server, A C & Shooter, E M, Adv protein chem 31 (1977) 339.
80. Bocchini, V & Angeletti, P U, Proc natl acad sci US 64 (1969) 787.
81. Cohen, S & Levi-Montalcini, R, Proc natl acad sci US 42 (1956) 571.
82. Aloe, L, Cozzari, C & Levi-Montalcini, R, Exp cell res 133 (1981) 475.
83. Harper, G P, Barde, Y A, Burnstock, G, Carstairs, J R, Dennison, M E, Suda, K & Vernon, C A, Nature 279 (1979) 160.
84. Harper, G P, Glanville, R W & Thoenen, H, J biol chem 257 (1982) 8541.
85. Calissano, P, Cattaneo, A, Aloe, L & Levi-Montalcini, R, Hormonal proteins and peptides (ed C H Li) vol. 12. Academic Press, New York. In press.
86. Young, M, Oger, J, Blanchard, M H, Asdourian, H, Amos, H & Arnason, B G W, Science 187 (1975) 361.

87. Longo, A M & Penhoet, E F, Proc natl acad sci US 71 (1974) 2347.
88. Lindsay, R M, Nature 282 (1979) 80.
89. Goldstein, L D, Reynolds, C P & Perez-Polo, J R, Neurochem res 3 (1977) 185.
90. Norrgren, G, Ebendal, T, Belew, M, Jacobson, C O & Porath, J, Exp cell res 130 (1980) 31.
91. Caramia, F, Angeletti, P U & Levi-Montalcini, R, Endocrinology 70 (1962) 915.
92. Levi-Montalcini, R & Angeletti, P U, Salivary glands and their secretions (ed L M Sreebny & J Meyer) p. 129. Pergamon Press, Oxford (1964).
93. Aloe, L & Levi-Montalcini, R, Exp cell res 125 (1980) 15.
94. Waterfield, M D, Scrace, G T, Whittle, N, Stroobant, P, Johnsson, A, Wasteson, A, Westermark, B, Heldin, C-H, Huang, J S & Deuel, T F, Nature 304 (1983) 35.
95. Downward, J, Yarden, Y, Mayes, E, Scrace, G, Totty, N, Stockwell, P, Ulrich, A, Schlessinger, J & Waterfield, M D, Nature 307 (1984) 521.
96. Francke, U, de Martinville, B H, Coussens, L & Ullrich, A, Science 222 (1983) 1248.

Brain Research, 133 (1977) 358–366

Mast cells increase in tissues of neonatal rats injected with the nerve growth factor

LUIGI ALOE and RITA LEVI-MONTALCINI

Laboratory of Cell Biology, Via Romagnosi 18, Rome (Italy) and Department of Biology, Washington University, St. Louis, Mo. (U.S.A.)

(Accepted June 2nd, 1977)

The discovery of the formaldehyde-induced fluorescence of biogenic amines in different cell lines, made available a most valuable marker for detecting cells endowed with the property of storing one or other of these amines[6]. Among amine-storing cells the mast cells, which in normal histological preparations are often overlooked in view of their small number and uneven distribution in peripheral tissues and organs, come sharply in relief in tissue sections prepared according to the Falck–Hillarp technique, thanks to the bright yellow fluorescence developed by 5-hydroxytryptamine stored together with other amines in the cell granules[7].

The present studies were prompted by the observation that tissues and organs in neonatal rats injected with nerve growth factor (NGF) and prepared for histofluorescence studies, revealed at the ultraviolet microscope the presence of a much larger number of fluorescent cells that the same tissues of control littermates. These findings were the starting point of more extensive and detailed investigations directed at the identification of these cells and at answering the question whether this NGF effect was specific of this protein molecule or was also elicited by other like or unlike effectors.

Litters of newborn rats (Sprague–Dawley strain), reduced to 8 pups each, were injected with: NGF, physiological solutions or other agents as indicated in Table I. The NGF was purified according to the method of Bocchini and Angeletti from the mouse salivary glands[4] and injected in the amount of 10 μg/g body weight (b.w.) as described in a previous article[8]. Bovine serum albumen (BSA), Cytochrome-c, lysozyme and insulin, were purchased from Sigma Co. St. Louis, U.S.A.; 6-hydroxydopamine (6-OHDA) was obtained from Fluka A.G., Chemische Fabrik, Switzerland; epidermal growth factor (EGF) was prepared according to the method of Cohen[5]. Ninety neonatal rats were divided in coded groups, injected daily with one or another agent listed in Table I and sacrificed by decapitation at the end of the second or third week.

Twenty young rats, 150 g b.w., were injected intraperitoneally for 14 consecutive days with 7 μg/g b.w. of NGF or physiological solution. Experimental and control rats were sacrificed one day after the last injection and tissues and organs were prepared for histofluorescence studies and examined with the UV microscope. Tissues (iris, ear pinna) and organs (digestive tract, heart, spleen, thymus, thyroid gland, superior cervical ganglia) were quickly removed and processed for histological, histofluorescence and ultrastructural studies. For observations at the light microscope, the tissues were fixed in alcoholic Bouin, embedded in Paraplast and stained with toluidine blue. For histofluorescence studies tissues and organs were freeze-dried, while the irides were air-dried and stretched on cover-slides for examination of the entire membrane in whole mount preparations.

The axial trunk, including the skin, subcutaneous tissue, vertebrae and adjacent sympathetic ganglia, were freeze-dried, exposed to formaldehyde vapors, embedded in Paraplast and sectioned at 8–10 μm. All tissues were examined at a Zeiss microscope equipped with a HBO 200 mercury lamp and EG 12 excitatory and 440–530 nm barring filters. Electron microscopic studies were performed on irides and superior cervical ganglia of NGF-treated pups, according to the technique described in a previous publication[1].

Light microscopic studies

Fig. 1 compares two sections cut through the ear pinna of a control (a) and a NGF-injected littermate pup (b), sacrificed at the end of the second post-natal week.

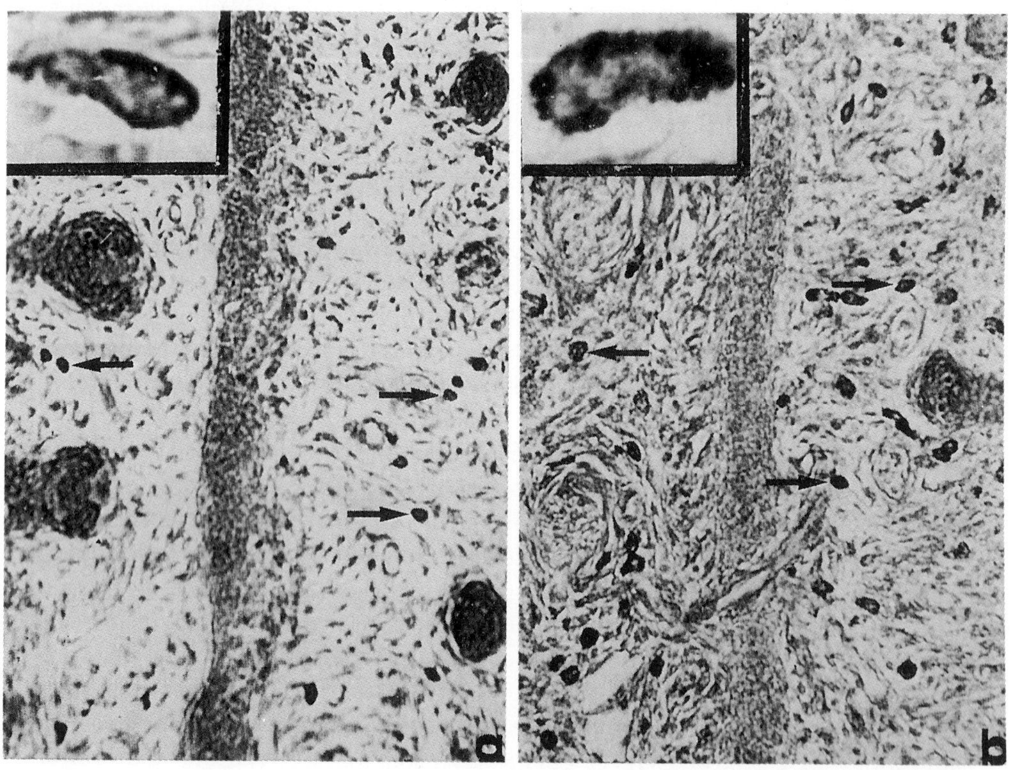

Fig. 1. Transverse sections of the ear pinna of littermate two-week old rats injected from birth with physiological solution (a) or NGF (b). Mast cells, indicated with arrows, are shown at higher magnification in the insets. Toluidine blue stain. Magnifications: a and b, $\times$ 179; insets $\times$ 1250.

Metachromatic cells indicated by arrows, are much more numerous in the experimental than in the control tissues and individual cells appear larger in the latter than in the former, as evident from a comparison of the two insets. The size difference is due to accumulation in the cytoplasm of cells in the NGF-injected pups of larger and darkly stained granules than in the cytoplasm of control cells. Metachromatic cells exhibiting the same morphological features as the cells shown in Fig. 1 are seen in great number in sections of superior cervical ganglia of NGF-injected neonatal rats. They are particularly abundant in the capsular layer surrounding the ganglia where they are oriented parallel to the collagen fiber bundles. In control ganglia these cells are exceedingly rare. Studies of sections of the organs listed above showed a marked increase in metachromatic cells in the experimental cases. The cells are dispersed in the loose intraparenchymal tissue with no preferential distribution, nor a tendency to aggregate in groups. Metachromatic cells are also present in control organs, but are much less nemerous than in sections of NGF-injected littermates.

Fluorescence microscopic studies

Fig. 2 shows two transverse sections through the skin and subepidermal layers

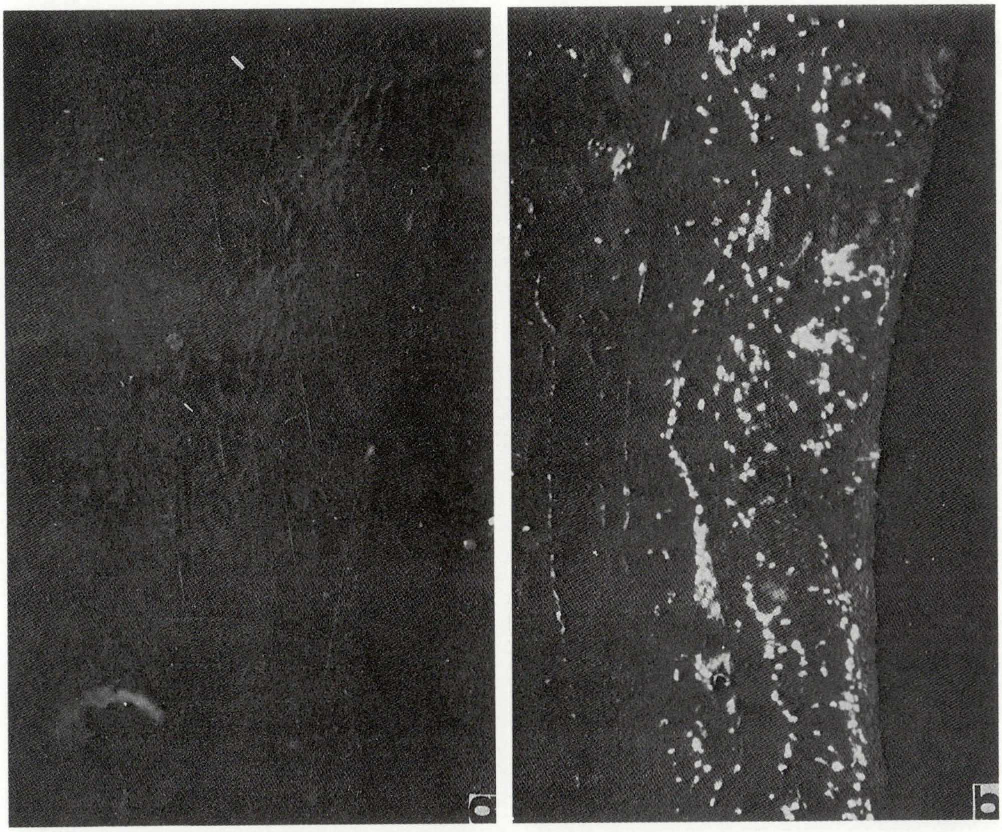

Fig. 2. Photomicrographs of histofluorescence preparations of skin and subepidermal layers of control (a) and NGF-injected (b) 17-day-old littermate rats. Large number of fluorescent cells are present in the experimental but not in control tissues. Magnification × 89.

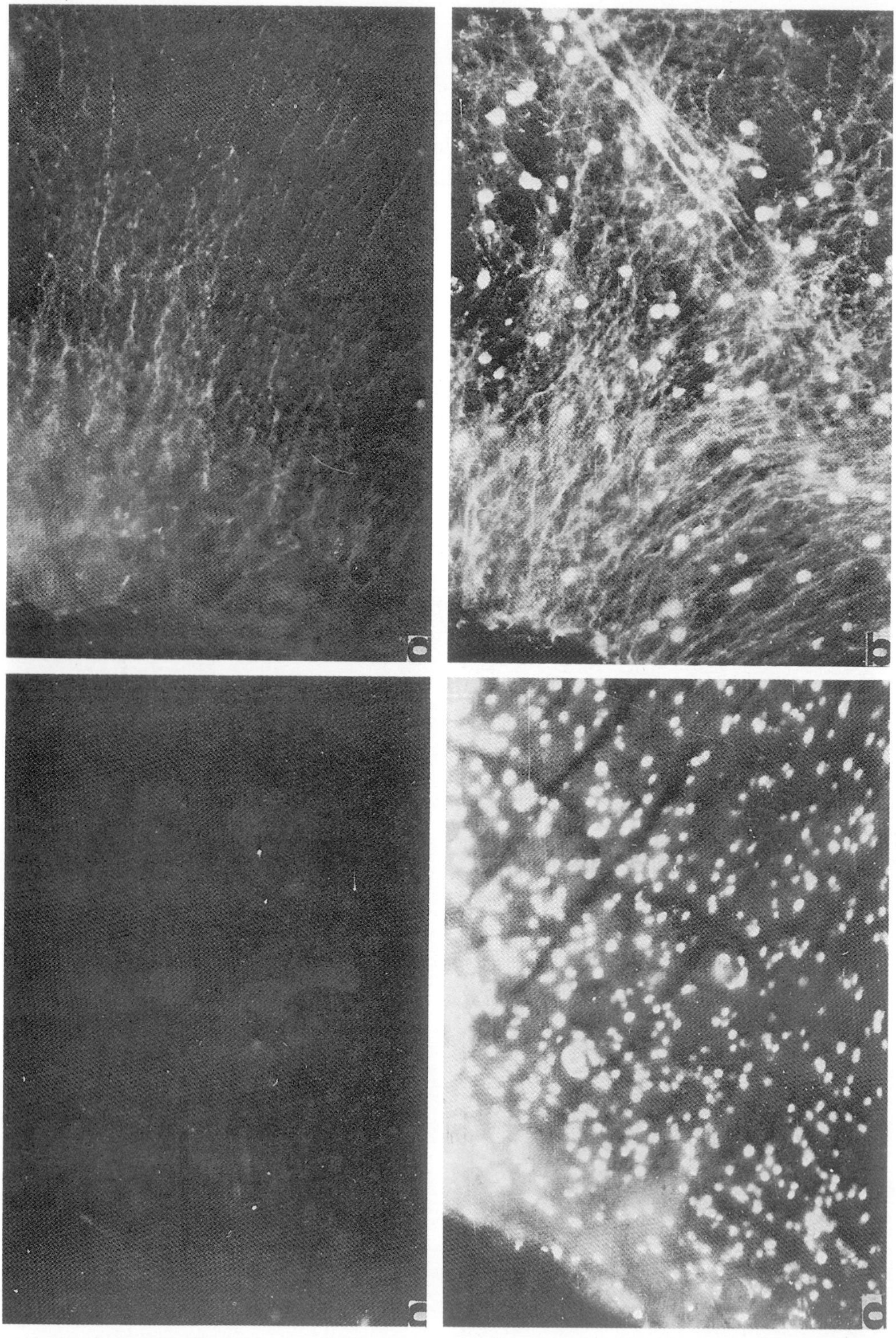

Fig. 3. Photomicrographs of histofluorescence stretch preparations of the irides of 20-day-old littermate rats injected since birth with physiological solution (a), NGF (b), 6-hydroxydopamine (c) and NGF + 6-hydroxydopamine (d). Magnifications: a and b, × 151; c and d, × 89.

of the dorso-lumbar region of the axial trunk in a control (a) and in a NGF-injected 17-day-old rat pup (b). The large number of fluorescent, unevenly scattered cells in the subepidermal tissue of the experimental section contrasts with the total lack of these cells in the control. Rostral and caudal sections of the axial trunk from the cervical to the sacral segment showed the same features as illustrated in Fig. 2. Fluorescent cells are also seen in large number in the bone marrow of the vertebral bodies; they are absent in the entire cerebro-spinal axis.

The results of these studies with light and fluorescent microscopy raised the question whether the increase in the number of metachromatic, fluorescent cells in NGF-injected rat pups could be due to the hyperinnervation of peripheral tissues and organs produced by NGF, or whether it resulted from a direct NGF-stimulating effect on these cells or their precursors. This problem was investigated by comparing the number of these cells in the irides of neonatal rats injected with physiological solution, NGF, 6-OHDA or NGF and 6-OHDA. In these last two experimental groups the pups were deprived of the sympathetic innervation of the peripheral tissues and organs by the toxic action of the dopamine derivative, 6-OHDA. In the treatment with 6-OHDA alone the sympathetic ganglia were irreversibly destroyed[2], while in pups submitted to the combined NGF and 6-OHDA treatment the NGF prevented the destruction of sympathetic nerve cells and even produced a more marked volume increase of the ganglia than injections of only NGF; but the peripheral tissues were entirely devoid of sympathetic innervation, due to the accumulation and toxic effects of 6-OHDA on the adrenergic nerve endings[9]. The effects of these four treatments are illustrated in Fig. 3a, b, c and d. The sympathetic innervation of the iris evident in the control (a) is markedly increased in that of the NGF-injected pup (b). No adrenergic fibers are seen in the irides of the 6-OHDA-injected pup (c) and in that of the littermate injected with NGF and 6-OHDA (d). Fluorescent cells are present in large number in the hyperinnervated as well as in the non-innervated irides (b and d), thus proving that their striking numerical increase upon NGF injection is not consequent to the hyperinnervation of peripheral tissues.

In order to see whether the same effect also materializes in older animals, one and a half-month-old rats were injected with NGF or with physiological solution. Studies with the UV microscope showed no difference in the number of fluorescent cells in the irides and in other tissues of control and NGF-injected rats.

Electron microscopic studies

The topographical distribution, morphological features and yellow fluorescence of cells present in large number in tissues and oigans of neonatal rats injected with NGF strongly suggested their identification as mast cells; definite evidence for such identity was obtained from ultrastructural studies.

Fig. 4 shows part of the cytoplasmic and nuclear compartments of one of these cells in the superior cervical ganglion of a 14-day-old rat injected with NGF. The cell exhibits an irregular contour, due to the presence of a large number of slender elongated projections, most likely sections of the folded plasmalemma membrane. The cytoplasm is crammed with large round-shaped granules delimited by a thin membrane; their

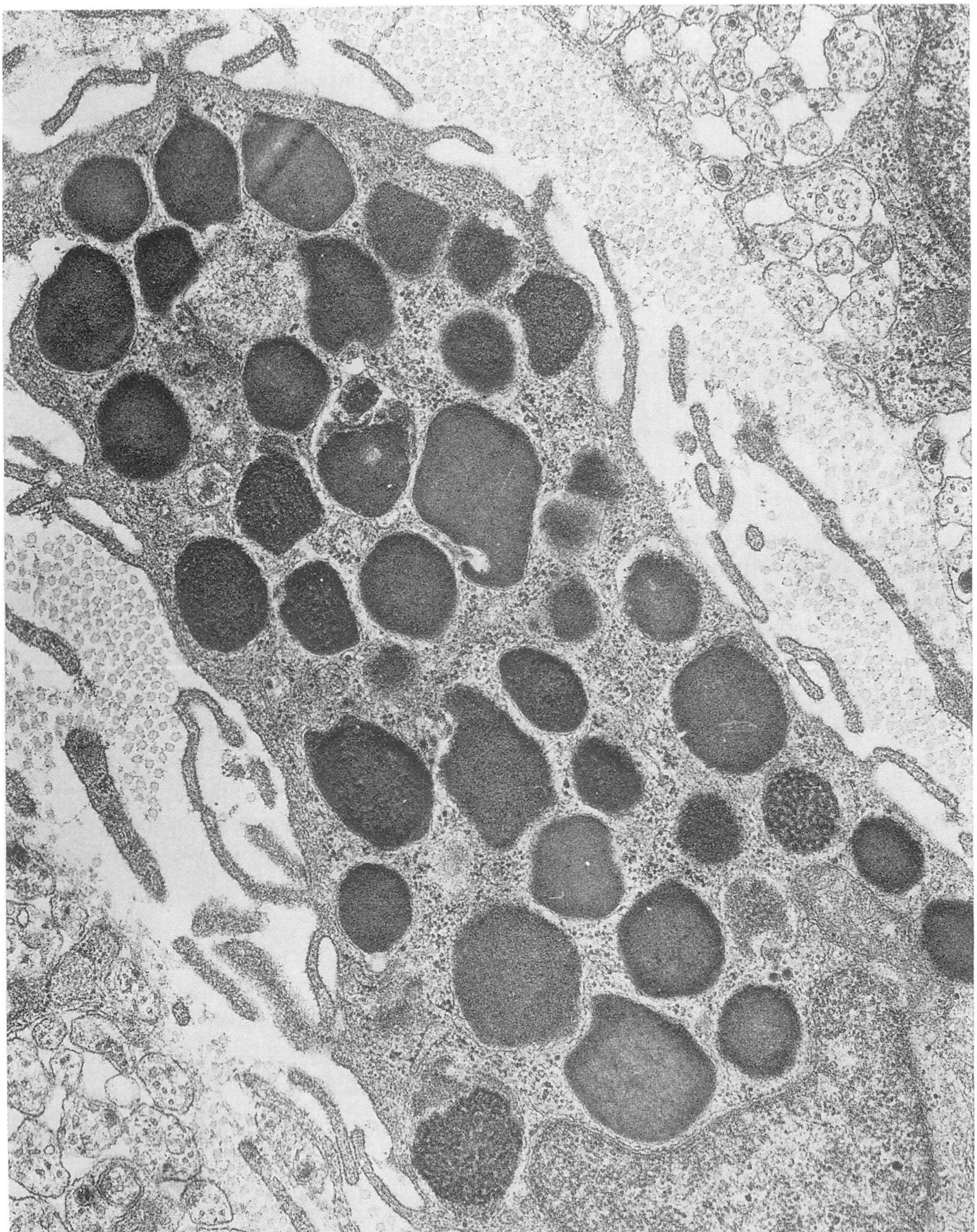

Fig. 4. Electron micrograph of part of the cytoplasmic and nuclear compartments of a mast cell in the superior cervical ganglion of a 14-day-old rat injected since birth with NGF. Magnification × 29,500.

TABLE I

Number of mast cells in histofluorescence stretch preparations of irides of 14-day-old rats treated for 13 consecutive days as indicated

	Normal saline (100 μl)	*NGF (10 μg/g)*	*BSA (200 μg/g)*	*Cytochrome c (10 μg/g)*	*Lysozyme (10 μg/g)*	*Insulin* (7×10^3 *U*)	*EGF (10 μg/g)*	*6-OHDA (100 μg/g)*	*6-OHDA (100 μg/g) + NGF (10 μg/g)*
Mast cell/iris	80	1250	85	80	72	75	90	15	1000
	70	1350	130	75	90	93	89	25	790
	100	1050	90	73	50	123	150	70	1190
	90	1180	140	98	115	160	127	50	1150
	135	1320	174	95	95	105	121	60	990
	108	980	104	108	112	87	94	55	1050
	88	1385	98	85	140	125	148	10	890
	130	1230	134	116	87	96	104	80	1110
	85	990	98	71	92	108	103	39	1020
	68	1260	108	90	74	115	67	48	1000
S.E.M.	96 ± 7.72	1200 ± 46.4	116 ± 8.5	89 ± 4.8	93 ± 7.39	109 ± 7.57	109 ± 8.48	46 ± 7.27	1019 ± 37.5

internal texture appears in most instances homogenous while a filamentous structure is apparent in a few others. Free ribosomes are irregularly scattered in the cytoplasm. The chromatin in the indented nucleus is denser at the periphery and more diffuse in the central core. Both the cytoplasmic and nuclear fine structures, conform to the criteria for definition of mast cells[3,12].

Quantitative studies

The problem whether the striking increase in mast cells in NGF-injected neonatal rats is a specific effect called forth by this protein molecule or whether it is also produced by other peptides was investigated by comparing the effect of NGF and that of these protein molecules on the total number of the mast cells in the irides of littermate pups submitted to one or the other treatments indicated in Table I. The peptides assayed share some properties in common with NGF, such as the electrical charge which is respectively 9.8–10.1 and 11.0–11.2 for cytochrome and lysozyme[10], or are endowed with biological activity (EGF and insulin) and antigenic properties (insulin and BSA). In the two last columns are given the results of cell counting in the irides of 6-OHDA and of the combined 6-OHDA and NGF treatments. The determinations were performed with the help of a digital counter on histofluorescence preparations.

As shown in Table I, the number of mast cells is about 10 times higher in the NGF- and NGF + 6-OHDA-injected pups than in those of other groups; a finding strongly in favor of a specific NGF effect on this cell type. It remains to be seen whether NGF promotes the differentiation of precursors of mast cells present in the loose connective of different tissues and organs, or whether it enhances the differentiation of already commited cells in specific sources such as thymus, spleen, bone marrow, and their subsequent mobilization toward other organs and peripheral tissues.

A localized or systemic increase in mast cells has been described in different conditions such as chronic inflammatory processes, systemic or cutaneous mastocytosis, allergic reactions, carcinogenesis[3] and traumatic nerve injuries[11]. The very marked increase in mast cells in neonatal rats injected with NGF differs from that described in these pathological disorders in at least two important respects: it occurs in newborn but not in fully differentiated rats and it is not associated with other localized or generalized changes, indicative of a response to a pathogenic factor. These findings suggest that the increase in mast cells in neonatal rats injected with NGF is a specific effect produced by this biologically and chemically well-defined growth factor on mast cells or their precursors, and not a neuronally mediated or a side effect related to antigenic or toxic properties of this protein molecule.

1 Aloe, L., Mugnaini, E. and Levi-Montalcini, R., Light and electron microscopic studies on the excessive growth of sympathetic ganglia in rats injected daily from birth with 6-OHDA and NGF, *Arch. ital. Biol.*, 113 (1975) 326–353.

2 Angeletti, P. U. and Levi-Montalcini, R., Sympathetic nerve cell destruction in newborn mammals by 6-hydroxydopamine, *Proc. nat. Acad. Sci. (Wash.)*, 65 (1970) 114–129.

3 Bloom, G. D., Structural and biochemical characteristic of mast cells. In B. W. Zweifach, I. Granit and R. T. McClustkey (Eds.), *Inflammatory processes*, Academic Press, New York, 1974, pp. 545–598.

4 Bocchini, V. and Angeletti, P. U., The nerve growth factor purification as a 30,000 molecular weight protein, *Proc. nat. Acad. Sci. (Wash.)*, 64 (1969) 787–794.

5 Cohen, S., The stimulation of epidermal proliferation by a specific protein (EGF), *Develop. Biol.*, 12 (1965) 394–407.

6 Falk, B., Observations on the possibilities of the cellular localization of monoamines by a fluorescence method, *Acta physiol. scand.*, 56 ,Suppl. 197, (1962) 1–25.

7 Fuxe, K., Hökfelt, T., Jonsson, G. and Ungerstedt, U., Fluorescence microscopy in neuroanatomy. In W. J. H. Nauta and S. O. E. Ebbesson (Eds.), *Contemporary Research Methods in Neuroanatomy*, 1970, pp. 275–314.

8 Levi-Montalcini, R., and Angeletti, P. U., Nerve growth factor *Physiological Rev.*, 48 (1969) 535–565.

9 Levi-Montalcini, R., Aloe, L., Mugnaini, E., Oesch, F. and Thoenen, H., Nerve growth factor induced volume increase and enhanced tyrosine hydroxylase synthesis in the chemically axotomized sympathetic ganglia of newborn rats, *Proc. nat. Acad. Sci. (Wash.)*, 72 (1975) 595–599.

10 Mahler, H. R. and Cordes, E. H., *Biological Chemistry*, Harper and Row, New York, 1966.

11 Olsoon, Y., Mast cells in the nervous system, *Int. Rev. Cytology*, 24 (1968) 27–70.

12 Padawer, J. and Gordon, A. S., Cellular elements in the peritoneal fluid of some mammals, *Anat. Rec.*, 124 (1956) 209–222.

Scientific American **240**, 68–77 (1979). Reprinted with permission.

The Nerve-Growth Factor

This protein plays a key role in the formation of the nervous system. It has also been employed as an "Ariadne's thread" to explore mechanisms of nerve-cell growth and differentiation

by Rita Levi-Montalcini and Pietro Calissano

The human nervous system is a vast network of several billion neurons, or nerve cells, endowed with the remarkable ability to receive, store and transmit information. In order to communicate with one another and with non-neuronal cells the neurons rely on the long extensions called axons, which are somewhat analogous to electrically conducting wires. Unlike wires, however, the axons are fluid-filled cylindrical structures that not only transmit electrical signals but also ferry nutrients and other essential substances to and from the cell body. Many basic questions remain to be answered about the mechanisms governing the formation of this intricate cellular network. How do the nerve cells differentiate into thousands of different types? How do their axons establish specific connections (synapses) with other neurons and non-neuronal cells? And what is the nature of the chemical messages neurons send and receive once the synaptic connections are made?

This article will describe some major characteristics and effects of a protein called the nerve-growth factor (NGF), which has made it possible to induce and analyze under highly favorable conditions some crucial steps in the differentiation of neurons, such as the growth and maturation of axons and the synthesis and release of neurotransmitters: the bearers of the chemical messages. The discovery of NGF has also promoted an intensive search for other specific growth factors, leading to the isolation and characterization of a number of proteins with the ability to enhance the growth of different cell lines.

The peripheral nervous system of vertebrate animals includes three kinds of nerve cells: sensory neurons, which transmit impulses from sensory receptor structures to the brain; motor neurons, which innervate the striated, or skeletal, muscles, and autonomic neurons, which regulate the functional activity of the circulatory system, the organs, the glands and the smooth muscles (such as those of the intestine). Autonomic neurons are of two kinds: sympathetic and parasympathetic. The sensory neurons and some of the sympathetic neurons are situated in chains of ganglia flanking the length of the spinal cord. Because these neurons are uniquely accessible to experimental manipulation much of the research on the development of the nervous system at the cellular level has focused on how the nerve fibers projecting from the sensory and sympathetic ganglia make connections with their corresponding target organs.

In the first half of this century the new science of experimental embryology seemed to offer the best way to study the intimate bond that interlocks the growth of the peripheral neurons and their target organs. Ross G. Harrison of Yale University challenged the nervous system of the larva of amphibians to solve problems it would ordinarily never confront, such as providing nerves for limbs and organs grafted from other species. He wanted to see how sensory and sympathetic ganglia, sending out their nerve fibers to these peripheral "fields of innervation," would adjust to the different dimensions and configurations of the foreign organs.

Harrison's results demonstrated that the developing amphibian nervous system is remarkably flexible in adapting to such novel situations, even to the point of accelerating the growth of nerve fibers in a host species to keep pace with the faster-growing limb of a smaller donor species. He concluded that the embryonic nervous system is highly receptive to influences exerted by the peripheral field. Such influences are not species-specific, however, since they can be evoked by organs or rudimentary limbs that have been transplanted from one species to another.

Viktor Hamburger of Washington University extended Harrison's work but chose to study the chick embryo because its nervous system, although more complex than that of an amphibian, lends itself better to experimental analysis: its nerve centers are more clearly delineated and their strong affinity for silver stain enables the experimenter to visually examine the nerve structures more easily. Hamburger grafted limb buds onto chick embryos at very early stages of development and observed how the modified peripheral field was innervated by sensory and sympathetic fibers. Unfortunately the responses were often so complex that they defied interpretation.

In 1948, in an effort to get more straightforward results, Elmer D. Bueker of Georgetown University modified Hamburger's experimental approach. He had the ingenious idea of replacing one limb bud of a chick embryo with a fragment of a bird or mammalian tumor. The tumor cells were all undifferentiated and hence provided a homogeneous peripheral field, in contrast to the cells of the normal limb bud, which were destined to differentiate into many types of tissue. Bueker's experiment was therefore expected to reveal how a homogeneous but fast-growing peripheral field would be innervated.

Of three different tumors implanted in the body wall of three-day-old chick embryos only one, a mouse tumor of connective-tissue cells called sarcoma 180, grew vigorously and was invaded by nerve fibers growing out from adjacent sensory ganglia. In embryos sacrificed after five days the sensory ganglia innervating the tumor were 33 percent larger than those innervating the normal limb bud on the opposite side of the embryo. At first these findings seemed to suggest that the size of a sensory ganglion depends on the size and rate of growth of its field of innervation. According to this hypothesis the rapidly growing tumor provided a more favorable peripheral field for the innervating sensory fibers than the slowly growing limb bud did.

A reexamination of Bueker's findings by our group at Washington University revealed new aspects of the phenomenon that obliged us to revise his conclusions. We found not only that the growth of the sensory ganglia innervating the sarcoma-180 tumors increased but also that the sympathetic ganglia increased enormously in volume, becom-

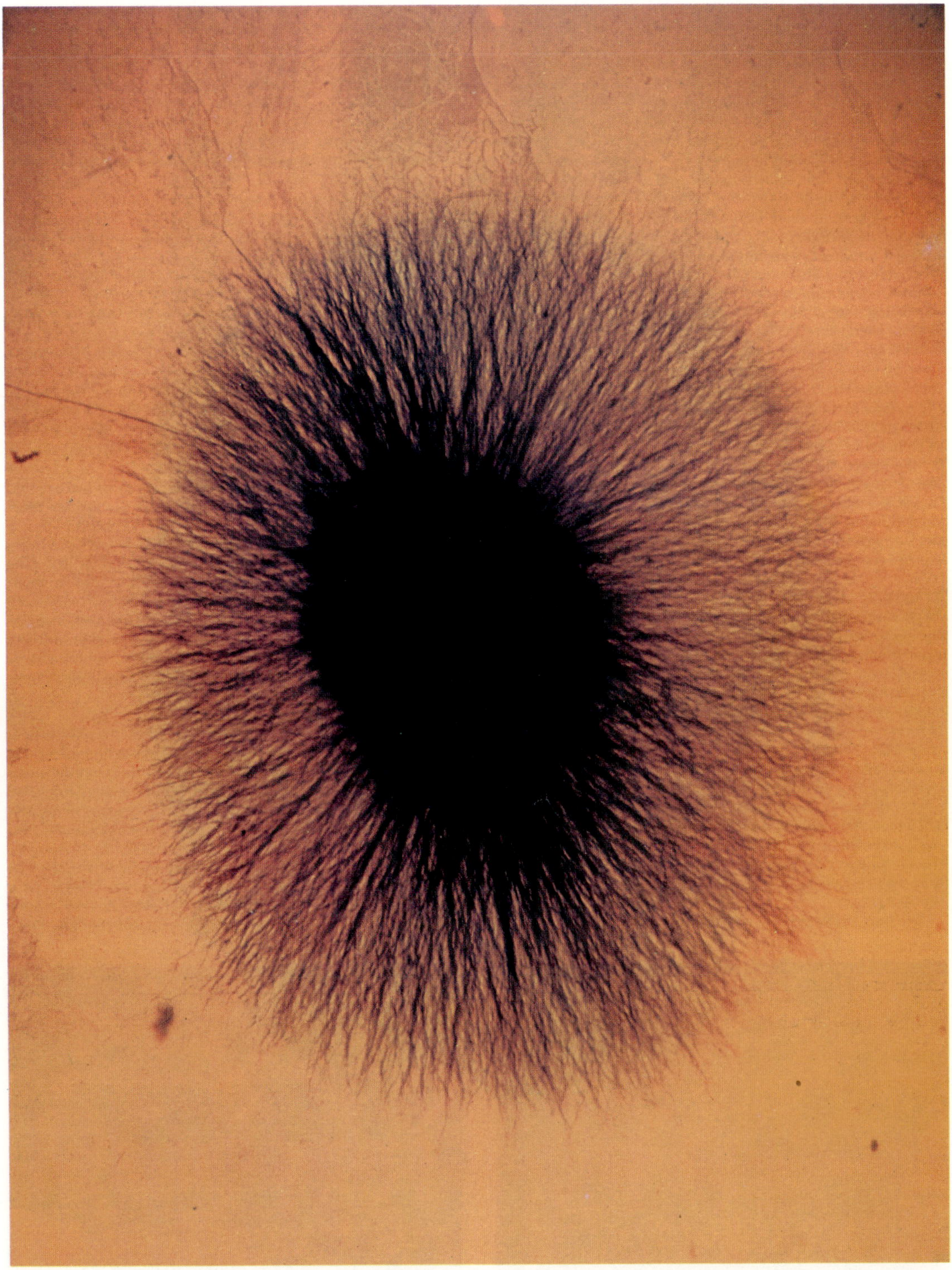

HALO OF NERVE FIBERS radiates from a sympathetic-nerve ganglion that was removed from a chick embryo and then incubated for 12 hours in a semisolid medium containing the nerve-growth factor (NGF). The factor was originally isolated from a mouse tumor designated sarcoma 180 but has subsequently been found to be secreted in trace amounts by a variety of normal and neoplastic (cancer) cells. In 1953 one of the authors (Levi-Montalcini) and Hertha Meyer found that NGF could induce the outgrowth of nerve fibers from an isolated sympathetic ganglion in tissue culture. This discovery led ultimately to the isolation of NGF and determination of its chemical structure. Micrograph, which magnifies ganglion 135 diameters, was made by J. S. Chen at Laboratory of Cell Biology in Rome.

ing five to six times larger than they were in control animals. This increase in size of the sympathetic ganglia was considerably greater than that exhibited by the sensory ganglia. Together with the sensory fibers the sympathetic fibers branched all over the peripheral field provided by the tumor but did not form any synapses with the tumor cells.

In addition, and indicating an even more striking departure from normality, the viscera of the embryos with the transplanted tumors were flooded with excessive numbers of sympathetic fibers long before the embryos of the control animals were even sparsely innervated. The sympathetic fibers forced their way into large and small veins, impeding and sometimes completely obstructing the flow of blood. These extraordinary effects suggested that the overgrowth of the sympathetic ganglia was more than simply a response to the rapidly growing peripheral field of innervation provided by the tumor, as Bueker had proposed. Rather it seemed the tumor was releasing some chemical factor that was in turn inducing the remarkable growth of the sympathetic ganglia and the exuberant branching of their nerve fibers.

This hypothesis was tested by transplanting sarcoma-180 tumors into the respiratory membranes of the chick egg, which are permeated with blood vessels from the embryo. The tumor and the developing chick embryo therefore shared the same blood supply, although they were not in direct contact. We found that when the tumor was transplanted into the respiratory membranes, it elicited the same growth-promoting effects on the sympathetic ganglia as it did when it was implanted in the embryo itself, providing convincing proof that the tumor was releasing a soluble factor that was carried in the bloodstream to the embryo.

The next challenge was to identify the postulated nerve-growth factor being released by the tumor. For this purpose we needed a system much less complex than the developing embryo. Tissue culture (which in the early 1950's had yet to become the universal biological tool it is today) seemed to offer a useful alternative. We reasoned that if sarcoma 180 was releasing a chemical factor with the ability to enhance nerve growth, the same effect should appear when an isolated sympathetic ganglion was incubated with the tumor in laboratory glassware. In 1953 one of us (Levi-Montalcini) and Hertha Meyer did this experiment at the Biophysics Institute in Rio de Janeiro. Sensory and sympathetic ganglia were dissected out of eight-day-old chick embryos and cultured in a semisolid medium in proximity to fragments of mouse sarcoma-180 tumors. Within 10 hours of incubation the isolated ganglion gave rise to a dense halo of nerve fibers, radiating out from the explant like rays from the sun. Control ganglia cultured for the same length of time in the absence of the sarcoma-180 cells displayed only a sparse and irregular outgrowth of nerve fibers.

The discovery that the tumor could exert its growth-enhancing effects on an isolated ganglion in tissue culture was the turning point of the investigation. Whereas our earlier experiments in developing chick embryos had required weeks of painstaking work, we could now in a few hours screen a large number of tissues, organic fluids and chemicals to determine if they were sources of the growth-promoting activity. Furthermore, it now became possible to attempt to isolate the nerve-growth factor from our simplified tissue-culture system.

Stanley Cohen, a biochemist, agreed to join our group at Washington University to undertake the task of identifying the active agent. He soon managed to restrict the growth factor to a fraction of the tumor cells containing both protein and nucleic acid. This finding opened the way to a more precise biochemical analysis, but we would not have made much progress if it had not been for a fortuitous event two years later. In order to determine whether the nerve-growth factor was a protein or a nucleic acid, Cohen and one of us (Levi-Montalcini) treated the tumor-cell extract with snake venom containing a high concentration of the enzyme phosphodiesterase, which degrades nucleic acids. To our surprise we found that adding minute amounts of the venom to the active fraction of the sarcoma-180 cells enhanced the growth-promoting effect of the fraction rather than reduced it. That the snake venom itself was the source of the increased activity was readily confirmed. The addition of a small amount of venom to the culture medium in the absence of the extract of sarcoma 180 elicited the growth of the same dense halo of nerve fibers around an isolated sensory or sympathetic ganglion.

In fact, the nerve-growth factor in the snake venom turned out to be much more abundant and potent than that in the sarcoma-180 cells. Cohen was therefore able to purify the factor from the venom and demonstrate that it was a protein. Injections of the purified snake-venom NGF into growing embryos resulted in the same excessive sympathetic innervation of the viscera and the blood vessels as that induced by the sarcoma-180 cells.

The discovery that two unrelated sources—mouse sarcoma and snake venom—harbor NGF suggested the possibility that still other tissues might secrete the factor. The search focused on the submaxillary salivary gland of rodents, which is similar in certain respects to the venom gland of snakes. Indeed, Cohen isolated an NGF from mouse salivary glands that was about 10,000 times more active than that puri-

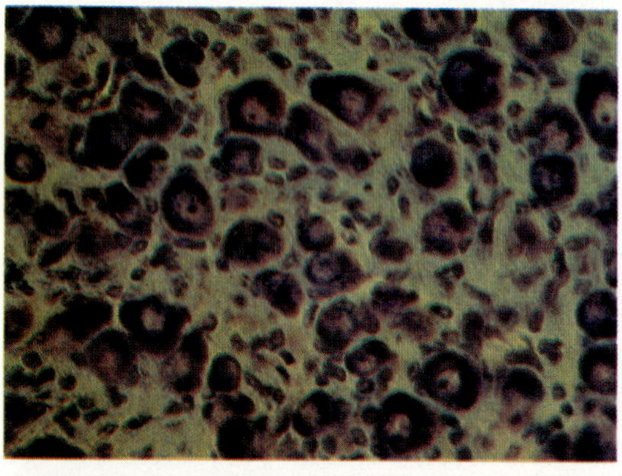

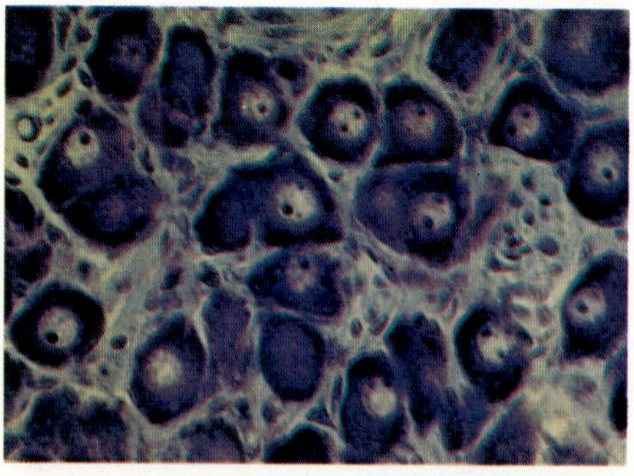

INCREASED SIZE AND NUMBER of neurons in the sympathetic ganglia of NGF-treated mice result in an increase in the volume of the ganglia. These micrographs are of sections through two ganglia, one ganglion from a three-week-old rat injected daily with saline solution only (*left*) and the other from a littermate mouse injected with NGF (*right*). Neurons in NGF-treated rat are substantially enlarged. Note also that cells in ganglia of animals treated with NGF exhibit a greater affinity for toluidine blue, the dye used to stain the cells.

fied from mouse sarcoma 180 and about 10 times more active than that purified from snake venom. Over the next two decades smaller amounts of NGF were found to be secreted by a wide variety of normal and neoplastic (cancer) cells.

In 1969 V. Bocchini and Pietro U. Angeletti at the Laboratory of Cell Biology in Rome devised a method for purifying NGF from mouse salivary glands. With the large quantities of NGF made available by their technique Ruth Hogue Angeletti and Ralph A. Bradshaw of the Washington University School of Medicine were able to determine the sequence of amino acids in the protein. Biologically active NGF is a dimer, or complex of two identical polypeptide chains, each of which has a molecular weight of 13,250 daltons. The NGF molecule has a marked predominance of positively charged amino acids and has a net positive charge at neutral *p*H. In addition the folded polypeptide chain is stabilized by three covalent sulfur-sulfur bridges between units of the amino acid cysteine at different positions along the chain. Such disulfide bonds are common in proteins that are actively secreted from cells, such as insulin and antibodies. The restraint these molecular bridges impose on the conformation of the polypeptide chain is apparently necessary to prevent its denaturation and inactivation in an environment that is much more subject to adverse conditions than the interior of the cell.

A comparison of the sequence of amino acids in NGF with that of several other polypeptides by William A. Frazier of Washington University revealed that NGF and insulin have certain sequences in common. This observation led to the hypothesis that the gene for NGF evolved from an ancestral gene for proinsulin, the large precursor polypeptide that is cleaved to yield the active molecule of insulin. Frazier suggested that the ancestral proinsulin gene had duplicated itself and the two copies had subsequently evolved divergently, giving rise to proinsulin and a postulated precursor polypeptide for NGF. Although this intriguing possibility remains theoretical, the similarities in the amino acid sequences of NGF and insulin are not great enough to result in any similarity of function: the two molecules have completely different target cells and biological activities.

In 1967 Silvio S. Varon and Eric M. Shooter of the Stanford University School of Medicine isolated NGF by a different procedure and found that the NGF dimer was present as a stable complex with two copies each of two other proteins, which they designated the alpha and gamma subunits. This finding was puzzling, since it was known that the NGF dimer by itself was biologically active. Further investigation revealed that the gamma subunit is a specific en-

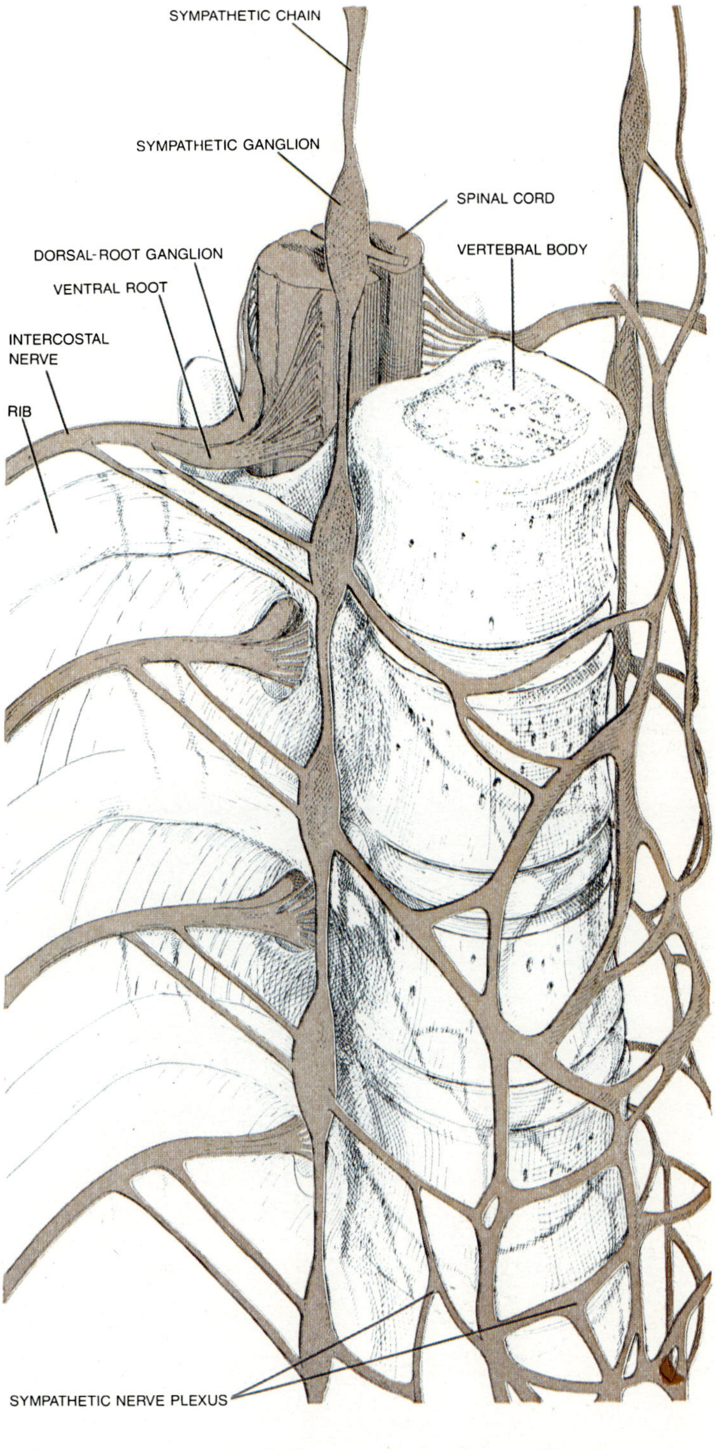

CHAIN GANGLIA flanking the spinal cord of mammals (those of a human being are shown here) contain the cell bodies of the sympathetic neurons that innervate the various organs, blood vessels, smooth muscles and glands of the body. The irregularly shaped nodules are held together by nerve fibers and are also connected to the roots of the spinal cord. Each ganglion innervates the blood vessels and organs in the corresponding segment of the trunk and legs. During the development of the nervous system the neurons in the ganglia send out fibers to innervate their specific target organs. Because the ganglia are readily accessible they are easy to study.

zyme that cleaves polypeptide chains only at a point adjacent to the amino acid arginine. The alpha subunit, on the other hand, has no detectable biological activity. This odd association of three disparate proteins demanded an explanation.

Shooter hypothesized that the alpha and gamma subunits serve to activate, store and protect the NGF molecule. According to his scheme the initial gene product in the manufacture of NGF is a large precursor polypeptide called proNGF, two copies of which form a dimer. Two gamma subunits then associate with the proNGF dimer and cleave the precursor chains to generate the active NGF dimer. Unlike the typical complex of an enzyme and its product, which rapidly dissociates, the gamma subunits remain bound to the NGF dimer after the cleavage of the proNGF chains. (The two alpha subunits may bind either to the proNGF dimer before cleavage or to the complex of the gamma subunit and the NGF dimer after cleavage.) The association of the gamma and alpha subunits with the NGF dimer apparently serves to protect it from further degradation by other protein-cleaving enzymes in the body fluids. Shooter's hypothesis recently received support when he and Edward A. Berger isolated the proNGF molecule and demonstrated that incubation of this precursor with the gamma subunit resulted in its complete conversion into active NGF.

The biological activity of a protein resides in its three-dimensional structure, which consists of not only its amino acid sequence but also the precise folding pattern of the polypeptide chain (the secondary structure) and the suprafolding of two or more such chains to form a unique globular entity (the tertiary structure). The only reliable way to determine the secondary and tertiary structure of a protein molecule is to crystallize it and mathematically analyze the crystal's X-ray-diffraction pattern. The recent crystallization of NGF by A. Wlodawer, Keith O. Hodgson and Shooter is therefore an important first step in determining the three-dimensional structure of the molecule by X-ray crystallography.

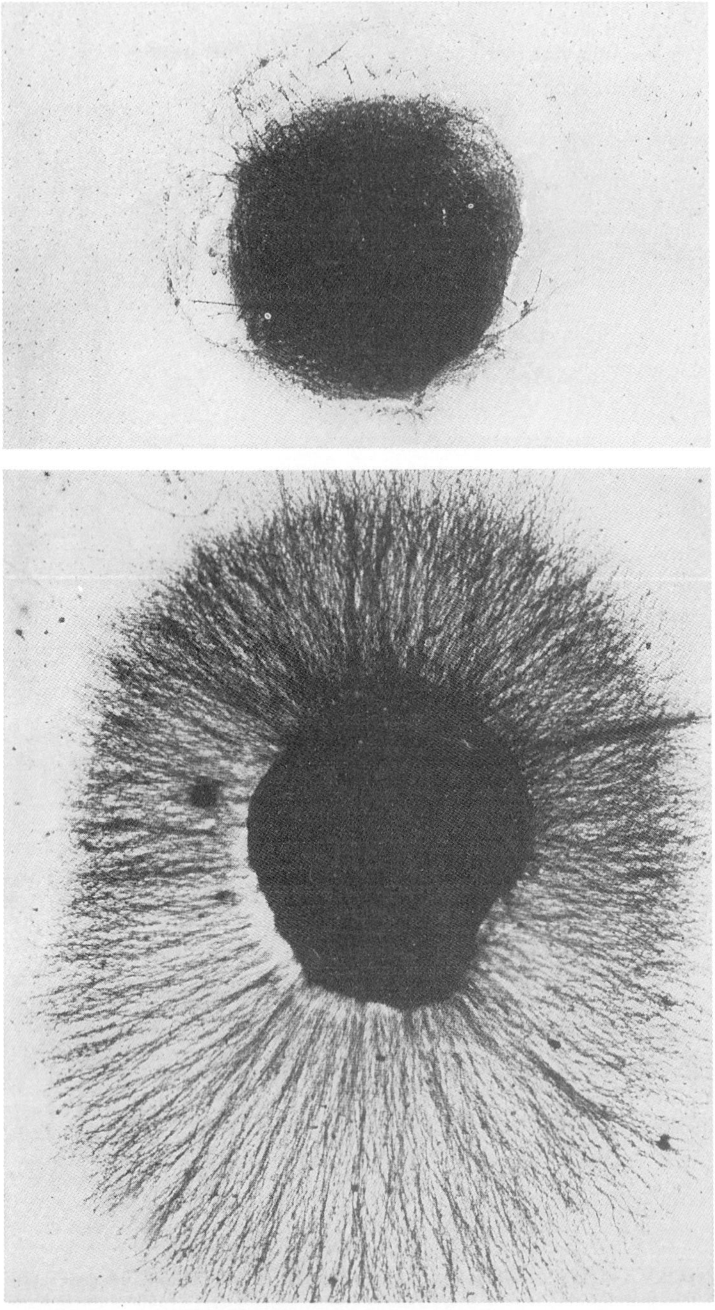

BIOLOGICAL ASSAY for NGF is based on the ability of the factor to induce the outgrowth of nerve fibers from an isolated ganglion. These micrographs show sensory ganglia dissected out of an eight-day-old chick embryo and cultured for 12 hours either in the absence (*top*) or the presence (*bottom*) of 10 micrograms of pure NGF. After incubation the fibers were stained.

The first real insight into the function of NGF in the living animal was provided by experiments done in our laboratory at Washington University in 1959. Cohen obtained specific antibodies to NGF by injecting purified mouse NGF into rabbits. Small amounts of rabbit serum containing the NGF antibodies were then injected into newborn mice. A month later the experimental and control mice were sacrificed. In the mice treated with the NGF antibodies the sympathetic ganglia were reduced to such diminutive size that they were barely visible in the dissecting microscope! Nevertheless, the NGF antibodies had no ill effect on any other organs or tissues and for unexplained reasons did not damage the peripherally located sympathetic ganglia that control the sex organs in both sexes. It has therefore been possible to raise to maturity entire colonies of mice entirely lacking in sympathetic nervous function but normal in all other respects. These animals have provided a most valuable model system for studying how the lack of sympathetic innervation interferes with a multiplicity of body functions.

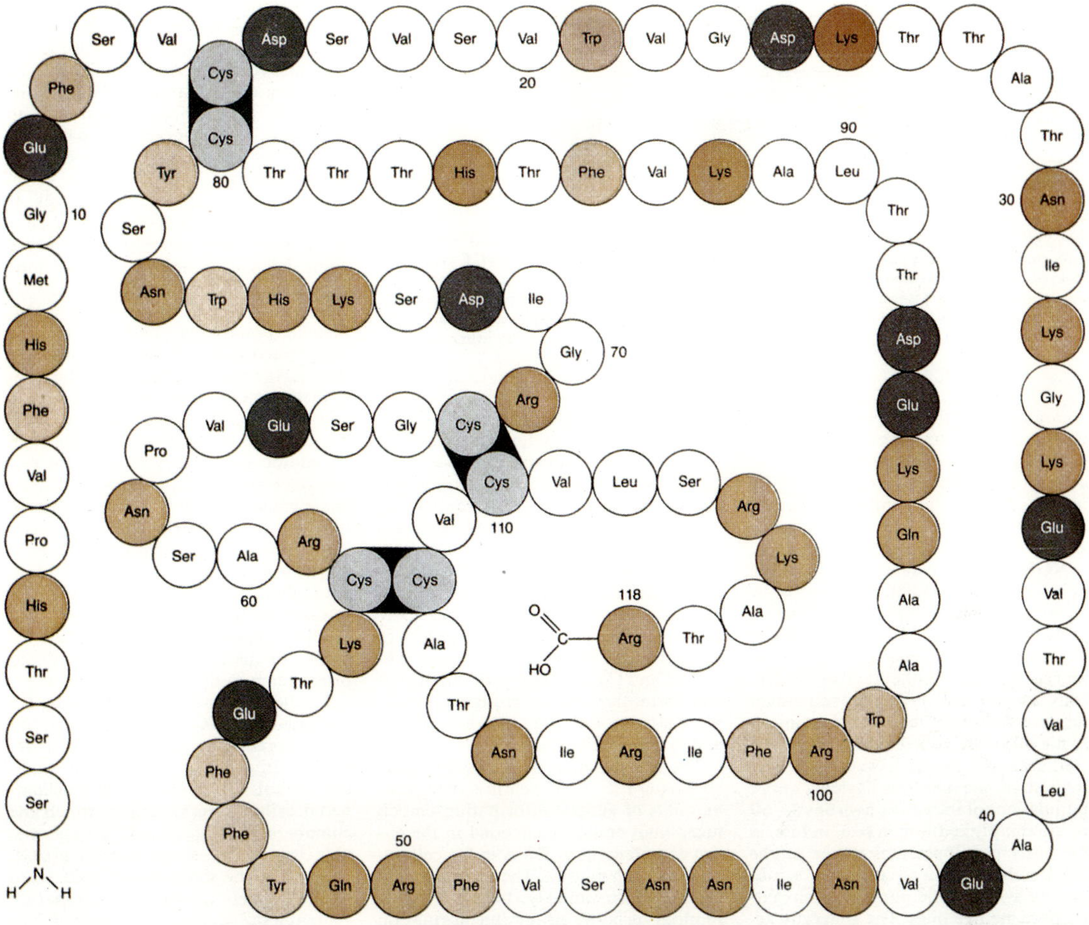

Ala	ALANINE	Leu	LEUCINE
Arg	ARGININE	Lys	LYSINE
Asn	ASPARAGINE	Met	METHIONINE
Asp	ASPARTIC ACID	Phe	PHENYLALANINE
Cys	CYSTEINE	Pro	PROLINE
Gln	GLUTAMINE	Ser	SERINE
Glu	GLUTAMIC ACID	Thr	THREONINE
Gly	GLYCINE	Trp	TRYPTOPHAN
His	HISTIDINE	Tyr	TYROSINE
Ile	ISOLEUCINE	Val	VALINE

PRIMARY STRUCTURE OF NGF is shown schematically as a necklace of amino acid units. The chain is bridged at three points by covalent bonds between the sulfur groups of cysteine units. As is true of all proteins, the sequence of amino acids in the chain gives rise to a unique three-dimensional structure, which is the result of weak interactions among the amino acid units. The precise folding pattern of NGF has not yet been determined. The various categories of amino acid are indicated by shading: positively charged (*dark color*), negatively charged (*dark gray*), incorporating an aromatic ring (*light color*), incorporating sulfur atoms (*light gray*) and uncharged (*white*). Above a critical concentration two monomeric chains of NGF, each having a molecular weight of 13,2[illegible]0 daltons, associate to form a dimer. It is not yet known whether the monomer or the dimer is the form of the NGF molecule that is biologically active.

It is not yet known whether the dramatic and selective effects of the NGF antibodies are due to a direct toxic action of the antibodies on immature sympathetic neurons or to an inactivation by the antibodies of circulating molecules of NGF, thereby indirectly causing the death of the sympathetic neurons by depriving them of the NGF they need in order to survive. This second alternative seems the more likely one; convincing evidence in its favor would provide tangible proof that NGF is an absolute requirement for the survival and growth of immature sympathetic neurons in the living animal.

That NGF is essential for the survival of sympathetic neurons in cell culture has been known for some time. In 1963 Pietro Angeletti and one of us (Levi-Montalcini), who were then at the Istituto Superiore di Sanità in Rome, dissected sensory and sympathetic ganglia into their cellular components: neurons, glial cells (which support and nourish the neurons) and fibroblasts (embryonic connective-tissue cells). After 24 hours of culture the glial cells and fibroblasts had survived and multiplied but the neurons had undergone a massive degeneration. The daily addition of minute amounts of NGF to the culture medium, however, enabled the neurons to survive for indefinite periods and to form a dense meshwork of nerve fibers that after a few days completely covered the surface of the culture dish. This ability of NGF to sustain sympathetic neurons in culture has since made possible some ingenious and revealing experiments on neuronal differentiation [see "The Chemical Differentiation of Nerve Cells," by Paul H. Patterson, David D. Potter and Edwin J. Furshpan; SCIENTIFIC AMERICAN, July, 1978].

The availability of milligram quantities of pure NGF has made it possible to test its effects in the living organism. All mammals respond to NGF the same way, although for practical reasons rodents are the experimental animals of choice. When 10 micrograms of NGF per gram of body weight is injected into newborn rodents for periods of up to three weeks, their sympathetic ganglia become 10 to 12 times larger than those of control animals. This excessive in-

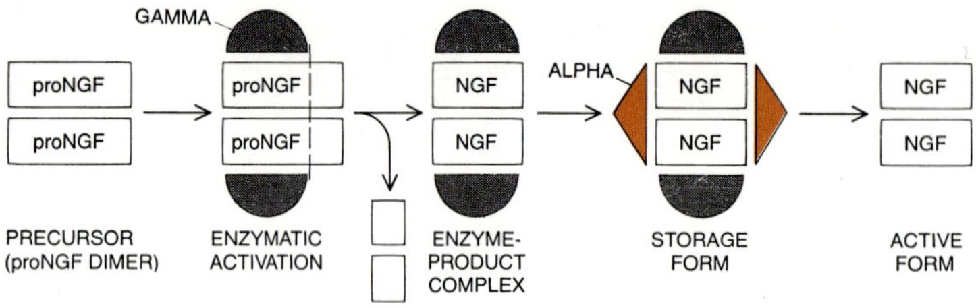

MANUFACTURE OF NGF requires a number of "processing" steps. The initial form of the molecule is a precursor chain called proNGF, which forms a dimer. The proNGF chains are cleaved by an enzyme known as the gamma subunit to yield the NGF dimer. The two gamma subunits remain associated with the dimer after it is generated, and two additional proteins, the alpha subunits, bind to the complex. The association of the alpha and gamma subunits with the NGF dimer apparently protects it from further degradation by other enzymes.

crease in size of the sympathetic ganglia has been traced to three separate processes: (1) an increased rate of differentiation of the sympathetic neurons, (2) an increase in the total number of neurons in the ganglia and (3) an increase in the size of the fully differentiated neurons.

NGF does not increase the number of neurons in the sympathetic ganglia by enhancing their multiplication. Instead, as was first proposed by I. A. Hendry of the Australian National University, the marked increase in the number of neurons is due to the survival of redundant immature neurons that would ordinarily die off in the course of development. Cell death is a common event in the formation of the nervous system: entire populations of immature neurons die off or are sharply reduced in size. In fact, in the early developmental stages of the chick embryo the dead cells in the sensory and sympathetic ganglia often outnumber the live ones. The generally accepted hypothesis is that the immature neurons that do not establish functional connections with their target cells are doomed to die; the redundant number of neurons ensures that all the appropriate connections are made. The effect of experimentally administered NGF is to enable the redundant neurons to survive and differentiate in spite of their failure to establish connections. It is this effect of NGF that gives rise to the substantially increased numbers of neurons in NGF-treated sympathetic ganglia.

In addition to apparently being essential to the survival of immature sympathetic neurons NGF seems to play a vital role in guiding nerve fibers toward their corresponding target organs. Three basic mechanisms have been proposed over the years to explain the formation of specific neural circuitry: (1) an elaborate predetermined program encoded genetically in each neuron that unfolds according to rigid and unmodifiable rules, (2) a random process of trial and error in which growing nerve fibers that make the right connections are consolidated and those that fail are reabsorbed and (3) a general program of circuit formation that is brought to completion by an interplay between genetic and extrinsic factors.

The first mechanism can be discounted because it would require very large amounts of genetic information—much more than could be encoded in the entire complement of DNA in the cell nucleus of each neuron. The second mechanism is also unlikely, because such a random process of circuit formation would be extremely time-consuming and wasteful of energy and resources. The third mechanism therefore seems the most probable: the circuitry of the nervous system is established through some combination of genetic and extrinsic factors.

The existence of such extrinsic factors was first suggested by the Spanish neurologist Santiago Ramón y Cajal, who visualized them as chemical signals emanating from peripheral tissues that would direct the growing nerve fibers toward their matching target cells. Cajal named the process neurotropism. His hypothesis was neglected for many years because the methodology for detecting such chemical factors in the living embryo was not yet available. The discovery and isolation of NGF, however, has provided an opportunity to reconsider the concept of neurotropism under more favorable experimental conditions.

Convincing evidence for the role of NGF in the formation of neural circuits in the sympathetic nervous system has come from both animal and test-tube experiments. At the Laboratory of Cell Biology in Rome, NGF was injected into the brain of newborn rodents. An unexpected effect of the treatment was that nerve fibers sprouted from the sympathetic ganglia and invaded the brain and spinal cord. Apparently the NGF injected into the brain diffused through the motor and sensory roots of the spinal cord and reached the sympathetic chain ganglia flanking the cord, where it induced the outgrowth of nerve fibers. This finding implies that the tip of a growing sympathetic nerve fiber elongates along a diffusion gradient of NGF released by the target organ. The NGF gradient does not entirely determine the course of the nerve fibers but helps to orient them in the right direction.

Further evidence for a neurotropism mediated by NGF was provided by some elegant experiments done by Robert B. Campenot of the Harvard Medical School. He devised a three-chambered cell-culture system in which the chambers were separated by impermeable barriers of sterile silicon grease, preventing the diffusion of fluid from one chamber to another. Immature sympathetic neurons were seeded into the central chamber in the presence of NGF and allowed to send out growing fibers under the silicon barriers along scratches made on the bottom of the culture dish. Campenot filled one side chamber with a nutrient solution containing NGF and the other chamber with the same nutrient solution devoid of NGF. He observed that the growing nerve fibers migrated from the central chamber only into the side chamber containing NGF. When the NGF-rich solution was withdrawn from that side chamber, the nerve fibers that had extended into it began to degenerate, even though the neuronal cell bodies in the central NGF-containing chamber persisted in good condition.

These experiments resolved the long-standing question of whether extrinsic chemical factors play a role in guiding the growth of nerve fibers. Since NGF is released in minute amounts from peripheral tissues receiving innervation from sympathetic ganglia, it now seems clear that a diffusion gradient of NGF directs the fibers toward their corresponding target tissues. When the growing nerve fiber and the target cell finally come in contact, the provisional adhe-

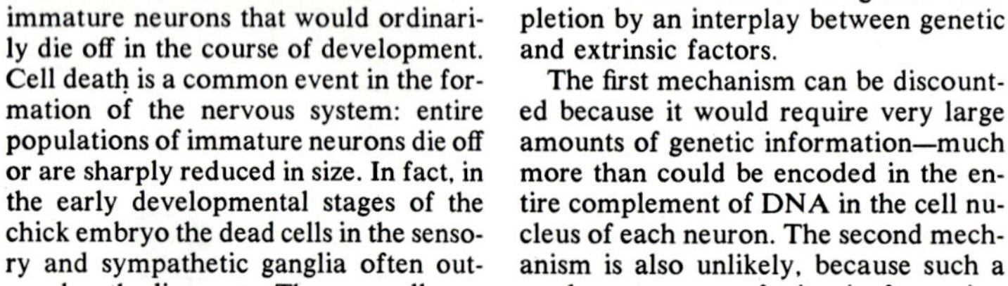

NGF ENHANCES THE GROWTH of sympathetic chain ganglia. The two sets of ganglia shown here were dissected from littermate three-week-old mice. The control mouse (*top*) was injected daily with a saline solution, whereas the test mouse (*bottom*) received 10 micrograms of NGF per gram of body weight. Chain ganglia showed a substantial increase in volume.

sion consolidates into the structural and functional organization of the synapse.

Once the elongating fibers have established the appropriate synaptic contacts with the target cells, the continued survival of the innervating cells in the ganglion appears to depend on the availability of NGF. Studies conducted by Hendry at the Australian National University and by K. Stöckel and H. Thoenen at the Basel Institute for Immunology have demonstrated that NGF is taken up at the terminal nerve endings of the sympathetic fibers and transported back to the neuronal cell body along the axon. This retrograde axonal transport of NGF is absolutely essential for the survival of the innervating neurons. When it is experimentally prevented (either by severing the projecting axons, by treating them with the drug vinblastine, which blocks axonal transport, or by administering 6-hydroxydopamine, which destroys the nerve endings), the innervating sympathetic neurons in the ganglion die off. The lethal effects of blocking the axonal transport of NGF can be completely overcome, however, by supplying the cell bodies with externally administered NGF. In this case the external NGF makes up for the NGF that would normally be transported back inside the axon to the cell body from the innervated cells.

Work in several laboratories has shown that the retrograde axonal transport of NGF follows its interaction with specific receptor sites on the nerve terminals of the newly established fibers. Receptors are proteins that are usually located on the external surface of the cell membrane; they provide specific recognition sites for messenger substances such as hormones, neurotransmitters and growth factors. The existence of such specific receptors on the neuronal surface makes it possible for NGF to exert its effects at exceedingly low concentrations (about 2.8 micrograms per liter). The binding of NGF to its receptors triggers a chain of biochemical events that leads ultimately to the outgrowth of the nerve fiber.

Immature sympathetic neurons respond to NGF with a burst of metabolic activity that provides the material necessary for the growth of the nerve fiber and the manufacture of molecules of neurotransmitter. The cells make more proteins and lipids, take up amino acids from the surrounding medium and burn glucose and other energy-rich compounds at a faster rate. These effects can be seen as a general response to the primitive signal conveyed by NGF through its specific receptors on the cell surface.

Within a short time after the binding of NGF to its receptors the protein constituents of the cytoplasm of the immature sympathetic neuron are profoundly rearranged. In particular the filamentous structures called microtubules and microfilaments come to fill all the available space between the cell nucleus and the cell membrane. These filaments play a key role in the growth of the nerve fiber by providing a structural framework and the propulsive force for its elongation.

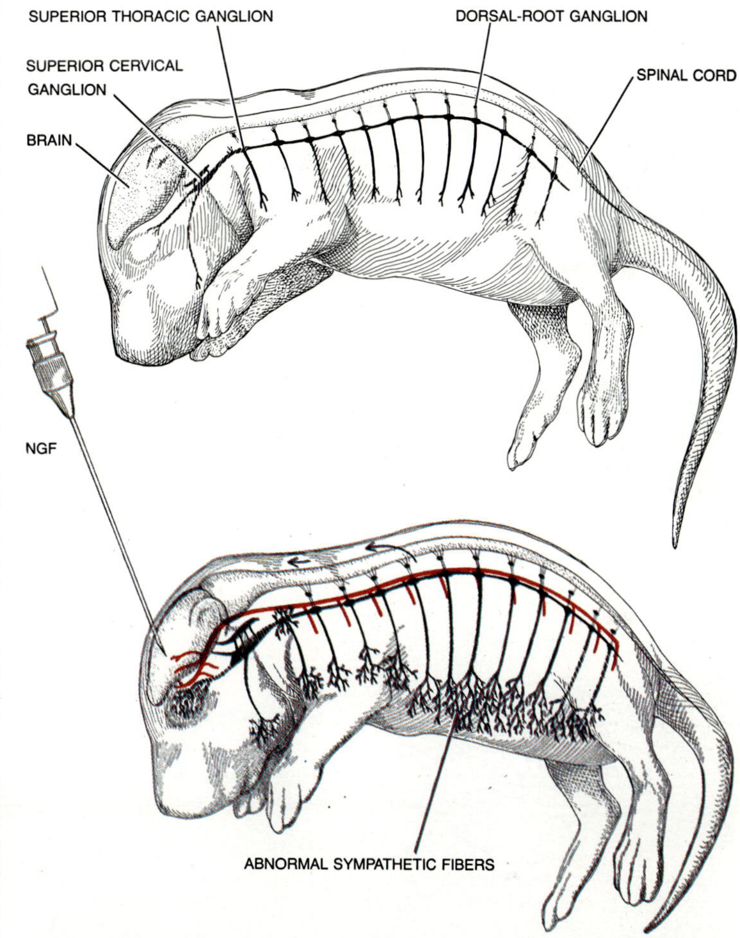

NEUROTROPIC EFFECT of NGF was demonstrated by injecting the factor into the brain of newborn rats for 10 days; the control animals were injected with saline solution only. In the NGF-treated animals (*bottom*) the growth factor diffused from the injection site to the spinal cord and reached the adjacent sympathetic ganglia through the motor and sensory roots. The NGF induced the outgrowth of sympathetic nerve fibers (*color*), which gained access to the spinal cord through the spinal roots and extended as far as the site of injection in the brain. This abnormal projection of sympathetic fibers (which usually innervate peripheral tissues) in response to NGF provided evidence for the hypothesis that a gradient of NGF guides the direction of growing sympathetic fibers in the periphery. In addition injection of NGF into either the brain or the systemic circulation enhanced the peripheral branching of the sympathetic fibers. Experiment was done by M. L. Menesini Chen and J. S. Chen in authors' laboratory.

How does NGF control the assembly of these filamentous proteins? One possibility is that it acts directly to enhance the polymerization of tubulin and actin, the monomeric proteins that give rise respectively to microtubules and microfilaments. Working at the Laboratory of Cell Biology in Rome, we tested the hypothesis by measuring the rate of assembly of tubulin and actin in the test tube in both the presence and the absence of NGF. In the absence of NGF dilute solutions of the proteins were unable to polymerize into filaments (or did so at a very low rate) because the thermodynamic tendency of the monomers to stay far apart in solution was greater than their tendency to aggregate. The addition of NGF to the solution, however, induced a rapid and massive polymerization reaction. Subsequent investigation revealed that NGF joins together the minimum number of monomers needed to initiate polymerization, after which the reaction proceeds spontaneously to completion. On the basis of these test-tube findings we have hypothesized that the massive formation of mi-

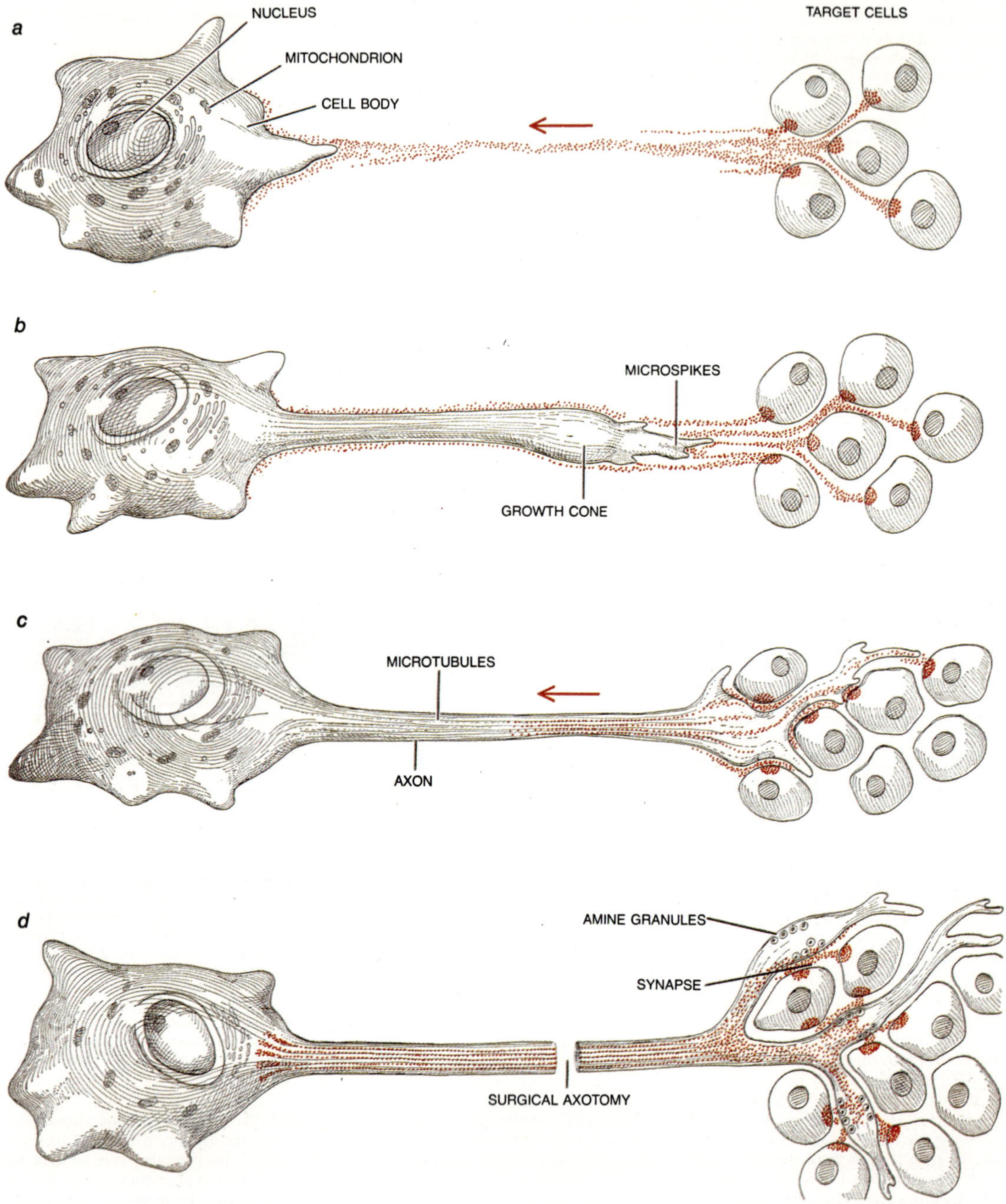

EXTENSION OF THE NERVE FIBER from an immature sympathetic neuron is induced and guided by NGF. Under normal conditions the target cells that are innervated by the sympathetic neurons or that are present along the path of the growing axon manufacture and secrete small amounts of NGF. The growth factor diffuses into the intercellular space and binds to specific receptor molecules on the surface of the immature sympathetic neuron (*a*). The interaction of NGF and its receptor triggers (by a still unknown mechanism) the assembly within the neuron of filamentous proteins called microtubules and microfilaments. These structures are essential constituents of the growth cone, the locomotory organelle of the growing axon. Situated at the tip of the extending fiber, the growth cone possesses motile, exploratory fibers called microspikes, which confer movement and directionality to the growing axon (*b*). Once the axon has established definitive connections with its target cells, the contact consolidates into the physical and functional link known as the synapse. NGF supplied by the target cells is then channeled directly along the axon to the cell body by a mechanism called retrograde axonal transport, which is most likely achieved by means of filamentous structures within the axon (*c*). Blockage of retrograde axonal transport (by cutting the axon, destroying the synapses with 6-hydroxydopamine or preventing axoplasmic flow with vinblastine) results in the degeneration of the sympathetic neuron in the ganglion (*d*). These different treatments all exert their lethal effects by preventing the normal access of NGF to the sympathetic neurons by way of retrograde axonal transport from the peripheral innervated tissues. Antibodies to NGF also kill immature sympathetic neurons by inactivating NGF so that it can no longer bind to neurons and ensure their continued survival.

crotubules and microfilaments in the developing sympathetic neuron is triggered by NGF, and that this effect may lead directly to the growth and elongation of the nerve fiber.

An entirely different function of NGF came to light recently with the discovery that certain non-neuronal cells are able to respond to the protein. Lloyd A. Greene and Arthur S. Tischler of the Harvard Medical School demonstrated in 1977 that cells of the line designated *PC*12 (which are derived from a rat tumor) respond to NGF by acquiring properties characteristic of sympathetic neurons. The NGF-treated cells send out fibers, become electrically excitable and store and release neurotransmitters of the catecholamine type. When NGF is withdrawn from the culture medium, the cells retract their fibers, lose their other neuronal properties and resume the uncontrolled proliferation characteristic of neoplastic cells.

Shortly after this discovery K. Unsicker and his colleagues at Johns Hopkins University found that immature chromaffin cells obtained from the medulla (inner part) of the adrenal gland and cultured in the presence of NGF acquire the biochemical and morphological properties of sympathetic neurons. Subsequent experiments carried out at the Laboratory of Cell Biology in Rome by Luigi Aloe and one of us (Levi-Montalcini) demonstrated that this remarkable transformation can also take place in the intact animal: repeated injections of NGF into rat fetuses, continued for two to three weeks after birth, resulted in the differentiation of chromaffin cells into sympathetic neurons in the core of the adrenal gland! These findings indicate that NGF plays a much broader role in the living organism than had been supposed.

It should be obvious to the reader that the investigation of NGF, which began more than two decades ago, is far from over. Among the many unanswered questions is why NGF is manufactured and secreted by the venom gland of snakes and the salivary gland of rodents, even though neither of these glands is essential for the life of the organism and the sympathetic neurons that depend on NGF for their survival. The molecule is also manufactured and released in minute amounts by a wide range of normal and neoplastic cells.

The discovery of NGF has made it possible to experimentally manipulate sympathetic neurons with considerable ease, both in tissue culture and in the living animal. With the aid of purified NGF and NGF antibodies the neurobiologist is now able to increase the number of sympathetic neurons in an animal tenfold or to eliminate them altogether. Moreover, the catecholamine-secreting neurons in the brain are strikingly similar to the sympathetic neurons in the peripheral ganglia in both their morphological and biochemical properties. Already information gained from the study of sympathetic neurons has proved applicable to the treatment of disorders of the central catecholamine-secreting neurons, such as Parkinson's disease. A. Björklung, B. Bierre and U. Stenevi of the University of Lund recently obtained preliminary evidence that the catecholamine-secreting neurons in the brain respond to NGF with a profuse branching of their nerve fibers. If these findings are confirmed, a powerful tool will become available for modulating the function of such brain circuits, which play a crucial role in many kinds of behavior.

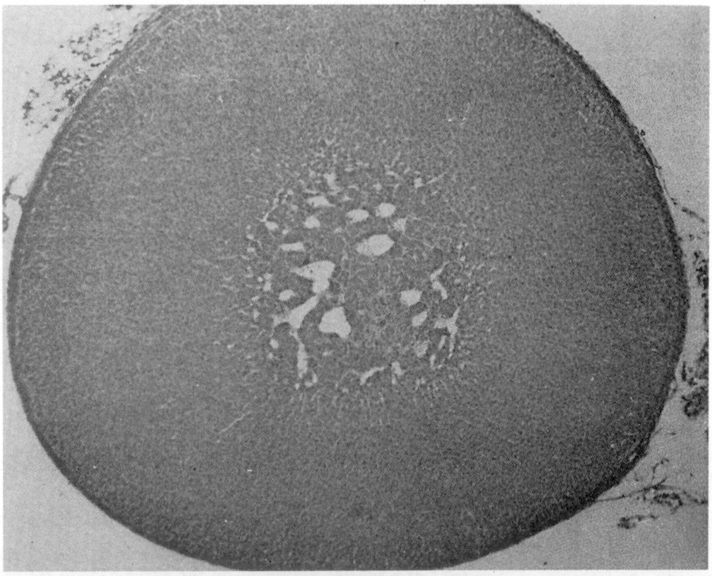

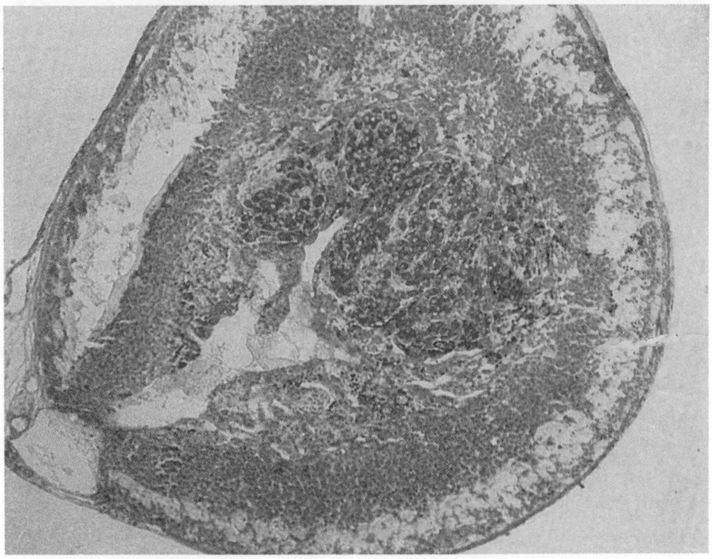

NGF TRANSFORMS ADRENAL CELLS into sympathetic neurons when it is injected into fetal rats before and after birth. These equatorial cross sections of the adrenal glands from two-day-old rats show the effect of injections of saline solution alone (*top*) and an NGF solution (*bottom*). The normal adrenal gland consists of a dense cortical layer of cells surrounding a central medullary area made up of adrenal cells and blood vessels. NGF treatment results in a total rearrangement of the gland. The central medulla dramatically increases in volume as a result of the transformation of the adrenal cells into sympathetic neurons, which sprout a large number of profusely branching fibers. These fibers form a large bundle that replaces the inner cortical layer and cuts across the cortex, running in a circular fashion around the external zone. The experiments were performed by Luigi Aloe of the Laboratory of Cell Biology in Rome.

Printed by permission from *Nature* Vol. 291, pp. 413–415.

Somatic and behavioural postnatal effects of fetal injections of nerve growth factor antibodies in the rat

L. Aloe, C. Cozzari, P. Calissano & R. Levi-Montalcini
Laboratory of Cell Biology, CNR, Via Romagnosi 18/A, 00196 Rome, Italy

Previous work has shown that injections to neonatal mammals of a specific antiserum to the nerve growth factor (AS-NGF) produce irreversible destruction of sympathetic ganglia[1,2]. The animals used in these studies were mice and rats. Because this treatment does not damage other nerve cells or produce adverse effects in other tissues and organs, the immunosympathectomized young and adult rodents do not exhibit—at gross inspection—somatic or behavioural differences compared with controls except for a marked ptosis[3]. Differences become apparent, however, when the treated rodents are submitted to physiological, pharmacological and behavioural tests[4]. We report here the dramatic effects that are produced by injections of purified antibodies to NGF in rat fetuses. A preliminary account of this work has already been presented[5,6], and similar effects have been demonstrated in another laboratory.

2.5 NGF was purified by the method of Bocchini and Angeletti[7]. After two successive carboxymethyl cellulose chromatographies, NGF was run in SDS-polyacrylamide gel electrophoresis and the corresponding band cut out, eluted from the gel matrix, mixed with complete Freund's adjuvant and injected into rabbit to produce antibodies. These were purified

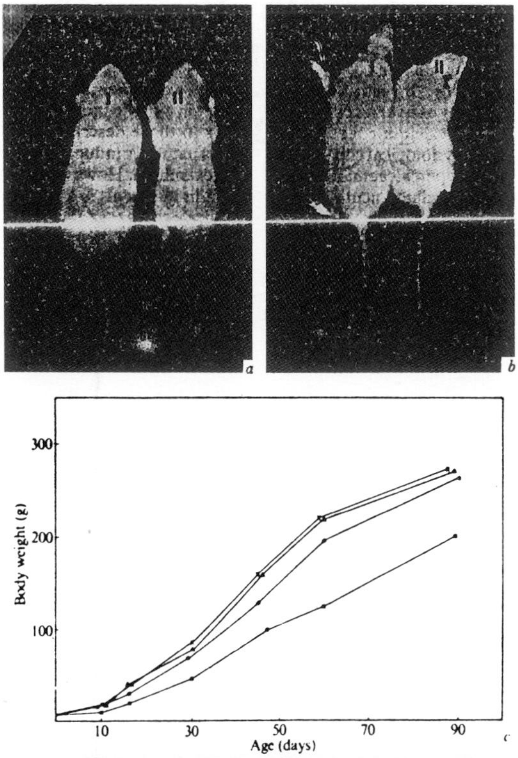

Fig. 1 Control (I) and experimental (II) rats killed respectively at 3 weeks (*a*) and 2 months (*b*) of age. Control rats were injected pre- and postnatally with 0.15 M NaCl for their first 3 weeks. Experimental rats were injected with 40 μg of purified NGF antibodies dissolved in 0.15 M NaCl. Further details are included in the text. *c*, Comparison of body weight increase in rats injected pre- and postnatally with NGF antibodies (■), 6-hydroxydopamine (●), rabbit IgG (▲) or 0.15 M NaCl (×). Each curve represents the average weight of 20 rats per group.

using the procedure of Stoeckel *et al.*[8]—affinity chromatography of the IgG fraction on Sepharose 4B with covalently linked 2.5 NGF. After absorption and washes as described, NGF antibodies were eluted with 4.5 M $MgCl_2$, dialysed extensively against twice distilled water and lyophilized. These purified preparations of NGF antibodies contained no detectable IgG fraction cross-reacting with the renin-like activity present in mouse salivary gland extract[9] and gave one single precipitin band in double-diffusion agar plates.

Female Sprague–Dawley rats were mated and at 16–17 days of gestation divided into four experimental groups. Pregnant rats were anaesthetized with ether, a middle incision made on the abdomen and the left and right uterine tubes exposed in sterile conditions. Each fetus of the first group (I) was injected subcutaneously with 40 μg of purified NGF antibodies dissolved in 20 μl of 0.15 M NaCl, using a 30-gauge needle inserted through the intact tube. The other groups served as controls and were injected with 20 μl of 0.15 M NaCl (group II), 40 μg of rabbit IgG (group III) or (group IV) with the dopamine derivative, 6-hydroxydopamine (6-OHDA), which in neonatal rodents produces massive destruction of para and pre-vertebral sympathetic ganglia[10,11]. Of the 29 pregnant rats, one of group I and one of group II died. In group I, two rats underwent abortion and a third had to have its litter delivered through caesarian intervention 2 days after the end of the gestation time. All neonatal rats of the four groups appeared vital and compared favourably at birth with untreated litters. The only noticeable difference to the untreated gestant rats was that the delivery of the litters was delayed by 24 h and in two cases by 48 h, with no apparent consequence for the mothers and the neonatal rats. The studies reported here are based on the analysis of 126 pups of the first group, 124 of the second, 48 of the third and 45 of the fourth group. In all neonatal rats the treatment was resumed immediately after birth at progressively higher doses according to the increase in body weight, as described previously[12].

Towards the end of the first postnatal week, pups injected with purified NGF antibodies (group I) were markedly smaller than those of the other three groups. 18 neonatal rats of group I died at the end of the first week and 22 before the end of the second week. In all instances death was preceded by progressive decrease in vitality and refusal to suckle even if they were taken care of by the mothers. To ensure that the weakest pups had access to nipples, the largest litters were reduced to nine pups by making available foster mothers deprived of their offspring. Six pups injected with saline (group II) and two treated with rabbit IgG (group III) died before the end of the second week. Three infants injected with 6-OHDA (group IV) died during the second postnatal week. The pups of groups II, III and IV developed normally although the latter were vital but slightly smaller than those of groups II and III (Fig. 1*c*). The 86 surviving rats treated with NGF antibodies were markedly smaller than all age-matched rats from the other three groups, as shown in Fig. 1*a–c*. Moreover, they exhibited sluggish, apathetic behaviour and moved only a few steps when exposed to strong stimuli, immediately resuming a static position, and did not show the exploratory behaviour characteristic of normal 2–4-week-old pups. On external inspection, their leg muscles appeared severely reduced in size. The behaviour of group I pups during resting of motility phases was strikingly similar to that of hypothyroid rats. Furthermore, they differed from those of other groups in their very weak reaction to pressure, prickling, burning or other noxious stimuli. Their body temperature to the hand was markedly lower than that of controls.

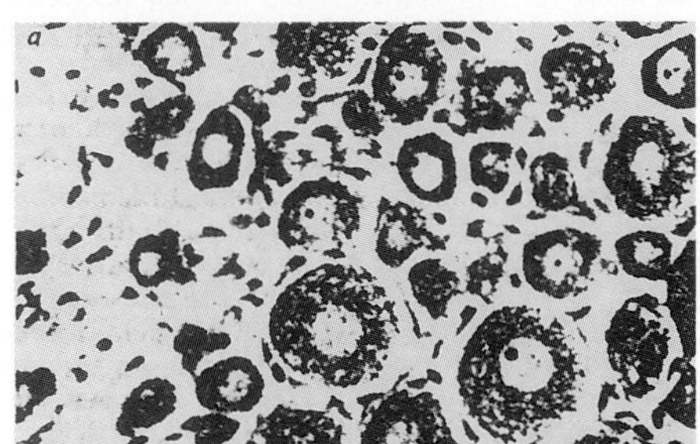

Fig. 2 Photomicrographs of sections through spinal sensory ganglia at the lumbar level of 9-day-old rats pre- and postnatally treated with 0.15 M NaCl (*a*) or NGF antibodies (*b*). Notice hypotrophic sensory nerve cells in the NGF antibody-treated rats. ×350.

Histological studies of serially sectioned sympathetic, sensory ganglia, adrenal and extra-adrenal chromaffin cells gave the following results. At birth, sympathetic ganglia were already reduced to sclerotic nodules consisting of glial cells, fibroblasts and few small atrophic sympathetic neurones. The effect was similar but more pronounced and much more precocious than that elicited by AS-NGF in neonatal rats and mice[4]. The massive degeneration of chromaffin adrenal and extra-adrenal cells has been discussed previously[13]. Sensory ganglia, which do not exhibit morphological or functional alterations in postnatal AS-NGF-injected rodents[14], appeared markedly reduced in size in 2–3 week-old group I pups. Figure 2 compares histological sections of sensory neurones of spinal ganglia of groups I and II. Nerve cells of the first group show various degrees of atrophy, ranging from decrease in cell size to evident signs of degeneration and piknosis. These findings provide evidence of severe damage to sensory neurones in rats treated with purified NGF antibodies during pre- and postnatal periods.

We also studied the accumulation of purified NGF antibodies in different organs and tissues. Fetuses, neonatal and 12-day-old rats were injected with unlabelled NGF antibodies and trace amounts of ^{125}I-NGF antibodies (1×10^6 c.p.m. per fetus or infant rat, specific activity 30,000 c.p.m. ng^{-1}). Fetuses and pups were killed 5 and 24 h later. Brain segments, hypophysis, kidneys, adrenals, salivary glands, liver, placenta, sensory and sympathetic ganglia were dissected out and the TCA-precipitable radioactivity counted. In fetuses the distribution of labelled antibodies was comparable in all organs tested, except for sympathetic and sensory ganglia, adrenal glands and hypophysis where the radioactivity was markedly higher. In a typical experiment, the specific radioactivity (counts per mg wet weight) 24 h after injections was 722, 222, 240 and 219 in sympathetic ganglia, adrenal glands, sensory ganglia and hypophysis, respectively, while the average value in all other organs, including brain regions, was 60 ± 20. In animals injected after birth, the overall level of ^{125}I-NGF antibody distribution did not differ significantly from one organ to another, except in all brain segments where the radioactivity level was about one order of magnitude lower than in animals injected before birth, probably as a result of the development of the brain–blood barrier.

Our findings suggest that neutralization of NGF or a NGF-like molecule by NGF antibodies during a critical prenatal period produces damages to sensory nerve cells, as well as to sympathetic or chromaffin cell precursors[13], and results in a syndrome reminiscent of human dysautonomic neuropathies. These findings support the hypothesis that this syndrome may result from a genetically produced NGF deficiency[15–20]. Decrease in body weight and size, and sluggish and apathetic behaviour of infant rats injected during fetal life with NGF antibodies could be due to a direct or indirect damage to one or more components of the neuroendocrine system. A NGF role in the central nervous system has been previously hypothesized[21–24]. Similar results were reported[25,26] while this paper was in preparation.

Received 7 November 1980; accepted 25 March 1981.

1. Levi-Montalcini, R. & Booder, B. *Proc. natn. Acad. Sci. U.S.A.* **42,** 384–391 (1960).
2. Cohen, S. *Proc. natn. Acad. Sci. U.S.A.* **46,** 302–311 (1960).
3. Levi-Montalcini, R. & Angeletti, P. U. *Pharmac. Rev.* **18,** 619–628 (1966).
4. Steiner, G. & Schönbaum, E. *Immunosympathectomy* (Elsevier, Amsterdam, 1972).
5. Levi-Montalcini, R. *Nerve Cells, Transmitters and Behaviour,* 84 (Elsevier, Amsterdam, 1980).
6. Levi-Montalcini, R., Aloe, L., Calissano, P. & Cozzari, C. *Abstr. 3rd Meet. int. Soc. dev. Neurosci.*, Strasbourg, 5–6 (1980).
7. Bocchini, V. & Angeletti, P. U. *Proc. natn. Acad. Sci. U.S.A.* **64,** 787–794 (1969).
8. Stoeckel, K., Gagnon, C., Guroff, G. & Thoenen, H. *J. Neurochem.* **26,** 1207–1221 (1976).
9. Cozzari, C., Angeletti, P. U., Lazar, J., Orth., H. & Cross, F. *Biochem. Pharmac.* **22,** 1321–1327 (1973).
10. Angeletti, P. U. & Levi-Montalcini, R. *Proc. natn. Acad. Sci. U.S.A.* **65,** 114–121 (1970).
11. Angeletti, P. U. & Levi-Montalcini, R. *Archs ital. Biol.* **108,** 213–221 (1970).
12. Levi-Montalcini, R., Aloe, L., Mugnaini, E., Oesch, F. & Thoenen, H. *Proc. natn. Acad. Sci. U.S.A.* **72,** 595–599 (1975).
13. Aloe, L. & Levi-Montalcini, R. *Proc. natn. Acad. Sci. U.S.A.* **76,** 1246–1250 (1979).
14. Levi-Montalcini, R. *Harvey Lecture* **60,** 217–259 (1966).
15. Riley, C. M. & Moore, R. H. *Pediatrics* **37,** 435–446 (1966).
16. Pearson, J., Pytel, B. A., Grover-Johnson, N., Axerod, F. & Dancis, J. *J. neurol. Sci.* **35,** 77–92 (1978).
17. Dolkart Gorin, P. & Johnson, E. M. *Proc. natn. Acad. Sci. U.S.A.* **76,** 5382–5386 (1979).
18. Schwartz, J. P. & Breakefield, X. O. *Proc. natn. Acad. Sci. U.S.A.* **77,** 1154–1158 (1980).
19. Otten, U., Goodert, H., Mayer, N. & Lembeck, F. *Nature* **287,** 158–159 (1980).
20. Crain, S. M., Peterson, E. R., Leibam, M. & Schulman, H. *Expl Neurol.* **67,** 205–214 (1980).
21. Freed, W. J. *Brain Res. Bull.* **1,** 393–412 (1976).
22. Benowitz, L. I. & Shashona, V. E. *Brain Res.* **172,** 561–565 (1979).
23. Walker, P., Weichsel, M. E., Fisher, D. A., Guo, S. M. & Fisher, D. A. *Science* **204,** 127–129 (1979).
24. Otten, U., Baumann, J. B. & Girard, J. *Nature* **282,** 413–414 (1979).
25. Gorin, P. D. & Johnson, E. M. *Proc. natn. Acad. Sci. U.S.A.* **76,** 5382–5386 (1980).
26. Johnson, E. M. *et al. Science* **210,** 916–918 (1980).

0028-0836/81/220415—04$01.00

Proc. Natl. Acad. Sci. USA
Vol. 82, pp. 7111–7115, October 1985
Neurobiology

Differentiating effects of murine nerve growth factor in the peripheral and central nervous systems of *Xenopus laevis* tadpoles

RITA LEVI-MONTALCINI AND LUIGI ALOE

Istituto di Biologia Cellulare, Consiglio Nazionale delle Ricerche, Via Romagnosi 18/A, 00196 Roma, Italy

Contributed by Rita Levi-Montalcini, May 31, 1985

ABSTRACT The present investigation was directed toward exploration of the spectrum of action of murine nerve growth factor (m-NGF) in peripheral cells and neurons in the central nervous system (CNS) of *Xenopus laevis* tadpoles. It was found that systemic m-NGF injections elicit growth and differentiative effects not only on sensory and sympathetic nerve cells but also on several populations in the CNS. The finding that aminergic and peptidergic neurons in brain centers are highly receptive to m-NGF activity provides evidence for the broad spectrum of action of this molecule in lower vertebrates and calls for a systematic search for these and other putative target cells in the CNS of higher vertebrates.

Two of the most interesting areas in the field of hormone research are those concerned with the molecular events underlying their mechanisms of action and evolutionary history. Studies in this past decade have resulted in considerable progress in elucidating some of these mechanisms and in tracing back several vertebrate hormones to unicellular eukaryotic organisms and even to prokaryotes (1). In the class of hormone-like peptides, designated as specific growth factors because of their properties of enhancing proliferative or differentiative processes in their target cell lines, extensive work has been performed to gain information on their mode of action, but few attempts have been made to trace their origin and biological significance in lower vertebrates. Among the biologically active peptides listed in this class, nerve growth factor (NGF) holds a privileged position as the first to have been identified (2–4) and to have lent itself to extensive studies pursued in living organisms, in organ cultures, and on dissociated nerve cells (5). Surprisingly, very little is known about the response elicited by NGF in the nervous system of fishes and amphibians. Although early studies (6–9) described proliferative and differentiative effects on sensory and sympathetic cells of specimens of these lower vertebrates, these reports received only little attention. The belief that information on the NGF range of activity and spectrum of action in these simpler organisms would be of considerable interest prompted a more detailed study on its effects in tadpoles of the amphibian *Xenopus laevis*. Our results provide evidence that in this species NGF not only enhances differentiative processes of sensory and sympathetic nerve cells but also is an important factor for growth and differentiation of neuronal cell types in the central nervous system (CNS). This last finding, discussed previously (10), is the main focus of this presentation.

MATERIALS AND METHODS

Adult male and female *X. laevis* were treated with chorionic gonadotropin and induced artificially to mate following the method described by Gurdon (11). After removing the jelly coat, the fertilized eggs were transferred to a large jar containing charcoal-inactivated tap water thermoregulated at 23°C. During the first week, tadpoles were fed homogenized dried urtica or spinach and in later stages were fed Tubifex (Euraquarium, Bologna, Italy).

Age- and stage-matched tadpoles, varying from stage 40 to stage 52 according to the Nieuwkoop and Faber table (12), were divided prior to each experiment into pairs of 10–20 specimens and kept in separate water containers throughout the time of experimentation.

NGF was prepared from male mouse salivary glands (m-NGF) according to the method of Bocchini and Angeletti (13). NGF was iodinated following the procedure of the chloramine-T method (14). The reaction yields 95% ^{125}I-labeled NGF (^{125}I-NGF) precipitable with trichloroacetic acid, with a specific activity of 150,000–200,000 cpm/ng of protein. More than 600 tadpoles at stages between 40 and 52 were used in these studies. They were anesthetized by immersion for 2–3 min in ice-cold water and placed in Petri dishes containing moist filter paper. Under a dissecting stereomicroscope, tadpoles were injected with NGF or vehicle solution in the right periorbital area by means of a calibrated glass micropipette with a tip of about 10–15 μm in diameter. They were injected daily for periods of 3–7 days with 2 μg (stages 40–44), 4 μg (stages 45–48), and 10 μg (stages 49–52) of NGF in 1–2 μl of vehicle solution. Mortality in the NGF- and vehicle-injected tadpoles was about 20% and was slightly higher in the former than in the latter set.

Histological Studies. Age- and stage-matched tadpoles were fixed in alcoholic Bouin's fluid, embedded in Paraplast, sectioned serially (7 μm thick), and stained in hematoxylin/eosin or toluidine blue. Tadpoles of all premetamorphic stages were also stained by a silver impregnation technique (unpublished method) to better visualize cells and neurites of the peripheral and central nervous systems. Morphometric studies were carried out as described (6, 15).

Radioautographic Studies. For the *in vivo* binding assays, tadpoles were injected in the right periorbital area with 2 μl of ^{125}I-NGF with or without a 1000-fold excess of unlabeled NGF. Tadpoles were also injected with ^{125}I-labeled IgG (^{125}I-IgG). The bound labeled NGF or IgG was measured in isolated brains with a γ counter at various time intervals. Tadpoles were fixed in Bouin's fluid or 4% paraformaldehyde at time intervals for 15 min to 48 hr after injection. Tadpoles embedded in Paraplast were sectioned (5 μm thick), deparaffinized, dipped in K-2 Ilford emulsion, and 4 weeks later developed in D-19 Kodak solution. The sections were lightly stained with hematoxylin/eosin and examined with a Zeiss microscope.

Histofluorescence Studies. Saline- and NGF-treated tadpoles were exposed to the monoamine oxidase inhibitor

The publication costs of this article were defrayed in part by page charge payment. This article must therefore be hereby marked "*advertisement*" in accordance with 18 U.S.C. §1734 solely to indicate this fact.

Abbreviations: NGF, nerve growth factor; m-NGF, murine NGF; CNS, central nervous system; SP, substance P; Som, somatostatin.

iproniazide phosphate (Sigma) for 48 hr to better visualize the monoamine-containing cells. The specimens were then fixed in isopentane cooled by liquid nitrogen, freeze-dried for 3–4 days, and exposed to paraformaldehyde vapors at 70% relative humidity for 1 hr at 80°C (16). After embedding in pure Paraplast, tadpoles were serially sectioned (7 μm thick) and examined under a Zeiss standard fluorescence microscope equipped with the appropriate filters.

Immunofluorescence Studies. Tadpoles were fixed by immersion in 4% paraformaldehyde in 0.1 M phosphate-buffered saline (pH 7.6) (P_i/NaCl) for 2 hr, washed several times in P_i/NaCl mounted on a tissue supporter, and frozen with dry ice. Serial sections (10–15μm) were cut with a cryostat at −20°C, collected on gelatin-coated glass slides, and processed for immunofluorescence according to the technique of Coons (17). Briefly, sections were treated for 12 hr with monoclonal antibodies against substance P (SP), enkephalin, or somatostatin (Som), rinsed in P_i/NaCl for 30 min, and incubated for 1 hr in fluorescein isothiocyanate-conjugated antibodies (Miles), diluted 1:20 at room temperature. After a brief rinse in P_i/NaCl, sections were mounted in glycerine and P_i/NaCl and examined with a Zeiss epi-fluorescence microscope. To test the specificity of the immunocytochemical reaction, alternative sections were incubated with antibodies previously preabsorbed for 60 min at 37°C with pure antigen.

***In Vitro* Studies.** Groups of 50 tadpoles ranging from stage 46 to stage 48 were washed several times with sterile distilled water, immersed in 0.1% H_2O_2 for 3 min, and immediately washed in Ca^{2+}/Mg^{2+}-free P_i/NaCl. Brains were then removed, cleaned, incubated for 30 min at 25°C in the same buffer containing 0.1% trypsin and 0.2% collagenase, washed in culture medium, and mechanically dissociated with a Pasteur pipette. Dispersed brain cells were plated in collagen-coated plastic dishes in a culture medium consisting of BEM/RPM, inactivated horse and fetal calf sera, hypoxanthine, and NGF or vehicle solution (to be reported elsewhere). Cultures were observed every 24 hr for 20 days under an inverted Leitz microscope.

RESULTS

m-NGF Effects on Sensory, Sympathetic, and Chromaffin Stem Cells. m-NGF daily injections in *Xenopus* tadpoles for a 7-day period at stages 42–46 produced a massive volume increase of sensory cephalic and dorsal root ganglia. The latter, which in stage-46 control specimens consisted of three pairs of small aggregates of 12–20 nerve cells innervating the cervicothoracic trunk muscles, underwent an 8-fold increase upon NGF daily administration as a result of a striking increase in size and number of individual cells. This effect is illustrated in Figs. 1 and 2 *A* and *B*. Hyperinnervation of trunk muscles was very evident in silver-stained preparations. A marked increase in synthesis of the peptidergic transmitter SP was detected by immunocytochemical studies with monoclonal antibodies. The response of sympathetic and chromaffin stem cells to NGF administration occurred according to the same pattern as in higher vertebrates (18): sympathetic nerve cells undergo precocious differentiation while chromaffin cell precursors are diverted toward the neuronal rather than the glandular phenotype upon NGF treatment. These effects are shown in Fig. 2 *C–F*.

m-NGF Effects in the CNS. Brain and spinal cord of tadpoles between stage 42 and stage 50 are very visible in living specimens due to the dense texture of the CNS that stands out against the background of soft, sponge-like surrounding tissue. Stereomicroscopic inspection of size-matched, stage-48 tadpoles submitted for a 5- to 7-day period to daily injection of m-NGF or vehicle solution showed that brain and rostral spinal segments are slightly but consistently larger in the experimental than control specimens (not

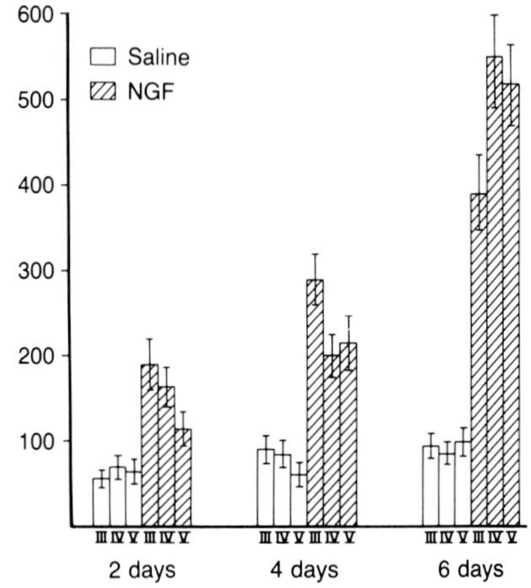

FIG. 1. Comparative volume (expressed in arbitrary units) of the third, fourth, and fifth spinal sensory ganglia of *Xenopus* tadpoles injected for 2, 4, and 6 days with saline (▭) or NGF (▭). Values are means ± SEM of four determinations.

shown). Camera lucida drawings and microphotographs of the brain and neural tube in serially sectioned larvae provided additional strong evidence for this effect (Fig. 2 *A* and *B*).

In toluidine blue-stained preparations, cells such as

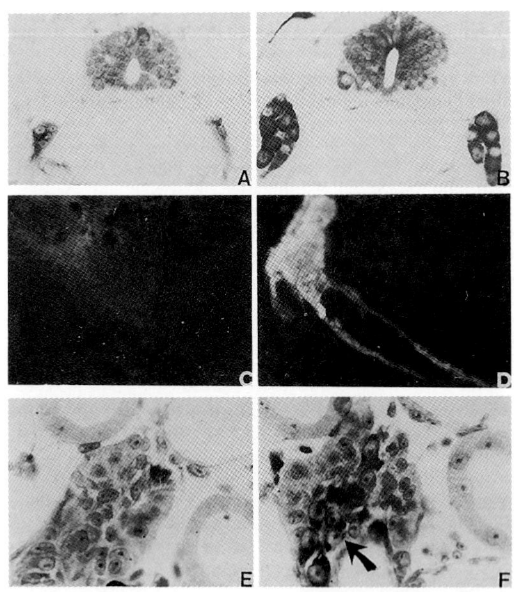

FIG. 2. (*A* and *B*) Comparative sections through the spinal cord of the third dorsal root ganglia of stage-48 tadpoles treated for 5 days with vehicle solution (*A*) and NGF (*B*). (Toluidine blue stain; ×125.) (*C* and *D*) Comparative sections at the level of the thoracic paravertebral ganglia in stage-50, vehicle-injected (*C*) and NGF-injected (*D*) tadpoles. The precocious differentiation of sympathetic ganglia in the NGF specimen (*D*) is very apparent. (Formaldehyde-induced fluorescence preparations; ×140.) (*E* and *F*) Comparative sections of the retroaortic area of stage-48 tadpoles treated for 6 days with vehicle solution (*E*) and NGF (*F*). Immature chromaffin cells are visible in *C*. In *F*, the arrow points to a group of sympathetic-like nerve cells. (Toluidine blue stain; ×185.)

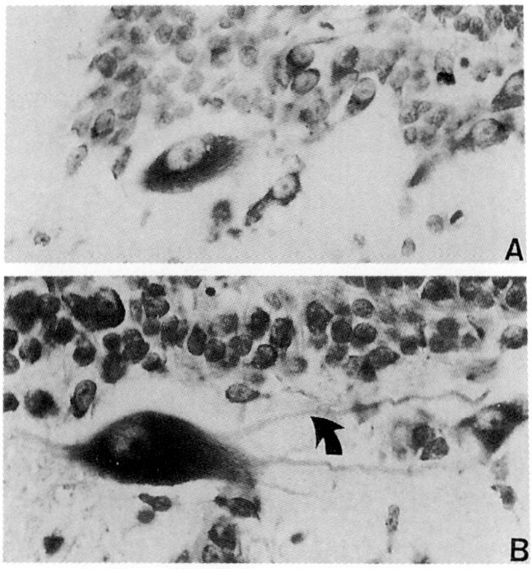

FIG. 3. Mauthner cells in stage-48 tadpoles injected for 6 days with vehicle (*A*) and NGF (*B*). The arrow in *B* points to well-developed dendritic processes. (Toluidine blue stain; ×320.)

Mauthner cells (Fig. 3 *A* and *B*), Miller pontotegmental neurons in the brainstem, motor neurons in the spinal cord, as well as associative and other still unidentified nerve cells exhibited a larger size and a more intense basophylia in NGF-injected than in vehicle-injected tadpoles. The hypothesis that these effects resulted from specific binding and interaction of m-NGF with a large number of nerve cells at an incipient stage of their differentiation received support from studies on ^{125}I-NGF uptake in the brains of stage-46 tadpoles. Fig. 4 compares the accumulation of radioactivity in brain tissue of specimens injected with ^{125}I-NGF or ^{125}I-IgG. More than 95% of the ^{125}I-NGF injected into the periorbital area and incorporated in the CNS was trichloroacetic acid-precipitable. Its localization in the spinal cord and brain sections of tadpoles fixed 2 hr after NGF injection is shown in microphotographs in Fig. 5. Most of the ^{125}I-NGF grains were concentrated on dorsal funiculi in the spinal cord (Fig. 5*A*), dorsolateral sensory fiber tracts originating from cephalic ganglia (Fig. 5*B*), and round-shaped cells lining the third ventricle (Fig. 5 *C–E*). Individual cells and cell clusters throughout the whole cerebrospinal axis and longitudinal fiber tracts running from the spinal cord to the rostral section of the diencephalon, illustrated in Fig. 5*G*, were covered with silver grains, although less densely than the above-mentioned fiber and cell systems. Injection of ^{125}I-NGF with an excess

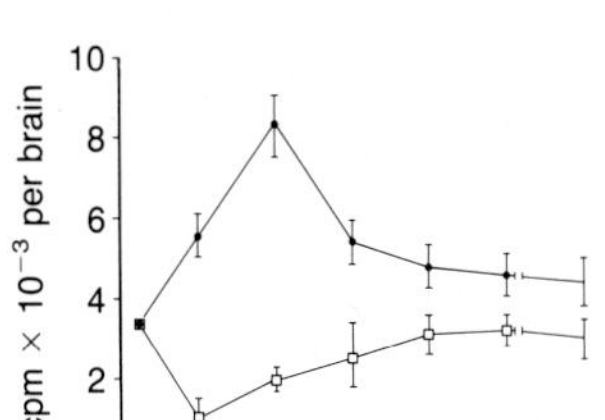

FIG. 4. Time-course accumulation of radioactive NGF in the isolated brain of *X. laevis* tadpoles at stage 46 injected in the right periorbital area with 2 μl of ^{125}I-NGF (●) or 2 μl of ^{125}I-IgG (□). Values are means ± SEM of four determinations.

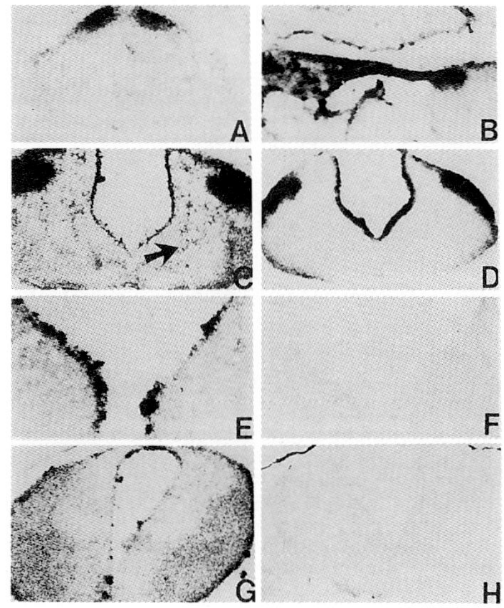

FIG. 5. Radioautographic histological sections of the CNS of *X. laevis* tadpoles at stage 46 following injection with ^{125}I-NGF (50 ng per tadpole) alone or with an excess of unlabeled NGF; fixation was 2 hr later. (*A*) Transverse section through the lower spinal cord showing the dorsal funicular area heavily labeled. (×140.) (*B*) Transverse section through the caudal medulla oblongata showing the intense labeling of the postotic ganglion, its nerve, and the afferent dorsal fiber bundles projecting into the medulla. (×140.) (*C* and *D*) Caudal and rostral brainstem sections showing heavily labeled fiber tracts afferent to the brainstem from the pre- and postotic ganglia. In *C*, intrinsic fiber tracts (arrow) and scattered cell clusters are also labeled. (*C*, ×180; *D*, ×140.) (*E*) Round-shaped heavily labeled cells bordering the third ventricle. (×106.) (*F*) Comparative histological section injected with ^{125}I-NGF and an excess of unlabeled NGF. (×106.) (*G* and *H*) Sections through rostral diencephalon injected with ^{125}I-NGF alone (*G*) and an excess of unlabeled NGF (*H*). Intrinsic fiber bundles in the ventrolateral area are also covered with iodinated NGF. (×73.)

of unlabeled NGF prevented completely the labeling of intracerebral nerve cells and fibers, as shown in Fig. 5 *F* and *H*.

Catechol/Indoleaminergic Systems. In stage-48 tadpoles, aminergic neurons are at the incipient phase of their differentiation. In histofluorescent transverse-sectioned preparations, small green fluorescent cell clusters—identified as belonging to the preoptic recess organ, the paraventricular organ, and the nucleus infundibularis dorsalis—are assembled around or at a short distance from (as in the case of the nucleus infundibularis dorsalis) the third ventricle. Upon m-NGF treatment, the same centers consist of markedly more fluorescent nerve cells that cover a larger area than in untreated specimens (Fig. 6 *A–F*). The number and cell size of individual units in NGF-injected tadpoles are markedly increased in pontomesencephalic reticular cells as compared to those of vehicle-injected matched specimens (Fig. 6 *G* and *H*).

Preliminary Neuropeptide Immunofluorescence Studies. Serial sections of tadpoles in stages 46–50 submitted for a 5-day period to NGF or vehicle solution injections were processed for immunoreactivity to enkephalin, SP, and Som. The weak histofluorescence reactivity to enkephalin monoclonal antibodies did not permit an evaluation of its distribution in brain sections of experimental and control specimens. Instead, immunoreactivity was quite evident in neuronal somata and

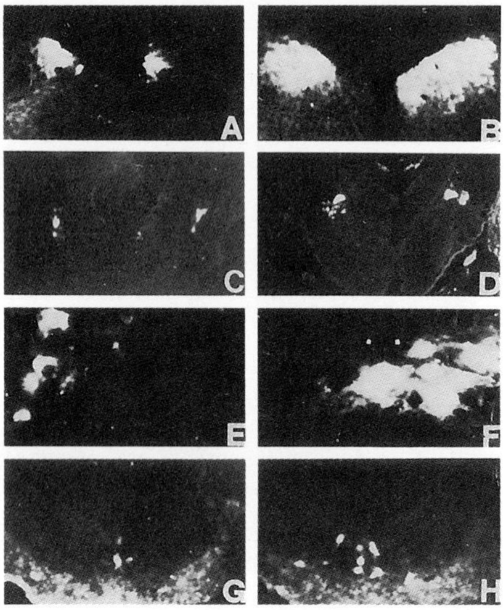

FIG. 6. Formaldehyde-induced fluorescence of tadpole brains at stage 48 injected for 6 days with vehicle (*A, C, E, G*) or NGF (*B, D, F, H*). (*A* and *B*) Comparative brain sections showing aminergic neurons in the paraventricular organs. (×50.) (*C* and *D*) Comparative brain sections showing the aminergic cells of the nucleus infundibularis dorsalis. (×15.) (*E* and *F*) Cells of nuclei infundibularis dorsalis at higher magnification. (×122.) (*G* and *H*) Comparative brain sections showing fluorescent nerve cells and fiber tracts in the pontomesencephalic reticular area. (×43.)

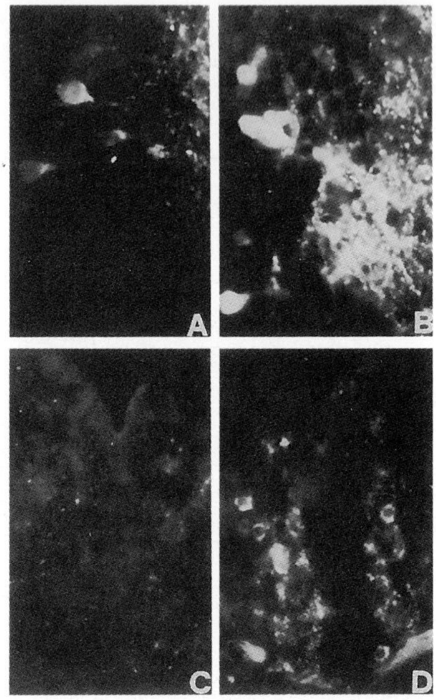

FIG. 7. (*A* and *B*) SP immunoreactive cells and fiber tracts in the mesencephalic area of tadpole brain at stage 48 injected for 6 days with vehicle (*A*) or NGF (*B*). (×86.) (*C* and *D*) Som immunoreactive nerve cells in the pontomesencephalic area of tadpole brains injected for 6 days with vehicle (*C*) or NGF (*D*). (×75.)

fiber network in diencephalic and pontomesencephalic segments processed to visualize SP and Som. Although in these preliminary studies no attempts were made to investigate the respective regional distributions of these two neuropeptides, the reactivity to SP was stronger in diencephalic than in pontomesencephalic centers, whereas the opposite was true for immunoreactivity to Som. In both experimental series, immunohistofluorescence was more pronounced in nerve cells, their dendrites, and axonal processes in sections of NGF-injected specimens than in control specimens. The more vivid fluorescence, larger somata size, and denser fibrillar network around cells immunoreactive to SP and Som is illustrated in Fig. 7 *A–D*.

***In Vitro* Studies.** Nerve cells dissociated from brain tissue of tadpoles between stage 46 and stage 48 were cultured for a 2- to 3-week period in the presence or absence of 10–50 ng of NGF per ml. The finding that these cells survived and produced axonal processes only upon addition of NGF to the culture medium provided evidence for its vital trophic role. Autoradiographic studies showed that the biological effect was mediated by NGF binding to specific receptors present in a fairly large number of cells; labeling of nerve cell somata in the presence of ^{125}I-NGF did not, in fact, materialize upon addition of excess unlabeled NGF to the incubation medium.

DISCUSSION

On the basis of previous reports (6–9), we anticipated that sensory and sympathetic amphibian nerve cells would be receptive to the growth and differentiation effects of m-NGF. The results, however, greatly exceeded our expectations with respect to the magnitude of the response elicited by NGF in early differentiated sensory neurons and to its range of action, which, in this species, also extends to several neuronal cell populations of the CNS. We shall here consider only the latter aspect of the biological activity of this protein molecule. In fact, with the exception of three reports on the enhanced regeneration elicited by m-NGF on surgically transected fish optic nerves (7–9), no attempts have been made to investigate whether other neuronal systems in lower vertebrates are receptive to the NGF differentiative effect. This effect is now experimentally demonstrated in several nerve centers in the amphibian CNS; the list of these cells is likely to increase as these studies are extended to other neuronal systems. Favoring this hypothesis is the finding that NGF selectively binds to a large number of nerve cells in the CNS of *Xenopus* tadpoles.

These observations call for some comment. The strikingly intense labeling of dorsal funiculi in the spinal cord and the dorsolateral fiber bundle collecting the central processes of sensory neurons of cephalic ganglia could result from (*i*) an anterograde ^{125}I-NGF transport, (*ii*) a direct binding to nerve fibers, or (*iii*) a retrograde ^{125}I-NGF transport along these fiber bundles. It is not possible at present to assess which of these mechanisms is responsible for the ^{125}I-NGF distribution. Previous studies performed on the mammalian nervous system would favor the latter hypothesis (19–21), which is also supported by a recent report (22, 23) showing that ^{125}I-NGF injected into the spinal cord of neonatal rats undergoes retrograde transport from the nerve endings to the cell compartments of sensory neurons in the dorsal root ganglia. The recent demonstration of a surprisingly high level of NGF mRNA in several mammalian tissues, including the brain, shows that the CNS shares a common property of most organs and vertebrate cell lines—synthesis of NGF (24). Also in favor of this hypothesis is a report based on immunohistochemical studies in goldfish brain, indicating that ependymal cells or astrocytes are the putative source of NGF in the CNS of these lower vertebrates (25). A local NGF production does not, however, rule out the possibility that during early ontogenetic stages, when nerve cells are in their most active growth and differentiative phase, other extracerebral NGF sources supply additional amounts of the growth factor.

The results reported here lend support to this view. Exogenous NGF injected into the periorbital area gains access to the CNS, as shown from the high uptake of labeled NGF in brain tissue. Autoradiographic studies with ^{125}I-NGF gave evidence of the extraordinarily dense accumulation of silver grains in round-shaped cells lining the third ventricle and cerebral aqueduct. They were identified as cerebrospinal fluid contacting neurons, although an alternative hypothesis that cannot at present be definitely discarded is that they are specialized ependymal cells or tanycytes. Both cell types belong to the complex of midline structures designated as the circumventricular organs, which have recently aroused great interest in view of the very high content of the peptides thyrotropin-releasing hormone, luteinizing hormone-releasing hormone, and Som in some of these cellular aggregates and in the cerebrospinal fluid (26). Lack of the blood–brain barrier in most of these circumventricular organs allows the free circulation through these structures, viewed as "windows in the brain" (27), of these and other peptides as well as of steroid hormones, which would, according to some authors, impart a functional organization upon neurons in the hypothalamus (28). The present finding of the strikingly high uptake of NGF in cells lining the third ventricle suggests the possibility that, in these lower vertebrates, NGF may be endowed with an important physiological role in brain organization and function, at least during these premetamorphic stages.

After the present study was completed, the widespread expression of NGF receptors in neuronal and nonneuronal cell populations in chicken embryos and their marked decrease in late developmental stages was reported (29).

We thank Dr. P. Calissano for helpful and stimulating discussions during preparation of the manuscript. This work was supported in part with Grants Progetto Finalizzato Medicina Preventiva e Riabilitativa from Consiglio Nazionale delle Ricerche and Progetto Finalizzato Chimica Fine e Secondaria from Consiglio Nazionale delle Ricerche to R.L.-M. We also acknowledge support from a Medical Research Council Wellcome Trust grant to Dr. C. Cuello (Department of Anatomy, Oxford Medical School, Oxford, England), who generously made available to us monoclonal antibodies.

1. Niall, H. D. (1982) *Annu. Rev. Physiol.* **44,** 615–624.
2. Levi-Montalcini, R. & Booker, B. (1960) *Proc. Natl. Acad. Sci. USA* **46,** 373–384.
3. Cohen, S. & Levi-Montalcini, R. (1960) *Proc. Natl. Acad. Sci. USA* **46,** 571–574.
4. Levi-Montalcini, R. (1952) *Ann. N.Y. Acad. Sci.* **55,** 330–343.
5. Levi-Montalcini, R. & Angeletti, P. U. (1963) *Dev. Biol.* **7,** 653–659.
6. Weis, J. S. (1969) *J. Exp. Zool.* **170,** 471–488.
7. Mathieu, C., Duprat, A. M., Zalta, J. P. & Beetschen, J. C. (1971) *Exp. Cell Res.* **68,** 25–32.
8. Vogel, D. L. & Model, P. G. (1977) *Dev. Biol.* **56,** 76–96.
9. Turner, J. E. & Delaney, R. C. (1979) *Brain Res.* **171,** 197–212.
10. Calissano, P., Cattaneo, A., Aloe, L. & Levi-Montalcini, R. (1984) in *Hormonal Proteins and Peptides*, ed. Li, C. H. (Academic, London), Vol. 12, pp. 1–56.
11. Gurdon, J. B. (1967) in *Methods in Developmental Biology*, eds. Wilt, F. H. & Wessells, N. K. (Crowell, New York), pp. 75–84.
12. Nieuwkoop, P. D. & Faber, J. (1956) *Normal Table of* Xenopus laevis (North-Holland, Amsterdam).
13. Bocchini, U. & Angeletti, P. U. (1969) *Proc. Natl. Acad. Sci. USA* **64,** 787–794.
14. Young, M. Y., Blanchard, M. H. & Saide, J. D. (1978) in *Methods of Radioimmunoassay*, eds. Jaffe, B. H. & Behrman, H. R., (Academic, New York), pp. 941–957.
15. Aloe, L., Mugnaini, E. & Levi-Montalcini, R. (1975) *Arch. Ital. Biol.* **113,** 326–353.
16. Falck, B., Hillarp, N. Å., Thieme, G. & Thorp, A. (1962) *Histochem. Cytochem.* **10,** 348–342.
17. Coons, A. H. (1958) in *General Cytochemical Methods*, ed. Danielli, J. F. (Academic, New York), pp. 399–422.
18. Aloe, L. & Levi-Montalcini, R. (1979) *Proc. Natl. Acad. Sci. USA* **76,** 1246–1250.
19. Hendry, I. A., Stockel, K., Thoenen, H. & Iversen, L. L. (1974) *Brain Res.* **68,** 103–121.
20. Seiler, M. & Schwab, M. E. (1984) *Brain Res.* **300,** 33–39.
21. Gnahn, H., Hejti, F., Heumann, R., Schwab, M. E. & Thoenen, H. (1983) *Dev. Brain Res.* **9,** 45–52.
22. Yip, H. K. & Johnson, E. M. (1984) *Proc. Natl. Acad. Sci. USA* **81,** 6245–6249.
23. Ayer-Lelievre, C. S., Ebendal, T., Olson, L. & Seiger, A. (1983) *Med. Biol.* **61,** 296–304.
24. Shelton, D. L. & Reichardt, L. F. (1984) *Proc. Natl. Acad. Sci. USA* **81,** 7951–7955.
25. Benowitz, L. I. & Greene, L. A. (1979) *Brain Res.* **162,** 164–168.
26. Oksche, A. & Ueck, M. (1976) in *Physiology of the Amphibia*, ed. Lofts, B. (Academic, New York), Vol. 3, pp. 314–419.
27. Jackson, I. M. D. (1984) in *Neuroendocrine Perspectives*, eds. Müller, E. E. & MacLeod, R. M. (Elsevier, Amsterdam), Vol. 3, pp. 121–159.
28. Bruni, J. E., Clattenberg, R. E. & Millar, E. (1983) *Anat. Anz. Jena* **153,** 53–68.
29. Raivich, G., Zimmermann, A. & Sutter, A. (1985) *EMBO J.* **4,** 637–644.

Proc. Natl. Acad. Sci. USA
Vol. 83, pp. 6184–6187, August 1986
Neurobiology

Aggressive behavior induces release of nerve growth factor from mouse salivary gland into the bloodstream

(submaxillary salivary gland/adrenal gland)

LUIGI ALOE*, ENRICO ALLEVA†, ARIELA BÖHM*, AND RITA LEVI-MONTALCINI*

*Istituto di Biologia Cellulare, Consiglio Nazionale delle Ricerche, Via Romagnosi 18/A, Roma 00196, Italy; and †Section of Neurobehavioral Pathophysiology, Laboratorio di Fisiopatologia di Organo e di Sistema, Istituto Superiore di Sanità, Viale Regina Elena 299, Roma 00161, Italy

Contributed by Rita Levi-Montalcini, March 26, 1986

ABSTRACT Intraspecific fighting induced by 6–8 weeks of social isolation results in massive release of nerve growth factor (NGF) into the bloodstream of adult male mice. The amount of circulating NGF is highly correlated with the number of fighting episodes. Biological, radioimmunological, immunohistochemical, and ultrastructural studies show that NGF is discharged from the salivary gland into the blood within minutes after fighting and reaches the highest level 3–4 hr later. Adrenergic innervation of the salivary gland or adrenalectomy does not abolish the NGF release. Corticotropic hormones do not induce NGF increase in the blood. Daily administrations of highly purified NGF (3 μg per g of body weight) result in a considerable increase in the volume of adrenal glands. These findings are unequivocable evidence for a physiological role of the mouse salivary glands as a major source of blood NGF.

Almost 3 decades ago, it was reported that mouse submaxillary salivary glands (SSGs) synthesize and release an extraordinarily large amount of a nerve growth factor (NGF) molecule (1, 2) endowed with a potent nerve growth-promoting activity on sympathetic and sensory nerve cells (3–6). This finding raised the question whether this molecule serves any specific function and, if so, whether it is discharged into the saliva, into the blood, or into both. All the results obtained so far seem to favor the hypothesis that this salivary product, the NGF, is routed from the gland secretory ducts into the saliva, where it plays some still uncertain function (7, 8). This hypothesis springs from the almost generally agreed upon observation that the injection of pharmacological agents shows NGF not to be present in mouse blood, whereas it is present in large amounts in the saliva (9–11). Structural, ultrastructural, and immunohistochemical studies have also favored this hypothesis and shown that the granular products are localized in the apical part of the cell lining in the gland tubular portion where NGF has been identified (11, 12). However, results from this laboratory provided convincing evidence that, upon pharmacological stimulation (13, 14), NGF from SSG can also be found in the bloodstream.

A recent report that aggressive behavior results in the release into the blood of a large amount of another biologically active protein synthesized in the gland tubular portion, renin (15), prompted us to investigate the effects of aggressiveness induced by social isolation on NGF content in the blood. The present study was aimed at the investigation of this problem and, more importantly, at the possible functional significance of such a large amount of NGF synthesis and release by mouse salivary glands. The results of these studies showed that aggressive behavior induces massive NGF discharge and that the adrenal gland might be a primary target for this endogenous NGF release.

The publication costs of this article were defrayed in part by page charge payment. This article must therefore be hereby marked "*advertisement*" in accordance with 18 U.S.C. §1734 solely to indicate this fact.

MATERIALS AND METHODS

Behavioral Studies. Male mice of a Swiss-derived (CD-1) strain (25–27 g), purchased from Charles River Italy, were used. They were either individually housed for 6–8 weeks or maintained in groups of three (nonisolated mice) in opaque Plexiglas boxes (33 × 13 × 14 cm) with metal tops. Water and standard pellet food were freely available. The animals were kept in an air-conditioned room at 21 ± 1°C and 50% ± 10% relative humidity, using a 9:30 a.m. to 9:30 p.m. lighting schedule. Isolated and nonisolated mice were randomly assigned to the control or fighting condition. Fighting mice were introduced in pairs in a clean box of the same type as the home box, and the number of fighting episodes was recorded. A pilot study showed a high consistency of the scores assigned by four different observers during testing or by the same observer in repeated examinations of videotape records. All scores in the present study were assigned by the same observer. Tests were carried out between 10:00 a.m. and 1:00 p.m. Unless otherwise indicated, the fighting sessions lasted 20 min. (Under these conditions, no significant sign of dominance was observed.) Blood was collected from the retro-orbital plexus with a glass pipette at the end of the fighting sessions (with the exception of the time-course study). All samples were used for biological (5) and radioimmunological (16) assays. Each mouse was used only once.

Surgical and Chemical Treatments. Mice were sialoadenectomized or adrenalectomized under deep sodium pentobarbital (Nembutal) anesthesia 2 days before fighting. For immunosympathectomy, neonates were injected daily during the first postnatal week with antibodies to NGF (17). At the age of 10 weeks they were isolated, and 6 weeks later they were used in fighting sessions. For chemical sympathectomy (18), 6-hydroxydopamine (50 μg per g of body weight) obtained from Fluka was injected intraperitoneally (i.p.) into isolated mice for 2 consecutive days; these animals underwent a fighting session 2 days later. Eye ptosis and the absence of adrenergic nerve terminals in the iris were used as confirmation that complete sympathectomy was achieved (19). Porcine corticotropin (ACTH) purchased from Sigma was dissolved in 0.09% saline [270 international units (IU) in 1 ml of saline]. Adrenal gland (30 in 4 ml) and hypophysis (20 in 2 ml) of adult male mice were homogenized in phosphate-buffered saline in an ice bath using a glass and Teflon homogenizer. The insoluble material was removed by centrifugation at 3000 rpm for 15 min at 4°C. Isolated nonfighting male mice were injected i.p. with (*i*) 0.5 ml of extract of adrenal gland, (*ii*) 0.2 ml of extract of hypophysis, or (*iii*) 27 IU of ACTH. Blood was collected 20–60 min and 120 min after the injection, and serum was assayed for the presence of NGF.

Abbreviations: NGF, nerve growth factor; SSG, submaxillary salivary gland; ACTH, corticotropin.

NGF was prepared by the method of Bocchini and Angeletti (20) and further purified by an additional passage through CM-52 column chromatography resulting in a NGF with undetectable renin activity (21). Adult male mice were injected i.p. for 10 consecutive days with NGF or vehicle solution (saline, 0.09%) (3 μg per g of body weight) and sacrificed on the 11th day for adrenal gland histological examination.

Histological and Histometrical Studies. Adrenal glands and SSGs were fixed in alcoholic Bouins' solution, washed, dehydrated, embedded in Paraplast, sectioned (7 μm thick), and stained in hematoxylin/eosin or toluidine blue. To determine the relative volume of adrenals, serial sections were projected with a camera lucida, and outlines of the cortical and medullary zones were traced on paper. The outlines of all sections were cut out and then weighed on a torsion balance (22). SSGs were also fixed and processed for immunohistochemical localization of NGF (14).

Ultrastructural Studies. Fighting and nonfighting mice were perfused with buffered paraformaldehyde (4%) and glutaraldehyde (1%) fixative. SSGs were removed, washed, and postfixed with 1% OsO_4. For light microscopic observation, tissues were dehydrated, embedded in Epon 812, sectioned (1 μm), and stained with toluidine blue. Ultrathin sections were stained with uranyl acetate and lead citrate and observed with a Philyps 200 electron microscope.

RESULTS

There is no detectable NGF in the whole (100%) blood serum of mice according to our biological and radioimmunological assays (Table 1; see also refs. 4–6). Likewise, blood serum collected from mice isolated for several weeks (control) contains no bioassayable NGF. On the contrary, blood samples from isolated animals that experience fighting (fighting mice) contain high levels of NGF ($\approx$100 ng/ml) and stimulate the outgrowth of nerve fibers in cultured chicken sympathetic and sensory ganglia when they are added to the medium in appropriate dilutions (Fig. 1). This finding indicates that biologically active NGF has been released into the bloodstream. A time-course study showed that the level of NGF released into the blood increases gradually up to 300 ng/ml during the first 2–3 hr and then decreases steadily, reaching basal levels 12–24 hr later (Fig. 2). As shown in Fig. 3, the concentration of released NGF is positively correlated with the frequency of fighting episodes, indicating some relationship between aggressive behavior and serum NGF concentration.

To determine how closely the amount of NGF found in the blood after fighting is related to the release of the NGF present in the SSG, we made histological, immunoperoxidase, and electron microscopic comparisons of the

Table 1. Relationship between number of fighting episodes and NGF serum concentrations of male mice subjected to different experimental conditions

Condition	*n*	Fighting episodes	Serum NGF, ng/ml
Control	6	—	ND
Fighting	10	32.9 ± 3.2	103.5 ± 20.1
Nonisolated	7	31.6 ± 3.3	88.4 ± 12.3
Sialoadenectomized	5	37.2 ± 4.4	ND
Immunosympathectomized	6	35.7 ± 5.1	101.7 ± 23.6
6-Hydroxydopamine	6	38.2 ± 5.9	93.3 ± 35.5
Adrenalectomized	5	32.2 ± 8.9	52.5 ± 13.5
Adrenal extract	8	—	ND
Hypophysis extract	8	—	ND
ACTH	6	—	ND

Values are means ± SEM. ND, not detectable.

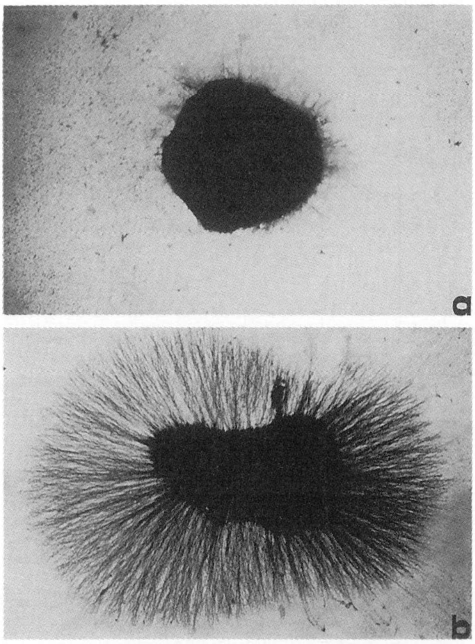

FIG. 1. Neurite outgrowth from chicken sensory ganglia cultured for 24 hr in the presence of 20 μl of blood serum collected from control (*a*) and fighting (*b*) mice. Silver stain. (×60.)

submaxillary granular convoluted tubules of fighting and nonfighting mice. The results of the three morphological studies clearly indicate that there is a depletion of NGF-containing granules in the SSG of fighting mice (Fig. 4). Furthermore, mice sialoadenectomized two days before isolation retained high levels of fighting behavior, but no increase in their blood NGF concentration was found. As shown in Table 1, blood samples collected from fighting mice that underwent chemical or surgical sympathectomy still

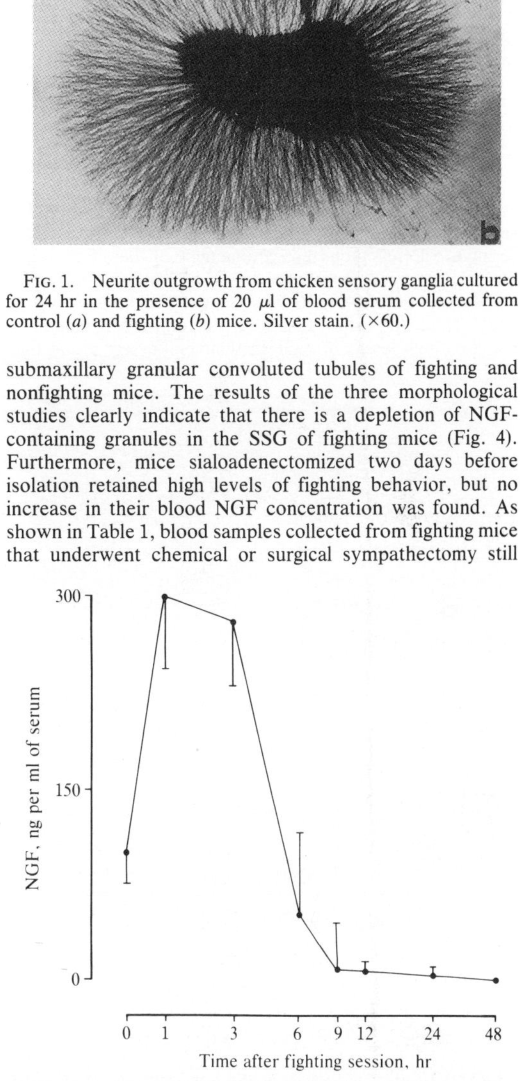

FIG. 2. Time course of NGF release in the bloodstream of fighting mice. Each point represents the mean ± SEM of six mice. Different animals were used for each time point.

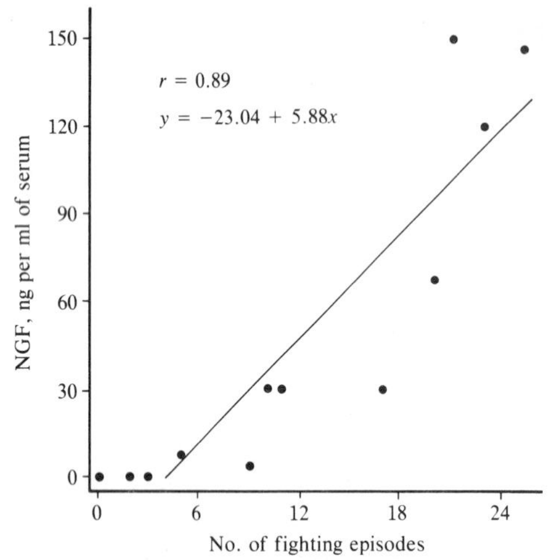

FIG. 3. NGF levels in blood serum of fighting mice as a function of aggressive behavior scored during a 6-min session. NGF values refer to mean levels of the two fighting animals.

contain large amounts of NGF. To assess the role of ACTH and adrenal gland hormones on NGF release, the blood serum of adrenalectomized fighting mice or that of nonfighting mice previously injected with ACTH, hypophysis extract, and adrenal gland extracts was assayed for the presence of NGF. The results show that adrenalectomized fighting mice maintain high levels of NGF in their blood serum and that

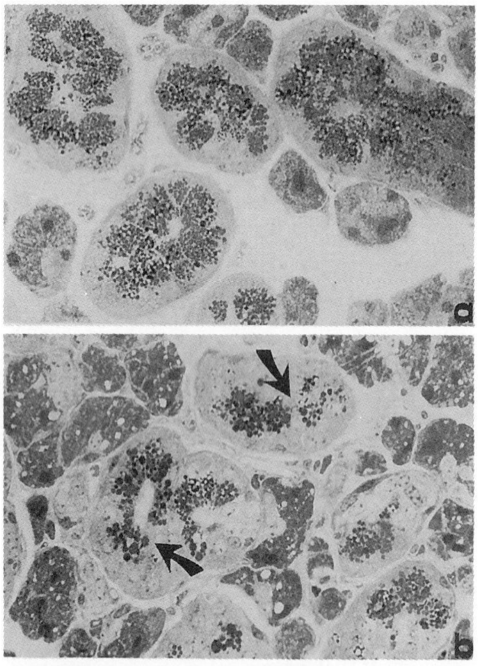

FIG. 4. Histological sections of SSG from control (*a*) and fighting (*b*) mice stained with toluidine blue. Note the depletion of NGF-containing granules in the convoluted tubules of the gland in the fighting animal (arrows). (×320.)

ACTH injection does not cause any NGF release into the serum. Moreover, injection of adrenal gland or hypophysis extract in adult mice did not result in NGF release into the blood (Table 1). It appears, therefore, that neither corticosteroid excess nor ACTH is involved in the NGF release into the blood serum.

As shown in Fig. 5, daily injections of NGF (3 μg per g of body weight) for 10 consecutive days result in marked weight and volume increases of these glands. Both the cortical and medullary zones are larger in weight and volume than those of vehicle-injected and unhandled controls. This effect is more pronounced in the medullary than in the cortical zones of the glands. In all cases, adrenals of NGF-treated mice differed significantly (paired t tests, $P < 0.01$ or less upon Bonferroni's adjustment) from those of the other groups, while glands of saline-injected and unhandled groups did not differ from each other ($P < 0.05$).

DISCUSSION

The results demonstrate that intraspecific fighting induced by social isolation (23, 24) causes massive NGF release from mouse SSG. Nonisolated male mice paired with isolated ones engage in aggressive behavior and likewise show high levels of NGF release. Thus, fighting *per se*, rather than a combined effect of isolation and fighting, produces the NGF release. This release appears to be specifically induced by intraspecific fighting, a form of psychosocial stress (25). An extensive series of experiments using a different stressful condition—i.e., inescapable footshock treatment (26, 27) (number of shocks, 5–100; intensity, 0.1–5 mA; duration, 1–60 sec; intershock interval, 15–600 sec; 1 or 3 days of treatment)—failed to evidence any NGF release into the blood. Moreover, another series of tests using stresses such as cold water swimming, forced immobilization, and forced biting yield similarly negative results.

The release of NGF does not appear to be mediated by the adrenergic innervation of the salivary glands. In fact, blood samples collected from fighting mice immunosympathectomized with antibody to NGF since birth, or chemically sympathectomized with 6-hydroxydopamine as adults, still contain large amounts of NGF. These findings suggest that under the present conditions NGF release into the bloodstream from SSG is not mediated by the adrenergic nerves innervating their tubular portion or, alternatively, by an enhancing of norepinephrine release from postganglionic nerves and the consequent potentiation, as first hypothesized by Cannon and Rosenblueth (28), of the defensive mechanism and/or aggressiveness. Since adrenalectomy did not block NGF release in fighting mice, and injections of commercially available ACTH or adrenal gland and hypophysis extracts in isolated adult male mice did not result in NGF release into the bloodstream (Table 1), it is unlikely that circulating corticosteroids or ACTH could account for this effect.

While the nature and mechanism triggering the NGF release into the bloodstream remain unknown, increase of circulating NGF results in adrenal gland hypertrophy, following administration of NGF, as demonstrated in these studies and suggested in a previous report (22). It is also known that repeated fighting causes an increase in size and weight of adrenal glands (25, 29–31), an effect that in light of present results could be elicited by circulating NGF. The hypertrophic hyperplastic effect, as well as the selective induction of tyrosine hydroxylase caused by NGF treatment in rodent adrenal medullary cells (32–34), lend support to the hypothesis of the role of NGF in the stress syndrome accompanying the aggressive behavior. It remains to be seen whether the adrenal cortical hypertrophy produced by NGF administration to adult mice is a direct NGF-specific corticotropic effect.

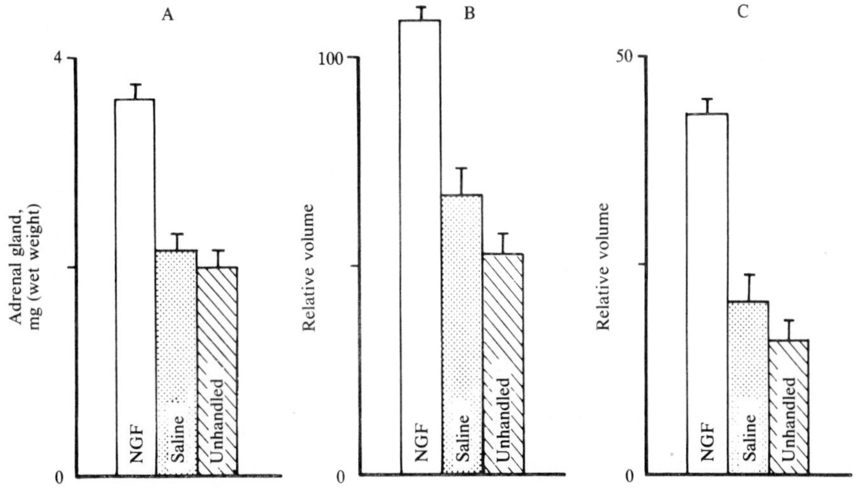

FIG. 5. Histogram of adrenal gland (wet weight) (*A*) and relative volumes of cortical (*B*) and medullary (*C*) zones of mice injected i.p. for 10 consecutive days with purified mouse NGF or vehicle (Saline) solution (3 μg per g of body weight) or left undisturbed (Unhandled). Relative volumes refer to area of paper covered by microscopic projections (×40) of adrenal sections. Values represent weight of paper (mg). At least 12 glands in each treatment group were used.

The results of these studies demonstrate that in adult male mice aggressive behavior, but not other kinds of stressful events, leads to a massive release of NGF from SSG into the blood. Concomitant evidence that exogenous NGF produces a marked increase in the medullary zone of the adrenal gland suggests an effect similar to that of the endogenous release of NGF. To what extent the latter effect may be related to the well-known aggressive behavior of males of this species (24, 35) remains to be investigated.

We thank Dr. C. Cozzari for preparing the NGF used in this study and Ms. Anna Sebastiano for typing the manuscript. This work was supported in part by grants Progetto Finalizzato Medicina Preventiva e Riabilitativa and Progetto Finalizzato Chimica Fine e Secondaria (from the Consiglio Nazionale delle Ricerche to R.L.-M.), and Progetto Patologia Non-Infettiva, Sottoprogetto Fisiopatologia Neurotessutale e Comportamentale (from the Istituto Superiore di Sanitá to E.A.).

1. Levi-Montalcini, R. (1958) in *Chemical Basis of Development*, eds. McElroy, W. D. & Glass, B. (Johns Hopkins Press, Baltimore), pp. 646–664.
2. Cohen, S. (1958) in *Chemical Basis of Development*, eds. McElroy, W. D. & Glass, B. (Johns Hopkins Press, Baltimore), pp. 665–676.
3. Levi-Montalcini, R. & Hamburger, V. (1951) *J. Exp. Zool.* **116,** 321–362.
4. Levi-Montalcini, R. (1952) *Ann. N.Y. Acad. Sci.* **55,** 330–343.
5. Levi-Montalcini, R. & Angeletti, P. U. (1968) *Physiol. Rev.* **48,** 534–565.
6. Thoenen, H. & Barde, Y. A. (1980) *Physiol. Rev.* **60,** 1284–1335.
7. Levi-Montalcini, R. & Angeletti, P. U. (1964) in *Salivary Glands and Their Secretions*, eds. Screebny, L. N. & Meyer, J. (Pergamon, New York), pp. 129–141.
8. Barka, T. (1980) *J. Histochem. Cytochem.* **28,** 836–859.
9. Murphy, R. A., Saide, J. D., Blanchard, M. & Young, M. (1977) *Proc. Natl. Acad. Sci. USA* **74,** 2330–2333.
10. Wallace, L. J. & Partlow, L. M. (1976) *Proc. Natl. Acad. Sci. USA* **73,** 4210–4214.
11. Murphy, R. A., Watson, A. Y., Metz, J. & Forssmann, W. J. (1980) *J. Histochem. Cytochem.* **28,** 890–902.
12. Schwab, M. E., Stoekel, K. & Thoenen, H. (1978) *Cell Tissue Res.* **169,** 289–299.
13. Levi-Montalcini, R. & Aloe, L. (1981) *Horm. Cell Regul.* **5,** 53–72.
14. Aloe, L., Cozzari, C. & Levi-Montalcini, R. (1985) *Brain Res.* **332,** 259–265.
15. Bing, J. & Poulsen, K. (1979) *Acta Physiol. Scand.* **105,** 64–72.
16. Suda, K., Barde, Y. A. & Thoenen, H. (1978) *Proc. Natl. Acad. Sci. USA* **75,** 4042–4046.
17. Levi-Montalcini, R. & Angeletti, P. U. (1966) *Pharmacol. Rev.* **18,** 619–628.
18. Tranzer, J. P. & Thoenen, H. (1968) *Experientia* **24,** 155–156.
19. Malmfors, T. & Thoenen, H. (1971) *6-Hydroxydopamine and Catecholamine Neurons* (North-Holland, Amsterdam).
20. Bocchini, V. & Angeletti, P. U. (1969) *Proc. Natl. Acad. Sci. USA* **64,** 787–794.
21. Cozzari, C., Angeletti, P. U., Lazar, J., Orth, H. & Gross, F. (1973) *Biochem. Pharmacol.* **22,** 1321–1327.
22. Zaimis, E. (1971) in *Nerve Growth Factor and Its Antiserum*, eds. Zaimis, E. & Knight, J. (Atlone, London), pp. 59–70.
23. Valzelli, L. (1973) *Psychopharmacologia* **31,** 305–320.
24. Valzelli, L. (1969) in *Aggressive Behavior*, eds. Garattini, S. & Sigg, E. B. (Excerpta Medica, Amsterdam), pp. 70–76.
25. Axelrod, J. & Reisine, T. D. (1984) *Science* **224,** 452–459.
26. Alleva, E., De Acetis, L., Amorico, L. & Bignami, G. (1983) *Behav. Neural. Biol.* **39,** 78–104.
27. Anisman, H., Kokkinidis, L. & Sklar, L. S. (1981) in *Theory in Psychopharmacology*, ed. Cooper, S. J. (Academic, London), Vol. 1, pp. 65–102.
28. Cannon, W. B. & Rosenblueth, T. (1937) *Autonomic Neuro-Effector Systems* (Macmillan, New York).
29. Christian, J. J. (1955) *Am. J. Physiol.* **82,** 292–300.
30. Axelrod, J., Mueller, R. A., Henry, J. P. & Stephens, P. M. (1970) *Nature (London)* **225,** 1059–1060.
31. Hucklebridge, F. H., Gamal-el-Din, L. & Brain, P. F. (1981) *Behav. Neural. Biol.* **33,** 345–363.
32. Eranko, O., Soinila, A. & Paivarinta, H. (1980) *Histochemistry and Cell Biology of Autonomic Neurons, SIF Cells and Paraneurons* (Raven, New York).
33. Lillien, L. E. & Claude, P. (1985) *Nature (London)* **317,** 632–634.
34. Otten, U., Schwab, M., Gagnon, C. & Thoenen, H. (1977) *Brain Res.* **133,** 291–303.
35. Mackintosh, J. H. (1970) *Anim. Behav.* **18,** 177–183.

Proc. Natl. Acad. Sci. USA
Vol. 86, pp. 8555–8559, November 1989
Neurobiology

Nerve growth factor mRNA and protein increase in hypothalamus in a mouse model of aggression

(trophic interactions/gene transcription/*in situ* hybridization)

M. G. SPILLANTINI*, L. ALOE†, E. ALLEVA‡, R. DE SIMONE†, M. GOEDERT*, AND R. LEVI-MONTALCINI†

*Medical Research Council Laboratory of Molecular Biology, Hills Road, Cambridge CB2 2QH, United Kingdom; †Istituto di Neurobiologia, Consiglio Nazionale delle Ricerche, Viale Marx 15, 00156 Rome, Italy; and ‡Laboratorio di Fisiopatologia di Organo e Sistema, Istituto Superiore di Sanità, Viale Regina Elena 299, 00161, Rome, Italy

Contributed by R. Levi-Montalcini, July 6, 1989

ABSTRACT **The effects of intermale aggressive behavior induced by social isolation on the level of nerve growth factor (NGF) mRNA and protein were investigated in central and peripheral mouse tissues. A large increase in NGF mRNA and protein was observed in hypothalamus, with no changes in cerebral cortex, hippocampus, and cerebellum. No change in NGF mRNA levels was found in heart, spleen, vas deferens, and submaxillary salivary gland. The cellular localization of NGF mRNA in the central nervous system was investigated by *in situ* hybridization. Numerous nerve cells were specifically labeled in preoptic and ventrolateral nuclei of the hypothalamus, as well as in the cornu ammonis region of the hippocampus and throughout all layers of the cerebral cortex, with the highest concentration in layer III. The present results firmly establish that nerve cells constitute the major source of NGF in the brain. They also open the way to understanding the regulation of NGF biosynthesis in the central nervous system.**

Nerve growth factor (NGF) is a protein required for the development and maintenance of sympathetic neurons and the majority of neural crest-derived sensory nerve cells (1). It functions as a retrograde trophic messenger between target tissues and their innervating nerve cells. Accordingly, NGF mRNA has been detected in peripheral tissues of various mammalian species (2, 3), whereas NGF receptor mRNA is mostly produced by the corresponding sensory and sympathetic ganglia (4). The actions of NGF are not limited to the peripheral nervous system. Both NGF and its receptor are produced throughout the central nervous system of various mammalian species (4–13) and NGF is retrogradely axonally transported from neocortex and hippocampus to basal forebrain magnocellular cholinergic nerve cells (14, 15), where it exerts trophic actions (16–18). Moreover, the degeneration of basal forebrain neurons observed following transection of the septohippocampal pathway is markedly reduced by NGF (19–21). These results suggest that NGF may function as a retrograde trophic messenger for basal forebrain cholinergic nerve cells. However, the widespread occurrence of NGF mRNA throughout the central nervous system suggests additional functions.

Whereas a great deal of information is available about the sites of synthesis of NGF in the central nervous system and about some of its biological actions, relatively little is known about the factors regulating NGF synthesis. Aloe *et al.* (22) showed that intermale aggressive behavior in mice induced by social isolation leads to a marked release of biologically active NGF from the submaxillary salivary gland into the bloodstream. We now report that the same behavioral paradigm leads to a large and specific increase in NGF mRNA and protein in the hypothalamus, where NGF mRNA was found in nerve cells by *in situ* hybridization. Part of this work has been presented in abstract form.§

MATERIALS AND METHODS

Behavioral Studies. Male mice of a Swiss-derived (CD-1) strain, with previous breeding experience, were purchased from Charles River Italy. They were either housed individually in Plexiglas boxes (33 × 13 × 14 cm) with metal tops for 6–8 weeks or maintained in groups of four or five (nonisolated mice); tap water and standard pellet food were freely available. The animals were kept in an air-conditioned room at 21 ± 1°C and 50 ± 10% humidity, using a 0930–2130 lighting schedule (red lights on from 0930 to 2130). Isolated and nonisolated male mice were randomly assigned to the nonfighting or fighting group. Fighting mice were introduced in pairs into a clean box of the same type as the home box, and the number of fighting episodes was recorded during a single 20-min session. A pilot study showed a high consistency in the scores assigned by four different observers during testing or by the same observer in repeated examinations of videotape records. In the present study all scores were assigned by the same observer. Nonfighting controls were treated similarly to fighting mice but did not undergo the fighting session. Tests were carried out between 1000 and 1300, i.e., during the first part of the red-light period (under these conditions no significant dominance was observed). Blood was collected from the retroorbital plexus with a glass pipette at the end of the fighting or nonfighting session. All samples were used for biological assays. Each mouse was used only once.

Surgical Procedures and Preparation of Tissue Samples. The animals were divided into the following groups: (*i*) isolated fighting; (*ii*) isolated nonfighting; (*iii*) nonisolated fighting; (*iv*) nonisolated nonfighting. For the NGF assay mice were decapitated, their brains were removed, and cerebral cortex, hippocampus, and hypothalamus (2 mm lateral and 2 mm frontal to the third ventricle) were homogenized in distilled water (10% wt/vol). Tissues were then centrifuged at 40,000 × *g* and supernatants were used for biological assays. For RNA blotting and *in situ* hybridization, brains were removed, dissected, quick-frozen in liquid nitrogen, and stored at −70°C until use.

Preparation of 2.5S NGF and of NGF Antibodies. NGF was isolated as the 2.5S subunit from the submaxillary glands of adult male Swiss albino mice (23) and further purified by an additional passage through a carboxymethylcellulose (CM-52, Whatman) column, resulting in NGF with undetectable renin activity (24). Polyclonal antibodies were raised in rabbits by multiple intradermal and intramuscular injections

The publication costs of this article were defrayed in part by page charge payment. This article must therefore be hereby marked "*advertisement*" in accordance with 18 U.S.C. §1734 solely to indicate this fact.

Abbreviations: NGF, nerve growth factor; SCG, superior cervical ganglion.
§Aloe, L. & Alleva, E., 5th Capo Boi Conference on Neuroscience, Villasimius, Sardinia, Italy, May 24–31, 1987 (abstr.).

of NGF, as described (25), and purified by affinity chromatography (26). The titer of the antiserum was evaluated in a bioassay employing 8-day-old chicken embryo dorsal root ganglia (27).

Biological Assays for NGF. The biological activity of NGF was estimated by a chicken dorsal root ganglion bioassay (27) or a bioassay using sympathetic neurons from fetal rat superior cervical ganglia (SCG). Briefly, nerve cells from SCG of 17- to 18-day rat fetuses were mechanically isolated, dissociated, and cultured in collagen-coated Falcon tissue culture dishes containing Dulbecco's modified Eagle's medium alone or with various amounts of NGF. In the absence of NGF only a small number of neurite-bearing cells appeared, while the addition of as little as 0.02 ng of NGF per ml yielded a detectable increase in the number of neurites. With increasing amounts of NGF the number of responding cells increased, resulting in a dose–response curve that was used to estimate the amount of NGF present in the tissue samples to be tested.

RNA Blot Analysis. A 629-base-pair *Pst* I fragment from the mouse NGF cDNA clone 12E4 (28) was subcloned into bacteriophage M13mp18 and single-stranded DNA probes were prepared as described (6). RNA was extracted by a modification of the guanidinium isothiocyanate/hot phenol method (29) and enriched for poly(A)$^+$ RNA by using oligo(dT)-cellulose. Poly(A)$^+$ RNA was electrophoresed in a 2.2 M formaldehyde-containing 0.9% agarose gel and directly transferred to a nylon membrane (Hybond-N, Amersham). Blots were prehybridized for at least 3 hr at 42°C in 50% (vol/vol) formamide/5× SSC/50 mM sodium phosphate buffer, pH 7/0.2% Ficoll/0.2% polyvinylpyrrolidone/0.2% bovine serum albumin/0.1% SDS containing acid/base-cleaved salmon sperm DNA (250 μg/ml) and heparin (200 μg/ml). (SSC is 150 mM sodium chloride/15 mM sodium citrate.) Hybridization was carried out for 16 hr at 42°C in prehybridization solution containing 10% dextran sulfate and 10^6 cpm of denatured labeled probe per ml. Filters were washed twice at room temperature in 2× SSC/0.1% SDS and twice at 60°C in 0.1× SSC/0.1% SDS. After autoradiography and dehybridization, blots were rehybridized at 42°C with a mouse actin cDNA clone (30) labeled by the random-primer technique (31). In some experiments rehybridization was performed using the cDNA clone H1 for the human NGF receptor (32), labeled by random priming.

***In Situ* Hybridization.** Coronal mouse brain sections were cut on a cryostat at 10–15 μm and processed (30). The sections were prehybridized for 2 hr at room temperature and then hybridized at 42°C in prehybridization solution containing 10% dextran sulfate and denatured, ^{35}S-labeled probe in the anti-mRNA-sense orientation (5 × 10^5 cpm per section). In control experiments probes in the mRNA-sense orientation (5 × 10^5 cpm per section) were used. The sections were washed at room temperature in three changes of 2× SSC and then washed overnight at 42°C in 50% formamide/0.6 M sodium chloride/10 mM Tris, pH 7.4/1 mM EDTA. Dithiothreitol (20 mM) was included in all hybridization and washing solutions. The slides were dehydrated in 70% ethanol, air-dried, dipped in Ilford K-5 emulsion, and kept at −20°C for 1 month. They were developed in safe-light conditions with Kodak D19 developer, postfixed in Carnoy's solution, stained with thionin, dehydrated, cleared in Histoclear (National Diagnostic, Manville, NJ), and mounted with Depex (BDH).

RESULTS

In isolated nonfighting animals, mRNA encoding NGF was found in high amounts in submaxillary gland and in lower amounts in peripheral tissues such as vas deferens, spleen, and heart. In the central nervous system the largest amounts

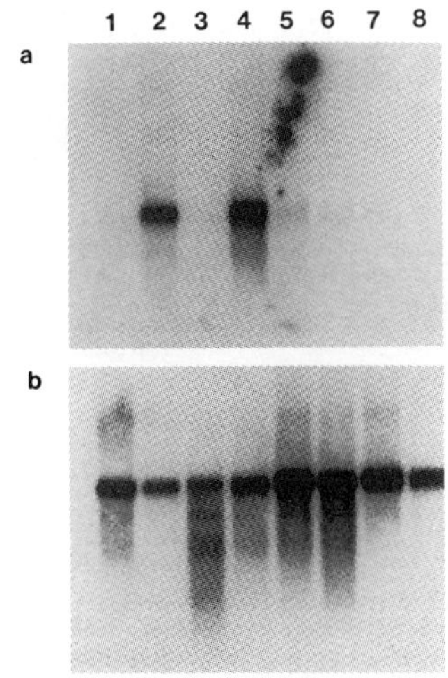

FIG. 1. Northern blot analysis of poly(A)$^+$ RNA from hypothalamus of isolated nonfighting and isolated fighting mice. ^{32}P-labeled NGF DNA (*a*) or β-actin DNA (*b*) was used as a probe. The fighting mice were killed at various times after the end of the fighting session, and the nonfighting mice served as controls. Lanes: 1, nonfighting mice, 30 min; 2, fighting mice, 30 min; 3, nonfighting mice, 3 hr; 4, fighting mice, 3 hr; 5, nonfighting mice, 6 hr; 6, fighting mice, 6 hr; 7, nonfighting mice, 12 hr; 8, fighting mice, 12 hr.

were present in cerebral cortex and hippocampus, with lower levels in hypothalamus and cerebellum (Figs. 1–3). A single 1.3-kilobase band was observed in all cases. The levels of mRNA for NGF were investigated in mice that had been kept isolated for 6–8 weeks, put together in pairs for a single 20-min fighting session, and killed at various times thereafter. All the results were normalized by rehybridizing the same blot with a mouse actin cDNA probe. A single fighting session resulted in a marked increase in NGF mRNA levels in the hypothalamus, when compared with the hypothalamic levels in controls that had been kept isolated for the same period of time but were not subjected to a fighting session

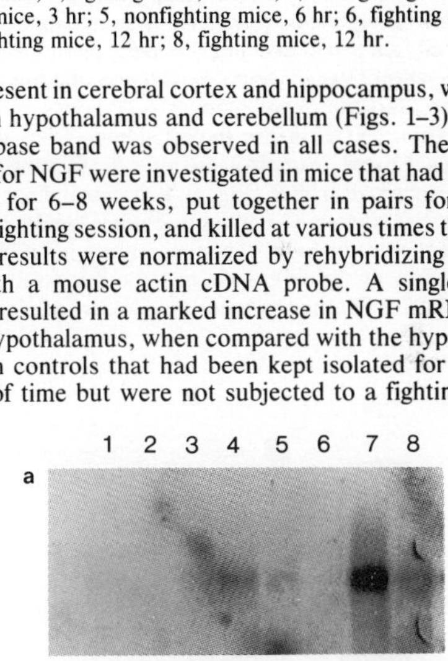

FIG. 2. Northern blot analysis of poly(A)$^+$ RNA from different brain regions of isolated nonfighting and isolated fighting mice. ^{32}P-labeled NGF DNA (*a*) or β-actin DNA (*b*) was used as a probe. The fighting mice were killed 30 min after the end of the fighting session, and the nonfighting mice served as controls. Lanes: 1, cerebellum, nonfighting mice; 2, cerebral cortex, nonfighting mice; 3, hypothalamus, nonfighting mice; 4, hippocampus, nonfighting mice; 5, cerebellum, fighting mice; 6, cerebral cortex, fighting mice; 7, hypothalamus, fighting mice; 8, hippocampus, fighting mice.

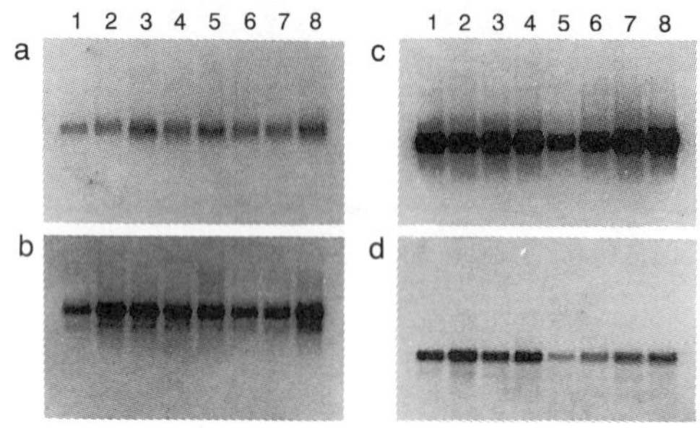

FIG. 3. Northern blot analysis of poly(A)$^+$ RNA from heart (*a* and *b*) and submaxillary gland (*c* and *d*) of isolated nonfighting and isolated fighting mice. ^{32}P-labeled NGF DNA (*a* and *c*) or β-actin DNA (*b* and *d*) was used as a probe. The fighting mice were killed at various times after the end of the fighting session, and the nonfighting mice served as controls. Lanes: 1, nonfighting mice, 30 min; 2, fighting mice, 30 min; 3, nonfighting mice, 3 hr; 4, fighting mice, 3 hr; 5, nonfighting mice, 6 hr; 6, fighting mice, 6 hr; 7, nonfighting mice, 12 hr; 8, fighting mice, 12 hr.

(Figs. 1 and 2). The increase in NGF mRNA was apparent after 30 min, peaked after 3 hr, and declined to reach control levels after 6 hr and 12 hr (Fig. 1). No change in NGF receptor mRNA levels was observed in hypothalamus (data not shown).

Biological assays demonstrated that hypothalamic extracts from isolated mice induced neurite outgrowth from both chicken dorsal root ganglia and rat SCG and that the fighting episode increased the hypothalamic NGF levels. After the fighting session, NGF in the hypothalamus from isolated mice increased 7-fold; in nonisolated mice the fighting session led to a 60% increase in NGF (Fig. 4). Isolation did not appear to raise the NGF levels in nonfighting mice, while it significantly potentiated the hypothalamic increase in NGF levels observed following fighting. This may be due to the much higher number of fighting episodes that occurs as a result of

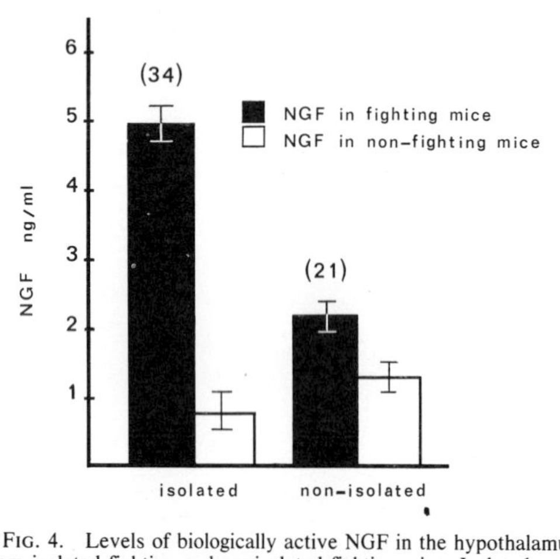

FIG. 4. Levels of biologically active NGF in the hypothalamus from isolated fighting and nonisolated fighting mice. Isolated nonfighting and nonisolated nonfighting mice served as controls. Numbers in parentheses refer to the means of the total number of fighting episodes occurring during the fighting session. Bars represent the means ± SEM of five animals. Two-way analysis of variance showed a signficant interaction between the isolation and the fighting factor ($P < 0.001$) for the NGF levels. Numbers of fighting episodes were significantly higher in the isolated groups than in the nonisolated groups ($P < 0.001$).

isolation. Fighting induced a large increase in biologically active NGF in serum from isolated mice, when compared with the isolated nonfighting group, in agreement with previous results (22). The neurite outgrowth induced by both serum and hypothalamic extracts from chicken dorsal root ganglia and rat SCG was completely inhibited by NGF antibodies.

The increase in NGF mRNA observed in isolated mice after fighting was restricted to the hypothalamus; no such effect was seen in cerebral cortex, hippocampus or cerebellum, with the 30-min time point shown in Fig. 2. Similarly, NGF mRNA levels were unchanged in heart (Fig. 3) as well as spleen and vas deferens (data not shown) from isolated fighting mice. Despite the massive release of NGF from the submaxillary gland of isolated fighting mice, no effect on NGF mRNA levels was observed in the submaxillary gland over the time period studied (Fig. 3).

The cellular localization of NGF mRNA was investigated by *in situ* hybridization in the hypothalamus from isolated fighting mice; the time points chosen were 30 min or 3 hr after the fighting session. When a probe in the anti-mRNA-sense orientation was used numerous nerve cells were labeled; only background labeling was observed with a probe in the mRNA-sense orientation (Fig. 5). The labeled cells were located in the magnocellular preoptic nucleus and the vent-

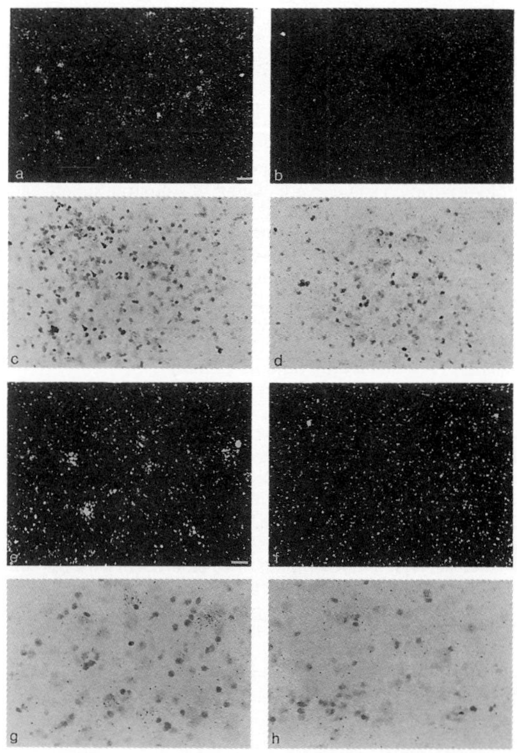

FIG. 5. Dark- and light-field photomicrographs of the cellular localization of NGF mRNA in hypothalamus from isolated fighting mouse. Hybridization probe was ^{35}S-labeled NGF DNA in the anti-mRNA-sense orientation (*a*, *c*, *e*, and *g*) or mRNA-sense orientation (*b*, *d*, *f*, and *h*). The fighting mice were killed 3 hr after the end of the fighting session. Note the presence of hybridization-positive cells in *a* and *c* (arrowheads) and in *e* and *g*, but not in *b* and *d* or in *f* and *h*. [Bar in *a* (for *a–d*) = 120 μm; bar in *e* (for *e–h*) = 80 μm.]

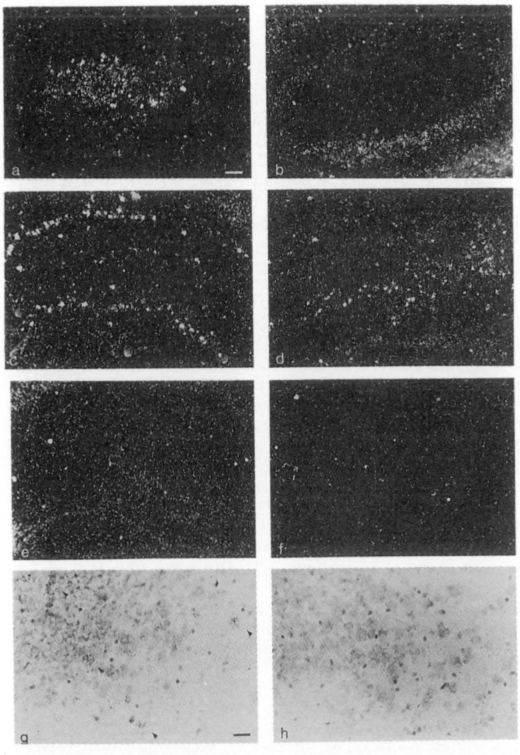

FIG. 6. Dark- and light-field photomicrographs of the cellular localization of NGF mRNA in serial coronal sections (frontal to rostral) of the hippocampal formation from isolated fighting mouse. Hybridization probe was ^{35}S-labeled NGF DNA in the anti-mRNA-sense orientation (*a–d* and *g*) (arrows) or mRNA-sense orientation (*e, f,* and *h*). Note the presence of numerous hybridization-positive cells in the cornu ammonis region (*a–c* and *g*) and in the subiculum (*d*) and their absence in the dentate gyrus (*a–c*); *e* represents the control to *c*, *f* the control to *a*, and *h* the control to *g*. [Bar in *a* (for *a–f*) = 120 μm; bar in *g* (for *g* and *h*) = 70 μm.]

rolateral nucleus of the hypothalamus. No specific labeling of glial cells was observed. Besides the hypothalamus, hybridization-positive cells were observed in hippocampal formation (Fig. 6) and cerebral cortex (Fig. 7). Strong specific labeling of nerve cells in the subiculum and in the cornu ammonis region was observed at all levels of the hippocampus; in the cornu ammonis region the strongest labeling was seen in layers CA1 and CA3 (Fig. 6). Granule cells in the dentate gyrus remained hybridization-negative (Fig. 6). In the cerebral cortex, numerous hybridization-positive cells were observed through all layers, with the largest number of positive cells located in layer III (Fig. 7). The size and distribution of these cells indicate that they represent nerve cells, possibly pyramidal cells. Only background labeling was observed with a probe in the mRNA-sense orientation (Fig. 7).

DISCUSSION

Previous studies (22) showed that intraspecific fighting induced by social isolation results in a massive release of biologically active NGF from the submaxillary gland of adult male mice into the bloodstream and that the amount of circulating NGF is highly correlated with the number of fighting episodes. In this report we show that the same experimental paradigm leads also to a marked increase in NGF mRNA in the hypothalamus. This increase was found to be specific, as it was not found in other central nervous system areas investigated, such as cerebral cortex, hippocampus, and cerebellum. No change in NGF mRNA levels was observed in vas deferens, spleen, heart, or submaxillary gland. The lack of change in the latter tissue indicates that the massive release of NGF into the bloodstream in fighting mice is not accompanied by a rapid change in NGF gene transcription. The large increase in NGF mRNA in the hypothalamus was accompanied by an increase in biologically active NGF, as evidenced by the ability of NGF antibody to block neurite outgrowth elicited from sensory and sympathetic ganglia by hypothalamic extracts from isolated fighting mice.

The sites of synthesis of NGF in the central nervous system are poorly understood. Earlier observations (33) indicated that NGF is produced by astrocytes in primary cultures taken from various rat brain regions. However, many cell types start to produce NGF once put into tissue culture (34).

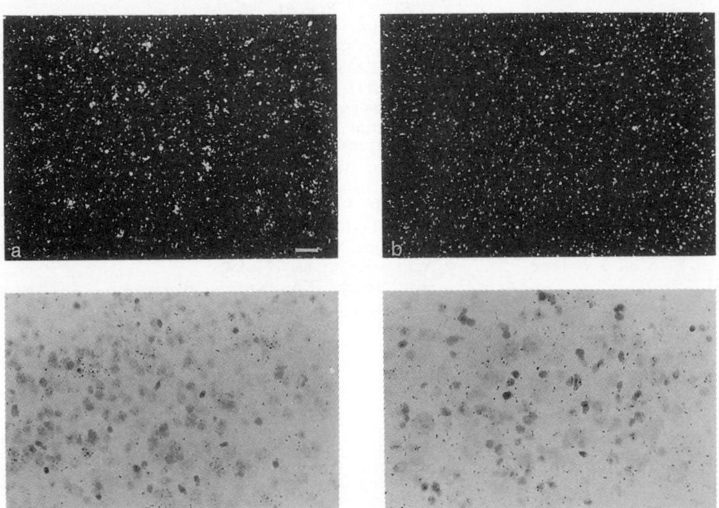

FIG. 7. Dark- and light-field photomicrographs of the cellular localization of NGF mRNA in cerebral cortex from isolated fighting mouse. Probe was ^{35}S-labeled NGF DNA in the anti-mRNA-sense orientation (*a* and *c*) or mRNA sense orientation (*b* and *d*). Note the presence of hybridization-positive cells in *a* and *c*. [Bar in *a* (for *a* and *b*) = 130 μm; bar in *c* (for *c* and *d*) = 60 μm.]

Subsequent immunohistochemical studies performed on rat brain with NGF antibodies did not distinguish whether the immunoreactive material was present in nerve cells or glial cells (35). By *in situ* hybridization on mouse and rat hippocampus, NGF mRNA has been localized to pyramidal and granule cells (36–38); however, the specificity of some of these results has been questioned (39). In the present study we found NGF mRNA in pyramidal cells at all levels of the mouse hippocampal formation, whereas granule cells were consistently hybridization-negative. The specificity of the hybridization was ascertained by washing the tissue sections at high stringency; unlike other studies (36–38), the present study used a probe in the mRNA-sense orientation as a control, and this probe always gave only background labeling. Besides pyramidal cells in the hippocampus, nerve cells in the cerebral cortex were also specifically labeled; the size and distribution of these cells suggested that they were pyramidal cells. *In situ* rehybridization detected numerous specific hybridization-positive cells in the hypothalamus from isolated aggressive mice, mainly in the preoptic and ventrolateral nuclei. No specific labeling of glial cells was observed in cerebral cortex, hippocampus, or hypothalamus. Taken together, these results thus strongly suggest that nerve cells constitute the major source of NGF in the central nervous system.

The physiological significance of the hypothalamic increase in NGF mRNA remains to be established. Exogenous NGF enhances the secretion of adrenocorticotropin from the pituitary gland (40). At present, nothing is known about the factors that regulate NGF biosynthesis in the central nervous system. Previous studies performed on the basal forebrain magnocellular cholinergic system and its projections have failed to find a change in NGF mRNA levels following denervation (6, 41). In the peripheral nervous system, sciatic nerve transection leads to a large increase in NGF mRNA in nerve segments proximal and distal to the lesion (42); this effect is the direct consequence of the release of interleukin 1 from macrophages (43). The lack of change in NGF mRNA levels in peripheral tissues from fighting mice makes it seem unlikely that circulating interleukin 1 could be responsible for the observed increase in NGF mRNA in the hypothalamus. At present, we cannot exclude the possibility that this increase could be mediated by the interleukin 1 that is intrinsic to the hypothalamus (44).

The identification of the substance(s) responsible for the increase in hypothalamic NGF mRNA and protein levels in fighting mice will be important for an understanding of the physiological function of NGF in the central nervous system and could be of practical value in view of a possible therapeutic use of NGF.

1. Levi-Montalcini, R. (1987) *EMBO J.* **6,** 1145–1154.
2. Shelton, D. L. & Reichardt, L. F. (1984) *Proc. Natl. Acad. Sci. USA* **81,** 7951–7955.
3. Heumann, R., Korsching, S., Scott, J. & Thoenen, H. (1984) *EMBO J.* **3,** 3183–3189.
4. Buck, C. R., Martinez, H. J., Black, I. B. & Chao, M. V. (1987) *Proc. Natl. Acad. Sci. USA* **84,** 3060–3063.
5. Korsching, S., Auburger, G., Heumann, R., Scott, J. & Thoenen, H. (1985) *EMBO J.* **4,** 1389–1393.
6. Goedert, M., Fine, A., Hunt, S. P. & Ullrich, A. (1986) *Mol. Brain Res.* **1,** 85–92.
7. Shelton, D. L. & Reichardt, L. F. (1986) *Proc. Natl. Acad. Sci. USA* **83,** 2714–2718.
8. Whittemore, S. R., Ebendal, T., Lärkfors, L., Olson, L., Seiger, Å., Strömberg, I. & Persson, H. (1986) *Proc. Natl. Acad. Sci. USA* **83,** 817–821.
9. Large, T. H., Bodary, S. C., Clegg, D. O., Weskamp, G., Otten, U. & Reichardt, L. F. (1986) *Science* **234,** 352–355.
10. Taniuchi, M., Schweitzer, J. B. & Johnson, E. M. (1986) *Proc. Natl. Acad. Sci. USA* **83,** 1950–1954.
11. Richardson, P. M., Verge Issa, V. M. K. & Riopelle, R. J. (1986) *J. Neurosci.* **6,** 2312–2321.
12. Hefti, F., Hartikka, J., Salvatierra, A., Weiner, W. J. & Mash, D. C. (1986) *Neurosci. Lett.* **69,** 37–41.
13. Goedert, M., Fine, A., Dawbarn, D., Wilcock, G. K. & Chao, M. V. (1989) *Mol. Brain Res.* **5,** 1–7.
14. Schwab, M. E., Otten, U., Agid, Y. & Thoenen, H. (1979) *Brain Res.* **168,** 473–483.
15. Seiler, M. & Schwab, M. E. (1984) *Brain Res.* **300,** 33–39.
16. Honegger, P. & Lenoir, D. (1982) *Dev. Brain Res.* **3,** 229–238.
17. Gnahn, H., Hefti, F., Heumann, R., Schwab, M. E. & Thoenen, H. (1983) *Dev. Brain Res.* **9,** 45–52.
18. Hefti, F., Hartikka, J., Eckenstein, F., Gnahn, H., Heumann, R. & Schwab, M. (1985) *Neuroscience* **14,** 55–68.
19. Hefti, F. (1986) *J. Neurosci.* **6,** 2155–2162.
20. Kromer, L. F. (1987) *Science* **235,** 214–216.
21. Williams, L. R., Varon, S., Peterson, G. W., Wictorin, K., Fischer, W., Björklund, A. & Gage, F. H. (1986) *Proc. Natl. Acad. Sci. USA* **83,** 9231–9235.
22. Aloe, L., Alleva, E., Böhm, A. & Levi-Montalcini, R. (1986) *Proc. Natl. Acad. Sci. USA* **83,** 6184–6187.
23. Bocchini, G. & Angeletti, P. U. (1969) *Proc. Natl. Acad. Sci. USA* **64,** 787–794.
24. Cozzari, C., Angeletti, P. U., Lazar, J., Orth, H. & Gross, F. (1973) *Biochem. Pharmacol.* **22,** 1321–1327.
25. Aloe, L., Cozzari, C. & Levi-Montalcini, R. (1985) *Brain Res.* **332,** 259–265.
26. Stoeckel, K., Gagnon, C., Guroff, G. & Thoenen, H. (1976) *J. Neurochem.* **26,** 1207–1211.
27. Greene, L. A. (1974) *Neurobiology* **6,** 286–292.
28. Ullrich, A., Gray, A., Berman, C. & Dull, T. J. (1983) *Nature (London)* **303,** 821–825.
29. Feramisco, J. R., Smart, J. E., Burridge, K., Helfman, D. M. & Thomas, G. P. (1982) *J. Biol. Chem.* **257,** 11026–11031.
30. Goedert, M. (1987) *EMBO J.* **6,** 3627–3632.
31. Feinberg, A. P. & Vogelstein, B. (1983) *Anal. Biochem.* **132,** 6–13.
32. Johnson, D., Lanahan, A., Buck, C. R., Sehgal, A., Morgan, C., Mercer, E., Bothwell, M. & Chao, M. (1986) *Cell* **47,** 545–554.
33. Lindsay, R. M. (1979) *Nature (London)* **282,** 80–82.
34. Harper, G. P., Al-Saffar, A. M., Pearce, F. L. & Vernon, C. A. (1980) *Dev. Biol.* **77,** 379–390.
35. Ayer-Le Lievre, C. S., Ebendal, T., Olson, L. & Seiger, Å. (1983) *Med. Biol.* **61,** 296–304.
36. Rennert, P. D. & Heinrich, G. (1986) *Biochem. Biophys. Res. Commun.* **138,** 813–818.
37. Ayer-Le Lievre, C. S., Olson, L,, Ebendal, T., Seiger, Å. & Persson, H. (1988) *Science* **240,** 1389–1391.
38. Whittemore, S. R., Friedman, P. L., Larhammer, D., Persson, H., Gonzalez-Carvajal, R. & Holets, V. R. (1988) *J. Neurosci. Res.* **20,** 403–410.
39. Thoenen, H., Bandtlow, C. & Heumann, R. (1987) *Rev. Physiol. Biochem. Pharmacol.* **109,** 145–178.
40. Otten, U., Baumann, J. B. & Girard, J. (1979) *Nature (London)* **282,** 413–415.
41. Korsching, S., Heumann, R., Thoenen, H. & Hefti, F. (1986) *Neurosci. Lett.* **66,** 175–180.
42. Heumann, R., Korsching, S., Bandtlow, C. & Thoenen, H. (1987) *J. Cell. Biol.* **104,** 1623–1631.
43. Lindholm, D., Heumann, R., Meyer, M. & Thoenen, H. (1987) *Nature (London)* **330,** 658–659.
44. Breder, C. D., Dinarello, C. A. & Saper, C. B. (1988) *Science* **240,** 321–324.

Reprinted from
TINS – October 1986

473

Nerve growth factor as a paradigm for other polypeptide growth factors

Rita Levi-Montalcini and Pietro Calissano

The last decade has witnessed a veritable explosion in the number of studies on polypeptide growth factors (PGFs) because of their ability to promote the proliferation and/or differentiation of different cell types[1]. A PGF's typical effect is to cause 'growth' in its target cells, where growth may be defined as an increase in the number (hyperplasia), in the size (hypertrophy) of cells, or, in the case of neurons in the extension of axons and dendrites. By the end of this growth process, cells have usually undergone terminal differentiation. The object of this article is to review recent developments in the study of one of the PGFs – nerve growth factor (NGF). This factor was initially considered a specific effector molecule for two neural crest derivatives, sensory and sympathetic nerve cells. Work during the last decade has shown that NGF activity also extends to CNS neurons and to non-neuronal cells such as chromaffin and mast cells.

Mapping and biological response of NGF target cells

If we define an NGF target cell as one that undergoes morphological changes and/or an altered functional activity in response to this factor, cells listed below belong to this category. It was the long sympathetic neurons that first revealed the existence of this diffusible factor[2], whose powerful differentiative activity was shown by precocious axonal production and elongation of nerve fibers in chick embryos and in an *in-vitro*-devised bioassay[3], leading to the name NGF[4]. A mitogenic effect of NGF was also postulated, following description of NGF-induced increase in the volume of the sensory and sympathetic ganglia[5], although later studies showed that the increase was merely the result of a reduction in natural cell death during the early stages of differentiation[6]. However, more recent experiments have shown that in some cells, such as chromaffin cells, NGF does have a mitogenic activity[7]. It has also been shown that NGF plays an essential trophic role, given the devastating effects of its removal by immunological[8,9], pharmacological[10] or surgical means[11]. The massive cell death that follows any one of these experimental manipulations, also known as sympathectomy, occurs only during the early ontogenetic stages. However, NGF continues to play a supporting role in fully differentiated cells. To our knowledge, the absolute dependency on NGF of immature sensory and sympathetic cells remains the most dramatic experimental evidence to date of a PGF playing an essential trophic role in its target cells. The same long sympathetic neurons have also revealed a third important property of NGF – its ability to direct growing or regenerating axons along its concentration gradient. This tropic activity, clearly seen in both *in-vivo*[12] and *in-vitro* studies[13], can be experimentally dissociated from the trophic effect of NGF, and was the first evidence of a diffusible factor having a chemotactic influence on nerve fibers. As *in-vitro* studies performed on individual neurons cultured under different conditions have shown, the type of neurotransmitter expressed by a particular neuron is determined by another diffusible factor released in the medium by non-neuronal cells, and does not depend on NGF[14].

The short sympathetic neurons in the ganglia innervating the male and the female genito-urinary tracts are only weakly responsive to NGF and only slightly damaged by its antibodies[15]. A unique property that they share with SIFs (see below) and adrenal chromaffin cells is their marked sensitivity to steroid hormones, which in the case of these neurons, belong to the sex gland class[16].

SIF cells (small intensely fluorescent cells) exhibit all the intermediate properties ranging between chromaffin and sympathetic nerve cells. Evidence has been obtained suggesting that they are precursors of the latter. These cells were first considered to be unresponsive to NGF, but more recently have been shown to respond to this factor by neurite outgrowth and conversion into neurons[17].

Adrenal chromaffin cells are the most precocious NGF target cells in terms of their ontogenic response: injection of the factor into rat fetuses results in their differentiation into chromaffin cells[18]. When added *in vitro* to dissociates of neonatal rodent adrenal medulla, NGF first causes the immature chromaffin cells to undergo several replicative cycles[7] and then to differentiate into neurite-bearing cells[19]. An even more impressive effect has been evidenced in a clonal cell line derived from a rat pheochromocytoma tumor that, when grown in the presence of NGF, undergoes mitotic arrest and differentiation into nerve cells[20]. Thus, NGF can exert opposite effects on the same type of cell, acting as a mitogen on normal chromaffin cells and as an antimitotic agent on their neoplastic counterpart. We can imagine this dual, opposite effect resulting from NGF's differentiating action. Whereas differentiation into neuronal-like cells necessarily involves mitotic arrest, the same event in normal chromaffin cells may require a certain number of rounds of replication, as is the case with most of the target cells of other PGFs.

A similar effect may be at work when NGF acts upon mast cells. *In-vivo* injections of the factor into newborn rats results in a very large increase in the number of mast cells in various organs and tissues[21]. Under proper culture conditions, the *in-vitro* incubation of newborn rat spleen dissociates in the presence of 25–50 ng of NGF leads to the formation of cell colonies of which 60–70%, at the end of a one month period, are composed of mast cells[22]. It is not yet clear whether this NGF effect, which extends the factor's spectrum of action to non-neuronal cells, is the result of a generalized effect on mast cell precursors or of an action exerted on a limited number of clones by way of a clonal selection mechanism not unlike the one by which lymphocytes produce antibodies.

Sensory neurons form two cell populations – dorso-medial and ventro-lateral (DM and VL)[5] – which differ in their response to NGF. The dependence of DM cells, and, to a lesser extent, of the VL cells on the factor is very similar to that of the long sympathetic neurons; the trophic, differentiative and trophic action of NGF on these cells is thoroughly documented[23–25].

CNS neurons have only recently come to the forefront as NGF target cells. The most extensive studies in this regard have been on cholinergic neurons. NGF, when locally injected in the brain of adult rodents, is carried via retrograde axonal transport[26] to the cell perikaryon and causes: (1) an increase of neurotransmitter synthesis[27,28]; (2) activation of cholinergic pathways involved in behavioural activity[29]; and (3) *in-vitro* neuron survival and neurite outgrowth[30]. The presence of endogenous NGF in adult rat CNS, in particular in the hippocampus and cortex, and of NGF-specific receptors in several rat CNS areas, strongly suggests that NGF also plays a role centrally[31–34]. This possibility is supported by the finding that injections of NGF antibodies in rat fetuses induce a severe post-natal neuroendocrine syndrome[35].

Studies on lower vertebrates have shown that the administration of NGF results in an enhanced regeneration of surgically transected optic nerves in fish[36] and in *Xenopus laevis* tadpoles, and in the precocious differentiation and hypertrophy of monoaminergic and peptidergic neurons, as well as cells that line the third ventricle and belong to the periventricular organs[37].

Evolutionary conservation of NGF and its receptor

The NGF genes have now been cloned in four different species – chick, mouse, cattle and man – and have been shown to have 80–90% sequence homology[38–40]. The NGF gene codes for a larger polypeptide precursor, 307 amino acids in length[38] in mouse, while the actual biologically active protein is made up of two identical, non-covalently associated, 118-amino-acid-long proteins.

In the mouse submandibular gland, which was the first large source of NGF to provide the factor for structural and sequence studies, the 118-amino-acids dimer can be isolated in the form of a 7S complex[41] made of three distinct proteins: the NGF (which in this complex is named the β-subunit) and two other proteins – a γ-subunit with a serine protease activity, which is likely to be involved in some step(s) of the processing of the 307 amino acid precursor, and a α-subunit with no known biological function. NGF may lose an arginine at the N-terminus and eight amino acids at C-terminus during its isolation, but it retains its full biological activity. This latter form, which was actually the first to be isolated[42] and sequenced[43] is known as 2.5S NGF.

What do other base pairs that constitute the NGF gene code for? They do not code for any known protein nor for the α- and γ-subunits[38]. Given that the NGF precursor possesses different potential cleavage sites, it is possible that it is processed in different ways in distinct types of cells, in a manner analogous to other hormones and neuromodulators. This possibility lengthens the list of potential neuronal and non-neuronal targets of NGF gene product(s). Whatever the nature of the processing of NGF precursor(s), the 118-amino-acid NGF dimer has a structure and biological activity that is highly conserved, especially if one considers the domain(s) responsible for interaction with the receptor. Thus, the similarity of the NGFs from different species, as evaluated by cross-reaction studies using polyclonal and monoclonal antibodies, is less pronounced than the differences in the specific activities of their corresponding proteins, as assessed by the classical bioassay using either chick embryo sensory or sympathetic ganglia, or rat cell lines such as PC12 cells. The most plausible hypothesis that could explain this fact is that the structure of the domain(s) that interacts in a specific way with the receptor has been more strongly conserved than other portions of the molecule[25,40]. This, in turn, suggests the hypothesis (now amenable to experimental verification thanks to the availability of cDNA clones of the NGF receptor[44]) that the receptor has also been conserved throughout evolution, or that at least the sequence coding for the region responsible for its interaction with the NGF has. The molecular cloning of the human NGF receptor, in addition to providing essential information on its chromosomal location[45], opens the way to settle a long-lasting controversy over the number of NGF-receptor types. Hybridization studies employing a 5-kb Ba HI genomic fragment have revealed a single 4-kb mRNA coding for the NGF receptor in various cell lines, suggesting once again, as kinetic studies had done previously[46,47], that high or low affinity and high or low molecular weight forms of NGF receptors are coded for by the same gene. Furthermore, the isolation of a cosmid DNA clone that spans the NGF-receptor gene, and produces NGF-receptor expression in mouse L cell lines, will allow one to introduce the NGF receptor into various cell types and to study the properties that a cell requires in order to respond to NGF.

NGF's mechanism of action

The mechanism by which NGF elicits its numerous and diversified responses has been, and to a large extent still is, one of the most elusive aspects of this PGF. Some information has been gained using suitable model systems such as the PC12 cell line (Ref. 20), and more will soon be available as recombinant DNA technology allows the analysis of genes that are expressed[48] or suppressed[49] in response to the NGF/receptor interaction. For the sake of simplicity, let us consider NGF's most impressive effect, namely that of inducing neurite outgrowth. No one doubts that this event is absolutely dependent on the concerted action of cytoskeletal components, such as microtubules (MTs) and microfilaments (MFs), which are essential to axonal growth and transport and to growth cone motility as well. Neurofilaments (NFs), the third major component of the cytoskeleton, appear to be less involved in the process, at least in its first phases; they intervene only to consolidate, with their more rigid structure, the growth work performed by MTs and MFs. However, all of these filamentous structures are only 'work generating structures'[25], which must be continuously directed to perform their task. What kind of message(s) does NGF send them?

The temporal sequence of events may be visualized as follows. NGF binds to receptors (70–80 kDa) that are present on the perikaryon and on the growth cone of target cells. Binding is followed by some sort of physicochemical change that results in an altered affinity of the ligand for its receptors[46], the clustering of these complexes[50], and their internalization[51]. At the terminal endings, the complexes are transported in a retrograde manner to the perikaryon[11], which is vitally dependent on this process since discontinuation will lead to the cell's death. A small proportion of the NGF-receptor complexes associate with certain cytoskeletal elements[52], while another small but detectable pool is found in proximity to or inside the nuclear compartment[53–55].

During this sequence of events, control of cytoskeletal function by NGF could be achieved via two temporally and spatially distinct mechanisms. In target cells that have never before been exposed to NGF, such as naive PC12 cells, or sensory or sympathetic neurons at early ontogenetic stages, NGF may stimulate the expression of genes coding for proteins such as tau[56] or other MAPs (microtubule- and microfilament-associated proteins), whose function seems to be that of triggering or modulating the assembly and organization of cytoskeletal elements. After prolonged exposure to NGF, and once the cells have accumulated a critical amount of these proteins, a second, more rapid and easily reversible mechanism could control the function of MAPs via post-translational modifications such as phosphorylation[57]. Such control may be achieved locally by the pool of NGF/receptor complexes associated with cytoskeletal[52] elements and, as has been suggested on the basis of *in-vitro* findings, might even involve the direct action of NGF[58]. Neosynthesis of MAPs may correspond to the so-called 'priming period'[59], while the faster, secondary response could be expressed by the regenerative events of previously primed cells. Oncogene products such as the pp60$^{v\text{-}src}$; v-Ki-*ras* or H-*ras* proteins (see below) might intervene at different points along this train of events, the former at the level of phosphorylation of MAPs and the other proteins involved in the modulation of cytoskeletal elements, and the latter two along the pathway that operates as a transduction system of NGF-receptor interaction.

Since NGF exerts a multiplicity of diversified actions, the simplest conclusion one might draw is that its function, once bound to the receptor, consists in selecting one out of many cellular patterns of response. This selection would, in the case of a developing or neurite-regenerating neuron, switch on the molecular mechanisms that control the cytoskeletal elements and thus manifest itself as neurite outgrowth or regeneration. A similar mechanism may be functioning in the tropic response of growth cones. The control of adhesivity and rate of motility might vary according to NGF-receptor occupancy by a gradient of NGF molecules released by target cells. In a neuron that is already differentiated, a different cell response would be selected, which might, for example, regulate the transport of essential nutrients[60], neurotransmitter synthesis, etc. The selection of this last program would be devoted to the regulation of the trophism proper to the neuron. Thus, NGF/receptor interaction could elicit an entire repertoire of responses, whose number depends on the sets of genes that are programmed to respond to NGF.

The role of the submandibular gland and other large NGF sources

The male mouse submandibular glands synthesize NGF and release it into the saliva in large quantities. In the blood, the factor is normally almost undetectable, a fact that appeared to rule out the possibility that salivary NGF could play a hormone-like role. However, such a role has recently been suggested by the discovery that the amount of NGF present in the blood increases by several orders of magnitude during episodes of intraspecific fighting of male mice[61]. Since injections of exogenous NGF induce both a marked increase in the volume of the adrenal glands and an enhanced activity of tyrosine-hydroxylase[62], the key enzyme in catecholamine synthesis, the emission of salivary NGF into the blood might play a functional part in the intraspecific fighting stress syndrome.

No hint is presently available as to the functional significance, if any, of the large amount of NGF in snake venom, guinea-pig prostrate and bull seminal fluid. A unifying hypothesis is that whatever the type of cell, the NGF-receptor complexes are endowed with the property of modulating the function of cytoskeletal elements, in the way similar to that postulated for neurite outgrowth. Monoclonal antibodies directed against NGF receptors have revealed the existence of cross-reacting antigens in different organs and tissues, even far away from terminal endings of sympathetic or sensory origin. Since the cytoskeleton operates as a transmission belt between signals impinging on the cell membrane and intracellular events (mitosis, exo- and endo-cytosis, control of gene expression, etc.), NGF could activate or inhibit any one of these processes in different, NGF-receptor-bearing cells.

NGF, PGFs and oncogenes

Most of the proteins coded for by oncogenes are involved to some extent in the train of events leading to cell proliferation. One of the most striking features of PGFs, which has often aided in their identification, is that of triggering or increasing the rate of this cellular event. It was thus only reasonable to suppose that there was some sort of link between certain oncogenes and PGFs. But the actual extent of this link turned out to be far greater than anyone expected: it was discovered that platelet-derived growth factor (PDGF), the receptors for the epidermal growth factor (EGF), the granulocyte macrophage colony stimulating factor (GM-CSF) and steroid hormones all share strong sequence homologies and functional activities with retroviral oncogenes such as v-*sis*, v-*erb*-B, c-*fms* and v-*erb*-A[64–67]. In this startling connection between PGFs and oncogenes, NGF occupies a peculiar position. It has been demonstrated that infection of PC12 cells with retrovirus carrying the *src* oncogene of the Rous sarcoma virus, which codes for a phosphotyrosine kinase known as pp60$^{v\text{-}src}$, causes the cells to differentiate in a way that bears marked resemblances to NGF-induced neuronal differentiation[68]. Moreover, the cellular homologue of this oncogene, c-*src*, is expressed in post-mitotic neurons in concentrations that are 15–20-fold higher, and with levels of specific activity that are 6–12-fold greater, than those in fibroblasts[69]. Other experiments have shown that PC12 cells infected with Kirsten or Harvey murine sarcoma viruses undergo morphological differentiation and mitotic arrest[70]. Finally, an effect analogous to that induced by v-*src* or by Kirsten sarcoma virus is also obtained when the protein product of H-*ras*, an oncogene whose cellular counterpart is thought to be involved in the amplification pathways of ligand/receptor interactions, is microinjected into PC12 cells[71]. Surprisingly though, the same effect does not take place upon microinjection of the protoncogene form of the *ras* protein, a fact that leads one to the puzzling conclusion that a differentiation event such as neurite outgrowth can be triggered by an oncogene product but not by its cellular counterpart. The simplest conclusion that may be drawn from these facts is that the products of these three oncogenes mimick or modulate the functions of proteins that play an essential role in neuronal differentiation, as the high levels of c-*src* present in post-mitotic neurons seem to suggest, and that NGF-receptor complexes act in the same train of events.

A crucial task will now be to prove that analogous effects occur in non-neoplastic NGF target cells, such as sensory and sympathetic cells. Whereas a positive answer would lead one to conclude that a step of the molecular mechanism that underlies promotion of neurite outgrowth has been uncovered, a negative finding would warn one against extrapolating from an otherwise excellent model system such as the neoplastic PC12 cells to neuronal NGF target cells, and imply that axonal growth and elongation are the result of more than one molecular pathway.

Together these findings add credence to the conclusion that there is a close link between various PGFs and oncogenes. Now that the genes that code for NGF and its receptor have been cloned, it will be possible to assess whether a mutated or overexpressed version of the NGF receptor can lead to the transformation of the recipient cells, as has been proven to be the case for other PGFs. Such a finding, in turn, would add further strength to the more and more widely accepted notion that not only are PGFs to differentiation what oncogene products are to transformation, but that the opposite is also true, i.e. that somewhat altered versions of PGFs or of their receptors may cause transformation while some products of oncogenes may induce differentiation.

Final remarks and perspectives

The discovery of NGF and the growing list of other, subsequently identified PGFs, has brought the existence of a new class of polypeptides to the fore. PGFs differ from classical hormones by virtue of: (1) their multifocal sites of synthesis; (2) the limited number of identified target cells; and (3) their actions being exerted mainly during final stages of differentiation. Among this ever-growing class of polypeptides, NGF can be visualized as a paradigm since it embodies most of the properties of PGFs. Its discovery has anticipated the existence of other similar PGFs acting on neuronal and satellite cells. The future isolation and characterization of these PGFs should have a tremendous impact on basic and applied neuroscience, since it will then be possible to grow different neuronal populations *in vitro*, to modulate their phenotypic expression and, if necessary, to repair damaged functions by applying one of the many procedures now already under investigation.

Acknowledgements

Research in the author's laboratory was supported by grants from Progetto Finalizzato Chimica Fine and from Progetto Finalizzato Oncologia.

Selected references

1 James, R. and Bradshaw, R. A. (1984) *Annu. Rev. Biochem.* 53, 259–292

2 Levi-Montalcini, R. (1952) *Ann. N.Y. Acad. Sci.* 55, 330–343

3 Levi-Montalcini, R., Meyer, H. and Hamburger, V. (1954) *Cancer Res.* 16, 29–57

4 Cohen, S., Levi-Montalcini, R. and Hamburger, V. (1954) *Proc. Natl Acad. Sci. USA* 40, 1014–1018

5 Levi-Montalcini, R. and Hamburger, V. (1953) *J. Exp. Zool.* 123, 233–288

6 Hendry, I. A. (1975) *J. Neurocytol.* 5, 337–349

7 Lillien, L. E. and Claude, P. (1985) *Nature* 317, 632–634

8 Levi-Montalcini, R. and Booker, B. (1960) *Proc. Natl Acad. Sci. USA* 46, 384–391

9 Cohen, S. (1960) *Proc. Natl Acad. Sci. USA* 46, 302–311

10 Angeletti, P. U. and Levi-Montalcini, R. (1970) *Proc. Natl Acad. Sci. USA* 65, 114–121

11 Hendry, I. A., Stockel, K., Thoenen, H. and Iversen, L. L. (1974) *Brain Res.* 68, 103–121

12 Levi-Montalcini, R. (1976) *Prog. Brain Res.* 45, 235–258

13 Gundersen, R. W. and Barrett, J. N. (1979) *Science* 206, 1079–1080

14 Patterson, P. H. (1978) *Annu. Rev. Neurosci.* 1, 1–17

15 Bjerre, B. and Rosengren, E. (1974) *Cell Tissue Res.* 150, 299–322

16 Owman, C., Sjoberg, N-O. and Sjostrand, N. O. (1974) in *Amine Fluorescence Histochemistry* (Fujiwara, M. and Tanaca, C., eds), pp. 47–66, Igaku Shoin

17 Doupe, A. J., Patterson, P. H. and Landis, S. C. (1985) *J. Neurosci.* 5, 2143–2160

18 Aloe, L. and Levi-Montalcini, R. (1979) *Proc. Natl Acad. Sci. USA* 76, 1246–1250

19 Unsicker, K., Kirsch, B., Otten, U. and Thoenen, H. (1978) *Proc. Natl Acad. Sci. USA* 75, 3498–3502

20 Greene, L. A. and Tischler, A. S. (1976) *Proc. Natl Acad. Sci. USA* 73, 2424–2428

21 Aloe, L. and Levi-Montalcini, R. (1977) *Brain Res.* 133, 358–366

22 Böhm, A. and Aloe, L. (1986) *Acad. Naz. dei Lincei* 80, 1–6

23 Johnson, E. M., Gorin, P. D., Brandeis, L. D. and Pearson, J. (1980) *Science* 210, 916–918

24 Hamburger, V. and Yip, J. W. (1984) *J. Neurosci.* 4, 767–774

25 Calissano, P., Cattaneo, A., Aloe, L. and Levi-Montalcini, R. (1984) in *Hormonal Proteins and Peptides* (Vol. XII) (Li, C. H., ed.), pp. 1–56, Academic Press

26 Seiler, M. and Schwab, M. (1984) *Brain Res.* 30, 33–39

27 Korsching, S., Auburger, G., Heumann, R., Scott, J. and Thoenen, H. (1985) *EMBO J.* 4, 1389–1393

28 Mobley, W. C., Rutkowski, J. L., Tennekoon, G. I., Buchanan, K. and Johnston, M. V. (1985) *Science* 229, 284–287

29 Alleva, E., Aloe, L. and Laviola, G. (1986) *Dev. Brain Res.* 28, 278–281

30 Martinez, H. J., Dreyfus, C. F., Jonakait, G. M. and Black, I. (1985) *Proc. Natl Acad. Sci. USA* 82, 7777–7781

31 Heifti, F., Hartikka, J., Eckenstein, F., Gnahn, H., Heuman, R. and Schwab, M. (1985) *Neuroscience* 14, 55–68

32 Whittemore, S. R., Ebendal, T., Larkfors, L., Olson, L., Seiger, A., Stromberg, I. and Persson, H. (1986) *Proc. Natl Acad. Sci. USA* 83, 817–821

33 Shelton, D. L. and Reichardt, L. F. (1986) *Proc. Natl Acad. Sci. USA* 83, 2714–2718

34 Taniuchi, M., Schweitzer, J. B. and Johnson, E. M., Jr (1986) *Proc. Natl Acad. Sci. USA* 83, 1950–1954

35 Aloe, L., Cozzari, C., Calissano, P. and Levi-Montalcini, R. (1981) *Nature* 291, 413–415

36 Turner, J. E., Delaney, R. K. and Johnson, J. E. (1980) *Brain Res.* 197, 319–330

37 Levi-Montalcini, R. and Aloe, L. (1985) *Proc. Natl Acad. Sci. USA* 82, 7111–7115

38 Scott, J., Selby, M., Urdea, M., Quiroga, M., Bell, G. and Rutter, W. J. (1983) *Nature* 302, 538–540

39 Ullrich, A., Gray, A., Berman, C. and Dull, T. J. (1983) *Nature* 303, 821–823

40 Meier, R., Becker-Andre, M., Gotz, R., Heuman, R., Shaw, A. and Thoenen, H. *EMBO J.* (in press)

41 Varon, S., Nomura, J. and Shooter, E. M. (1967) *Biochemistry* 6, 2202–2215

42 Bocchini, V. and Angeletti, P. U. (1967) *Proc. Natl Acad. Sci. USA* 64, 787–792

43 Hogue Angeletti, R. and Bradshaw, R. A. (1971) *Proc. Natl Acad. Sci. USA* 68, 2417–2420

44 Chao, M. V., Bothwell, M. A., Ross, A. H., Koprowski, H., Lanahan, A. A., Buck, C. R. and Sehgal, A. (1986) *Science* 232, 518–521

45 Huebner, K., Isobe, M., Chao, M., Bothwell, M., Ross, A. H., Finan, J., Hoxie, J. A., Sehgal, A., Buck, C. R., Lanahan, A., Nowell, P. C., Koprowski, H. and Croce, C. (1986) *Proc. Natl Acad. Sci. USA* 83, 1403–1407

46 Landreth, G. and Shooter, E. M. (1980) *Proc. Natl Acad. Sci. USA* 77, 4751–4755

47 Cattaneo, A., Biocca, S., Nasi, S. and Calissano, P. (1983) *Eur. J. Biochem.* 135, 285–290

48 Levi, A., Eldridge, J. D. and Paterson, B. M. (1985) *Science* 229, 393–395

49 Biocca, S., Cattaneo, A. and Calissano, P. (1984) *Proc. Natl Acad. Sci. USA* 81, 2080–2084

50 Levi, A., Schechter, Y., Neufeld, E. J. and Schlessinger, J. (1980) *Proc. Natl Acad. Sci. USA* 77, 3469–3473

51 Calissano, P. and Shelanski, M. L. (1980) *Neuroscience* 5, 1033–1039

52 Vale, R. D., Ignatius, M. J. and Shooter, E. M. (1985) *J. Neurosci.* 5, 2762–2770

53 Yankner, B. A. and Shooter, E. M. (1979) *Proc. Natl Acad. Sci. USA* 76, 1269–1273

54 Marchisio, P. C., Naldini, L. and Calissano, P. (1980) *Proc. Natl Acad. Sci. USA* 77, 1656–1660

55 Bernd, P. and Greene, L. A. (1983) *J. Neurosci.* 3, 631–643

56 Drubin, D. G., Feinstein, S. C., Shooter, E. M. and Kirschner, M. W. (1985) *J. Cell Biol.* 101, 1799–1807

57 Burstein, D. E., Seeley, P. J. and Greene, L. A. (1985) *J. Cell Biol.* 101, 862–870

58 Nasi, S., Cirillo, D., Naldini, L., Marchisio, P. C. and Calissano, P. (1982) *Proc. Natl*

Acad. Sci. USA 79, 820–825
59 Burnstein, D. and Greene, L. A. (1978) *Proc. Natl Acad. Sci. USA* 75, 6059–6064
60 Horii, Z. I. and Varon, S. (1977) *Brain Res.* 124, 121–133
61 Aloe, L., Alleva, E., Böhm, A. and Levi-Montalcini, R. *Proc. Natl Acad. Sci. USA* (in press)
62 Otten, U., Schwab, M., Gagnon, C., and Thoenen, H. (1977) *Brain Res.* 133, 291–303
63 Bruni, A., Bigon, E., Boarato, E., Leon, A. and Toffano, G. (1982) *FEBS Lett.* 138, 190–192
64 Doolittle, R., Hunkapiller, M. W., Hood, L. E., DeVare, S. G., Robbins, K. L., Aaronson, S. A. and Antoniades, H. N. (1983) *Science* 221, 275–276
65 Downward, J., Yraden, Y., Mayes, E., Scrace, G., Totty, N., Stockwell, P., Ullrich, A., Schlessinger, J. and Waterfield, M. D. (1984) *Nature* 307, 521–527
66 Sherr, C. J., Rettenmier, C. W., Sacca, R., Roussel, M. F., Look, T. A. and Stanley, E. R. (1985) *Cell* 41, 665–676
67 Weinberger, C., Hollenberg, S. M., Rosenfeld, M. G. and Evans, R. M. (1985) *Nature* 318, 670–673
68 Alemà, S., Casalbore, P., Agostini, E. and Tatò, F. (1985) *Nature* 316, 557–559
69 Brugge, J. S., Cotton, P. C., Queral, A. E., Barrett, J. N., Nonner, D. and Keane, R. (1985) *Nature* 316, 554–557
70 Noda, M., Ko, M., Ogura, A., Liu, D., Amano, T., Takano, T. and Ikawa, Y. (1985) *Nature* 318, 73–75
71 Bar-Sagi, D. and Feramisco, J. R. (1985) *Cell* 42, 841–848

Rita Levi-Montalcini and Pietro Calissano are at the Istituto di Biologia, CNR, Via Romagnosi 18/A, 00196 Rome, Italy.

Science **237**, 1154–1162 (1987).

The Nerve Growth Factor 35 Years Later

RITA LEVI-MONTALCINI

Embryogenesis is in some way a model system. It has always been distinguished by the exactitude, even punctilio, of its anatomical descriptions. An experiment by one of the great masters of embryology could be made the text of a discourse on scientific method. But something is wrong, or has been wrong. There is no theory of development in the sense in which Mendelism is a theory that accounts for the results of breeding experiments. There has therefore been little sense of progression or timeliness about embryological research. Of many papers delivered at embryological meetings, however good they may be in themselves, one too often feels that they might have been delivered 5 years beforehand without making anyone much the wiser, or deferred for 5 years without making anyone conscious of a great loss.—P. MEDAWAR

THIS FEELING OF FRUSTRATION SO INCISIVELY CONVEYED by these considerations by P. Medawar (*1*), in the passage quoted above, pervaded, in the forties, the field of experimental embryology, which had been enthusiastically acclaimed in the mid-thirties when the upper lip of the amphibian blastopore brought this area of research to the forefront of the biological stage. The side branch of experimental neuroembryology, which had stemmed from the common tree and was entirely devoted to the study of the trophic interrelations between neuronal cell populations, and between these and the innervated organs and tissues, was then in its initial vigorous growth phase. It, in turn, suffered from a sharp decrease in the enthusiasm that had inflamed the pioneers in this field, ever since R. G. Harrison delivered his celebrated lecture on this topic at the Royal Society in London in 1935 (*2*). Although the alternate "wax and wane" cycles are the rule rather than the exception in all fields of human endeavor, in that of biological sciences the "wane" is all too often indicative of a justified loss of faith in the rational and methodical approach that had at first raised so much hope.

A brief account of the state of the art of experimental neuroembryology in the forties, when interest in this approach to the study of the developing nervous system was waning, is a prerequisite for understanding the sudden unforeseeable turn of events which resulted in the discovery of the nerve growth factor.

Experimental Neuroembryology in the Early Forties

The replacement, in 1934 by Viktor Hamburger, of the chick embryo for the amphibian larva as object of choice for the analysis of the effects of limb bud extirpation on spinal motor neurons and sensory nerve cells innervating the limbs (*3*) signaled the beginning of a long series of investigations centered on the analysis of this and related experimental systems in avian embryos. Here I shall list only the major advantages offered by the chick embryo over amphibian larvae as objects of neurological investigations.

The author is a professor at the Institute of Cell Biology, National Research Council, Rome, Italy. This article is the lecture she delivered in Stockholm, Sweden, 8 December 1986, when she received the Nobel Prize in Physiology or Medicine, which she shared with Stanley Cohen. It is published here with the permission of the Nobel Foundation and will also be included in the complete volume of *Les Prix Nobel en 1986* as well as in the series *Nobel Lectures* (in English) published by Elsevier Publishing Company, Amsterdam and New York. The lecture by Dr. Cohen will appear in a later issue.

The avian nervous system is built according to a more elaborate design than that of amphibians, and it lends itself to a more rigorous analysis of its nerve centers than does the nervous system of lower vertebrates. Extensive fundamental studies on the nervous system of the chick embryo, with use of the invaluable silver-specific techniques of Ramón y Cajal and his co-workers, extended recently by myself and other investigators (*4, 5*), provided a very accurate blueprint of most nerve centers and their developmental history during neurogenesis. This allowed the detection of even small infractions of normal developmental rules in experimentally manipulated embryos. Unlike the ontogenetic processes in amphibians, these processes in chick embryos unfold according to a rigidly scheduled time sequence that never departs from the anticipated. It is therefore possible to compare the central and peripheral nerve centers of experimental and control specimens in embryos incubated under the same temperature and other environmental conditions. The analysis, in amphibian larvae, was extended to the brain, spinal cord, and peripheral nervous system in various experimental situations. In the chick embryo, it was mainly confined to the study of the effects that extirpation of limb primordia or implantation of additional wing or leg buds had on innervating motor and sensory nerve centers.

In 1934, Viktor Hamburger published an article (*3*) on the effects of wing bud extirpation on the development of the brachial spinal motor segment and sensory dorsal root ganglia innervating the wing. He came to the conclusion that the hypoplasia of motor nerve cells of the ventral horn and of other nerve cells of the same hemisection of the spinal cord resulted from lack of stimuli centripetally transmitted by nerve fibers of the first differentiated neurons. These normally exert a regulatory effect on proliferation and differentiation of neighboring nerve cells. A reinvestigation of the effects produced by limb bud extirpation suggested a different control mechanism of the developing nerve centers by peripheral tissues. Through serial studies of silver-stained embryos, the conclusion was reached that the severe hypoplasia of nerve centers deprived of their fields of innervation resulted from death of differentiated neurons and not from failure of recruitment of neurons from a pool of still uncommitted nerve cell precursors (*6, 7*). In 1947, Hamburger invited me to join him for the purpose of reinvestigating this problem. This invitation marked the beginning of a 30-year period that I spent at Washington University and of my lifelong friendship with Viktor. Our 1949 article (*7*) confirmed the hypothesis previously submitted by G. Levi and myself. The satisfaction of this confirmation of an important theoretical issue and the successful analysis of other neuroembryological problems (*8, 9*) was, however, perturbed by the realization of the low resolution power of the techniques in our possession for in-depth exploration of the tremendously complex neurogenetic processes. The temptation to abandon the experimental analysis of the developing nervous system and move into the phage field, which was in full blossom in the forties, did not, however, take hold, thanks to unpredictable and fortunate events that occurred at the same time and opened a new era in developmental neurobiology.

The Unexpected Break: A Gift from Malignant Tissues

In a 1948 article, a former student of Viktor Hamburger, Elmer Bueker, reported the results of a bold and imaginative experiment consisting in the grafting of fragments of mouse sarcoma 180 into the body wall of 3-day chick embryos. The histological study of the embryos fixed 3 to 5 days later showed that sensory nerve fibers emerging from adjacent dorsal root ganglia had gained access into the neoplastic tissue, whereas no motor nerve fibers entered the tumor (*10*). The author concluded that histochemical properties of the fast-growing mouse sarcoma offered a favorable field for growth of sensory fibers. This condition, in turn, resulted in a slight but consistent increase in the volume of these ganglia as compared to that of homologous ganglia innervating the wing on the contralateral side. Viktor and I reinvestigated this remarkable phenomenon adopting the method I had developed during my first neuroembryological studies—namely, that of a daily inspection of control and experimental embryos serially sectioned and impregnated with a specific silver technique. Our results confirmed those reported by Bueker but at the same time uncovered other effects elicited by grafts of the mouse tumor which hardly fit in with the hypothesis that they were in the same range and of the same nature as those called forth by transplants of normal embryonic tissues. They differed from the latter in the following significant respects: sympathetic fibers as well as sensory fibers entered the neoplastic tissues, where they built a network of extraordinarily high density. Nerve fibers branched at random between tumor cells without, however, establishing synaptic connections with them. Sensory and sympathetic ganglia innervating the tumor underwent a progressive increase in volume, attaining—in the case of sympathetic ganglia—a size about six times larger than that of same control ganglia (*11*).

Subsequent experiments uncovered another astonishing deviation from the norm in embryos bearing transplants of mouse sarcoma 180 or of another tumor of identical origin, known as sarcoma 37. Embryonic viscera, which in normal specimens are devoid of innervation—such as the mesonephroi—or which become scarcely innervated only in late developmental stages—such as the sex glands, the thyroid, the parathyroid, and the spleen—were loaded with sympathetic nerve fibers at early embryonic stages (*12*). A patent infraction of all developmental rules came to light with the finding of thick sympathetic fiber bundles inside the veins of the host where they protruded in the form of large neuromas obstructing blood circulation (Fig. 1). All sympathetic chain ganglia, and not only ganglia adjacent to or in direct connection with neoplastic tissues, were enormously enlarged. The hypothesis that these anomalous effects could result from the release by neoplastic cells of a soluble, diffusible agent that altered the differentiating and growth properties of its target cells received full confirmation from experiments in which one or the other mouse sarcoma was transplanted onto the

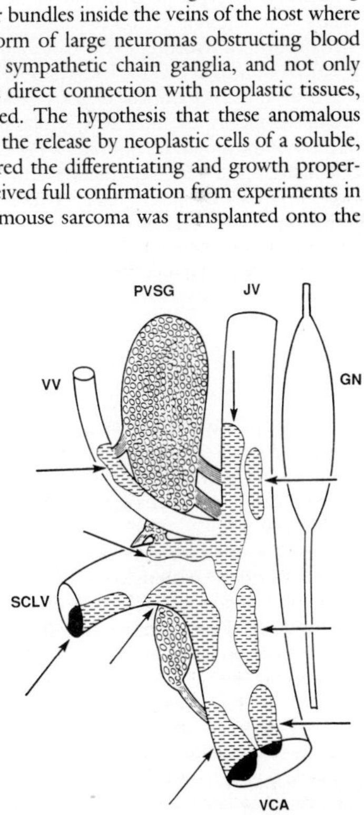

Fig. 1. Sixteen-day chick embryo with intra-embryonic tumor (sarcoma 180). Ingrowth of sympathetic nerve fibers into the jugular, vertebral, subclavian, and anterior caval veins. Abbreviations: GN, ganglium nodosum; JV, jugular vein; PVSG, paravertebral sympathetic ganglion; SCLV, subclavian vein; VCA, anterior caval vein; and VV, vertebral vein. Arrows point to nerve agglomerations [from (*12*)].

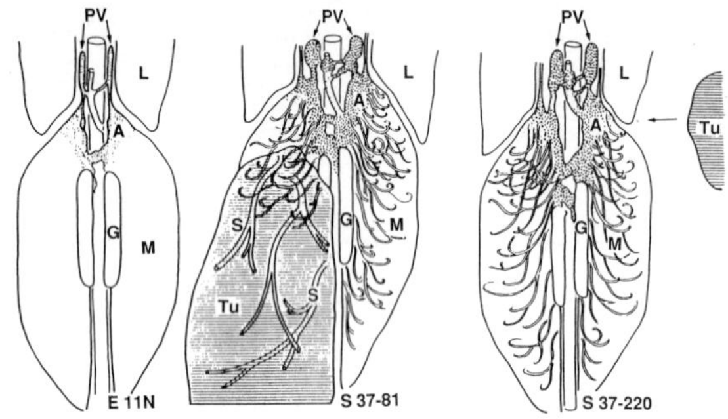

Fig. 2. Semidiagrammatic reconstruction of a normal 11-day chick embryo (E 11N), of an 11-day embryo carrying an intra-embryonic transplant of mouse sarcoma (S 37-81), and of an 11-day embryo with a transplant of sarcoma 37 on the chorioallantoic membrane (S 37-220). Note the hyperplastic growth of the prevertebral ganglia in embryos carrying tumor transplants. Visceral nerve fibers from these ganglia invade the nearby mesonephroi. The abbreviations are A, adrenal; G, gonad; L, lung; M, mesonephros; PV, prevertebral ganglia; S, sensory nerves; and Tu, tumor [from (*12*)].

chorioallantoic membrane of 4- to 6-day chick embryos in such a position as to prevent direct contact between embryonic and neoplastic tissues (Fig. 2). Embryonic and tumor tissues were, however, in reciprocal connection through the circulatory system. The finding that these extraembryonic transplants elicited the same effects as intraembryonic grafts gave definite evidence for the diffusible nature of the tumor nerve growth–promoting factor (*12, 13*).

Attempts to replicate these effects by implanting dried tumor pellets or by injecting extract of either sarcoma were unsuccessful. I then thought of resorting to the tissue culture technique that I had practiced with G. Levi at the University of Turin. Lack of facilities in this field in the department of zoology at Washington University prompted me to ask hospitality from Professor Carlos Chagas, director of the Biophysic Institute of the University of Brazil in Rio de Janeiro. There, a friend of mine, Hertha Meyer, had built and was director of a most efficient tissue culture unit. Upon approval and invitation from Professor Chagas, I boarded a plane for Rio de Janeiro, carrying in my handbag two mice bearing transplants of mouse sarcomas 180 and 37.

The Nerve Growth Factor at Its Early in Vitro and in Vivo Debut

"The tumor had given a first hint of its existence in St. Louis but it was in Rio de Janeiro that it revealed itself, and it did so in a theatrical and grand way, as if spurred by the bright atmosphere of that explosive and exuberant manifestation of life that is the Carnival in Rio" (*14*).

The discovery of the growth response elicited by a soluble tumor agent revealed the receptivity of developing nerve cells to hitherto unknown humoral factors and in this way opened a new area of investigation. The in vitro bioassay offered a practical and invaluable tool for uncovering the identity of this factor and paved the way for the study of its mechanism of action. Ink drawings that I had enclosed in several letters mailed from Rio to Viktor give an eloquent account of the spectacular way in which this still unknown agent revealed itself. Sensory and sympathetic ganglia explanted from 8-day chick embryos in a semisolid medium in proximity to, but not in contact with, fragments of mouse sarcoma 180 or 37, produced, in a 24-hour period, a halo of nerve fibers of maximal density on the side facing the tumor (*15*) (Fig. 3). The euphoric state elicited by this discovery was, however, dampened shortly afterward by the discovery that normal mouse tissues, at variance with those of chick embryos, induce a milder but not substantially different effect from that of mouse sarcomas. In retrospect, this should have alerted us to a novel and even more significant aspect of these in vitro experiments—namely, the widespread presence of the factor endowed with nerve growth–promoting activity in normal and neoplastic tissues. The failure to realize the significance of this "mouse effect" was beneficial rather than detrimental since for the next 2 years our attention was entirely focused on the study of the chemical nature of the factor released from the two mouse sarcomas in much larger quantities than from normal mouse tissues.

A young biochemist, Stanley Cohen, who joined our group shortly before my return from Rio, isolated from the two tumors a nucleoprotein fraction endowed with the in vitro nerve growth–promoting activity (*16*). Chance, rather than calculated search, signaled a new, most fortunate turn of events. In order to degrade the nucleic acids present in this active fraction, Stan made use of snake venom, which contains among other enzymes the nucleic acid–degrading enzyme phosphodiesterase. Its addition in minute amounts to the nucleoprotein tumor fraction was expected to suppress the formation of the fibrillar halo if nucleic acids rather than the protein were responsible for the nerve growth–promoting effect elicited by this fraction. The startling result was a marked increase in the density of the fibrillar halo around the ganglia incubated in the presence of the tumor fraction treated with snake venom. Since a dense fibrillar halo was also produced around ganglia cultured in the presence of minute amounts of snake venom alone, it became apparent that the venom itself was a potent source of nerve growth–promoting activity. On the basis of biochemical studies, Cohen was in fact able to show that equivalent growth-stimulating effects were obtained by 15,000 μg of a sarcoma 180 homogenate and 6 μg of the moccasin snake venom. From the latter, after several purification steps, he isolated a nondialyzable heat-labile particle endowed with nerve growth–promoting activity, identified in a protein molecule with a molecular weight on the order of 20,000 (*17, 18*). Microgram quantities of the purified snake venom fraction endowed with nerve growth–promoting property, injected daily into the yolk of 6- to 8-day chick embryos for a 3- to 5-day period, resulted in the overgrowth of sensory and sympathetic ganglia and excessive production of their fibers. Sympathetic nerve bundles branched profusely into the viscera and protruded into the cavity of the veins, mimicking in all details the effects elicited by grafts of mouse sarcomas (*19*).

If chance brought to our attention the unforeseeable presence of two nerve growth–promoting sources in mouse sarcomas and in snake venom, the subsequent findings that mouse submandibular salivary gland extract added in a minute quantity to the culture medium elicits an even denser and more compact fibrillar halo was the result of a calculated search. These glands, as the homolog of the snake venom glands, were chosen by Stanley Cohen (*20*) as more likely than other organs screened with the in vitro bioassay, to store the nerve growth factor (NGF). These results were soon followed by purification and identification by Cohen of the salivary factor in a protein molecule with an evaluated molecular weight of 44,000 (*20*). Its availability in larger quantities than the venom NGF, and its moderate toxicity when injected in a highly purified form, made possible the exploration of its biological activity in neonatal, young, and adult mammals (*21*). The results of these investigations signaled the beginning of an ever more extensive and systematic in vivo and in vitro analysis of the salivary NGF, of its chemical structure, mechanism, and spectrum of action. Only the most significant findings reported from several laboratories in original and review articles will be considered in the following pages.

The Vital Role of NGF in the Life of Its Target Cells

In spite of, or perhaps because of, its unusual and almost extravagant deeds in living organisms and in vitro systems, NGF did not at first find enthusiastic reception by the scientific community, as also indicated by the reluctance of other investigators to engage in this line of research. The finding that a protein molecule from such diverse and unrelated sources as mouse sarcomas, snake venom, and mouse salivary glands elicited such a potent and disrupting action on normal neurogenetic processes did not fit into any conceptual preexisting schemes, nor did it seem to bear any relation to normal control mechanisms at work during ontogenesis. It was in this skeptical atmosphere that NGF asserted, in a most forceful way, its vital role in the life of its target cells. Previous in vitro experiments had shown that incubation of snake venom with its antiserum inhibited the fiber outgrowth induced by the venom NGF. A specific antiserum to salivary NGF likewise abolished the formation of the in vitro fibrillar halo. These results suggested testing the effect of daily injections of small amounts of this antiserum (AS-NGF) in neonatal rodents. The inspection of treated mice, performed at the end of the first month with the stereo- and optic microscopes, revealed the near total disappearance of sympathetic para- and prevertebral chain ganglia (*22–24*). This dramatic effect—the loss of the sympathetic system in newborn rodents and other neonatal mammals injected with antiserum to salivary NGF, without interfering with their normal development and vitality—became known as immunosympathectomy (*25, 26*). The same treatment produces much less damaging effects in young and adult animals.

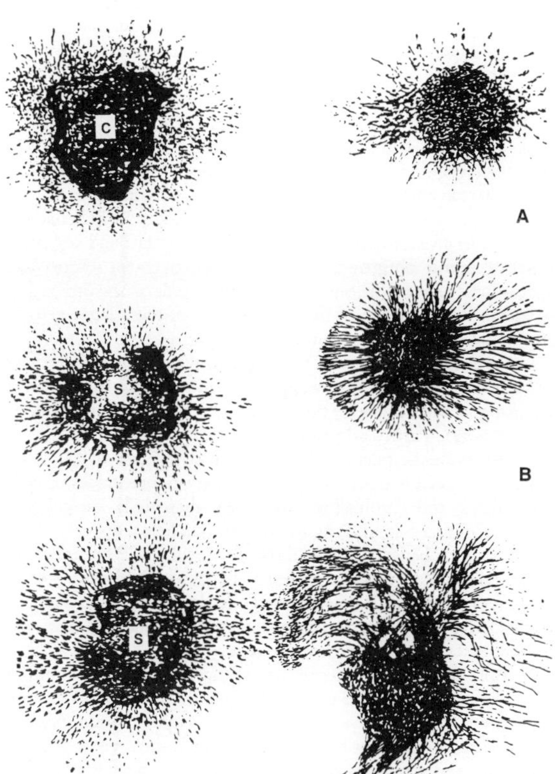

Fig. 3. Drawings illustrating the in vitro "halo" effect on 8-day chick embryo sensory ganglia cultured in the presence of fragments of mouse sarcoma 180 for 24 hours (**B**) or 48 hours (**C**). In (**A**), the ganglion, which faces a fragment of chick embryonic tissue (c) shows fibroblasts but few nerve fibers. In (B) and (C), the ganglia, facing fragments of sarcoma 180 (s) show the typical "halo" effect elicited by the growth factor released from the sarcoma. Note in (C) the first evidence of a neurotropic effect of the growth factor.

Two alternative hypotheses were submitted to explain the mechanism underlying the destructive effects of the antiserum: (i) a complement-mediated cytotoxic effect, or (ii) the inactivation of NGF or of an NGF-like protein essential for differentiation and survival of sympathetic nerve cells. Although the first hypothesis was favored in early articles, the second progressively gained support and is now generally accepted on the basis of the results with antiserum and an in vitro experimental approach, which provided additional, unequivocal evidence of the essential role NGF plays in the early differentiation stages of its target cells. The in vitro experiments consisted of the dissociation of sensory and sympathetic nerve cells from ganglia of 8- to 11-day chick embryos and their incubation in minimal essential media. Nerve cells failed to survive unless nanogram quantities of NGF were added daily to the culture medium (*27*). The in vitro evidence for the role of NGF in the early phases of development of sensory nerve cells received confirmation from subsequent experiments, which proved that administration of NGF antiserum to rodent fetuses (*28, 29*) and autoimmunization of pregnant rodents against endogenous NGF result in failure of sensory ganglia to undergo normal development (*30*).

NGF as a Retrograde Trophic Messenger and Tropic Factor

The evidence in favor of the hypothesis that immunosympathectomy results from removal, by AS-NGF (*31*), of circulating endogenous NGF opened the questions of how NGF reached its target cells and what were its sources of production. Subsequent experimental pharmacological and surgical approaches provided satisfactory answers to both questions, and in view of the interest in these problems, techniques and main findings are briefly reported here.

Administration to neonatal rodents of drugs such as 6-hydroxydopamine, which destroys adrenergic nerve endings (*32*), or vinblastine, which blocks axonal transport (*33*), results in death of the majority of sympathetic nerve cells in their most active phase of differentiation and growth. The degenerative effects produced by these pharmacological components compare in magnitude to those produced by administration of AS-NGF and result in the destruction of para- and prevertebral sympathetic ganglia through a process which became known as chemical sympathectomy (*32, 33*). A third experimental manipulation, consisting of the surgical transectomy of postganglionic axons of the superior cervical ganglion, performed in neonatal rodents, results in the death of about 90 percent of immature sympathetic cells in this ganglion (*34*). The experimental evidence that in all instances nerve cell death is prevented by supply of exogenous NGF (*30, 35–37*), demonstrates the vital role played by this molecule in the life and differentiation of these cells. The subsequent demonstration that labeled NGF is taken up by nerve endings of sympathetic (*38*) or sensory fibers (*39*) and is retrogradely transported to the cell perikarya lent strong support to the concept of NGF as a trophic messenger, conveyed through nerve fibers from peripheral cells to the innervating neurons. Disconnection of the partners by chemical or surgical axotomy results in death of differentiating nerve cells deprived of this essential molecule.

At the same time as the vital role of NGF in developing sympathetic and sensory nerve cells was assessed and its retrograde transport from peripheral tissues was well documented, another important property of NGF—its ability to direct growing or regenerating axons of sensory and sympathetic fibers along its concentration gradient (neurotropism)—was definitely established through different in vivo and in vitro experimental approaches (*40–45*).

The first strong evidence for an NGF neurotropic effect was obtained from experiments of daily microinjections of NGF into the floor of the fourth ventricle. A 7-day treatment resulted in the penetration of fiber bundles originating from sympathetic ganglia inside the neural tube and in their ending at the level of experimentally produced NGF pools (*40, 41*) (Fig. 4). In vitro experimental approaches gave more rigorous proof that neurites of NGF target cells grow along a NGF concentration gradient and deflect their route according to the changed position of the NGF-releasing pipette (*42*). While these studies unequivocally establish the NGF neurotropic effect as independent from its trophic action, they leave unanswered the question of whether this effect is exerted through a local control of growth cone motility (*43*), altered adhesion of this locomotor organelle to the substratum (*44, 45*), or via other mechanisms (*46, 47*).

Neuronal and Nonneuronal Target Cells

As indicated in Table 1, targets of NGF action that have been well characterized up to now, can be classified under three main categories: (i) neural-crest derivatives, (ii) central nervous system (CNS) neurons, and (iii) cells of nonneuronal origin. For a thorough analysis of the many and diversified effects exerted by NGF on each one of these cells, the reader is referred to review articles on this specific matter (*47–52*). I wish to make in this context only some general considerations.

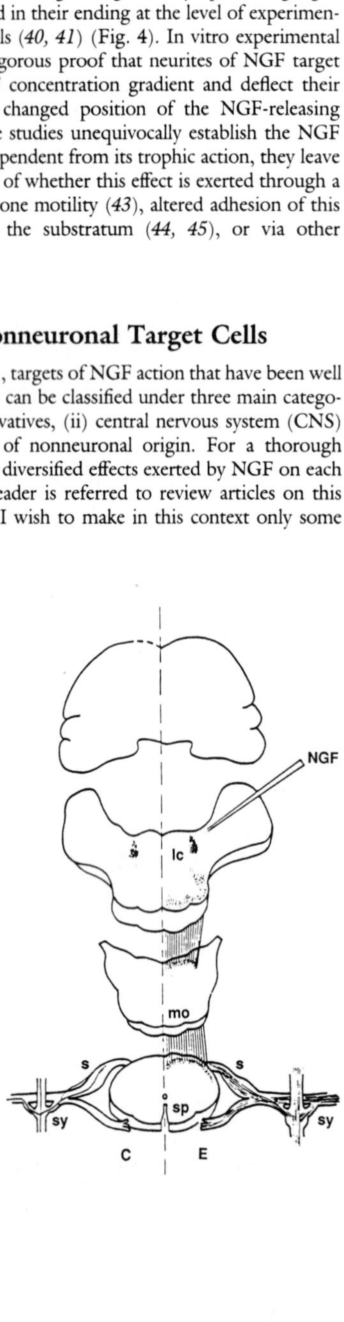

Fig. 4. Diagrammatic representation of sympathetic fiber bundles that enter the spinal cord and medulla oblongata from adjacent sympathetic ganglia in infant rats given intracerebral injections of NGF. Left half, control (C); right half, experimental (E) case. Abbreviations: NGF, site of injection of NGF into the floor of the fourth ventricle; lc, locus coeruleus; mo, medulla oblongata; sp, spinal cord; s, sensory ganglia; and sy, sympathetic ganglia. Sympathetic fibers run across the sensory ganglion and enter into the neural tube with the dorsal roots [from (*40*)].

A generally valid rule is that all cells are maximally responsive to the action of NGF during their early differentiation; the response undergoes progressive restriction in the adult without, however, being totally effaced. Long sympathetic neurons and sensory neurons, with particular reference to those of the dorsomedial quadrant of spinal ganglia in chick embryo (*12*), provided a valuable system for demonstrating the three main activities of NGF—that is, (i) its vital trophic role during the early developmental stages; (ii) its property of enhancing differentiative processes such as neurite outgrowth; and (iii) its property of guiding the growing or regenerating neurites along its own concentration gradient (*43, 44*). These same cells offered an in vivo model system to study the induction of enzymes involved in neurotransmitter synthesis (*53*) and were also instrumental in providing the first demonstration of NGF retrograde transport (*54*) and its role as a trophic messenger (*55*). If sensory and sympathetic cells played a key role in revealing these properties of NGF, chromaffin cells and their neoplastic counterpart, the clonal cell line PC12, became the model of choice for studying the capacity of NGF to modulate phenotypic expression and the molecular mechanism subserving this process (*56*). The phenotypic shift induced by NGF both in chromaffin (*57, 58*) and PC12 cells (*56, 59*), resulting in their neuronal differentiation and accompanied by a plethora of chemical, ultrastructural, and morphological changes characteristic of the neuronal rather than glandular phenotype, is too well known to warrant a detailed description (*50*). These cells, moreover, uncovered the startling capacity of NGF to act as either a mitogenic (*60*) or antimitotic agent (*56*), even within the context of the clonal cell line PC12 and a mutated version of it (*61*). This, in turn, clearly pointed to the "versatility" of NGF receptors and of their transduction machinery, whose message is evidently read and interpreted in different ways according to the cell type and previous cell history. The "priming model," which had the potential to give a molecular account of the very fast and very slow onset of neurite outgrowth occurring, respectively, in sensory and sympathetic cells on the one hand and PC12 cells on the other (*50*), is an excellent example of the contribution of these latter cells to studies on the mode of action of NGF.

Other examples of the wide and at the same time diversified effects of NGF are illustrated by other sympathoadrenal derivatives such as paraganglia, small intensely fluorescent cells (SIF), and carotid body cells (*62–64*). Particularly impressive evidence of the capacity of NGF to modulate phenotypic expression is the case of SIF cells, which have been hypothesized to be immediate precursors of both sympathetic and chromaffin cells. When these cells are cultured under appropriate conditions, they can be channeled toward the first or the second phenotype in media supplied with NGF or dexamethasone (*63, 64*). Such an interplay between NGF and steroid hormones, even in fully differentiated cells, is also indirectly suggested by in vivo studies on the short adrenergic neurons that innervate, in both sexes, the genitourinary systems (*65*).

In more recent years, two new populations came to the forefront of studies on NGF target cells: CNS neurons and cells originating from the hematopoietic system. Small and large neuronal populations located in different brain areas have been shown to exhibit all properties and responses typical of sensory and sympathetic cells, such as: (i) the presence of specific receptors (*66*), (ii) retrograde transport of NGF (*67*), (iii) increased neurotransmitter synthesis with special reference to acetylcholine (*68–70*), and (iv) trophic response manifested as protection by exogenous NGF administration to selective noxious treatments or surgical transections otherwise leading to cell death (*71, 72*). A role for NGF in the development of as yet unidentified hypothalamic brain centers has been suggested by the finding that injections of affinity-purified polyclonal antibodies against NGF in rat fetuses induce a severe postnatal neuroendocrine syndrome (*29*). The loop of an unquestionable NGF role in brain is completed by the demonstration that other nerve cells, especially those located in the hippocampus and cortical areas, manufacture large quantities of NGF messenger RNA and of NGF protein, thus closing the functional link between NGF-producing and NGF-responding cells (*73–75*). As indicated in the last section of this article, although the range of NGF action in the CNS is qualitatively comparable to that previously observed in peripheral neurons, the actual extent of the NGF role in brain is far from foreseeable because of the vast repertoire of possible responses from nerve cells in the CNS.

An analogous general consideration holds for the effect exerted by NGF on mast cells and possible on other cells of the immune system. The increased number of mast cells in vivo and in vitro following NGF treatment (*76, 77*), as well as the effect of this growth factor on histamine release (*78–80*), point to an unquestionable role for NGF in the physiology of these cells. It is not yet clear, however, whether such an effect is exerted through a generalized action on all mast cell precursors or through a sort of clonal selection mechanism. The more recent reports of an NGF effect on other spleen cells such as mononuclear cells (*81*) and the existence of NGF receptors on thymocytes (*82*) clearly suggest that the NGF action extends to cells belonging to a network of enormous functional significance. The role played by histamine as an immunomodulator and the obvious involvement of spleen cells in the immune response of the organism suggest new scenarios in which NGF may gain access, not through a back door but through the main entrance.

The ID Card of NGF

Sequencing of mouse submandibular gland NGF, achieved in 1971 (*83*), not only provided invaluable information on its primary structure but was recently instrumental in the preparation of synthetic oligonucleotides which resulted in the identification of NGF complementary DNA (*84*). The cloning of mouse (*84*), human (*85*), bovine (*86*), and chick (*87*) genes, which followed in rapid succession, demonstrated their high degree of homology. The NGF gene, located in the human species on the proximal short arm of chromosome 1 (*88*), codes for a large polypeptide of 307 amino acid residues which, upon cleavages, gives rise to the 118–amino acid mature NGF subunit protein and, possibly, to other peptides of unknown function and with no sequence homology with presently identified proteins (*84*). NGF is a dimer of two identical subunits held together by noncovalent bonds. The dimer can be isolated as such (*89*) or in the form of a complex consisting of two additional proteins, one with an esteropeptidase activity, probably involved in processing an NGF precursor, and the other with an as yet unknown function (*90–92*). Although it remains to be established whether each NGF subunit is biologically active, it has been demonstrated that a covalently cross-linked form of the dimer maintains full activity (*91, 92*). Between the two well-identified molecular entities of NGF and of its coding gene, which can be visualized as the summit and the base of an iceberg, are several other possible intermediate forms of unknown nature and biological properties. Their identification would answer important questions such as: Are other biologically active peptides coded for by the NGF gene? What is the significance of different splicing in different cells of NGF messenger RNA (*93*)? Is the processing of pre-pro-NGF identical in all neuronal and nonneuronal cells or, as in other peptides (*94*), do alternative processing pathways result in the production of peptides endowed with different biological functions? Since the same peptides may undergo post-transcriptional or post-translational modification, the submerged areas of the NGF iceberg loom very large.

Studies on the immunological and biological relatedness of NGFs purified from different species strongly support the hypothesis that the site (or sites) of interaction with their receptors has remained structurally more constant than is the case for other epitopes, probably free to mutate in view of their less fundamental biological functions (*95*).

NGF Growth Factor and Oncogenes

The discovery of NGF, soon followed by that of epidermal growth factor (EGF), led to the biological identification of an ever-growing list of polypeptide growth factors (*48*). In the seventies, another apparently unrelated area of biology came to the forefront of research with the discovery of single gene products (oncogenes) causing transformation. Polypeptide growth factor and oncogene research, pursued at first independently of each other, converged when homology between some oncogenes and growth factors or their receptors was shown by sequence analysis. Evidence is steadily increasing that excessive synthesis or an altered version of polypeptide growth factors or of their receptors may result in transformation of recipient cells (*95–98*). More recently, the demonstration that the opposite is also true—namely, that certain oncogene products may induce differentiation of recipient cells, called attention to another facet of this intricate interplay between differentiative and transforming processes. The case of H-*ras* and that of v-*src*, whose expression in PC12 cells (*99, 100*) result in mitotic arrest and neuronal differentiation comparable to those elicited by NGF, provide instances of a list most likely to grow. The obvious conclusion is that a given polypeptide growth factor, or intracellular proteins playing essential roles in the cell cycle or in differentiation of some cells, may exert markedly different actions in distinct cell types. In the case of NGF, one wonders if and how other actions are elicited by this versatile molecule. For instance, is an altered version of NGF or of its receptors capable of causing transformation of some recipient cells, as has been shown for other polypeptide growth factors? If this is the case, could NGF in a modified version or its receptors be implicated in neoplasia in the central and peripheral nervous systems?

NGF in Exocrine Glands: A Fortuitous Presence or a Biological Function?

The early discovery that mouse submandibular glands synthesize

Table 1. NGF target cells.

Neural crest derivatives
- Sympathoadrenal
  - Long sympathetic neurons
  - Short sympathetic neurons
  - Paraganglia (carotid and abdominal paraganglia) cells
  - SIF (small, intensely fluorescent) cells
  - Chromaffin cells
    - Normal
    - Neoplastic (PC12)
- Sensory neurons

Central nervous system
- Cholinergic neurons:
  - Corpus striatum, basal forebrain, septum
  - Nucleus diagonal band of Broca
- Adrenergic, indoleaminergic, and peptidergic neurons
  - *Xenopus laevis* tadpoles

Nonneuronal origin
- Mast cells

and release into the saliva large quantities of NGF, that the synthesis of this protein molecule is under the control of testosterone and of thyroxine (*101, 102*), and that the NGF protein content is about tenfold higher in male than in female mice, remained a puzzling and unexplained finding for about three decades. The conflicting but altogether negative attempts to reveal the presence of this molecule in the circulating blood (*49, 51*), and the lack of any adverse effects on sympathetic and sensory cells by removal of these glands, which deprived these rodents of such a large NGF source, militated against the hypothesis that salivary NGF gains access to their target cells. An alternative biological function for salivary NGF was first hypothesized by our group (*103*) and recently proved by us (*104*) and another investigator (*105*). We demonstrated that intraspecific fighting, experimentally induced in adult male mice by 6 to 8 weeks of social isolation, results in massive NGF release into the bloodstream, an event prevented by previous sialoadenectomy. Since injections of NGF induce weight and size increase of the adrenal glands (*105*) and stimulate the synthesis of the catecholamine key enzyme, TH (*106*), we suggested that such a massive discharge into the blood circulation of endogenous salivary NGF may be instrumental in defense or offense mechanisms (or both) of vital significance for male mice that engage in intraspecific fighting among individuals of the same sex. Supporting this hypothesis is a recent report that aggressive behavior results in the release into the blood of another biologically active protein, renin, synthesized in the same tubular portions of these glands (*107*). The mechanism triggering this NGF release is not yet understood, nor whether other stations are activated and play a role in this specific stress syndrome.

As for the presence of large NGF sources in snake venom (*18*) and male genital organs (*108, 109*), they may be conceived as instances of bizarre evolutionary gene expression. Alternatively, in these cases NGF may subserve other functions that may somehow be linked with the poisonous action of snake venom or the reproductive activity of the genital apparatus. In the case of snake venom, one can envisage the possibility that a highly specific neurotropic molecule such as NGF is utilized by reptiles as a carrier of other neurotoxins devoid of specific receptors in the central and peripheral nervous systems. For instance, enzymes such as phospholipases, phosphodiesterases, and proteases of various nature, which by themselves may lack specific recognition sites in target cells, may exploit NGF as a carrier to gain access inside cells wherever there are specific NGF receptors. The widespread distribution of these specific molecules also in several nonneuronal cells could offer some toxins or enzymes a better access to their target organs.

In the reproductive tract, NGF could participate in fertilization mechanisms by cytoskeletal-mediated activation of spermatozoa locomotion much in the same way as in neurite outgrowth, or by favoring egg implantation via inhibition of rejection through the immune system. This latter hypothesis is presently under investigation (*110*) by assessing the effect of NGF on uterine mast cells which, through histamine release, are postulated to prevent the local immune reaction (*111*).

Foreseeable Approaches and Predictions of the Unpredictable

The most obvious among the foreseeable approaches is the search for other NGF target cells with the use of the ever more sophisticated in vivo and in vitro techniques that became available in these last decades from the molecular to the supercellular levels. It was this multimodal approach that in recent years led to the discovery of NGF target cells in the CNS of lower and higher vertebrates and in cell lineages playing a role in the immune system. This list is likely to increase, as the search extends to other neuronal and nonneuronal cell populations. Furthermore, one should take into account the fact that some of these populations are receptive to NGF mainly during developmental stages in prenatal life. This was already demonstrated in sensory cells of avian and mammalian species (*49, 51, 52*), and in cells lining the third ventricle in amphibian tadpoles (*112*) and prenatal and neonatal rodents (*113*). Likewise, the systematic screening of neuroendocrine and hematopoietic cell lines in both in vitro and in vivo systems may reveal other as yet uncovered roles of this growth factor.

Another foreseeable approach now in progress in many laboratories is the search and characterization of NGF-like factors active on other neuronal populations. These factors may be subdivided into two major classes: (i) those coded by the NGF gene itself but processed through alternate post-transcriptional or post-translational pathways leading to polypeptide growth factors with a somewhat different structure and function and (ii) other proteins or peptides having one or more of the trophic, chemotactic, and differentiative activities of NGF but encoded by other genes.

The search and identification of factors belonging to the first group will take advantage of the techniques of molecular biology and immunology. These should provide valuable information on some of the still unexplored, submerged areas of the NGF iceberg, dealing with the processes of the NGF gene transcription or translation. Of particular importance would be the identification of the NGF sequence responsible for its binding to receptors that may presumably trigger a given cellular response (*47*). As previously surmised (*95*), this portion has possibly been better conserved than other parts of the molecule. Once identified it will be feasible to introduce, in its synthetic counterpart, amino acid substitutions or chemical modifications (or both) and explore the biological potency of the newly manufactured peptide. This approach should not only provide invaluable information on the nature and properties of the NGF active center but, it is hoped, will result in the synthesis of peptides endowed with an activity even higher than that of NGF itself, so brilliantly achieved in the field of other biologically active peptides (*114, 115*).

Within this category of studies on NGF and its coding gene, one can conceive a strategy aimed at exploiting the property of nonneuronal cells in peripheral tissues and of neurons and satellites in the CNS to manufacture and release NGF by resorting to pharmacological agents that modify NGF gene expression or processing. The well-established finding that NGF synthesis is increased after transection of nerve fibers connecting NGF-receptive nerve cells to their targets (*116*) or via hormonal action (*101, 102*) is an additional indication of the remarkable plasticity of the mechanisms controlling NGF gene expression. It is conceivable that this property might be modulated by pharmacological agents acting on the same pathway as those involved in regulation of the synthesis and release of NGF.

The search for neurotrophic factors coded for by genes other than the NGF gene could take advantage, at least in its main lines, of the classical approach so successfully applied in the isolation and identification of NGF. Two main causes may explain why extensive work invested in this attempt has not been so successful in providing evidence for the existence of other polypeptide growth factors activating different neuronal cell lines: (i) the lack of fast and reliable bioassays such as those developed for NGF and (ii) the failure to detect large sources of these factors comparable to those fortuitously discovered in early NGF studies. The realization of rapid, highly reliable bioassays can, however, now be achieved by resorting to the use of most stringent, chemically defined media, permitting survival and differentiation of only given cell types, upon addition to the medium of putative growth factors extracted from different sources and screened with the in vitro bioassay for their potential specific growth-enhancing activity. The problem of finding by sheer chance large sources of NGF-like peptides, such as those which played a key role in the discovery of NGF, can be solved by resorting to techniques of protein chemistry and recombinant DNA technology. A few micrograms of purified protein—in order to decipher the sequence, prepare the corresponding complementary DNAs, identify the gene of the polypeptide growth factor in question, and express it in bacteria—thus replace a search once guided by unpredictable strikes of luck with a rational and systematic strategy.

Predictions of the unpredictable are encouraged by the same history of NGF, which may be defined as a long sequence of unanticipated events, which each time resulted in a new turn in the NGF unchartered route and opened new vistas on an ever-changing panorama. This trend, which became manifest from the very beginning and in fact alerted me to the existence of NGF, is perhaps the most attractive, even though elusive, trait of this 35-year-long adventure. One can at present only predict where future developments are most likely to occur. The main causes of unpredictability of the findings reside in the intricacy of the new surroundings where NGF is moving—the CNS and the immune system—rather than in NGF itself. The enormous complexity of these two networks, which on the basis of recent findings are closely interrelated and influence each other through bidirectional signals (*117, 118*), opens endless possibilities for NGF activation of distinct repertoires of cells belonging to one or the other system. How many indirect effects can be elicited by direct NGF action on cholinergic, adrenergic, and peptidergic neurons interlinked via fiber pathways and humoral channels or through short-distance diffusion? Likewise, how many effects could follow the simple histamine release by NGF-activated mast cells, considering the well-established role of this amine as an immunomodulator or an immunosuppressor? These considerations hold also for the potential utilization of NGF in brain and immune system disorders. For instance, whenever cell death or specific neuronal populations may be linked to a decreased local availability of neurotrophic factors, such as NGF, its exogenous supply or stimulation of its endogenous production via pharmacological agents may offer a promising approach to presently incurable diseases.

I shall end this account of the unfolding of the NGF story with a remark made more than a decade ago by Viktor Hamburger: "... the fact that this discovery, which grew out of a seemingly peripheral problem (peripheral in every sense of the word), has blazed so many new trails is its greatest contribution to neuroembryology" (*119*). Studies in this last decade have not only provided new strong evidence of the most important contributions of NGF in the field of neuroembryology, but brought to the fore its significance in the more general field of neuroscience and also foretell its role in that of the immune system.

REFERENCES AND NOTES

1. P. B. Medawar, *The Art of the Soluble* (Methuen, New York, 1967), pp. 106–107.
2. R. G. Harrison, *Proc. R. Soc. London Ser. B* **118**, 155 (1935).
3. V. Hamburger, *J. Exp. Zool.* (1934), p. 449.
4. J. F. Tello, *Trab. Lab. Invest. Biol. Univ. Madrid* **21**, 1 (1922).
5. R. Levi-Montalcini, *Prog. Brain Res.* **4**, 1 (1964).
6. ——— and G. Levi, *Arch. Biol. Liège* **54**, 183 (1943).
7. V. Hamburger and R. Levi-Montalcini, *J. Exp. Zool.* **111**, 457 (1949).
8. R. Levi-Montalcini, *J. Comp. Neurol.* **91**, 209 (1949).
9. ———, *J. Morphol.* **86**, 256 (1950).
10. E. D. Bueker, *Anat. Rec.* **102**, 369 (1948).
11. R. Levi-Montalcini and V. Hamburger, *J. Exp. Zool.* **116**, 321 (1951).
12. R. Levi-Montalcini, *Ann. N.Y. Acad. Sci.* **55**, 330 (1952).
13. ——— and V. Hamburger, *J. Exp. Zool.* **123**, 233 (1953).
14. R. Levi-Montalcini, in *The Neurosciences: Paths of Discovery*, F. G. Worden, J. P. Swayzey, G. Adelman, Eds. (MIT Press, Cambridge, MA, 1975), pp. 245–265.
15. ———, H. Meyer, V. Hamburger, *Cancer Res.* **14**, 49 (1954).
16. S. Cohen, R. Levi-Montalcini, V. Hamburger, *Proc. Natl. Acad. Sci. U.S.A.* **40**, 1014 (1954).
17. S. Cohen and R. Levi-Montalcini, *ibid.* **42**, 571 (1956).
18. S. Cohen, *J. Biol. Chem.* **234**, 1129 (1959).
19. R. Levi-Montalcini and S. Cohen, *Proc. Natl. Acad. Sci. U.S.A.* **42**, 695 (1956).
20. S. Cohen, *ibid.* **46**, 302 (1960).
21. R. Levi-Montalcini and B. Booker, *ibid.*, p. 373.
22. ———, *ibid.*, p. 384.
23. R. Levi-Montalcini, *Science* **143**, 105 (1964).
24. ———, *Harvey Lect.* **60**, 217 (1966).
25. ——— and P. U. Angeletti, *Pharmacol. Rev.* **18**, 819 (1966).
26. G. Steiner and E. Schönbaum, Eds., *Immunosympathectomy* (Elsevier, Amsterdam, 1972).
27. R. Levi-Montalcini and P. U. Angeletti, *Dev. Biol.* **7**, 653 (1963).
28. R. Levi-Montalcini, L. Aloe, P. Calissano, C. Cozzari, in *First Meeting of the International Society for the Development of Neuroscience,* Strasbourg (1980), vol. 1, p. 5.
29. L. Aloe, C. Cozzari *et al.*, *Nature (London)* **291**, 413 (1981).
30. E. M. Johnson, Jr., P. D. Gorin, L. D. Brandeis, J. Pearson, *Science* **210**, 916 (1980).
31. M. Goedert, U. Otten, T. Schaefer, M. Schwab, H. Thoenen, *Brain Res.* **201**, 399 (1980).
32. P. U. Angeletti and R. Levi-Montalcini, *Proc. Natl. Acad. Sci. U.S.A.* **65**, 114 (1970).
33. P. Calissano, G. Monaco, M. G. Menesini-Chen, J. S. Chen, R. Levi-Montalcini, in *Contractile Systems in Non-Muscle Tissue,* S. W. Perry, A. Margret, R. S. Adelstein, Eds. (Elsevier, Amsterdam, 1976), pp. 201–211.
34. I. A. Hendry, *Brain Res.* **90**, 235 (1975).
35. R. Levi-Montalcini, L. Aloe, E. Mugnaini, F. Oesch, H. Thoenen, *Proc. Natl. Acad. Sci. U.S.A.* **72**, 595 (1975).
36. I. A. Hendry, *Brain Res.* **94**, 87 (1975).
37. L. Aloe, E. Mugnaini, R. Levi-Montalcini, *Arch. Ital. Biol.* **113**, 326 (1975).
38. K. Stöckel, U. Paravicini, H. Thoenen, *Brain Res.* **76**, 413 (1974).
39. V. Hamburger, V. J. K. Brunso-Bechtold, J. W. Yip, *J. Neurosci.* **1**, 60 (1981).
40. R. Levi-Montalcini, *Prog. Brain Res.* **45**, 235 (1976).
41. M. L. Menesini-Chen, J. S. Chen, R. Levi-Montalcini, *Arch. Ital. Biol.* **116**, 53 (1978).
42. R. W. Gundersen and J. N. Barrett, *Science* **206**, 1079 (1979).
43. ———, *J. Cell Biol.* **87**, 546 (1980).
44. R. B. Campenot, *Dev. Biol.* **93**, 1 (1982).
45. ———, *ibid.*, p. 13.
46. K. H. Pfenninger and M. P. Johnson, *Proc. Natl. Acad. Sci. U.S.A.* **78**, 7797 (1981).
47. P. Calissano, A. Cattaneo, L. Aloe, R. Levi-Montalcini, in *Hormonal Proteins and Peptides,* C. H. Li, Ed. (Academic Press, New York, 1984), vol. 12, pp. 1–56.
48. R. A. Bradshaw, *Annu. Rev. Biochem.* **47**, 191 (1978).
49. H. Thoenen and Y. A. Barde, *Physiol. Rev.* **60**, 1284 (1980).
50. L. A. Greene and E. M. Shooter, *Annu. Rev. Neurosci.* **3**, 353 (1980).
51. P. Calissano *et al.*, *Exp. Cell Res.* **154**, 1 (1984).
52. R. Levi-Montalcini and P. Calissano, *Trends Neurosci.* **9**, 473 (1986).
53. H. Thoenen, P. U. Angeletti, R. Levi-Montalcini, R. Kettler, *Proc. Natl. Acad. Sci. U.S.A.* **68**, 1598 (1971).
54. U. Paravicini, K. Stoeckel, H. Thoenen, *Brain Res.* **84**, 279 (1975).
55. E. M. Johnson, Jr., and H. K. Yip, *Nature (London)* **314**, 751 (1985).
56. L. A. Greene and A. S. Tischler, *Proc. Natl. Acad. Sci. U.S.A.* **73**, 2424 (1976).
57. K. Unsicker, B. Krisch, U. Otten, H. Thoenen, *ibid.* **75**, 3498 (1978).
58. L. Aloe and R. Levi-Montalcini, *ibid.* **76**, 1246 (1979).
59. L. A. Greene, R. K. H. Liem, M. L. Shelanski, *J. Cell Biol.* **96**, 76 (1983).
60. L. E. Lillien and P. Claude, *Nature (London)* **317**, 632 (1985).
61. D. E. Burnstein and L. A. Greene, *Dev. Biol.* **94**, 477 (1982).
62. R. Levi-Montalcini and L. Aloe, *Adv. Biochem. Psychopharmacol.* **25**, 3 (1980).
63. A. J. Doupe, P. H. Patterson, S. C. Landis, *J. Neurosci.* **5**, 2143 (1985).
64. A. J. Doupe, S. C. Landis, P. H. Patterson, *ibid.*, p. 2119.
65. C. Owman and N. O. Sjoberg, in *Proceedings of the 5th International Congress of Endocrinology*, V. H. T. James, Ed. (Excerpta Medica, Amsterdam, 1977), vol. 1, pp. 205–209.
66. A. Szutowicz, W. A. Frazier, R. A. Bradshaw, *J. Biol. Chem.* **251**, 1516 (1976).
67. M. Seiler and M. Schwab, *Brain Res.* **30**, 33 (1984).
68. H. Gnahn, F. Hefti, R. Heumann, M. E. Schwab, H. Thoenen, *Dev. Brain Res.* **9**, 45 (1983).
69. F. Hefti, A. Dravid, J. Hartikka, *Brain Res.* **293**, 305 (1984).
70. W. C. Mobley, J. L. Rutkowskį, G. I. Tennekoon, K. Buchanan, M. V. Johnston, *Science* **229**, 284 (1985).
71. L. R. Williams *et al.*, *Proc. Natl. Acad. Sci. U.S.A.* **83**, 9231 (1986).
72. L. F. Kromer, *Science* **235**, 214 (1987).
73. S. Korsching, G. Auburger, R. Heumann, J. Scott, H. Thoenen, *EMBO J.* **4**, 1389 (1985).
74. D. L. Shelton and L. F. Reichardt, *Proc. Natl. Acad. Sci. U.S.A.* **83**, 2714 (1986).
75. S. R. Whittemore *et al.*, *ibid.*, p. 817.
76. L. Aloe and R. Levi-Montalcini, *Brain Res.* **133**, 358 (1977).
77. A. Böhm and L. Aloe, *Accad. Naz. Lincei* **80**, 1 (1986).
78. A. Bruni, E. Bigon, E. Boarato, A. Leon, G. Toffano, *FEBS Lett.* **138**, 190 (1982).
79. K. Sugiyama, Y. Suzuki, H. Furuta, *Arch. Oral. Biol.* **30**, 93 (1985).
80. N. Mazurek, G. Weskamp, P. Erne, U. Otten, *FEBS Lett.* **198**, 315 (1986).
81. L. W. Thorpe, K. Werrbach-Perez, J. R. Perez-Polo, *Proceedings, Second International Workshop on Neuroimmunomodulation,* Dubrovnik, 1986 (abstr.), p. 151.
82. A. Cattaneo and D. S. Secher, *Exp. Cell Res.*, in press.
83. R. Hogue-Angeletti and R. A. Bradshaw, *Proc. Natl. Acad. Sci. U.S.A.* **68**, 2417 (1971).

84. J. Scott *et al.*, *Nature (London)* **302**, 538 (1983).
85. A. Ullrich, A. Gray, C. Berman, T. J. Dull, *ibid.* **303**, 821 (1983).
86. R. Meier *et al.*, *EMBO J.* **5**, 1489 (1986).
87. T. Ebendal, D. Larhammar, H. Persson, *ibid.*, p. 1483.
88. U. Francke, B. de Martinville, L. Coussens, A. Ullrich, *Science* **222**, 1248 (1983).
89. V. Boccini and P. U. Angeletti, *Proc. Natl. Acad. Sci. U.S.A.* **64**, 787 (1969).
90. S. Varon, J. Normura, E. M. Shooter, *Biochemistry* **6**, 2202 (1967).
91. ——, *ibid.* 7, 1296 (1968).
92. R. W. Stach and E. M. Shooter, *J. Biol. Chem.* **249**, 6668 (1974).
93. R. H. Edwards, M. J. Selby, W. J. Rutter, *Nature (London)* **319**, 784 (1986).
94. B. A. Eipper, R. E. Mains, E. Herbert, *Trends Neurosci.* **9**, 463 (1986).
95. R. F. Doolittle *et al.*, *Science* **221**, 275 (1983).
96. J. Downward *et al.*, *Nature (London)* **307**, 521 (1984).
97. C. J. Sherr *et al.*, *Cell* **41**, 665 (1985).
98. C. Weinberger, S. M. Hollenberg, M. G. Rosenfeld, R. M. Evans, *Nature (London)* **318**, 670 (1985).
99. D. Bar-Sagi and J. R. Feramisco, *Cell* **42**, 841 (1985).
100. S. Alemà, P. Casalbore, E. Agostini, F. Tatò, *Nature (London)* **316**, 557 (1985).
101. R. Levi-Montalcini and P. U. Angeletti, in *Salivary Glands and their Secretions,* L. M. Sreebny and J. Meyer, Eds. (Pergamon, Oxford, 1964), pp. 129–141.
102. L. Aloe and R. Levi-Montalcini, *Exp. Cell Res.* **125**, 15 (1980).
103. L. Aloe, C. Cozzari, R. Levi-Montalcini, *Brain Res.* **332**, 259 (1985).
104. L. Aloe *et al.*, *Proc. Natl. Acad. Sci. U.S.A.* **83**, 6184 (1986).
105. J. Lakshmanan, *Am. J. Physiol.* **250**, E386 (1986).
106. U. Otten, M. Schwab, C. Gagnon, H. Thoenen, *Brain Res.* **133**, 291 (1977).
107. J. Bing, K. Poulsen, E. Hackenthal, E. Rix, R. Taugner, *J. Histochem. Cytochem.* **28**, 874 (1980).

28, 874 (1980).

108. G. P. Harper *et al.*, *Nature (London)* **279**, 160 (1979).
109. G. P. Harper and H. Thoenen, *J. Neurochem.* **34**, 893 (1980).
110. Geraci, Cocchiara, Calissano, in preparation.
111. D. J. Beer, S. M. Matloff, R. E. Rocklin, *Adv. Immunol.* **35**, 209 (1984).
112. R. Levi-Montalcini and L. Aloe, *Proc. Natl. Acad. Sci. U.S.A.* **82**, 7111 (1985).
113. L. Aloe and R. Levi-Montalcini, unpublished observations.
114. E. T. Kaiser and D. S. Lawrence, *Science* **226**, 505 (1984).
115. B. Rajashekhar and E. T. Kaiser, *J. Biol. Chem.* **261**, 13617 (1986).
116. T. Ebendal, L. Olson, Å. Seiger, K. O. Hedlund, *Nature (London)* **286**, 25 (1980).
117. T. L. Roszman *et al.*, *J. Immunol.* **135**, 769 (1985).
118. N. R. Hall, J. P. McGillis, B. L. Spangelo, A. L. Goldstein, *ibid.*, p. 806s.
119. V. Hamburger, *Perspect. Biol. Med.* **18**, 162 (1985).
120. I dedicate this article to Viktor Hamburger, who promoted and took part in this search, and to whom I am forever indebted for invaluable suggestions and generosity. Without him, the nerve growth factor would never have come to our attention. To my dear friends, Pietro Calissano and Luigi Aloe, I wish to express my deepest gratitude for their fundamental contributions. In this 35-year-long investigation, a large number of colleagues, research associates, and graduate students took part in this scientific adventure. I am particularly indebted and I very gratefully acknowledge the most important work performed by two of them: Drs. Piero Angeletti and Vincenzo Bocchini. To Professor Carlos Chagas, for his generous hospitality in the Biophysic Institute of the University of Brazil, and to Dr. Hertha Meyer, who helped me in devising the tissue culture bioassay of NGF, my warmest thanks.

PART IV

THE MODULATORY ROLE OF NGF IN HOMEOSTATIC PROCESSES IN THE NEURO-ENDOCRINE-IMMUNE TRIAD

Results reported in Part III of this volume provided evidence of a much broader spectrum of action of NGF than had been attributed to it at the time of its discovery and in the two and a half decades that ensued.

In these last years great progress has been made toward a better understanding of the role of NGF not only in neuronal and non-neuronal cell populations, but also in the homeostatic interplay between the nervous, immune and endocrine systems.

The presence of sympathetic fibers in lymphoid tissues in a highly dense array suggests, furthermore, that there exists a structural link between the nervous and immune systems that could be utilized for translating neural messages into chemical signals on specific cellular elements of the immune system. An immunomodulatory role of NGF is suggested by the higher density and fluorescence of varicose adrenergic fiber end terminals in immunocompetent organs and by the large increase in the catecholamine content of these tissues. Sympathetic nerve fibers also form dense plexuses in endocrine organs (paper 33).

This hypothesis, advanced for the first time in 1990 and illustrated in paper 33, received further support from the finding of NGF high affinity TrkA receptors on the membrane of cells belonging to these three systems.[1–4]

Additional evidence of such a role came from the demonstration of a marked local increase of NGF in experimental animal models and clinical cases of autoimmune diseases resulting from a deregulation of homeostatic functions. Enhanced NGF levels are also found in all inflammatory processes, and in other instances of homeostatic function deregulation consequent to psychologically stressful events.

Studies reported in Part IV of this volume present data in favor of the concept that NGF plays an important role in all these processes. They prospect the hypothesis that this molecule is endowed with a physiological function, particularly evident in

---

[1]Ehrhard P.B., Erb P., Graumann U. and Otten U., Expression of nerve growth factor and nerve growth factor receptor tyrosine kinase Trk in activated CD4-positive T-cell clones, *Proc. Natl. Acad. Sci. USA* **90**, 10984–10988 (1990).

[2]Meakin S.O., Suter U., Drinkwater C.C., Welcher A.A. and Shooter E.M., The rat trk protooncogene product exhibits properties characteristic of the slow NGF receptor, *Proc. Natl. Acad. Sci. USA* **89**, 2374–2378 (1992).

[3]Meakin S. and Shooter E.M., Molecular investigations on the high-affinity nerve growth factor receptor, *Neuron.* **6**, 153–163 (1991).

[4]Kaplan D.R., Hempstead B.L., Martin-Zanca D., Chao M.V. and Parada L.F., The trk proto-oncogene product: a tranducing receptor for nerve growth factor, *Science* **252**, 554–558 (1991).

conditions which threaten the welfare of the organism through deregulation of its defense mechanisms.

The holistic approach now followed in our and other laboratories would, not have been possible, however, without the information gathered from past studies, as well as from those in progress at the cellular, subcellular and molecular levels which are being pursued by an ever-increasing number of investigators. An important step forward has been accomplished as a result of the exploitation of powerful new molecular techniques and in the wake of the disappearance of sectarian loyalties. A typical example of the results of this holistic approach is the finding that NGF, as well as other growth factors, hormones, cytokines, mast cells and a plurality of other types of cells are all involved in altered systemic homeostatic functions.[5–7]

Studies reported in papers 34–40 provided evidence of marked increases in mast cell number in the course of autoimmune diseases, which as a result became the main object of our recent investigations. Since such conditions are typically associated with rapid mast cell activation and NGF accumulation, we at first hypothesized, and then demonstrated, that these cells play the most important role indeed in autoimmune processes (paper 42).

NGF levels were found to be elevated in inflammatory cells, inflammatory exudates and inflamed skin. These effects can be seen within hours after the beginning of inflammation. The rapid appearance and high levels of NGF in acute inflammatory exudates is compatible with the release of NGF from resident tissue cells in which the protein is stored. In these conditons, NGF exerts a profound modulatory role on sensory nociceptive nerve physiology during adulthood, which appears to correlate with hyperalgesic phenomena occurring in the course of tissue inflammation (paper 35).[8–11]

There is strong evidence indicating that the action of NGF on sensory neurons might be aimed at specific functional groups of neurons. Small sensory neurons containing neuropeptides, such as substance P and calcitonin gene related protein (CGRP), die when deprived of NGF during development. In adults, the levels of

[5]Johnson D. and Krenger W., Interactions of mast cells with the nervous system, recent advances, *Neurochem. Res.* **17**, 939–951 (1992).

[6]Purcell W.M. and Atterwill C.K., Mast cells in neuroimmune function: neurotoxicological and neuropharmacological perspectives, *Neurochem. Res.* **20**, 521–532 (1995).

[7]Gordon J.R., Burd P.R. and Galli S.J., Mast cells as a source of multifunctional cytokines, *Immunol. Today*, **11**, 12, 458–464 (1990).

[8]Donnerer J., Schligoi R. and Stein C., Increased content and transport of substances P and Calcitonin gene-related peptide in sensory nerves innervating inflammed tissues: evidence for a regulatory function of nerve growth factor in vivo, *Neurosci.* **49**, 693–698 (1992).

[9]Silver R., Silverman A.J., Vitkovic L. and Lederhenderler I.I., Mast cells in the brain: evidence and functional significance, *TINS* **19**, 25–31 (1996).

[10]Tuveri M.A., Passiu G., Mathieu A. and Aloe L., Nerve growth factor and mast cell distribution in the skin of patients with systemic sclerosi, *Clin. Exp. Rheumatol.* **11**, 319–322 (1993).

[11]Dray A. and Bevan S., Inflammation and hyperalgesia: highlighting the team effort, *TIPS* **14**, 287–290 (1993).

these peptides are down regulated upon NGF deprivation. These peptide-containing neurons are nociceptive sensory neurons with unmyelinated or thinly myelinated axons (paper 39). Animals receiving pre- or post-natal treatment with NGF antibodies show a decreased sensitivity to mechanical and heat stimuli. It has been suggested that the loss of putative nociceptive neurons is responsible for the hypoalgesia that has been observed.[12]

As has been illustrated in Part III of this volume, NGF is also a potent stimulator of mast cells, which not only possess high affinity NGF receptors, but also synthesize, store and release NGF as may be seen from that reported in paper 44. Interestingly, cytokines such as interleukin 1β (IL-1) and Tumor Necrosis Factor-∂ (TNF∂), are not only important regulatory factors in immune and inflammatory processes, but also strong inducers of NGF synthesis.[13]

Similarities between NGF and cytokines are not restricted to an overlap in their cellular targets, but involve a more complex series of biological functions that are of critical importance for the integrity and survival of multicellular organisms.[14,15]

The concept of an NGF modulation of homeostatic processes has received further support from recent studies which demonstrate that NGF blood levels are significantly increased in humans both prior to and in the wake of psychologically stressful and anxiety-causing events.[16,17] It is of interest that serum levels of interleukin-1, tumor necrosis factor-∂, cortisol and adrenocorticotropic hormones are, however, not modified. Together, the data suggest that NGF may play an "alerting" role in regulatory processes involving homeostatic adaptive mechanisms.

Studies performed in other laboratories have shown that NGF and its receptors are expressed in the developing rat hypothalamus and that NGF is also synthesized by the pituitary gland. Thyroid and parathyroid glands in rats express NGF, its mRNA and the precursor protein.[18] NGF stimulates the pituitary-adrenocorticotropic hormone and the concentration of plasma glucocorticoids. Plasma NGF levels in women during labor and lactation are increased in comparison with concentrations found during the

[12]Lewin R. and Mendell L.M., Nerve growth factor and nociception, *TINS* **16**, 353–359 (1993).

[13]Keffer J., Probert L., Cazlaris E., Kiooussis D. and Kollias G., Transgenic mice expressing human tumour necrosis factor: a predictive genetic model of arthritis, *Embo J.* **10**, 13, 4025–4031 (1991).

[14]Rothwell N.J. and Hopkins S.J., Cytokines and the nervous system II: actions and mechanisms of action, *TINS* **18**, 130–136 (1995).

[15]Luheshi G.N., Hammond E. and Van Dam A.M., Cytokines as messengers of neuroimmune interactions, *TINS* **19**, 46–47 (1996).

[16]Aloe L., Bracci-Laudiero L., Alleva E., Lambiase A., Micera A. and Tirassa P., Emotional stress induced by parachute jumping enhances blood nerve growth factor levels and the distribution of nerve growth factor receptors in lymphocytes, *Proc. Natl. Acad. Sci. USA* **91**, 10440–10444 (1994).

[17]Black P.H., Central nervous system-immune system interactions: psychoneuroendocrinology of stress and its immune consequences, *Antimicrobial Agents and Chemotherapy* **38**, 1, 1–7 (1994).

[18]Dicou E., Lee J. and Brachet P., Synthesis of nerve growth factor mRNA and precursor protein in the thyroid and parathyroid glands of the rat, *Proc. Natl. Acad. Sci. USA* **83**, 7084–7088 (1986).

term of pregnancy and in non-pregnant women. Furthermore, intravenous injection of oxytocin in virgin rats increases hypothalamic NGF content (paper 41).[19]

The ensemble of these results suggests that NGF acts as a modulator of the neuro-endocrine-immune triad. In the context of a network of interconnecting systems, circulating NGF represents a molecule whose activities are not restricted to either single organs or systems, but which instead assumes a pivotal role in the global homeostasis of the organism. NGF can thus act as a molecule able to recruit and prime both local tissues and systemic defense processes.

Within this framework, the mast cell can be viewed as a gatekeeper between the immune and nervous systems.[20,21] Mast cells are not only involved in hypersensitivity, they also actively participate in autoimmune diseases and in inflammatory processes of various types, especially those of a neurogenic nature. Not only do mast cells synthesize and secrete NGF, but NGF itself has both survival and sensitizing effects on mast cells (paper 43).

Paper 42 outlines a natural endogenous defense mechanism that protects the organism against the consequences of uncontrolled mast cell activation as occurs in inflammatory processes.[22]

Paper 45 lists some of the most crucial aspects of NGF, from the molecular level to that of the organism, which have been brought to light in recent years.

The functional role of this protein molecule, as viewed almost half a century after its discovery, is prospected in paper 46.

The early discovery of the neurotrophic function of NGF in two cell lines belonging to the PNS has been fully confirmed and extended to an ever-increasing number of nerve cell populations in the CNS. The recent finding of NGF-responsive cells in the immune and endocrine systems indeed suggests that this protein molecule plays a much broader role than had been initially conceived.

NGF appears now, rather, as a pleiotropic agent with an unexpectedly broad spectrum of target cells, something which could not have been foreseen on the basis of earlier studies which established its neurotrophic function. In reality, NGF is able to modulate neuro-endocrine-immune functions of vital importance to the regulation of homeostatic interactions.

NGF may help to co-ordinate components of systems involved in peripheral tissue defense mechanisms, as a result of its ability to interact with most of their

---

[19]Richard P., Moos F. and Freund-Mercier M.J., Central effects of oxytocin, *Physiol. Rev.* **71**, 2, 331–355 (1991).

[20]Orr E.L., Nervous-system-associated mast cells-gatekeepers of neuronal and immune interactions, *Drug Dev. Res.* **15**, 195–205 (1988).

[21]Galli S.J., Biology of disease, new insights into 'The Riddle of the Mast Cells': microenvironmental regulation of mast cell development and phenotypic heterogeneity, *Lab. Invest.* **62**, 1, 5–33 (1990).

[22]Facci L., Dal Toso R., Romanello S., Buriani A., Skaper S.D. and Leon A., Mast cells express a peripheral cannabinoid receptor with differential sensitivity to anandamide and palmitoylethanolamide, *Proc. Natl. Acad. Sci. USA* **92**, 3376–3380 (1995).

elements. Such defense mechanisms include both the local activation of tissue-specific cells and the recruitment at the site of injury of elements of the homeostatic systems (immune and nervous). NGF-induced sensitization of mast cells, as well as that of sensory nerves via increase of neuropeptide synthesis and their release from central and peripheral terminals, can contribute to an enhanced tissue responsiveness toward noxious stimuli.

NGF can also recruit and activate, together with other cytokines, cellular elements of the immune system. These cells, by releasing cytokines, stimulate NGF synthesis in fibroblasts, Schwann cells and others — thereby helping to maintain the elevated NGF levels found in inflamed tissue and inflammatory exudates. NGF does not directly induce hyperalgesia or pain; rather, it reduces the threshold of mechanical and thermal nociceptors and thus generates, wherever its local concentration is increased, areas of heightened sensitivity to pressure and heat.

In the context of the interconnected neuro-endocrine-immune triad, circulating NGF represents a molecule that acts on any cell type which expresses functional NGF receptors belonging to any one of these systems. Such a role suggests a new link between neuronally-controlled mediators and the immune system. At physiological concentrations, NGF alone might not directly activate cells of immune lineage, but instead modulate their sensitivity to other triggering stimuli. All of the above findings support the hypothesis that circulating NGF might function as a general 'alerting' signal utilized by the brain in conditions of stress and anxiety to 'prime' the immune system against external noxious perturbations. They greatly expand the range of biological actions attributable to NGF.

NGF is therefore beginning to emerge as a crucial coordinator of the body's major homeostatic systems and defense strategies.[23]

The elucidation of autoregulatory mechanisms in inflammatory and autoimmune diseases will undoubtedly forward our understanding of the role of mast cells and NGF under both physiological and altered homeostatic conditions.

[23] Levi-Montalcini R., Skaper S.D., Aloe L. and Leon A., Nerve growth factor, *Encyclopedia of Neuroscience* (1996), in press.

Progress in NeuroEndocrinImmunology Vol. 3, No. 1, 1990

PNEI REVIEW

A Role for Nerve Growth Factor in Nervous, Endocrine and Immune Systems

Rita Levi-Montalcini, Luigi Aloe and Enrico Alleva*

Institute of Neurobiology, Consiglio Nazionale delle Ricerche, Rome, Italy
**Section of Behavioral Pathophysiology, Laboratorio di Fisiopatologia O.S., Istituto Superiore di Sanità, Rome, Italy*

SYNOPSIS: A major shift in nerve growth factor (NGF) studies has occurred with the realization that the spectrum of action of this protein molecule is not restricted, as previously believed, to enhancing differentiative processes of two peripheral neuronal cell lines (the sensory and sympathetic neurons), but extends also to several other neuronal and non-neuronal cell types. The newly detected NGF-responsive cells belong to the hemopoietic-immune system, to the cholinergic system of the basal forebrain nuclei, and to other cell populations of the central nervous system involved in neuroendocrine functions. It is the object of this article to report on findings that project a much wider role for this protein molecule than initially had been envisaged. Additional evidence promoting this concept came from results of other investigations showing that NGF synthesis in some hypothalamic nuclei, and its level in the bloodstream of mice displaying aggressive behavior, are markedly increased, as compared to controls. These and previous findings support the hypothesis that this protein molecule may exert a modulatory role on neuro-immuno-endocrine functions of vital importance in the regulation of homeostatic processes. (*Progress in NeuroEndocrinImmunology* **3**:1-10, 1990)

"One of the most gratifying things about science is that it is impossible to predict just where a particular trail of research will lead. Indeed, an extended investigation may wind up shedding light on a long-standing puzzle in a field that initially appeared to be quite distant." (1)

In all sciences, but perhaps especially in biological sciences, investigations pursued with different techniques in apparently unrelated areas of research suddenly merge together, opening up new vistas in the ever-widening scenario of control mechanisms in living organisms. This trend is evidenced by the realization of the tight interactions among the three main homeostatic systems: the nervous, immune and endocrine systems (2-6). Such interaction, and its far-reaching implications in both basic research and clinically applied science, has been documented (7).

A new dimension of the interactions taking place among cell populations belonging to these three systems came to light with the realization that the role of the protein molecule known as nerve growth factor (NGF) extends far beyond that formerly attributed to it (8,9). It is the object of this article to re-examine the past in the context of new findings that support this hypothesis.

Address correspondence to: Professor Rita Levi-Montalcini, Institute of Neurobiology, Consiglio Nazionale delle Ricerche, Viale Marx 15, I-00156 Rome, Italy.

Discovery of the NGF polypeptide in 1954 was foretold by its property of enhancing dramatically the growth and differentiative processes of two peripheral nerve cell types, the sensory and sympathetic cells (10). This molecule is synthesized and released in varying quantities from different animal sources: malignant mouse tumor (11), snake venom (12), submaxillary male mouse salivary gland (13), and accessory male sex glands (14,15). It also is synthesized in trace amounts in cells of tissues and organs innervated by sensory and sympathetic nerves (16). The physiological role of NGF was demonstrated by the finding that administration to neonatal animals with specific antibodies to this molecule results in the destruction of sympathetic nerve cells, a process known as immunosympathectomy (17). Additional evidence in support of this role came from the finding that NGF produced in peripheral tissues and organs is transported in a retrograde manner to the innervating neurons. Impedance of this transport system results in death of these cells (18-20).

These studies signaled the opening of an entirely new area of research, leading to the discovery of other polypeptides endowed with properties similar to those of NGF, but acting on different cell lines. Further investigations of NGF in the following two decades confirmed early findings and centered on the study of this molecule and its mechanism of action on the two receptive nerve cell types (8,9).

However, studies beginning in 1977-78 identified unforeseen properties of NGF, properties which demonstrated that this molecule exerts identical biological activity on a cell line belonging to the immune system. Subsequent studies showed that many other cell types belonging to the nervous, immune and endocrine systems are likewise receptive to the biological activity of NGF, suggesting a much broader role for this molecule than was first conceived. A list of these findings is given in the following section.

NGF-Peripheral Nerve Cells

Ever since the discovery that NGF, whatever its source, elicits hypertrophy of sensory and sympathetic ganglia in chick embryos, it has been accepted that not all nerve cells in these ganglia undergo enhanced differentiation and hypertrophy upon NGF administration (8,9). The number and physiological properties of NGF-receptive sensory and sympathetic cells differs in the two neuronal populations, as will be reported below.

Dorsal root ganglia (DRG). Although the NGF effect also extends to sensory nerve cells in some cephalic sensory ganglia, most of the investigations centered on DRG in which NGF-receptive and -non-receptive cells form two topographically distinct subpopulations in avian species (21,22). During early embryonic stages, NGF-receptive cells represent about 50% of the total DRG population. Their number sharply decreases in late embryonic stages as a result of natural cell death, which reduces their total population approximately by half (21,23). Studies performed on adult rodents and other species provide evidence that the number of NGF-receptive cells amounts to about 20% of the total DRG nerve cell population (23,25). The identification of these cells as substance P (SP)-synthesizing and -releasing neurons (26,27) is predicated on several criteria, which will be listed below.

NGF treatment markedly increases SP content of DRG in both neonatal and adult rats and also causes hypertrophy of a percentage of postnatal DRG sensory neurons (24-28). Exposure to NGF antiserum leads to a significant reduction in the SP level in ganglia of adult rodents (24,25). NGF exposure protects the SP-containing DRG neurons from the toxic effects of capsaicin (29).

SP-synthesizing and -releasing neurons belong to the category of small sensory neurons. Their thin, unmyelinated axons, designated as C fibers, terminate in the superficial layers of the dorsal horn of the spinal cord. Through an internuncial neuron, they convey nociceptive signals to the contralateral spinothalamic tract. However, about 90% of the SP synthesized in the cell bodies of these neurons is antidromically transported to their peripheral terminals (30,31). Release of SP from the nerve endings can be elicited by either electrical or physiological stimuli (32). SP discharge results in local plasma leakage, mast cell degranulation (33) and vasodilation, resulting in hyperemia and other effects typical of neurogenic inflammation (34,35) (Figure 1).

Another neuropeptide, somatostatin, is also released from these fibers. Both SP and somatostatin modify diverse activities of

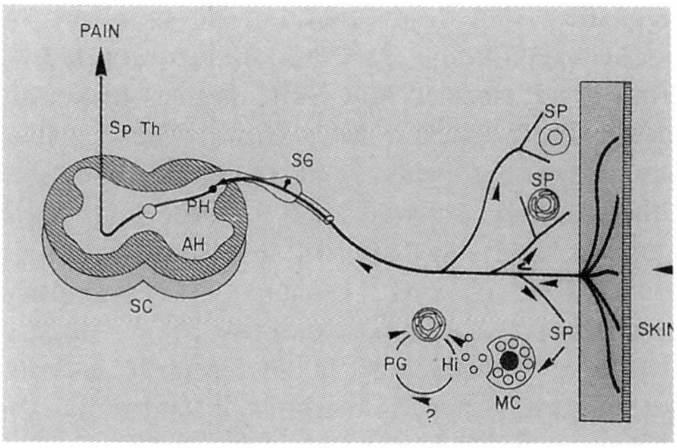

Figure 1. This drawing illustrates the mechanism, suggested by Pernow (34), underlying the axon reflex in the skin. Noxious stimuli (indicated by arrow) elicit afferent impulses in a sensory neuron, giving a sensation of pain, probably via a release of SP in the dorsal horn of the spinal cord. Other impulses travel antidromically, leading to a release of SP at the sensory nerve. **SP** = substance P; **M** = mast cells; **Hi** = histamine; **PG** = prostaglandin; **SG** = spinal ganglia; **SC** = spinal cord; **PH** = posterior horn; **AH** = anterior horn; **SpTh** = spinothalamic tract.

lymphocytes, PNM leukocytes and monocytes, resulting in the local modulation of the immune response. Thus, as stated by Payan *et al.*, "The receptor mediated interactions that evoke lymphocyte responses provide one mechanism through which the PNS may influence localized immune responses." (36). Daily NGF injections greatly enhance this response (26).

Sympathetic ganglia. Since the 1960s, two distinct classes of adrenergic neurons, the long and the short adrenergic neurons, have been described (37,38). They differ from each other in possessing a long or short axon (a morphological feature that correlates with their location either distant from or close to their effectors), and also in many other physiological and pharmacological properties. The finding that the short adrenergic neurons of the ganglionic complexes innervating the urogenital organs in both sexes are receptive to the action of male and female steroid hormones (39,40) led to the hypothesis that these neurons belong to a peripheral neuroendocrine system endowed with important motor functions in these organs. Early studies gave evidence for the striking differential response elicited by NGF in the long and short adrenergic neurons. The former, but not the latter, are highly receptive to this polypeptide (41,42). The long adrenergic neurons are assembled in the paraventricular chain ganglia at the two sides of the spinal cord and in the

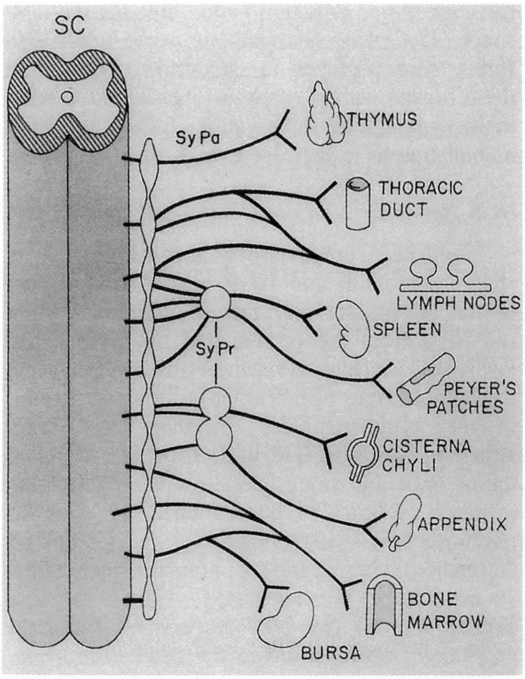

Figure 2. Innervation by long para- and prevertebral sympathetic noradrenergic nerve fibers of organs and viscera belonging to the immune system. **SC** = spinal cord; **SyPa** = sympathetic paraventricular ganglia; **SyPr** = sympathetic preventricular ganglia.

prevertebral ganglia, which consist of large nerve cell aggregates located at the origin of the vessels of the abdominal aorta. They innervate immune-mediating organs and structures such as the thymus, spleen, lymph nodes, gut-associated lymphoid tissue and bone marrow (Figure 2). In all these organs, bundles of sympathetic fibers enter in close association with blood vessels, travel in the parenchymal tissue, and branch out among single cells; these fibers form dense tangles among lymphocytes and several other cell types without, as a rule, establishing synaptic connections with individual cells. Their presence in these lymphoid organs in such a dense array suggests a structural link between the nervous and immune systems that could be utilized for translating neural messages into chemical signals that interact with specific cellular elements of the immune system (43,44). The immunomodulatory role of NGF was shown by the higher density and fluorescence of varicose adrenergic fibers and terminals in the immunocompetent organs, by the total increase in the catecholamines in these organs, and by the catecholamine diffusion into the intercellular space (41,42,45). Sympathetic nerve fibers also form dense plexuses in endocrine organs. In these organs, as in lymphoid organs and tissues, their sympathetic innervation following NGF administration is increased markedly (41,42).

NGF-Nerve Cells in the Central Nervous System

Evidence compiled over the last few years shows that NGF and NGF mRNA are present in the developing and mature vertebrate brain (46-50). Both the protein and the mRNA are synthesized in high concentrations in the cortex, the hippocampus (48,49), and, to a lesser extent, in the hypothalamus. In this latter brain structure, the NGF and its mRNA are found in the preoptic and ventrolateral nucleus, where the synthesis appears to be regulated by stressful psychosocial events (50-52). Combined autoradiographic and immunohistochemical studies have demonstrated NGF receptor expression on the NGF-responsive forebrain cholinergic neurons and in the cerebellum, both in developing and in adult males (53,54). Radiolabeled NGF injected into the rat hippocampus or cerebral cortex is taken up selectively by cholinergic terminals and is transported axonally in a retrograde manner to the forebrain NGF-responsive magnocellular cholinergic neurons (46,47). The biological significance of NGF in the central nervous system is supported by findings that exogenous NGF administration causes an increase of choline acetyltransferase activity (55-57) and locomotory activity upon injection of scopolamine, a specific muscarinic blocker (58). Moreover, this molecule enhances the survival of lesioned basal forebrain neurons following transection of the septo-hippocampal pathways (55-59).

NGF-Cell Lines of the Immune System

The first evidence showing that cells belonging to the immune system lineage are receptive to the action of NGF came from our laboratory in 1977 (60). In these studies, it was shown that *in vivo* administration of NGF to neonatal rodents caused a widespread increase in the size and number of mast cells (MCs) in several peripheral tissues. In a subsequent report, we demonstrated that spleen cells cultured in the presence of physiological amounts of NGF are phenotypically converted into MCs (61), suggesting that NGF exerts a direct effect on rodent spleen cells. In 1982, studies reported by Bruni *et al.* showed that NGF induces histamine release from isolated MCs (62). These results later were confirmed and extended by other authors, who showed that intradermal NGF injection results in a specific release of histamine from MCs (63,64), causing a concomitant increase of vascular permeability (65). Since it has been shown that, in the central nervous system, MCs act as cellular gatekeepers for interaction between the immune and the nervous systems (66), it would be of interest to explore the effect of NGF on this mechanism.

Recently, numerous reports have been published which provide evidence that lymphocytes and accessory cells express cell surface receptors for NGF (67), and that NGF is involved in the modulation of such lymphocytic activities as proliferation and/or differentiation (45,68-70). It was shown that lymphocytes (71), as well as activated macrophages (72), can synthesize and release NGF. More recently, it was reported that during development, the Bursa of Fabricius, the spleen and the thymus express high levels of NGF-mRNA receptors (54). Collectively, these studies support the notion of an NGF role in cell lines of the immune system.

Our group recently has carried out studies aimed at elucidating the effect of NGF in the developing and adult thymus gland. The results indicate that, both in young and in adult rodents, thymic cells are responsive to NGF and to NGF antibodies (Levi-Montalcini, Aloe, Santoni, Vigneti, unpublished results).

NGF Antibodies: The Neuroendocrine Immune Deficiency Syndrome

As previously reported by our laboratory, injection of rat fetuses with specific antibodies to NGF (AB-NGF) at the 16th to 17th day of gestation results in a marked neuroendocrine deficit in postnatal life (73). A high percentage of offspring were severely affected and did not survive beyond a few days of post-natal life. After birth, AB-NGF-injected animals showed a progressive decrease in vitality, exhibiting sluggish, apathetic behavior and moving only a few steps when exposed to strong stimuli. They also differed from controls in their very weak reaction to pressure, pricking, burning and other noxious stimuli. Their body temperature was markedly lower than that of vehicle-injected rats.

Histological studies showed that sympathetic ganglia were reduced to sclerotic nodules consisting of glial cells, fibroblasts and a few small atrophic sympathetic neurons. Sensory ganglia were reduced markedly in size. Among endocrine organs, the most extensively examined gland, the thyroid, appeared severely reduced in size. Using a different methodology, deleterious effects of NGF deprivation also were described in rabbits and guinea pigs (24). These authors showed that maternal exposure to AB-NGF during fetal development causes, in newborn pups, loss of body weight, as well as sensory deficits including insensitivity to heat, cold, pinpricks, pinching and corneal irritation, resulting in a high percentage of animal death (25).

The mechanism(s) of action leading to this lethal effect is not yet clear. A plausible explanation for this dramatic effect is that AB-NGF injected during prenatal life inactivates the endogenous NGF co-localized in neuroendocrine structures, thus causing the widespread neuroendocrine immune deficiency syndrome observed in these studies. Recent reports showing that NGF is synthesized and stored in the hypothalamus (51,52) and hypophysis (74), and that NGF stimulates the pituitary-adrenocortical axis (75), would support this hypothesis.

NGF Activation of Aggressive and Defensive Behavior

An entirely new feature of the biological activity of NGF became evident in 1986, when we reinvestigated one of the most intriguing unexplained findings in NGF studies: the synthesis of large quantities of this polypeptide in the convoluted tubules of the adult male mouse submaxillary salivary glands (76). Results accumulated during the past three decades showed that this salivary product is routed through the gland's excretory duct into the saliva in untreated, as well as in pharmacologically stimulated, mice (77). The failure to detect NGF in the blood serum seemed to militate against the hypothesis that this molecule served some physiological function, unrelated to that thought to exist in the digestive system. The potent nerve growth stimulating effect of this polypeptide then was tentatively listed in the large category of bizarre "side effects" of molecules produced for entirely different purposes in organisms of the vegetal and animal kingdoms.

Unwilling to accept this interpretation, we resorted to a different experimental approach, suggested by considerations based on the specific behavior of wild mice. Under natural conditions, male mice exhibit aggressive behavior against males of the same colonies. They engage in fighting behavior that results in severe biting and even killing of one of the two combatants. Similarly aggressive behavior is displayed by laboratory-reared male mice maintained for a 4 to 6 week period in social isolation and then paired with another male in the same cage (78). The finding of high NGF levels in the blood of both contenders immediately after each aggressive episode hinted that this NGF released from the salivary glands could be endowed with a definite physiological role activated and/or greatly enhanced by specific stressful conditions (79).

Previous studies, indicating that administration of exogenous NGF cells elicits hypertrophy of the adrenal gland and enhanced functional activity of the medullar and cortical sections of this gland, suggested that this organ would be the most likely target of the polypeptide (80,81). This hypothesis received confirmation from studies demonstrating that prolonged fighting episodes indeed result in hypertrophy of both cortical and medullary components of the adrenal (79).

An additional significant finding in the studies of the NGF role in agonistic interaction in mice was that the level of this molecule in the circulating blood consistently reaches a higher peak in the subordinate rather than in the dominant partner (81,82). Thus, the massive NGF discharge from the salivary glands into the bloodstream results not only from activation of aggressive behavior but also from anxiety, which obviously is more acute in the subordinate mouse. The significance of these findings will be considered further later in this article.

NGF: From a Specific Growth Factor to a Modulator of the Three Main Homeostatic Systems

Neither NGF administration nor daily injections of any specific antibodies to this polypeptide in neonatal and young rodents results in any marked changes in their somatic development and behavior. These findings seemed consistent with the generally accepted idea that NGF, although its role was broader than first conceived, did not serve any specific biological function (8,9). Results (reported in previous sections) revealed other facets of the NGF spectrum of action which suggest that it may, on the contrary, be endowed with a most important general role consisting of the synergistic modulation of the nervous, endocrine and immune systems.

Initial support for this hypothesis came from results, reported above, showing that NGF-target cells are not randomly distributed throughout the peripheral and central nervous systems, nor in other organs or tissues. Rather, it became apparent that they are concentrated in distinct areas: the nervous, endocrine and immune systems. Since each of these consists of billions of cells endowed with diversified activities, it was of great interest to determine whether the NGF-responsive cells, irrespective of their specific properties, share any trait in common that would benefit from the NGF property of stimulating their differentiation and function. We suggest that this common trait is the role these systems play in homeostasis (2-4) (Figure 3).

Each of the NGF effects on target cells (listed above) calls for some comment. We reported that not all sensory neurons, but only those belonging to a large subpopulation of these cells, are highly receptive to NGF biological activity. The identification of these cells is based on morphological and biochemical criteria reported in detail in previous articles (8,9,50). Among the most revealing features of their morphological and functional properties are their small size, and also the synthesis and release of SP from their central and peripheral axons. The central projections of these neurons connect through an interneuron with the contralateral spinothalamic fiber tract and convey pain-related impulses to the thalamus. Noxious and other stimuli travel antidromically in the peripheral axons (31,32). The SP released from their peripheral nerve terminals triggers multiple responses typical of neurogenic immune-related inflammation, such as hyperemia consequent to local vasodilation, mast cell activation resulting in histamine release, plasma leakage, and smooth muscle contraction (34-36). Substance P release from central axons also is enhanced by NGF administration, as indicated by increased

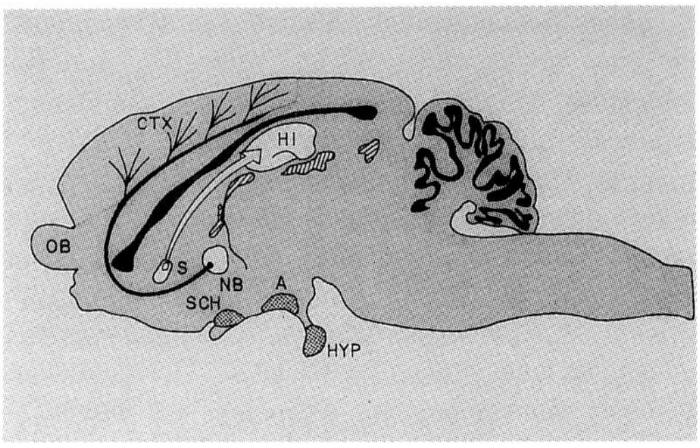

Figure 3. Nerve growth factor (NGF) synthesizing and receptive cholinergic forebrain centers and pathways as described in text. The same diagram illustrates hypothalamic centers and hypophysis involved in NGF synthesis. **CTX** = cortex; **HI** = hippocampus; **OB** = olfactory bulb; **S** = septum; **NB** = nucleus basalis; **SCH** = suprachiasmatic nucleus; **A** = arcuate nucleus; **HYP** = hypophysis.

sensitivity toward nociceptive stimuli (26,28,65) in rodents injected with NGF, and the reduced response of these animals to pain following administration of specific antibodies to NGF (25,26). We have observed recently that, following intravenous NGF administration, adult mice show a hyperalgesic response in a hot-plate test (Alleva, Aloe and Levi-Montalcini, unpublished results).

Medium-sized and large neurons in spinal and cephalic sensory ganglia are non-responsive, or weakly responsive, to NGF (8,9,50). The cell size and large diameter of the myelinated axons of these neurons set them in the class of proprioceptive neurons and of neurons endowed with functional properties different from those of small NGF target nerve cells. Only these latter cells would be involved in activities related to control of homeostatic processes.

As described above, the long adrenergic neurons, but not the short adrenergic neurons or the mature small intensely fluorescent cells, are highly receptive to NGF (50), which plays an essential role in promoting their differentiation and in enhancing their function throughout their life (Figure 4).

A Guide for Future Investigation

In few instances does the statement by Washburn that "A theory is only a licence for research" (83) better apply than to the actual state of research on NGF. The concept formulated almost four decades ago, of a polypeptide which became known as NGF on the

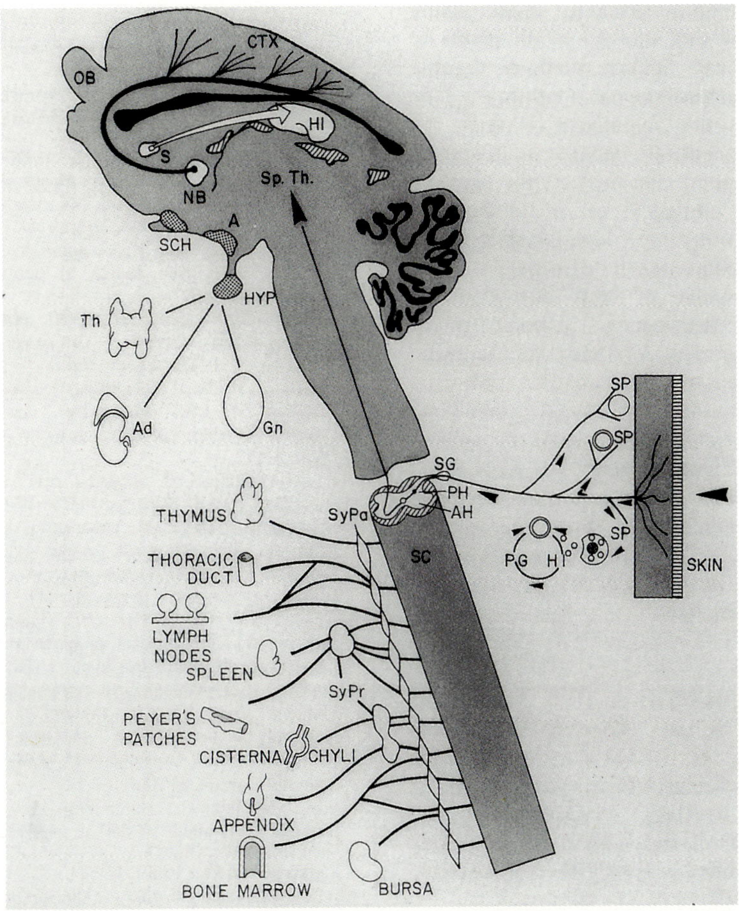

Figure 4. Illustration of the hypothesized nerve growth factor role in nervous, endocrine and immune homeostatic systems shown in detail in Figures 1-3. **A** = arcuate nucleus; **AD** = adrenal gland; **AH** = anterior horn; **CTX** = cortex; **GN** = gonad; **HI** = hippocampus; **Hi** = histamine; **Hyp** = hypophysis; **M** = mast cells; **NB** = nuclear basalis; **OB** = olfactory bulb; **PH** = posterior horn; **PG** = prostaglandin; **S** = septum; **SC** = spinal cord; **SCH** = suprachiasmatic nucleus; **SG** = spinal ganglia; **SP** = substance P; **SpTh** = spinothalamic tract; **SyPa** = sympathetic paraventricular ganglia; **SyPr** = sympathetic preventricular ganglia; **Th** = thyroid gland.

basis of its growth-differentiating and function-enhancing activity on two peripheral nerve cell types, requires expansion. The hypothesis submitted in this article does not in any way conflict with or negate the results which first brought this polypeptide to our attention. Innumerable studies performed in our own and in a large number of other laboratories certify the validity of early observations. At the same time, however, these studies reveal the property of NGF to influence an ever-increasing number of cells belonging to the immune, neuroendocrine and central nervous systems. A detailed analysis of NGF-receptive cells highlighted a startling finding: all of them in one way or another activate homeostatic processes and reinforce the physiological defense mechanisms of the organism. The most impressive of these findings, and one that in no way could be listed as a function attributable to a specific growth factor, was the discovery of the massive release of NGF into the blood circulation during episodes of aggressive behavior, which suggested to us that NGF is endowed with a far more general and vital role in the organism than that previously assigned to it. This hypothesis found further support in the now generally acknowledged concept that the nervous, endocrine and immune systems share many signals in common and interact at all levels to protect the organism against noxious organic agents or hostile environmental conditions. The heuristic value of this hypothesis is dual: it offers a guideline, entirely lacking in the past, for future investigation, and projects the possible utilization of this molecule, or of its specific antibodies, in restoring the functions of systems suffering from pathological disorders. The potential clinical value of NGF and/or of its antibodies in the treatment of a long list of disorders affecting any of these homeostatic mechanisms is obvious.

The ever-expanding NGF scenario before us in these past several decades warns us against believing that we have at long last discovered the entire spectrum of action and pinpointed the only biological function of this polypeptide. New findings may yet reveal other properties of NGF which may or may not fit into the role projected in this article.

ACKNOWLEDGMENTS: This work was supported in part by the "Progetto Finalizzato, Biotechnologia e Biostrumentazione: Sottoprogetto, Biofarmaci (Consiglio Nazionale delle Ricerche)" to Rita Levi-Montalcini and Luigi Aloe. The authors wish to thank Linda Delucchi for her help in typing the manuscript.

REFERENCES

1. Edelman GM 1989 Topobiology. Sci Am 260:76
2. Kaltenbach JC 1988 Endocrine aspects of homeostatis. Am Zool 28:761
3. Bernard C 1858 Leçons sur la Physiologie et la Pathologie du Systeme Nerveux. Balliere, Paris
4. Cannon WB 1929 Bodily Changes in Pain, Hunger, Fear and Rage. Appleton-Century-Crofts, New York
5. Cotman CW, Brinton R, Galaburda A, Israel B, McEwen B, Schneider DM 1987 (eds) The Neuro-Immune-Endocrine Connection. Raven Press, New York
6. Blalock JE 1988 Neuroendocrine peptide hormones in the immune system. Prog NeuroEndocrinImm 1(1):9
7. Kordon C, Bihoreau C 1989 Integrated communication between the nervous, endocrine and immune systems. Horm Res 31:100
8. Levi-Montalcini R, Angeletti PU 1968 The nerve growth factor. Physiol Rev 48:534
9. Thoenen H, Barde YA 1980 Physiology of nerve growth factor. Physiol Rev 60:1284
10. Levi-Montalcini R, Meyer H, Hamburger Y 1954 *In vitro* experiments on the effects of mouse sarcoma 180 and 37 on the spinal and sympathetic ganglia of the chick embryo. Cancer Res 14:49
11. Levi-Montalcini R 1952 Effects of mouse tumor transplantation on the nervous system. Ann NY Acad Sci 55:330
12. Cohen S 1959 Purification and metabolic effects of a nerve growth-promoting protein from snake venom. J Biol Chem 234:1129
13. Levi-Montalcini R, Cohen S 1960 Effects of the extract of the mouse salivary glands on the sympathetic system of mammals. Ann NY Acad Sci 85:324
14. Harper GP, Barde YA, Burnstock G, Carstairs JR, Dennison ME, Suda K, Vernon CA 1979 Guinea pig prostate is a rich source of nerve growth factor. Nature 279:160
15. Harper GP, Thoenen H 1980 The distribution of nerve growth factor in the male sex organs of mammals. J Neurochem 77:391
16. Thoenen H, Barde YA, Edgar D, Hatanaka H, Otten U, Schwab M 1979 Mechanism of action and possible sites of synthesis of nerve growth factor. Prog Brain Res 51:95
17. Levi-Montalcini R, Angeletti PU 1966 Immuno-sympathectomy. Pharmacol Rev 18:619
18. Stockel K, Schwab M, Thoenen H 1975 Comparison between the retrograde axonal transport of nerve growth factor and tetanus toxin in motor sensory and adrenergic neurons. Brain Res 99:1
19. Hendry TA, Campbell J 1976 Horphometric analysis of rat superior cervical ganglion after axotomy and nerve growth factor treatment. J Neurocytol 5:351
20. Aloe L, Levi-Montalcini R 1979 Nerve growth factor induced overgrowth of axotomized superior cervical ganglion in neonatal rats: similarities and differences with NGF effects in chemically axotomized sympathetic ganglia. Arch Ital Biol 117:287
21. Levi-Montalcini R, Hamburger V 1951 Selective growth stimulating effects of mouse sarcoma on the sensory and sympathetic nervous system of the chick embryo. J Exp Zool 116:321
22. Levi-Montalcini R 1962 Analysis of a specific nerve growth factor and of its antiserum. Sci Repts Ist Super Sanità 2:345
23. Landmesser L, Pilar G 1976 Fate of ganglionic synapses and ganglion cell axons during normal and induced cell death. J Cell Biol 68:357
24. Johnson EM Jr, Osborne PA, Rydel RE, Schmidt RE, Pearson J 1983 Characterization of the effects of autoimmune nerve growth factor deprivation in the developing guinea pig. Neuroscience 8:631
25. Yip HK, Rich KM, Lampe PA, Johnson EM Jr 1984 The effects of nerve growth factor and its antiserum on the postnatal develpment and survival after injury of sensory neurons in rat dorsal root ganglia. J Neurosci 4:2986
26. Otten U, Baumann JB, Girard J 1985 Nerve growth factor induces plasma extravasation in rat skin. Eur J Pharmacol 106:199
27. Schwartz JP, Costa E 1979 Nerve growth factor-mediated increase of the substance P content of chick embryo dorsal root ganglia. Brain Res 170:198
28. Goedert M, Stoeckel K, Otten U 1981 Biological importance of the retrograde axonal transport of nerve growth factor in sensory neurons. Proc Natl Acad Sci USA 78:5895
29. Miller MS, Buck SH, Sipes IG, Yamamura HI, Burks

TF 1982 Regulation of substance P by nerve growth factor: disruption by capsaicin. Brain Res 250:193

30. Brimjoin S, Lundberg JM, Brodin E, Hokfelt T, Nilsson G 1980 Axonal transport of substance P in the vagus and sciatic nerves of the guinea pig brain. Brain Res 191:443
31. Mandahl A, Bill A 1981 Ocular responses to antidromic trigeminal stimulation, intracameral prostaglandin E_1 and E_2, capsaicin and substance P. Acta Physiol Scand 112:331
32. Jessell TM 1985 Cellular interactions at the central and peripheral terminals of primary sensory neurons. J Immunol 135:746
33. Light AR, Perl ER 1979 Spinal termination of functionally identified primary afferent fibres with slowly conducting myelinated fibres. J Comp Neurol 186:133
34. Pernow B 1985 Role of tachykinins in neurogenic inflammation. J Immunol 135:812
35. Bynke G 1983 Capsaicin pretreatment prevents disruption of the blood-aqueous barrier in the rabbit eye. Invest Ophthalmol Vis Sci 24:744
36. Payan DG, Goetzl EJ 1985 Modulation of lymphocyte function by sensory neuropepetides. J Immunol 135:783
37. Sjoestrand NO 1962 Inhibition by ganglionic blocking agents of the motor response of the isolated guinea-pig vas deferens to hypogastric nerve stimulation. Acta Physiol Scand 54:306
38. Sjostrand NO 1965 The adrenergic innervation of the vas deferens and the accessory male genital glands. Acta Physiol Scand 65:1
39. Owman D, Sjoberg N-O 1977 Influence of pregnancy and sex hormones on the system of short adrenergic neurons in the female reproductive tract. In James VHT (ed) Endocrinology, Vol 1, Proceedings of the 5th International Congress of Endocrinology. Excerpta Medica, Amsterdam, p 305
40. Owman C, Alm P, Sjoberg N-O 1983 Pelvic autonomic ganglia: structure, transmitters, function and steroid influence. In Elfvin L-G (ed) Autonomic Ganglia. John Wiley & Sons, New York, p 125
41. Levi-Montalcini R 1972 The morphological effects of immunosympathetectomy. In Steiner G, Schönbaum E (eds) Immunosympathectomy. North-Holland Publishing Company, Amsterdam, p 55
42. Levi-Montalcini R, Angeletti PU 1966 Immunosympathectomy. Pharmacol Rev 18:619
43. Martin JB 1988 Neural effector mechanisms that modify or modulate the immune response Prog NeuroEndocrinImm 1(1):5.
44. Felten DL, Felten SY, Carlson SL, Olschowka JA, Livnat S 1985 Noradrenergic and peptidergic innervation of lymphoid tissue. J Immunol 135:755
45. Abramichik GV, Termakova SS, Kaliunov YN, Tanina RM, Tumilovich MK 1988 The immunomodulatory effect of nerve growth factor. J Neurosci Res 19:349
46. Gnahn H, Hefti F, Haumann R, Schwab ME, Thoenen H 1983 NGF-mediated increase of choline acetyltransferase (ChAT) in the neonatal rat forebrain: evidence for a physiological role of βNGF in the brain? Dev Brain Res 3:229
47. Seiler M, Schwab ME 1984 Specific retrograde transport of nerve growth factor from the neocortex to nucleus basalis in the rat. Brain Res 300:33
48. Korsching S, Auburger G, Heumann R, Scott J, Thoenen H 1985 Levels of nerve growth factor and its mRNA in the central nervous system of the rat correlate with cholinergic innervation. EMBO J 4:1009
49. Shelton DL, Reichart LF 1986 Studies on the expression of the β nerve growth factor (NGF) gene in the central nervous system: level and regional distribution of NGF mRNA suggest that NGF functions as a trophic factor for several distinct populations of neurons. Proc Natl Acad Sci USA 83:2714
50. Levi-Montalcini R 1987 The nerve growth factor: thirty-five years later. Science 237:1154
51. Spillantini MG, Aloe L, Alleva E, De Simone R, Goedert M, Levi-Montalcii R 1989 Nerve growth factor mRNA and protein increase in hypothalamus in a mouse model of aggression. Proc Natl Acad Sci USA 87:8555
52. Aloe L, Alleva E, De Simone R 1989 Changes of NGF level in mouse hypothalamus following intermale aggressive behavior: biological and immunohistochemical evidence. Behav Brain Res (in press)
53. Pioro EP, Cuello AC 1988 Purkinje cells of adult rat cerebellum express nerve growth factor receptor immunoreactivity: light microscopic observations. Brain Res 455:182
54. Ernfors P, Hallbook F, Ebendal T, Shooter EM, Radeke MJ, Misko TP, Persson H 1988 Developmental and regional expression of β-nerve growth factor receptor mRNA in the chick and rat. Neuron 1:983
55. Thoenen H, Bendtlow C, Heumann R 1987 The physiological function of nerve growth factor in the central nervous system: comparison with the periphery. Physiol Biochem Pharmacol 109:146
56. Auburger G, Heumann R, Hellweg R, Korsching S, Thoenen H 1987 Developmental changes of nerve growth factor and its mRNA in the rat hippocampus: comparison with choline acetyltransferase. Dev Biol 120:322
57. Whittemore SR, Seiger A 1987 The expression, localization and functional significance of β-nerve growth factor in the central nervous system. Brain Res Rev 12:439
58. Alleva E, Aloe L, Laviola G 1986 Pretreatment of young mice with nerve growth factor enhances scopolamine-induced hyperactivity. Dev Brain Res 28:278
59. Korsching S, Heumann R, Thoenen H, Hefti R 1986 Cholinergic denervation of the rat hippocampus by fimbrial transection leads to a transient accumulation of nerve growth factor (NGF) without change in mRNA NGF content. Neurosci Lett 66:175
60. Aloe L, Levi-Montalcini R 1977 Mast cells increase in tissues of neonatal rats injected with the nerve growth factor. Brain Res 133:358
61. Bohm A, Aloe L, Levi-Montalcini R 1986 Nerve growth factor enhances precocious differentiation and numerical increase in mast cells in cultures of rat splenocytes. Accad Naz Lincei 80:1
62. Bruni A, Bigon E, Boarato E, Mietto L, Leon A, Toffano G 1982 Interaction between nerve growth factor and lysophosphatidylserine on rat peritoneal mast cells. FEBS Lett 138:190
63. Sugiyama K, Suzuki Y, Furuta H 1985 Histamine release induced by 7S nerve growth factor of mouse submandibular salivary glands. Arch Oral Biol 30:93
64. Mazurek N, Weskamp G, Erne P, Otten U 1986 Nerve growth factor induces mast cell degranulation without changing intracellular calcium levels. FEBS Lett 198:31

65. Otten U 1984 Nerve growth factor and the peptidergic sensory neurons. Trends Pharmacol Sci 5:307
66. Orr EL 1988 Nervous system-associated mast cells: gatekeepers of neural and immune interactions. Drug Dev Res 15:195
67. Thorpe LW, Stach RW, Hashim CA, Marchetti D, Perez-Polo JR 1987 Receptors for nerve growth factor on rat spleen mononuclear cells. J Neurosci Res 17:128
68. Thorpe LW, Stach RW, Morgan B, Perez-Polo JR 1988 The influence of nerve growth factor on the *in vitro* proliferative response of rat spleen lymphocytes. J Neurosci Res 18:134
69. Bienenstock J, Denburg J, Scicchitano R, Stead R, Perdue M, Stanisz A 1988 Role of neuropeptides, nerves and mast cells in intestinal immunity and physiology. In Dukor P *et al.* (eds) Monographs in Allergy. Karger, Basel, p 134
70. Matsuda H, Coughlin MD, Bienenstock J, Denburg JA 1988 Nerve growth factor has human hemopoietic colony stimulating activity. Proc Natl Acad Sci USA 85:6508
71. Gozes Y, Moskowitz MA, Strom TB, Gozes I 1983 Conditioned media from activated lymphocytes maintain sympathetic neurons in culture. Dev Brain Res 6:93
72. Lindholm D, Heumann R, Meyer M, Thoenen H 1987 Interleukin-1 regulates synthesis of nerve growth factor in non-neuronal cells of rat sciatic nerve. Nature 330:658
73. Aloe L, Cozzari C, Calissano P, Levi-Montalcini R 1981 Somatic and behavioral postnatal effects of fetal injections of nerve growth factor antibodies in the rat. Nature 291:413
74. Lahtinen T, Soinila S, Lakshmanan J 1989 Biological demonstration of nerve growth in the rat pituitary gland. Neuroscience 30:165
75. Otten U, Baumann JB, Girard J 1979 Stimulation of the pituitary-adrenocortical axis by nerve growth factor. Nature 282:413
76. Levi-Montalcini R, Angeletti PU 1964 Hormonal control of the NGF content in the submaxillary glands of mice. In Sreebny LM, Meyer J (eds) Salivary Glands and Their Secretions, Vol 3. International Series of Monographs on Oral Biology, Pergamon Press, London, p 129
77. Wallace LJ, Partlow LM 1976 α-Adrenergic regulation of secretion of mouse saliva rich in nerve growth factor. Proc Natl Acad Sci USA 73:4210
78. Valzetti L 1973 The "isolation syndrome" in mice. Psychopharmacologia (Berl) 31:305
79. Aloe L, Alleva E, Bohm A, Levi-Montalcini R 1986 Aggressive behavior induces release of nerve growth factor from mouse salivary gland into blood stream. Proc Natl Acad Sci USA 83:6187
80. Zaimis E 1971 Nerve growth factor: the target cells. In Zaimis E, Knight J (eds) Nerve Growth Factor and Its Antiserum. Athlone Press, London, p 59
81. Maestripieri D, De Simone R, Aloe L, Alleva E 1989 Social status and nerve growth factor serum levels after agonistic encounters in mice. Physiol Behav 47:1
82. Alleva E, Aloe L 1990 Physiological role of nerve growth factor in adult rodents: a biobehavioral perspective. Int J Comp Psychol (in press)
83. Washburn SL 1980 Human behavior and behavior of other animals. In Montague A (ed) Sociobiology Examined. Oxford University Press, Oxford, p 254

Clinical and Experimental Rheumatology ***10:*** *203-204, 1992*

Letter to the Editor

Nerve growth factor and distribution of mast cells in the synovium of adult rats

Sir,

Previous studies have shown that the administration of nerve growth factor (NGF) in young rodents results in an increase in the number and size of peripheral mast cells (MCs) (1). These findings were later confirmed (2, 3) and extended by showing that NGF also induces degranulation and histamine release from MCs (4).

More recently, we have reported that the synovium and synovial fluid of patients affected by rheumatoid arthritis (RA) contain high levels of NGF and the hypothesis was prospected that this specific growth factor might be involved in the event of the early onset of this disease (5). Other studies have shown that the distribution of MCs increases significantly in the synovium of patients affected by RA and in the synovium of an animal model of RA (results to be published), suggesting that MCs play a role as homeostatic regulators in the pathophysiology of this disease (6, 7). Moreover, IgE rheumatoid factor, which is frequently associated with RA (8, 9), could also be involved in the histamine released from MCs, which induces inflammation of the synovium and the proliferation of connective tissue. At present, however, nothing is known about the correlation between the expression of NGF and the distribution of MCs in RA. We present here new findings showing that NGF causes an increase in MCs in the synovium, and examine the hypothesis that NGF might be involved in the process of synovial inflammation.

One μg of highly purified 2.5S NGF (2) dissolved in physiological solution was injected into the ankle synovium of adult Sprague Dawley rats (n = 6). Arthritis was induced in the adult rats (n = 14) by injecting 50 μl of 1% carrageenann (Sigma Chemical Co., St. Louis, USA) into the ankle synovium. The injection of 50 μl of physiological solution was used as control. Twenty-four hours later, the animals were sacrificed with an overdose of nembutal and the synovial tissues were removed, weighed, fixed in alcoholic Bouin, embedded in paraplast, sectioned at a 10 μm thickness and stained with toluidine blue to visualize the MCs, as previously described (2). Sections were examined under a Zeiss Axiophot microscope at a 400X magnification, the MC distribution counted in 6 consecutive comparable sections, and the values expressed as the mean ± S.E.M.

The criteria for MC identification were based on the presence of clearly visible round or oval nuclei and of cytoplasmic basophilic granules, stained metachromatically. Figure 1 (A, B) illustrates the distribution of MCs in the synovium of control (a) and NGF (b) rats, 24 hrs after treatment. The number of MCs per mm^2 increased significantly in the NGF-treated rats (52 ± 4) compared to the control rats (23 ± 3), $p < 0.001$.

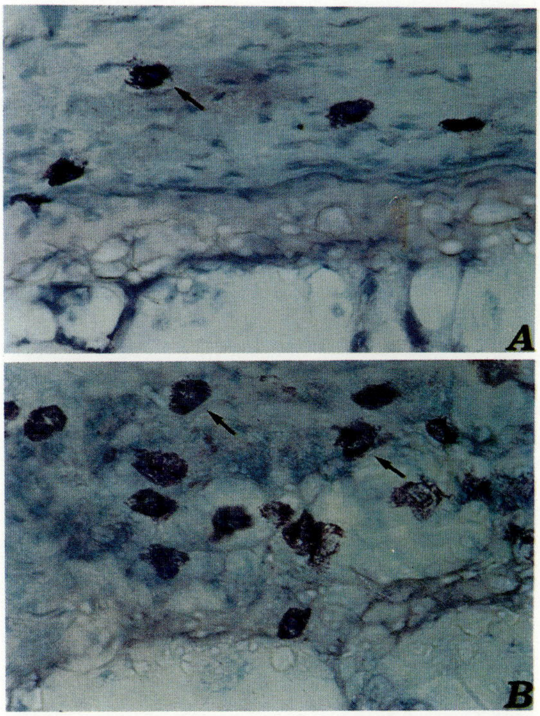

Fig. 1. Photomicrographs of comparable sections of synovium of adult rats injected 24 hours before with saline (***A***), or with 1 μg of NGF (***B***) in the synovium. Note the numerical increase in mast cells (arrows) in the NGF-treated rats as compared to control tissues (toluidine blue; X-400).

The results of our studies indicate that the synovial injection of NGF, as well as of carrageenann, which induces arthritis and increases the NGF level, results in an augmentation in the local distribution of MCs. The potential involvement of MCs in the pathophysiology of arthritis has been reported, and their homeostatic regulatory role in joint inflammation has also been prospected (6, 7). Our findings, therefore, raise the question whether NGF might be a promoter or a contributing factor in RA. Since the synovium from humans affected by RA as well as that from experimentally-induced models expresses high levels of NGF, we hypothesize that the increase in MCs occurring in RA (6) can be mediated by the presence of NGF (2, 4, 5). Moreover, it is also possible that NGF, by inducing the release of histamine (3), might also influence the formation of the synovial pannus. Other studies in progress are aimed at establishing whether a correlation exists between MC activation by NGF and RA.

L. ALOE[1,3]
M.A. TUVERI[2]
R. LEVI-MONTALCINI[1]

[1]*Institute of Neurobiology, Consiglio Nazionale delle Ricerche, viale Marx 15, 00137 Rome, Italy;* [2]*II Chair of Rheumatology, University "La Sapienza", Rome, Italy.*
[3]*To whom correspondence should be sent.*

This work was supported by the "Progetto Finalizzato Invecchiamento" grant no. 912087, CNR to Dr. R. Levi-Montalcini and L. Aloe.

References

1. ALOE L, LEVI-MONTALCINI R: Mast cells increase in tissues of neonatal rats injected with the nerve growth factor. *Brain Res* 1977; 133: 358-66.
2. ALOE L: The effect of nerve growth factor and its antibody on mast cells *in vivo*. *J Immunol* 1988; 18: 1-12.
3. TOMIOKA M, STEAD RH, NIELSEN L, COUGHLIN MD, BIENENSTOCK J: Nerve growth factor enhances antigen and other secretagogue-induced histamine release from rat peritoneal mast cells in the absence of phosphatidylserine. *J Allergy Clin Immunol* 1988; 82: 599-607.
4. PEARCE FL, THOMPSON HL: Some characteristics of histamine secretion from rat peritoneal mast cells stimulated with nerve growth factor. *J Physiol* 1986; 372: 379-93.
5. ALOE L, TUVERI MA, CARCASSI U, LEVI-MONTALCINI R: Nerve growth factor in synovial fluid of patients with chronic arthritis. *Arthritis Rheum* 1992; 35.
6. GODFREY HP, ILARDI C, ENGBER W, GRAZIANO FM: Quantitation of human synovial mast cells in rheumatoid arthritis and other rheumatic diseases. *Arthritis Rheum* 1984; 27: 852-6.
7. WASSERMAN SI: The mast cell and synovial inflammation (or, what's a nice cell like you doing in a joint like this?). *Arthritis Rheum* 1984; 27: 841-4.
8. ZURAW BL, O'HAIR CH, VAUGHAN JH, MATHISON DA, CURD JG, KATZ DH: Immunoglobulin E-rheumatoid factor in the serum of patients with rheumatoid arthritis, asthma and other diseases. *J Clin Invest* 1981; 68: 1610-3.
9. PERMIN H, EGESKJOLD EM: IgE anti-IgG antibodies in patients with juvenile and adult rheumatoid arthritis including Felty's syndrome. *Allergy* 1982; 37: 421-7.

Reprinted from ARTHRITIS & RHEUMATISM
Vol. 35, No. 3, March 1992

J. B. Lippincott Company
Printed in U.S.A.

NERVE GROWTH FACTOR IN THE SYNOVIAL FLUID OF PATIENTS WITH CHRONIC ARTHRITIS

LUIGI ALOE, MARCO A. TUVERI, UGO CARCASSI, and RITA LEVI-MONTALCINI

Cytokines regulate nerve growth factor (NGF) synthesis during inflammatory processes. Since cytokines are also involved in the inflammatory processes of autoimmune rheumatic diseases, we examined levels of NGF in patients with rheumatoid or other types of chronic arthritis. NGF was present in the synovial fluid and synovium of patients with chronic arthritis, but was undetectable in control fluids. We conclude that NGF might be involved in the pathogenesis of arthritis.

Nerve growth factor (NGF) is a well-characterized neurotrophic protein which plays an essential role in the development and differentiation of sympathetic and sensory nerve cells and the cholinergic neurons of the central nervous system (1). Studies published in recent years have shown that lymphocytes express NGF receptors (2) and that during inflammatory processes, cells of the immune system lineage are able to synthesize and/or release NGF (3,4). More recently, it has been shown that cytokines, such as interleukin-1 (IL-1) and tumor necrosis factor (TNF), elicit a significant increase in levels of NGF in the peripheral and central nervous systems during inflammation (2,5). This observation prompted us to examine levels of NGF in synovial fluid from patients with rheumatoid arthritis (RA) and other forms of chronic arthritis, and to determine whether NGF is present in their synovial tissues.

The synovium is a type of connective tissue that is densely vascularized and innervated by fibers that originate in the sympathetic and sensory nervous systems. These fibers release neuropeptides, including substance P (SP) and calcitonin gene–related peptide, the levels of which increase during inflammatory disease (6). Since it has been reported that the synthesis and/or release of these neuropeptides is also regulated by NGF (7), the possibility of NGF involvement in the pathogenesis of joint inflammation was considered. We present here our findings regarding the levels of NGF and SP in synovial fluid and synovium from patients with various forms of chronic inflammatory arthritis.

From the Institute of Neurobiology, Consiglio Nazionale delle Ricerche, and the II Chair of Rheumatology, University La Sapienza, Rome, Italy.

Supported by a grant from the Consiglio Nazionale delle Ricerche, Progetto Finalizzato Invecchiamento (no. 912085) to Dr. Levi-Montalcini and Dr. Aloe.

Luigi Aloe, PhD: Institute of Neurobiology, Consiglio Nazionale delle Ricerche; Marco A. Tuveri, MD: II Chair of Rheumatology, University La Sapienza; Ugo Carcassi, MD, PhD: II Chair of Rheumatology, University La Sapienza; Rita Levi-Montalcini, MD, PhD: Institute of Neurobiology, Consiglio Nazionale delle Ricerche.

Address reprint requests to Luigi Aloe, Institute of Neurobiology, CNR, via Carlo Marx, 15, 00156 Rome, Italy.

Submitted for publication June 3, 1991; accepted in revised form October 10, 1991.

PATIENTS AND METHODS

Synovial fluids. Synovial fluid samples were collected from the knees of 22 outpatients of the Department of Rheumatology, University "La Sapienza" (Rome, Italy). Six of these 22 patients (3 women and 3 men) had osteoarthritis (OA). Their mean age was 66 years (range 59–76), and their mean duration of disease was 10.5 years (range 1–20 years). Eight patients (6 women and 2 men) had RA. Their mean age was 49 years (range 24–83), and their mean duration of disease was 7.5 years (range 1–20 years). Eight patients had other types of chronic arthritis: 4 had psoriatic arthritis (PA), 3 had ankylosing spondylitis (AS), and 1 had juvenile chronic arthritis (JCA). Their mean age was 33 years (range 14–43), and their mean duration of disease was 6.3 years (range 3–10 years). The diagnoses of OA and RA were made according to the criteria of the American College of Rheumatology (formerly, the American Rheumatism Association) (8,9). PA, AS, and JCA were diagnosed according to established standards reported elsewhere (10–12).

Two synovial fluid samples were taken from the knees of patients who did not have a rheumatic disease. These 2 samples served as controls.

Synovial tissues. Synovial tissues were obtained at the time of surgical synovectomy of the knees of 2 patients. One was a 65-year-old woman with a 12-year history of RA, which was in the acute phase at the time of synovectomy. The other was a 55-year-old man with a 2-year history of asymptomatic OA. For histopathologic analysis, tissues were fixed in 4% paraformaldehyde, and sections were stained with hematoxylin and eosin, and examined with a Zeiss Axiophot microscope.

Biologic assays. Synovial tissues were homogenized and centrifuged as previously described (13). Synovial fluids were centrifuged at 10,000*g* for 20 minutes, and the supernatants were removed and used for the NGF bioassays. NGF levels were determined with the in vitro bioassay developed in our laboratory and others (14,15). A few samples were also tested by enzyme-linked immunosorbent assay, as described elsewhere (16). SP levels in synovial fluid samples were measured with a radioimmunoassay, following the methodology described previously (17).

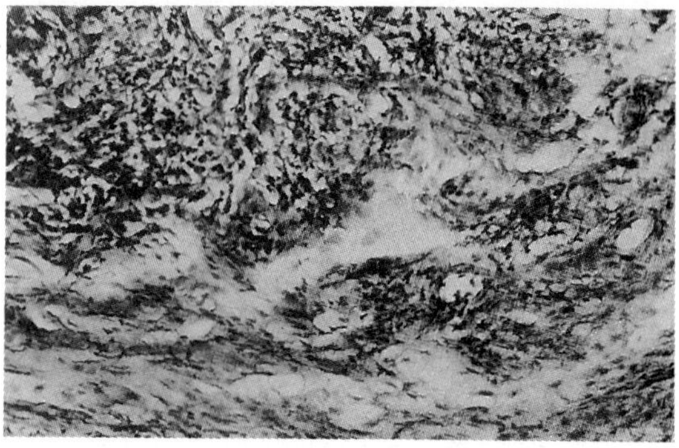

Figure 1. Hematoxylin and eosin–stained section of knee joint synovium from a patient with rheumatoid arthritis, showing increased infiltration by lymphocytes and plasma cells. Capillary proliferation is also present (original magnification × 114).

Table 1. NGF and SP levels in human knee joint synovial fluid*

Diagnostic group	NGF (ng/ml)	SP (pg/ml of synovial fluid)
Osteoarthritis		
1	ND	90
2	2	80
3	6	38
4	6	35
5	ND	45
6	ND	110
Rheumatoid arthritis		
7	3	50
8	9	260
9	6	125
10	2	190
11	3	90
12	5	150
13	1	125
14	ND	90
Other chronic arthritis		
15	4	98
16	6	95
17	9	120
18	4	90
19	6	35
20	12	190
21	2	90
22	12	235
Non–rheumatic disease control		
23	ND	60
24	ND	80

* Nerve growth factor (NGF) levels were assayed in vitro, using 8-day-old chick embryo sensory ganglia, as described in Patients and Methods; substance P (SP) levels were determined by radioimmunoassay. Patients 15–17 had ankylosing spondylitis; patients 18–21 had psoriatic arthritis; patient 22 had juvenile chronic arthritis. See Patients and Methods for a description of control patients. ND = not detectable.

Immunohistochemistry. For NGF immunohistochemistry studies, synovial tissues were quickly removed and immediately fixed by immersion in 2% paraformaldehyde/0.4% *p*-benzoquinone/50 m*M* phosphate buffer, pH 7.4, for 6 hours at room temperature and processed as reported elsewhere (13). Sections were incubated overnight with either affinity-purified polyclonal NGF antibodies, diluted 1:600–1:3,000 (initial concentration 20 mg/ml), or monoclonal NGF antibodies raised against mouse NGF (at the Institute of Neurobiology, Rome, Italy) (unpublished methodology), diluted 1:30 (initial concentration 1.0 mg/ml), and then processed for immunoperoxidase staining as described (13). Staining specificity was assessed by 1) preabsorption of the first antibody with an excess of NGF, 2) omission of the primary antibody, and 3) incubation of sections with preimmune rabbit serum.

RESULTS

Synovial fluid NGF. The level of NGF in synovial fluid from patients with OA, RA, and other types of chronic arthritis varied greatly. Table 1 shows NGF and SP levels in the 24 synovial fluid samples examined. Levels of NGF in the arthritis patients ranged from nondetectable to 12 ng/ml of synovial fluid. Control patients' synovial fluid samples showed undetectable amounts of NGF.

The highest levels of SP were found in synovial fluids from RA patients and patients with other forms of chronic arthritis (Table 1), while low levels were found in OA patients.

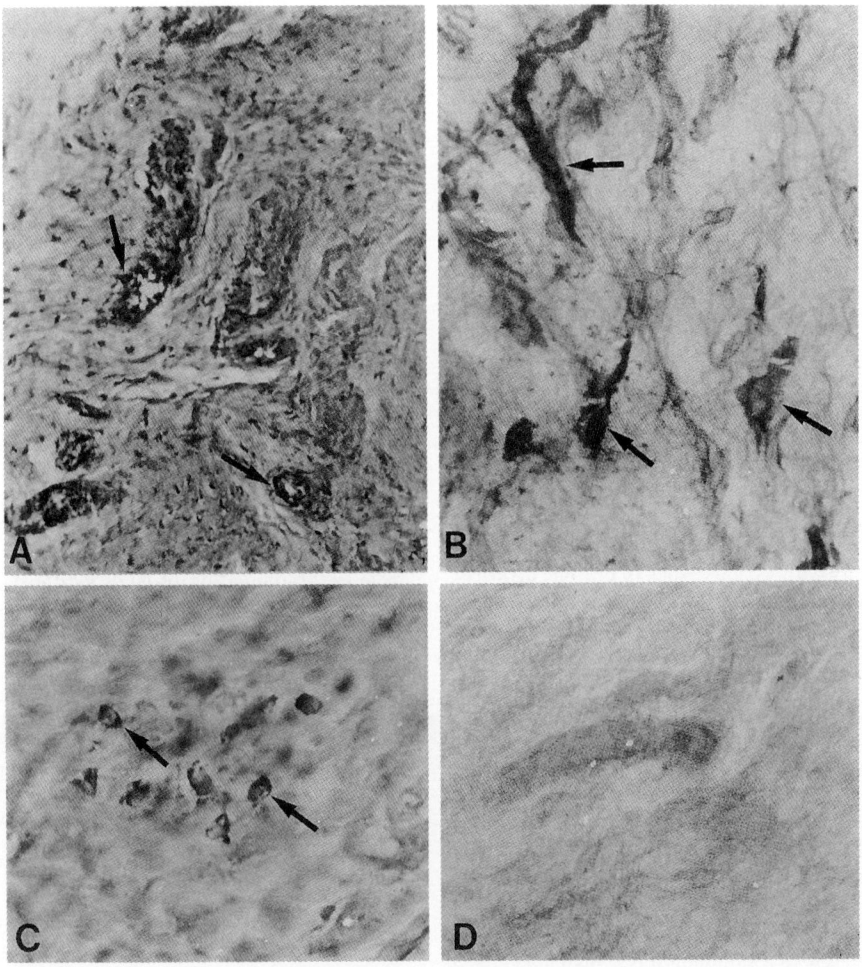

Figure 2. Immunostained section of knee joint synovium from a patient with rheumatoid arthritis, showing nerve growth factor (NGF)–positive cells (**arrows**) in perivascular areas (**A**), the synovial space (**B**), and within a synovial mass (**C**). **D,** Control synovial tissue exposed to preimmune serum. (Original magnification × 114 for **A** and **D**, × 330 for **B**, and × 250 for **C**.)

Synovial membrane NGF. Two tissue samples, 1 from an RA patient and 1 from an OA patient, were screened for histopathologic changes and for NGF immunoreactivity. The synovial membrane of the RA patient showed diffuse infiltration by lymphocytes and plasma cells, and capillary proliferation (Figure 1). No sign of inflammation was present in the synovium of the OA patient.

Immunohistochemical studies showed NGF-positive cells, either singly or in small groups, mainly near the perivascular region (Figure 2) in the RA synovial tissue. These were identified as lymphocyte-like cells. In the synovium from the patient with asymptomatic OA, there were no NGF-positive cells; no staining was observed when the primary antibodies were omitted.

DISCUSSION

The first evidence that inflammation causes an increase in NGF levels was reported by Levi-Montalcini and Angeletti in 1960 (18). Their research showed that a dramatic increase in NGF, far above the level found in sarcoma 180, occurred in granuloma tissue induced by carrageenan (18). This seminal work led to the study of NGF levels in carrageenan-induced arthritis (unpublished) and in chronic arthritis in humans. One of the major mechanisms underlying these diseases is the activation of immunocompetent cells that release protein mediators such as IL-1 and TNF (19). These cytokines also act as growth-promoting substances. The findings reported in the last few years, showing that synthesis of NGF can be enhanced during inflammation through the activation of macrophages and the release of IL-1, are of particular interest (2,5,20).

The present study shows that, in the synovial fluid of patients with various types of chronic arthritis, there is an increase in levels of NGF. The highest NGF levels were found in RA patients and patients with other chronic arthritis, while low levels were found in OA patients. In noninflammatory joint fluids (control fluids) NGF was undetectable.

Immunohistochemical studies showed that lymphocyte-like cells localized to the perivascular region in the synovium of RA patients exhibit NGF positivity, thus suggesting that NGF is synthesized by the lymphocyte-like cells. However, the possibility that other cells in the synovium are able to synthesize or store NGF cannot be excluded. In the synovium of an OA patient, NGF-immunoreactive cells were absent. Since histologic examination revealed no signs of inflammation in this OA synovium specimen, it appears that the presence of NGF in RA synovium is the result of an inflammatory process.

Several studies have implicated SP, which is released from the terminal field of the peripheral nervous system, in the inflammatory process of RA (6). In fact, SP released from afferent nerves causes a series of events, such as vasodilatation, increased permeability of small blood vessels, stimulation of phagocytosis by polymorphonuclear leukocytes, and degranulation of mast cells (2), which lead to inflammatory conditions. Thus, it has been hypothesized that there is a direct correlation between the release of SP and the extent of inflammation in joint synovium. Furthermore, it has been suggested that activation and infiltration of immune cells, particularly lymphocytes, plasma cells, and macrophages, contribute greatly to the pathogenesis of RA through the release of cytokines and growth factors (19). Other studies have shown that macrophages are important in the regulation of NGF synthesis through IL-1 production (5). Thus, the possibility is suggested that activated macrophages present in the synovial tissue are responsible for the induction of NGF synthesis found in the inflamed synovium.

Recent studies have shown that other growth factors, such as transforming growth factor β, platelet-derived growth factor, fibroblast growth factor, and epidermal growth factor, are released during chronic arthritis (20–22). However, the precise role of these diverse growth factors in the pathophysiology of chronic synovitis has not yet been elucidated. Our studies represent the first evidence that NGF is found in synovial fluids and synovial tissues of patients with inflammatory joint diseases, and these findings support the hypothesis that cells of the immune system play a role in the regulation of NGF synthesis. Whether the increase of NGF in synovial fluid and synovium is a cause or a result of these immune processes is, at present, a matter for investigation.

REFERENCES

1. Levi-Montalcini R: The nerve growth factor 35 years later. Science 237:1154–1162, 1987
2. Levi-Montalcini R, Aloe L, Alleva E: A role for nerve growth factor in nervous, endocrine and immune systems. Prog Neurol Endocrinol Immunol 1:1–10, 1990
3. Matsuda H, Coughlin MD, Bienenstock J, Denburg JA: Nerve growth factor has human hemopoietic colony stimulating activity. Proc Natl Acad Sci USA 85:6508–6512, 1988
4. Gozes Y, Moskowitz MA, Strom TB, Gozes I: Conditioned media from activated lymphocytes maintain sympathetic neurons in culture. Dev Brain Res 6:93–97, 1983
5. Friedman WJ, Larkfors L, Ayer-LeLievre C, Ebendal T, Olson L, Persson H: Regulation of β-nerve growth factor expression by inflammatory mediators in hippocampal cultures. J Neurosci Res 27:374–382, 1990
6. Levine JD, Goetzl EJ, Basbaum AI: Contribution of the nervous system to the pathophysiology of rheumatoid arthritis and other poliarthritides. Rheum Dis Clin North Am 2:369–383, 1987
7. Lindsay RM, Lockett C, Sternberger J, Winter J: Neuropeptide expression in cultures of adult sensory neurons: modulation of substance P and calcitonin gene-related peptide levels by nerve growth factor. Neuroscience 1:53–65, 1989
8. Altman R, Asch E, Bolch D, Bole G, Borenstein D,

Brandt K, Christy W, Cooke TD, Greenwald R, Hochberg M, Howell D, Kaplan D, Koopman W, Longley S III, Mankin H, McShane DJ, Medsger T Jr, Meenan R, Mikkelsen W, Moskowitz R, Murphy W, Rothschild B, Segal M, Sokoloff L, Wolfe F: Development of criteria for the classification and reporting of osteoarthritis: classification of osteoarthritis of the knee. Arthritis Rheum 29:1039–1049, 1986

9. Arnett FC, Edworthy SM, Bloch DA, McShane DJ, Fries JF, Cooper NS, Healey LA, Kaplan SR, Liang MH, Luthra HS, Medsger TA Jr, Mitchell DM, Neustadt DH, Pinals RS, Schaller JG, Sharp JT, Wilder RL, Hunder GG: The American Rheumatism Association 1987 revised criteria for the classification of rheumatoid arthritis. Arthritis Rheum 31:315–324, 1988
10. Roberts MET, Wright V, Hill AGS, Mehra AC: Psoriatic arthritis: follow-up study. Ann Rheum Dis 35:206–212, 1976
11. Calin A: Diagnostic criteria (new and old) for the spondyloarthropathies (editorial). Clin Exp Rheum 5:101–102, 1987
12. Cassidy JT, Levinson JE, Bass JC, Baum J, Brewer EJ Jr, Fink CW, Hanson V, Jacobs JC, Masi AT, Schaller JG, Fries JF, McShane D, Young D: A study of classification criteria for a diagnosis of juvenile rheumatoid arthritis. Arthritis Rheum 29:274–281, 1986
13. Aloe L: Adrenalectomy decreases nerve growth factor in young adult rat hippocampus. Proc Natl Acad Sci USA 86:5636–5640, 1989
14. Levi-Montalcini R: In vivo and in vitro experiments on the effect of mouse sarcoma 180 and 37 on the sensory and sympathetic system of the chick embryo, Fourteenth International Congress of Zoology, Copenhagen, 1953
15. Stephani U, Sutter A, Zimmermann A: Nerve growth factor (NGF) in serum: an evaluation of serum NGF levels with a sensitive bioassay employing embryonic sensory neurons. J Neurosci Res 17:25–35, 1987
16. Larkfors L, Ebendal T: Highly sensitive enzyme immunoassays for nerve growth factor. J Immunol Methods 97:41–47, 1987
17. McMahon SB, Lewin GR, Anand P, Ghatei MA, Bloom SR: Quantitative analysis of peptide levels and neurogenic extravasation following regeneration of afferents to appropriate and inappropriate targets. Neuroscience 1:67–73, 1989
18. Levi-Montalcini R, Angeletti PU: Biological properties of a nerve-growth promoting protein and its antiserum, Regional Neurochemistry: The Fourth International Neurochemical Symposium. Edited by SS Kety, J Elkes. New York, Pergamon Press, 1960
19. Zvaifler NJ: Etiology and pathogenesis of rheumatoid arthritis, Arthritis and Allied conditions. Edited by DJ Mc Carty. Philadelphia, Lea & Febiger, 1989
20. Lindholm D, Hengerer B, Zafra F, Thoenen H: Transforming growth factor-1 stimulates expression of nerve growth factor in the rat CNS. NeuroReport 1:9–12, 1990
21. Bucala R, Ritchlin C, Winchester R, Cerami A: Constitutive production of inflammatory and mitogenic cytokines by rheumatoid synovial fibroblasts. J Exp Med 173:569–574, 1991
22. Hamerman D, Taylor S, Kirshenbaum I, Klagsbrun M, Raines EW, Ross R, Thomas KA: Growth factors with heparin binding affinity in human synovial fluid. Proc Soc Exp Biol Med 186:384–389, 1987

Rheumatol Int (1992) 12:213–216

Rheumatology International
Clinical and Experimental Investigations

Studies on carrageenan-induced arthritis in adult rats: presence of nerve growth factor and role of sympathetic innervation

L. Aloe, M. A. Tuveri, and R. Levi-Montalcini

Institute of Neurobiology, CNR, Viale K. Marx 15, I-00137 Rome, Italy

Received February 6, 1992/Accepted June 23, 1992

Summary. Recent studies effected by our Institute indicate that various forms of human arthritis express both immunohistochemically and biologically active nerve growth factor (NGF) in the synovium. In the present study, we used a model of carrageenan-induced arthritis to further evaluate the effects of joint inflammation on NGF level. These studies showed that experimentally-induced arthritis in rats caused a significant increase in NGF in the perivascular area of the synovium. We also showed that injection into the synovium of purified NGF did not cause inflammation per se and that the destruction of peripheral sympathetic innervation significantly reduced both the inflammation and the level of NGF following carrageenan injection.

Key words: Nerve growth factor (NGF) – Carrageenan – Arthritis – Sympathetic innervation

Introduction

Rheumatoid arthritis (RA) is a disease characterized by chronic joint inflammation, cell proliferation, synovium enlargement, pannus formation and destruction of joint cartilage [1]. Various studies have focused attention on the relationship between the invading immune cells and the local production of immuno-mediators such as cytokines [2] and growth factor [3, 4]. We have recently reported that the synovial fluid and synovium of patients effected by various types of arthritis express higher levels of nerve growth factor (NGF) than the synovium of unaffected patients [5]. The physiological significance of NGF in joint inflammation is not known.

NGF, a polypeptide capable of promoting the development and differentiation of nerve cells of the peripheral and central nervous systems [6], is also an essential tropic factor for cells originating outside the nervous system [7]. For example, NGF has been shown to effect mast cell distribution [8], histamine release [9], the modulation of lymphocyte activity [10], and vascular permeability [7]. It has also been reported that a variety of different stimuli can induce cells to release stored and newly synthesized NGF [7]. Thus, NGF synthesis is altered by hormones [11], cytokines [12, 13], and aggressive behaviour [14]. Recent studies have shown that the level of NGF is also altered by autoimmune diseases [5, 15–17]. The present studies extended this latter investigation by using an animal model of arthritis induced by carrageenan. This agent first induces a chemical-type inflammation, and later an immunological-like inflammation characterized by the release of tissue-specific antigens [18]. Data showing that NGF was increased in joint inflammation in our animal model are presented here. Since synovium is richly innervated by the sympathetic nervous system [2], we also sought to determine whether sympathectomy could affect joint disease in our animal model.

Correspondence to: L. Aloe

Materials and methods

Adult male and female Sprague Dawley rats were taken from our colony. NGF was extracted from mouse submaxillary salivary glands according to the method of Bocchini and Angeletti [19], and further purified to remove all renin activity as described by Cozzari et al. [20]. NGF has a sedimentation coefficient of 2.5 *s*, a molecular weight of about 26,000 daltons and is free from contaminants as indicated by its migration as a single band in gel (pH 4.3) in the presence of detergent. NGF antibodies (AS-NGF) were raised in rabbits [21] and purified by affinity chromatography as described by Stoeckel et al., [22]. Carrageenan was purchased from Sigma Chem. Co., St. Louis, Mo., USA. We obtained 6-hydroxydopamine (6-OHDA), a pharmacological compound that acts as a false neurotransmitter leading to destruction of sympathetic nerve terminals [23, 24] from Fluka Chemist Co, FRG.

Thirty-two adult rats were divided into four experimental groups and injected in the ankle joint as follows: (a) eight with 50 µg carrageenan dissolved in 50 µl of physiological saline (p. s., i.e. 0.9% NaCl); (b) eight with 100 mg of AS-NGF; (c) eight with 50 µg of purified NGF; and (d) eight with 50 µl of p. s. Two rats from each group were sacrified with an overdose of nembutal 24 h, and 3, 5, or 7 days after treatment. The synovium was then removed and used for routine histological studies. Another group of rats was injected with the same dose of carrageenan ($n = 12$) or p. s. ($n = 12$) and used for the evaluation of NGF content in the synovium.

To induce sympathectomy, 6-OHDA was injected in a dose of 100 µg/gm of body weight in ten newborn rats [23, 24]. In adult life

(90 days old), these animals were used to assess synovial inflammation following carrageenan injection. The rats were sacrificed 48 h after the injection and synovium was removed and used for NGF immunohistochemistry and NGF biological assays.

Morphometric studies. Tissues were fixed in 4% parabenzoquinone plus 2% paraformaldehyde in 0.05 *M* phosphate buffer, pH 7.4, cryoprotected with 20% sucrose, sectioned at 20 μm thickness and immunostained for NGF localization, according to the method described previously [11, 25]. Alternate sections were stained with hematoxyline-eosine, and lymphocyte infiltrates were counted under light microscopy with a 40 × objective. These different fields were evaluated for each experimental group and values expressed as the mean ± SEM. For immunohistochemistry, sections were incubated overnight with affinity-purified antibodies, diluted 1:600–1:3000 (initial concentration 20 mg/ml) and processed for immunoperoxidase staining using the Vectastain Kit (Vector Laboratories Inc., Burlingame, USA). Staining specificity was assessed by (1) preabsorption of the first antibody with an excess of NGF, (2) omission of the primary antibody, or (3) incubation of sections with preimmune rabbit serum.

Biological assay. NGF biological activity in the synovium was estimated by using sympathetic neurons from fetal superior cervical ganglia (SCG), which are highly responsive to NGF [11], but not to other growth factors of the NGF family [26]. Briefly, nerve cells from SCG of 17- to 18-day-old rat fetuses were mechanically isolated, dissociated, and cultured in collagenated Falcon tissue culture dishes containing Dulbecco's basal medium only or the medium with the addition of increasing amounts of NGF. In the absence of NGF, only a small number of neurite-bearing cells appeared in the culture, while the addition of as little as 0.02 ng of NGF yielded a detectable increase in the number of neurites. By adding increasing amounts of NGF, the number of responding cells also increased, resulting in a dose-response curve which was used to evaluate the amount of NGF present in the tissue samples to be tested. Samples were also tested with the enzyme-linked immunoassay (ELISA), according to the methodology developed by others [27, 28].

Statistical analysis. Data was evaluated using the Student's *t*-test and values expressed as the mean ± SEM.

Results

Injection of carrageenan into the ankle synovium induced inflammation, similar to that described in other studies [18]. Histological examination of the synovium after injection of carrageenan showed progressive proliferation of fibroblasts and infiltration of lymphocytes (Fig. 1). Injection of NGF (Fig. 1 C) or NGF-antibody did not cause joint inflammation.

Immunohistochemical studies of the synovium carried out 24 h after carrageenan injection showed the presence of NGF in the perivascular cells and in the lymphocytes (Fig. 2). NGF immunoreactivity lasted for 2–3 days after which it decreased steadily and 1 week later no NGF positivity was observed in the synovium. As shown in Fig. 3, the level of NGF in the synovium of control rats was undetectable, while in that of carrageenan-induced arthritis the level of NGF was more than 0.5 ng/mg of wet tissue. Time-course studies showed that the highest level of NGF biological activity occurred 24 h after carrageenan injection and remained quite stable for up to 72 h. Thereafter, it began to decline, reaching a baseline level by 7 days after treatment.

In order to assess the contribution of the peripheral sympathetic nervous system in RA, sympathectomized

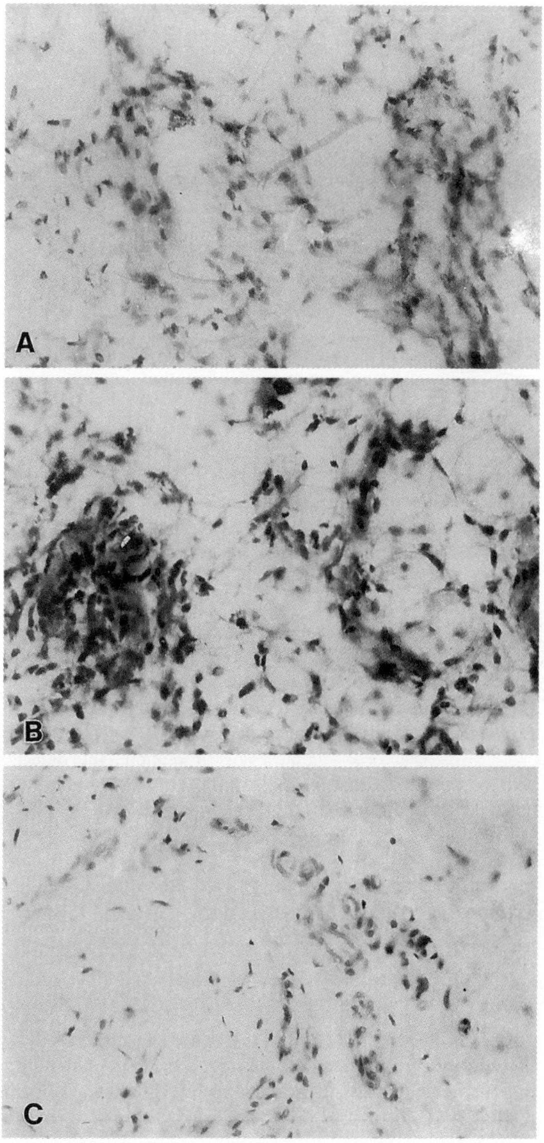

Fig. 1. Sections of ankle joint synovium of adult rats injected into that joint **A** with physiological solution, **B** with carrageenan or **C** with NGF, and sacrificed 24 h later. Note infiltrated lymphocytes in the synovium of carrageenan-treated rats (**B**) as compared to **A** and **C**. Hematoxyline-eosine stained. × 116

rats were injected in the ankle joint with carrageenan. The results showed that in the absence of sympathetic innervation there was remarkably less synovial inflammation following carrageenan injection. The morphometric examination of the synovium, stained with hematoxyline-eosine, indicated a mild infiltration of lymphocytes, as compared to the effect of carrageenan non-sympathectomized rats. Moreover, both the level of NGF (Fig. 3) and NGF-immunoreactivity in the synovium of sympathectomized rats were very low or negative.

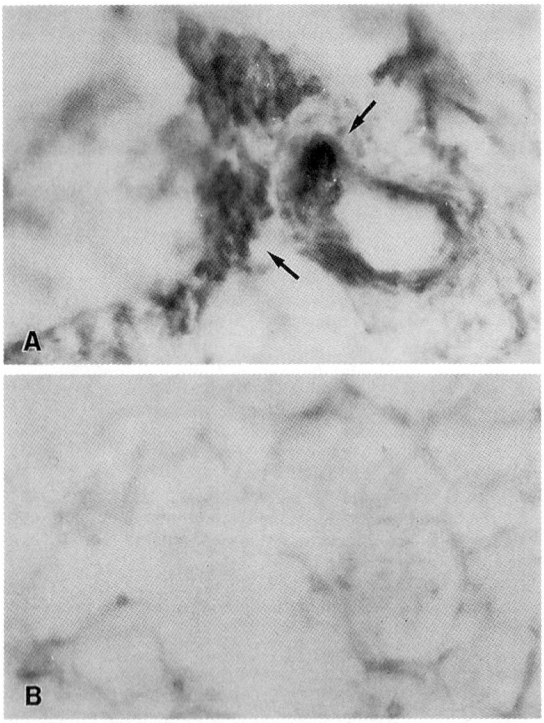

Fig. 2. NGF immunoreactivity in the synovium of rats treated **A** with carrageenan or **B** with physiological solution. Note numerous NGF-positive cells in **A** and their near absence in **B**. Stained cells around the perivascular region (shown by arrow in **A**). × 270

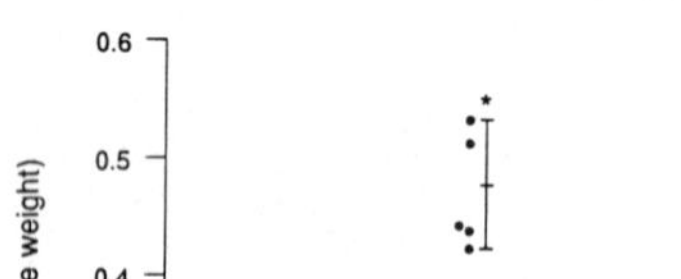

Fig. 3. NGF content in the synovium of adult rats injected into the synovium with **A** 50 μl of physiological solution, **B** with 50 μl of 1% of carrageenan or into the synovium of sympathectomized adult rats with **C** 50 μl of 1% carrageenan and sacrificed 24 h later. Note the significant increase ($P<0.01$) in NGF level after carrageenan injection (**B**) as compared to control synovium (**A**, **C**)

Discussion

The experiments were designed to evaluate three issues: (1) whether carrageenan-induced arthritis in rats caused a local increase of NGF; (2) if NGF was the single cause, or one amongst several causes of the induction of joint inflammation; and (3) whether sympathetic innervation modulated the synovial inflammation caused by carrageenan. The answer to the first question was clearly positive; carrageenan injection in the ankle joint of adult rats caused inflammation of the synovium and increased NGF content from non-detectable levels in untreated rats to 0.5 ng/mg of wet tissue in treated rats, supporting previous findings showing that human joint RA inflammation is associated with local increases in the synthesis and/or release of NGF [5]. In addition, the results obtained using animal models would exclude the possibility that the increased NGF level in patients suffering from RA might be related to the therapeutic treatment of the patient.

It has been reported that a variety of agents can induce cells to release stores of preexisting or newly synthesized NGF. Some of these stimuli, such as lesions of the nervous and non-nervous systems, generate augmentation of local or circulating NGF [7, 14]. In a previous study reported by Levi-Montalcini in 1960, for example, it was shown that chemically-induced granuloma led to an increase in NGF synthesis [29]. Other more recent studies have confirmed and extended these findings by showing a close correlation between inflammation and NGF synthesis and/or release [7, 13, 15, 30].

Surprisingly, the results showed that injections of highly purified NGF did not cause joint inflammation. Thus, the possibility that NGF contributes directly to the pathology of such a disease is highly unlikely. The findings raise the question of what role NGF plays in local inflammation. One possible explanation is that the presence of large quantities of NGF at the site of inflammation may serve as a regulatory factor for a subpopulation of immunocompetent cells that are necessary for healing processes. Studies published in recent years, demonstrating that the presence of bioactive compounds released in the microenvironment are associated with healing and nerve reparative action [3, 4, 12, 30, 31] suggest that NGF might be involved in these effects. The possibility, therefore, that the presence of NGF is necessary for healing rather than for induction of the disease is a hypothesis which should be further investigated. The evidence presented here and the findings showing that the level of NGF and/or its receptors is enhanced in multiple sclerosis [15] and in diabetes [16, 17], not only suggest that NGF is associated with nerve repair after injury [6], but also support the hypothesis that it plays a role in autoimmune diseases.

Our results showed that, in an animal model, sympathectomy drastically reduced the carrageenan-induced inflammation, as well as the level of NGF in the synovium. One possible explanation for the reduction of the inflammation after sympathectomy is a decreased facilitation of peripheral interaction between primary afferent nociceptors and sympathetic afferent neurons [2]. For example, it has been shown that the activity of sympathetic efferents excites sensory afferents in experimentally-induced neuromas [32, 33] and the interaction between primary afferents and sympathetic efferents may lead to a local increase of prostaglandins, which are potent mediators of inflammation [34]. The low level of NGF is most probably due to a local decrease of cytokines, which are known to be involved in NGF synthesis during inflammatory processes [7]. Our studies are in line with clinical studies showing that patients with sympathetic hyperactivity exhibit pain and swollen joints, which can be ameliorated by local injections of guanethidine [35, 36], a substance that interferes with the sympathetic terminals.

In conclusion, our findings indicated that NGF increased as a result of joint inflammation in an animal model. We have recently shown this to be the case in humans suffering from arthritis [5]. The physiological significance of such an increase remains, however, to be elucidated.

Acknowledgement. This study was supported by "Progetto Finalizzato Invecchiamento", CNR to R. Levi-Montalcini and L. Aloe.

References

1. Zvaifler NJ (1989) Etiology and pathogenesis of rheumatoid arthritis. McCorty DJ (ed) Arthritis and allied conditions. Lea and Febiger, Philadelphia, pp 659–673
2. Levine JD, Goetzl EJ, Basbaum AI (1987) Contribution of the nervous system to the pathophysiology of rheumatoid arthritis and other poliarthritides. Rheum Dis Clin North Am 13:369–383
3. Thornton SC, Por SB, Penny R, Richter M, Shelley L, Breit SN (1991) Identification of the major fibroblast growth factors released spontaneously in inflammatory arthritis as platelet derived growth factor and tumour necrosis factor-alpha. Clin Exp Immunol 86:79–86
4. Bucala R, Ritchlin C, Winchester R, Cerami A (1991) Constitutive production of inflammatory and mitogenic cytokines by rheumatoid synovial fibroblasts. J Exp Med 173:569–574
5. Aloe L, Tuveri MA, Carcassi U, Levi-Montalcini R (1992) Nerve growth factor in the synovial fluid of patient with chronic arthritis. Arthritis Rheum 35:351–355
6. Levi-Montalcini R (1987) The nerve growth factor: thirty-five years later. Science 237:1154–1162
7. Levi-Montalcini R, Aloe L, Alleva E (1990) A role for nerve growth factor in nervous, endocrine and immune system. Prog Neuroendocrinimmunol 1:1–10
8. Aloe L, Levi-Montalcini R (1977) Mast cells increase in tissues of neonatal rats injected with the nerve growth factor. Brain Res 133:358–366
9. Bruni A, Bigon E, Boarato E, Mietto L, Leon A, Toffano G (1982) Interaction between nerve growth factor and lysophosphatidylserine on rat peritoneal mast cells. FEBS Lett 138:190–193
10. Thorpe LW, Stach RW, Hashim GA, Marchetti D, Perez-Polo JR (1987) Receptors for nerve growth factor on rat spleen mononuclear cells. J Neurosci Res 17:128–134
11. Aloe L (1989) Adrenalectomy decreases nerve growth factor in young adult rat hippocampus. Proc Natl Acad Sci USA 86:5636–5640
12. Gadient RA, Cron KC, Otten U (1990) Interleukin-1 β and tumor necrosis factor-∞ synergistically stimulate nerve growth factor (NGF) release from cultured rat astrocytes. Neurosci Lett 117:335–340
13. Lindholm D, Heumann R, Meyer M, Thoenen H (1987) Interleukin-1 regulates synthesis of nerve growth factor in non-neuronal cells of rat nerve. Nature 330:658–659
14. Aloe L, Alleva E, De Simone R (1990) Changes of NGF level in mouse hypothalamus following intermale aggressive behaviour: biological and immunohistochemical evidence. Behav Brain Res 39:53–61
15. Otten U, Ehrhard P, Peck R (1989) Nerve growth factor induces growth and differentiation of human B lymphocytes. Proc Natl Acad Sci USA 86:10059–10063
16. Hellweg R, Hartung HD (1990) Endogenous levels of nerve growth factor (NGF) are altered in experimental diabetes mellitus: a possible role of NGF in the pathogenesis of diabetic neuropathy. J Neurosci Res 26:258–267
17. Faradji V, Sotelo J (1990) Low serum levels of nerve growth factor in diabetic neuropathy. Acta Neurol Scand 81:402–406
18. Muirden KD, Peace G (1969) Light and electron microscope studies in carrageenan, adjuvant, and tuberculin-induced arthritis. Ann Rheum Dis 28:392–401
19. Bocchini G, Angeletti PU (1969) The nerve growth factor: purification as a 30 000-molecular-weight protein. Proc Natl Acad Sci USA 64:787–794
20. Cozzari C, Angeletti PU, Lazar J, Orth H, Gross F (1973) Separation of isorenin activity from nerve growth factor (NGF) activity in mouse submaxillary gland extracts. Biochem Pharmacol 22:1321–1327
21. Aloe L, Cozzari C, Calissano P, Levi-Montalcini R (1981) Somatic and behavioral postnatal effects of fetal injections of nerve growth factor antibodies in the rat. Nature 291:358–366
22. Stoeckel K, Gagnon C, Guroff G, Thoenen H (1976) Purification of nerve growth factor antibodies by affinity chromatography. J Neurochem 26:1207–1211
23. Aloe L, Mugnaini E, Levi-Montalcini R (1975) Light and electron microscope studies on the excessive growth of sympathetic ganglia in rats injected daily from birth with 6-OHDA and NGF. Arch Ital Biol 113:326–353
24. Angeletti PU, Levi-Montalcini R (1970) Sympathetic nerve cell destruction in newborn mammals by 6-hydroxydopamine. Proc Natl Acad Sci USA 65:114–121
25. Sternberger LA, Hardy PH Jr, Cuculis JJ, Meyer HG (1970) The unlabeled antibody-enzyme method of immunohistochemistry. Preparation and properties of soluble antigen-antibody complex (horseradish peroxidase-antiperoxidase) and its use in identification of spirochetes. Histochem Cytochem 18:315–333
26. Alderson RF, Alterman AL, Barde Y-A, Lindsay RM (1990) Brain-derived neurotropic factors increase survival and differentiated functions of rat septal cholinergic neurons in culture. Neurons 5:297–306
27. Larkfors L, Ebendal T (1987) Highly sensitive enzyme immunoassays for nerve growth factor. J Immunol Methods 97:41–47
28. Weskamp G, Otten U (1987) An enzyme-linked immunoassay for nerve growth factor (NGF): A tool for studying regulatory mechanism involved in NGF production in brain and in peripheral tissues. J Neurochem 48:1779–1786
29. Levi-Montalcini R, Angeletti PU (1960) Biological properties of a nerve-growth promoting protein and its antiserum. 4th International Neurochemical Symposium. Pergamon, Oxford, pp 362–377
30. Boyle MDP, Lawman MJP, Gee AP, Young M (1985) Nerve growth factor: a chemotactic factor for polymorphonculear leukocytes in vivo. J Immunol 134:564–568
31. Alheim K, Andersson C, Tingsborg S, Ziolkowska M, Schultzberg M, Bartfai T (1991) Interleukin 1 expression is inducible by nerve growth factor in PC12 pheochromocytoma cells. Proc Natl Acad Sci USA 88:9302–9306
32. Devor M, Janig W (1981) Activation of myelinated afferents ending in a neuroma by simulation of the sympathetic supply in the rat. Neurosci Lett 24:43
33. Woll PD, Gutnik M (1974) Ongoing activity in peripheral nerves: the physiology and pharmacology of impulses originating from neuroma. Exp Neurol 43:580
34. Levine GD, Taiwo YO, Collins SD et al (1986) Noradrenaline hyperalgesia is mediated through interaction with sympathetic postganglionic neurone terminals rather than activation of primary afferent nociceptors. Nature 323:158
35. Herfort RA (1956) Extended sympathectomy in the treatment of advanced rheumatoid arthritis. NY J Med 56:1292–1297
36. Koizumi K, Collier R, Kaufman A et al (1970) Contribution of unmyelinated afferent excitation to sympathetic reflex. Brain Res 20:99–106

Neuroscience Letters, 147 (1992) 9–12

NSL 09077

Multiple sclerosis patients express increased levels of β-nerve growth factor in cerebrospinal fluid

L. Bracci Laudiero[a], L. Aloe[a], R. Levi-Montalcini[a], C. Buttinelli[b], D. Schilter[c], S. Gillessen[c] and U. Otten[c]

[a]*Institute of Neurobiology, CNR, Rome (Italy),* [b]*Department of Neurological Sciences, Università 'La Sapienza', Rome (Italy) and* [c]*Department of Physiology, University of Basel, Basel (Switzerland)*

(Received 27 July 1992; Revised version received 17 August 1992; Accepted 18 August 1992)

Key words: Multiple sclerosis; Cerebrospinal fluid; β-Nerve growth factor (NGF); Inflammation

We describe the measurement of β-nerve growth factor (NGF) content in cerebrospinal fluid (CSF) from multiple sclerosis (MS) patients compared with CSF from age-matched normal subjects using a specific sandwich immunoassay (ELISA). During acute attacks patients exhibit a significant increase of NGF content compared to controls. In contrast during remission the mean NGF levels in CSF markedly decrease. These results strongly indicate that increased NGF production in CSF is a characteristic feature of the MS inflammatory response.

Normal development and function of neurons depend on the presence of neurotrophins [16, 21]. The best characterized neurotrophin is β-nerve growth factor (NGF). NGF supports the survival of sympathetic neurons and neural crest-derived sensory neurons [16], as well as central cholinergic neurons [15, 30]. In addition to its neurotrophic actions NGF modulates inflammatory and immune responses [2, 17, 22–25, 27]. Human lymphocytes express functional NGF receptors [25] and NGF induces proliferation as well as the differentiation of B-cells [14, 25]. Accumulation of NGF at sites of lesion [18] and inflammation [29] suggests that inflammatory stimuli contribute to increasing NGF production. Thus NGF may act as an immunomodulator mediating cross talk between cells of the nervous and immune systems.

Neurodegenerative diseases which involve inflammation may provide valuable information about the pathophysiological role of NGF–cytokine interactions. Multiple sclerosis (MS) is a chronic recurrent inflammatory disease of the central nervous system (CNS) which may be an autoimmune disorder [12]. It is characterized by a patchy destruction of the myelin sheath during which lymphocytes and monocytes infiltrate white matter and release cytokines when stimulated. Recent studies indicate that cytokines, including interleukin-1 (IL-1), interleukin-6 (IL-6) and tumor necrosis factor-α (TNF-α), are potent inducers of NGF synthesis both in CNS [8, 9] and in peripheral tissues [18, 26]. We have used a specific sandwich immunoassay [ELISA] to measure developmental changes in NGF content in cerebrospinal fluid (CSF) from newborn to adult humans without neuronal inflammatory disease. Subsequently we investigated variation of NGF levels with the disease process in MS patients.

More than 80 human CSF samples from newborns and children were provided by the Department of Microbiology of Kinderspital Basel, and adult CSF was provided by Central Laboratories of Kantonsspital, Basel. We analysed 15 CSF samples of definite MS patients (followed at the 'Multiple Sclerosis Center' of the Department of Neurological Science, University 'La Sapienza', Rome, Italy), 7 males and 8 females, mean age 45 years, range 19–54. Twelve patients had a relapsing-remitting and 3 a chronic-progressive form of MS. The CSF was divided into 2 groups according to the clinical diagnosis (Table I) by parameters such as nuclear magnetic resonance (NMR) analysis, the IgG index (CSF IgG/plasma IgG):(CSF albumin/plasma albumin), oligoclonal IgG bands and absolute IgG values. Samples were taken (a) during acute attack (relapsing phase) defined as the occurrence of a new symptom or symptoms of neurological dysfunction lasting more than 24 h, and (b) during remission. In addition, CSFs from 8 patients (4 males and 4 females, mean age 43.3, range 21–78) with other neurological diseases (OND), (1 headache, 1 cerebellar neoplastic syndrome, 1 myelopathy, 2 polyneu-

Correspondence: U. Otten, Department of Physiology, University of Basel, Vesalgasse 1, 4051 Basel, Switzerland.

10

TABLE I

DIAGNOSTIC PARAMETERS DISTINGUISHING MS FROM OTHER NEUROLOGICAL DISORDERS (OND)

IgG index is equivalent/* to (CSF IgG/plasma IgG):(CSF albumin/plasma albumin). Values > 0.7 were taken as intrathecal IgG. NMR (nuclear magnetic resonance) imaging: 82–100% of patients with clinically defined MS show multifocal cerebral white matter lesions by NMR analysis.

	MS acute (n=4)	MS remission (n=11)	Other neurologic disorders (OND) (n=8)
Mononuclear cells (5×10^6/l)			
Range	3–8	1–9	1–7
Mean	4.8	3.8	3.1
% abnormality	25	18	25
IgG values (1.3–2.1 mg %)			
Range	1.3–10.6	1.5–7.3	0.9–7.9
Mean	4.6	3.7	3.4
% abnormality	75	81	62
IgG Index (0.6–0.7)			
Range	0.6–1.8	0.4–1.4	0.4–0.8
Mean	1.0	0.7	0.5
% abnormality	75	27	25
Oligoclonal IgG bands			
% abnormality	75	27	0
NMR			
% abnormality	100	82	0

ropathies, 1 leukodystrophy, 1 leukoencephalopathy, 1 cerebral ischemia) were used as controls.

The NGF content of human CSF was measured using a 2-site ELISA, modified from that previously described [1]. In brief: 96-microtiter plates (Nunc) were coated with 50 μl/well of 0.02 μg/ml of monoclonal mouse anti-NGF 21/27 (Boehringer) diluted in 0.05 M carbonate buffer (pH 9.6). To assess the non-specific binding, parallel wells were coated with equal amounts of purified mouse IgG (Zymed). After an overnight incubation at room temperature the coating buffer was replaced with a blocking buffer (coating buffer + 1% BSA), 100 μl/well for at least 4 h at room temperature. CSF samples were diluted 1:1 (v/v) with sample buffer 0.2% Triton X-100 (100 mM Tris-HCl pH 7.2, 400 mM NaCl, 4 mM EDTA, 0.2 mM PMSF, 0.2 mM benzethonium chloride, 2 mM benzamidine, 40 U/ml aprotinin, 0.05% sodium azide, 2% BSA and 0.5% gelatin). After washing (50 mM Tris-HCl pH 7.2, 200 mM NaCl, 0.25% gelatin, 0.05% sodium azide and 0.1% Triton X-100), 50 μl/well of standard and sample solution were distributed into the wells and left at room temperature (22°C) overnight. The NGF standard solutions ranged between 1–100 pg of NGF. After extensive washings (at least 1 h) and incubation for 2 h at 37°C with 4 mU/well anti-β-NGF galactosidase, 100 μl of substrate solution (4 mg of Chlorophenol red/ml substrate buffer: 100 mM HEPES, 150 mM NaCl, 2 mM $MgCl_2$, 0.1% sodium azide and 1% BSA) were added to each well and the plates incubated for 2 h at 37°C. The optical density was measured at 575 nm using an ELISA reader (Dynatech) and the values of standards and samples were corrected by subtraction of the background value due to non-specific binding. Specificity for NGF was assessed using recombinant human NGF (Genentech), biologically active NGF from conditioned media of neuroblastoma cells (SH-SY5Y), and conditioned media from cultures of human skin fibroblasts stimulated with IL-1β as standards. Recombinant brain-derived neurotrophic factor (BDNF) is not recognized in our assay at concentrations up to 20 ng/ml. Furthermore, biologically active BDNF protein in conditioned media from cultures of BDNF-transfected COS cells did not cross react with the antibody used in the ELISA. However, we cannot rule out that other neurotrophins, such as neurotrophin 3 (NT-3) and neurotrophin 4 (NT-4) [13], may be recognized by the antibodies used.

We first measured the NGF content in human CSF in newborns, children and adults. More than 80 CSFs were divided into groups according to age and sex. While no significant difference between males and females was detected, NGF levels do seem to be closely related to age. As shown in Fig. 1, the CSF from newborns and children

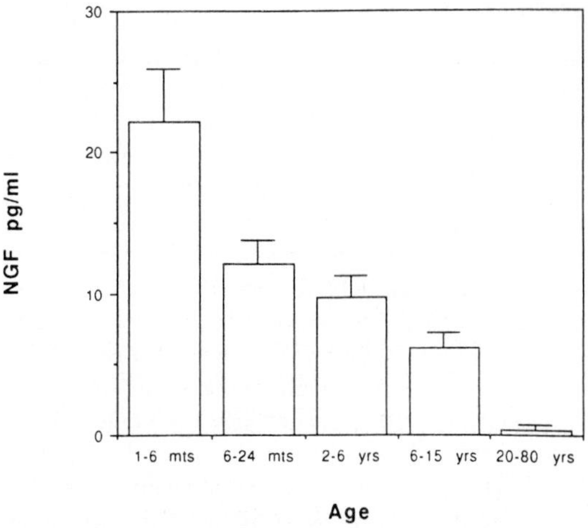

Fig. 1. NGF content in normal human CSF from newborns, children and adults without inflammatory disease. Early in postnatal development the NGF level peaks and then decreases until it is almost undetectable in adults. The values for each group represent the mean of at least 15 CSF samples tested in triplicate ± standard error of the mean (S.E.M.).

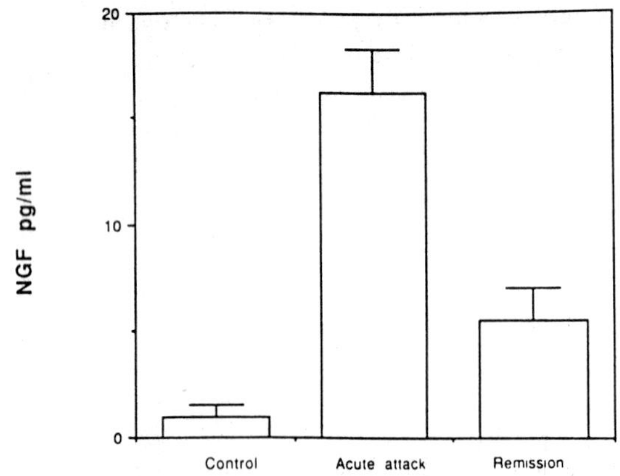

Fig. 2. NGF levels in the CSF of patients with MS. NGF content has been measured in patients in acute attacks (n=4), in remission (n=11) and in controls (n=8). All patients with acute attacks showed increased NGF levels compared with those of controls, while during remission the NGF content in CSF decreases. The values represent the mean of triplicate measures obtained in 3 independent experiments. The difference between the mean values is significant (P<0.001) according to Student's t-test.

of up to 6 months (mean age of 2.5 months) shows the highest measured NGF levels, with a mean value of 23 pg/ml. During postnatal development the NGF content gradually decreases in CSF and is almost undetectable in adults.

Because the NGF level in healthy adults seems to be extremely low, we wondered whether the NGF content in CSF would be altered by inflammation in the CNS. We have used CSF from 15 patients with MS. Since a characteristic feature of MS is recurrent attacks of inflammation we measured the NGF content in CSF in different clinical phases of the disease (acute attack and in remission). CSFs from 8 patients with other neurological diseases (OND) have been used as controls. NGF was detectable in all CSFs of MS patients analyzed (Fig. 2). Interestingly, NGF levels closely follow the phases of the disease. During acute attacks, patients exhibit a significant increase of NGF content (mean value 16.2 pg/ml) when compared to controls. In contrast during remission the mean NGF levels in the CSF markedly decrease (mean value 5.6 pg/ml). These results strongly indicate that the NGF content follows the inflammatory response. This is supported by our recent data showing that patients with viral and bacterial meningitis express elevated NGF levels of CSF (LBL and UO, unpublished results). The strict age-dependency of NGF content in human CSF confirms its importance in the development of CNS neurons. During early mammalian development NGF expression is maximal in certain brain areas [15, 21]; it then decreases or becomes undetectable but could be re-expressed at sites of lesion or after injury [10, 19]. Increased expression of NGF at sites of injury seems to be essential for neuronal recovery, and may also modulate the local immune response. Inflammatory stimuli enhance NGF expression, and NGF seems to act as an immunomodulatory cytokine stimulating lymphocyte proliferation and immunoglobulin production [25].

In the present study we have found increased NGF levels in the CSF of patients in acute phase MS, while during remission NGF in CSF markedly declines. Thus, there is a good correlation between the inflammatory episodes of the disease and increased NGF production. NGF could be used for clinical characterisation of the inflammatory phase of MS. It is worth noting that in experimental allergic encephalomyelitis (EAE), an animal model for MS, both NGF mRNA and protein increase in the inflamed brain tissues [5]. Although the mechanisms controlling NGF expression are unclear, recent reports demonstrate that increased levels of cytokines such as IL-1, IL-6 and TNF-α stimulate NGF expression. Increased levels of these cytokines have been found in the CSF of MS patients [6, 7, 20, 28] and they could be responsible for the enhanced NGF production we have observed.

The authors wish to thank Dr. M. Falcone (Department of Neurological Science, Università 'La Sapienza', Rome), Dr. R. Frei and Dr. J. Muser (Kantonsspital Basel) and Mrs. sc nat. G. Schläpfer (Kinderspital Basel) for providing CSF samples. The authors thank Dr. J. Scully for critical reading of the manuscript. This study was supported by 'Associazione Italiana Sclerosi Multipla', by CNR 'Progetto Finalizzato Biotecnologie e Biostrumentazione: Sottoprogetto Biofarmaci' to Prof. R. Levi-Montalcini and Dr. L. Aloe, by Swiss National Foundation for Scientific Research (Grant 31-2995490), by Deutsche Forschungsgemeinschaft (SFB 325) and from BMFT (Grant 01 KL 8904/6) to Prof. U. Otten.

1 Allen, S.J., MacGowan, S.H., Treanor, J.J.S., Feeney, R., Wilcock, G.K. and Dawbarn, D., Normal β-NGF content in Alzheimer's disease cerebral cortex and hippocampus, Neurosci. Lett., 131 (1991) 135–139.

2 Aloe, L. and Levi-Montalcini, R., Mast cells increase in tissue of neonatal rats injected with the nerve growth factor, Brain Res., 133 (1977) 358–366.

3 Aloe, L., Tuveri, M., Carcassi, U. and Levi-Montalcini, R., Nerve growth factor in the synovial fluid of patients with chronic arthritis, Arthritis Rheum., 35 (1992).

4 Calder, V., Owen, S., Watson, C., Feldmann, M. and Davidson, A., MS: a localized immune disease of the central nervous system, Immunol. Today, 10 (1989) 99–103.

5 De Simone, R., Micera, A. and Aloe, L., An analysis of the expression of NGF, NGF-r and their messengers in the experimental aller-

gic encephalomyelitis (EAE), Vth Congress of the Italian Society of Neuroscience, Modena, 1–4 december 1992, Abstract book.

6 Franciotta, D.M., Grimaldi, L.M.E., Martino, G.V., Piccolo, G., Bergamaschi, R., Citterio, A. and Melzi d'Eril, G.V., Tumor necrosis factor in serum and cerebrospinal fluid of patients with multiple sclerosis, Ann. Neurol., 26 (1989) 787–789.

7 Frei, K., Fredrikson, S., Fontana, A. and Link, H., Interleukin-6 is elevated in plasma in multiple sclerosis, J. Neuroimmunol., 31 (1991) 147–153.

8 Frei, K., Malipiero, U.V., Leist, T.P., Zinkernagel, R.M., Schwab, M.E. and Fontana, A., On the cellular source and function of interleukin-6 produced in the central nervous system in viral diseases, Eur. J. Immunol., 19 (1989) 689–694.

9 Gadient, R.A., Cron, K.C. and Otten, U., Interleukin-1β and tumor necrosis factor-α synergistically stimulate nerve growth factor (NGF) release from cultured rat astrocytes, Neurosci. Lett., 117 (1990) 335–340.

10 Gall, C., Murray, K. and Isackson, P.J., Kainic acid-induced seizures stimulate increased expression of nerve growth factor mRNA in rat hippocampus, Mol. Brain Res., 9 (1991) 113–123.

11 Gijbels, K., Van Damme, J., Proost, P., Put, W., Carton, H. and Billiau, A., Interleukin-6 production in the central nervous system during experimental autoimmune encephalomyelitis, Eur. J. Immunol., 20 (1990) 233–235.

12 Hafler, D.A. and Weiner, H.L., MS: a CNS and systemic autoimmune disease, Immunol. Today, 10 (1989) 104–107.

13 Ip, N.Y., Ibanez, C.F., Nye, S.H., Mc Clain, J., Jones, P.F., Gies, D.R., Belluscio, L., Le Bean, M.M., Espinosa, R., Squinto, S.P., Persson, H. and Yancopoulos, G.D., Mammalian neurotrophin 4: structure chromosomal localization, tissue distribution, and receptor specificity, Proc. Natl. Acad. Sci. USA, 89 (1992) 3060–3064.

14 Kimata, H., Yoshida, A., Ishioka, C., Kusunoki, T., Hosoi, S. and Mikawa, H., Nerve growth factor specifically induces human IgG4 production, Eur. J. Immunol., 21 (1991) 137–141.

15 Large, T.H., Bodary, S.C., Clegg, D.O., Weskamp, G., Otten, U. and Reichardt, L.F., Nerve growth factor gene expression in the developing rat brain, Science, 234 (1986) 352–355.

16 Levi-Montalcini, R., The nerve growth factor 35 years later, Science, 237 (1987) 1154–1162.

17 Levi-Montalcini, R., Aloe, L. and Alleva, E., A role for nerve growth factor in nervous, endocrine and immune systems, PNEI, 3 (1990) 1–10.

18 Lindholm, D., Heumann, R., Meyer, M. and Thoenen, H., Interleukin-1 regulates synthesis of nerve growth factor in non-neuronal cells of rat sciatic nerve, Nature, 330 (1987) 658–659.

19 Lorez, H.P., von Frankenberg, M., Weskamp, G. and Otten, U., Effect of bilateral decortication on nerve growth factor content in basal nucleus and neostriatum of adult rat brain, Brain Res., 454 (1988) 355–360.

20 Maimone, D., Gregory, S., Arnason, B.G.W. and Reder, A.T., Cytokine levels in the cerebrospinal fluid and serum of patients with multiple sclerosis, J. Neuroimmunol., 32 (1991) 67–74.

21 Maisonpierre, P.C., Belluscio, L., Friedman, B., Alderson, R.F., Wiegand, S.J., Furth, M.E., Lindsay, R.M. and Yancopoulos, G.D., NT-3, BDNF, and NGF in the developing rat nervous system: parallel as well as reciprocal pattern of expression, Neuron, 5 (1990) 501–509.

22 Matsuda, H., Coughlin, M.D., Bienenstock, J. and Denburg, J.A., Nerve growth factor has human hemopoietic colony stimulating activity, Proc. Natl. Acad. Sci. USA, 85 (1988) 6508–6512.

23 Mazurek, N., Weskamp, G., Erne, P. and Otten, U., Nerve growth factor induces mast cell degranulation without changing intracellular calcium levels, FEBS Lett., 198 (1986) 315–320.

24 Otten, U., Nerve growth factor: a signaling protein between the nervous and the immune systems, In A.I. Basbaum and J.-M. Besson (Eds.), Towards a new pharmacotherapy of pain, Wiley, Philadelphia, PA, 1991, p. 353–363.

25 Otten, U., Ehrhard, P. and Peck, R., Nerve growth factor induces growth and differentiation of human B lymphocytes, Proc. Natl. Acad. Sci. USA, 86 (1989) 10059–10063.

26 Steiner, P., Pfeilschifter, J., Boeckh, C., Radeke, H. and Otten, U., Interleukin-1β and tumor necrosis factor-α synergistically stimulate nerve growth factor synthesis in rat mesangial cells, Am. J. Physiol., 261 (1991) F792–F798.

27 Thorpe, L.W. and Perez-Polo, J.R., The influence of nerve growth factor on the in vitro proliferative response of rat spleen lymphocytes, J. Neurosci. Res., 18 (1987) 134–139.

28 Tsukada, N., Miyagi, K., Matsuda, M., Yanagisawa, N. and Yone, K., Tumor necrosis factor and interleukin-1 in the CSF and sera of patients with multiple sclerosis, J. Neurol. Sci., 102 (1991) 230–234.

29 Weskamp, G. and Otten, U., An enzyme-linked immunoassay for nerve growth factor (NGF): a tool for studying regulatory mechanism involved in NGF production in brain and peripheral tissues, J. Neurochem., 48 (1987) 1779–1786.

30 Whittemore, S.R. and Seiger, A., The expression, localization and functional significance of beta nerve growth factor in the central nervous system, Brain Res. Rev., 12 (1987) 439–464.

INT. J. TISS. REAC. **XV**(4) 139–143 (1993)

LEVEL OF NERVE GROWTH FACTOR AND DISTRIBUTION OF MAST CELLS IN THE SYNOVIUM OF TUMOUR NECROSIS FACTOR TRANSGENIC ARTHRITIC MICE

ALOE L.,[1*] PROBERT L.,[2] KOLLIAS G.,[2] BRACCI-LAUDIERO L.,[1] MICERA A.,[1] MOLLINARI C.,[1] LEVI-MONTALCINI R.[1]

1) Institute of Neurobiology, Viale Marx 15, I-00137 Rome, Italy.
2) Laboratory of Molecular Genetics, Hellenic Pasteur Institute, 127 Av Vas Sofias, Athens 11521, Greece.

Summary: *We have recently reported that nerve growth factor (NGF) increases in the synovium of patients affected by arthritis, as well as in animal models. We report here that the synovium of transgenic arthritic mice expressing human tumour necrosis factor (TNF) contains numerous mast cells (MC) and that their appearance is a phenomenon which was correlated to the local increase in NGF level. These findings provide further evidence that NGF plays a role in inflammation and suggest a functional link between NGF and MC.*

Introduction

Mast cells (MC) are granule-containing cells which are widely distributed throughout the connective tissues of vertebrates (1). They are characterized by an ability to release numerous biologically active molecules following immunological and non-immunological stimuli (2) and have been implicated in biological responses such as angiogenesis, wound healing and other inflammatory conditions (3).

We have previously reported that nerve growth factor (NGF), a well-characterized neurotrophic factor which plays an essential role in the development and survival of certain neurons of the peripheral and central nervous system (4, 5), causes an increase in the number and size of MC (6, 7). More recently, we have shown that NGF increases in the synovium of patients affected by rheumatoid arthritis (RA) (8), in the synovium of rats with pharmacologically-induced arthritis (9) and in the synovium of transgenic arthritic mice expressing human tumour necrosis factor (10, 11). We have also reported that, in rat synovium, the high levels of NGF correlate to an increased number of MC (9, 12).

To further characterize the link between NGF and MC, we have studied the distribution of MC in the synovia of transgenic arthritic mice (TAM). The results are consistent with the hypothesis that MC are NGF-target cells (6, 7, 9, 12).

Materials and methods

A transgenic mouse line (Tg197) expressing human tumour necrosis factor and developing inflammatory polyarthritis has been described (10). A

* Author to whom correspondence should be addressed.

0250-0868/93/0400139 + 5 $02.00/0

total of forty-one Tg adult mice (eighteen male Tg197, eighteen non-transgenic males of the same strain and five Tg 197 females) were used for these studies. Twelve mice were sacrificed at 2 weeks of age and twenty-nine at 8 weeks. All animals were maintained two per cage on a 12-h light and dark cycle with food and water provided *ad libitum*. All animal procedures were carried out in accordance with the guidelines approved by the Italian Bioethical Commission on Animal Care and Use.

For MC localization, Tg197 males ($n = 9$), Tg females ($n = 5$) and non-transgenic males ($n = 9$), were sacrificed with an overdose of nembutal, and the knee joints were fixed in alcoholic Bouin, embedded in paraffin, sectioned at 10 μm and stained with Toluidine blue. Coded stained sections were examined under a Zeiss microscope, using a magnification of 400×. All MC in at least 20 random fields were counted from each synovium, and MC numbers were expressed in square millimetres per tissue. Submaxillary salivary glands which are known to store large amounts of NGF (4), as well as spleen and tongue which express very low or undetectable levels of NGF, were also fixed and examined for MC distribution. Knee joints of nine TAM mice (three 2 weeks old and six 8 weeks old) and nine non-transgenic mice of the same age were used for NGF bioassay. NGF levels were measured using a highly sensitive two-site enzyme-linked immunoassay as described in a previous paper (13).

The data were evaluated statistically by analysis of variance and Student's t test.

Results

After being stained with Toluidine blue, MC were easily distinguishable from other cells by their pale blue nuclei and cytoplasm containing small, round, dark purple granules (1). At 2 weeks of age (the arthritis had not yet developed), the number of MC in the synovium of TAM was very low (less than 4/mm^2) and the level of NGF was undetectable. Differences between TAM and control mice were not observed. At 8 weeks of age, when mice were clearly arthritic, the number of MC in the synovium in TAM mice had increased about five- to six-fold as compared to the number of MC found in the synovium of non-transgenic mice (18 ± 2 versus 96 ± 8 MC/mm^2 of TAM mice), see Table I. In TAM mice, MC were particularly numerous in the dermal tissues (not shown). The level of NGF in the knee joint rose from 400 ± 36 pg/g wet weight (control) to 1.900 ± 109 pg/g wet weight in the joints of TAM mice. In the salivary gland, a tissue which synthesizes and stores large amounts of NGF, no difference in NGF levels or in MC distribution was observed (Table II). Likewise, in tongue and spleen no differences in NGF levels and MC distribution between control and TAM mice were detected.

Discussion

We have previously reported that *in vivo* administration of NGF into neonatal rodents results in an increase in the number and size of MC, and that spleen cells cultured in the presence of physiological amounts of NGF are phenotypically con-

Table I *Number of mast cells/mm^2 in tissue of control and transgenic arthritic mice*[a]

Mouse	Mast cells per mm^2			
	Tongue	Spleen	Synovium	Salivary gland
Control	130 ± 11	24 ± 6	18 ± 2	9 ± 1
TG197	138 ± 13	28 ± 7	96 ± 8[b]	12 ± 1

[a] Animals ($n = 6$ for each experimental group) were sacrificed at 2 months of age and the tissues indicated were fixed and stained with Toluidine blue. Mast cells were counted and expressed as number/mm^2 of tissue.

[b] Statistically significant ($p < 0.001$) and correlated to the increased NGF level.

verted into MC (6, 7). More recently, we have shown that the synovium of both humans and rodents affected by arthritis also expresses high levels of NGF (8, 9, 11) and that the increase in NGF correlates to the local number of MC (9). To further characterize the effect of NGF on MC in synovia, we used transgenic mice which develop arthritis 6 to 8 weeks after birth (10, 11). The results indicate that the number of MC in the synovium of TAM mice is higher than that of control non-arthritic

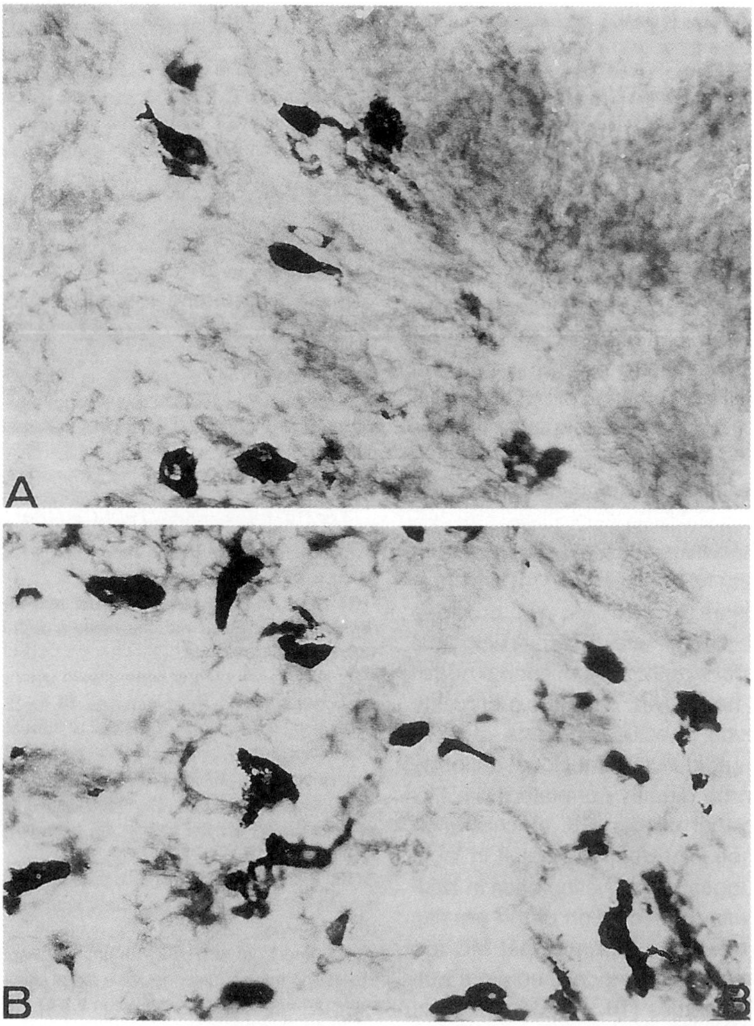

Fig. 1 Sections of synovia from knee joints of normal (A) and transgenic arthritic mice (B), showing the distribution of mast cells (darkly stained with Toluidine blue (×320).

Table II *Level of NGF[a] in tissues of control and transgenic arthritic mice*

Mouse	NGF (pg/mg wet tissue)			
	Tongue	Spleen	Synovium	Salivary gland[b]
Control	<1	31.5 ± 21	400 ± 36	340 ± 24
Tg197	<1	26.5 ± 19	1900 ± 109	380 ± 31

[a] Values represent mean of 6 different samples.
[b] NGF is expressed in ng/mg of wet tissue for the salivary gland.

mice, whereas in tissues where no differences in NGF levels were seen between arthritic and non-arthritic mice (salivary gland, spleen, tongue), no significant differences in MC distribution were observed. Thus, the present findings confirm and extend previous observations indicating that the level of NGF affects the distribution of MC in arthritis (12). The fact that at 2 weeks of age (when arthritis has not yet developed) no changes can be seen in either NGF levels or in MC distribution favours the hypothesis of a link between NGF and MC in TAM. Moreover, the observation that MC express both low- and high-affinity NGF-receptors (14) and proliferate after exposure to NGF (6, 15) provides further support to this hypothesis. The physiological or pathophysiological significance of such a correlation is at present not known. Further experiments are needed in which specific blockade of NGF function (*e.g.* through *in-vivo* monoclonal antibody administration) can be directly correlated with MC status. The finding that exogenous administration of NGF into the knee joint does not result in local joint disease (9) suggests that the increase in NGF is not associated with the induction of the arthritic disorder. Since it has been reported that MC are able to release numerous biological mediators with anti-inflammatory properties (16, 17), it is possible that NGF, along with other biological mediators released by MC, is involved in tissue remodelling and nerve reparative processes. Thus our results, together with those from other laboratories (8, 9, 11, 15–22), point to a functional link between NGF and MC during joint inflammatory processes.

Acknowledgement

This investigation was supported by P.F. 'Invecchiamento' CNR no. 93/2/352 to R. Levi-Montalcini and L. Aloe.

References

(1) Padawer J. *Quantitative studies with mast cells.* Ann. N.Y. Acad. Sci., **103**, 87–137, 1963.

(2) Galli S.J., Gordon J.R., Wershil B. *Cytokine production by mast cells and basophils.* Curr. Opinion Immunol., **3**, 865–873, 1991.

(3) Uvnäs B. *The mast cell and its biogenic amines.* In: Bloom F.E., Acheson G.H. (eds.) "Pharmacology and the future of man." Vol. 4: "Brain, nerves and synapses." S. Karger, Basel/New York, pp. 306–330, 1973.

(4) Levi-Montalcini R. *The nerve growth factor: its mode of action on sensory and sympathetic nerve cells.* Harvey Lect., **60**, 217–259, 1966.

(5) Levi-Montalcini R. *The nerve growth factor 35 years later.* Science, **237**, 1154–1162, 1987.

(6) Aloe L., Levi-Montalcini R. *Mast cells increase in tissues of neonatal rats injected with the nerve growth factor.* Brain Res., **133**, 358–366, 1977.

(7) Aloe L. *The effect of nerve growth factor and its antibody on mast cells* in vivo. J. Neuroimmunol., **18**, 1–12, 1988.

(8) Aloe L., Tuveri M.A., Carcassi U., Levi-Montalcini, R. *Nerve growth factor in the synovial fluid of patients with chronic arthritis.* Arth. Rheum., **35**, 351–355, 1992.

(9) Aloe L., Tuveri M.A., Levi-Montalcini R. *Nerve growth factor and distribution of mast cells in the synovium of adult rats.* Clin. Exp. Rheum., **10**, 203–204, 1992.

(10) Keffer J. et al. *Transgenic mice expressing human tumour necrosis factor: a predictive genetic model of arthritis.* EMBO J., **10**, 4025–4031, 1991.

(11) Aloe L. et al. *The synovium of transgenic arthritic mice expressing human tumor necrosis factor contains a high level of nerve growth factor.* Growth Factor, **9**, 149–155, 1993.

(12) Aloe L., Tuveri M.A., Levi-Montalcini R. *Studies on the carrageenan-induced arthritis in adult rats. Presence of nerve growth factor and role of sympathetic innervation.* Rheum. Int., **12**, 213–216, 1992.

(13) Bracci Laudiero L. et al. *Multiple sclerosis patients express increased levels of β-nerve growth factor in cerebrospinal fluid.* Neurosci. Lett., **147**, 9–12, 1992.

(14) Horigome, K., Pryor C., Bullock E.D., Johnson E.M. *Mediator release from mast cells by nerve growth factor: Neurotrophin specificity and receptor mediation.* J. Biol. Chem., in press.

(15) Marsall J.S., Stead R.S., McSharry C., Nielsen L., Bienestock J. *The role of mast cell degrabulation in mast cell hyperplasia: mechanism of action of nerve growth factor.* J. Immunol., **144**, 1886–1892, 1990.

(16) Levi-Montalcini R., Angeletti P.U. *Biological properties of a nerve growth factor promoting protein and its antiserum.* In: Kety S.S., Elkes J. eds. "Fourth Int. Neurochem. Symp." Pergamon, New York, 1960.

(17) Weskamp G., Otten U. *An enzyme-linked immunoassay for nerve growth factor (NGF): a tool for studying regulatory mechanisms involved in NGF production in brain and peripheral tissues.* J. Neurochem., **48**, 1779–1786, 1987.

(18) Lewin G.R., Mendell L.M. *Nerve growth factor and nociception.* Trends Neurosci., **16**, 353–359, 1993.

(19) Malone D.G., Wilder R.L., Saavedra-Delgado A.M., Metcalfe D.D. *Mast cell numbers in rheumatoid synovial tissues.* Arth. Rheum., **30**, 130–137, 1987.

(20) Levi-Montalcini R., Aloe L., Alleva E. *A role for nerve growth factor in nervous, endocrine and immune systems.* Prog. Neurol. Endocrinol. Immunol., **1**, 1–10, 1990.

(21) Mican J.A.M., Metcalfe D.D. *Arthritis and mast cell activation.* J. Allergy Clin. Immunol., **86**, 677–683, 1990.

(22) Bischoff S.C., Dahinden C.A. *Effect of nerve growth factor on the release of inflammatory mediators by mature human basophils.* Blood, **79**, 2662–2699, 1990.

Growth Factors, 1993, Vol 9, pp. 149–155
Reprints available directly from the publisher
Photocopying permitted by license only

The Synovium of Transgenic Arthritic Mice Expressing Human Tumor Necrosis Factor Contains a High Level of Nerve Growth Factor

L. ALOE, L. PROBERT[1], G. KOLLIAS[1], L. BRACCI-LAUDIERO, M.G. SPILLANTINI[2] and R. LEVI-MONTALCINI

Institute of Neurobiology, CNR, Viale Marx 15, 00156, Rome, Italy; [1]Laboratory of Molecular Genetics, Hellenic Pasteur Institute, 127 Av. Vas Sofias, Athens 11521, Greece; [2]Institute of Cell Biology, CNR, Viale Marx 15, 00156, Rome, Italy

(Received January 26 1993, Accepted April 19 1993)

We have recently reported that nerve growth factor (NGF) increases in the synovium of patients affected by rheumatoid arthritis and in the synovium of pharmacologically-induced arthritis in animal models. In the present study, we demonstrate that arthritic transgenic mice which carry and express the human TNF gene (Tg197) also express elevated levels of NGF, and that subcutaneous injection of NGF-antibodies attenuates the loss of body weight caused by the developement of disease in these mice. Along with our previous findings, which show an increase in the level of NGF during the acute phase of other autoimmune diseases, these results suggest a role of NGF in these pathologies. The functional significance of NGF in rheumatoid arthritis (RA) is currently under study.

KEYWORDS: transgenic mice, arthritis, tumor necrosis factor alpha, nerve growth factor, synovium

INTRODUCTION

Rheumatoid arthritis (RA) is an inflammatory autoimmune disease in which chronic activation of the resident cells causes alteration of joint cartilage and bone, leading to fibrosis (Brackertz et al., 1977; Godfrey et al., 1984; Wasserman, 1984; Thornton et al., 1991; Young et al., 1992). The autoimmune mechanism that induces joint damage is not known, but numerous studies indicate that cytokines, particularly Interleukin-1 beta (IL-1β) and tumor necrosis factor alpha (TNF-α) may play a major role in the pathogenesis of RA (Palombella et al., 1987; Hopkins and Meafer, 1988; Brennan et al., 1989). TNF is a cytokine with a wide range of biological effects on lymphoid and non-lymphoid cell types (Levine et al., 1987; Schall et al., 1990; Soliven and Albert, 1992). It has also been shown that in both rats and humans affected by arthritis, the level of TNF in synovial tissue increases significantly (Brennan et al., 1989; Schall et al., 1990). The hypothesis that TNF plays a part in arthritis is firmly supported by recent studies (Keffer et al., 1991) showing that transgenic mice which express the human TNF-α (Tg197) gene develop arthritis which can be blocked by exogenous administration of anti-TNF antibodies (Keffer et al., 1991).

Other studies have shown that IL-1β (Heumann et al., 1987; Lindholm et al., 1987), either alone or in combination with TNF-α (Gadient et al., 1990; Steiner et al., 1991; Thornton et al., 1991), is a potent inducer of nerve growth factor (NGF). NGF is a well-characterized neurotrophic factor which plays a central role in the development, survival and repair of peripheral and central nervous tissue (Levi-Montalcini et al., 1954; Levi-Montalcini and Angeletti, 1960; Thoenen and Barde, 1980; Levi-Montalcini, 1987; Koliatos et al., 1990; Garofalo et al., 1992), as well as in cells of the immune and endocrine systems (Levi-Montalcini et al., 1990). NGF induces mast cell and B lymphocyte proliferatiation and differention (Otten et al., 1989; Kimata et al., 1991) and exerts a growth-promoting activity on myeloid progenitor cells (Matsuda et al., 1988a; Matsuda et al., 1988b). NGF is known to increase following lesion of the nervous system (Heumann et al., 1987; Lindholm et al., 1987) and in

Corresponding author: L. Aloe, Institute of Neurobiology, CNR, Viale Marx 15, 00156, Rome, Italy.

experimentally-induced inflammation (Levi-Montalcini and Angeletti, 1960). The fact that NGF has such a large spectrum of action, both in the nervous and immune systems, raises the question if, and to what extent, NGF is involved in the pathophysiology of autoimmune inflammatory diseases.

We have recently reported that the synovium of patients affected by RA (Aloe et al., 1992a) and of adult rats with experimentally-induced RA (Aloe et al., 1992b) expresses high levels of NGF, while in non-inflamed synovium the level of NGF was undetectable. Since NGF accumulates at the sites of inflammation, it was hypothesized that NGF might be involved in the induction of the inflammatory state (Heumann et al., 1987; Lindholm et al., 1987). However, highly purified NGF injected in the joint synovium of normal rats does not induce knee joint inflammation, suggesting that NGF per se is not the cause of the disease.

We now report findings showing that arthritic transgenic mice which express high levels of TNF-α contain large amounts of NGF in the synovium and tha the injection of NGF antibodies in young Tg197 mice attenuates the TNF-wasting effect on these mice.

MATERIALS AND METHODS

Tg197 mice which express the human TNF-α gene were obtained from the Hellenic Pasteur Institute of Athens, Greece. As reported (Keffer et al., 1991), between 4 and 6 weeks of age these mice develop swelling of the ankles, impaired movement and, frequently, progressive weight loss. At 8 weeks of age, when the swelling of the ankles was clearly evident and the mice weighed 50% less than the controls, they were sacrificed with an overdose of Nembutal, and ankle joints, kidney, spleen, thymus, heart and brain regions and blood samples were removed and used for morphometric, biological and biochemical evaluation. Tissue samples of non-transgenic mice of the same age and strain were used as controls. 2.5S NGF (Bocchini and Angeletti, 1969; Cozzari et al., 1973) and NGF-antibodies were prepared as previously described (Aloe, 1989).

NGF was evaluated by a two-site immunoassay performed as described (Bracci-Laudiero et al., 1992) with minor modifications. Tissues were homogenized with a polytron system PT 2000 (Italscientifica, Italy) in Tris buffer (see below), centrifuged at 10,000 RPM and supernatant was used for NGF determination. In brief, 96-microtiter plates were coated with 50 μl/well of 0.02 μg/ml of monoclonal mouse anti-NGF diluted in 0.05 M carbonate buffer (pH 9.6). In order to assess the non-specific binding, parallel wells were coated with equal amounts of purified mouse IgG. After overnight incubation at room temperature, the coating buffer was replaced with a 100 μl/well blocking buffer (coating buffer+1% BSA) and left for at least 4 h at room temperature. Tissue samples were diluted 1:1(v/v) with sample buffer containing 0.2% Triton X-100 (100 mM Tris-HCl, ph 7.2, 400 mM NaCl, 4 mM EDTA, 0.2 mM PMSF, 0.2 mM benzethonium chloride, 2 mM benzamidine, 40 U/ml aprotinin, 0.05% sodium azide, 2% BSA and 0.5% gelatin). After washing (50 mM Tris-HCl pH 7.2, 200 mM NaCl, 0.25% gelatin, 0.05% sodium azide and 0.1% Triton X-100), 50 μl/well of standard and sample solution were distributed among the wells and left overnight at room temperature.

Plates were washed and incubated for 2 h at 37°C with 4 mU/well anti B-NGF galactosidase and 100 μl of substrate solution (4 mg of clorophenol red/ml substrate buffer: 100 mM HEPES, 150 mM NaCl, 2 mM $MgCl_2$) were added to each well and then incubated for an additional 2 h at 37°C. The optical density was measured at 575 nm using an ELISA reader (Dynatec) and the results corrected by subtraction of the background values (for further details see Bracci-Laudiero et al., 1992a).

Twenty Tg197 mice were divided into 2 groups and from the second postnatal week were injected intraperitoneally every other day with rabbit polyclonal NGF antibodies (10 mice), or with preimmune rabbit serum (10 mice) for 5 consecutive weeks. At the end of the experimental session, mice were weighed, blood was collected and, after an overdose of Nembutal, the ankle joint was removed as described above for TNF-α determination using an enzyme-linked immuno-assay (ELISA) (Genzyme, Cambridge, MA, USA). To assess whether TNF cross-reacts with AS-NGF, extracts from ankle joints of Tg197 mice were also preincubated with purified AS-NGF and the level of TNF evaluated.

For morphometric studies, the ankle joints of control and experimental mice were fixed in alcohol Bouin containing 2% nitric acid and 0.2% EDTA, embedded in paraplast, sectioned at 8 μm thickness and stained with haematoxilin-eosin. Sample sections were immunostained for NGF as described previously (1).

The data are presented as the mean of ±SEM. The significance of different groups was assessed by the Student's t-test.

RESULTS

An ELISA system was utilized to determine the levels of NGF in the joints and other tissues of Tg197 mice. As shown in Fig. 1, the arthritic joint of transgenic mice contains about 5 times more NGF as compared to the NGF levels measured in the joint of non-arthritic mice. NGF also increases in the spleen (about threefold) and to a lesser extent in the heart, but not in blood sera or the brain, which remain at baseline levels (data not shown). Likewise, the levels of NGF in the salivary gland, which contains the largest amount of tissue-bound NGF, is not affected. The NGF found in the joints of Tg197 mice is biologically active since it stimulates the outgrowth of neurites from sympathetic neurons, while the addition of NGF antibodies completely inhibits neuronal outgrowth. Using this bioassay, the NGF level in the joints of Tg197 is 210 ± 23 pg/g of fresh tissue (for methodological details, see Aloe, 1989).

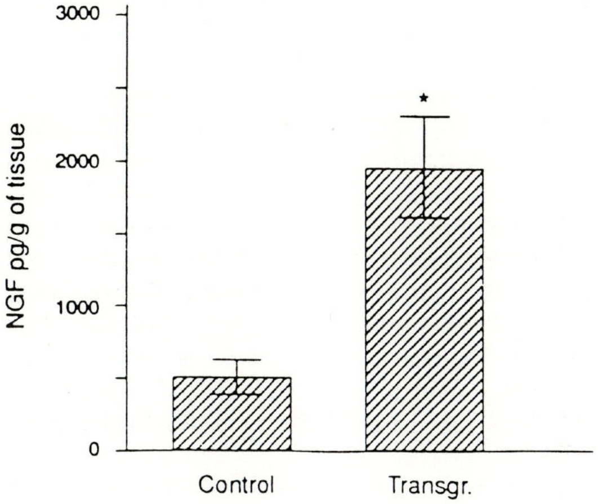

FIGURE 1. NGF levels in the synovium of 45-day-old Tg197 (seven) and of wild strain (seven) mice. All synovia of arthritic mice showed increased NGF levels compared with those of wild strain. The values represent the mean of triplicate measures obtained in three independent experiments (each sample tested 3 times). The difference between the mean values is significant *(P<0.001) according to Student's t-test.

Injection of 800 U of human TNF-α into the joint synovium of adult non-transgenic mice results in a small increase of NGF (256 ± 16 pg/g versus 200 ± 18 pg/g/wet weight of control synovium, P<0.02 Student's t-test). This weak effect suggests that factors other than TNF might be responsible for the increased NGF level found in the synovium of Tg197 mice.

To determine whether NGF antibodies would interfere with TNF-induced arthritis, polyclonal NGF antibodies (4 mg, activity: 1/5000) were Intraperitoneally injected in 2-week-old Tg197 mice (n = 10) 3 times a week for 6 consecutive weeks. The results showed that AS-NGF treatment does not prevent but reduces knee joint swelling (Fig. 2) and the loss of body weight in Tg197 mice as compared to non-transgenic controls. Thus, at 45 days of age, the mean body weight of Tg197 mice treated with preimmune serum is 16.0 ± 1 g; those treated with AS-NGF being about 22.0 ± 2 g, while those treated with antibodies against TNF-α about 25.0 ± 2 g (P<0.01 Student's t-test).

Figure 2a–c shows the morphological organization of the knee joint of 45-day-old Tg197 mice treated for 6 weeks, starting from the second postnatal life of age with pre-immune serum (a), or with AS-NGF (b). Figure 1c illustrates the joint of wild strain mice. As shown by arrows, numerous darkly stained cells (lymphocyte-like cells) have invaded the synovium of untreated Tg197 mice (Fig. 1a) while following injection with AS-NGF the cell layer lining the synovium is greatly reduced in thickness (Fig. 2b).

The mice treated with AS-NGF, unlike those treated with preimmune serum, reached sexual maturity, since they were able to breed. The effect of anti-NGF antibodies in these mice, however, is not comparable to that of anti-TNF-treated Tg197 mice. Thus, this latter treatment permits a normal lifespan, as long as TNF-antibodies are provided, whereas treatment with AS-NGF leads to weight loss after 2 months and death shortly thereafter. Moreover, the effect of AS-NGF on Tg197 mice was more pronounced in males than in females.

To assess whether NGF-antibodies interfere with endogenous levels of TNF-α, the latter was then estimated in the joints of untreated (n = 4) and AS-NGF-treated (n = 4) Tg197 mice. Likewise, no changes were observed in TNF levels extracted from synovium of non-injected Tg197 mice preincubated with purified AS-NGF.

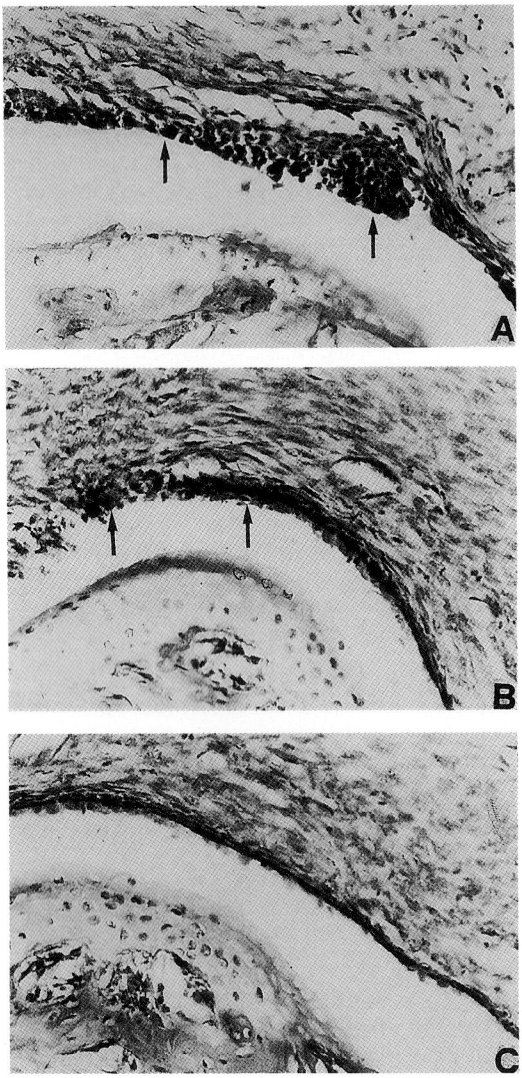

FIGURE 2. (**A–C**) Photomicrographs of hematoxiline-eosine stained sections of knee joint of 45-day-old Tg197 mice treated for 3 weeks, starting from the tenth postnatal day of life with pre-immune serum (a) or with NGF antibodies (b). Figure 1c: knee joint of 45-day-old wild mouse. Arrows point to synovial lining layer and lymphocyte-like cell aggregate. Note a decrease in cell layer in Fig. (b) as compared to Fig. 1a. Original magnification X-180.

DISCUSSION

We have recently reported that both the synovium of patients affected by RA (Aloe et al., 1992a) and the synovium of rats with experimentally-induced arthritis (Aloe et al., 1992b) express higher than baseline levels of NGF. We have also shown that the NGF is synthesized locally, since numerous cells in the knee joint are immunohistochemically positive for this factor and that highly purified NGF injected in the knee joint of normal rats does not cause inflammation (Aloe et al., 1992b). Moreover, when injected into the rat synovium of experimentally-induced arthritis, NGF does not enhance joint inflammation (unpublished results). These observations suggest that NGF is not involved in the induction of the inflammation observed in arthritis.

In the present study, we have demonstrated that the synovium of arthritic transgenic mice expressing human TNF gene contains higher NGF levels than the synovium of non-transgenic mice (Fig. 1). A smaller but statistically significant increase occurs in the spleen and heart, while in the brain and blood circulation NGF remains nearly at baseline, levels. As reported in a previous study, the ELISA used is highly specific for NGF (Bracci-Laudiero et al., 1992a) and does not recognize recombinant brain-derived neurotropic factor (DiStefano et al., 1992) at a concentration of up to 20 ng/ml.

To assess whether TNF-α caused the augmentation of synovial NGF in Tg197 mice, exogenous TNF-α was injected in the knee joint of wild strain mice. The results showed that administration of TNF induces only a slight increase (less than 20%) of NGF in the synovium, suggesting that other mediators and/or different mechanisms besides TNF are involved in the regulation of NGF in the pathogenesis of RA.

Although the mechanism(s) leading to arthritis is not fully understood, numerous studies have indicated that the presence of cytokines, particularly TNF-α and IL-1β released by activated immunocompetent cells and by activated mast cells, plays a major role in the pathogenesis of the disease (Brackertz et al., 1977; Lindau and Fernandez, 1986; Palombella et al., 1987; Brennan et al., 1989; Keffer et al., 1991; Thornton et al., 1991). Mast cells have also been indicated as markers of inflammatory disorders playing a role in homeostatic regulation of the joint (Wasserman, 1984; Levi-Montalcini et al., 1990).

Studies reported in recent years have shown that NGF induces proliferation and release of mediators from mast cells (Levi-Montalcini et al., 1990), exerts proliferating and differentiating activity on B lymphocytes (Otten et al., 1989; Kimata et al., 1991; Brodie and Gelfand, 1992) and induces growth-promoting activity on myeloid progenitor cells (Matsuda et al., 1988a; Matsuda et al., 1988b). It has also been reported that NGF, but not brain-derived neurotrophic factor (BDNF), a neurotrophic protein structurally related to NGF (Leibrock et al., 1989; Di Stefano et al., 1992), causes release of inflammatory mediators from mature human basophilic cells (Bischoff and Dahinden, 1992).

It is known that the peripheral nervous system contributes to the pathophysiology of RA (Levine et al., 1987) and that bioactive products released from nerve terminals might enhance or reduce the severity of RA. For example, Substance P (SP) released from sensory nerve terminals enhances the severity of joint destruction (Gronblad et al., 1988), while B-adrenergic receptor blockers and sympathectomy markedly reduce joint inflammation (Levine et al., 1987). Since synthesis of SP and sympathetic neurite outgrowth are regulated by NGF (Thoenen and Barde, 1980; Levi-Montalcini, 1987; Levi-Montalcini et al., 1990), it is possible that NGF, either alone or synergistically, induces release of biologically active compounds from nerve terminals (Donnerer et al., 1992), mast cells (Wasserman, 1984) and/or immuno cells (Kimata et al., 1991; Bischoff and Dahinden, 1992; Brodie and Gelfand, 1992), which might be involved in local reparative processes. More recently a functional link between NGF and neuropeptides during inflammation (see below) has been suggested (Donnerer et al., 1992). The observation that NGF administration does not induce inflammation supports this hypothesis. Also consistent with this hypothesis is the observation that the level of NGF in human cerebrospinal fluid (CSF) is high at birth and decreases steadily with age, and that animal models affected by experimentally induced Allergic Encephalomyelitis (De Simone et al., 1992), or Lupus Erythematosus (Bracci-Laudiero et al., 1992a) and Multiple Sclerosis patients (Bracci-Laudiero et al., 1992b) show high levels of NGF in the acute phase of the disease and low levels during remission. Thus, although the precise functional significance of the increased level of NGF in the synovium remains to be defined, the possibility that NGF is associated with a transient protective mechanism modulating inflammatory and helaing processes cannot be excluded. Evidence of a regulatory function of NGF during prolonged inflammatory process has been reported recently (Donnerer et al., 1992).

We found that subcutaneous AS-NGF administration significantly attenuates the loss of body weight and the decrease in reproductive capacity in these mice. This observation raises the question as to whether TNF and NGF might act on common target cells, which might indirectly affect the biological activity of TNF-α. Therefore, NGF antibodies, either directly or indirectly, might interfere with the biological activity of circulating and/or tissue-bound TNF. The fact that the level of TNF decreases in the synovium of Tg197 mice following NGF antibody administration and that the gene coding for the low-affinity NGF-receptor shares a high degree of homology with TNF-receptors (Schall et al., 1990) does not exclude this hypothesis.

To summarize, we have shown that (a) NGF increases in the joint of hTNF-transgenic arthritic mice; (b) injection of NGF into the synovium of normal mice neither induces nor exacerbates inflammation; (c) chronic subcutaneous injection of NGF antibodies attenuates some of the wasting effect induced by TNF-α in Tg197 mice. Based on these observations, it is hypothesized that high levels of circulating NGF antibodies might directly interfere with or indirectly affect the biological activity of TNF-α, while the increased NGF levels might be correlated to a local cascade of events, such as activation of immunocompetent mast cell hyperplasia and/or degranulation, and neuropeptide synthesis (Wasserman, 1984; Levi-Montalcini et al., 1990; Donnerer et al., 1992). However, the mechanism(s) through which NGF is up-regulated in plasma and tissue affected by inflammatory autoimmune disorders and the role that this molecule plays in early and later phases of the disease remain to be elucidated. The identification of cells which synthesize NGF and/or TNF within the knee joint of both healthy and arthritic mice, as well as the distribution of low and/or high forms of NGF-receptor in inflamed synovia might lead to a better understanding of the role NGF plays in this inflammatory disease.

ACKNOWLEDGMENTS

This study was supported in part by P.F, CNR Invecchiamento, Subproject Neuroendocrinology Gerontobiology to R. Levi-Montalcini and L. Aloe (Grant C.P. 93/2/290).

REFERENCES

Aloe, L. (1989) Adrenalectomy decreases nerve growth factor in young adult rat hippocampus. *Proc. Natl Acad. Sci. USA* **86**, 5636–5640.

Aloe, L. and Levi Montalcini R. (1977) Mast cell increase in tissues of neonatal rats injected with the nerve growth factor. *Brain Res* **133,** 358–366.

Aloe, L., Tuveri, M. A., Carcassi, U. and Levi-Montalcini, R. (1992a) Nerve growth factor in the synovial fluid of patients with chronic arthritis. *Arth. Rheum.* **35,** 351–355.

Aloe, L., Tuveri, M. A. and Levi-Montalcini, R. (1992b) Nerve growth factor and distribution of mast cells in the synovium of adult rats. *Clin. Exp. Rheum.* **10,** 203–204.

Aloe, L., Tuveri, M. and Levi-Montalcini, R. (1992c) Studies on carrageenan-induced arthritis in adult rats: presence of nerve growth factor and role of sympathetic innervation. *Rheum. Int.* **12,** 213–216.

Bischoff, S. C. and Dahinden, C. A. (1992) Effect of nerve growth factor on the release of inflammatory mediators by mature human basophils. *Blood* **79,** 2662–2669.

Bocchini, G. and Angeletti, P. U. (1969) The nerve growth factor: purification as a 30,000-molecular-weight protein. *Proc. Natl Acad. Sci USA* **64,** 787–794.

Bracci-Laudiero, L., Aloe, L., Levi-Montalcini, R., Buttinelli, C., Schilter, D., Gillessen, S. and Otten, U. (1992a) Multiple sclerosis patients express increased levels of B-nerve growth factor in cerebrospinal fluid. *Neurosci. Lett.* **147,** 9–12.

Bracci-Laudiero, L., Levi-Montalcini, R., Aloe, L., Galeazzi, M., Shilter, D., Gillessen, S. and Otten, U. (1992b) Increased levels of NGF in sera of Systemic Lupus Erytematous patients. VIII International Symposium on Cellular Endocrinology, 31 August-3 September 1992, Lake Placid (NY), USA.

Brackertz, D., Mitchell, G. F. and Mackay, I. R. (1977) Antigen-induced arthritis in mice. *Arth. Rheum.* **20,** 841–850.

Brennan, F. M., Chantry, D., Jackson, A., Maini, R. and Feldmann, M. (1989) Inhibitory effect of TNF-α antibodies on synovial cell Interleukin-1 production in rheumatoid arthritis. *Lancet* **ii,** 244–247.

Brodie, C. and Gelfand, E. W. (1992) Functional nerve growth factor receptors on human B Lymphocytes. Interaction with IL-2. *J. Immunol.* **148,** 3492–3497.

Cozzari, C., Angeletti, P. U., Lazar, J., Orth, H. and Gross, F. (1973) Separation of isorenin activity from nerve growth factor (NGF) activity in mouse submaxillary gland extracts. *Biochem. Pharmacol.* **22,** 1321–1327.

De Simone, R., Micera, A. and Aloe, L. (1992) An analysis of the expression of NGF, NGF-r and their messengers in the experimental allergic encephalomyelitis (EAE). *Neurosci. Lett. Supp.* **43,** S.37.

DiStefano, P. S., Friedman, B., Radziejewski, C., Alexander, C., Boland, P., Schick, C.M., Lindsay, R. M. and Wiegand, S. J. (1992) The neurotrophins BDNF, NT-3, and NGF display distinct paaterns of retrograde axonal transport in peripheral and central neurons. *Neuron* **8,** 983–993.

Donnerer, J. Schligoi, R. and Stein, C. (1992) Increased content and transport of substance P and Calcitonin gene-related peptide in sensory nerves innervatang inflammed tissues: evidence for a regulatory function of nerve growth factor in vivo. *Neurosci.* **49,** 693–698.

Gadient, R. A., Cron, K. C. and Otten, U. (1990) Interleukin-1 B and tumor necrosis factor-synergistically stimulate nerve growth factor (NGF) release from cultured rat astrocytes. *Neurosci. Lett.* **117,** 335–340.

Garofalo, L., Ribeiro-da-Silva, A. and Cuello, A. C. (1992) Nerve growth factor-induced synaptogeneis and hypertrophy of cortical cholinergic terminals. *Proc. Natl Acad. Sci. USA* **89,** 2639–2643.

Godfrey, H. P., Ilardi, C., Engber, W. and Graziano, F. M. (1984) Quantitation of human synovial mast cells in rheumatoid arthritis and other rheumatic diseases. *Arth. Rheum.* **27,** 852–856.

Gronblad, M., Konttinen, Y. T., Korkala, O., Liesi, P., Hukkanen, M. and Polak, J. M. (1988) Neuropeptides in synovium of patients with rheumathoid arthritis and osteoarthritis. *J. Rheum.* **15,** 1807–1811.

Heumann, R., Korsching, S., Bandtlow, C. and Thoenen, H. (1987) Changes of nerve growth factor synthesis in nonneuronal cells in response to sciatic nerve transection. *J. Cell Biol.* **104,** 1623–1631.

Hopkin S. J. and Meafer, A. (1988) Cytokines in synovial fluid: II. The presece of tumour necrosis factor and interferon. *Clin. Exp. Immunol.* **73,** 88–92.

Keffer, J., Probert, L., Cazlaris, H., Georgopoulos, S., Kaslaris, E., Kioussis, D. and Kollias, G. (1991) Transgenic mice expressing human tumour necrosis factor: a predictive genetic model of arthritis. *EMBO J.* **10,** 4025–4031.

Kimata, H., Yoshida, A., Ishioka, C., Kusonoki, T., Hosoi, S. and Mikawa, H. (1991) Nerve growth factor specifically induces human IqG4 production. *Eur. J. Immunol.* **21,** 137–142.

Koliatos, V., Haring, E., Nauta, J. W., Clatterbuck, R. E., Holtzman, D. M., Mobley, W. C. and Price, D. L. (1990) Mouse nerve growth factor prevents degeneration of axotomized basal forebrain cholinergic neurons in the monkey. *J. Neurosci.* **10,** 3801–3813.

Leibrock, J., Lottspeich, F., Hohn, A., Hofer, M., Hengerer, B., Masiakowski, P., Thoenen, H. and Barde, Y. A. (1989) Molecular cloning and expression of brain-derived neurotrophic factor. *Nature* **341,** 149–152.

Levi-Montalcini, R. (1987) The nerve growth factor 35 years later. *Science* **237,** 1154–1162.

Levi-Montalcini, R., Aloe, L. and Alleva, E. (1990) A role of nerve growth factor in nervous, endocrine and immune systems. *Prog. Neuro. Endocrin. Immunol.* **3,** 1–10.

Levi-Montalcini, R. and Angeletti, P. U. (1960) Biological properties of a nerve growth factor promoting protein and its antiserum. *Fourth International Neurochemical Symposium.* S. S. Kety and J. Elkes, eds., Pergamon Press, New York.

Levi-Montalcini, R., Meyer, H. and Hamburger, V. (1954) *In vitro* experiments on the effects of mouse sarcomas 180 and 37 on the spinal and sympathetic ganglia of the chick embryo. *Cancer Res.* **14,** 49–57.

Lindau, M. and Fernandez, J. M. (1986) IgE-mediated degranulation of mast cells does not require opening of ion channels. *Nature* **319,** 150–154.

Levine, J. D., Goetzl, E. J. and Basbaum, A. I. (1987) Contribution of the nervous system to the pathophysiology of rheumatoid arthritis and other polyarthritides. In *Rheumatic Disease Clinics of North America* **Vol. 13,** University of California, San Francisco, 1987.

Lindholm, D., Heumann, R., Meyer, M. and Thoenen, H. (1987) Interleukin-1 regulates synthesis of nerve growth factor in non-neuronal cell of rat sciatic nerve. *Nature* **330,** 658–661.

Matsuda, H., Coughlin, M. D., Bienenstock, J. and Denburg, J. A. (1988a) Nerve growth factor promotes human hemopoietic colony growth and differentiation. *Proc. Natl Acad. Sci. USA* **85,** 6508–6512.

Matsuda, H., Switzer, J., Coughlin, M. D., Bienenstock, J. and Denburg, J. A. (1988b) Human basophilic cell differentiation promoted by 2.5S nerve growth factor. *Int. Arch. Allergy Appl. Immunol.* **86,** 453–460.

Otten, U., Ehrhard, P. and Peck, R. (1989) Nerve growth factor induces growth and differentiation of human B lymphocytes. *Proc. Natl Acad. Sci. USA* **86,** 10059–10063.

Palombella, V. J., Yamashiro, D. J., Maxfield, F. R., Decker, S. J. and Vilcek, J. (1987) Tumor necrosis factor increases the number of epidermal growth factor receptors on human fibroblasts. *J. Biol. Chem.* **262,** 1950–1954.

Schall, T. J., Lewis, M., Koller, K. J., Lee, A., Rice, G. C., Wong, G. H. W., Gatanaga, T., Granger, G. A., Lentz, R., Raab, H., Kohr, W. J. and Goeddel, D. V. (1990) Molecular cloning and expression of a receptor for human tumor necrosis factor. *Cell* **61,** 61–370.

Soliven, B. and Albert, J. (1992) Tumor necrosis factor modulates the inactivation of catecholamine secretion in cultured sympathetic neurons. *J. Neurochem.* **58,** 1073–1078.

Steiner, P., Pfeilschifter, J., Boeckh, C., Radeke, H. and Otten, U. (1991) Interleukin-1β and tumor necrosis factor-alpha synergistically stimulate nerve growth factor synthesis in rat mesangial cells. *Am. J. Physiol.* **261,** F792-F798.

Thoenen, h. and Barde, Y. A. (1980) Physiology of nerve growth factor. *Physiol. Rev.* **60,** 1284–1335.

Thornton, S. C., Por, S. B., Penny, R., Richter, M., Shelley, L. and Breit, S. N. (1991) Identification of the major fibroblast growth factor released spontaneously in inflammatory arthritis as platelet growth factor and tumor necrosis factor-alpha. *Clin. Exp. Immunol.* **86,** 79–86.

Wasserman, S. I. (1984) The mast cell and synovial inflammation. *Arth. Rheum.* **27,** 841–846.

Young, L. H.Y., Joag, S. V., Lin, P. Y., Luo, S-F., Zheng, L. M., Liu, C-C. and Young, J.D-E. (1992) Expression of cytolytic mediators by synovial fluid lmphocytes in rheumatoid arthritis. *Am. J. Pathol.* **140,** 1261–1268.

Clinical Neuroscience and Neuropathology

neuroReport

NeuroReport 4, 563–565 (1993)

USING a specific enzyme-linked immunosorbent assay (ELISA) for human nerve growth factor (NGF), serum levels in patients with systemic lupus erythematosus (SLE) were measured. We found a consistent increase in NGF levels in SLE patients compared with controls. A good correlation exists between serum NGF level and severity of clinical manifestation. We hypothesize that NGF might play a role in the pathogenesis of autoimmune disorders such as SLE.

Key words: Systemic lupus erythematosus; Nerve growth factor; ELISA; Autoimmunity

Increased levels of NGF in sera of systemic lupus erythematosus patients

L. Bracci-Laudiero,[1] L. Aloe,[1] R. Levi-Montalcini,[1] M. Galeazzi,[2] D. Schilter,[3] J. L. Scully[3] and U. Otten[3,CA]

[1]Institute of Neurobiology, CNR, Rome; [2]Institute of Rheumatology, University of Siena, Siena, Italy; [3]Department of Physiology, University of Basel, Vesalgasse 1, 4051 Basel, Switzerland

[CA] Corresponding Author

Introduction

Nerve growth factor (NGF) exerts a biological action which is not restricted to cells of neuronal origin but also affects cells of the endocrine and immune systems.[1] Cells involved in the humoral immune response are responsive to NGF as demonstrated by numerous studies. *In vivo* administration of NGF to neonatal rats causes a widespread increase in size and number of mast-cells in several peripheral tissues.[2] *In vitro* these cells degranulate and release histamine in response to NGF.[3,4] NGF is a chemoattractant *in vitro* and *in vivo*[5] for polymorphonuclear leucocytes, has growth-promoting activity on myeloid progenitor cells,[6,7] prolongs *in vitro* survival of neutrophils and enhances phagocytosis.[8] In addition NGF has been shown to enhance lipid mediator release in basophils[9] without influencing eosinophil function.[10]

The biological effects of NGF are directly dependent on initial binding of NGF to specific cell surface receptors[11] whose presence has been demonstrated on rat splenocytes[12] and human T- and B-cells.[13] Human resting B-cells from peripheral blood and tonsil express NGF-r, and this expression increases upon activation of the cells.[13,14] Treatment of B-cells with NGF causes an increase in proliferation[13,14] and differentiation of B-cells into immunoglobulin-secreting plasma cells.[13] Cytokines involved in inflammation and immune responses such as IL-1, TNFα and IL-6 are potent inducers of NGF synthesis in peripheral tissues[15,16] and the central nervous system.[17,18] Thus NGF seems to play a central role in the cascade of events that leads to immune cell activation and inflammatory response.

Autoimmune diseases, which are characterized by an abnormal activation of the immune system, could provide useful models for studying the immunomodulatory role of NGF and its interaction with cytokines. In the present study we investigate the NGF content of the serum of systemic lupus erythematosus (SLE) patients. SLE is a multi-system autoimmune disease characterized by B-cell hyperactivity, aberrant immune regulation and an excessive production of immunoglobulins and autoantibodies. Continuous overproduction of specific cytokines[19,20] seems to contribute to B-cell hyperactivity in SLE. Using a two-site enzyme-linked immunosorbent assay (ELISA) specific for human NGF we found a significant increase in NGF levels in the sera of SLE patients, which appears to be related to the inflammatory episodes of the disease. This and previous studies[21,22] showed that the baseline level of NGF increases significantly in autoimmune states, suggesting that this factor may play a role in these diseases.

Materials and Methods

We used sera from 16 patients with SLE (12 female and 4 male, mean age 34 years, range 18–55 years, mean disease duration 5.4 years) diagnosed according to the American Rheumatism Association (ARA) 1982 revised criteria for the classification of SLE.[23] Disease activity was assessed with an arbitrary scoring system:[24] a value was assigned to 19 clinical signs and laboratory findings of SLE (Table 1). The patient index is the sum of these values. Using this index the SLE patients were divided into two groups: patients with less active disease (score 1–4) and patients with more active disease (score > 4). Serum from 15 healthy donors (9 female and 6 male, mean age 35.4 years, range 22–50 years) was used as control. In addition, we tested sera from a SLE family group consisting of a father (44 years) with a mild form of SLE (score < 4), one daughter (12 years) with active SLE (score > 4) and two healthy daughters (14 and 18 years). All sera were aliquoted to prevent repeated thawing and freezing and stored at – 70°C until used.

Before being tested the sera were acidified in order to reduce nonspecific binding as follows: 25 μl of 2 N HCl were added to 400 μl of serum and then, after 1 h at 20°C, 25 μl of 2 N NaOH and 12 μl of 1 M Na-Hepes were added. The acidified sera were then diluted with sample buffer and the NGF content of SLE sera was measured using a specific two-site immunoassay for human NGF as previously described.[22] As standards for the immunoassay we used human biologically active NGF from conditioned media of neuroblastoma cells (SH-SY5Y) and conditioned media from cultures of human skin fibroblasts stimulated with IL-1β.

Recombinant brain-derived neurotrophic factor (BDNF), that shares an identity of more than 50% with NGF, is not recognized in our assay at concentrations up to 20 ng ml^{-1}. Furthermore, conditioned medium from cultures of BDNF-transfected COS cells, which contains biologically active BDNF protein, did not crossreact with the antibody used in the ELISA.

Results

In this study we measured the NGF content in the sera of SLE patients using a specific two-site immunoassay for human NGF. NGF was detectable in all the sera analysed (Fig. 1) and the 16 SLE patients showed a higher NGF content (mean value 17.1 ± 1.7 pg ml^{-1}) than healthy controls (5.5 ± 1.15 pg ml^{-1}) ($p < 0.001$). According to the diagnosis, the patients were divided into two groups characterized by the degree of disease activity. As shown in Fig. 2, patients with a more active disease exhibit a higher amount of serum NGF (mean value 22.3 ± 2.1 pg ml^{-1}) than less active SLE patients (mean value 13.9 ± 1.9) but both SLE groups showed a significant increase of NGF content over basal levels found in healthy donors (mean value 5.6 ± 1.2 pg ml^{-1}). Thus NGF levels in the sera of SLE patients closely follow the activity of the disease. In addition, we tested sera from a SLE family group mentioned in Materials and Methods. The daughter with SLE showed higher serum NGF levels (mean value 21 ± 2.5 pg ml^{-1}) than her sisters (mean value

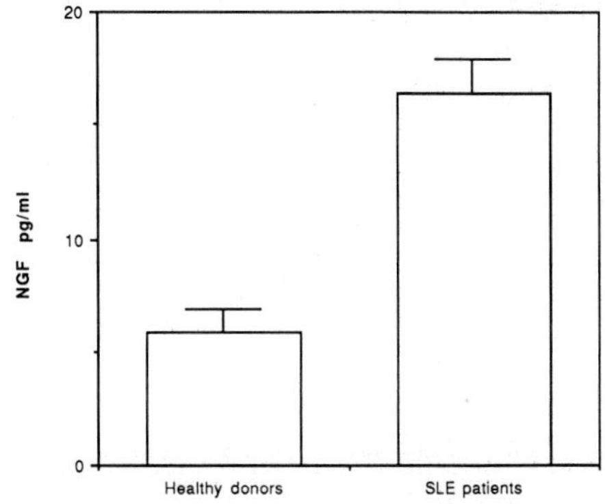

FIG. 1. NGF content of sera of SLE patients. NGF was detectable in all the SLE patients ($n = 16$), where the amount of this growth factor is higher than the baseline NGF levels found in the sera of healthy donors ($n = 15$). Serum samples were tested in triplicate in three different experiments and the values represent the mean of the samples + the standard error of the mean (s.e.m.). The difference between the mean values is significant ($p < 0.001$) according to Student's *t*-test.

Table 1. Systemic lupus erythematosus activity scoring system (according to G. D. Sebastiani, G. Passiu, M. Galeazzi *et al.*, 1991)

	Score
Vasculitis	3
Neurological	
seizures	3
psychosis	3
cranial nerve paralysis	3
chorea	3
headache	1
Renal	
proteinuria > 0.5 g 24 h^{-1}	2
haematuria > 5 RBC/HPF*	1
cellular casts	2
Arthralgias/arthritis	1
Skin and mucous membrane involvement	1
Serositis	2
Haematological	
anaemia < 10 g l^{-1}	2
leucocytopenia < 4.000 ul^{-1}	2
thrombocytopenia < 100.000 ul^{-1}	2
Eye involvement	2
Fever	1
Antibodies to double stranded DNA	1
Half-reduction in complement factors	1

*RBC = red blood cells; HPF = high power field.

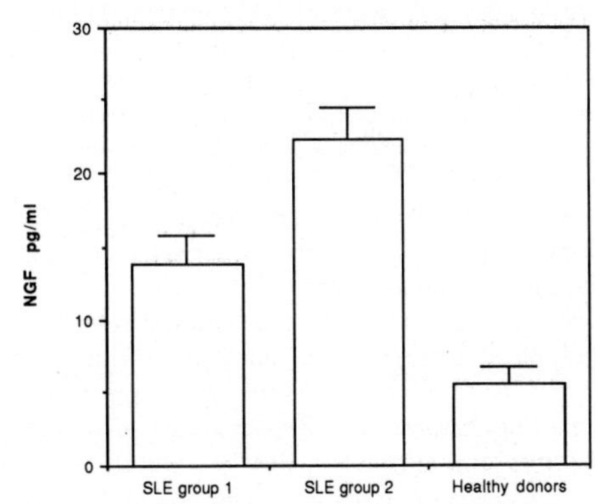

FIG. 2. NGF levels in sera of inactive and active SLE patients. SLE patients were divided in two groups according to disease activity. The patients in group 1 ($n = 8$) had an inactive disease (< 4) while patients in group 2 ($n = 8$) showed a more active disease (> 4). Sera from healthy donors ($n = 15$) were used as control. The patients of both SLE groups showed a significant increase of NGF content in the serum ($p < 0.001$). The values for each group represent the mean of the samples tested in three independent experiments, each done in triplicate + s.e.m.

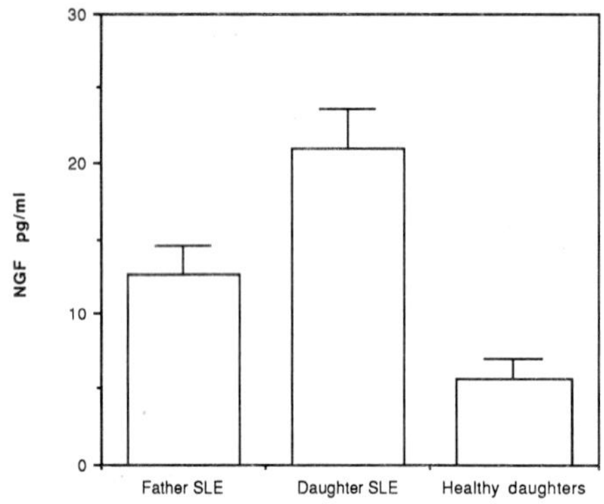

FIG. 3. NGF levels in sera of a SLE family. In this family the father and one daughter were affected by SLE. The SLE daughter was in the active phase (> 4) of the disease while the father showed a less active disease (< 4). The other two daughters were healthy and were used as controls. The NGF level of the active phase daughter was higher than that of her father and her healthy sisters. All the samples were tested in triplicate and the values represent the mean of three independent experiments + s.e.m.

5.7 ± 1.3 pg ml^{-1}). As shown in Fig. 3, the SLE daughter expressed higher NGF levels than her father (mean value 12.5 ± 2 pg ml^{-1}). These results indicate that there is a strong correlation between increased levels of serum NGF and the activity of SLE.

Discussion

In the present study we have analysed NGF levels in serum from patients affected by SLE, a multisystemic autoimmune disease whose principal feature is B-cell hyperactivity accompanied by abnormalities in immunoregulation. We detected NGF in all the sera tested and we observed that NGF levels increased in serum of SLE patients. The patients with more active disease showed higher levels of NGF when compared with the patients with less active SLE, although in both cases the NGF content was higher than baseline levels found in healthy donors.

A study on a family group affected by SLE indicated that there is a strong correlation between the activity of the disease and the serum NGF content. We have recently reported that in autoimmune diseases, such as multiple sclerosis[22] and rheumatoid arthritis,[21] there is an increase in NGF levels in the cerebrospinal and synovial fluid respectively. Recent studies demonstrated that cytokines such as IL-1β, TNFα and IL-6 stimulate NGF expression in the periphery and in the CNS.[15-18] Since higher amounts of these cytokines are produced during inflammation it is possible that enhanced serum NGF levels, found in the active phase of SLE, are regulated by these cytokines. However, the mechanisms involved in this increased NGF expression remain to be elucidated.

It is possible that the NGF found at sites of inflammation,[25] in CSF of MS patients[22] and sera of SLE patients may act as a regulatory factor for subpopulations of immune cells. NGF may be part of a cytokine cascade which can lead to immune cell activation, including the B-cell hyperactivity observed in autoimmune pathogenesis. Recent reports, which indicate that NGF interacts with specific cytokines to enhance their effect on immune cells,[6,7,9,14] are in line with this hypothesis.

Conclusion

Using a two-site ELISA specific for human NGF, we found a significant increase in NGF levels in the sera of patients affected by systemic lupus erythematosus (SLE), a multisystemic autoimmune disease. This study also showed a correlation between inflammatory episodes of the disease and increase in NGF levels. Since NGF accumulates at sites of inflammation, in the joints of patients affected by rheumatoid arthritis and in cerebrospinal fluids of multiple sclerosis patients, it is suggested that NGF is involved in the modulation of the immune response in these states.

References

1. Levi-Montalcini R, Aloe L and Alleva E. *PNEI* **3**, 1–10 (1990).
2. Aloe L and Levi-Montalcini R. *Brain Res* **133**, 358–366 (1977).
3. Bruni B, Bigon E, Boarato E *et al. FEBS Lett* **138**, 190–192 (1982).
4. Mazurek N, Weskamp G, Erne P *et al. FEBS Lett* **198**, 315–320 (1986).
5. Boyle M, Lawman M, Gee A *et al. J Immunol* **134**, 564–568 (1985).
6. Matsuda H, Coughlin M, Bienenstock J *et al. Proc Natl Acad Sci USA* **85**, 6508–6512 (1988).
7. Tsuda T, Wong D, Dolovich J *et al. Blood* **77**, 971–979 (1991).
8. Kannan Y, Ushio H, Koyama H *et al. Blood* **77**, 1320–1325 (1991).
9. Bischoff S and Dahinden C. *Blood* **79**, 2662–2669 (1992).
10. Takafuji S, Bischoff S, De Weck A *et al. Eur J Immunol* **22**, 969–974 (1992).
11. Bothwell M. *Nerve Growth Factors.* Chichester: J Wiley & Sons Ltd, 1989: 299–313.
12. Thorpe L, Stach R, Hashim G *et al. J Neurosci Res* **17**, 128–134 (1987).
13. Otten U, Ehrhard P and Peck R. *Proc Natl Acad Sci USA* **86**, 10059–10063 (1989).
14. Brodie C and Gelfand E. *J Immunol* **148**, 3492–3497 (1992).
15. Lindholm D, Heumann R, Meyer M *et al. Nature* **330**, 658–659 (1987).
16. Steiner P, Pfeilschifter J, Boeckh C *et al. Am J Physiol* **261**, F792–F798 (1991).
17. Frei K, Malpiero UV, Leist TP *et al. Eur J Immunol* **19**, 689–694 (1989).
18. Gadient RA, Cron KC and Otten U. *Neurosci Lett* **117**, 335–340 (1990).
19. Huang YP, Perrin L, Miescher P *et al. J Immunol* **141**, 827–833 (1988).
20. Tanaka Y, Saito K, Shirakawa F *et al. J Immunol* **141**, 3043–3049 (1988).
21. Aloe L, Tuveri M, Carcassi U *et al. Arth Rheum* **35**, 351–355 (1992).
22. Bracci-Laudiero L, Aloe L, Levi-Montalcini R *et al. Neurosci Lett* **147**, 9–12 (1992).
23. Tan EM, Cohen AS, Fries JF *et al. Arth Rheum* **25**, 1271–1277 (1982).
24. Sebastiani GD, Passiu G, Galeazzi M *et al. Clinic Rheumatol* **10**, 289–293 (1991).
25. Weskamp G and Otten U. *J Neurochem* **48**, 1779–1786 (1987).

ACKNOWLEDGEMENTS: We thank Dr G. D. Sebastiani, University of Catanzaro, for providing SLE sera. This study was supported by Target Projects CNR:FATMA (Stress) and BTBS (Biopharmaceutics) to Professor Levi-Montalcini and Dr Aloe, Swiss National Foundation (Grant 31-2995490), Deutsche Forschungsgemeinschaft (SFB 325), and from BMFT (Grant 01KL8904/6) to Professor Otten.

Received 30 January 1993; accepted 8 March 1993

Neuroendocrinology

neuroreport

NeuroReport 4, 1063–1065 (1993)

NGF is released into plasma during human pregnancy: an oxytocin-mediated response?

Patrizia Luppi,[1] Rita Levi-Montalcini,[1] Luisa Bracci-Laudiero,[1] Alfio Bertolini,[3] Rossana Arletti,[3] Donatella Tavernari,[4] Eliana Vigneti[2] and Luigi Aloe[1,CA]

[1]Institute of Neurobiology and [2]Cell Biology, C.N.R., Viale C. Marx 15, I-00156, Roma; [3]Institute of Pharmacology, University of Modena, Viale Campi 287, I-41100, Modena; [4]Department of Obstetrics and Gynaecology, Hospital of Mirandola, Via Fogazzaro 1, I-41037 Mirandola, Modena, Italy

[CA] Corresponding Author

THE presence of biologically active nerve growth factor (NGF) in the peripheral circulation of women during pregnancy, labour and lactation was investigated. Using a sensitive immunoenzymatic assay (ELISA), we found an approximately five-fold increase in plasma NGF levels during labour and lactation compared with the concentrations found at the term of gestation or in control healthy women. Since labour and lactation are characterized by activation of the hypothalamo-pituitary-adrenal axis and by high plasma levels of the neurohypophyseal hormone oxytocin, and since the intravenous injection of oxytocin in female rats causes a 176% increase in the hypothalamic levels of NGF, it is possible that the increased amount of circulating NGF is correlated with one or both of these events.

Key words: Labour; Stress; Lactation; NGF; Oxytocin; Two-site enzyme immunoassay

Introduction

Pregnancy, parturition and lactation of the reproductive processes represent the paradigm of maternal functions. During these states, female mammals exhibit a number of neuroendocrine morphological and behavioural changes which prepare for the onset of maternal behaviour and interaction with the pups. Ovarian steroids play a central role in these reproductive-related behaviours and together with oxytocin, a nonapeptide synthesized principally in magnocellular neurones of paraventricular and supraoptic nuclei of the hypothalamus and released from the neurohypophisis in the systemic circulation in response to appropriate physiological stimuli, constitute the hormonal bases of sexual,[1] maternal[2,3] and other social behaviours.[4]

The possibility of an involvement of nerve growth factor (NGF) in these processes arises from recent studies showing that this protein, well-known for its role in the survival and development of peripheral sympathetic and sensory neurones[5] and for its neurotrophic effects on magnocellular cholinergic neurones[6] takes part in the regulation of neuronal and cellular populations involved in the control of specific neuroendocrine functions[7] and in the acquisition of male and female reproductive capacity.[8,9] Moreover, it has been shown that both NGF and its mRNA increase in the hypothalamus following stressful events.[7,10]

Other recent reports provide evidence that the gene encoding NGF and its receptor are expressed in the developing female rat hypothalamus[11] and that NGF is also synthesized by the anterior, intermediate and posterior lobes of the pituitary gland throughout the postnatal period as well as in adult rats.[12] These findings, together with the evidence that NGF induces activation of the pituitary-adrenocortical axis, enhancing the secretion of ACTH and the concentration of plasma glucocorticoids[13] point to a role for NGF in the modulation of certain neuroendocrine functions.[13,14]

The aim of the present study was to investigate whether the level of NGF in the human bloodstream changes during the third trimester of pregnancy, labour and lactation. The possibility that plasma NGF levels might be correlated to circulating oxytocin is suggested.

Materials and Methods

Samples: Blood samples were collected from individual women (age 18–40 years) at 36–38 weeks of gestation ($n = 17$), during uncomplicated labours and vaginal deliveries ($n = 10$) and at the fourth day of lactation ($n = 19$). Blood from healthy women from the same age-range was used as a control ($n = 18$). NGF was also measured in paired arterial and venous umbilical cord plasma samples from vaginal delivery ($n = 10$). At delivery, the umbilical cord was double clamped near the foetus and close to the vaginal introitus, and this portion of the cord was cut off; puncture of the umbilical artery and vein was then performed. All samples were centrifuged as quickly as possible and the sera were stored at – 70°C until used for NGF determination. In another set of experiments, adult virgin Sprague–Dawley female rats ($n = 4$) were injected in

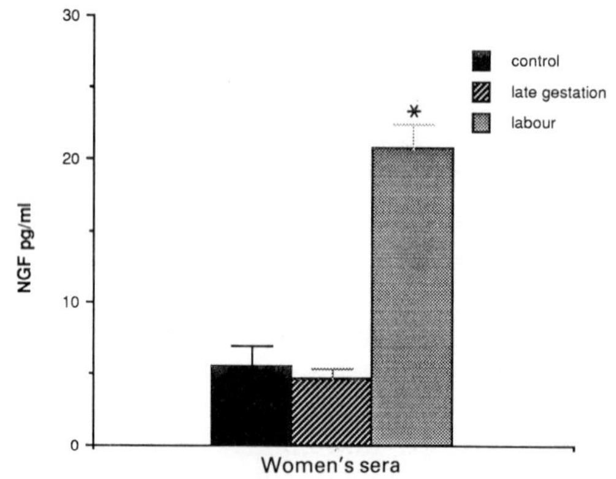

FIG. 1. NGF levels in women's sera during gestation and labour. NGF content has been measured in women at 36–38 weeks of gestation (n = 17, during uncomplicated labour (n = 10) and in controls (n = 18). A marked increase of NGF levels was measured in the sera of women in several phases of dilatation during labour compared with the values measured at the term of gestation and in the controls. The values represent the mean ± s.e.m. of triplicate measures obtained in 3 independent experiments. *$p < 0.001$ (Student's t-test).

the tail vein with 10 μg of oxytocin and sacrificed after 3 h with an overdose of nembutal, following the guideline of the Ethical Committee of the Institute of Neurobiology. A virgin female rat injected with a saline solution was used as a control. Hypothalamus and other brain tissues were then removed and used for NGF assay.

NGF determination: NGF was prepared from mouse salivary glands following the procedure by Bocchini and Angeletti[15] and a monoclonal anti-NGF antibody which recognizes rodent and human NGF was prepared in our laboratory. The NGF content in human sera and in the rat hypothalamic area was measured by a two-site immunoenzymatic assay (ELISA): briefly, polystyrene 96 well microtiter immunoplates (Nunc) were coated with affinity purified polyclonal goat anti-NGF antibody. Parallel wells were coated with pre-immune goat IgG for evaluation of the non-specific signal. After 5 h at 20°C the plates were washed (as in subsequent steps) with PBS-0.05% Tween 20 and then incubated for 2 hours at 20°C with PBS-0.05% Tween 20, 1% FCS to block non-specific binding sites. After washing, the NGF standard solutions, ranging from 0 to 1 ng ml^{-1}, were distributed in each plate. The samples were ultrasonicated at 4°C in the sample buffer and centrifuged at 10,000 rpm for 30 min at 4°C. 100 μl of supernatant was added to each well and the plates incubated overnight at 4°C. The plates were then washed and 100 μl of monoclonal antibody against NGF, diluted 1:100 in PBS, was added to each well. After incubation for 4 h at 20°C, the plates were washed and incubated with biotinylated rat immunoglobulins (1:8,000, Zymed). The subsequent incubation with perioxidase-conjugated streptavidin (Zymed) and addition of orto-phenyldiamine resulted in a colourimetric reaction whose optical density was measured at 490 mn using a Dynatech MR 5000 microplate reader.[17] Specificity for NGF was also assessed by using a recombinant human NGF (Genetech, USA). Recombinant brain-derived neurotrophic factor (BDNF)[18] is not recognized in the ELISA at concentrations up to 20 ng ml^{-1}.[17] The biological activity of NGF was evaluated using sympathetic neurones from foetal rat superior cervical ganglion ((SCG) which are receptive to NGF but not to other neurotrophins.[18] The specific NGF neurite outgrowth-promoting effect of human plasma was assessed by adding NGF antibodies to tissue culture.[19]

Data analysis: All the samples were tested in triplicate and at two dilutions, and the data represent the mean ± s.e.m. of different tests. Statistical differences between our samples were determined by Student's t-test.

Results

Using a sensitive immunoassay for detecting human NGF, we determined the level of NGF in women's sera during late pregnancy, labour and lactation. Sera of healthy women were also tested to assess basal NGF levels. While plasma NGF concentration during late pregnancy (4.71 ± 0.76 pg ml^{-1}, n = 17) was not significantly different from that of our controls (5.53 ± 1.41 pg ml^{-1}, n = 18), a marked increase was found in the sera of pregnant women in several phases of dilatation during labour (20.77 ± 1.62 pg ml^{-1}, n = 10), which represents an approximately five-fold increase compared with the values measured at late gestation (Fig. 1).

A similar significant enhancement in NGF content was also observed in the sera of women at the fourth day of lactation (20.17 ± 1.85 pg ml^{-1}, n = 19) (Fig. 2). Blood serum collected from venous and arterial umbilical cord (= foetal circulation) contained baseline levels of NGF. The addition of 2 μ ml^{-1} of human plasma from parturient women, stimulated neurite outgrowth from sympathetic neurones and this neurite outgrowth was inhibited by NGF antibodies, suggesting that NGF was biologically active.

In another experiment, virgin female rats were injected in the tail vein with 10 μg of oxytocin or saline solution. A significant increase in NGF levels was found in the hypothalamus of the oxytocin-injected rats (876 ± 110 pg g^{-1} w.w.) compared with those injected with saline (317 ± 23 pg g^{-1} w.w.).

Discussion

This study shows that NGF, a neurotrophic polypeptide, increases in the plasma of women during labour and lactation. Plasma NGF levels at 36–38 weeks of gestation were not significantly different

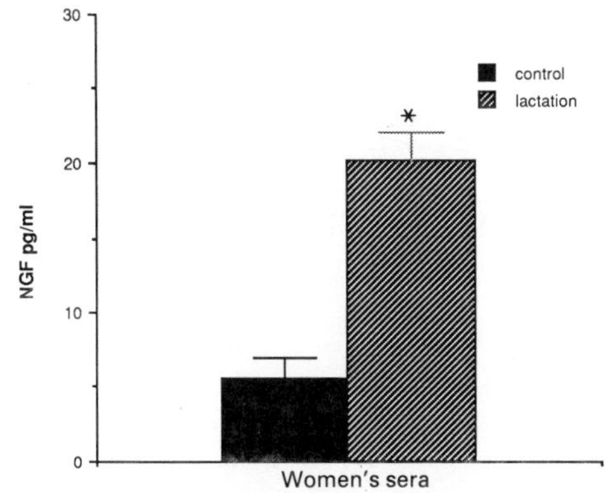

FIG. 2. NGF levels in women's sera during lactation. NGF has been measured in women at the fourth day of lactation ($n = 19$). Blood of healthy women was used as controls ($n = 18$). All lactating women showed increased NGF levels compared with those of controls. The values represent the mean ± s.e.m. of triplicate measures obtained in four independent experiments. $*p < 0.001$ (Student's t-test).

from those found in control healthy women, while during labour the maternal NGF concentration was about five-fold higher than in late gestation and the same rise in NGF levels was also detected in the sera of women during lactation (see Figs 1 and 2).

The NGF in human plasma was biologically active since it was able to stimulate neurite outgrowth of cultured neurones from superior cervical ganglion (SCG). This neurite outgrowth was inhibited by NGF antibodies, suggesting that the activity was mostly due to NGF rather than other neurotrophins.[18] Thus, the present data provide evidence that plasma NGF concentration, like several hormones,[20,21] increases during human labour and lactation. The presence of increased plasma NGF levels in several phases of human labour and during *post-partum* lactation raises the question as to whether NGF released into the circulation during these stressful situations[20] might take part in the physiological activation of the classical and typical hypothalamo-pituitary-adrenal axis response to stress, perhaps in co-operation with oxytocin.[22,23]

Although this hypothesis is supported by our findings, the physiological role(s) played by NGF in the peripheral circulation and the mechanism(s) involved in the increase of NGF in human plasma during these reproductive-related behaviours remain, however, to be determined. It is known that the neurohypophyseal hormone oxytocin is released into the systemic circulation during several reproductive functions such as ovulation and coitus[24] and that it promotes fertilization by enhancing sperm and ova transport after mating.[21] Oxytocin also plays an important role in the induction of labour, leading to contractions of the uterine wall.[22] Moreover, after delivery, this neurohormone stimulates milk ejection from the mammary gland, caused by the suckling reflex, and affects the release of prolactin (PRL) from the anterior lobe of the pituitary gland, which itself promotes milk production.[25] It is, therefore, possible that at the end of gestation oxytocin may be involved not only in the phenomenon of parturition, but also in the adrenocortical response that characterizes the stress of labour.[20,22] The participation of oxytocin in the regulation of NGF secretion is supported by our observation that intravenous injection of oxytocin in female virgin rats elicits a significant enhancement of the hypothalamic NGF content compared with saline-injected virgin rats.

In conclusion, the data presented here support the hypothesis that NGF plays a role in neuroendocrine functions both in pregnancy and lactation and that oxytocin might be involved in the regulation of brain and blood NGF levels. Nevertheless, the functional role of NGF in these physiological situations remains to be clarified.

Conclusions

The results of our study demonstrate that during human labour and lactation there is an increase of biologically active NGF in women's sera. These two states are also characterized by an extensive synthesis of oxytocin and its release into the peripheral circulation, suggesting that during labour and lactation the increased levels of NGF and oxytocin might be correlated. This hypothesis is supported by the observation that after oxytocin injection there is an enhancement of NGF in the hypothalamic region of female virgin rats.

References

1. Arletti R and Bertolini A. *Neuropeptides* **6**, 247–253 (1985).
2. Arletti R, Benelli A and Bertolini A. Oxytocin involvement in male and female sexual behavior. In: Pedersen CA, Caldwell JD, Jirikowski GF, Insel T, eds. *Oxytocin in maternal, sexual and social behaviors.* A.N.A.Y.S., 1991: **652**, 180–193.
3. Pedersen CA and Prange AJ. *PNAS* **76**, 6661–6665 (1979).
4. Carter SC Jr, William DM, Witt TR *et al.* Oxytocin and social bonding. In: Pedersen CA, Caldwell JD, Jirikowski GF, Insel T, eds. *Oxytocin in maternal, sexual and social behaviors.* A.N.A.Y.S., 1991: **652**, 204–211.
5. Levi-Montalcini R. *Science* **237**, 1154–1162 (1987).
6. Gnahn H, Hefti F, Heumann R *et al. Dev Brain Res* **9**, 45–52 (1983).
7. Aloe L, Alleva E and De Simone R. *Behav Brain Res* **39**, 53–61 (1990).
8. Persson H, Ayer-Le-Lievre C, Soder O *et al. Science* **247**, 704–707 (1990).
9. Lara E, McDonald JK and Ojeda SR. *Endocrinol* **126**, 364–375 (1990).
10. Spillantini MG, Aloe L, Alleva E *et al. PNAS* **86**, 8555–8559 (1989).
11. Ojeda RS, Hill DF and Katz KH. *Mol Brain Res* **9**, 47–55 (1991).
12. Lahtinen T, Soinila S and Lakshmanan J. *Neuroscience* **30**, 1: 165–170 (1989).
13. Otten U, Beaumann JB and Girard J. *Nature* **282**, 413–414 (1979).
14. Aloe L, Cozzari C, Calissano P *et al. Nature* **291**, 358–366 (1981).
15. Bocchini G and Angeletti PU. *PNAS* **64**, 787–794 (1969).
16. Vigneti E, Bracci-Laudiero L and Aloe L. In: *New generation of monoclonal antibodies in diagnosis and therapy.* In press (1993).
17. Bracci-Laudiero L, Aloe L, Levi-Montalcini R *et al. Neurosci Lett* **147**, 9–12 (1992).
18. Alderson RF, Alterman AL, Barde YA *et al. Neuron* **5**, 297–306 (1990).
19. Aloe L. *PNAS* **88**, 5636–5640 (1989).
20. Campbell EA, Linton EA, Wolfe CDA *et al. J Clin Endocrinol Metab* **64**, 1054–1059 (1987).
21. Swaab DF. *Prog Brain Res* **38**, 225–244 (1972).
22. Fuchs AR and Saito S. *Endocrinol* **88**, 574–578 (1971).
23. Lang RE, Heil WEJ, Ganten D. *et al. Neuroendocrinol* **37**, 314–316 (1983)
24. Murphy MR, Seckl JR, Burton S *et al. J Clin Endocrinol Metab* **65**, 738–741, (1987).
25. Arey BJ and Foreman ME. *Endocrinol* **130**, 126–132 (1992)

ACKNOWLEDGEMENTS: This study was supported by P. F. FATMA, Subproject Stress to Dr Luigi Aloe.

Received 12 May 1993; accepted 26 May 1993

Agents Actions *39, Special Conference Issue* (1993)

0065-4299/93/03C145-03 $ 1.50 + 0.20/0

A proposed autacoid mechanism controlling mastocyte behaviour

Luigi Aloe, Alberta Leon [1] and Rita Levi-Montalcini

Institute of Neurobiology, National Research Council, Rome, Italy and [1]Researchlife, Padua, Italy

Abstract

Evidence is provided here supporting the existence of a novel autacoid mechanism negatively modulating mast cell behaviour in response to noxious stimuli *in vivo*; hence, the denomination "autacoid local inflammation antagonism" (ALIA). In particular, as lipid amides of the *N*-acylethanolamine type have been reported to accumulate in tissues in degenerative inflammatory conditions, we examined whether these *N*-acylated lipids could exert regulatory effects on mast cell activation *in vivo*. The results reported show that both long- and short-chain *N*-acylethanolamines, when systemically administered, are effective in reducing mast cell degranulation induced by local injection of substance P in the ear *pinna* of developing rats. These and other data suggest that the endogenous production of *N*-acylethanolamines may constitute a local autocrine/paracrine response for the negative feedback control of mast cell responses to various activating signals. Such a process may be of physio-pathological relevance in the regulation of functional neuro-immune-mast cell interactions.

Introduction

Mast cells, involved in hypersensitivity and inflammatory processes, have now been shown to be in functional bidirectional communication with both the nervous and immune systems [1]. This indicates that neuro-immune interactions can occur via mast cells and that mast cells may be critical gate cells in the regulation of these interactions [2]. Thus, abnormal mast cell numbers or phenotypes, as well as uncontrolled mast cell activity, may underlie alterations in neuro-immune interactions and *vice versa*, with significant implications for several disease states.

To date, the mechanisms capable of regulating mast cell properties in normal and pathological conditions remain largely obscure. Mast cell phenotypic heterogeneity has been documented in normal conditions and an increase in mast cell numbers has been reported in a variety of immuno-inflammatory conditions. Interestingly, the latter conditions have been shown to be associated with increased tissue levels of nerve growth factor (NGF), a well-characterized neuronotrophic factor, which, when administered to young rodents, causes mast cell activation and proliferation [2–5]. Furthermore, mast cells have been reported to produce and respond to a variety of cytokines [6]. As cytokines are known to affect NGF synthesis [2], differences in tissue cytokine and NGF expression, either alone or in combination, may provide a positive feedback loop for autocrine/paracrine regulation of mast cell properties. However, despite this knowledge, no information is yet available concerning the negative control mechanisms for these cell populations. Such mechanisms, in view of

Correspondence to: Luigi Aloe, Institute of Neurobiology, National Research Council, Viale Marx, 15, 00137 Rome, Italy

Agents Actions, *Special Conference Issue* (1993)

the potential hazards of uncontrolled mast cell behaviour, may be postulated to exist naturally and to be involved in the regulation of neuro-immune-mediated inflammatory responses. We herein report evidence supporting the possibility that lipid amides of the *N*-acylethanolamine type are potential prototypes of naturally ocurring molecules capable of modulating mast cell activation *in vivo*.

Materials and methods

A long-chain (C16) *N*-acylethanolamine (LG 2110/1) and a short (C4)-chain analogue (LG 2130/2) (provided by Lifegroup, Padua, purity >99%) or saline were administered s.c. in the back of 2-week-old rats (Sprague–Dawley, Charles River) followed, 20 min later, by s.c. administration of substance P (1 μl of a $10^{-6} M$ solution or saline in the ear pinna). After 5 min, the rats were sacrificed and the pinna removed for histological assessment of mast cell degranulation, by counting the number of mast cells showing membrane disruption and surrounded by at least five extracellular metachromatic granules.

Results and discussion

Tissue accumulation of *N*-acylated glycerophospholipids and free *N*-acylethanolamines has been reported to occur in pathological degenerative conditions [7]. As such conditions are known to be associated with inflammatory reactions, one attractive hypothesis is that the production of these lipid metabolites may play an autacoid role in controlling mast cell behavior in pathological conditions. To test this hypothesis, we examined whether systemically administered short (C4)- and/or long (C16)-chain *N*-acylethanolamines were effective in decreasing mast cell degranulation induced by local injection of substance P in the ear pinna of developing rats.

As shown in Table 1 and exemplified in Fig. 1, the local administration of substance P in the ear pinna of the developing rats rapidly resulted in massive mast cell degranulation. This degranulation was significantly reduced in animals pretreated, 20 min earlier, with both *N*-acylethanolamines, with the short chain (LG 2130/2) being more effective than the long-chain (LG 2110/1) analogue at a dose of 20 mg/kg s.c. and showing significant effects already at a dose of 0.1 mg/kg s.c. As this

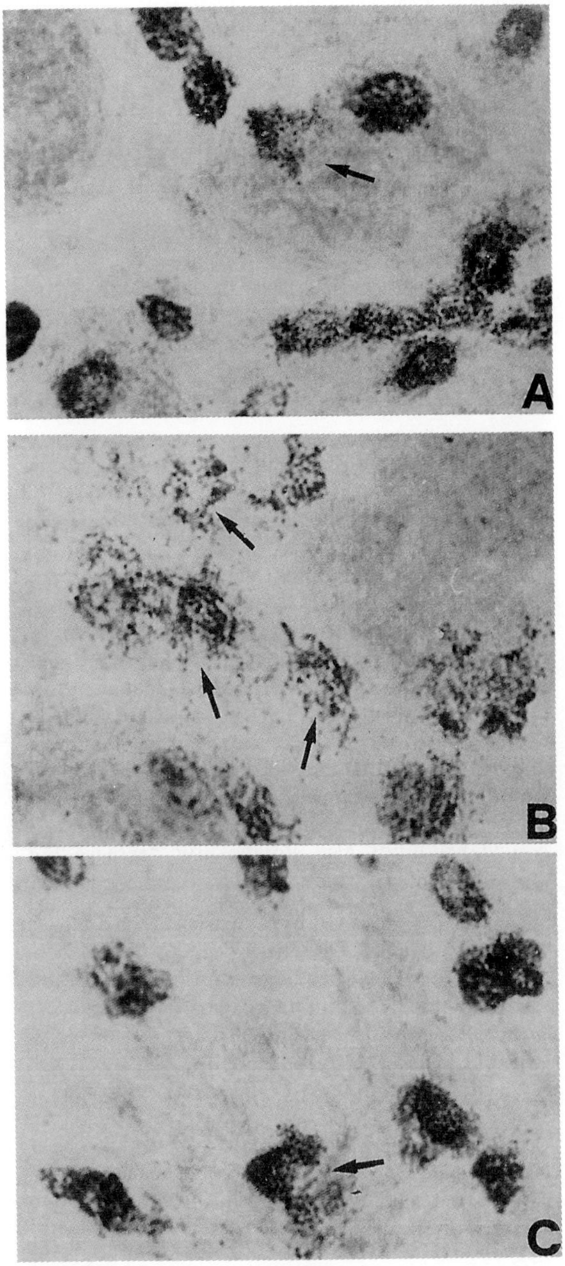

Figure 1
Cross section of the ear pinna of 12-days-old rats showing mast cell morphology: (a) untreated; (b) injected in the pinna with 1 μl of SP $10^{-6} M$; (c) injected with SP as in b, preceded 20 min earlier by subcutaneous administration of 20 mg/kg of G2130/2. Note decrease of mast cell degranulation (arrows) in c as compared to b. Toluidine blue stained. X-780.

Agents Actions, *Special Conference Issue* (1993) C147

Table 1
Effects of *N*-acylethanolamine administration on substance-P-induced mast cell degranulation in the rat ear pinna

Treatment	Degranulated mast cells (%)
Saline + saline	6
Substance P + saline	91
+ LG 2110/1 (20 mg/kg)	65
+ LG 2130/2 (20 mg/kg)	25
(5 mg/kg)	24.6
(1 mg/kg)	32.5
(0.1 mg/kg)	68.8

Results are the mean of duplicate experiments, each conducted by counting at least 700–800 cells in a total surface area of approx. 3 mm^2 per pinna. SEM was at all times < 10% (see Materials and Methods section for further relevant information).

dose–response diversity may be due to differences in bioavailability, these results suggest that both lipid amides are capable of exerting rapid and potent local anti-inflammatory effects *in vivo*.
The above results are in accord with previous reports concerning anti-inflammatory effects of *N*-acylethanolamines [8] and further indicate that such effects may occur via their action on mast cells. Since mast cell activation *in vivo* was here induced utilizing substance P, the compounds are effective in negatively controlling mast cell responses correlated with neurogenic inflammatory processes. In addition, we have preliminary evidence that both *N*-acylated lipids also significantly decrease the passive cutaneous anaphylaxis reaction, indicating that they may control mast cells independently of the type of activating stimuli. Thus, it may be speculated that endogenous local production of lipid amides of the *N*-acylethanolamine type may be an adaptive response for the regulation of mast cell activation and consequent expression of inflammatory processes. The acronym ALIA (autacoid local inflammation antagonism) is proposed for this novel mast cell modulation mechanism. Further studies concerning the cellular and molecular processes involved in the ALIA mechanisms are thus warranted and may well lead to the elucidation of novel pathways regulating neuro-immune interactions, mediated via the mast cells.

References

[1] J. S. Marshall and J. Bienenstock, *Mast cells*. Semin. Immunopathol. *12*, 191–202 (1990).

[2] R. Levi-Montalcini, L. Aloe and E. Alleva, *A role for nerve growth factor in nervous, endocrine and immune system*. Prog. Neurol. Endocrinol. Immunol. *1*, 1–10 (1990).

[3] L. Aloe, M. Tuveri, U. Carcassi and R. Levi-Montalcini, *Nerve growth factor in the synovial fluid of patients with chronic arthritis*. Arth. Rheum. *35*, 351–355 (1992).

[4] L. Aloe and R. Levi-Montalcini, *Mast cell increase in tissues of neonatal rats injected with nerve growth factor*. Brain Res. *133*, 358–366 (1977).

[5] J. S. Marshall, R. H. Stead, C. McSharry, L. Nielsen and J. Bienenstock, *The role of mast cell degranulation in mast cell hyperplasia, I. Mechanism of action of nerve growth factor*. J. Immunol. *144*, 1886–1892 (1990).

[6] S. J. Galli, J. R. Gordon and B. Wershil, *Cytokine production by mast cells and basophils*. Current Opinion Immunol. *3*, 865–873 (1991).

[7] V. Natarajan, V. P. Reddy, P. C. Schmid and H. H. O. Schmid, *N-Acylation of ethanolamine phospholipids in canine myocardium*. Biochim. Biophys. Acta *712*, 342–355 (1982).

[8] O. H. Ganley, O. Graessle and H. J. Robinson, *Anti-inflammatory activity of compound obtained from egg yolk, peanut oil and soybean lecithin*. Lab. Clin. Med. *51*, 709–714 (1958).

Proc. Natl. Acad. Sci. USA
Vol. 91, pp. 3739–3743, April 1994
Neurobiology

Mast cells synthesize, store, and release nerve growth factor

(plasticity/inflammation/hyperalgesia/neuroimmune interactions)

A. LEON*†, A. BURIANI*, R. DAL TOSO*, M. FABRIS*, S. ROMANELLO*, L. ALOE‡, AND R. LEVI-MONTALCINI‡

*Researchlife, c/o Centro di Ricerca Biomedica, Ospedale Civile, 31033 Castelfranco Veneto, Italy; and ‡Institute of Neurobiology, National Research Council, 00137 Rome, Italy

Contributed by R. Levi-Montalcini, December 20, 1993

ABSTRACT Mast cells and nerve growth factor (NGF) have both been reported to be involved in neuroimmune interactions and tissue inflammation. In many peripheral tissues, mast cells interact with the innervating fibers. Changes in the behaviors of both of these elements occur after tissue injury/inflammation. As such conditions are typically associated with rapid mast cell activation and NGF accumulation in inflammatory exudates, we hypothesized that mast cells may be capable of producing NGF. Here we report that (*i*) NGF mRNA is expressed in adult rat peritoneal mast cells; (*ii*) anti-NGF antibodies clearly stain vesicular compartments of purified mast cells and mast cells in histological sections of adult rodent mesenchymal tissues; and (*iii*) medium conditioned by peritoneal mast cells contains biologically active NGF. Mast cells thus represent a newly recognized source of NGF. The known actions of NGF on peripheral nerve fibers and immune cells suggest that mast cell-derived NGF may control adaptive/reactive responses of the nervous and immune systems toward noxious tissue perturbations. Conversely, alterations in normal mast cell behaviors may provoke maladaptive neuroimmune tissue responses whose consequences could have profound implications in inflammatory disease states, including those of an autoimmune nature.

Mast cells are involved in inflammatory and hypersensitivity reactions (1) and occur in many peripheral tissues, in perivascular regions in close apposition to innervating sensory or autonomic nerve fibers (2), and also within the peripheral and central nervous systems (3). Nervous and immunological mediators such as neuropeptides or IgE can affect the state of mast cell activation (4). Secretory products of activated mast cells can stimulate or facilitate axon reflexes, thereby inducing positive feedback loops (5). Activated mast cells also secrete a wide array of pluripotent cytokines and other inflammatory mediators (6) and may thus act as bidirectional carriers of information between the nervous and immune systems, suggesting profound implications for tissue homeostatic mechanisms.

Neurotrophic and/or tropic cues could be produced directly by mast cells. Nerve growth factor (NGF) reportedly accumulates in inflammatory sites or exudates (7, 8) caused by various noxious stimuli, including those of autoimmune origin (9, 10). NGF is the prototype of target-derived neurotrophic factors critical for development and maintenance of specific peripheral and central neuronal populations (11, 12). Evidence also points to an action for NGF on cells of the immune system (13, 14), including mast cells (15). NGF may thus play a key role not only within the nervous system but, more importantly, in cross-talk between cells of the nervous and the immune systems (16).

NGF accumulation in acute inflammatory exudates has been attributed to cellular infiltrates or to the induction of tissue NGF expression. However, the rapid appearance (within 2 h) and degree of NGF accumulation in rat skin blister fluid (8) is more compatible with release of NGF from resident tissue cells containing the stored protein. Rapid induction of preprotachykinin mRNA expression, which is a NGF-sensitive process, in dorsal root ganglia (DRG) of rats with adjuvant monoarthritis (17) supports this hypothesis. In addition, the remodeling of intestinal mucosal nerve fibers during intestinal inflammation is correlated with changes in mast cell density (18). Mast cells and sympathetic neurons in culture form contacts (19), suggesting a mast cell–nerve ending chemotactic NGF-like gradient effect.

Given the potential involvement of mast cells and NGF in neuroimmune interactions and the close microanatomical associations between mast cells and sensory or autonomic fibers in several tissues (20, 21) mast cells may, in fact, produce NGF. This study investigates the capability of purified rat peritoneal mast cells to synthesize and release biologically active NGF.

MATERIALS AND METHODS

Immunocytology and Histochemistry. Rat peritoneal mast cells (RPMCs) were prepared from male Wistar rats (150–200 g) (Charles River Breeding Laboratories) as described (22). Over 90% cell purity was indicated by toluidine blue or safranin staining (23). Immunocytochemistry for NGF was performed with an affinity-purified rabbit polyclonal antibody to mouse NGF (1–0.1 μg/ml) and a peroxidase-conjugated goat anti-rabbit polyclonal antibody (Vectastain, Vector Laboratories). Cryostat sections (15 μm) from adult rat ear pinna were similarly processed. To control for staining specificity, primary antibody was preincubated with excess NGF, substituted with nonspecific rabbit IgG (2 μg/ml; Sigma), or omitted.

Biological Assays and Quantitation of NGF Activity. NGF activity in RPMC conditioned medium was tested on dissociated cultures of chicken embryonic day 8 (E8) DRG and E10 sympathetic neurons (11). Conditioned medium was obtained by incubating 10^6 RPMCs per ml in Dulbecco's modified Eagle's medium (DMEM) plus 10% fetal calf serum (FCS) for 24 h, at which time cells were still viable by microscopic appearance. Sympathetic and DRG neurons were cultured (24) for 24 and 48 h, respectively, without or with various amounts of the conditioned medium. The percentage of cells bearing neurites of >4 somal diameters was determined in random fields, and NGF was quantitated as described (24). Biospecificity was evaluated by adding a goat anti-mouse NGF polyclonal antibody (10 μg/ml) or a mouse

The publication costs of this article were defrayed in part by page charge payment. This article must therefore be hereby marked "*advertisement*" in accordance with 18 U.S.C. §1734 solely to indicate this fact.

Abbreviations: NGF, nerve growth factor; DRG, dorsal root ganglia; RPMC, rat peritoneal mast cell; E8, embryonic day 8; BDNF, brain-derived neurotrophic factor.
†To whom reprint requests should be addressed.

monoclonal anti-NGF antibody (1 μg/ml; clone 27/21; Boehringer Mannheim). A two-site ELISA was used to quantitate NGF in appropriate dilutions of mast cell conditioned media versus mouse 2.5S NGF as described (25). Blanks consisted of samples added to microwells coated with mouse myeloma IgG (Calbiochem) instead of anti-NGF antibody.

RNA Extraction and PCR. Total RNA was extracted as described (26) from RPMCs, rat basophilic leukemia (RBL-2H3) cells (P. Ghezzi, Mario Negri Institute, Milan), mouse neuroblastoma neuro-2a cells (ATCC-CCL 131), or male mouse submaxillary glands. The RNA was then precipitated with 4 M sodium acetate (4°C overnight) and separated by centrifugation (8000 × *g* for 15 min). PCR primers and an internal hybridization probe for rat NGF were from Severn Biotech (Kidderminster, U.K.). Their sequences were as follows: 5′ primer, 5′-TCA TCC ACC CAC CCA GTC TTC-3′, 5′ corresponding to residue 654; 3′ primer, 5′-GGC AGC CTG TTT GTC GTC TGT-3′, 3′ corresponding to residue 946; internal probe, 5′-CGC CTT GAC AAA GGT GTG AGT CGT-3′, corresponding to residues 896–919 of the NGF gene sequence (27). Primer sequence identity with other members of the NGF gene family was <50% and PCR of a brain-derived neurotrophic factor (BDNF) cDNA clone with the chosen primers gave negative results. First-strand cDNA synthesis was performed with the 3′ antisense NGF primer and Moloney murine leukemia virus reverse transcriptase (United States Biochemical). The volume was then increased 5-fold with water and 25 μl of the reverse transcriptase products were used for PCR in a final vol of 100 μl containing 4 mM $MgCl_2$, 0.5 mM dNTPs, 0.5 μM primers, and 10 units of *Taq* DNA polymerase Stoffel fragment (Perkin–Elmer/Cetus). The reaction cycle consisted of 1 min each at 94°C, 50°C, and 72°C. Mast cell, neuro-2a, and submaxillary gland cDNAs were amplified for 23 cycles, while 35 cycles were used for RBL-2H3 cDNA. Samples were then subjected to electrophoresis and a 4% agarose gel and transferred to a nylon filter. The internal probe was 3′-end-labeled with [α-^{32}P]dCTP (3000 Ci/mmol; 1 Ci = 37 GBq) (New England Nuclear) using terminal deoxynucleotidyl transferase (United States Biochemical). Hybridization was carried out in standard solutions

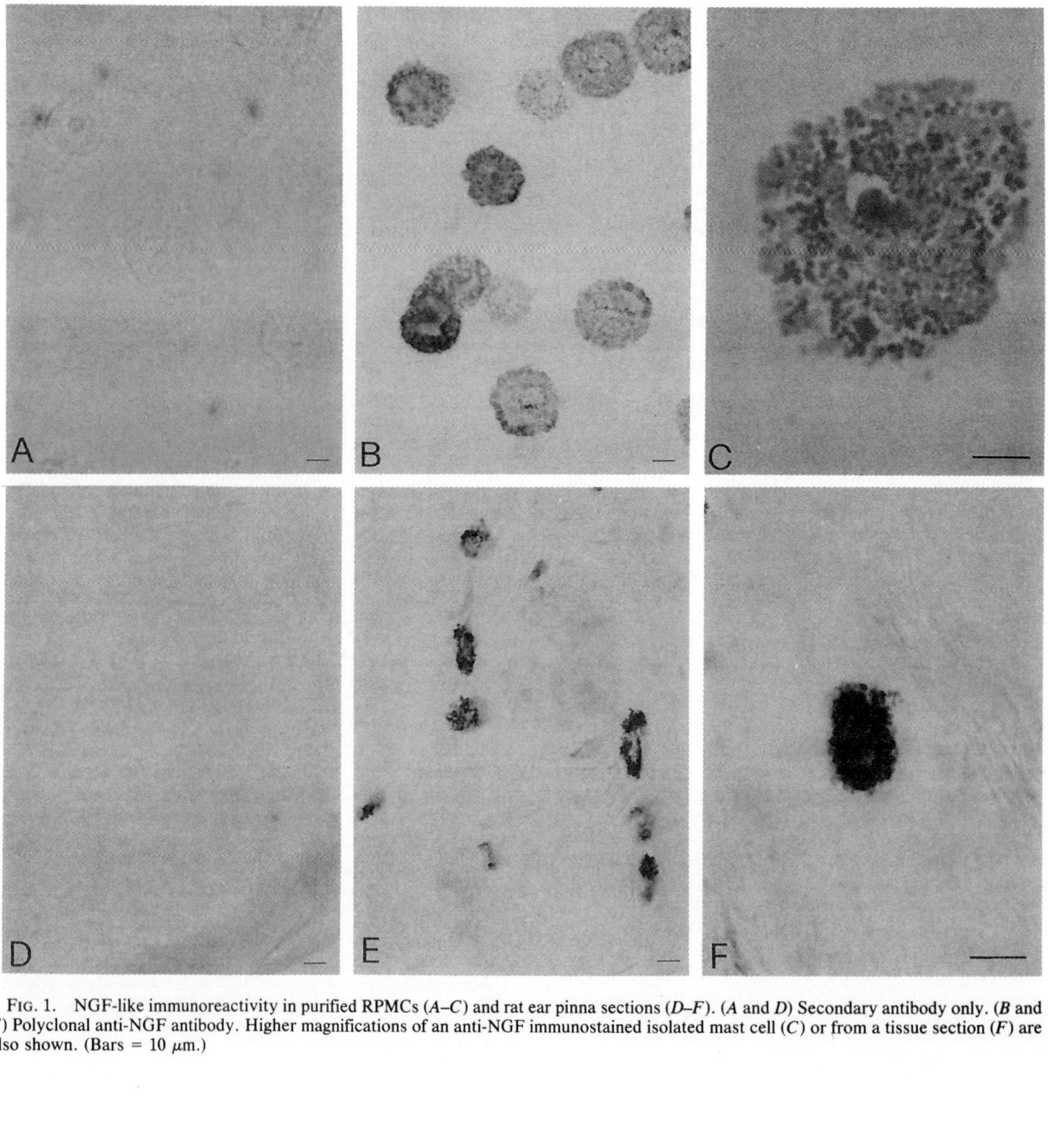

FIG. 1. NGF-like immunoreactivity in purified RPMCs (*A–C*) and rat ear pinna sections (*D–F*). (*A* and *D*) Secondary antibody only. (*B* and *E*) Polyclonal anti-NGF antibody. Higher magnifications of an anti-NGF immunostained isolated mast cell (*C*) or from a tissue section (*F*) are also shown. (Bars = 10 μm.)

(28) with 10% formamide (25°C overnight) and filters were washed in 1× standard saline citrate at room temperature before autoradiography (Hyperfilm-MP; Amersham).

RESULTS

NGF-Like Immunoreactivity in RPMCs. All mast cells, identified by safranin or toluidine blue staining, were clearly labeled by affinity-purified anti-NGF antibodies (Fig. 1*B*); highest levels of immunoreactivity appeared to be localized within granules (Fig. 1*C*). Omission of primary antibody or absorption with excess NGF yielded no or barely visible staining (Fig. 1*A*). Another primary rabbit anti-NGF polyclonal antibody (Sigma) gives positive labeling. No staining was seen in the non-mast cell resident cell populations present in the peritoneal cavity washes before mast cell purification.

NGF immunoreactivity was also observed in the mast cell line RBL-2H3 but with a lower intensity of staining (data not shown). Specific NGF-like immunoperoxidase staining was detected in mast cells of histological sections from rat ear pinna (Fig. 1 *D–F*).

RPMCs Contain NGF mRNA. The presence of NGF mRNA in RPMCs was examined by PCR, as no NGF hybridizing band was observed by Northern blot analysis. PCR amplification of mast cell reverse-transcribed total RNA (1 μg) with NGF-specific primers yielded a 292-bp band, which hybridized with a NGF-specific internal probe (Fig. 2), indicating the expression of the NGF gene in RPMCs. Mouse submaxillary gland total RNA served as a positive control (Fig. 2). No hybridizing band was found in PCR-processed mast cell RNA without the reverse transcription step. Although only a very minor proportion of non-mast cells was present in the purified RPMC preparation, PCR amplification could have resulted in a contribution to the NGF signal by the non-RPMCs. When total RNA (1 μg) from the rat mast cell line RBL-2H3 was reverse transcriptase PCR amplified, a NGF hybridizing band was clearly detected (Fig. 2), albeit at higher amplification values. No NGF hybridizing signal was evident in the PCR amplification products of mouse neuroblastoma neuro-2a cells (Fig. 2).

Biological Activity of Spontaneously Released NGF from Mast Cells. Chicken embryo DRG and sympathetic neurons in *in vitro* culture were used to measure NGF in medium conditioned for 24 h with RPMCs (Fig. 3). Mast cell viability was assessed by quantifying the histamine content of randomly selected conditioned medium using an enzyme immunoassay kit (Immunotech, Marseilles, France). Mast cells released 1.65 ± 0.2 μg of histamine per 10^6 cells (mean ± SD; $n = 4$) over the same 24-h period, being equivalent to 9.6% ± 1.2% of total histamine content and only slightly above routinely determined levels of basal histamine release (22).

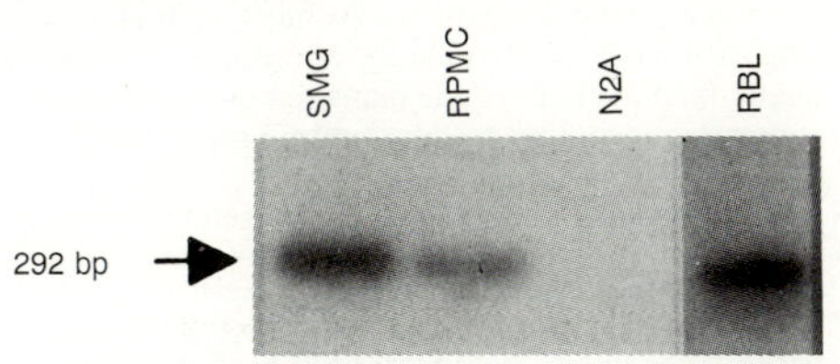

FIG. 2. NGF oligonucleotide probe hybridization of PCR amplified products from male mouse submaxillary glands (SMG), RPMCs, mouse neuro-2a cells (N_2A), and rat basophil leukemia cells (RBL). Autoradiograms of hybridization bands of SMG (positive control), N_2A (negative control), and RPMCs were obtained after 23 cycles of amplification. The RBL cDNA was amplified 35 cycles. No NGF hybridization band was observed in N_2A cDNA amplification products even after 35 cycles.

Sympathetic (E10) and DRG (E8) neurons were cultured 24 and 48 h, respectively, with conditioned medium (3–50 μl in 100-μl final vol). Maximal neurite-promoting activity for DRG neurons was obtained with 12 μl (Fig. 3*B*), the quantitation of which was biologically equivalent to ≈2 ng of NGF per ml per 10^6 mast cells over 24 h. Preincubation of RPMC conditioned medium with a goat polyclonal anti-NGF antibody (Fig. 3*B*) largely reduced this biological activity, excluding a significant contribution by factors other than those of the NGF neurotrophin family. RPMC conditioned medium was also active for sympathetic neurons (Fig. 3*A*), which respond neither to BDNF nor to neurotrophin 4 (29). Preincubation of conditioned medium with a monoclonal anti-NGF antibody (clone 27/21), which does not cross-react with other neurotrophins (30), significantly diminished biological activity for sympathetic neurons (Fig. 3*A*). Higher amounts of RPMC conditioned medium (≥25 μl) showed some cytotoxicity.

In agreement with the bioassay data, significant levels of β-NGF protein were detected in RPMC conditioned medium (2.56 ± 0.65 ng of β-NGF per 10^6 cells per ml per 24 h; mean ± SEM; $n = 10$), as measured by a specific two-site ELISA.

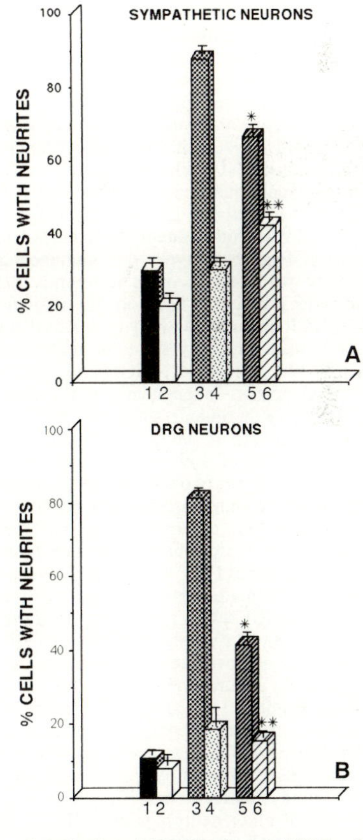

FIG. 3. Quantitation of RPMC-released NGF trophic activity. Histograms represent mean ± SEM of the percentage of neurite-bearing cells (*A*, $n = 6$; *B*, $n = 10$). Bars: 1, control culture medium; 2, anti-NGF antibody; 3, NGF (1 ng/ml); 4, NGF (1 ng/ml) and anti-NGF antibody; 5, RPMC conditioned medium; 6, RPMC conditioned medium and anti-NGF antibody. (*A*) Monoclonal anti-NGF antibody 27/21. (*B*) Polyclonal anti-NGF antibodies. Statistical analysis was conducted by Student's–Newman–Keuls *t* test for multiple comparisons. *, $P < 0.001$, control vs. mast cell conditioned medium; **, $P < 0.001$, mast cell conditioned medium alone vs. addition of anti-NGF antibody.

DISCUSSION

The present study shows that RPMCs express the gene transcript for NGF and display a NGF-like immunoreactivity largely associated with granule-containing compartments. The RBL-2H3 cell line behaved similarly, indicating that cells of the mastocytic–basophilic lineage are NGF-producing cells. Furthermore, a two-site ELISA detected abundant β-NGF protein levels in RPMC conditioned medium. NGF-like neurotrophic activity was also observed when the same conditioned medium was added to cultured avian embryonic sensory and sympathetic neurons, the latter of which respond to neither BDNF nor neurotrophin 4 (29). A monoclonal antibody recognizing NGF but not BDNF or neurotrophin 3 (30) effectively blocked mast cell-derived trophic bioactivity. Together with the occurrence of NGF-like immunopositive mast cells in histological sections of adult rat ear pinna, these data show that rodent connective tissue-type mast cells synthesize, store, and release biologically active NGF. The possibility that other mast cell types or basophils have similar properties, or that mast cells produce neurotrophic factors other than NGF, remains to be investigated.

Numerous cell types synthesize NGF including other hematopoietic cells (31). NGF expression and/or production has been observed *in vivo* either during development (32, 33) or during tissue injury (34) and *in vitro* following stimulation (e.g., with inflammatory agents) (31, 34, 35). Despite evidence of NGF in peripheral adult tissues such as skin, the cellular element(s) contributing to NGF production in normal or pathological conditions remains to be clearly identified (36, 37). Our results indicate NGF-like immunoreactivity in mast cells of mesenchymal adult rodent tissues or RPMCs. Although some cross-reactivity of the polyclonal antibody toward other neurotrophins cannot be excluded (38), the expression of NGF mRNA favors this immunostaining to reflect reactivity with one or more differently processed forms of the NGF protein. In addition, the vesicular localization of anti-NGF immunoreactivity indicates that connective tissue-type mast cells may store, in analogy to tumor necrosis factor (39), preformed β-NGF and/or specialized precursor storage forms. Independent of the intracellular NGF form(s), these experiments show that RPMCs release relatively large amounts of authentic biologically active β-NGF in the absence of any overt cellular stimulation. Preliminary studies with two-site ELISA indicate that β-NGF also occurs within unstimulated RPMCs (unpublished data). Given their defensive role, mast cells may represent a continuous source of NGF for tissue homeostatic functions and/or a prompt NGF supply for their reequilibration after tissue inflammatory insults; other cell types may contribute to tissue NGF levels at later stages of the inflammation process (7, 8). Mast cells could, by releasing cytokines such as tumor necrosis factor, also indirectly upregulate NGF production in surrounding cells (35).

Inflammatory insults to a variety of tissues, including peripheral nerves, are often accompanied by mast cell activation, hyperalgesia (40), and extracellular NGF accumulation (7, 8). Mast cells are frequently found closely apposed to small caliber or unmyelinated fibers (20), which may contain substance P and/or other neuropeptides (21). Small peptidergic fibers are generally assumed to be nociceptive sensory fibers, whose dependence on NGF for survival and differentiation is developmentally restricted. However, in the adult these neurons continue to express high-affinity NGF receptors and to respond to NGF with enhanced neuropeptide expression (12), while anti-NGF antibodies modify their morphology (41). Induction of neuropeptide synthesis in sensory neurons is also found in many models of tissue inflammation. In some instances, neuropeptide induction occurs within hours of initiation of inflammation (17, 40) and is abolished by anti-NGF antibodies (42). The increased levels of neuropeptides that accompany tissue inflammation may, upon primary afferent stimulation, result in their increased release from the sensory terminals in the spinal cord (43) and likely underlie centrally mediated prolonged behavioral hyperalgesia (40). Interestingly, a single high systemic dose of NGF in adult rats caused long-lasting thermal and mechanical hyperalgesia (44). These data, together with ours, support the notion that activated mast cells may, by releasing NGF, directly control plastic modifications in the central processing of sensory nociceptive inputs after tissue injury and inflammation. At the same time, NGF-induced neuropeptide upregulation may cause increased neuropeptide release at peripheral nociceptor terminals, and proteolytic products of NGF in inflamed tissues may indirectly modulate their excitation (45). Given the potential NGF effects on collateral sprouting or growth of sensory and sympathetic fibers (11, 36), mast cells, via release of NGF, could control peripheral short- and long-term modifications in the sensitivity and/or amplitude of sensory terminal fields toward persistent or subsequent, even subthreshold, noxious tissue stimuli. The degree of mast cell activation may thus contribute significantly to the hyperalgesic and pathological pain states frequently accompanying tissue inflammation, especially in peripheral nerves.

Mast cell progenitors migrate from the bloodstream to those tissues where they undergo terminal phenotypic differentiation, proposing that these cells represent potential mobile quanta for replenishment of tissue NGF. NGF itself can increase tissue mast cell number *in vivo* (15, 46) and affect mast cell survival, differentiation, and mediator release *in vitro* (47–50). RPMCs express the functional NGF receptor tyrosine kinase (51), indicating possible autocrine and paracrine actions of NGF on mast cells. Given that NGF can affect not only neurons but also hematopoietic cells, mast cells may, through the release of NGF, convey information to the nervous system as well as modulate their own behavior and the reactivity of tissue-infiltrating cells.

The capacity of at least connective tissue-type mast cells to produce NGF points to their having a pathophysiological potential far beyond that currently recognized. The ability of mast cells to synthesize and release NGF may represent a well-integrated tissue defense mechanism for maintaining and/or restoring homeostatic functions after noxious perturbations (16). On the other hand, mast cell hyperplasia and nerve remodeling are found in some chronic inflammatory states (18, 52–54). Such tissue modifications may be triggered and/or sustained by mast cell-derived NGF and lead to long-lasting changes in tissue reactivity and behavioral responses toward persistent or recurrent inflammatory stimuli. Accordingly, alterations in mast cell properties—e.g., in the entity and duration of their activation—could play a critical role in the progression and/or secondary complications of inflammatory tissue responses. A more complete understanding of their local stimulatory and inhibitory regulation (55) might open avenues to the management of inflammatory disease states, including those of autoimmune origin.

The authors thank Dr. S. D. Skaper for useful discussions and comments, Dr. L. Facci for the histamine measurements, and Ms. P. Lentola for expert secretarial assistance. A special note of thanks to the late Prof. Angelo Burlina, whose continuous effort and initiative in biomedical research helped to make this work possible.

1. Galli, S. J., Dvorak, A. M. & Dvorak, H. F. (1984) *Prog. Allergy* **34,** 1–141.
2. Bienenstock, J., Tomioka, M., Stead, R. M., Quinonez, G., Simon, G. T., Coughlin, M. D. & Denburg, J. A. (1987) *Int. Arch. Allergy Appl. Immunol.* **82,** 238–243.
3. Johnson, D. & Krenger, W. (1992) *Neurochem. Res.* **17,** 939–951.

4. Olsson, Y. (1968) *Int. Rev. Cytol.* **24,** 27–70.
5. Ratzlaff, R. E., Cavanaugh, V. J., Miller, G. W. & Oakes, S. G. (1992) *J. Neuroimmunol.* **41,** 89–96.
6. Gordon, J. R., Burd, P. R. & Galli, S. J. (1990) *Immunol. Today* **11,** 458–464.
7. Aloe, L., Tuveri, M. A., Carcassi, U. & Levi-Montalcini, R. (1992) *Arthritis Rheum.* **35,** 351–355.
8. Weskamp, G. & Otten, U. (1987) *J. Neurochem.* **118,** 1779–1786.
9. Bracci-Laudiero, L., Aloe, L., Levi-Montalcini, R., Buttinelli, C., Schilter, D., Gillessen, S. & Otten, U. (1992) *Neurosci. Lett.* **147,** 9–12.
10. Bracci-Laudiero, L., Aloe, L., Levi-Montalcini, R., Galeazzi, M., Schilter, D., Scully, J. L. & Otten, U. (1993) *Neuroreport* **4,** 563–565.
11. Levi-Montalcini, R. (1987) *Science* **237,** 1154–1162.
12. Lindsay, R. M. & Hamar, A. J. (1989) *Nature (London)* **337,** 362–364.
13. Otten, U., Ehrhard, P. & Peck, R. (1989) *Proc. Natl. Acad. Sci. USA* **86,** 10059–10063.
14. Bischoff, S. C. & Dahinden, C. A. (1992) *Blood* **79,** 2662–2669.
15. Aloe, L. & Levi-Montalcini, R. (1977) *Brain Res.* **133,** 358–366.
16. Levi-Montalcini, R., Aloe, L. & Alleva, E. (1990) *Prog. Neurol. Endocrinol. Immunol.* **3,** 1–10.
17. Donaldson, L. F., Harmar, A. J., McQueen, D. S. & Seckl, J. R. (1992) *Mol. Brain Res.* **16,** 143–149.
18. Stead, R. H., Kosecka-Janiszewska, U., Oestreicher, A. B., Dixon, M. F. & Bienenstock, J. (1991) *J. Neurosci.* **11,** 3809–3821.
19. Blennerhassett, M. G., Tomioka, M. & Bienenstock, J. (1991) *Cell Tissue Res.* **265,** 121–128.
20. Newson, B., Dahlström, A., Enerbäck, L. & Ahlman, H. (1983) *Neuroscience* **10,** 565–570.
21. Skofitsh, G., Savitt, J. M. & Jacobowitz, D. M. (1985) *Histochemistry* **82,** 5–8.
22. Mousli, M., Bronner, C., Bueb, J. L., Tschirhart, E., Gies, J. P. & Landry, Y. (1989) *J. Pharmacol. Exp. Ther.* **250,** 329–335.
23. Mayrhofer, G. (1980) *Histochem. J.* **12,** 513–526.
24. Skaper, S. D., Facci, L., Milani, D., Leon, A. & Toffano, G. (1990) in *Methods and Neurosciences*, ed. Conn, P. M. (Academic, New York), Vol. 2, pp. 17–33.
25. Näher-Noé, M., Gnahn, H., Grundler, A., Klingelhöfer, J., Weindl, A. & Conrad, B. (1993) *Eur. J. Clin. Chem. Clin. Biochem.* **31,** 375–380.
26. Chomczynski, P. & Sacchi, N. (1987) *Anal. Biochem.* **162,** 156–159.
27. Whittemore, S. R., Friedman, P. L., Larhammer, D., Persson, H., Gonzales-Carvajal, M. & Holets, V. R. (1988) *J. Neurosci. Res.* **20,** 403–410.
28. Sambrook, J., Fritsch, E. F. & Maniatis, T. (1989) *Molecular Cloning: A Laboratory Manual* (Cold Spring Harbor Lab. Press, Plainview, NY), Vols. 1–3.
29. Korsching, S. (1993) *J. Neurosci.* **13,** 2739–2748.
30. Södeström, S., Hallböök, F., Ibáñez, C. F., Persson, H. & Ebendal, T. (1990) *J. Neurosci. Res.* **27,** 665–677.
31. Mallat, M., Houlgatte, R., Brachet, P. & Prochiantz, A. (1989) *Dev. Biol.* **133,** 309–311.
32. Davies, A. M., Bandtlow, C., Heumann, R., Korsching, S., Rohrer, H. & Thoenen, H. (1987) *Nature (London)* **326,** 353–358.
33. Whittemore, S. R., Lärkfors, L., Ebendal, T., Holets, V. R., Ericsson, A. & Persson, H. (1987) *J. Neurosci.* **7,** 244–251.
34. Friedman, W. S., Lärkfors, L., Ayer-Le Lievre, C., Ebendal, T., Olson, L. & Persson, H. (1990) *J. Neurosci. Res.* **27,** 374–382.
35. Hattori, A., Tanaka, E., Murase, K., Ishida, N., Chatani, Y., Tsujimoto, M., Hayashi, K. & Kohno, M. (1993) *J. Biol. Chem.* **268,** 2577–2582.
36. Diamond, J., Coughlin, M., Macintyre, L., Holmes, M. & Visneau, B. (1987) *Proc. Natl. Acad. Sci. USA* **84,** 6596–6600.
37. Scarisbrick, I. A., Jones, E. G. & Isackson, P. J. (1993) *J. Neurosci.* **13,** 875–893.
38. Murphy, R. A., Acheson, A., Hodges, R., Haskins, J., Richards, C., Reklow, E., Chlumecky, V., Barker, P. A., Alderson, R. F. & Lindsay, R. M. (1993) *J. Neurosci.* **13,** 2853–2862.
39. Young, J. D., Liu, C., Butler, G., Cohn, Z. A. & Galli, S. J. (1987) *Proc. Natl. Acad. Sci. USA* **84,** 9175–9179.
40. Dubner, R. & Ruda, A. (1992) *Trends Neurosci.* **15,** 96–103.
41. Gold, B. G., Mobley, W. C. & Matheson, S. F. (1991) *J. Neurosci.* **11,** 943–955.
42. Donnerer, J., Schuligoi, R. & Stein, C. (1992) *Neuroscience* **49,** 693–698.
43. Garry, M. C. & Margreaves, K. M. (1992) *Brain Res.* **582,** 139–142.
44. Lewin, G. R., Ritter, A. M. & Mendell, L. M. (1993) *J. Neurosci.* **13,** 2136–2148.
45. Taiwo, Y. O., Levine, J. D., Burch, R. M., Woo, J. E. & Mobley, W. C. (1991) *Proc. Natl. Acad. Sci. USA* **88,** 5144–5148.
46. Marshall, J. S., Stead, R. H., McSharry, C., Nielsen, L. & Bienenstock, J. (1990) *J. Immunol.* **144,** 1886–1892.
47. Aloe, L. & De Simone, R. (1989) *Int. J. Dev. Neurosci.* **7,** 565–573.
48. Bruni, A., Bigon, E., Boarato, E., Mietto, L., Leon, A. & Toffano, G. (1982) *FEBS Lett.* **138,** 140–193.
49. Tomioka, M., Stead, R. H., Nielsen, L., Coughlin, M. D. & Bienenstock, J. (1988) *J. Clin. Immunol.* **82,** 599–607.
50. Matsuda, H., Kannah, Y., Ushio, H., Kiso, Y., Kenemoto, T., Suzuki, H. & Kitamura, Y. (1991) *J. Exp. Med.* **174,** 7–14.
51. Horigome, K., Pryor, J. C., Bullock, E. D. & Johnson, E. M. (1993) *J. Biol. Chem.* **268,** 14881–14887.
52. Nennesmo, I. & Reinholt, F. (1986) *Neurosci. Lett.* **69,** 296–301.
53. Dvorak, A. M. & Silen, W. (1985) *Ann. Surg.* **201,** 53–63.
54. Naukkarinen, A., Harvima, I., Paukkonen, K., Aalto, M.-L. & Horsmanheimo, M. (1993) *Arch. Dermatol. Res.* **285,** 341–346.
55. Aloe, L., Leon, A. & Levi-Montalcini, R. (1993) *Agents Actions* **39,** C145–C147.

Autoimmunity, 1994, Vol. 19, pp. 141–150
Reprints available directly from the publisher
Photocopying permitted by license only

Review

NERVE GROWTH FACTOR AND AUTOIMMUNE DISEASES

LUIGI ALOE*, STEPHEN D. SKAPER, ALBERTA LEON AND RITA LEVI-MONTALCINI*

Institute of Neurobiology, Consiglio Nazionale delle Ricerche, Rome and Researchlife Castelfranco Veneto (TV), Italy*

(Received 15 June 1994; in final form 15 September 1994)

The initiation of a humoral immune response to a foreign antigen is a complex biologic process involving the interaction of many cell types and their secreted products. Autoimmune diseases, which are characterized by an abnormal activation of the immune system, probably result from the failure of normal self-tolerance mechanisms. The etiology of such illnesses, however, is far from being understood. While there have been extensive studies on the participation of the immune and endocrine systems in autoimmune diseases, few have dealt with nervous system-mediated immunoregulation in such situations. Evidence continues to grow suggesting that nerve growth factor (NGF), first identified for its activity in promoting the growth and differentiation of sensory and sympathetic neurons, may exert a modulatory role on neuroimmunoendocrine functions of vital importance in the regulation of homeostatic processes. Newly detected NGF-responsive cells belong to the hemopoietic-immune system and to populations in the brain involved in neuroendocrine functions. NGF levels are elevated in a number of autoimmune states, along with increased accumulation of mast cells. NGF and mast cells both appear to be involved in neuroimmune interactions and tissue inflammation. Moreover, mast cells themselves synthesize, store, and release NGF, proposing that alterations in normal mast cell behaviors may provoke maladaptive neuroimmune tissue responses whose consequences could have profound implications in inflammatory disease states, including those of an autoimmune nature. This review focuses on these cellular events and presents a working model which attempts to explain the close interrelationships of the neuroendocrinoimmune triade via a modulatory action of NGF.

KEY WORDS: Nerve growth factors, mast cell, inflammation, autoimmune disease, homeostasis.

INTRODUCTION

Although the pathology of most autoimmune diseases has been well defined, the mechanisms responsible for the breakdown of self-tolerance and which lead to the development of systemic and organ-specific autoaggression are still unclear. Soluble mediators and receptors shared by both the neuro-endocrine and immune systems have been identified in increasing numbers. Most of the classical endocrine factors are considered to have immunomodulatory properties, rather than to be essential for immunity. For instance, estrogens interfere with the balance between cell-mediated and humoral immunity, which may perhaps explain the higher prevalence among women of a variety of immune-mediated diseases, including multiple sclerosis (MS)[1], a chronic inflammatory demyelinating disease of the central nervous system (CNS) caused by immunological mechanisms involving T lymphocytes and macrophages. The pituitary-adrenal axis appears to play a role in autoimmune diseases, in particular in animal models of arthritis, Hashimoto's disease and MS[2]. In experimental allergic encephalomyelitis (EAE), an animal model widely used for demyelinating diseases, particularly MS, treatment with glucocorticoid receptor antagonists enhances neurological manifestations of the disease[3]. In addition, the susceptibility of Lewis rats to inflammatory diseases including EAE is due to a genetically determined dysfunction of a regulatory feedback loop between the immune system and the brain[4]. Lewis rats show an inherited defect in the pituitary-adrenal response to inflammatory stimuli and immune mediators such as interleukin-1 (IL-1), apparently due to a defect in the regulatory elements of the gene for corticotrophin-releasing factor in the hypothalamus[5].

Cytokines are important regulatory factors in immune processes. Among these potent inflammatory agents are IL-1β, interleukin-6 (IL-6), interferon-g (IFN-γ), and tumor necrosis factor-a (TNF-α)[6]. The cytokine network represents an important bidirectional communication link between the nervous and

Address correspondence to: Dr Luigi Aloe, Institute of Neurobiology, Consiglio Nazionale delle Ricerche, Via Carlo Marx, 15, 00156 Rome, Italy.

immune systems. An imbalance in the normally carefully regulated cytokine system may affect the expression of autoimmune pathology. There are several possible ways in which cytokines might play a part in autoimmune disease processes. IL-1β is synthesized by macrophages, microglia and astrocytes. Its pleiotropic effects include chemotaxis, induction of increased adherence, enhanced vascular permeability, and release of prostaglandins and thromboxanes by neutrophils, macrophages, microglia and astrocytes[7]. This cytokine also increases expression of adhesion molecules on immunocompetent cells[8]. IL-1β immunoreactivity has been demonstrated in active MS plaques and elevated IL-1β levels in brains of EAE rats[9]. Interferon-γ, predominantly produced by CD4^{+} T lymphocytes of the T_H1 inflammatory phenotype, also exerts a multitude of inflammatory effects. It upregulates MHC class II and adhesion molecule expression on macrophages, astrocytes[10,11], and microglia. TNF-α, synthesized predominantly by macrophages, but also by T cells, mast cells, astrocytes and microglial cells[12–15], has been linked to the inflammatory demyelinating process of MS, EAE and experimental allergic neuritis[15].

Nerve growth factor (NGF) is a well-characterized neurotrophic protein which plays an essential role in the development and differentiation of sympathetic and neural crest-derived sensory nerve cells and the basal forebrain cholinergic neurons of the CNS[16,17]. In addition to its neurotrophic actions NGF also affects cells of the endocrine and immune systems[18]. Cells involved in the humoral immune response are sensitive to NGF (this will be discussed in more detail later). Accumulation of NGF at sites of lesion[19] and inflammation[20] suggests that inflammatory stimuli contribute to increasing NGF production. Interestingly, cytokines involved in inflammation and immune responses such as IL-1β and TNF-α are potent inducers of NGF synthesis in peripheral tissues[21,22] and in the CNS[23,24]. Thus, NGF may act as an immunomodulator mediating cross-talk between cells of the nervous and immune systems.

Autoimmune diseases, which are characterized by an abnormal activation of the immune system, could provide useful models for studying the immunomodulatory role of NGF and its interaction with cytokines. A number of autoimmune diseases with a potential involvement of cytokines in the inflammatory process appear, in fact, to be characterized by alterations in the levels of NGF. For example, NGF has been found in the synovium of patients with rheumatoid and other types of chronic arthritis (but not in the control fluids)[25], as well as in the synovium of pharmacologically-induced arthritis in animal models[26]. In the latter instance, destruction of peripheral sympathetic innervation reduced both the inflammation and the level of NGF following carrageenan injection. More recently, arthritis transgenic mice which carry and express the human TNF gene have been reported to show elevated levels of NGF[27]. Subcutaneous injections of NGF antibodies in these transgenic animals attenuated the wasting effect of TNF-α[27]. A significant increase in NGF levels has been found in the sera of patients with systemic lupus erythematosus (SLE)[28], which appeared to be related to the inflammatory episodes of the disease. SLE is a multi-system autoimmune disease characterized by B-cell hyperactivity, aberrant immune regulation and an excessive production of immunoglobulin and autoantibodies. Continuous overproduction of specific cytokines seems to contribute to B-cell hyperactivity in SLE[29]. Furthermore, during acute attacks MS patients exhibited a sigificant increase of NGF content in cerebrospinal fluid compared to age-matched normal subjects[30]. Interestingly, NGF levels closely followed the phases of the disease, markedly decreasing during remission[30]. Such studies showing that the baseline level of NGF increases in autoimmune state suggest that this factor may play a role in these diseases. This review will focus on the cellular regulatory elements in which NGF may participate, involving interactions between the nervous, immune and endocrine systems. The investigations summarized here will propose that loss of homeostatic control in the neuro-immuno-endocrine axis and aberrant NGF expression, perhaps mediated at the level of the mast cell, may be a critical component in the development and/or progression of autoimmune disorders.

NGF AND THE NERVOUS SYSTEM

Nerve growth factor is thought of as a target-derived factor responsible for the survivial and phenotypic maintenance of specific sets of peripheral and central neurons during development and maturation[16,17]. Many sympathetic neurons continue to depend on NGF for survival into adulthood, whereas sensory neurons cease to rely on this molecule in the postnatal period[17]. This change in NGF dependence for survival occurs despite the fact that many small-diameter peptide-containing neurons in the adult have high-affinity NGF receptors and retrogradely transport NGF from their target tissues.[31,32].

Several lines of evidence indicate that the action of NGF on sensory neurons might be directed to specific functional groups of nociceptive neurons. For example, small sensory neurons containing neuropeptides such as substance P do not survive NGF deprivation during development. High-affinity NGF receptors are present primarily on this same population of neuropeptide-containing neurons[32]. It is widely assumed that peptide-containing neurons are nociceptive sensory neurons with unmyelinated or thinly myelinated axons, designated as C fibers. Substance P-synthesizing and -releasing neurons belong to this category. Their axons terminate in the superficial layers of the dorsal horn of the spinal cord. Through an internuncial neuron, they convey nociceptive signals to the contralateral spinothalamic tract. However, about 90% of the substance P synthesized in the cell bodies of these neurons is antidromically transported to their peripheral terminals[33]. Release of substance P from the nerve endings can be elicited by either electrical or physiological stimuli[34]. NGF treatment increases substance P and calcitonin gene-related peptide gene expression in adult sensory neurons[35]. Induction of neuropeptide synthesis in sensory neurons is also found in many models of tissue inflammation. In some instances, neuropeptide induction occurs within hours of inflammation[36,37] and is abolished by anti-NGF antibodies[38]. Substance P discharge results in local plasma leakage, mast cell degranulation[39] and vasodilation, resulting in hyperemia, infiltration of leukocytes and other effects typical of neurogenic inflammation[40,41]. Interestingly, the hyperalgesia that follows an NGF injection[42] is qualitatively very similar to that which often accompanies inflammatory tissue injury[43].

NGF AND THE IMMUNE SYSTEM

In addition to its neurotrophic actions, there is ample evidence that the NGF/NGF receptor system is able to exert a wide variety of effects on cells of the immune system. In the first such study showing that cells belonging to the immune system lineage are receptive to the action of NGF, *in vivo* administration of NGF to neonatal rats caused a widespread increase in the size and number of mast cells in several peripheral tissues[44]. This finding was subsequently confirmed by others[45]. NGF-primed spleen cells transplanted into the brain ventricles of donor rats differentiate into mast cells[46]. Furthmore, peritoneal mast cells exposed to NGF antibodies show cytological alterations and decreased numbers. NGF also induces degranulation and histamine release from isolated peritoneal MCs[48,49].

Nerve growth factor is involved in the modulation of several activities of lymphocytes and accessory cells. Treatment of young rats with NGF prior to and after immunization with sheep erythrocytes results in an enhancement of T-lymphocyte-dependent antibody synthesis[50]. NGF has been shown to enhance lymphocyte proliferation in both B- and T-cell populations[51–53] and to stimulate production of IgM, IgA[52] and IgG4[54]. NGF is capable of inducing high-affinity interleukin-2 receptors on human peripheral blood mononuclear cells[55] and of promoting human hemopoietic colony growth and differentiation[56]. Additionally, NGF is chemotactic for neutrophilic leukocytes both *in vitro*[57] and *in vivo*[58].

The multiple effects of NGF on target cells are directly dependent on initial binding of NGF to specific cell surface receptors. Two classes of NGF receptors can be differentiated by their relative affinities: a major class of low-affinity NGF receptors (Kd ª10^{-9} M), and relatively small numbers of high-affinity receptors (Kd ª10^{-11}M)[59]. High-affinity NGF receptors are known to transduce functional responses to NGF[60]. Recent studies have established that the *trk* protoncogene product, a receptor tyrosine kinase, is the signal-transducing receptor unit for NGF[61]. Following the early demonstration of low-affinity NGF binding and receptor immunoreactivity by human peripheral blood mononuclear cells[62], several reports have now described the presence of functional Trk protein (but not low-affinity NGF receptors) by human monocytes[63], activated mouse $CD4^+$ T-cell clones[64], and rat peritoneal mast cells[65]. The finding that activated $CD4^+$ T-cell clones and resting rat peritoneal mast cells not only express Trk but also synthesize and release biologically active NGF[64,66] implicates NGF as an autocrine and/or paracrine factor in the development and regulation of immune responses.

From an anatomical perspective, the long adrenergic (and highly NGF-receptive) neurons of the para- and prevertebral sympathetic ganglia innervate organs and structures such as the thymus, spleen, lymph nodes, gut-associated lymphoid tissue and bone marrow. The presence of sympathetic fibers in these lymphoid tissues, in a highly dense array, suggests a structural link between the nervous and immune systems that could be utilized for translating neural messages into chemical signals that interact with specific cellular elements of the immune system[67]. An immunomodulatory role of NGF is suggested by the higher density and fluorescence of varicose adrenergic fibers and terminals in the immunocompetent organs, and by the large increase in catecholamines in these organs[68]. In fact, both the developing spleen and thymus express high levels of NGF receptor mRNA[69]. Sympathetic nerve fibers also form dense plexuses in endocrine organs. In these organs, as in lymphoid organs and tissues, their sympathetic innervation after NGF administration is markedly increased[70].

NGF AND THE NEUROENDOCRINE SYSTEM

Apart from its well-known roles in the survival and development of peripheral sympathetic and sensory neurons and in magnocellular cholinergic neurons, NGF also appears to take part in the regulation of neuronal and cellular populations involved in the control of specific neuroendocrine functions[71] and in the acquisition of male and female reproductive capacity[72,73]. Moreover, it has been shown that both

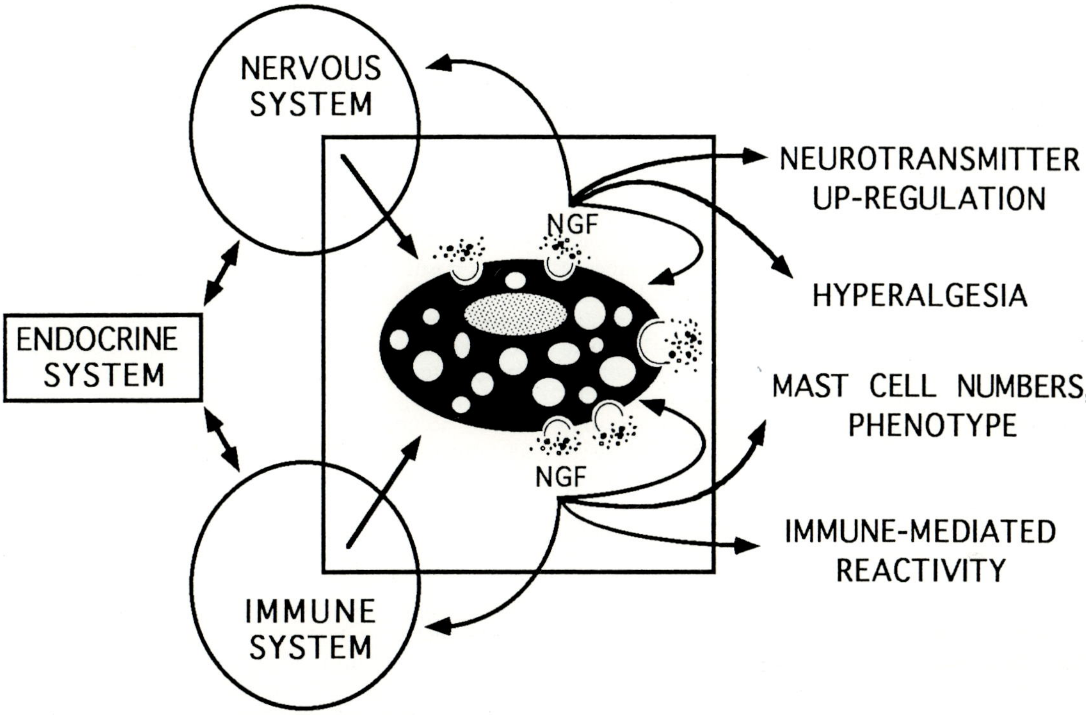

Figure 1 Schematic illustration of a working hypothesis in which NGF functions as a modulator of the neuroendocrinoimmune triade. NGF released from tissue mast cells as a consequence of inputs from the nervous or immune systems (and indirectly from the endocrine system) can, in turn, influence the nervous and immune systems. Mast cell-released NGF could also function in an autocrine manner. Possible actions of masst cell-derived NGF are indicated on the right-hand side.

NGF and its mRNA increase in the hypothalamus following stressful events[71,74]. There is evidence that the gene encoding NGF and its receptor are expressed in the developing female rat hypothalamus[75] and that NGF is also synthesized by the pituitary[76]. Thyroid and parathyroid glands of the rat express NGF mRNA and precursor protein[77]. NGF stimulates the pituitary-adrenocortical axis, enhancing the secretion of ACTH and the concentration of plasma glucocorticoids[78]. Plasma NGF levels in women during labor and lactation are increased compared with the concentrations found at the term of gestation or in control women[79], a time when plasma levels of the neurohypophyseal hormone oxytocin are high. Furthermore, intravenous injection of oxytocin in female virgin rats significantly raises hypothalamic NGF content[79]. NGF suppresses the transforming phenotype of human prolactinomas, reverting them to more differentiated, less malignant, dopamine-sensitive lactotroph-like cells[80]. Prolactinoma cells express both low-affinity NGF receptors and Trk[80]. In addition GH-3 cells, a clonal strain related to the bipotential somatomammotroph, express NGF receptors and are committed to generating mammortroph-like cells by NGF[81]. These data thus suggest a physiological role for NGF during pituitary development.

Antibodies to NGF, injected into rat fetuses, results in a marked neuroendocrine deficit in postnatal life[82]. Such offspring have severely atrophied sympathetic ganglia and sensory ganglia greatly reduced in size. Among endocrine organs, the thyroid appeared considerably smaller. Deleterious effects of NGF deprivation have also been described in rabbits and guinea pigs[83]. Maternal exposure to NGF antibodies during fetal development causes, in newborn pups, loss of body weight, sensory deficits, and lethality[84]. One explanation for this death is that exposure to NGF antibodies during prenatal life inactivates the endogenous NGF co-localized in neuroendocrine structures, thus causing a widespread neuroendocrine immune deficiency syndrome. The above-cited studies pointing to a modulatory role for NGF in the hypothalamo-pituitary-adrenal axis would support this idea. Intriguingly, natural autoantibodies to NGF have been detected in the sera of patients with SLE, autoimmune thyroditis and rheumatoid arthritis[85], and with herpes simplex virus infection[86].

In adult male mice, NGF is produced and stored in

large amounts in the submaxillary salivary glands[16], although the biological significance of this is not fully understood. Initially, a massive release of NGF into the bloodstream was reported to occur after exposure to psychosocially stressful conditions, after repeated aggressive interactions with conspecifics[87]. Intermale aggressive behavior also induces a large increase in NGF mRNA and protein in hypothalamus[74]. NGF release produced by psychosocial stress is not mimicked by physical stresses. Serum levels of NGF were found to correlate positively with the total number of fighting episodes in a mouse pair[87]. Prolonged fighting episodes resulted in hypertrophy of both cortical and medullary components of the adrenal[87]. The ability of exogenous NGF to elicit hypertrophy of the adrenal gland and enhanced functional activity of the medullar and cortical sections of this gland suggests this organ to be the most likely target of NGF[88]. NGF levels in serum consistently reach higher peaks in the subordinate rather than in the dominant partner[88], indicating the massive NGF discharge from the salivary glands into the bloodstream results not only from activation of aggressive behavior but also from anxiety, which would be more acute in the subordinate mouse. Thus, NGF release may contribute to the mediation of adrenal function in a situation-specific fashion.

NGF AND MAST CELLS IN INFLAMMATION

Mast cells are necessary for the development of immediate, and possibly delayed, hypersensitivity reactions[89] and occur in many peripheral tissues, in perivascular regions in close opposition to innervating sensory or autonomic nerve fibers[90], and also within the peripheral and central nervous systems[91]. Nervous and immunological mediators such as neuropeptides (e.g. substance P) or IgE can affect the state of mast cell activation[92]. Activated mast cells also secrete a wide array of cytokines and others inflammatory mediators[93]. This indicates that neuro-immune interactions can occur via mast cells and that mast cells may be pivotal in the regulation of these interactions[18].

Abnormal mast cells numbers or phenotypes, as well as uncontrolled mast cell activity, may underlie alterations in nerve-immune system interactions. Mast cell phenotypic heterogeneity has been observed in normal conditions, and increased mast cell numbers in a variety of immuno-inflammatory conditions. For example, mast cell numbers increase in the dermas of patients with early systemic sclerosis[94] and in the synovium of rats with rheumatoid synovitis[95]. In the latter two instances, NGF was elevated in the dermas[94] and synovium[95], respectively. Given that NGF causes mast cell activation and proliferation[44–49], a possible correlation between mast cells and NGF may exist in certain inflammatory diseases. In fact, intradermal application of NGF reportedly induces plasma extravasation in skin[96]. Moreover, NGF enhances histamine release and strongly modulates the formation of lipid mediators by basophils in response to several stimuli[97].

Mast cells both produce and respond to a variety of cytokines[98]. As cytokines are known to affect NGF synthesis[18], differences in tissue cytokine and NGF expression, either alone or in combination, may provide a positive feedback loop for autocrine/paracrine regulation of mast cell properties. This idea has recently taken on a new dimension with the demonstration that mast cells themselves are capable of synthesizing, storing, and releasing biologically active NGF[66].

NGF accumulation in acute inflammatory exudates has usually been attributed to cellular infiltrates or to the induction of tissue NGF expression. The rapid appearance and level of NGF accumulation in rat skin blister fluid[20], however, seems more compatible with NGF being released from resident tissue cells containing the stored protein. This notion receives support from the rapid induction of preprotachykinin mRNA expression (an NGF-sensitive process) in sensory ganglia of rats with adjuvant monoarthritis[36]. The remodelling of intestinal mucosal nerve fibers during intestinal inflammation correlates with changes in mast cell density[99]. Mast cells and sympathetic neurons *in vitro* form contacts[100], suggestive of a chemotactic NGF-like gradient effect between the mast cell and nerve ending. The above observations are now easier to rationalize in the context of the mast cell as a readily available source of NGF.

NGF AND MAST CELLS IN AUTOIMMUNE DISEASES

The etiology and cellular mechanisms underlying autoimmune diseases remain largely unknown. Certain autoimmune disorders such a MS[1], irritable bowel disease and interstitial cystitis[101] show a high prevalence in women, suggesting that a hormonal component could be implicated in the pathophysiology. Indeed, Theoharides *et al.*[102] have shown estradiol and myelin basic protein to act synergistically in inducing mast cell secretion. Myelin basic protein, a major suspected immunogen in MS by itself, activates mast cells and induces peripheral[103] and central[104] demyelination. Mast cells, probably originating from the neural crest[105], have also been identified intracranially, where they show a strict perivascular location[104] and where they secrete vasoactive amines in responses to mediators[106].

MS results in penetration of blood-borne immune cells within the brain parenchyma and subsequent destruction of myelin. Most studies investigating the cellular infiltrate in areas of demyelination have focused on involvement of T-lymphocytes and monocytes[107,108] and their role in EAE[109] or experimental allergic neuritis. In addition, mast cells have been localized in MS plaques[110,111]. In a preliminary report, brains from monkeys with EAE were claimed to show secreting mast cells in areas of demyelination[112]. Most cells also secrete during experimental allergic neutritis[113]. Evidence indicates that a delayed T-cell response may depend on early release of mast cell mediators[114] and that T-cell products can cause mast cell activation[115]. Mast cells themselves have the capacity to present antigen, in MHC class II-restricted manner, to T cells[116]. Exposure to myelin basic protein from demyelinated axons may in turn trigger mast cells, leading to a feedforward reaction. Conceivably brain mast cell-derived NGF, through its known actions on lymphocytes and monocytes, could contribute to such a cycle.

A number of studies suggest that MS may be caused by the immune response to some latent viral infection[117,118]. It is of interest that mast cells are involved in virus-induced inflammation in the CNS[119], and that mast cell secretion is triggered by certain viruses[120].

Evidence for a functional dichotomy among mouse $CD4^+$ T-cell clones suggests the existence of two subsets among fully differentiated helper T cells, namely T_H1 and T_H2 cells[121], which can be subdivided on the basis of their cytokine profile[122], denominated type-1 or type-2. Production of type-1 cytokines (IFN-g, interleukins-2 and -12) is characteristic of cell-mediated immune responses, while the so-called type-2 cytokines (interleukins-4, -5, -6, -10) are produced primarily in response to humoral immune activation. Activation of type-1 and type-2 responses is reciprocally exclusive. For example, secretion of IFN-γ during a type-1 response induces a functional suppression of the type-2 response and, conversely, production of interleukins-4 and -10 in the course of type-2 immune responses impedes activation of type-1 responses[123]. Human immunodeficiency virus infection would appear to be a likely example of a pathology with an altered T_H1-T_H2 balance[124]. The fact that mast cells display a predominantly type-2 cytokine pattern[93,125] could, conceivably, implicate their participation in immunodeficiency syndromes.

Autoimmune diseases of extraneural origin may also involve the participation of mast cells. As discussed earlier, at least in the case of rheumatoid arthritis mast cells accumulate in the synovial fluid[95,126,127]. In several types of chronic arthritis and in SLE, increased quantities of NGF occur[25–28] (Table 1). In the latter instance, the source(s) of the expressed NGF remains to be identified. Interestingly, intrasynovial injection of NGF is able to induce an increase in the local distribution of mast cells[95]. Considering that mast cells are frequently found at sites of tissue inflammation, and their potential capacity for elaborating substantial amounts of NGF it is not unreasonable to presume that mast cells will be found to contribute, at both acute and later times, to NGF expression in inflammatory conditions.

CONCLUSIONS AND FUTURE PERSPECTIVES

The nervous, immune and endocrine systems can no longer be considered as autonomous units, but as continually communicating elements integrated through anatomical and functional relationships. These elements thus comprise a unique homeostatic network regulating mutual communication between the three systems. Within the pathway of such interconnections, a highly important role is played by molecules produced and released by components of this network. Such substances include cytokines, hormones, neurotransmitters and neuropeptides, that function to mediate intersystemic communication by modulating the activation, proliferation or differentiation of cellular elements in the network.

Table 1 Autoimmune and others inflammatory conditions with increased NGF levels

Condition	Species	NGF Accumulation	Reference [no.]
Rheumatoid synovitis	rat	synovium	Aloe *et al.*, 1992 [95]
Carrageenan-induced arthritis	rat	synovium	Aloe *et al.*, 1992 [26]
Transgenic TNF overexpression (arthritis)	rat	synovium	Aloe *et al.*, 1994 [27]
Chronic arthritis	human	synovium	Aloe *et al.*, 1992 [25]
Systemic lupus erythematosus	man	serum	Bracci-Laudiero *et al.*, 1993 [28]
Multiple sclerosis	man	cerebrospinal fluid	Bracci-Laudiero *et al.*, 1992 [30]
Noxious thermal stimulus	rat	skin blister fluid	Weskamp and Otten, 1987 [20]
Carrageenan-induced pleurisy	rat	pleural exudate	Weskamp and Otten, 1987 [20]
Systemic sclerosis	human	dermis	Tuveri *et al.*, 1993 [94]

Nerve growth factor, originally discovered for its capacity to promote the growth and differentiation of peripheral ganglionic neurons, today emerges as a single modulator of the neuroendocrinoimmune triade. In the context of a network of interconnecting systems, circulating NGF represents a molecule without limitation to either organ or system — assuming a key role in the global homeostasis of the organism.

Within the above framework, the mast cell can be viewed as a gatekeeper between the immune system and the nervous system[128]. Mast cells are not only involved in hypersensitivity but also actively participate in inflammatory phenomena of various types, above all those of a neurogenic nature. Mast cells release and respond to numerous inflammatory mediators, as well as to NGF. Several autoimmune diseases are characterized by increased numbers of mast cells and NGF levels, although whether the elevation in NGF is a cause or a result of these immune processes is, at present, a matter for investigation. In at least one case, however, anti-NGF antibodies prevented the wasting induced by TNF-a in transgenic arthritic mice over expressing TNF-a. Recent evidence also indicates that endocrine influences, in particular glucocorticoids, are important in negatively modulating mast cells[129,130]. Mast cells can thus be considered as constituting an essential element of the neuroendocrinoimmune network.

By their very nature, mast cell activation and the relative effects of degranulation are highly localized phenomena, yet potentially dangerous for homeostatic equilibrium overall. The hyperinducibility of mast cells, a key element in rheumatoid arthritis and in many other pathologies of an inflammatory character (e.g. multiple sclerosis) has been hypothesized to reside in a deficit of antagonistic autacoid control, that is, by substances locally produced at the site of inflammation and having a regulatory effect on the state of mast cell activation[66,131,132]. Elucidating these autoregulatory mechanisms may hold possibilities for understanding the neuroendocrinoimmune circuits coinvolved in homeostatic equilibrium of the organism, their role and that of NGF in autoimmune diseases, and for designing rational pharmacological strategies to overcome such pathologies.

Acknowledgement

These studies were supported by Target Projects. Aging and BTBS, CNR, Italy, to R. Levi-Montalcini e Luigi Aloe.

References

1. Waksman B.H. Autoimmunity in demyelinating diseases. *Ann. NY. Acad. Sci.* 1988; **540**: 13–24.
2. Derijik R. and Berbenbosch F. The immuno-hypothalamo-pituitary-adrenal axis and autoimmunity. *Int. J. Neurosci.* 1991; **59**: 91–100.
3. Bolton C. and Flower R.J. The effects of the anti-glucocorticoid RU38486 on steroid-mediated suspression of experimental allergic encephalomyelitis (EAE) in the Lewis rat. *Life Sci.* 1989; **45**: 97–104.
4. Sternberg E.M., Hill J.H., Chrousos G.P., Kamilaris T., Listwak S.J., Gold P.W., *et al.* Inflammatory mediator-induced hypothalamic-pituitary-adrenal axis activation is defective in streptococcal cell wall arthritis-susceptible Lewis rats. *Proc. Natl Acad. Sci. USA* 1989; **86**: 2374–2378.
5. Sternberg E.M., Young III W.S., Bernardini R., Calogero A.E., Chrousos G.P., Gold P.W., *et al.* A central nervous system defect in biosynthesis of corticotropin-releasing hormone is associated with susceptibility to streptococcal cell wall-induced arthritis in Lewis rats. *Proc. Natl Acad. Sci. USA* 1989; **86**: 4771–4775.
6. Arai K., Lee F., Miyajima A., Miyatake S., Arai N. and Yokata T. Cytokines: coordinators of immune and inflammatory responses. *Ann. Rev. Biochem.* 1990; **59**: 783–836.
7. Hartung H.P., Schäfer B, Heininger K. and Toyka K.V. Recombinant interleukin-1b stimulates eicosanoid production in rat primary culture astroycytes. *Brain Res.* 1989; **489**: 113–119.
8. Frohman E.M., Frohman T.C., Dustin M.L., Vayuvegula B., Choi B., Gupta A., *et al.* The induction of intercellular adhesion molecules 1 (ICAM-1) expression on human fetal astrocytes by interferon-g, tumor necrosis factor a, lymphotoxin, and interleukin-1: relevance to intracerebral antigen presentation. *J. Neuroimmunol.* 1989; **23**: 117–124.
9. Fontana A., Bodmer S. and Frei K. Immunoregulatory factors secreted by astrocytes and glioblastoma cell. *Lymphokines* 1987; **14**: 91–121.
10. Fierz W., Endler B., Reske K., Wekerle H. and Fontana A. Astrocytes as antigen-presenting cells. I. Induction of Ia antigen expression on astrocytes by T cells via immune interferon and its effect on antigen presentation. *J. Immunol.* 1985; **134**: 3785–3793.
11. Sun D. Enhanced interferon-g-induced la-antigen expression by glial cells after previous exposure to this cytokine. *J. Neuroimmunol.* 1991; **34**: 205–214.
12. Lieberman A.P., Pitha P.M., Shi H.S. and Shin M.L. Production of tumor necrosis factor and other cytokines by astrocytes stimulated with lipopolisaccharide or a neurotropic virus. *Proc. Natl Acad. Sci. USA* 1989; **86** 6348–6352.
13. Sawada M., Kondo N., Suzumura A. and Marunouchi T. Production of tumor necrosis fator-alpha by microglia and astrocytes in culture. *Brain Res.* 1989; **491**: 394–397.
14. Gordon J.R. and Galli S.J. Mast cells as a source of both preformed and immunologically inducible TNF-a/cachetin. *Nature* 1990; **346**: 274–276.
15. Hartung H.P., Jung S., Stoll G., Zielasek J., Schmidt B., Archelos J.J., *et al.* Inflammatory mediators in demyelinating disorders of the CNS and PNS. *J. Neuroimmunol.* 1992; **40**: 197–210.
16. Levi-Montalcini R. The nerve growth factor 35 years later. *Science* 1987; **237**: 1154–1162.
17. Snider W.D. and Johnson E.M. Jr. Neurotrophic molecules. *Ann. Neutral.* 1989; **26**: 489–506.
18. Levi-Montalcini R., Aloe L. and Alleva E. A role for nerve growth factor in nervous, endocrine and immune systems. *Prog. NeuroEndocrinImmunology* 1990; **3**: 1–10.
19. Gasser U.E., Weskamp G., Otten U. and Dravid A.R. Time course of the elevation of nerve growth factor (NGF) content in the hippocampus and septum following lesions

of the septohippocampal pathyway in rats. *Brain Res.* 1986; **376**: 351–356.
20. Weskamp G. and Otten U. An enzyme-linked immunoassay for nerve growth factor (NGF): at tool for studying regulatory mechanisms involved in NGF production in brain and peripheral tissues. *J. Neurochem.* 1987; **48**: 1779–1786.
21. Lindholm D., Heumann R., Meyer M. and Thoenen H. Interleukin-1 regulates synthesis of nerve growth factor in non-neuronal cells of rat sciatic nerve. *Nature* 1987; **330**: 658–659.
22. Steiner P., Pfeilschifter J., Boeckh C., Radeke H. and Otten U. Interleukin-1b and tumor necrosis factor-a synergistically stimulate nerve growth factor synthesis in rat mesangial cells. *Am. J. Physiol.* 1991; **261**: F792–F798.
23. Gadient R.A., Cron K.C. and Otten U. Interleukin-1b and tumor necrosis factor-a synergistically stimulate nerve growth factor (NGF) release from cultured rat astrocytes. *Neurosci. Lett.* 1990; **117**: 335–340.
24. Friedman W.J., Lärkfors L., Ayer-LeLievre C., Ebendal T., Olson L. and Persson H. Regulation of b-nerve growth factor expression by inflammatory mediators in hippocampal cultures. *J. Neurosci. Res.* 1990; **27**: 374–382.
25. Aloe L., Tuveri M.A., Carcassi U. and Levi-Montalcini R. Nerve growth factor in the synovial fluid of patients with chronic arthritis. *Arth. Rheum.* 1992; **35**: 351–355.
26. Aloe L., Tuveri M.A. and Levi-Montalcini R. Studies on carrageenan-induced arthritis in adult rats; presence of nerve growth factor and role of sympathetic innervation. *Rheumatol. Int.* 1992; **12**: 213–216.
27. Aloe L., Probert L., Kollias G., Bracci-Laudiero L., Spillantini M.G. and Levi-Montalcini R. The synovium of transgenic arthritic mice expressing human tumor necrosis factor contains a high level of nerve growth factor. *Growth Factors*, in press.
28. Bracci-Laudiero L., Aloe L., Levi-Montalcini R., Galeazzi M., Schilter D., Scully J.L., *et al.* Increased levels of NGF in sera of systemic lupus erythematosus patients. *NeuroReport* 1993; **4**: 563–565.
29. Huang Y.P., Perrin L.H., Miescher P.A. and Zubler R.H. Correlation of T and B cell activities *in vitro* and serum IL-2 levels in systemic lupus erythematosus. *J. Immunol.* 1988; **141**: 827–833.
30. Bracci-Laudiero L., Aloe L., Levi-Montalcini R., Buttinelli C., Schilter D., Gillessen S., *et al.* Multiple sclerosis patients express increased levels of b-nerve growth factor in cerebrospinal fluid. *Neurosci. Lett.* 1992; **147**: 9–12.
31. Goedert M., Stoeckel K. and Otten U. Biological importance of the retrograde transport of nerve growth factor in sensory neurons. *Proc. Natl Acad. Sci. USA* 1981; **78**: 5895–5898.
32. Verge V.M.K., Richardson P.M., Benoit R. and Riopelle R.J. Histochemical characterization of sensory neurons with high-affinity receptors for nerve growth factor. *J. Neurocytol.* 1989; **18**: 583–591.
33. Brimjoin S., Lundberg J.M., Brodin E., Hokfelt T. and Nilsson G. Axonal transport of substance P in the vagus and sciatic nerves of the guinea pig brain. *Brain Res.* 1980: **191**: 443–457.
34. Jessell T.M. Central interactions at the central and peripheral terminals of primary sensory neurons. *J. Immunol.* 1985; **135** (Suppl. 2): 746s–749s.
35. Lindsay R.M. and Harmar A.J. Nerve growth factor regulates expression of neuropeptide genes in adult sensory neurons. *Nature* 1989; **337**: 362–364.
36. Donaldson L.F., Harmar A.J., McQueen D.S. and Seckl J.R. Increased expression of preprotachykinin, calcitonin gene-related peptide, but not vasoactive intestinal polypeptide messenger RNA in dorsal root ganglia during the development of adjuvant monoarthritis in the rat. *Mol. Brain Res.* 1992; **16**: 143–149.
37. Dubner R. and Ruda M.A. Activity-dependent neuronal plasticity following tissue injury and inflammation. *Trends Neurosci.* 1992; **15**: 96–103.
38. Donnerer J., Schuligoi R. and Stein C. Increased content and transport of substance P and calcitonin gene-related peptide in sensory nerves innervating inflamed tissue: evidence for a regulatory function of nerve growth factor *in vivo. Neuroscience* 1992; **49**: 693–698.
39. Light A.R. and Perl E.R. Spinal termination of functionally identified primary afferent fibres with slowly conducting myelinating fibres. *J. Comp. Neurol.* 1979; **186**: 133–143.
40. Pernow B. Role of tachykinins in neurogenic inflammation. *J. Immunol.* 1985; **135** (Suppl. 2): 812s–815s.
41. Bynke G. Capsaicin pretreatment prevents disruption of the blood-aqueous barrier in the rabbit eye. *Invest. Ophthalmol. Vis. Sci.* 1983; **24**: 744–749.
42. Lewin G.R., Ritter A.M. and Mendell L.M. Nerve growth factor-induced hyperalgesia in the neonatal and adult rat. *J. Neurosci.* 1993; **13**: 2136–2148.
43. Treede R.D., Meyer R.A., Raja S.N. and Campbell J.N. Peripheral and central mechanisms of cutaneous hyperalgesia. *Prog. Neurobiol.* 1992; **38**: 397–421.
44. Aloe L. and Levi-Montalcini R. Mast cells increase in tissues of neonatal rats injected with the nerve growth factor. *Brain Res.* 1977; **133**: 358–366.
45. Stead R.H., Tomioka M., Quinonez G., Simon G.T., Felten S.Y. and Bienenstock J. Intestinal mucosal mast cells in normal and nematode-infected intestines are in intimate contact with peptidergic nerves. *Proc. Natl Acad. Sci. USA* 1987; **84**: 2975–5979.
46. Aloe L. and De Simone R. NGF primed spleen cells injected in brain of developing rats differentiate into mast cells. *Int. J. Devl. Neurosci.* 1989; **7**: 565–573.
47. Aloe L. The effect of nerve growth factor and its antibody on mast cells *in vivo. J. Neuroimmunol.* 1988; **18**: 1–12.
48. Bruni A., Bigon E., Boarato E., Mietto L., Leon A. and Toffano G. Interaction between nerve growth factor and lysophosphatidylserine on rat peritoneal mast cells. *FEBS. Lett.* 1982; **138**: 190–192.
49. Pearce F.L. and Thompson H.L. Some characteristics of histamine secretion from rat peritoneal mast cells stimulated with nerve growth factor. *J. Physiol.* 1986; **372**: 379–393.
50. Manning P.T., Russell J.H., Simmons B. and Johnson E.M. Jr. Protection from guanethidine-induced neuronal destruction by nerve growth factor: effects of NGF on immune function. *Brain Res.* 1985; **340**: 61–69.
51. Thorpe L.W. and Perez-Polo J.R. The influence of nerve growth factor on the *in vitro* proliferative response of rat spleen lymphocytes. *J. Neurosci. Res.* 1987; **18**: 134–139.
52. Otten U., Ehrhard P. and Peck R. Nerve growth factor induces growth and differentiation of human B lymphocytes. *Proc. Natl Acad. Sci. USA* 1989; **86**: 100 59–100 63.
53. Brodie C. and Gelfand E.W. Functional nerve growth factor receptors on human B lymphocytes. Interaction with IL-2. *J. Immunol.* 1992; **148**: 3492–3497.
54. Kimata H., Yoshida A., Ishioka C., Kusunoki T., Hosoi S. and Mikawa H. Nerve growth factor specifically induces human IgG4 production. *Eur. J. Immunol.* 1991; **21**: 137–141.
55. Thorpe L.W., Werrbach-Perez K. and Perez-Polo J.R. Effects of nerve growth factor expression on interleukin-2 receptors on cultured human lymphocytes. *Ann. NY. Acad. Sci.* 1987; **496**: 310–311.
56. Matsuda H., Coughlin M.D., Bienestock J. and Denburg J.A. Nerve growth factor promotes human hemopoietic

colony growth and differentiation. *Proc. Natl Acad. Sci. USA* 1988; **85**: 6508–6512.
57. Gee A.P., Boyle M.D.P., Munger K.L., Lawman M.J.P. and Young M. Nerve growth factor: stimulation of polymorphonuclear leukocyte chemotaxis *in vitro*. *Proc. Natl Acad. Sci. USA* 1983; **80**: 7215–7218.
58. Boyle M.D.P., Lawman M.J.P., Gee A.P. and Young M. Nerve growth factor: a chemotactic factor for polymorphonuclear leukocytes *in vivo*. *J. Immunol.* 1985; **134**: 564–568.
59. Meakin S.O. and Shooter E.M. The nerve growth factor family of receptors. *Trends Neurosci.* 1992; **15**: 323–331.
60. Green S.H., Rydel R.E., Connolly J.L. and Greene L.A. PC 12 mutants that possess low but not high-affinity nerve growth factor receptors neither respond to nor internalize nerve growth factor. *J. Cell Biol.* 1986; **102**: 830–843.
61. Barbacid M. Nerve growth factor: a tale of two receptors. *Oncogene* 1993; **8**: 2033–2042.
62. Morgan B., Thorpe L.W., Marchetti D. and Perez-Polo J.R. Expression of nerve growth factor receptors by human peripheral blood mononuclear cells. *J. Neurosci. Res.* 1989; **23**: 41–45.
63. Enrhard P.B., Ganter U., Bauer J. and Otten U. Expression of functional *trk* protooncogene in human monocytes. *Proc. Natl Acad. Sci. USA* 1993; **90**: 5423–5427.
64. Ehrhard P.B., Erb P., Graumann U. and Otten U. Expression of nerve growth factor and nerve growth factor receptor tyrosine kinase Trk in activated CD4-positive T-cell clones. *Proc. Natl Acad. Sci. USA* 1990; **90**: 109 84–109 88.
65. Horigome K., Pryor J.C., Bullock E.D. and Johnson E.M. Jr. Mediator release from mast cells by nerve growth factor. Neurotrophin specificity and receptor mediation. *J. Biol. Chem.* 1993; **268**: 148 81–148 87.
66. Leon A., Buriani A., Dal Toso R., Fabris M., Romanello S., Aloe L., *et al.* Mast cells synthesize, store, and release nerve growth factor. *Proc. Natl Acad. Sci. USA* 1994; **91**: 3739–3743.
67. Felten D.L., Feltein S.Y., Carlson S.L., Olschowka J.A. and Livnat S. Noradrenergic and peptidergic innervation of lymphoid tissue. *J. Immunol.* 1985; **135**: 755s–765s.
68. Abramchik G.V., Yermakova S.S., Kaliunov V.N., Tanina R.M. and Tumilovich M.K. The immunomodulatory effect of nerve growth factor. *J. Neurosci. Res.* 1988; **19**: 349–356.
69. Ernfors P., Hallbook F., Ebendal T., Shooter E.M., Radeke M.J. and Misko T.P., *et al.* Developmental and regional expression of b-nerve growth factor receptor mRNA in the chick and rat. *Neuron.* 1988; **1**: 983–996.
70. Levi-Montalcini R. and Angeletti P.U. Immunosympathectomy. *Pharmacol. Rev.* 1966; **18**: 619–628.
71. Aloe L., Alleva E. and De Simone R. Changes of NGF level in mouse hypothalamus following intermale aggressive behavior: biological and immunohistochemical evidence. *Behav. Brain Res.* 1990; **39**: 53–61.
72. Persson H., Ayer-Le Lievre C., Söder O, Villar M.J., Madis Metsis M., and Olson L., *et al.* Expression of b-nerve growth factor receptor mRNA in Sertoli cells downregulated by testosterone. *Science* 1990; **247**: 704–707.
73. Lara H.E., McDonald J.K. and Ojeda S.R. Involvement of nerve growth factor in female sexual development. *Endocrinology* 1990; **126**: 364–375.
74. Spillantini M.G., Aloe L., Alleva E., De Simone R., Goedert M. and Levi-Montalcini R. Nerve growth factor mRNA and protein increase in hypothalamus in a mouse model of aggression. *Proc. Natl Acad. Sci. USA* 1989; **86**: 8555–8559.
75. Ojeda S.R., Hill D.F. and Katz K.H. The genes encoding nerve growth factor and its receptor are expressed in the developing female rat hypothalamus. *Mol. Brain Res.* 1991; **9**: 47–55.
76. Lahtinen T., Soinila S. and Lakshmanan J. Biological demonstration of nerve growth factor in the rat pituitary gland. Neuroscience 1989; **30**: 165–170.
77. Dicou E., Lee J. and Brachet P. Synthesis of nerve growth factor mRNA and precursor protein in the thyroid and parathyroid glands of the rat. *Proc. Natl Acad. Sci. USA* 1986; **83**: 7084–7088.
78. Otten U., Baumann J.B. and Girard J. Stimulation of the pituitary-adrenocortical axis by nerve growth factor. *Nature* 1979; **282**: 413–414.
79. Luppi P., Levi-Montalcini R., Bracci-Laudiero L., Bertolini A., Arletti R., Tavernari D., *et al.* NGF is released into plasma during human pregnancy: an oxytocin-mediated response? *NeuroReport* 1993; **4**: 1063–1065.
80. Missale C, Boroni F., Losa M., Giovanelli M., Zanellato A., Dal Toso R., *et al.* Nerve growth factor suppresses the transforming phenotype of human prolactinomas. *Proc. Natl Acad. Sci. USA* 1993; **90**: 7961–7965.
81. Missale C., Boroni F., Sigala S., Zanellato A., Dal Toso R., Balsari A., *et al.* Nerve growth factor directs differentiation of the bipotential cell line GH-3 into the mammotroph phenotype. *Endocrinology* 1994; **135**: in press.
82. Aloe L., Cozzari C., Calissano P. and Levi-Montalcini R. Somantic and behavioral postnatal effects of fetal injections of nerve growth factor antibodies in the rat. *Nature.* 1981; **291**: 413–415.
83. Johnson E.M. Jr., Osborne P.A., Rydel R.E., Schmidt R.E. and Pearson J. Characterization of the effects of autoimmune nerve growth factor deprivation in the developing guinea pig. *Neuroscience* 1983; **8**: 631–642.
84. Yip H.K., Rich K.M., Lampe P.A. and Johnson E.M. Jr. The effects of nerve growth factor and its antiserum on the postnatal development and survival after injury of sensory neurons in rat dorsal root ganglia. *J. Neurosci.* 1984; **4**: 2986–2992.
85. Dicous E., Hurez D., Nerrière. V. Natural autoantibodies against nerve growth factor in autoimmune diseases. *J. Neuroimmunol.* 1993; **47**: 159–168.
86. Dicou E., Nerrière V. and Labropoulou V. Naturally occurring antibodies against nerve growth factor in human and rabbit sera: comparison between control and herpes simplex virus-infected patients. *J. Neuroimmunol.* 1991; **34**: 153–158.
87. Aloe L., Alleva E., Böhm A. and Levi-Montalcini R. Aggressive behavior induces release of nerve growth factor from mouse salivary gland into the bloodstream. *Proc. Natl Acad Sci. USA* 1986; **83**: 6184–6187.
88. Maestripieri D., De Simone R., Aloe L. and Levi-Montalcini R. Social status and nerve growth factor serum levels after agonistic encounters in mice. *Physiol. Behav.* 1990; **47**: 161–164.
89. Galli S.J., Dvorak A.M. and Dvorak H.F. Basophils and mast cells: morphologic insights into their biology, secretory patterns and function. *Prog. Allergy* 1984; **34**: 1–141.
90. Bienenstock J., Tomioka M., Matsuda H., Stead R.H., Quinonez G., Simon G.T., *et al.* The role of mast cells in inflammatory processes: evidence for nerve/mast cell interactions. *Int. Arch. Allergy Apply Immunol.* 1987; **82**: 238–243.
91. Johnson D. and Krenger W. Interactions of mast cells with the nervous system-recent advances. *Neurochem. Res.* 1992; **9**: 939–951.
92. Olsson Y. Mast cells in the nervous system. *Int. Rev. Cytol.* 1968; **24**: 27–70.
93. Gordon J.R., Burd P.R. and Galli S.J. Mast cells as a source of multifunctional cytokines. *Immunol. Today* 1990; **11**: 458–464.

94. Tuveri M.A., Passiu G., Mathieu A. and Aloe L. Nerve growth factor and mast cell distribution in the skin of patients with systemic sclerosis. *Clin. Exp. Rheumatol.* 1993; **11**: 319–322.
95. Aloe L., Tuveri M.A. and Levi-Montalcini R. Nerve growth factor and distribution of mast cells in the synovium of adult rats. *Clin. Exp. Rheumatol.* 1992; **10**: 203–204.
96. Otten U., Baumann J.B. and Girard J. Nerve growth factor induces plasma extravasation in rat skin. *Eur. J. Pharmacol.* 1984; **106**: 199–201.
97. Bischoff S.C. and Dahinden C.A. Effect of nerve growth factor on the release of inflammatory mediators by mature human basophils. *Blood* 1992; **79**: 2662–2669.
98. Galli S.J., Gordon J.R. and Wershil B.K. Cytokine production by mast cells and basophils. *Curr. Opin. Immunol.* 1991; **3**: 865–872.
99. Stead R.H., Kosecka-Janiszewska U., Oestreicher A.B., Dixon M.F. and Bienenstock J. Remodelling of B-50 (GAP-43)-and NSE-immunoreactive mucosal nerves in the intestines of rats with *Nippostrongylus brasiliensis. J. Neurosci.* 1991; **11**: 3809–3821.
100. Blennerhasset M.G, Tomioka M. and Bienenstock J. Formation of contacts between mast cells and sympathetic neurons *in vitro. Cell Tissue Res.* 1991; **265**: 121–128.
101. Sant G.R. Interstitial cystitis: pathophysiology, clinical evaluation and treatment. *Urol. Ann.* 1989; **3**: 172–179.
102. Theoharides T.C., Dimitriadou V., Letourneau R., Rozniecki J.J., Vliagoftis H. and Boucher W. Synergistic action of estradiol and myelin basic protein on mast cell secretion and brain myelin changes resembling early stages of demyelination. *Neuroscience* 1993; **57**: 861–871.
103. Johnson D., Seeldrayers P.A. and Weiner H.L. The role of mast cells in demyelination. 1. Myelin proteins are degraded by mast cell proteases and myelin basic protein and P_2 can stimulate mast cell degranulation. *Brain Res.* 1988; **444**: 195–198.
104. Theoharides T.C., Baloyannis S.J. and Manolidis L.S. Activated rat peritoneal mast cells can cause syngeneic brain demyelination *in vitro. Int. J. Immunopath. Pharmacol.* 1991; **4**: 137–144.
105. Lambracht-Hall M., Dimitriadou V. and Theoharides T.C. Migration of mast cells in the developing rat brain. *Dev. Brain Res.* 190; **56**: 151–159.
106. Lambracht-Hall M., Konstantinidou A.D. and Theoharides T.C. Serotonin release from rat brain mast cells *in vitro. Neuroscience* 1990; **30**: 199–207.
107. Hauser S.L., Bhan A.K., Gilles F., Kemp M., Kerr C. and Weiner H.L. Immunohistochemical analysis of the cellular infiltrate in multiple sclerosis lesions. *Ann. Neurol.* 1986; **19**: 578–587.
108. Prineas J.W. and Wright R.G. Macrophages, lymphocytes, and plasma cells in the perivascular compartment in chronic multiple sclerosis. *Lab. Invest.* 1978; **38**: 409–421.
109. Sobel R.A., Blanchette B.W., Bhan A.K. and Colvin R.B. The immunopathology of experimental allergic encephalomyelitis: quantitative analysis of inflammatory cells *in situ. J. Immunol.* 1984; **132**: 2393–2401.
110. Olsson Y. Mast cells in plaques of multiple sclerosis. *Acta Neurol. Scand.* 1974; **50**: 611–618.
111. Toms R., Weiner H.L. and Johnson D. Identification of IgE-positive cells and mast cells in frozen sections of multiple sclerosis brains. *J. Neuoimmunol.* 1990; **30**: 169–177.
112. Theoharides T.C., Dimitriadou V., Massacesi L. and Hauser S.L. Ultrastructural evidence of mast cell involvement in monkey experimental allergic encephalomyelitis (EAE). In *Proc International Congress on Inflammation*, Barcelona, 1990; p 282.
113. Brosnan C.F., Lyman W.D., Tansey F.A. and Carter T.H. Quantitation of mast cells in experimental allergic neuritis. *J. Neuropath. Exp. Neurol.* 1985; **44**: 196–203.
114. Askenase P.W. and Van Loveren H. Delayed-type hypersensitivity activation of mast cells by antigen-specific T-cell factors initiates the cascade of cellular interactions. *Immunol. Today* 1983; **4**: 259–264.
115. Kaplan A.P., Reddigari S., Baeza M. and Kuna P. Histamine releasing factors and cytokine-dependent activation of basophils and mast cells. *Adv. Immunol.* 1991; **50**: 237–260.
116. Frandji P., Oskéritzian C., Cacaraci F., Lapeyre J., Peronet R., David B., *et al.* Antigen-dependent stimulation by bone marrow-derived mast cells of MHC class II-restricted T cell hybridoma. *J. Immunol.* 1993; **151**: 6318–6328.
117. Reddy E., Sanberg-Wollheim M., Mettus R.V., Ray P.E., De Freitas E. and Koprowski H. Amplification and molecular cloning of HTLV-1 sequences from DNA of multiple sclerosis patients. *Science* 1989; **243**: 529–533.
118. Vandvik B. and Norrby E. Paramyxovirus SV5 and multiple sclerosis. *Nature* 1989; **338**: 769–771.
119. Mokhtarian F. and Griffin D.E. The role of mast cells in virus-induced inflammation in the murine central nervous system. *Cell Immunol.* 1984; **86**: 491–500.
120. Sugiyama K. Histamine release from rat mast cells induced by Sendai virus. *Nature* 1977; **270**: 614–615.
121. Mosmann T.R. and Coffman R.I. Two types of mouse T helper cell clone: implication for immune regulation. *Immunol. Today* 1987; **8**: 223–226.
122. Mosmann T.R. and Coffman R.I. T_H1 and T_H2 cells: different patterns of lymphokine secretion lead to different functional properties. *Annu. Rev. Immunol.* 1989; **7**: 145–168.
123. Fiorentino D.F., Bond M.W. and Mosmann T.R. Two types of mouse T helper cells. IV. The clones secrete a factor that inhibits cytokines production by T_h1 clones. *J. Exp. Med.* 1989; **170**: 2081–2098.
124. Clerici M., Shearer G.M. A TH1 ÆTH2 switch is a critical step in the etiology of HIV infection. *Immunol. Today* 1993; **14**: 107–111.
125. Costa J.J., Burd P.R. and Metcalfe D.D. Mast cell cytokines. In: Kaliner M.A., Metcalfe D.D. The Mast Cell in Health and Disease, Vol. 62. New York: M. Dekker, 1993; 443–466.
126. Malone D.G. and Metcalfe D.D. Mast cells and arthritis. *Ann. Allergy* 1988; **61**: 27–30.
127. Mican J.M. and Metcalfe D.D. Arthritis and mast cell activation. *J. Allergy Appl. Immunol.* 1990; **86**: 677–683.
128. Orr R.L. Nervous system-associated mast cells: gatekeepers of neural and immune interactions. *Drug. Dev. Res.* 1988; **15**: 195–205.
129. Lavker R.M., Schechter N.M., Guzzo C. and Lazarus G.S. Aggressive topical corticosteroid therapy: a novel approach to mast cell-dependent cutaneous disorders. *Dermatologica* 1987; **175**; 213–216.
130. Bachelet C.M., Bernaudin J.F. and Fleury-Feith J. Histological study of mast cells in the actively sensitized guinea pig lung and after challenge: effect of a corticosteroid. *Int. Arch. Allergy Appl. Immunol.* 1990; **91**: 171–174.
131. Aloe L., Leon A. and Levi-Montalcini R. A proposed autacoid mechanism controlling mastocyte behaviour. *Agents Actions* 1993; **39**: C145-C147.
132. Mazzari S., Canella R., Marcolongo G. and Leon A, N-(α-hydroxyethyl) hexadecanamide given orally reduces hyperalgesia by down-modulating mast cell activation. *Soc. Neurosci Abstr.* 1994. (in press).

Neuroscience Year, Supplement 2, *Encyclopedia of Neuroscience*, 110–114 (1992).

Nerve Growth Factor

Rita Levi-Montalcini and Pietro Calissano

The term *nerve growth factor*, or NGF, was introduced to define the effect of a macromolecular agent endowed with the property of inducing, in an in vitro test, rapid and exuberant nerve fiber outgrowth from sensory and sympathetic embryonic neurons (Fig. 1). This fast, unique action offered a simple and reliable test to assess its presence and to uncover its putative biological sources.

Historically, the effect of the still-unknown NGF was first called to attention when a mouse tumor, implanted into the body of 3-day chick embryos, was innervated by nerve fibers outgrowing from the adjacent spinal root ganglia of the host. The original interpretation that this effect was due to the rapid growth and histochemical properties of neoplastic cells was later revised and replaced with the hypothesis that the implanted tumor released a humoral factor that selectively promoted fiber outgrowth from sympathetic and sensory ganglia. This hypothesis was tested and confirmed by in vitro bioassay.

NGF effects in neurogenesis

The identified NGF target cells are a) the large majority of sympathetic, long adrenergic neurons, b) the small, late differentiated neurons that make up 50% to 60% of the root spinal ganglia, c) neural crest derivatives normally committed to become chromaffin cells, and d) the mast cells. Recent investigations suggest that several neuronal cell populations in the central nervous system are also receptive to this protein. Well-identified target cells respond to NGF action according to a pattern that is similar for all cells: it is maximal during early ontogenetic stages and gradually decreases but still persists in the adult.

During early ontogenetic phases of sympathetic cells, three experimentally distinguishable NGF effects have been described: trophic, differentiative, and neurotropic. The trophic role played by NGF in early ontogenetic stages is evident in in vivo and in vitro studies. In the developing organism, deprivation of this molecule through immunological, pharmacologic, or surgical procedures results in death of sympathetic NGF receptive cells. In vitro, the same cells neither produce axons nor survive if NGF is not present in the culture medium. The differentiative NGF action is apparent by commitment to or acceleration of expression of the cell phenotype upon injections of this protein. This action is manifested by cell size enlargement, cytoskeletal organization, axonal growth and branching, and synthesis of neurotransmitters via stimulation of enzymes such as tyrosine hydroxylase, the rate-limiting enzyme in the biosynthesis of catecholamines. The in vivo neurotropic effect is shown by the channeling of nerve fiber growth toward an experimentally produced NGF pool. In vitro, the same effect has been unequivocally demonstrated by directional neurite outgrowth toward a calibrated NGF concentration gradient.

Phylogenetic features of the NGF receptor system and its target cells

Two distinct in vitro tests employing cells of avian or mammalian origin gives unequivocable evidence of the existence, from fishes to humans, of a protein with the nerve fiber growth-promoting activity of mouse NGF. Furthermore, it was found that a large number of nerve cells of the central nervous system in *Xenopus laevis*

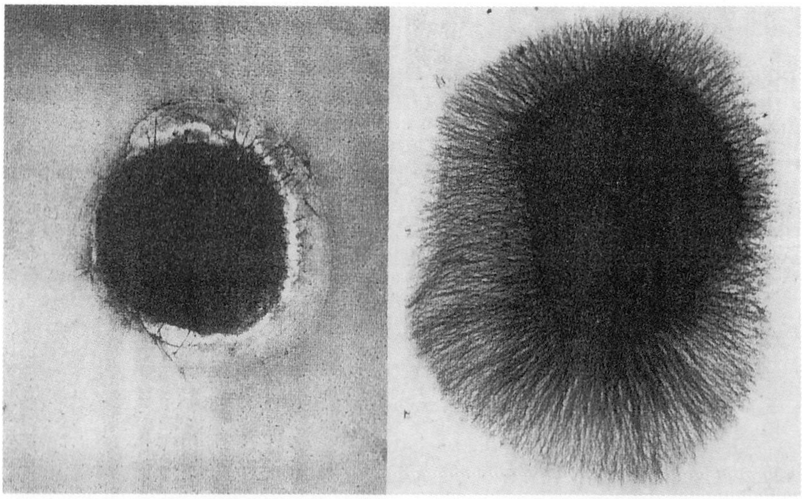

Figure 1. Photomicrographs of two sensory ganglia from an 8-day chick embryo cultured for 24 hours at 37°C in a medium with (right) or without (left) 10 ng/ml of the protein nerve growth factor.

tadpoles are responsive to administration of murine NGF. The same molecule enhances regenerative processes in the optic system of teleost fish. These findings, and the detection of the NGF gene in amphibia, clearly demonstrate the evolutionary preservation not only of the gene, but also of its product (NGF) and of the recognition system (receptor) involved in the transduction of its message. In addition, the spectrum of action and the plurality of target cells to mouse NGF would suggest that the NGF-receptor system is operative in a larger number of nerve cells in lower than in higher vertebrates.

The NGF molecule and its gene

Through the availability of the in vitro bioassay, the fast screening of putative NGF sources became possible. Two of these were discovered in the venom of several species of poisonous snakes and in the mouse submandibular salivary gland. Other sources were subsequently uncovered in the guinea pig prostate and in the bull seminal fluid. While these glands or their secretory products contain NGF in the order of micrograms per milligram total proteins, a large number of normal and neoplastic cells from different vertebrate species synthesize minute but measurable quantities of NGF.

NGF is a protein whose primary structure (amino acid sequence) was first elucidated from the molecule purified from the mouse submandibular gland. It is a dimer made of two identical subunits, each of 13.250 daltons, held together by noncovalent bonds. Similar primary structures have been reported for NGF isolated from snake venom, guinea pig prostate, and bull seminal fluid. Recent studies identified the mRNA for the mouse salivary gland NGF. It codes for a much larger precursor of 307 amino acids (the biologically active NGF contains 118 amino acids), a finding that demonstrates that the NGF protein is part of a precursor, as is the case for several hormones and specific growth factors. Its processing is mediated by specific proteases that cleave it at specific sites. In the mouse salivary gland, it has been hypothesized that the processor enzyme remains associated to NGF to form a large, multimolecular complex defined as 7S NGF. The complex consists of three different molecular species—the alpha subunit, of unknown function; the gamma subunit, endowed with protease activity; and the beta subunit, which is biologically active NGF. Although this complex is detectable both in salivary gland homogenate and mouse saliva, its presence in other small or large NGF sources has not yet been established. Upon removal of eight amino acids at the C-terminal and of an N-terminal arginine, β-NGF is converted to a slightly smaller, biologically identical molecule, named 2.5S NGF on the basis of its sedimentation co-efficient.

The isolation of NGF mRNA in the mouse was followed by the cloning and isolation of its gene not only in rodents, but also in humans. Preliminary studies demonstrate that the NGF gene is present in fish and amphibians. Its structure in mammals, from rodents to primates, has been highly conserved and—as inferred from sequence studies of the protein product—this is also true for nonmammalian species.

Mechanisms of action

NGF action on target cells is mediated by specific, high-affinity receptors present on the external surface of plasma membranes. These receptors have a K_d for NGF which varies, in different target cells, between 10^{-11} and 10^{-9} M. The optimal NGF concentration for the in vitro bioassay employed for NGF screenings in biological sources is 5 to 10 ng/ml, that is, 2.5 to 5.0×10^{-10} M, a value of the same magnitude as the K_d of NGF receptors in these nerve cells. NGF receptors are present in the plasma membrane of undifferentiated target cells and on the nerve endings of axons and dendrites of the same fully differentiated cells. Binding of NGF to its receptors is followed by the internalization of their complexes and intracellular distribution into different cellular compartments, including lysosomes that eventually destroy the factor. NGF internalization from nerve endings is followed by retrograde transport along nerve fibers up to the cell perikaryon. Evidence that this transport is crucial, at least for the vital, trophic effect of NGF, is provided by the demonstration that prevention of retrograde transport by different manipulations leads to death of target cells, an effect counteracted by an exogenous supply of NGF. It is not yet clear whether there are one or more distinct types of receptors possibly mediating different NGF effects, nor whether these receptors operate as transducers of NGF action via as yet unidentified second messengers. An alternative mechanism to the amplification of the NGF message through second messengers (cyclic adenosine monophosphate, Ca^{++}, etc.) is a direct action of NGF or of NGF receptor complexes on some intracellular structures or components (e.g., the cell's genome and the cytoskeleton), which, because of their vital function or strategic position, control several other properties of the cells.

Morphological, ultrastructural, and chemical changes of target cells are detectable minutes after NGF interaction with its receptors and become more pronounced and of a broader range within hours and days. They involve increased uptake of amino acids and other nutrients across the plasma membrane, stimulation of the synthesis of specific classes of mRNAs and of their corresponding protein products, and organization of cytoskeletal elements providing the force-generating structure for axonal growth and elongation. These anabolic changes are accompanied by an increased uptake of glucose and its consumption through oxidative pathways. This pleiotypic response of target cells to NGF is particularly evident in certain cells, such as those from the clonal cell line PC12, derived from a rat pheochromocytoma tumor. PC12 cells, in the presence of NGF, change their phenotype from a neoplastic, undifferentiated, proliferating cell type to a mitotically arrested, neurite-bearing cell population markedly resembling normal sympathetic nerve cells.

NGF as a paradigm of neuronal and nonneuronal growth factors

The discovery of NGF prompted the search for other peptide molecules endowed with the property of selectively stimulating proliferation and differentiation of specific cell populations. Among these factors, NGF holds a privileged position as the first to have been identified, sequenced, and shown to play a key physiological role not only in in vitro systems, but also in the intact organism. Although at present it is the only well-identified neuronotrophic factor, others are likely to be identified.

The discovery of NGF and of other growth factors acting on cell populations such as epidermal cells, fibroblasts, and ematopoietic cells is at the same time opening new avenues to the understanding of the biological basis of neoplastic transformation. Thus, there are malignant cells which produce factors that stimulate their own proliferation and invasion of the host (alpha tumor growth factor, angiogenesis growth factor, etc.). The epidermal growth factor receptor shares strong sequence homologies with gene products causing some types of sarcomas. Finally, a factor released by platelets and postulated to play a role in wound healing after injury has the same, identical structure of a cellular protein (protoncogene) causing transformation of several cell types. Among the few cellular proteins inducing transformation, the growth factors appear to hold a key position in normal and deviant biological processes. However, at variance with mitogenic factors that are identical to or share common properties with oncogenes, differentiating growth factors (having their paradigm in NGF) have up to now been shown to play a role only in differentiation. Cell differentiation or cancer are the end result of a proper or improper expression or function of these few proteins.

A role for NGF in nervous, endocrine, and immune systems

A major shift in the NGF studies has recently occurred with the realization that the spectrum of action of this protein molecule is not restricted, as previously conceived, to enhancing differentiative processes of two peripheral neuronal cell lines (the sensory and sympathetic neurons), but extends also to several other neuronal and nonneuronal cell types. The newly detected NGF-responsive cells belong to the hemopoietic-immune system, to the cholinergic system of the basal forebrain nuclei, and to other cell populations of the central nervous system involved in neuroendocrine functions. Additional investigation showed that NGF is synthesized in some hypothalamic nuclei, and its level in the bloodstream of mice displaying aggressive behavior is markedly increased. These and previous findings support the hypothesis that this protein molecule may exert a modulatory role on neuro-immunoendocrine functions of vital importance in the regulation of homeostatic processes.

Acknowledgments. This work has been supported by two grants from the National Council of Research (CNR), Italy: Progetto Finalizzato Chimica Fine to R.L.-M. and Progetto Finalizzato Oncologia to P.C.

Further reading

Buecker ED (1948): Implanation of tumors in the hind limb field of the embryonic chick and the developmental response of the lumbo-sacral nervous system. *Anat Rec* 102:369–390

Calissano P, Cattaneo A, Aloe L. Levi Montalcini R (1984): The nerve growth factor. In: *Hormonal Proteins and Peptides. vol. 12.* Li CH, ed, pp 2–47, New York: Academic Press

Greene LA, Shooter EM (1980): The nerve growth factor: Biochemistry, synthesis, and mechanism of action. *Annual Review of Neuroscience* 3: 353–402

James R, Bradshaw RA (1984): Polypeptide growth factors. *Annual Review of Biochemistry* 53:259–293

Levi-Montalcini R (1952): Effects of mouse tumor transplantation on the nervous system. *Annals of the Academy of Science* 55:330–343

Levi-Montalcini R, Aloe L, Alleva E (1990): A role for nerve growth factor in nervous, endocrine and immune systems. *Prog in Neuroendocrinimmunology* 3:1–10

Thoenen H, Barde YA (1980): Physiology of nerve growth factor. *Physiology Review* 60:1284–1335

ELSEVIER

Journal of the Neurological Sciences 130 (1995) 119–127

JOURNAL OF THE
NEUROLOGICAL SCIENCES

Update of the NGF saga

Rita Levi-Montalcini [a,*], Roberto Dal Toso [b], Federica della Valle [c], Stephen D. Skaper [b], Alberta Leon [b]

[a] *Institute of Neurobiology, National Research Council, Viale Marx 15, 00156 Rome, Italy*
[b] *Researchlife S.c.p.A., Castelfranco Veneto (Treviso), Italy*
[c] *Lifegroup S.p.A., Padua, Italy*

Received 25 October 1994; accepted 4 December 1994

Abstract

Nerve growth factor (NGF), initially characterized for its survival and differentiating actions on embryonic sensory and sympathetic neurons, is now known to display a greatly extended spectrum of biological functions. NGF exerts a profound modulatory role on sensory nociceptive nerve physiology during adulthood which appears to correlate with hyperalgesic phenomena occurring in tissue inflammation. Other newly detected NGF-responsive cells belong to the hematopoietic-immune and neuroendocrine systems. In particular, mast cells and NGF both appear to be involved in neuroimmune interactions and tissue inflammation, with NGF acting as a general "alert" molecule capable of recruiting and priming both local tissue and systemic defense processes following stressful events. NGF can thus be viewed as a multifactorial mediator modulating neuroimmune-endocrine functions of vital importance to the regulation of homeostatic interactions, with potential involvement in pathological processes deriving from dysregulation of either local or systemic homeostatic balances.

Keywords: Nerve growth factor; Neuroimmune interactions; Mast cells; Inflammation

1. Introduction

Advances in the neurosciences have been taking place at an extraordinary pace over the last several decades. Neuroscientists are making use today of the latest technological advancements in molecular biology, genetics and biophysics to unravel the processes underlying morphofunctional organization of the nervous system, and are rapidly uncovering the sophisticated molecular mechanisms regulating neuronal functions and behavior. We no longer view the adult nervous system as a static entity but rather as a mutable plastic structure undergoing continuous morphofunctional adjustments in response to perturbations both from within and without. Recent developments tell us that this property of the nervous system is brought about in cooperation with both the immune and endocrine systems and vice-versa, indicating that these three major homeostatic systems are in constant communication with each other to maintain a balance to meet the necessities of the organism.

Nerve growth factor (NGF), while initially characterized as a neurotrophic factor for specific neuronal populations in the peripheral nervous system (PNS) and, more recently, in the central nervous system (CNS), clearly is capable of exerting a far broader spectrum of biological activities. As early as 1977, effects of NGF were being reported on mast cells (Aloe and Levi-Montalcini, 1977), a cell type belonging to the immune lineage. NGF was later found to act also on other immunocompetent cells, as well as on cells of the endocrine system. A hypothesis has now gained ground in which NGF is viewed as playing a key role not only within the nervous system but, more importantly, in coordinating cross-talk between systems involved in the control of homeostatic functions. In the following article we will re-examine this issue, especially in view of a number of new findings that lend further credence to this notion.

* Corresponding author. Tel.: (+39-6) 86 090 510; Fax: (+39-6) 86 090 269.

SSDI 0022-510X(95)00007-0

2. NGF and plasticity of the peripheral nervous system

A functional role for NGF in the development of peripheral sympathetic and some neural crest-derived sensory neurons is now well-established from in vitro and in vivo studies.

Early experiments made use of NGF-specific antibodies applied either in utero or to newborn rodents, which resulted in a profound reduction in the number of sympathetic neurons (Levi-Montalcini and Booker, 1960; Gorin and Johnson, 1979; Goedert et al., 1980), cholinergic sympathetic neurons innervating sweat glands (Landis et al., 1985), adrenal medullary cells (Aloe and Levi-Montalcini, 1979), and small diameter spinal sensory neurons (Johnson et al., 1980). NGF is also important for the phenotypic traits of myelinated nociceptors (Ritter et al., 1991; Lewin et al., 1992a). More recent studies relying on the generation of homologous recombinant mice lacking either NGF or its signal transducing receptor Trk have resulted in predictable and dramatic phenotypes. Behaviors in these animals are very similar to those exposed prenatally to NGF antibodies, having severe losses of sympathetic and sensory ganglion neurons (Crowley et al., 1994; Smeyne et al., 1994). Animals bearing a null mutation for the low-affinity neurotrophin receptor $p75^{LNGFR}$ displayed largely cutaneous sensory deficiencies (Lee et al., 1992). Some neural crest-derived cells (e.g. enteric neurons and cranial cartilage) apparently are not affected by "in vivo" anti-NGF treatment, although yet other derivatives such as melanocytes and thymic stromal cells may be sensitive to NGF treatment.

Common features of both immunologic and genetic experimental models of NGF deprivation include sympathectomy and severe sensory neuropathy which is expressed as a markedly reduced pain sensitivity and responsiveness to both mechanical and thermal stimuli. This is in keeping with the observation that, in Trk-deficient (Smeyne et al., 1994) or NGF-deficient (Crowley et al., 1994) transgenic rodents, a selective loss occurs during development in substance P- and calcitonin gene-related peptide (CGRP)-containing small sensory neurons. These are mainly unmyelinated or thinly myelinated neurons thought to mediate peripheral afferent nociceptive stimuli to the dorsal horn of the spinal cord (Leah et al., 1985).

Evidence also points to an important pathophysiological role of NGF in sensory function during adulthood. Although NGF is not required for neuron survival (unlike the case for adult sympathetic neurons) (Snider and Johnson, 1989), NGF induces a series of phenotypical and morphofunctional modifications of responsive adult sensory neurons, both in their peripheral and central terminal fields. In fact, small peptidergic sensory neurons retain functional high-affinity NGF receptors in the adult (Verge et al., 1989). NGF upregulates CGRP mRNA expressed in adult dorsal root ganglion neurons both in vitro (Lindsay and Harmar, 1989) and in vivo (Donnerer et al., 1992), increases the density of sensory nociceptive endings in dermis and epidermis (Albers et al., 1994), modulates distal collateral sprouting (Diamond et al., 1987,1992) and upregulates central connectivity of nociceptive afferents (Lewin et al., 1992b). Administration of NGF to adult rodents rapidly induces a decrease in the threshold of nociceptors to heat and mechanical stimuli, which then slowly recovers (Lewin et al., 1993). Hyperalgesic responses have also been observed in transgenic animals overexpressing NGF in the skin. In contrast, genetically modified rodents expressing antisense NGF mRNA exhibited hypoalgesia to mechanical stimuli (Davis et al., 1993), consistent with reduced behavior responses towards noxious stimuli in animals exposed to anti-NGF antibodies (Aloe et al., 1981).

The precise mechanism(s) of hyperalgesia induction by NGF in adult rats is not yet clear, although high levels of NGF appear able to strengthen the effects of both orthodromic and antidromic sensory synaptic release by modifying amplitude and/or density of peripheral and central terminal fields, and by increasing peptidergic transmitter content and reducing the activation threshold to noxious heat and mechanical stimuli. The two latter effects occur very rapidly. The initial, rapid phase of NGF-induced thermal hyperalgesia seems also to be peripherally mediated by mast cells and to involve a central NMDA receptor-dependent mechanism (Lewin and Mendell, 1993). Thus, NGF or other NGF-derived peptides do not directly activate nociceptive neurons (Taiwo et al., 1991), but rather modulate a set of cellular parameters and functional conditions, which coherently act to increase or facilitate sensory responsiveness. This is borne out by a recent report in which healthy human subjects who had received NGF subcutaneously experienced long-lasting hyperalgesia at the injection site, but no spontaneous pain (Petty et al., 1994). Nerve growth factor-induced hyperalgesia is qualitatively analogous to that occurring in tissue inflammation. Local pain is one of the hallmarks associated with inflammatory tissue injury. Induction of neuropeptide synthesis in sensory neurons occurs within hours of onset of inflammation (Donaldson et al., 1992; Dubner and Ruda, 1992), an effect which can be blocked by anti-NGF antibodies (Donnerer et al., 1992). Sensory peptides such as tachykinins also play a role in mediating neurogenic inflammatory responses and in regulating immune system function (e.g. mast cell activation) (Holzer, 1988). Sympathetic post-ganglionic neurons as well contribute notably to tissue inflammation and hyperalgesia by producing peptide mediators such as interleukin-1 (IL-I) (Freidin et al., 1992), substance P (Jonakait, 1993), and prostaglandin E2 (Gonzales et al., 1989). The involvement of sympathetic neurons m processes such as plasma extravasation appears to be closely regulated by mast cell-derived signals and by epinephrine released from the adrenal medulla (Basbaum and Levine, 1991). These observations may help to assign a physiological meaning to the high levels of NGF which, in association with inflammatory cytokines accumulate in a variety of inflamed tissues and body fluids (Aloe et al., 1994; Weskamp and Otten, 1987). As such, NGF could represent a key player in coordinating the defensive responses of both the immune and nervous systems to external noxious stimuli (Levi-Montalcini et al., 1990).

3. NGF and immune cell function

NGF is now recognized to influence a variety of bone marrow-derived cell types. For example, NGF can induce shape changes in rabbit platelets (Gudat et al., 1981). Furthermore, NGF stimulates the proliferation of both B and T lymphocytes (Thorpe and Perez-Polo, 1987; Otten et al., 1989; Brodie and Gelfand, 1992) and the production of IgM, IgA (Otten et al., 1989) and IgG4 antibodies (Kimata et al., 1991). Exposure to NGF induces high-affinity IL-2 receptors on human peripheral blood mononuclear cells (Thorpe et al., 1987) and promotes human hemopoietic cell growth and proliferation (Matsuda et al., 1988). NGF is also involved in chemotaxis, viability and functional properties of human polymorphonuclear neutrophils in vitro (Gee et al., 1983; Kannan et al., 1991) and in vivo (Boyle et al., 1985), and in the differentiation of thymic stromal non-lymphoid cells (Screpanti et al., 1992). These effects of NGF are consistent with the recent finding that human monocytes (Ehrhard et al., 1993a) and activated $CD4^+$ T-cell clones of the Th2-type (Ehrhard et al. 1993b) express the functional NGF receptor Trk. These latter cells also elaborate biologically active NGF, suggesting a possible autocrine or paracrine action of this neurotrophic factor in the development and regulation of immune cell responses.

Mast cells were the first cells of immune lineage to be shown responsive to NGF, both in vivo (Aloe and Levi-Montalcini, 1977) and in vitro (Böhm et al., 1986). Neonatal rats which received NGF over the first 14 days of life experienced a massive mast cell hyperplasia (increase in both number and size) involving connective tissue mast cells (CTMC) in several peripheral tissues, an effect later extended also to mucosal mast cells (Stead et al., 1987; Marshall et al., 1990). In addition, anti-NGF antibodies reduce rat peritoneal mast cell numbers (Aloe, 1988). These data suggest a specific and direct NGF effect on the differentiation and maturation of mast cells and/or their precursors and are supported by the fact that NGF can induce the development of CTMC from mouse bone marrow cells (Matsuda et al., 1991) and from spleen cells of new-born rats treated with NGF (Aloe and De Simone, 1989). Interestingly, NGF appears to act as a cofactor with interleukin-3 in the development of basophils or mast cells from human umbilical cord blood cells (Richard et al., 1992).

Nerve growth factor is an extremely potent degranulating agent for cultured rat peritoneal mast cells in the presence of phosphatidylserine or its lyso derivative (Bruni et al., 1982; Sugiyama et al., 1985), being far more potent than compound 48/80 or anti-IgE on a molar basis (Pearce and Thompson, 1986). NGF-induced non-cytotoxic mast cell degranulation appears to require the presence of extracellular calcium (Mazurek et al., 1986), does not depend on the presence of cell-bound IgG or IgE, differs from the action of polycationic secretagogues (Pearce and Thompson, 1986), and can be blocked by anti-allergic compounds (which also inhibit NGF-induced hyperplasia) (Marshall et al., 1990). By itself, NGF at physiological concentrations is a very poor secretagogue for rat peritoneal mast cells, while markedly enhancing antigen and other secretagogue-induced histamine release from such cells (Tomioka et al., 1988). NGF may thus act as an immunomodulator in the inflammatory response by regulating mediator release from mast cells. A very similar sensitizing or priming effect of NGF on histamine release has been observed in mature human basophils (Bischoff and Dahinden, 1992). In keeping with such findings, subcutaneous injection of NGF produces plasma extravasation (Otten et al., 1984), while intravenous administration induces hypotension (Yan et al., 1991). Both responses are consistent with histamine release caused or facilitated by NGF.

Mast cells express the functional (high-affinity) NGF receptor Trk, but neither other members of this receptor tyrosine kinase family nor the p75 (low-affinity) neurotrophin receptor (Horigome et al., 1993). Thus, NGF appears to be the only neurotrophin capable of affecting mast cell mediator release. A survival-promoting effect of NGF on rat peritoneal mast cells in vitro may involve prevention of an apoptotic death mechanism (Horigome et al., 1994). In keeping with other reports (Marshall et al., 1990), this NGF effect is suggested to be mediated by the increased expression and release of one or several autocrine growth factors. We have recently shown that rat peritoneal mast cells and cells of the basophilic lineage synthesize, store and release biologically active NGF in physiologically relevant amounts (Leon et al., 1994), raising the possibility that at least one autocrine agent inducing mast cell survival and sensitization may be NGF itself.

The ability of mast cell degranulation products, including NGF, to induce cell hyperplasia in an autocrine fashion has clear implications for atopic disease states or other chronic inflammatory pathologies where repeated challenge with antigens, neurogenically derived secretagogues or naturally-occurring IgE may cause on-going mast cell degranulation, whereby the resulting products could contribute to the observed hyperplasia and chronicity of the disease state.

4. NGF and endocrine function

NGF synthesis and maturation in the mouse submaxillary gland (SMG) as well as in other tissues is controlled by a number of hormonal mediators. Thyroid hormones and androgens increased NGF gene expression in the SMG of both male and female mice (Aloe and Levi-Montalcini, 1980; Black et al., 1992; Siminoski et al., 1993) while castration of male mice marked reduced NGF mRNA content in the SMG (Caramia et al., 1962). Testosterone increased NGF content (Ishii and Shooter, 1975) and modified the post-translational processing of NGF in female mice (Lakshmanan et al., 1988). Adrenal steroids, while reported to positively regulate NGF content in the mouse SMG (Siminoski et al., 1993) and NGF biosynthesis in the rat brain (Aloe, 1989; Fabrazzo et al., 1991) inhibited lesion-induced increases of NGF mRNA levels in transected sciatic nerves (Lindholm et al., 1990), suggesting the observed regulatory activity for single hormones to differ according to cell type, tissue or physiological condition. Thyroxine markedly increased NGF content in the cerebral cortex, brainstem and cerebellum (Walker et al., 1979), potentiated NGF effects in the CNS (Patel et al., 1988) and regulated gene expression of the p75 NGF receptor (Figueiredo et al., 1993); 17β-estradiol increased synthesis and release of NGF from cultured glioma cells (Perez-Polo et al., 1977).

NGF mRNA and protein are present in a number of tissues and areas involved in endocrine control of both local and systemic homeostasis. Both NGF and its receptor components have been detected in rat male reproductive organs (Harper et al., 1982; Persson et al., 1990; Graham et al., 1992; Parvinen et al., 1992) and ovary (Lara et al., 1990, Dissen et al., 1991). In these tissues NGF content is under the control of both pituitary and sympathetic innervation, where it appears to play a central role in paracrine interactions mediating spermatogenesis, prostatic growth, and acquisition of female reproductive capacity.

Thyroid and parathyroid glands of the rat express NGF mRNA and NGF precursor protein (Dicou et al., 1986). NGF immunoreactivity has been detected in the pineal gland (Garcia-Mauriño et al., 1992), where it likely maintains the high density sympathetic innervation.

The role of NGF in the hypothalamus-pituitary-adrenal (HPA) endocrine axis is of particular interest, given the strategic importance of this hormonal system in the control of systemic homeostasis and fundamental regulatory activities of the nervous system. NGF and the p75 receptor are present in male and female rats in various hypothalamic nuclei (Talamini and Aloe, 1993), with NGF gene expression being regulated during development (Ojeda et al., 1991) and after behaviorally stressful conditions (Spillantini et al., 1989; Aloe et al., 1990) – implying a hormonal and neuro-mediated control of NGF synthesis. It is not clear if hypothalamic NGF can be released into the surrounding extracellular space or into the portal blood vessels.

Biologically active NGF is also found in the rat pituitary (Lathinen et al., 1989), where levels of expression appear to be under neuronal control (Soinila et al., 1988). Transgenic mice overexpressing the NGF gene in lactotrophs showed a marked hyperplasia of this cell type and the ability to release NGF (Borrelli et al., 1992). Furthermore, it would seem that NGF may be a differentiating factor for pituitary prolactin secreting cells. Upon activation of Trk, NGF directs the differentiation of GH3 cells from a prevalent growth hormone secreting somatotroph phenotype to prolactin producing lactotrophs (Missale et al., 1994), at the same time suppressing the tumorigenicity of human prolactinomas in vitro and in vivo (Missale et al., 1993). NGF may also be present in the pituitary in a stored form and secreted into the circulation in vivo (Lathinen et al., 1989). This appears to be likely since cell lines of pituitary origin, when transfected with the NGF gene via a replication-defective retroviral vector, are able to store mature β-NGF in secretory granules and to release it upon stimulation (Wolf et al., 1988).

The presence of NGF in both hypothalamus and pituitary may contribute to serum levels of NGF and other hormones present in a variety of physiological and behavioural conditions. NGF is known to stimulate the HPA axis, enhancing the secretion of adrenocorticotrophic hormone (ACTH) and glucocorticoid concentrations (Otten et al., 1979), an effect probably mediated through the hypothalamus (Scaccianoce et al., 1993). Following aggressive behaviour, high NGF serum levels in rodents are associated with hypertrophy of cortical and medullary components of the adrenal (Aloe et al., 1986). Plasma NGF levels in women during labor and lactation are increased when compared to concentrations found in control women or at term (Luppi et al., 1993), a time when plasma levels of oxytocin are high. Oxytocin itself increases hypothalamic NGF content in virgin female rats (Luppi et al., 1993). Our recent results indicate that NGF blood levels are significantly increased in humans both prior to and after psychologically stressful and anxious situations, while interleukin-1 (IL-I) and tumor necrosis factor-α (TNF-α) serum levels are not modified (Aloe et al., 1994). Together, these data suggest that NGF may be a crucial component of regulatory processes involving general homeostatic mechanisms (Levi-Montalcini et al., 1990).

5. NGF as a short-distance mediator in the control of tissue defense strategies

Peripheral tissue defense mechanisms, for example inflammation, invoke a complex series of molecular and cellular events encompassing local activation of tissue-specific elements (fibroblasts, mast cells, keratinocytes, Schwann cells) and recruitment at the site of injury of elements of general homeostatic systems, such as the immune system (macrophages, neutrophils, basophils, lymphocytes) and the nervous system (sympathetic and sensory neurons). Properly regulated, these components work together to limit the extent and duration of the damage.

NGF may help to coordinate these systems, given its ability to interact with most of the elements involved. The NGF-induced sensitization of mast cells, as well as that of sensory nerves via a long-term increase of neuropeptide synthesis and release from both central and peripheral terminals, would contribute to an increased tissue responsiveness towards exogenous and endogenous noxious stimuli. The release of stored NGF from mast cells may represent a local and readily available source of this trophic factor to rapidly induce nociceptive C fiber sensitization and to recruit and activate, together with other cytokines, cellular elements of the immune system. These cells, by releasing IL-1β and TNF-α, can induce NGF synthesis in fibroblasts, Schwann cells and others (Lindholm et al., 1987; Hattori et al., 1993) – thereby helping to maintain elevated tissue NGF levels known to occur in both acute and chronic inflammatory exudates (Aloe et al., 1994). This latter point may be of importance in the case of wound healing (Li et al., 1980), possibly through NGF induction of transforming growth factor β-1 (TGF-β1) (Kim et al., 1994) – a multifunctional peptide playing a key role in tissue repair processes.

Thermal hyperalgesia induced by NGF administration (Lewin et al., 1993) involves a very rapid and early mast cell-dependent sensitization of nociceptive neurons, followed by longer-lasting phases probably associated with changes in central processing and neuropeptide release (Lewin et al., 1992b), as occurs after inflammatory stimuli (Garry and Hargreaves, 1992). As discussed earlier, NGF does not directly induce hyperalgesia or pain, but rather reduces the threshold of mechanical and thermal nociceptors thus generating, in situations of locally elevated NGF levels (e.g. inflammatory or injured sites) an area of increased sensitivity to pressure and heat, which might serve to limit damage of the lesioned tissue and facilitate healing.

Tissue defense mechanisms, particularly inflammation-related, should be strictly limited in time and extent to avoid involving healthy tissue and a possible prolongation which generates a chronic inflammatory state. Given that NGF may be one mediator in this coordinated activation, its action may be restricted in one of several ways. Sympathetic and sensory neurons possess an efficient uptake system for NGF which could greatly reduce NGF levels, giving rise to the increased density of peripheral nerve fibers seen in inflamed tissue. NGF diffusion may also be limited by the presence of specific cell surface receptors, or by interaction with components of the extracellular matrix. Proteases at inflammatory sites may help to terminate NGF trophic effects, while protracting actions directed to nerve sensitization (Taiwo et al., 1991). Together, these processes may contribute to a short-distance action of NGF. Controlling inflammation and NGF levels by pharmacologic regulation of excessive mast cell neurogenic or immunogenic activation may be one approach to locally limit the effects of tissue defense processes.

6. NGF as a long-distance modulator in the control of body defense strategies?

The greatly increased NGF serum levels in rodents after behaviorally and psychologically stressful events are intriguing. The increase in serum NGF concentration apparently is not a matter of physical stress alone but seems rather to be associated with psychological stress status and, perhaps, anxiety (Maestripieri et al., 1990). Levels of HPA axis-derived hormones such as ACTH, prolactin, corticosteroids and epinephrine are modified by psychological stress. The presence of NGF in the hypothalamus and hypophysis, the sensitivity of NGF synthesis to endocrine and neuronal mediators, and the activation of the HPA axis by exogenous NGF together propose an integrated role for NGF in HPA axis function. Several points, however, need to be established to validate this hypothesis.

First, one needs to consider the tissue or cellular elements that secret NGF into the bloodstream. A critical requirement is that they have the capacity to synthesize and release biologically relevant amounts of NGF under strict control of cognitive and neuromediated actions, and hence be sensitive to neurotransmitters or in proximity to neuronal structures. The hypothalamic-pituitary system and, at least in male mice, also the submaxillary glands appear to meet these conditions, although other sources cannot be excluded. A corollary to this point is the identification of positive and negative neuromediators of NGF endocrine release.

A second point concerns possible targets of physiological significance for serum NGF. Any cell type expressing high-affinity NGF receptors within range of serum proteins could respond to this NGF. Endocrine organs such as the adrenal glands, particularly the medullary portion, with its neural crest derivation and responsiveness to NGF could be likely targets. Cellular elements of the immune system itself, such as monocytes, lymphocytes and basophils, however, cannot be excluded. This idea, if verified by other experimental observations, would define an important new link between neuronally controlled mediators and the immune system, thereby adding to the known neuroimmune interactions which occur locally in central and peripheral lymphoid organs between immune-derived cellular elements and peripheral innervation. One must note that at physiological concentrations NGF alone does not appear to directly activate cells of the immune lineage, but rather modulates their sensitivity to other triggering stimuli. Circulating NGF could function as a general "alerting" signal utilized by the CNS in settings of stress and anxiety to "prime" the immune system towards external noxious perturbations, greatly expanding the range of biological activities attributable to NGF. This neurotrophic molecule begins to emerge as a crucial coordinator for major homeostatic systems and body defense strategies.

The transport and clearance of NGF from blood represents yet another point in need of study, and which may be of importance in minimizing local actions of systemic NGF. It is known that NGF in serum, like other cytokines reversibly binds to α-2 macroglobulins (Ronne et al., 1979; Koo and Stach, 1989; Crookston et al., 1994), and that monoamine activated α-2 macroglobulin specifically binds to and functionally inhibits Trk (Koo and Qiu, 1994).

7. Future goals

Nerve growth factor now appears to be a complex pleiotropic agent active on an unexpectedly wide array of cell types and biological functions, which was not envisioned on the basis of earlier work establishing its trophic role for sympathetic and sensory neurons during development and adulthood. One important future goal will be to more fully evaluate the possible involvement of NGF in pathological processes resulting from dysregulation of local or systemic homeostatic processes which may occur, for example, in chronic inflammatory diseases (for a review see Aloe et al., 1994). High levels of NGF have been detected in the inflammatory exudates and biological fluids of patients suffering from autoimmune diseases. Not only do mast cells, a crucial component in virtually all inflammatory processes, synthesize and secrete NGF, but NGF itself has both survival and sensitizing activities on mast cells. These findings raise the possibility of autocrine actions that, if not properly controlled, could have deleterious effects on surrounding tissues and lead to chronic inflammatory processes – such as those present in many autoimmune-based pathologies. Should such an autocrine loop exist for mast cells, the question arises as to how tissues normally regulate and control their NGF activity. Evidence is now emerging suggesting the presence of local antagonistic autacoid mechanisms which depend on the synthesis, at the site of inflammation, of substances having negative regulatory effects on the state of mast cell activation. Elucidation of these autoregulatory mechanisms holds enormous potential for our understanding of local tissue defense and homeostatic processes, which may one day open up novel strategies for pharmacologic intervention.

Recently, NGF has been shown to induce HIV-1 gene expression and replication in both neuronal and glial cell lines (Ensoli et al., 1994), carrying major implications in the neuropathological manifestations associated with the acquired immunodeficiency syndrome (AIDS). This observation opens a completely new field of interest in NGF research and further implicates NGF regulation as a potential target for therapeutic manipulation.

References

Albers, K.M., Douglas, E.W. and B.M. Davis (1994) Overexpression of nerve growth factor in epidermis of transgenic mice causes hypertrophy of the peripheral nervous system. J. Neurosci., 14: 1422–1432.

Aloe, L. (1988) The effect of nerve growth factor and its antibody on mast cells in vivo. J. Neuroimmunol. 18: 1–12.

Aloe, L. (1989) Adrenalectomy decreases nerve growth factor in young adult rat hippocampus. Proc. Natl. Acad. Sci. USA, 86: 5636–5640.

Aloe, L. and De Simone, R. (1989) NGF primed spleen cells injected in brain of developing rats differentiate into mast cells. Int. J. Dev. Neurosci., 7: 565–573.

Aloe, L. and R. Levi-Montalcini (1977) Mast cells increase in tissues of neonatal rats injected with the nerve growth factor. Brain Res., 133: 358–366.

Aloe, L. and R. Levi-Montalcini (1979) Nerve growth factor in vivo induced transformation of immature cromaffin cells in sympathetic neurons: effect of antiserum to the nerve growth factor. Proc. Natl. Acad. Sci. USA, 76: 1246–1250.

Aloe, L. and R. Levi-Montalcini (1980) Comparative studies on testosterone and L-thyroxine. Effects on the synthesis of nerve growth factor in mouse submaxillary salivary glands. Exp. Cell Res., 125:15–22.

Aloe, L., Cozzari, C., Calissano, P., and R. Levi-Montalcini (1981) Somatic and behavioural postnatal effects of fetal injections of nerve growth factor antibodies in the rat. Nature, 291: 413–415.

Aloe, L., Alleva, E., Böhm, A. and R. Levi-Montalcini (1986) Aggressive behavior induces release of nerve growth factor from mouse salivary gland into the bloodstream. Proc. Natl. Acad. Sci. USA, 83: 6184–6187.

Aloe, L., Alleva, E. and R. De Simone (1990) Changes of NGF level in mouse hypothalamus following intermale aggressive behavior: biological and immunohistochemical evidence. Behav. Brain. Res., 39: 53–61.

Aloe, L., Bracci-Laudiero, L., Lambise, A., Tirassa, P., Micera A. and R. Levi-Montalcini (1994) Emotional stress induced by parachute jumping enhances blood NGF levels and the distribution of NGF-receptors in Lymphocytes. III. Int. Conf. on NGF and Related Molecules. Lake Lewis, Alberta, Canada, April 26th–May 1st, 1994.

Aloe, L., Skaper, S.D., Leon, A. and R. Levi-Montalcini (1995) Nerve growth factor and autoimmune disease. Autoimmunity, in press.

Basbaum, A.I. and J.D. Levine (1991) The contribution of the nervous system to inflammation and inflammatory disease. Can. J. Physiol. Pharmacol., 69: 647–651.

Bischoff, S.C. and C.A. Dahinden (1992) Effect of nerve growth factor on the release of inflammatory mediators by mature human basophils. Blood, 79: 2662–2669.

Black, M.A., Lefebvre, F.A., Pope, L., Lefrebvre, Y.A. and P. Walker (1992) Thyroid hormone and androgen regulation of nerve growth factor gene expression in the mouse submandibular gland. Mol. Cell. Endocrinol., 84: 145–154.

Böhm, A., Aloe, L., and R. Levi-Montalcini (1986) Nerve growth factor enhances precocious differentiation and numerical increase in mast cells in cultures of rat splenocytes. Accad. Naz. Lincei, 80: 1–6.

Borrelli, E., Sawchenko, P.E. and R.M. Evans (1992) Pituitary hyperplasia produced by ectopic expression of nerve growth factor. Proc. Natl. Acad. Sci. USA, 89: 2764–2768.

Boyle, M.D.P., Lawman, M.J.P., Gee, A.P. and M. Young (1985) Nerve growth factor: a chemotactic factor for polymorphonuclear leukocytes in vivo. J. Immunol., 134: 564–568.

Brodie, C. and E.W. Gelfand (1992) Functional nerve growth factor receptors on human B lymphocytes. Interaction with IL-2. J. Immunol., 148: 3492–3497.

Bruni, A., Bigon, E., Boarato, E., Mietto, L., Leon, A. and G. Toffano (1982) Interaction between nerve growth factor and lysophosphatidylserine on rat peritoneal mast cells. FEBS Lett., 138: 190–192.

Caramia, F., Angeletti, P.U. and R. Levi-Montalcini (1962) Experimental analysis of the mouse submaxillary salivary gland in relationship to its nerve growth factor content. Endocrinology, 70: 915–922.

Crookston, K.P., Webb, D.J., Wolf, B.B. and S.L. Gonias (1994) Classification of the α-2-macroglobulin-cytokine interactions based on affinity of noncovalent association in solution under apparent equilibrium conditions. J. Biol. Chem., 269: 1533–1540.

Crowley, C., Spencer, S.D., Nishimura, M.C., Chen, K.S., Pitis-Meek, S., Armanini, M.P., Ling, L.H., McMahon, S.B., Shelton, D.L., Levinson, A.D. and H.S. Philips (1994) Mice lacking nerve growth factor display perinatal loss of sensory and sympathetic neurons yet develop basal forebrain cholinergic neurons. Cell, 76: 1001–1011.

Davis, B.M., Lewin, G.R., Mendell, L.R., Jones, M.E. and K.M. Albers (1993) Altered expression of nerve growth factor in the skin of transgenic mice leads to changes in response to mechanical stimuli. Neuroscience, 56: 789–792.

Diamond, J., Coughlin, M., Macintyre, L., Holmes, M. and B. Visheau (1987) Evidence that endogenous β-nerve growth factor is responsible for the collateral sprouting, but not the regeneration, of nociceptive axons in adult rats. Proc. Natl. Acad. Sci. USA, 84: 6576–6600.

Diamond, J., Holmes, M. and Coughlin, M. (1992) Endogenous NGF and nerve impulses regulate the collateral sprouting of sensory axons in the skin of the adult rat. J. Neurosci., 12: 1454–1466.

Dicou, E., Lee, J. and P. Brachet (1986) Synthesis of nerve growth factor mRNA and precursor protein in the thyroid and parathyroid glands of the rat. Proc. Natl. Acad. Sci. USA, 83: 7084–7088.

Dissen, C.A., Hill, D.F., Costa, M.E., Ma, Y.J. and S.R. Ojeda (1991) Nerve growth factor receptors in the peripuberal rat ovary. Mol. Endocrinol., 5: 1642–1650.

Donaldson, L.F., Harmar, A.J., McQueen, D.S. and J.R. Seckl (1992) Increased expression of preprotachykinin, calcitonin gene-related peptide, but not vasoactive intestinal polypeptide messenger RNA in dorsal root ganglia during the development of adjuvant monoarritis in the rat. Mol. Brain Res., 16: 143–149.

Donnerer, J., Schuligoi, R. and C. Stein (1992) Increased content and transport of substance P and calcitonin gene-related peptide in sensory nerves innervating inflamed tissue: evidence for a regulatory function of nerve growth factor in vivo. Neuroscience, 49: 693–698.

Dubner, R. and Ruda, M.A. (1992) Activity-dependent neuronal plasticity following tissue injury and inflammation. Trends Neurosci., 15: 96–103.

Ehrhard, P.B., Ganter, U., Bauer, J. and U. Otten (1993a) Expression of functional *trk* protooncogene m human monocytes. Proc. Natl. Acad. Sci. USA, 90: 5423–5427.

Ehrhard, P.B., Erb, P., Graumann, U. and U. Otten (1993b) Expression of nerve growth factor and nerve growth factor receptor tyrosine kinase Trk in activated CD4-positive T-cell clones. Proc. Natl. Acad. Sci. USA, 90: 10984–10988.

Ensoli, F., Ensoli, B. and C.J. Thiele (1994) HIV-1 gene expression and replication in neuronal and glial cell lines with immature phenotype: effects of nerve growth factor. Virology, 200: 668–676.

Fabrazzo, M., Costa, E. and I. Mocchetti (1991) Reserpine stimulates nerve growth factor biosynthesis in developing rat brain: steroids and possible mediators. Mol. Pharmacol., 39: 144–149.

Figueiredo, B.C., Otten, U., Strauss, S., Volk, B. and Maysinger D. (1993) Effects of perinatal hypo- and hyperthyroidism on the levels of nerve growth factor and its low-affinity receptor in cerebellum. Dev. Brain Res., 72: 237–244.

Freidin, M., Bennett, M.V.L. and J.A. Kessler (1992) Cultured sympathetic neurons synthesize and release the cytokine interleukin 113. Proc. Natl. Acad. Sci. USA, 89: 10440–10443.

Garcia-Mauriño, J.E., Boya, J., Lopez-Muñoz, F. and J.L. Calvo (1992) Immunohistochemical localization of nerve growth factor in the rat pineal gland. Brain Res., 585: 255–259.

Garry, M.G. and K.M. Hargreaves (1992) Enhanced release of immunoreactive CGRP and substance P from spinal dorsal horn slices occurs during carrageenan inflammatioin. Brain Res., 582: 139–142.

Gee, A.P., Boyle, M.D.P., Munger, K.L., Lawman, M.J.P. and M. Young (1983) Nerve growth factor: stimulation of polymorphonuclear leukocyte chemotaxis in vitro. Proc. Natl. Acad. Sci. USA, 80: 7215–7218.

Goedert, M., Otten, U., Schafer, T., Schwab, M. and H. Thoenen (1980) Immunosympathectomy: lack of evidence for a complement-mediated cytotoxic mechanism. Brain Res., 210: 399–409.

Gonzales, R., Goldyne, M.E., Taiwo, Y.O. and J.D. Levine (1989) Production of hyperalgesic prostaglandins by sympathetic postganglionic neurons. J. Neurochem., 53: 1595–1598.

Gorin, P.D. and E.M. Johnson (1979) Experimental autoimmune model of nerve growth factor deprivation: effect of developing sympathetic and sensory neurons. Proc. Natl. Acad. Sci. USA, 76: 5382–5386.

Graham, C.W., Lynch, J.H. and D. Djakiew (1992) Distribution of nerve growth factor like protein and nerve growth factor receptor in human benign prostatic hyperplasia and prostatic adenoma. J. Urol., 147: 1444–1447.

Gudat, F., Laubscher, A., Otten, U. and A. Pletscher (1981) Shape changes induced by biologically active peptides and nerve growth factor in blood platelets of rabbits. Br. J. Pharmacol. 74: 533–538.

Harper, G.P., Glanville, R.W. and H. Thoenen (1982) The purification of nerve growth factor from bovine seminal plasma: biochemical characterization and partial amino acid sequence. J. Biol. Chem., 257: 8541–8548.

Hattori, A., Tanaka, E., Murase, K., Ishida, N., Chatani, Y., Tsujimoto, M., Hayashi, K. and M. Kohno (1993) Tumor necrosis factor stimulates the synthesis and secretion of biologically active nerve growth factor in non-neuronal cells. J. Biol. Chem., 268: 2577–2582.

Holzer, P. (1988) Local effector functions of capsaicin-sensitive sensory nerve endings: involvement of tachykinins, calcitonin gene-related peptide and other neuroppetides. Neuroscience, 24: 739–768.

Horigome, K., Pryor, J.C., Bullock, E.D. and E.M. Johnson, Jr. (1993) Mediator release from mast cells by nerve growth factor. Neurotrophin specificity and receptor mediation. J. Biol. Chem., 268: 14881–14887.

Horigome, K., Bullock, E.D. and E.M. Johnson, Jr. (1994) Effects of nerve growth factor on rat peritoneal mast cells. Survival promotion and immediate-early gene induction. J. Biol. Chem., 269: 2695–2702.

Ishii, D.N. and E.M. Shooter (1975) Regulation of nerve growth factor synthesis in mouse submaxillary glands by testosterone. J. Neurochem., 25: 843–851.

Johnson, E.M., Gorin, P.D., Brandeis, L.D. and J. Pearson (1980) Dorsal root ganglion neurons are destroyed by exposure in utero to maternal antibody to nerve growth factor. Science, 210: 916–918.

Jonakait, J.M. (1993) Neural-immune interactions in sympathetic ganglia. Trends Neurosci., 16: 419–423.

Kannan, Y.H., Ushuo, H., Koyama, H., Okada, M., Oikawa, M., Yoshihara, T., Kaneko, M. and H. Matsuda (1991) 2.55 Nerve growth factor enhances the survival, phagocytosis, and superocide production of murine neutrophils. Blood, 77: 1320–1325.

Kim, S.J., Park, K., Rudkin, B.B., Dey, B.R., Sporn, M.B. and A.B. Roberts (1994) Nerve growth factor induces transcription of transforming growth factor β-1 through a specific promoter element in PC-12 cells. J. Biol. Chem., 269: 3739–3744.

Kimata, H. Yoshida, A. Ishioka, C. Kusunoki, T. Hosoi, S. and H. Mikawa (1991) Nerve growth factor specifically induces human IgG4 production. Eur. J. Immunol., 21: 137–141.

Koo, P.H. and R.W. Stach (1989) Interaction of nerve growth factor with murine α-macroglobulin. J. Neurosci. Res., 22: 247–261.

Koo, P.H. and W.S. Qiu (1994) Monoamine-activated α-2-macroglubulin binds trk receptor and inhibits nerve growth factor-stimulated trk phosphorylation and signal transduction. J. Biol. Chem., 269: 5369–5379.

Lakshmanan, J., Burns, C. and R.A. Smith (1988) Molecular forms of nerve growth factor in mouse submaxillary glands. Biochem. Biophys. Res. Commun., 152: 1008–1014.

Landis, S.C., Fredieu, J.R. and M. Yodlowski (1985) Neonatal treatment with nerve growth factor antiserum eliminates cholinergic sympathetic innervation of sweat glands. Dev. Biol., 112: 222–229.

Lara, H.E., Hill, D.F., Katz, K.H. and S.R. Ojeda (1990) The gene encoding nerve growth factor is expressed in the immature rat ovary: effect of denervation and hormonal treatment. Endocrinology, 126: 357–363.

Lathinen, T., Soinila, S. and J. Lakshanan (1989) Biological demonstration of nerve growth factor in the rat pituitary gland. Neuroscience, 30: 165–170.

Leah, J.D., Cameron, A.A. and P.J. Snow (1985) Neuropeptides in physiologically identified mammalian sensory neurones. Neurosci. Lett., 56: 257–264.

Lee, K.F., Li, E., Huber, L.J., Landis, S.C., Sharpe, A.H., Chao, M.V. and R. Jaenisch (1992) Targeted mutation of the gene encoding the low affinity NGF receptor p75 leads to dificits in the peripheral sensory nervous system. Cell, 69: 737–749.

Leon, A., Buriani, A., Dal, Toso, R., Fabris, M., Romanello, S., Aloe, L. and R. Levi-Montalcini (1994) Mast cells synthesize, store, and release nerve growth factor. Proc. Natl. Acad. Sci. USA, 91: 3739–3743.

Levi-Montalcini, R. and B. Booker (1960) Destruction of the sympathetic ganglia in mammals by an antiserum to a nerve-growth protein. Proc. Natl. Acad. Sci. USA, 46: 384–391.

Levi-Montalcini, R., Aloe, L. and E. Alleva (1990) A role for nerve growth factor in nervous, endocrine and immune systems. Prog. Neuroendocrine Immunol., 3: 1–10.

Lewin, O.R., Ritter, A.M. and L.M. Mendell (1992a) On the role of NGF in the development of myelinated nociceptors. J. Neurosci., 12: 1896–1905.

Lewin, G.R., Winter, J. and S.B. McMahon (1992b) Regulation of afferent connectivity in the adult spinal cord by nerve growth factor. Eur. J. Neurosci., 4: 700–707.

Lewin, O.R., Ritter, A.M. and L.M. Mendell (1993) Nerve growth factor-induced hyperalgesia in the neonatal and adult rat. J. Neurosci., 13: 2136–2148.

Lewin, G.R. and L.M. Mendell (1993) Nerve growth factor and nociception. Trends Neurosci., 16: 353–359.

Li, A.K.C., Koroly, M.J., Schattenkerk, M.E., Malt, R.A. and M. Young (1980) Nerve growth factor: acceleration of the rate of wound healing in mice. Proc. Natl. Acad. Sci. USA, 77: 4379–4381.

Lindholm, D., Heumann, R., Meyer, M. and H. Thoenen (1987) Interleukin-1 regulates synthesis of nerve growth factor in non-neuronal cells of rat sciatic nerve. Nature, 330: 658–659.

Lindholm, D., Hengerer, B., Heuman, R., Carroll, P. and H. Thoenen (1990) Glucocorticoid hormones negatively regulate nerve growth factor expression in vivo and in cultured rat fibroblasts. Eur. J. Neurosci., 2: 795–801.

Lindsay, R.M. and M. Harmar (1989) Nerve growth factor regulates expression of neuropeptide genes in adult sensory neurons. Nature, 337: 362–364.

Luppi, P., Levi-Montalcini, R., Bracci-Laudiero, L., Bertolini, A., Arletti, R., Tavernari, D., Vigneti, E. and L. Aloe (1993) NGF is released into plasma during human pregnancy: an oxytocin-mediated response? NeuroReport, 4: 1063–1065.

Maestripieri, D., De Simone, R., Aloe, L. and R. Levi-Montalcini (1990) Social status and nerve growth factor serum levels after agonistic encounters in mice. Physiol. Behav., 47: 161–164.

Marshall, J.S., Stead, R.H., McSharry, C., Nielsen, L. and J. Bienenstock (1990) The role of mast cell degranulation products in mast cell hyperplasia I. Mechanism of action of nerve growth factor. J. Immunol., 144: 1886–1892.

Matsuda, H., Coughlin, M.D., Bienestock, J. and J.A. Denburg (1988) Nerve growth factor promotes human hemopoietic colony growth and differentiation. Proc. Natl. Acad. Sci. USA, 85: 6508–6512.

Matsuda, H., Kannan, Y., Ushio, H., Kiso, Y., Kanemoto, T., Suzuki, H. and Y. Kitamura (1991) Nerve growth factor induces development of connective tissue-type mast cells in vitro from murine bone marrow cells. J. Exp. Med., 174: 7–14.

Mazurek, N., Weskamp, G., Erne, P. and U. Otten (1986) Nerve growth factor induces mast cell degranulation without changing intracellular calcium levels. FEBS Lett., 198: 315–320.

Missale, C., Boroni, F., Losa, M., Giovanelli, M., Zanellato, A., Dal, Toso, R., Balsari, A. and P. Spano (1993) Nerve growth factor suppresses the transforming phenotype of human prolactinomas. Proc. Natl. Acad. Sci. USA, 90: 7961–7965.

Missale, C., Boroni, F., Sigala, S., Zanellato, A., Dal, Toso, R., Balsari, A. and P. Spano (1994) Nerve growth factor directs differentiation of the bipotential cell line GH-3 into the mammotroph phenotype. Endocrinology, 135: 290–298.

Ojeda, S.R., Hill, D.F. and K.H. Katz (1991) The genes encoding nerve growth factor and its receptor are expressed in the developing female rat hypothalamus. Mol. Brain Res., 9: 47–55.

Otten, U., Baumann, S.B. and J. Girard (1979) Stimulation of the pituitary-adrenocortical axis by nerve growth factor. Nature, 282: 413–414.

Otten, U., Baumann, J.B. and J. Girard (1984) Nerve growth factor induces plasma extravasation in rat skin. Eur. J. Pharmacol., 106: 199–201.

Otten, U., Ehrhard, P. and R. Peck (1989) Nerve growth factor induces growth and differentiation of human B lymphocytes. Proc. Natl. Acad. Sci. USA, 86: 10059–10063.

Parvinen, F.L., Peito-Huikko, M., Soder, O., Schultz, R., Kaipia, A., Mali, P., Toppari, J., Hakovirta, H., Lonnerberg P. and E.M. Ritzen (1992) Expression of nerve growth factor and its receptor in rat seminiferous epithelium: specific function at the onset of meiosis. J. Cell. Biol., 117: 629–641.

Patel, A.J., Hayaschi, M. and A. Hunt (1988) Role of thyroid hormone and nerve growth factor in the development of choline acetyl-transferase and other cell-specific marker enzymes in the basal forebrain of the rat. J. Neurochem., 50: 803–811.

Pearce, F.L. and H.L. Thompson (1986) Some characteristics of histamine secretion from rat peritoneal mast cells stimulated with nerve growth factor. J. Physiol., 372: 379–393.

Perez-Polo, J.R., Hall, K., Livingston, K. and K. Westhind (1977) Steroid induction of nerve growth factor synthesis in cell culture. Life Sci., 21: 1535–1544.

Persson, H., Ayer-Le, Lievre, C., Söder, O., Villar, M.J., Madis, Metsis, M., Olson, L., Ritzen, M. and T. Hokfelt (1990) Expression of β-nerve growth factor receptor mRNA in Sertoli cells downregulated by testosterone. Science, 247: 704–707.

Petty, B.G., Cornblath, D.R., Adornato, B.T., Chaudhry, V., Flexner, C., Wachsman, M., Sinicropi, D., Burton, L.E. and S.J. Peroutka (1994) The effect of systemically administered recombinant human nerve growth factor in healthy human subjects. Ann. Neurol., 36: 244–246.

Richard, A., McColl, S.R. and G. Pelletier (1992) Interleukin-4 and nerve growth factor can act as cofactors for interleukin-3-induced histamine production in human umbilical cord blood in serum-free culture. Br. J. Haematol., 81: 6–11.

Ritter, A.M., Lewin, C.R., Kremer, N.E. and L.M. Mendell (1991) Requirement for nerve growth factor in the development of myelinated nociceptors in vivo. Nature, 350: 500–502.

Ronne, H., Anundi, H., Rask, L. and P.A. Peterson (1979) Nerve growth factor binds to serum α-2-macroglobulin. Biophys. Biochem. Res. Commun., 87: 330–336.

Scaccianoce, S., Cigliana, G., Nicolai, R., Muscolo, L.A., Porcu, A., Navarra, D., Perez-Polo, J.R. and L. Angelucci (1993) Hypothalamic involvement in the activation of the pituitary-adrenalcortical axis by nerve growth factor. Neuroendocrinology, 58: 202–209.

Screpanti, I., Meco, D., Scarpa, S., Morrone, S., Frati, L., Gulino, A. and A. Modesti (1992) Neuromodulatory loop mediated by nerve growth factor and interleukin 6 in thymic stromal cell cultures. Proc. Natl. Acad. Sci. USA, 89: 3209–3212.

Siminoski, K., Bernanke, J. and R.A. Murphy (1993) Nerve growth factor and epidermal growth factor in mouse submandibular glands: identical diurnal changes and rates of secretagogue-induced synthesis. Endocrinology, 132: 2031–2037.

Smeyne, R.J., Klein, R., Schnapp, A., Long, L.K., Bryant, S., Lewin, A., Lira, S.A. and M. Barbacid (1994) Severe sensory and sympathetic neuropathies in mice carrying a disrupted trk/NGF receptor gene. Nature, 368: 246–249.

Snider, W.D. and E.M. Johnson, Jr. (1989) Neurotrophic molecules. Ann. Neurol., 26: 489–506.

Soinila, S., Lakshnanan, J. and T. Lathinen (1988) Nerve growth factor-like activity in the rat pituitary intermediate lobe. Neurosci. Res., 6: 123–130.

Spillantini, M.G., Aloe, L., Alleva, E., De Simone, R., Goedert, M. and R. Levi-Montalcini (1989) Nerve growth factor mRNA and protein increase in hypothalamus in a mouse model of aggression. Proc. Natl. Acad. Sci. USA, 86: 8555–8559.

Stead, R.H., Bienenstock, J. and A.M. Stanisz (1987) Neuropeptide regulation of mucosal immunity. Immunol. Rev., 100: 333–359.

Sugiyama, K., Suzuki, Y. and H. Furuta (1985) Histamine-release induced by 7S nerve growth factor of mouse submandibular salivary glands. Arch. Oral. Biol., 30: 93–95.

Taiwo, Y.O, Levine, I.D., Burch, R.M., Woo, J.E. and W.C. Mobley (1991) Hyperalgesia induced in the rat by the amino-terminal octapeptide of nerve growth factor. Proc. Natl. Acad. Sci. USA, 88: 5144–5148.

Talamini, L.M. and L. Aloe (1993) Immunohistochemical localization of nerve growth factor (NGF) and NGF-receprot in the hypothalamus of adult rats. Arch. Ital. Biol., 131: 255–266.

Thorpe, L.W. and Perez-Rolo, (1987) The influence of nerve growth factor on the in vitro proliferative response of rat spleen lymphocytes. J. Neurosci. Res., 18: 134–139.

Thorpe, L.W., Werrbach-Perez, K. and J.R. Perez-Polo (1987) Effects of nerve growth factor expression on interleukin-2 receptors on cultured human lymphocytes. Ann. NY Acad. Sci, 496: 310–311.

Tomioka, M., Stead, R.H., Nielsen, L., Coughlin, M.D. and J. Bienenstock (1988) Nerve growth factor enhances antigen and other secretagogue-induced histamine release from rat peritoneal mast cells in the absence of phosphatidylserine. J. Allergy Clin. Immunol., 82: 599–607.

Verge, V.M.K., Richardson, P.M., Benoit, R. and R.J. Riopelle (1989) Histochemical characterization of sensory neurons with high-affinity receptors for nerve growth factor. J. Neurocytol., 18: 583–591.

Walker, C.D., Weichsel, M.E., Fisher, D., Guo, S.M. and D.A. Fisher (1979) Thyroxine increases nerve growth factor concentration in adult mouse brain. Science, 204: 427–429.

Weskamp, G. and U. Otten (1987) An enzyme-linked immunoassay for nerve growth factor (NGF): a tool for studying regulatory mechanisms involved in NGF production in brain and peripheral tissues. J. Neurochem., 48: 1779–1786.

Wolf, D., Richter-Landsberg, C., Short, M.P., Cepko, C. and X.O. Breakefield (1988) Retrovirus mediated transfer of beta-nerve growth factor into mouse pituitary line AcT-20. Mol. Biol. Med., 5: 43–59.

Yan, Q., Settle, S.L. and M.R. Wilkins (1991) Hypotension induced by intravascular administration of nerve growth factor in the rat. Clin. Sci., 80: 565–569.

BIBLIOGRAPHY

1. Levi-Montalcini R. and Sacerdote E., Ricerche quantitative sul sistema nervoso di Mus Musculus, *Monitore Zoologico* **XLV**, 163–172 (1934).
2. Levi-Montalcini and Sacerdote E., I caratteri e le modalità di sviluppo delle fibre collagene e reticolari in vitro, *Arch. Ital. Anat. Embriol.* **XLII**, 309–318 (1937).
3. Levi-Montalcini R. and Sacerdote E., Rapporti genetici fra la trama preesistente del tessuto e le fibrille collagene e reticolari neoformatesi in vitro, *Bull. d'histologie* **XIII**, 121–132 (1937).
4. Levi-Montalcini R. and Visintini F., Lesioni parziali dei centri nervosi nell'embrione di pollo dalla fine dal 4° all'8° giorno d'incubazione, *Boll. Soc. Ital. Biol. Sper.* **XIII**, 981–983 (1938).
5. Levi-Montalcini R. and Visintini F., Le prime vie nervose dimostrabili con metodi neurofibrillari nell'embrione di pollo all'epoca dei primi movimenti, *Boll. Soc. Ital. Biol. Sper.* **XIII**, 983–985 (1938).
6. Levi-Montalcini R. and Visintini F., Azione del curaro, della stricnina, dell'eserina, dell'acetilcolina sulla trasmissione dell'influsso nervoso nell'embrione di pollo dal 4° all'8° giorno d'incubazione, *Boll. Soc. Ital. Biol. Sper.* **XIII**, 979–981 (1938).
7. Visintini F. and Levi-Montalcini R., Relazione fra differenziazione strutturale e funzionale dei centri e delle vie nervose nell'embrione di pollo, *Arch. Suisses Neur. et Psyc.* **XLIII**, 44, 119–150 (1939).
8. Levi-Montalcini R., Origine ed evoluzione del nucleo accessorio del nervo abducente nell'embrione di pollo, *Acta Pontifica Accademia Scientiarum* **VI**, 335–347 (1942).
9. Levi-Montalcini R. and Levi G., Les conséquences de la destruction d'un territoire d'innervation périphérique sur le dévelopment des centres nerveux correspondants dans l'embryon de poulet, *Arch. Biol.* 537–545 (1942).
10. Levi-Montalcini R. and Levi G., Recherches quantitatives sur la marche du processus de differentiation des neurones dans le ganglions spinaux de l'embryon de poulet, *Arch. Biol.* **LIV**, 189–206 (1943).
11. Levi-Montalcini R. and Levi G., Correlazioni nello sviluppo tra varie parti del sistema nervoso, *Acta Pontifica Accademia Scientiarum* **VIII**, 527–575 (1944).
12. Levi-Montalcini R. and Levi G., Correlations dans le développement des différentes parties du système nerveux, *Arch. Biol.*, **56**, 71–81 (1945).
13. Levi-Montalcini R., Ricerche sperimentali sulla determinazione del placode otico nell'embrione di pollo, *Rend. Acc. Naz. Lincei* **1**, 443–448 (1946).
14. Levi-Montalcini R., Ricerca sperimentale sull'origine dei gangli del glossofaringeo e del vago nell'embrione di pollo, *Rend. Acc. Naz. Lincei* **1**, 1349–1352 (1946).

15. Levi-Montalcini R., Correlazioni nervose nello sviluppo embrionale, CNR, Conferenza Staz. Zool. Napoli 1–11 (1946).
16. Levi-Montalcini R. and Amprino R., Ricerche sperimentali sull'origine dei neuroni del ganglio ciliare nel pollo, *Rend. Acc. Naz. Lincei* **1**, 439–442 (1946).
17. Levi-Montalcini R., Ricerche sperimentali sull'origine del simpatico toraco-lombare nell'embrione di pollo, *Rend. Acc. Naz. Lincei*, Serie VIII, **III**, 140–144 (1947).
18. Levi-Montalcini R., Ricerche sulle correlazioni nello sviluppo del sistema nervoso. Regressione secondaria del ganglio ciliare dopo asportazione della vescicola mesencefalica in embrione di pollo, *Rend. Acc. Naz. Lincei* **III**, 144–146 (1947).
19. Levi-Montalcini R. and Amprino R., Recherches expérimentales sur l'origine du ganglion ciliare dans l'embryon de Poulet, *Arch. de Biol.*, 265–288 (1947).
20. Hamburger V. and Levi-Montalcini R., Proliferation, differentiation and degeneration in the spinal ganglia of the chick embryo under normal and experimental conditions, *J. Exp. Zool.* **111**, 457–501 (1949).
21. Levi-Montalcini R., The development of the acoustico-vestibular centers in the chick embryo in the absence of the afferent root fibers and of descending fiber tracts, *J. Comp. Neurol.* **91**, 209–242 (1949).
22. Levi-Montalcini R., The origin and development of the visceral system in the spinal cord of the chick embryo, *J. Morphol.* **86**, 253–284 (1950).
23. Hamburger V. and Levi-Montalcini R., Some aspects of Neuroembryology, in: Weiss P. ed., *Genetic Neurology* (Chicago Univ. Press) 127–160 (1950).
24. Levi-Montalcini R. and Hamburger V., Selective growth-stimulating effects of mouse sarcoma on the sensory and sympathetic system of the chick embryo, *J. Exp. Zool.* **116**, 321–336 (1951).
25. Levi-Montalcini R., Effects of mouse tumor transplantation on the nervous system, in: D.A. Karnofsky ed., The chick embryo in biological research, *Ann. N.Y. Acad. Sci.* **55**, 330–343 (1952).
26. Levi-Montalcini R., In vivo and in vitro experiments on the effect of mouse sarcoma 180 and 37 on the sensory and sympathetic system of the chick embryo, *Proc. XIV Int. Congress of Zool.*, Copenhagen, p. 309 (1953).
27. Levi-Montalcini R. and Hamburger V., A diffusible agent of mouse sarcoma, producing hyperplasia of sympathetic ganglia and hyperneurotiziation of viscera in the chick embryo, *J. Exp. Zool.* **123**, 233–287 (1953).
28. Levi-Montalcini R., Meyer H. and Hamburger V., In vitro experiments on the effects of mouse sarcomas 180 and 37 on the spinal and sympathetic ganglia of the chick embryo, *Cancer Res.* **14**, 49–57 (1953).
29. Cohen S., Levi-Montalcini R. and Hamburger V., A nerve growth-stimulating factor isolated from sarcomas 37 and 180, *Natl. Acad. Sci. USA* **40**, 1014–1018 (1954).
30. Levi-Montalcini R., In vivo and in vitro experiments on a nerve growth promoting

agent isolated from mouse sarcomas 180 and 37, *Pubbl. Staz. Zool.* **XXIX**, 1–14 (1955).

31. Cohen S. and Levi-Montalcini R., Stimulation of nerve fiber outgrowth in vitro by a heat-labile factor in snake venom, *Anat. Record* **125**, 630–631 (1956).
32. Levi-Montalcini R. and Cohen S., In vitro and in vivo effects of a nerve growth-stimulating agent isolated from snake venom, *Proc. Natl. Acad. USA* **42**, 695–699 (1956).
33. Cohen S. and Levi-Montalcini R., A nerve growth-stimulating factor isolated from snake venom, *Proc. Natl. Acad. Sci. USA* **42**, 571–574 (1956).
34. Levi-Montalcini R., In vivo and in vitro effect of a nerve growth stimulating agent isolated, *Proc. XIV Int. Congress of Zool.*, Copenhagen, 1953, p. 30 (1956).
35. Levi-Montalcini R., Some trends in research on cellular differentiation and morphogenesis, *Journal of the National Cancer Institute*, Vol. 19, **4**, 609–641 (1957).
36. Levi-Montalcini R., La biologie des homogreffes. Centre National de la recherche scientifique, 115–126 (1957).
37. Levi-Montalcini R., Chemical stimulation of nerve growth, *A Symp. on the Chemical Basis of Development*, 645–664 (1958).
38. Jones W. A. and Levi-Montalcini R., Patterns of differentation of the nerve centers and fiber tracts in the avian cerebral hemispheres, *Arch. Ital. Biol.* **96**, 231–284 (1958).
39. Levi-Montalcini R., Nerve growth controlling factors, *Proc. 1st Int. Conf. on Congenital Malformations*, 141–157, London (1960).
40. Levi-Montalcini R. and Booker B., Destruction of the sympathetic ganglia in mammals by an antiserum to a nerve growth protein, *Proc. Natl. Acad. Sci. USA* **46**, 384–391 (1960).
41. Levi-Montalcini R. and Angeletti P.U., Biological properties of a nerve growth promoting protein and its antiserum, in: Regional Neurochemistry, *Proc. 4th Int. Neurochem. Symposium*, S. S. Kety and J. Elkes eds., Pergamon Press, New York, 362–371 (1960).
42. Levi-Montalcini R. and Booker B., Excessive growth pf the sympathetic ganglia evoked by a protein isolated from mouse salivary glands, *Proc. Natl. Acad. Sci. USA* **46**, 373–384 (1960).
43. Levi-Montalcini R. and Cohen S., Effects of the extract of the mouse submaxillary salivary glands on the sympathetic system of mammals, *Ann. New York Acad. Sci. USA* **85**, 324–341 (1960).
44. Levi-Montalcini R. and Angeletti P.U., Growth control of the sympathetic system by a specific protein factor, *Quart. Rev. Biol.* **36**, 99–108 (1961).
45. Levi-Montalcini R., Analysis of a specific nerve growth factor and of its antiserum, *Sci. Repts. Ist. Sup. Sanità* **2**, 345–368 (1962).
46. Levi-Montalcini R., Angeletti P.U. and Lee R.H., Glycogen content in the heart of immunosympathectomized animals, *Int. J. Neurophar.* **1**, 161–164 (1962).

47. Levi-Montalcini R. and Angeletti P.U., Considerations on a nerve growth promoting activity of some tumors, in: M.J. Brennan and W.L. Simpson eds., *Biological Interactions in Normal and Neoplastic Growth. Henry Ford Hospital International Symposium*, Little, Brown & Co., 225–237 (1962).
48. Levi-Montalcini R. and Angeletti P.U., Growth and differentiation. *Ann. Rev. Physiol.* **24**, 11–56 (1962).
49. Caramia F., Angeletti P.U. and Levi-Montalcini R., Experimental analysis of the mouse submaxillary salivary gland in relationship to its nerve growth factor content, *Endocrinology* **70**, 915–922 (1962).
50. Levi-Montalcini R., Growth and differentiation in the nervous system, in: J.M. Allen ed., *The Nature of Biological Diversity*, McGraw-Hill, New York, 261–295 (1963).
51. Levi-Montalcini R. and Angeletti P.U., Essential role of the nerve growth factor in the survival and maintenance of dissociated sensory and sympathetic embryonic nerve cells in vitro, *Dev. Biol.* **7**, 653–655 (1963).
52. Levi-Montalcini R., Growth control of nerve cells by a protein factor and its antiserum, *Science* **143**, 105–110 (1964).
53. Levi-Montalcini R., Isolamento e purificazione di un fattore proteico essenziale alla differenziazione e sviluppo della cellula nervosa simpatica, *Boll. Atti Acc. Med. Roma* **1–3**, 71–75 (1964).
54. Levi-Montalcini R., The nerve growth factor, *Ann. New York Acad. Sci. USA* **118**, 149–170 (1964).
55. Levi-Montalcini R. and Angeletti P.U., Hormonal control of the NGF content in the submaxillary salivary glands of mice, in: M. Sreebny and J. Meyerm eds., *Salivary Glands and their Secretions*, Pergamon, Oxford, 129–141 (1964).
56. Levi-Montalcini R., Events in the developing nervous system, in: D.P. Purpura and J.P. Schade eds., *Progress in Brain Research*, Vol. 4, Growth and Maturation of the Brain, Elsevier, Amsterdam, 1–29 (1964).
57. Gandini Attardi D., Levi-Montalcini R., Wenger B.S. and Angeletti P.U., Submaxillary gland of mouse: effects of a fraction on tissues of mesodermal origin in vitro, *Science* **150**, 3701, 1307–1309 (1965).
58. Hamberger B., Levi-Montalcini R., Norberg K.W. and Sjöqvist F., Monoamines in immunosympathectomized rats, *Int. J. Neuropharmacol.* **4**, 91–95 (1965).
59. Levi-Montalcini R., Morphological and metabolic effects of the nerve growth factor, *Arch. Biol.* **76**, 387–417 (1965).
60. Levi-Montalcini R., The nerve growth factor: its mode of action on sensory and sympathetic nerve cells, *The Harvey Lectures*, Series **60**, Academic Press, 217–259 (1965).
61. Levi-Montalcini R., Tentori L., Vivaldi G., Angeletti P.U. and Marini Bettolo G.B., Composizione in amminoacidi del fattore di crescita del sistema nervoso (NGF), *Gazzetta Chim. Ital.* **95**, 333–337 (1965).

62. Angeletti P.U., Gandini Attardi D., Toschi G., Salvi M.I. and Levi-Montalcini R., Metabolic aspects of the effect of nerve growth factor on sympathetic and sensory ganglia: Protein and ribonucleic acid synthesis, *Biochim. Biophys. Acta* **95**, 111–120 (1965).
63. Levi-Montalcini R., Growth regulation of sympathetic nerve cells, *Arch. Ital. Biol.* **103**, 832–846 (1965).
64. Liuzzi A., Angeletti P.U. and Levi-Montalcini R., Metabolic effects of a specific nerve growth factor (NGF) on sensory and sympathetic ganglia: enhancement of lipid biosynthesis, *J. of Neurochem.* **12**, 705–708 (1965).
65. Levi-Montalcini R. and Angeletti P.U., The action of nerve growth factor on sensory and sympathetic cells, in: R.L. Dehaan and H. Ursprung eds., *Organogenesis*, New York, 187–198 (1965).
66. Levi-Montalcini R., Biological aspects of specific growth promoting factors, *Proc. Roy. Soc. Med.*, May 1965, **585**, 126–142 (1965).
67. Levi-Montalcini R., Problems and perspectives in neurobiology: a brief survey of old and new concepts, *Arch. Zool. Ital.* Parte **II**, 629–666 (1966).
68. Toschi G., Dore E., Angeletti P.U., Levi-Montalcini R. and De Haen Ch., Characteristics of labelled RNA from spinal ganglia of chick embryo and the action of a specific growth factor (NGF), *J. Neurochem.* **13**, 539–544 (1966).
69. Liuzzi A., Angeletti P.U., Levi-Montalcini R. and Tacchini G., Glucose oxidation in the heart and diaphragm of immunosympathectomized animals, *Il Farmaco* **3**, 191–195 (1966).
70. Sabatini M.T., Levi-Montalcini R. and Angeletti P.U., Effetti precoci del siero anti-nerve growth factor (anti-NGF) sulla cellule nervosa, *Ann. Ist. Super. Sanità* **2**, 349–355 (1966).
71. Levi-Montalcini R., Immunosympathectomy, *Molecular basis of some aspects of mental activity* **1**, 105–111 (1966).
72. Levi-Montalcini R. and Angeletti P.U., H. Immunosympathectomy, *Pharmacol. Rev.* **18**, 619–629, *Exp. Biol. Med.* **1**, 170–182 (1966).
73. Levi-Montalcini R., Differentiation and growth control mechanism in the nervous systems, in: E. Hagen, W. Wechsler and P. Ailliken eds., Experimental Biology and Medicine, Vol. **1**, *Morphological and Biochemical Aspects of Cytodifferentiation*, Karger, Basel, 170–182 (1967).
74. Angeletti P.U., Calissano P., Chen J.S. and Levi-Montalcini R., Multiple molecular forms of the nerve growth factor, *Biochim. Biophys. Acta* **147**, 180–182 (1967).
75. Levi-Montalcini R. and Angeletti P.U., Biological aspects of the nerve growth factor, in: G.E.W. Wolstenholme and M. O'Connor, eds., Ciba *Symp. on Growth of the Nervous System*, Churchill, London, 126–142 (1968).
76. Levi-Montalcini R., Caramia F., Luse S.A. and Angeletti P.U., In vitro effects

of the nerve growth factor on the fine structure of the sensory nerve cells, *Brain Res.* **8**, 347–362 (1968).

77. Levi-Montalcini R. and Angeletti P.U., Nerve growth factor, *Physiol. Rev.* **48**, 534–569 (1968).
78. Angeletti P.U., Levi-Montalcini R. and Calissano P., The nerve growth factor (NGF): chemical properties and metabolic effects, in: F.F. Nord ed., *Advances in Enzymology* Vol. **31**, Interscience, New York, 51–75 (1968).
79. Zanini A., Angeletti P.U. and Levi-Montalcini R., Immunochemical properties of the nerve growth factor, *Proc. Natl. Acad. Sci. USA* **61**, 835–842 (1968).
80. Levi-Montalcini R., Caramia F. and Angeletti P.U., Alterations in the fine structure of nucleoli in sympathetic neurons following NGF-antiserum treatment, *Brain Res.* **12**, 54–73 (1968).
81. Levi-Montalcini R., The nerve growth factor, in: *Molecular Basis of Some Aspects of Mental Activity*, Academic Press, London, N.Y., Vol. **I**, 385–388 (1968).
82. Levi-Montalcini R., Immunosympathectomy, *Giornale Accademia Medicina*, Torino, 7–12, 1–16 (1968).
83. Chen J.S. and Levi-Montalcini R., Axonal outgrowth and cell migration in vitro from nervous system of cockroach embryos, *Science* **166**, 631–632 (1969).
84. Levi-Montalcini R. and Chen J.S., In vitro studies of the insect embryonic nervous system, *Symp. of the Intl. Soc. Cell. Biol.* **8**, 277–298 (1969).
85. Chen J.S. and Levi-Montalcini R., Axonal outgrowth and cell migration in vitro from nervous system of cockroach embryos, *Science* **166**, 631–632 (1969).
86. Levi-Montalcini R., Fattori specifici di accrescimento, *Encicl. Sci. Tecnica Mondadori* **69**, 301–312 (1969).
87. Levi-Montalcini R., Meccanismi di controllo del sistema nervoso simpatico, *Accad. Naz. Lincei* **1**, 7, 137–150 (1970).
88. Angeletti P.U. and Levi-Montalcini R., Sympathetic nerve cell destruction in newborn mammals, *Proc. Natl. Acad. Sci. USA* **65**, 114–121 (1970).
89. Levi-Montalcini R. and Angeletti P., Action of nerve growth factor on synthesis of neurofilaments and neurotubules in the target nerve cells, *Proc. Natl. Acad. Sci. USA* **7A**, (1970).
90. Angeletti P.U. and Levi-Montalcini R., Specific cytotoxic effect of 6-hydroxydopamine on sympathetic neuroblasts, *Arch. Ital. Biol.* **108**, 213–221 (1970).
91. Chen J.S. and Levi-Montalcini R., Long term cultures of dissociated nerve cells from the embryonic nervous system of the cockroach periplaneta americana, *Arch. Ital. Biol.* **108**, 503–537 (1970).
92. Chen J.S. and Levi-Montalcini R., Axonal growth from insect neurons in glia-free cultures, *Proc. Natl. Acad. Sci. USA* **66**, 32–39 (1970).
93. Angeletti P.U. and Levi-Montalcini R., Cytolytic effect of 6-hydroxydopamine on neuroblastoma mast cells, *Cancer Res.* **30**, 2863–2869 (1970).

94. Angeletti P.U., Levi-Montalcini R. and Caramia F., Ultrastructural changes in sympathetic neurons of newborn and adult mice treated with nerve growth factor, *J. Ultrastruc. Res.* **36**, 24–36 (1971).
95. Angeletti P.U. and Levi-Montalcini R., Growth regulation of the sympathetic nervous system: immunosympathectomy and chemical sympathectomy, *Eur. J. Clin. Biol. Res.* **16**, 866–874 (1971).
96. Levi-Montalcini R. and Angeletti P.U., Ultrastructure and metabolic studies on sensory and sympathetic nerve cells treated with the nerve growth factor and its antiserum, in: M. Hamburgh and Barrington E.J.W. eds., *Hormones in Development* **56**, 719–730 (1971).
97. Angeletti P.U., Levi-Montalcini R. and Caramia F., Analysis of the effects of the antiserum to the nerve growth factor in adult mice, *Brain Res.* **27**, 343–355 (1971).
98. Levi-Montalcini R., Control mechanism in the sympathetic nervous system, in: R. Paoletti and A. Davidson, eds., *Chemistry Brain and Development*, 185–194 (1971).
99. Levi-Montalcini R., In vitro analysis of the insect nervous system, *Bull. Zool.* **38**, 385–399 (1971).
100. Seshan K.R. and Levi-Montalcini R., In vitro analysis of corpora cardiaca and corpora allata from nymphal and adult specimens of periplaneta americana, *Arch. Ital. Biol.* **109**, 81–109 (1971).
101. Thoenen H., Angeletti P.U., Levi-Montalcini R. and Kettler R., Selective induction by Nerve Growth Factor of tyrosine hydroxylase and dopamine ß hydroxylase in the rat superior cervical ganglia, *Proc. Natl. Acad. Sci. USA* **68**, 1598–1602 (1971).
102. Levi-Montalcini R. and Chen R.S., Selective outgrowth of nerve fibers in vitro from embryonic ganglia of periplaneta americana, *Arch. Ital. Biol.* **109**, 307-337 (1971).
103. Caramia F., Angeletti P.U., Levi-Montalcini R. and Carratelli L., Mitochondrial lesions of developing sympathetic neurons induced by bretylium tosylate, *Brain Res.* **43**, 237–246 (1972).
104. Angeletti P.U., Levi-Montalcini R. and Caramia F., Structural and ultrastructural changes in developing sympathetic ganglia induced by guanethidine, *Brain Res.* **43**, 515–525 (1972).
105. Levi-Montalcini R. and Angeletti P.U., Nerve Growth Factor: evaluation and perspectives, in: Zaimis E. and Knight J. eds: *Nerve Growth Factor and Its Antiserum*, 46–58 (1972).
106. Angeletti P.U., Levi-Montalcini R. and Vigneti E., Localization of the nerve growth factor in subcellular fractions of peripheral tissues, in: Zaimis E. and Knight J. eds., *Nerve Growth Factor and Its Antiserum*, 39–45 (1972).
107. Aloe L. and Levi-Montalcini R., Interrelation and dynamic activity of visceral

muscle and nerve cells from insect embryos in long-term cultures, *J. Neurobiol.* **3**, 3–23 (1972).
108. Aloe L. and Levi-Montalcini R., In vitro analysis of the frontal and ingluvial ganglia from nymphal specimens of the cockroach periplaneta americana, *Brain Res.* **44**, 147–163 (1972).
109. Angeletti R.H., Angeletti P.U. and Levi-Montalcini R., Selective accumulation of [121] labelled nerve growth factor in sympathetic ganglia, *Brain Res.* **46**, 421–425 (1972).
110. Levi-Montalcini R., The nerve growth factor, in: G. Steiner and E. Schonbaum eds., *Immunosympathectomy*, Elsevier, Amsterdam, 25–45 (1972).
111. Levi-Montalcini R. and Aloe L., Neuronal nets and nerve cell interactions in vitro insect system, *In vitro* **8**, 178–193 (1972).
112. Angeletti P.U., Levi-Montalcini R., Kettler R. and Thoenen H., Comparative studies on the effect of the nerve growth factor on sympathetic ganglia and adrenal medulla in newborn rats, *Brain Res.* **44**, 197–206.
113. Levi-Montalcini R., Angeletti R.H. and Angeletti P.U., The nerve growth factor, in: G.H. Bourne ed., *The Structure and Function of Nervous Tissue*, Vol. **V**, 1–38 (1972).
114. Angeletti P.U. and Levi-Montalcini R., Growth inhibition of sympathetic cells by some adrenergic blocking agents, *Proc. Natl. Acad. Sci. USA* **69**, 1, 86–88 (1972).
115. Levi-Montalcini R., J.S. Chen, Seshan K.R. and Aloe L., An in vitro approach to the insect, in: D. Young ed., *Development Neurobiology of Arthropods*, Cambridge Press, 5–36 (1973).
116. Levi-Montalcini R., Excessive glial cell proliferation in sympathetic rat ganglia by combined nerve growth factor (NGF) and 6-hydroxydopamine (6-OHDA) treatment, in: *Society for Neuroscience 3rd Annual Meeting* (1973).
117. Levi-Montalcini R., Aloe L. and Johnson E.M., Interaction between the nerve growth factor (NGF) and adrenergic blocking agents in sympathetic neurons, in: E. Usdin and S. Snyder, eds, *Front. Catechol. Res.*, 267–276 (1973).
118. Levi-Montalcini R. and Seshan K.R., Long term cultures of embryonic and mature insect nervous and neuroendocrine systems, in: Gordon Sato ed., *Tiss. Cult. Nervous System*, Plenum Press, 1–33 (1973).
119. Seshan K.R. and Levi-Montalcini R., Neuronal properties of nymphal and adult insect neurosecretory cells in vitro, *Science* **182**, 291–293 (1973).
120. Levi-Montalcini R., Dervenwachstums-factor, *Umschau* **73**, 700 (1973).
121. Angeletti P.U., Angeletti R.H. and Levi-Montalcini R., The nerve growth factor, in: J. Lobue and A.S. Gordon eds., *Humoral Control of Growth and Differentiation*, Vol. **I**, 229–247 (1973).
122. Levi-Montalcini R., Revoltella R. and Calissano P., Microtubule proteins in the nerve growth factor mediated response, *Rec. Progr. Horm. Res.* **38**, 635–669 (1974).

123. Seshan K.R., Provine R.R. and Levi-Montalcini R., Structural and electophysiological properties of nymphal and adult insect medial neurosecretory cells: an in vitro analysis, *Brain Res.* **78**, 359–376 (1974).

124. Levi-Montalcini R., Control mechanism of the adrenergic neuron, in: K. Fuxe, L. Olson and Y. Zotterman eds., *Dynamics of Degeneration and Growth in Neurons*, 297–314 (1974).

125. Levi-Montalcini R., Aloe L., Mugnaini E., Oesch F. and Thonen H., Nerve growth factor induces volume increase and enhances tyrosine hydroxylase synthesis in chemically axotomized sympathetic ganglia of newborn rats, *Proc. Natl. Acad. Sci. USA* **72**, 595–599 (1974).

126. Levi-Montalcini R., Regulatory processes in the sympathetic adrenergic neuron, *Golgi Centennial Symp. Proc.* in: M. Santini ed., Raven Press, New York, 437–447 (1975).

127. Levi-Montalcini R., NGF: an uncharted route, *The Neuroscience Paths of Discovery*, MIT Press, 245–265 (1975).

128. Calissano P., Levi A., Alema S. and Levi-Montalcini R., Studies on the interaction of the nerve growth factor with tubulin and actin, 26° Colloquium Mosbach, Molecular Basis of Motility, L. Heilmeyer and Th. Wieland, eds, Springer-Verlag, 186–201 (1975).

129. Aloe L., Mugnaini E. and Levi-Montalcini R., Light and electron microscopic studies on the excessive growth of sympathetic ganglia in rats injected daily from birth with 6-OHDA and NGF, *Arch. Ital. Biol.* **113**, 326–353 (1975).

130. Levi-Montalcini R., One of Hans Spemann's pupils, in: W.M. Cowan ed., *Studies in Developmental Neurobiology. Essays in Honor of Viktor Hamburger*, 22–43 (1975).

131. Levi-Montalcini R., New developments in neurobiological research, *Pont. Acad. Sci.* **3**, 1–24 (1976).

132. Levi-Montalcini R., Milestones, unanswered questions and current studies of nerve growth factor, *Proc. 6th Int. Congr. Pharmacol.* **2**, 221–230 (1976).

133. Levi-Montalcini R., The nerve growth factor: its widening role and place in neurobiology, *Adv. Biochem. Psych.* **15**, 237 (1976).

134. Levi-Montalcini R., The nerve growth factor: its role in growth, differentiation and function of the sympathetic adrenergic neuron, in: Corner M.A. and Scwaab D.F. eds., *Perspectives in Brain Research*, 235–256 (1976).

135. Calissano P., Monaco G, Levi A., Menesini Chen J.S. and Levi-Montalcini R., New developments in the study of NGF-tubulin interaction, in: S.V. Perry *et al.* eds., *Contrac. System Non-Muscle Tiss.*, Elsevier, North-Holland, p. 201 (1976).

136. Levi-Montalcini R., Nerve growth factor in familial dysautonomia, *New England J. Med.* **295**, 12, 671–672 (1976).

137. Levi-Montalcini R., Regulatory processes in the sympathetic adrenergic neuron,

Atti dei Seminari Biologici della Facoltà di Medicina e Chirurgia, 1973–1974, Vita e Pensiero, (Ed. Università Cattolica, Milano) 243–262 (1976).

138. Menesini Chen M.G., Chen J.S., Calissano P. and Levi-Montalcini R., Nerve growth factor prevents vinblastine destructive effects on sympathetic ganglia in newborn mice, *Proc. Natl. Acad. Sci. USA* **74**, 11, 5559–5563 (1977).
139. Aloe L. and Levi-Montalcini R., Mast cells increase in tissues of neonatal rats injected with the nerve growth factor, *Brain Res.* **133**, 358–366 (1977).
140. Liuzzi A., Foppen F.H., Saavedra J.M., Levi-Montalcini R. and Kopin I.J., Gas chromatographic-mass spectrometric assay of serotonin in rat superior cervical ganglia. Effects of nerve growth factor and 6-hydroxydopamine, *Brain Res.* **133**, 354–357 (1977).
141. Levi-Montalcini R., ed., Nerve cells, neurotransmitters and behaviour, Introduction, *Proc. Study Week, Pontifical Academy of Sciences*, 9–14 October 1978, (Elsevier, North-Holland), 1–10 (1980).
142. Levi-Montalcini R., Aloe L., Menesini M.G. and Chen J.S., New features of the nerve growth factor-target cells interaction, *Proc. Study Week, Pontifical Academy of Sciences*, 9–14 October 1978, (Elsevier, North-Holland), 1-10 (1980).
143. Levi-Montalcini R., Menesini M.G. and Chen J.S., Neurotropic effects of the nerve growth factor in chick embryos and in neonatal rodents, in: Formshaping movements in neurogenesis *Zoom*, Vol. **8**, 201–212 (1978).
144. Menesini M.G., Chen J.S. and Levi-Montalcini R., Sympathetic nerve fibers ingrowth in the central nervous system of neonatal rodents upon intracerebral NGF injections, *Arch. Ital. Biol.* **116**, 53–84 (1978).
145. Aloe L. and Levi-Montalcini R., Nerve growth factor induced overgrowth of axotomized superior cervical ganglia in neonatal rats, *Arch. Ital. Biol.* **117**, 287-307 (1979).
146. Levi-Montalcini R., The sympathetic nerve cell as a model system of nerve cell growth and differentiation, *Int. Symp. Neuroblastoma Gaslini* **11**, 109–122 (1979).
147. Aloe L. and Levi-Montalcini R., Nerve growth factor-induced transformation of immature chromaffin cells in vivo into sympathetic neurons: effects of antiserum to nerve growth factor, *Proc. Natl. Acad. Sci. USA* **76**, 1246-1250 (1979).
148. Calissano P. and Levi-Montalcini R., Is NGF an enzyme?, *Nature* 280–359 (1979).
149. Levi-Montalcini R. and Calissano P., The nerve growth factor, *Scient. Amer.* **240**, 6, 67–77 (1979).
150. Levi-Montalcini R., Recent studies on the NGF-target cells interaction, *Differentiation* **13**, 51–53 (1979).
151. Levi-Montalcini R., Neurogenesi, *Enciclopedia del Novecento* Vol. **IV**, Istituto Enciclopedia Treccani, 637–661 (1979).
152. Aloe L. and Levi-Montalcini R., Comparative studies on testosterone and L-thyroxine effects on the synthesis of nerve growth factor in mouse submaxillary salivary glands, *Exp. Cell. Res.* **125**, 15–22 (1980).

153. Aloe L. and Levi-Montalcini R., Enhanced differentiation of sexually dimorphic organs in L-tyroxine treated TFM mice, *Cell Tiss. Res.* **205**, 15–29 (1980).

154. Levi-Montalcini R., Polydipsia after intracranial injections a property of NGF or a contaminant?, *Nature* **284**, 577 (1980).

155. Levi-Montalcini R. and Aloe L., Trophic, trophic, and transforming effects of nerve growth factor, in: Eranko *et al.* eds., *Histochem. Cell Biol. of Auton. Neurons* (Raven Press), *Acad. Roy. Med. Belgique* **134**, 217–227 (1980).

156. Aloe L. and Levi-Montalcini R., Comparative studies on the effects elicited by pre-and postnatal injections of anti-NGF, guanethidine and 6 hydroxydopamine in chromaffin and ganglion cells of the adrenal medulla and carotid body in infant rats, in: O. Eranko ed al. eds., *Histochem. Cell Biol. of Auton. Neurons*, 221 (1980).

157. Levi-Montalcini R., Aloe L., Menesini Chen J.S. and Chen J.S., Effects of the nerve growth factor in a new cell line: the chromaffin cell., in: C. di Benedetta *et al.*, eds., *Multidisciplinary Approach to Brain Development*, 467 (1980).

158. Levi-Montalcini R. and Aloe L., Mechanisms of action of nerve growth factor in intact and lethally injured sympathetic nerve cell in neonatal rodents, in: I.D. Bowen and R.A. Lockshin, eds., *Cell Death in Biology and Pathology*, (Chapman and Hall, London) 295–326 (1981).

159. Aloe L., Cozzari C., Calissano P. and Levi-Montalcini R., Somatic and behavioural postnatal effects of fetal injections of nerve growth factor antibodies in the rat, *Nature* **291**, 413–415 (1981).

160. Aloe L., Cozzari C. and Levi-Montalcini R., The submaxillary salivary glands of the African rodent Pramoys (mastomys) natalensis as the richest available source of the nerve growth factor, *Exp. Cell Res.* **133**, 475–480 (1981).

161. Levi-Montalcini R., Aggression and violence a physico/biological and clinical approach, in: L. Valzelli and L. Morgese eds., *Views on Human Aggressive Behaviour and Wars*, (Ed. Saint-Vincent) 21–30 (1981).

162. Levi-Montalcini R. and Aloe L., Synthesis and release of the nerve growth factor from the mouse submaxillary salivary glands: hormonal and neuronal regulatory mechanism, in: J.E. Dumont and J. Nunez eds., *Hormones and Cell Regulation* (Elsevier, North Biomedical Press), 53–72 (1981).

163. Levi-Montalcini R., Il Nerve Growth Factor. Cenni Storici sulla scoperta, risultati conseguiti e prospettive, *Cultura e Scuola* n. **80** (1981).

164. Levi-Montalcini R., Developmental neurobiology and the natural history of nerve growth factor, *Ann. Rev. Neurosci.* **5**, 341–362 (1982).

165. Aloe L., Calissano P. and Levi-Montalcini R., Effects of oral administration of nerve growth factor and of its antiserum on sympathetic ganglia of neonatal mice, Elsevier Biomedical Press, *Dev. Brain Res.* **4**, 31–34 (1982).

166. Levi-Montalcini R., Nerve Growth Factor — Cellule bersaglio — Un modello di elezione per lo studio della differenziazione neuronale, *Atti 3° Meeting of the*

Italian Branch of the European Tissue Culture Society, Fondazione Carlo Erba, Milano, 12 November 1981, 164–169 (1982).

167. Levi-Montalcini R. and Aloe L., The effect of nerve growth factor on autonomic ganglion cells, in: Lara-Gosta Elfvin eds., *Autonomic Ganglia*, 401–426 (1983).
168. Levi-Montalcini R., The nerve growth factor-target cell interaction: a model system for the study of directed axonal growth and regeneration, in: B. Haber, J.R. Perez-Polo, G.A. Mashin, A.M. Giuffrida Stella and A.R. Liss eds., *Nervous System Regeneration* Vol. **19**, 4, 3–22 (1983).
169. Calissano P., Cattaneo A., Biocca S., Aloe L., Mercanti D. and Levi-Montalcini R., The Nerve Growth Factor. Established Findings and Controversial Aspects, *Exp. Cell Res.* **154**, 1–9, Academic Press (1984).
170. Calissano P., Cattaneo A., Aloe L. and Levi-Montalcini R., The nerve growth factor (NGF) in: *Hormonal Proteins and Peptides*, XII, Academic Press **XII**, 1–56 (1984).
171. Levi-Montalcini R., Nerve Growth Factor activation of differentiation programs in neuronal and non-neuronal cell lines from amphibians to mammals, Atti Fifth International Meeting of the International Society for Developmental Neuroscience, Chieti, 24–28 giugno, *Int. J. Dev. Neurosci.* **3**, 4, 413 (1984).
172. Levi-Montalcini R. and Aloe L., Differentiating effects of murine nerve growth factor in the peripheral and central nervous systems of Xenopus laevis tadpoles, *Proc. Natl. Acad. Sci. USA* **82**, 7111–7115, Neurobiology (1985).
173. Aloe L., Cozzari C. and Levi-Montalcini R., Cyclocytidine-induced release of nerve growth factor from mouse submandibular glands enhances regeneration of sympathethic fibers in adult mice, *Brain Res.* **332**, 259–265 (1985).
174. Böhm A., Aloe L. and Levi-Montalcini R., Nerve growth factor enhances precocious differentiation and numerical increase in mast cells in cultures of rat splenocytes, *Rend. Acc. Naz. Lincei* **LXXX**, 1–2 (1986).
175. Aloe L., Alleva E., Böhm A. and Levi-Montalcini R., Aggressive behavior induces release of nerve growth factor mouse salivary gland into the bloodstream, Neurobiology, *Proc. Natl. Acad. Sci. USA* Vol. **83**, 6184–6187 (1986).
176. Levi-Montalcini R., Calissano P., Kandel E.R. and Maggi A., eds, *Molecular Aspects of Neurobiology*, Springer-Verlag, Berlin, 205 pp. (1986).
177. Levi-Montalcini R., The Nerve Growth Factor: thirtyfive years later in: The Nobel Foundation, Biochimica Clinica 11/87, VCH Verlagsgesellschaft MBH, D-6940 Weinheim; *Science* **237**, 1154–1162, *In vitro Cell. Dev. Biol.* **23**, 4, (1987).
178. Spillantini M.G., Aloe L., Alleva E., De Simone R., Goedert M. and Levi-Montalcini R., Nerve growth factor mRNA and protein increase in hypothalamus in a mouse model of aggression, Neurobiology, *Proc. Natl. Acad. Sci. USA* **86**, 8555–8559 (1989).
179. Levi-Montalcini R. and Aloe L., Culture medium and technique for long-term cultures of the nervous system and dissociated nerve cells of insects (Periplan-

eta americana), Evidence for specific growth factor?, in: *Manual of the Nervous Systems*, 215–218, Alan R. Liss, Inc. (1989).

180. Levi-Montalcini R. and Aloe L., In vitro bioassay of the nerve growth factor, in: *Manual of the Nervous Systems*, 313–316, Alan R. Liss, Inc. (1989).
181. Levi-Montalcini R. and Aloe L., Silver impregnation method for insect brain, ganglia, dissociated nerve cells in vitro and for the amphibian nervous system, in: *Manual of the Nervous System*, 280–282, Alan R. Liss, Inc. (1989).
182. Levi-Montalcini R., NGF: Apertura di una nuova frontiera nella neurobiologia, Theoria, Sonde, 20 (1989).
183. Aloe L., Bracci-Laudiero L., Tirassa P. and Levi-Montalcini R., NGF-receptor distribution and level of ChAT activity in forebrain of thymectomized rats and athymic nude mice, *2nd Eur. Cong. Endocrinol.*, Abstract, Kjubliana-Yugoslavia, July 1–6 (1990).
184. Aloe L. and Levi-Montalcini R., The effect of adrenalectomy on NGF level in the developing rat hippocampus, *Int. Symp. on Perinatal Biol.*, Abstract, Rome (1990).
185. Levi-Montalcini R., Aloe L. and Alleva E., A role for nerve growth factor in nervous, endocrine and immune systems, *Prog. Neuroendocrinimmunol.* **3**, 1–10 (1990).
186. Aloe L. and Levi-Montalcini R., Effect of early postnatal cocaine injection on brain cholinergic neurons, in Cocaina Oggi: Effetti sull'uomo e sulla società, *Proc. Int. Symp. Stop Droga*, Roma, June, 19–48 (1991).
187. Aloe L. and Levi-Montalcini R., Level of NGF in rat and human inflamed tissues, Abstract, *Int. Symp. Biotechnology and Growth Factors*: Vascular and Nervous Systems, Milan, May, 13–15 (1991).
188. Aloe L., Tuveri M.A., Vigneti E., Carcassi U. and Levi-Montalcini R., Expression of NGF-like activity in inflammatory synovial fluid, Abstract, *4th Interscience World Conf. on Inflammation: Antirheumatics, Analgesics, Immuno-modulators*, Geneva, Switzerland, April 15–18 (1991).
189. Aloe L., Tuveri M.A., Carcassi U. and Levi-Montalcini R., Nerve growth factor in the synovial fluid of patients with chronic arthritis, *Arthritis Rheum.* **35**, 3, 351–355 (1992).
190. Levi-Montalcini R. and Calissano P., Nerve growth factor, in: Barry Smith and George Adelman eds., *Neuroscience Year*, 110–114, Boston (1992).
191. Aloe L., Tuveri M.A. and Levi-Montalcini R., Nerve growth factor and distribution of mast cells in the synovium of adult rats, *Clin. Exp. Rheum.* **10**, 203–204 (1992).
192. Aloe L., Tuveri M.A. and Levi-Montalcini R., Studies on carrageenan-induced arthritis in adult rats: effect on nerve growth factor and role of sympathetic innervation, *Rheumatol. Int.* **12**, 213–216 (1992).
193. Bracci-Laudiero L., Aloe L., Levi-Montalcini R., Buttinelli C., Schilter D.,

Gillessen S. and Otten U., Multiple sclerosis patients express increased levels of B-nerve growth factor in cerebrospinal fluid, *Neurosci. Lett.* **147**, 9–12 (1992).

194. Micera A., Levi-Montalcini R., Alleva E. and Aloe L., Presence of antibodies against NGF in mouse serum following prolonged exposure to psychosocial stress, *Neurosci. Lett.* Suppl. **43**. Abstracts of the Italian society of Neuroscience Modena, Italy, 1–4 December (1992).
195. Aloe L., Leon A. and Levi-Montalcini R., A proposed autacoid mechanism controlling mastocyte behaviour, *Agents Action* **39**, Special Conference Issue, C146–C147 (1993).
196. Levi-Montalcini R., Il fattore di crescita neuronale nelle reti omeostatiche dell'organismo, *Reti, Scienza, Cultura, Economia* (Biologica **6**) 9–13 (1993).
197. Aloe L., Probert L., Kollias G., Bracci-Laudiero L., Spillantini M.G. and Levi-Montalcini R., The synovium of transgenic arthritic mice expressing human Tumor Necrosis Factor contains a high level of nerve growth factor, *Growth Factors* **9**, 149–155 (1993).
198. Aloe L., Probert L., Kollias G., Bracci Laudiero L., Micera A., Mollinari C. and Levi-Montalcini R., Level of nerve growth factor and distribution of mast cells in the synovium of tumour necrosis factor transgenic arthritic mice, *Int. J. Tiss. Reac.* **XV**, 4, 139–143 (1993).
199. Aloe L., Vigneti E., Cavalieri R., Didona B., Bracci Laudiero L., Mazzanti C., Papa G., Stasi R. and Levi-Montalcini R., Presence of nerve growth factor and NGF-receptors in human Schwannoma cells, *Clin. Chem. Enzym. Comms.* **5**, 309–313 (1993).
200. Bracci L., Aloe L., Levi-Montalcini R., Galeazzi M., Schilter D., J.L. Scully and Otten U., Increased levels of NGF in sera of systemic lupus erythematosus patients, *Neuroreport* **4**, 563–565 (1993).
201. Luppi P., Levi-Montalcini R., Bracci Laudiero L., Bertolini A., Arletti R., Tavernari D., Vigneti E. and Aloe L., NGF is released into plasma during human pregnancy: an oxytocin-mediated response?, *Neuroreport* **4**, 1063–1065 (1993).
202. Aloe L., Skaper S.D., Leon A. and Levi-Montalcini R., Nerve growth factor and autoimmune diseases, *Autoimmunity* **19**, 141–150 (1994).
203. Leon A., Buriani A., Dal Toso R., Fabris M., Romanello S., Aloe L. and Levi-Montalcini R., Mast cells synthesize, store, and release nerve growth factor, *Proc. Natl. Acad. Sci. USA* **91**, 3739–3743 (1994).
204. Levi-Montalcini R., Dal Toso R., Della Valle F., Skaper D. and Leon A., Update of the NGF saga, *J. Neurol. Sci.* **130**, 119–127 (1995).
205. Levi-Montalcini R., Skaper S.D., Aloe L. and Leon A., Nerve growth factor, *Encyclopedia of Neuroscience* (1996), in press.